Lecture Notes in Artificial Intelligence   8562

Subseries of Lecture Notes in Computer Science

LNAI Series Editors

Randy Goebel
University of Alberta, Edmonton, Canada
Yuzuru Tanaka
Hokkaido University, Sapporo, Japan
Wolfgang Wahlster
DFKI and Saarland University, Saarbrücken, Germany

LNAI Founding Series Editor

Joerg Siekmann
DFKI and Saarland University, Saarbrücken, Germany

T0183475

Stéphane Demri   Deepak Kapur
Christoph Weidenbach (Eds.)

Automated Reasoning

7th International Joint Conference, IJCAR 2014
Held as Part of the Vienna Summer of Logic, VSL 2014
Vienna, Austria, July 19-22, 2014
Proceedings

 Springer

Volume Editors

Stéphane Demri
New York University
Courant Institute of Mathematical Sciences
250 Mercer Street, New York, NY 10012, USA
E-mail: demri@lsv.ens.cachan.fr

Deepak Kapur
University of New Mexico
Department of Computer Science
Albuquerque, NM 87131-0001, USA
E-mail: kapur@cs.unm.edu

Christoph Weidenbach
Max Planck Institute for Informatics
Campus E1 4, 66123 Saarbrücken, Germany
E-mail: weidenbach@mpi-inf.mpg.de

ISSN 0302-9743   e-ISSN 1611-3349
ISBN 978-3-319-08586-9   e-ISBN 978-3-319-08587-6
DOI 10.1007/978-3-319-08587-6
Springer Cham Heidelberg New York Dordrecht London

Library of Congress Control Number: 2014941780

LNCS Sublibrary: SL 7 – Artificial Intelligence

Typesetting: Camera-ready by author, data conversion by Scientific Publishing Services, Chennai, India

Printed on acid-free paper

Springer is part of Springer Science+Business Media (www.springer.com)

logic n. **1** the science of reasoning.
– ORIGIN from Greek *logikē tekhnē*
  'art of reason'.

Foreword

In the summer of 2014, Vienna hosted the largest scientific conference in the history of logic. The Vienna Summer of Logic (VSL, http://vsl2014.at) consisted of twelve large conferences and 82 workshops, attracting more than 2000 researchers from all over the world. This unique event was organized by the Kurt Gödel Society and took place at Vienna University of Technology during July 9 to 24, 2014, under the auspices of the Federal President of the Republic of Austria, Dr. Heinz Fischer.

The conferences and workshops dealt with the main theme, logic, from three important angles: logic in computer science, mathematical logic, and logic in artificial intelligence. They naturally gave rise to respective streams gathering the following meetings:

Logic in Computer Science / Federated Logic Conference (FLoC)

- 26th International Conference on Computer Aided Verification (CAV)
- 27th IEEE Computer Security Foundations Symposium (CSF)
- 30th International Conference on Logic Programming (ICLP)
- 7th International Joint Conference on Automated Reasoning (IJCAR)
- 5th Conference on Interactive Theorem Proving (ITP)
- Joint meeting of the 23rd EACSL Annual Conference on Computer Science Logic (CSL) and the 29th ACM/IEEE Symposium on Logic in Computer Science (LICS)
- 25th International Conference on Rewriting Techniques and Applications (RTA) joint with the 12th International Conference on Typed Lambda Calculi and Applications (TLCA)
- 17th International Conference on Theory and Applications of Satisfiability Testing (SAT)
- 76 FLoC Workshops
- FLoC Olympic Games (System Competitions)

Mathematical Logic

- Logic Colloquium 2014 (LC)
- Logic, Algebra and Truth Degrees 2014 (LATD)
- Compositional Meaning in Logic (GeTFun 2.0)
- The Infinity Workshop (INFINITY)
- Workshop on Logic and Games (LG)
- Kurt Gödel Fellowship Competition

Logic in Artificial Intelligence

- 14th International Conference on Principles of Knowledge Representation and Reasoning (KR)
- 27th International Workshop on Description Logics (DL)
- 15th International Workshop on Non-Monotonic Reasoning (NMR)
- 6th International Workshop on Knowledge Representation for Health Care 2014 (KR4HC)

The VSL keynote talks which were directed to all participants were given by Franz Baader (Technische Universität Dresden), Edmund Clarke (Carnegie Mellon University), Christos Papadimitriou (University of California, Berkeley) and Alex Wilkie (University of Manchester); Dana Scott (Carnegie Mellon University) spoke in the opening session. Since the Vienna Summer of Logic contained more than a hundred invited talks, it would not be feasible to list them here.

The program of the Vienna Summer of Logic was very rich, including not only scientific talks, poster sessions and panels, but also two distinctive events. One was the award ceremony of the Kurt Gödel Research Prize Fellowship Competition, in which the Kurt Gödel Society awarded three research fellowship prizes endowed with 100.000 Euro each to the winners. This was the third edition of the competition, themed Logical Mind: Connecting Foundations and Technology this year.

The 1st FLoC Olympic Games formed the other distinctive event and were hosted by the Federated Logic Conference (FLoC) 2014. Intended as a new FLoC element, the Games brought together 12 established logic solver competitions by different research communities. In addition to the competitions, the Olympic Games facilitated the exchange of expertise between communities, and increased the visibility and impact of state-of-the-art solver technology. The winners in the competition categories were honored with Kurt Gödel medals at the FLoC Olympic Games award ceremonies.

Organizing an event like the Vienna Summer of Logic was a challenge. We are indebted to numerous people whose enormous efforts were essential in making this vision become reality. With so many colleagues and friends working with us, we are unable to list them individually here. Nevertheless, as representatives of the three streams of VSL, we would like to particularly express our gratitude to all people who helped to make this event a success: the sponsors and the Honorary Committee; the Organization Committee and the local organizers; the conference and workshop chairs and Program Committee members; the reviewers and authors; and of course all speakers and participants of the many conferences, workshops and competitions.

The Vienna Summer of Logic continues a great legacy of scientific thought that started in Ancient Greece and flourished in the city of Gödel, Wittgenstein and the Vienna Circle. The heroes of our intellectual past shaped the scientific world-view and changed our understanding of science. Owing to their achievements, logic has permeated a wide range of disciplines, including computer science, mathematics, artificial intelligence, philosophy, linguistics, and many more. Logic is everywhere – or in the language of Aristotle, πάντα πλήρη λογικῆς τέχνης.

July 2014   Matthias Baaz
                                                                Thomas Eiter
                                                                Helmut Veith

Preface

This volume contains the papers presented at IJCAR'14: 7th International Joint Conference on Automated Reasoning (IJCAR) held on July 19-22, 2014 in Vienna. This year's meeting was a merging of three leading events in automated reasoning – CADE (International Conference on Automated Deduction), FroCoS (International Symposium on Frontiers of Combining Systems) and TABLEAUX (International Conference on Automated Reasoning with Analytic Tableaux and Related Methods). IJCAR is the premier international joint conference on all topics in automated reasoning, including foundations, implementations, and applications. Previous IJCAR conferences were held at Siena (Italy) in 2001, Cork (Ireland) in 2004, Seattle (USA) in 2006, Sydney (Australia) in 2008, Edinburgh (UK) in 2010 and Manchester (UK) in 2012.

IJCAR 2014 is part of Federated Logic Conference (FLoC) that is itself part of Vienna Summer in Logic (VSL) and 24 workshops are affiliated with IJCAR. The Vienna Summer of Logic is a unique event organized by the Kurt Gödel Society at Vienna University of Technology from July 9 to 24, 2014.

The call for papers for IJCAR'14 invited authors to submit full papers (of 15 pages) and system descriptions (of 7 pages). There were 83 submissions (63 regular papers and 20 system descriptions) of which 37 were accepted (26 regular papers and 11 system descriptions). Each submission was assigned to at least three Program Committee members, who carefully reviewed the papers, with the help of 116 external referees. We wish to thank the Program Committee members and all their reviewers for their works and efforts in evaluating the submissions. It was a pleasure to work with all of them. The EasyChair conference management system was a great help in dealing with all aspects of putting our program and the proceedings together.

IJCAR 2014 had invited talks by Rajeev Goré (The Australian National University) and Ken McMillan (Microsoft Research). In addition, IJCAR together with other FLoC conferences, had two invited plenary talks by Véronique Cortier (Centre National de la Recherche Scientifique) and by Orna Kupferman (Hebrew University). These proceedings contain three papers and an abstract of these invited talks. We want to thank the invited speakers for contributing to the success of the IJCAR 2014.

Many people helped to make IJCAR 2014 a success. We want to thank the the conference co-chairs and the organizing committee consisting of Christian Fermüller, Stefan Hetzl and Giselle Reis, the publicity chair Morgan Deters and the workshop chair Matthias Horbach. We are also indebted to the FLoC and VSL organization committees.

Most importantly, we would like to thank all the authors for submitting their work to IJCAR 2014: we believe the outcome is an exciting technical program.

May 2014

Stéphane Demri
Deepak Kapur
Christoph Weidenbach

Organization

Program Committee

Franz Baader	TU Dresden, Germany
Peter Baumgartner	National ICT Australia
Bernhard Beckert	Karlsruhe Institute of Technology, Germany
Jasmin Christian Blanchette	TU München, Germany
Bernard Boigelot	University of Liège, Belgium
Maria Paola Bonacina	Universita' degli Studi di Verona, Italy
Agata Ciabattoni	TU Wien, Austria
Koen Claessen	Chalmers University of Technology, Sweden
Leonardo De Moura	Microsoft Research, USA
Stéphanie Delaune	CNRS, LSV, France
Stéphane Demri	CNRS, France and NYU, USA
Stephan Falke	Karlsruhe Institute of Technology (KIT), Germany
Christian Fermüller	TU Wien, Austria
Pascal Fontaine	Loria, INRIA, University of Nancy, France
Silvio Ghilardi	Università degli Studi di Milano, Italy
Jürgen Giesl	RWTH Aachen, Germany
Valentin Goranko	Technical University of Denmark
Radu Iosif	Verimag/CNRS/University of Grenoble, France
Deepak Kapur	University of New Mexico, USA
Boris Konev	The University of Liverpool, UK
Konstantin Korovin	Manchester University, UK
Daniel Kroening	Oxford University, UK
Viktor Kuncak	EPFL, Switzerland
Martin Lange	University of Kassel, Germany
Stephan Merz	Inria Lorraine, France
Aart Middeldorp	University of Innsbruck, Austria
Enric Rodríguez Carbonell	Technical University of Catalonia, Spain
Renate A. Schmidt	University of Manchester, UK
Carsten Schuermann	IT University of Copenhagen, Denmark
Roberto Sebastiani	DISI, University of Trento, Italy
Viorica Sofronie-Stokkermans	University Koblenz-Landau, Germany

Geoff Sutcliffe   University of Miami, USA
Cesare Tinelli   The University of Iowa, USA
Uwe Waldmann   MPI für Informatik, Germany
Christoph Weidenbach   MPI für Informatik, Germany
Jian Zhang   Institute of Software, Chinese Academy of
                           Sciences, China

Additional Reviewers

Alama, Jesse   Gimenez, Stéphane
Areces, Carlos   Gladisch, Christoph
Armas, Ana   Graham-Lengrand, Stéphane
Artale, Alessandro   Grebing, Sarah
Atkey, Robert   Greco, Giuseppe
Audemard, Gilles   Griggio, Alberto
Badban, Bahareh   Herda, Mihai
Baldi, Paolo   Heule, Marijn
Barrett, Clark   Hoder, Krystof
Bellodi, Elena   Horbach, Matthias
Bengtson, Jesper   Hou, Zhe
Benzmüller, Christoph   Huang, Guan-Shieng
Bezhanishvili, Nick   Hustadt, Ullrich
Bormer, Thorsten   Jacobs, Swen
Bresolin, Davide   Jovanović, Dejan
Brock-Nannestad, Taus   Kapur, Deepak
Brockschmidt, Marc   King, Timothy
Bruns, Daniel   Koopmann, Patrick
Bruse, Florian   Kop, Cynthia
Bucheli, Samuel   Kuraj, Ivan
Chen, Hong-Yi   Lammich, Peter
Conchon, Sylvain   Leitsch, Alexander
Cyriac, Aiswarya   Lellmann, Bjoern
De Nivelle, Hans   Lisitsa, Alexei
Della Monica, Dario   Liu, Jun
Demri, Stephane   Liu, Wanwei
Dietl, Werner   Lozes, Etienne
Dyckhoff, Roy   Ludwig, Michel
Eades Iii, Harley   Lutz, Carsten
Ehlers, Rüdiger   Madhavan, Ravichandhran
Enea, Constantin   Marchi, Jerusa
Erbatur, Serdar   Mccabe-Dansted, John
Ferreira, Francisco   Mclaughlin, Sean
Fiorino, Guido   Metcalfe, George
Franconi, Enrico   Momigliano, Alberto
Galmiche, Didier   Nagele, Julian

Neufeld, Eric
Nieuwenhuis, Robert
Papacchini, Fabio
Park, Sungwoo
Pelletier, Francis Jeffry
Peltier, Nicolas
Pelzer, Björn
Penczek, Wojciech
Perrussel, Laurent
Peñaloza, Rafael
Poggiolesi, Francesca
Popeea, Corneliu
Popescu, Andrei
Quaas, Karin
Ramanayake, Revantha
Reynolds, Andrew
Ringeissen, Christophe
Rubio, Albert
Schrammel, Peter
Seylan, Inanc
Simari, Gerardo
Spendier, Lara
Sternagel, Christian

Straccia, Umberto
Strassburger, Lutz
Szeider, Stefan
Tessaris, Sergio
Toman, David
Tsarkov, Dmitry
Ulbrich, Mattias
Vescovi, Michele
Walther, Dirk
Wand, Daniel
Wandelt, Sebastian
Wang, Kewen
Weller, Daniel
Wiedijk, Freek
Winkler, Sarah
Woltzenlogel Paleo, Bruno
Xu, Ke
Zankl, Harald
Zarrieß, Benjamin
Zhan, Naijun
Zhang, Heng
Zhang, Wenhui

Invited Talks
(Abstracts)

From Reachability to Temporal Specifications in Cost-Sharing Games

Guy Avni[1], Orna Kupferman[1], and Tami Tamir[2]

[1] School of Computer Science and Engineering, The Hebrew University, Jerusalem, Israel

[2] School of Computer Science, The Interdisciplinary Center, Herzliya, Israel

Abstract. Multi-agents cost-sharing games are commonly used for modeling settings in which different entities share resources. For example, the setting in which entities need to route messages in a network is modeled by a network-formation game: the network is modeled by a graph, and each agent has to select a path satisfying his reachability objective. In practice, the objectives of the entities are often more involved than reachability. The need to specify and reason about rich specifications has been extensively studied in the context of verification and synthesis of reactive systems. This paper suggests and analyzes a generalization of cost-sharing games that captures such rich specifications. In particular, we study network-formation games with regular objectives. In these games, the edges of the graph are labeled by alphabet letters and the objective of each player is a regular language over the alphabet of labels. Thus, beyond reachability, a player may restrict attention to paths that satisfy certain properties, referring, for example, to the providers of the traversed edges, the actions associated with them, their quality of service, or security. Our results show that the transition to regular objectives makes the game considerably less stable.

Electronic Voting: How Logic Can Help[*]

Véronique Cortier

LORIA - CNRS, France

Abstract. Electronic voting should offer at least the same guarantees than traditional paper-based voting systems. In order to achieve this, electronic voting protocols make use of cryptographic primitives, as in the more traditional case of authentication or key exchange protocols. All these protocols are notoriously difficult to design and flaws may be found years after their first release. Formal models, such as process algebra, Horn clauses, or constraint systems, have been successfully applied to automatically analyze traditional protocols and discover flaws. Electronic voting protocols however significantly increase the difficulty of the analysis task. Indeed, they involve for example new and sophisticated cryptographic primitives, new dedicated security properties, and new execution structures.

After an introduction to electronic voting, we describe the current techniques for e-voting protocols analysis and review the key challenges towards a fully automated verification.

[*] The research leading to these results has received funding from the European Research Council under the European Union's Seventh Framework Programme (FP7/2007-2013)/ERC grant agreement no 258865, project ProSecure.

And-Or Tableaux for Fixpoint Logics
with Converse: LTL, CTL, PDL and CPDL

Rajeev Goré

Logic and Computation Group
Research School of Computer Science
The Australian National University
rajeev.gore@anu.edu.au

Abstract. Over the last forty years, computer scientists have invented or borrowed numerous logics for reasoning about digital systems. Here, I would like to concentrate on three of them: Linear Time Temporal Logic (LTL), branching time Computation Tree temporal Logic (CTL), and Propositional Dynamic Logic (PDL), with and without converse. More specifically, I would like to present results and techniques on how to solve the satisfiability problem in these logics, with global assumptions, using the tableau method. The issues that arise are the typical tensions between computational complexity, practicality and scalability. This is joint work with Linh Anh Nguyen, Pietro Abate, Linda Postniece, Florian Widmann and Jimmy Thomson.

Structured Search and Learning

Kenneth L. McMillan

Microsoft Research

Abstract. Most modern Boolean satisfiability (SAT) solvers use conflict-driven clause learning (CDCL). In this approach, search for a model and search for a refutation by resolution are tightly coupled in a way that helps to focus search on relevant decisions and resolution on relevant deductions. Decision making narrows the search by applying arbitrary constraints. When a contradiction is reached, a "learned" fact is deduced in response. This fact generalizes the conflict and constrains future decisions. The learned fact can also be viewed as a Craig interpolant. As we will see this view allows us to generalize the notion of conflict learning in useful ways.

Satisfiability Module Theories (SMT solvers) of the lazy type apply the same paradigm to first-order decision problems with certain background theories, such as linear arithmetic or the theory of arrays. In this case, the interpolants may be validities of the theory generated by "theory solvers", but the basic conflict-driven mechanism remains the same.

A common shortcoming of these procedures, successful though they are, is that model search and conflict learning are essentially unstructured. That is, they do not take into account any modular structure that may be present in the decision problem. Decisions are made on variables regardless of their structural relationship, and consequently learned facts do not reflect the problem structure. This is in contrast to a saturation approach, in which we might order resolution so as to exploit, say, narrow tree width of the problem.

In this talk we will consider structured approaches to conflict learning. These techniques have been developed in the context of model checking, an area in which the need to exploit structure is acute. Structured learning can produce facts about reachable states of a system or summaries of procedures, which in turn can be combined to form inductive invariants. Examples of such techniques include IC3 [1] and Lazy Annotation [2].

These techniques have similar search strategies, differing primarily in their approach to computing interpolants. The approaches make different trade-offs between cost and generality, which in turn determine the usefulness of the resulting generalizations. We observe, for example, that more specialized decisions can make the learning problem easier, but possibly at the cost of reduced generality or relevance of the learned facts. Moreover, a substantial effort in generalizing the interpolants can be justified by the corresponding reduction in search.

The net effect of structured learning can be a dramatic improvement in performance, as we observe by comparing with unstructured SMT solvers on bounded software model checking problems.

References

1. Bradley, A.R.: SAT-based model checking without unrolling. In: Jhala, R., Schmidt, D. (eds.) VMCAI 2011. LNCS, vol. 6538, pp. 70–87. Springer, Heidelberg (2011)
2. McMillan, K.L.: Lazy annotation for program testing and verification. In: Touili, T., Cook, B., Jackson, P. (eds.) CAV 2010. LNCS, vol. 6174, pp. 104–118. Springer, Heidelberg (2010)

Table of Contents

Verification

Proof Theory

Modal and Temporal Reasoning

SMT and SAT

Modal Logic

Complexity

From Reachability to Temporal Specifications in Cost-Sharing Games

Guy Avni[1], Orna Kupferman[1], and Tami Tamir[2]

[1] School of Computer Science and Engineering, The Hebrew University, Jerusalem, Israel
[2] School of Computer Science, The Interdisciplinary Center, Herzliya, Israel

Abstract. Multi-agents cost-sharing games are commonly used for modeling settings in which different entities share resources. For example, the setting in which entities need to route messages in a network is modeled by a network-formation game: the network is modeled by a graph, and each agent has to select a path satisfying his reachability objective. In practice, the objectives of the entities are often more involved than reachability. The need to specify and reason about rich specifications has been extensively studied in the context of verification and synthesis of reactive systems. This paper suggests and analyzes a generalization of cost-sharing games that captures such rich specifications. In particular, we study network-formation games with regular objectives. In these games, the edges of the graph are labeled by alphabet letters and the objective of each player is a regular language over the alphabet of labels. Thus, beyond reachability, a player may restrict attention to paths that satisfy certain properties, referring, for example, to the providers of the traversed edges, the actions associated with them, their quality of service, or security. Our results show that the transition to regular objectives makes the game considerably less stable.

1   Introduction

The classical definition of a *computation* in computer science uses to the model of a Turing machine that recognizes a decidable language: once an input word is received, the machine operates on it, and eventually terminates, accepting or rejecting the word. Such a mode of operation corresponds to the use of computers for the solution of decidable problems, and there is no need to elaborate on the extensive research in theoretical computer science about this model and issues like decidability and complexity. The specification of a Turing machines is done by means of the language it recognizes. Indeed, the specification of hardware and software systems that are input-output transformers refers to the transformation they perform, for example "$z = x \cdot y$" or "the vector of strings is alphabetically sorted".

The classical definition of a computation does not capture the mode of operation of *reactive systems* [23]. Such systems maintain an on-going interaction with their environment. Operating systems, ATMs, elevators, satellites – these are all reactive systems. The computations of reactive systems need not terminate, and their specifications refer to the on-going interaction of the system with its environment, for example "every request is eventually granted" or "two requests are never granted simultaneously". Formal methods for specification, verification, and design of reactive systems have been a very active research area since the 80s.

S. Demri, D. Kapur, and C. Weidenbach (Eds.): IJCAR 2014, LNAI 8562, pp. 1–15, 2014.

The classical definition of a computation is prevalent in many areas in computer science, where users need to calculate a function or reach a certain desired goal. In particular, the classical setting in game theory is such that payoffs are being determined and paid after some finitely (often one) many rounds [32]. In recent years, we see exchange of ideas between formal methods and game theory. In one direction, the setting of a system interacting with its environment is naturally modeled by a game, where a correct system corresponds to a winning strategy for the system [35]. Beyond the relevancy of fundamental concepts from game theory, like partial observability [38] or different types of strategies [33], this gives rise also to the adoption of ideas like stability and anarchy in the context of reasoning about reactive systems. For example, [17] studies synthesis in the presence of rational environments, [7,11] study non-zero-sum games in formal methods. In the second direction, rich specification formalisms, especially quantitative ones, enables the extension of classical games to ones that consider on-going behavior. For example, [9] studies Nash Equilibria in games with ω-regular objectives, [3,10] introduce logics for specifying multi-agent systems, and [26] considers selfish on-going behaviors. Our work here belongs to this second direction, of lifting ideas from formal methods to game theory, and we focus on games corresponding to network design and formation.

Network design and formation is a fundamental well-studied problem that involves many interesting combinatorial optimization problems. In practice, network design is often conducted by multiple strategic users whose individual costs are affected by the decisions made by others. Early works on network design focus on analyzing the efficiency and fairness properties associated with different sharing rules (e.g., [24,31]). Following the emergence of the Internet, there has been an explosion of studies employing game-theoretic analysis to explore Internet applications, such as routing in computer networks and network formation [1,2,13,18]. In network-formation games (for a survey, see [40]), the network is modeled by a weighted graph. The weight of an edge indicates the cost of activating the transition it models, which is independent of the number of times the edge is used. Players have reachability objectives, each given by sets of possible source and target nodes. Players share the cost of edges used in order to fulfill their objectives. Since the costs are positive, the runs traversed by the players are simple. Under the common Shapley cost-sharing mechanism, the cost of an edge is shared evenly by the players that use it.

The players are selfish agents who attempt to minimize their own costs, rather than to optimize some global objective. In network-design settings, this would mean that the players selfishly select a path instead of being assigned one by a central authority. The focus in game theory is on the *stable* outcomes of a given setting, or the *equilibrium* points. A Nash equilibrium (NE) is a profile of the players' strategies such that no player can decrease his cost by an unilateral deviation from his current strategy, that is, assuming that the strategies of the other players do not change.[1]

Reachability objectives enable the players to specify possible sources and targets. Often, however, it is desirable to refer also to other properties of the selected paths. For example, in a *communication* setting, edges may belong to different providers,

---

[1] Throughout this paper, we focus on pure strategies and pure deviations, as is the case for the vast literature on cost-sharing games.

and a user may like to specify requirements like "all edges are operated by the same provider" or "no edge operated by AT&T is followed by an edge operated by Verizon". Edges may also have different quality or security levels (e.g., "noisy channel", "high-bandwidth channel", or "encrypted channel"), and again, users may like to specify their preferences with respect to these properties. In *planning* or in *production systems*, nodes of the network correspond to configurations, and edges correspond to the application of actions. The objectives of the players are sequences of actions that fulfill a certain plan, which is often more involved than just reachability [14]; for example "once the arm is up, do not put it down until the block is placed".

We extend network-formation games to a setting in which the players can specify regular objectives. This involves two changes of the underlying setting: First, the edges in the network are labeled by letters from a designated alphabet. Second, the objective of each player is specified by a *language* over this alphabet. Each player should select a path labeled by a word in his objective language. Thus, if we view the network as a *nondeterministic weighted finite automaton* (WFA) $\mathcal{A}$, then the set of strategies for a player with objective L is the set of accepting runs of $\mathcal{A}$ on some word in L. Accordingly, we refer to our extension as *automaton-formation games*. As in classical network-formation games, players share the cost of edges they use. Unlike the classical game, the runs selected by the players need not be simple, thus a player may traverse some edges several times. Edge costs are shared by the players, with the share being proportional to the number of times the edge is traversed. This latter issue is the main technical difference between automaton-formation and network-formation games, and as we shall see, it is very significant.

Many variants of cost-sharing games and congestion games have been studied. A generalization of the network-formation game of [2] in which players are weighted and a player's share in an edge cost is proportional to its weight is considered in [12], where it is shown that the weighted game does not necessarily have a pure NE. In a different type of congestion games, players' payments depend on the resource they choose to use, the set of players using this resource, or both [19,27,28,30]. In some of these variants a NE is guaranteed to exist while in others it is not. All these variants are different from automaton-formation games, where a player needs to select a *multiset* of resources (namely, the edges he is going to traverse) rather than a single one.

We study the theoretical and practical aspects of automaton-formation games. In addition to the general game, we consider classes of instances that have to do with the network, the specifications, or their combination. Recall that the network can be viewed as a WFA $\mathcal{A}$. We consider the following classes of WFAs: (1) *all-accepting*, in which all the states of $\mathcal{A}$ are accepting, thus its language is prefix closed (2) *uniform costs*, in which all edges have the same cost, and (3) *single letter*, in which $\mathcal{A}$ is over a single-letter alphabet. We consider the following classes of specifications: (1) *single word*, where the language of each player is a single word, (2) *symmetric*, where all players have the same objective. We also consider classes of instances that are intersections of the above classes.

Each of the restricted classes we consider corresponds to a real-life variant of the general setting. Let us elaborate below on single-letter instances. The language of an automaton over a single letter $\{a\}$ induces a subset of $\mathbb{N}$, namely the numbers $k \in \mathbb{N}$

such that the automaton accepts a^k. Accordingly, single-letter instances correspond to settings in which a player specifies possible lengths of paths. Several communication protocols are based on the fact that a message must pass a pre-defined length before reaching its destination. This includes *onion routing*, where the message is encrypted in layers [37], or *proof-of-work* protocols that are used to deter denial of service attacks and other service abuses such as spam (e.g., [16]).

We provide a complete picture of the following questions for various classes of the game (for formal definitions, see Section 2): (i) Existence of a *pure Nash equilibrium*. That is, whether each instance of the game has a profile of pure strategies that constitutes a NE. As we show, unlike the case of classical network design games, a pure NE might not exist in general automaton-formation games and even in very restricted instances of it. (ii) The complexity of finding the *social optimum* (SO). The SO is a profile that minimizes the total cost of the edges used by all players; thus the one obtained when the players obey some centralized authority. We show that for some restricted instances finding the SO can be done efficiently, while for other restricted instances, the complexity agrees with the NP-completeness of classical network-formation games. (iii) An analysis of *equilibrium inefficiency*. It is well known that decentralized decision-making may lead to solutions that are sub-optimal from the point of view of society as a whole. We quantify the inefficiency incurred due to selfish behavior according to the *price of anarchy* (PoA) [25,34] and *price of stability* (PoS) [2] measures. The PoA is the worst-case inefficiency of a Nash equilibrium (that is, the ratio between the worst NE and the SO). The PoS is the best-case inefficiency of a Nash equilibrium (that is, the ratio between the best NE and the SO). We show that while the PoA in automaton-formation games agrees with the one in classical network-formation games and is equal to the number of players, the PoS also equals the number of players, again already in very restricted instances. This is in contrast with classical network-formation games, where the PoS tends to *log* the number of players. Thus, the fact that players may choose to use edges several times significantly increases the challenge of finding a stable solution as well as the inefficiency incurred due to selfish behavior. We find this as the most technically challenging result of this work. We do manage to find structural restrictions on the network with which the social optimum is a NE.

The technical challenge of our setting is demonstrated in the seemingly easy instance in which all players have the same objective. Such *symmetric* instances are known to be the simplest to handle in all cost-sharing and congestion games studied so far. Specifically, in network-formation games, the social optimum in symmetric instances is also a NE and the PoS is 1. Moreover, in some games [21], computing a NE is PLS-complete in general, but solvable in polynomial time for symmetric instances. Indeed, once all players have the same objective, it is not conceivable that a player would want to deviate from the social-optimum solution, where each of the k players pays $\frac{1}{k}$ of the cost of the optimal solution. We show that, surprisingly, symmetric instances in AF-games are not simple at all. First, we answer negatively a question we left open in [5] and show that not only the social optimum might not be a NE, a symmetric instance need not have a NE at all. Also, the PoS is at least $\frac{k}{k-1}$, and for symmetric two-player AF games, we have that $PoS = PoA = 2$. We also show that the PoA equals the number of players already for very restricted instances.

The paper is based on our paper "Network-Formation Games with Regular Objectives" [5]. Due to the lack of space, some proofs and examples are missing and can be found in the full version.

2 Preliminaries

2.1 Automaton-Formation Games

A *nondeterministic finite weighted automaton* on finite words (WFA, for short) is a tuple $\mathcal{A} = \langle \Sigma, Q, \Delta, q_0, F, c \rangle$, where Σ is an alphabet, Q is a set of states, $\Delta \subseteq Q \times \Sigma \times Q$ is a transition relation, $q_0 \in Q$ is an initial state, $F \subseteq Q$ is a set of accepting states, and $c : \Delta \to \mathbb{R}$ is a function that maps each transition to the cost of its formation [29]. A *run* of $\mathcal{A}$ on a word $w = w_1, \ldots, w_n \in \Sigma^*$ is a sequence of states $\pi = \pi^0, \pi^1, \ldots, \pi^n$ such that $\pi^0 = q_0$ and for every $0 \le i < n$ we have $\Delta(\pi^i, w_{i+1}, \pi^{i+1})$. The run π is *accepting* iff $\pi^n \in F$. The *length* of π is n, whereas its size, denoted $|\pi|$, is the number of different transitions in it. Note that $|\pi| \le n$.

An *automaton-formation game* (AF game, for short) between k selfish players is a pair $\langle \mathcal{A}, O \rangle$, where $\mathcal{A}$ is a WFA over some alphabet Σ and O is a k-tuple of regular languages over Σ. Thus, the objective of Player i is a regular language L_i, and he needs to choose a word $w_i \in L_i$ and an accepting run of $\mathcal{A}$ on w_i in a way that minimizes his payments. The cost of each transition is shared by the players that use it in their selected runs, where the share of a player in the cost of a transition e is proportional to the number of times e is used by the player. Formally, The set of strategies for Player i is $\mathcal{S}_i = \{\pi : \pi \text{ is an accepting run of } \mathcal{A} \text{ on some word in } L_i\}$. We assume that $\mathcal{S}_i$ is not empty. We refer to the set $\mathcal{S} = \mathcal{S}_1 \times \ldots \times \mathcal{S}_k$ as the set of *profiles* of the game.

Consider a profile $P = \langle \pi_1, \pi_2, \ldots, \pi_k \rangle$. We refer to π_i as a sequence of transitions. Let $\pi_i = e_i^1, \ldots, e_i^{\ell_i}$, and let $\eta_P : \Delta \to \mathbb{N}$ be a function that maps each transition in Δ to the number of times it is traversed by all the strategies in P, taking into an account several traversals in a single strategy. Denote by $\eta_i(e)$ the number of times e is traversed in π_i, that is, $\eta_i(e) = |\{1 \le j \le \ell_i : e_i^j = e\}|$. Then, $\eta_P(e) = \sum_{i=1\ldots k} \eta_i(e)$. The *cost of player i in the profile P* is

$$cost_i(P) = \sum_{e \in \pi_i} \frac{\eta_i(e)}{\eta_P(e)} c(e). \tag{1}$$

For example, consider the WFA $\mathcal{A}$ depicted in Fig. 1. The label $e_1 : a, 1$ on the transition from q_0 to q_1 indicates that this transition, which we refer to as e_1, traverses the letter a and its cost is 1. We consider a game between two players. Player 1's objective is the language is $L_1 = \{ab^i : i \ge 2\}$ and Player 2's language is $\{ab, ba\}$. Thus, $\mathcal{S}_1 = \{\{e_1, e_2, e_2\}, \{e_1, e_2, e_2, e_2\}, \ldots\}$ and $\mathcal{S}_2 = \{\{e_3, e_4\}, \{e_1, e_2\}\}$. Consider the profile $P = \langle \{e_1, e_2, e_2\}, \{e_3, e_4\} \rangle$, the strategies in P are disjoint, and we have $cost_1(P) = 2 + 2 = 4, cost_2(P) = 1 + 3 = 4$. For the profile $P' = \langle \{e_1, e_2, e_2\}, \{e_1, e_2\} \rangle$, it holds that $\eta_1(e_1) = \eta_2(e_1)$ and $\eta_1(e_2) = 2 \cdot \eta_2(e_2)$. Therefore, $cost_1(P') = \frac{1}{2} + 2 = 2\frac{1}{2}$ and $cost_2(P') = \frac{1}{2} + 1 = 1\frac{1}{2}$.

We consider the following instances of AF games. Let $G = \langle \mathcal{A}, O \rangle$. We start with instances obtained by imposing restrictions on the WFA $\mathcal{A}$. In *one-letter* instances,

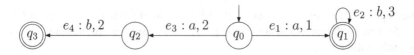

Fig. 1. An example of a WFA

$\mathcal{A}$ is over a singleton alphabet, i.e., $|\Sigma| = 1$. When depicting such WFAs, we omit the letters on the transitions. In *all-accepting* instances, all the states in $\mathcal{A}$ are accepting; i.e., $F = Q$. In *uniform-costs* instances, all the transitions in the WFA have the same cost, which we normalize to 1. Formally, for every $e \in \Delta$, we have $c(e) = 1$. We continue to restrictions on the objectives in O. In *single-word* instances, each of the languages in O consists of a single word. In *symmetric* instances, the languages in O coicide, thus the players all have the same objective. We also consider combinations on the restrictions. In particular, we say that $\langle \mathcal{A}, O \rangle$ is *weak* if it is one-letter, all states are accepting, costs are uniform, and objectives are single words. Weak instances are simple indeed – each player only specifies a length of a path he should patrol, ending anywhere in the WFA, where the cost of all transitions is the same. As we shall see, many of our hardness results and lower bounds hold already for the class of weak instances.

2.2 Nash Equilibrium, Social Optimum, and Equilibrium Inefficiency

For a profile P, a strategy π_i for Player i, and a strategy π, let $P[\pi_i \leftarrow \pi]$ denote the profile obtained from P by replacing the strategy for Player i by π. A profile $P \in \mathcal{S}$ is a *pure Nash equilibrium* (NE) if no player i can benefit from unilaterally deviating from his run in P to another run; i.e., for every player i and every run $\pi \in \mathcal{S}_i$ it holds that $cost_i(P[\pi_i \leftarrow \pi]) \geq cost_i(P)$. In our example, the profile P is not a NE, since Player 2 can reduce his payments by deviating to profile P'.

The (social) cost of a profile P, denoted $cost(P)$, is the sum of costs of the players in P. Thus, $cost(P) = \sum_{1 \leq i \leq k} cost_i(P)$. Equivalently, if we view P as a set of transitions, with $e \in P$ iff there is $\pi \in P$ for which $e \in \pi$, then $cost(P) = \sum_{e \in P} c(e)$. We denote by OPT the cost of an optimal solution; i.e., $OPT = \min_{P \in \mathcal{S}} cost(P)$. It is well known that decentralized decision-making may lead to sub-optimal solutions from the point of view of society as a whole. We quantify the inefficiency incurred due to self-interested behavior according to the *price of anarchy* (PoA) [25,34] and *price of stability* (PoS) [2] measures. The PoA is the worst-case inefficiency of a Nash equilibrium, while the PoS measures the best-case inefficiency of a Nash equilibrium. Formally,

Definition 1. *Let $\mathcal{G}$ be a family of games, and let $G \in \mathcal{G}$ be a game in $\mathcal{G}$. Let $\Upsilon(G)$ be the set of Nash equilibria of the game G. Assume that $\Upsilon(G) \neq \emptyset$.*

- *The price of anarchy of G is the ratio between the* maximal *cost of a NE and the social optimum of G. That is, $PoA(G) = \max_{P \in \Upsilon(G)} cost(P)/OPT(G)$. The price of anarchy of the family of games $\mathcal{G}$ is $PoA(\mathcal{G}) = sup_{G \in \mathcal{G}} PoA(G)$.*
- *The price of stability of G is the ratio between the* minimal *cost of a NE and the social optimum of G. That is, $PoS(G) = \min_{P \in \Upsilon(G)} cost(P)/OPT(G)$. The price of stability of the family of games $\mathcal{G}$ is $PoS(\mathcal{G}) = sup_{G \in \mathcal{G}} PoS(G)$.*

Uniform Sharing Rule: A different cost-sharing rule that could be adopted for automaton-formation games is the uniform sharing rule, according to which the cost of a transition e is equally shared by the players that traverse e, independent of the number of times e is traversed by each player. Formally, let $\kappa_P(e)$ be the number of runs that use the transition e at least once in a profile P. Then, the cost of including a transition e at least once in a run is $c(e)/\kappa_P(e)$. This sharing rule induces a potential game, where the potential function is identical to the one used in the analysis of the classical network design game [2]. Specifically, let $\Phi(P) = \sum_{e \in E} c(e) \cdot H(\kappa_P(e))$, where $H_0 = 0$, and $H_k = 1 + 1/2 + \ldots + 1/k$. Then, $\Phi(P)$ is a potential function whose value reduces with every improving step of a player, thus a pure NE exists and BRD is guaranteed to converge[2]. The similarity with classical network-formation games makes the study of this setting straightforward. Thus, throughout this paper we only consider the proportional sharing rule as defined in (1) above.

3   Properties of Automaton-Formation Games

In this section we study the theoretical properties of AF games: existence of NE and equilibrium inefficiency. We show that AF games need not have a pure Nash equilibrium. This holds already in the very restricted class of weak instances, and is in contrast with network-formation games. There, BRD converges and a pure NE always exists. We then analyze the PoS in AF games and show that there too, the situation is significantly less stable than in network-formation games.

Theorem 1. *Automaton-formation games need not have a pure NE. This holds already for the class of weak instances.*

Proof. Consider the WFA $\mathcal{A}$ depicted in Fig. 2 and consider a game with $k = 2$ players. The language of each player consists of a single word. Recall that in one-letter instances we care only about the lengths of the objective words. Let these be ℓ_1 and ℓ_2, with $\ell_1 \gg \ell_2 \gg 0$ that are multiples of 12. For example, $\ell_1 = 30000, \ell_2 = 300$. Let C_3 and C_4 denote the cycles of length 3 and 4 in $\mathcal{A}$, respectively. Let D_3 denote the path of length 3 from q_0 to q_1. Every run of $\mathcal{A}$ consists of some repetitions of these cycles possibly with one pass on D_3.

We claim that no pure NE exists in this instance. Since we consider long runs, the fact that the last cycle might be partial is ignored in the calculations below. We first show that the only candidate runs for Player 1 that might be part of a NE profile are $\pi_1 = (C_4)^{\frac{\ell_1}{4}}$ and $\pi'_1 = D_3 \cdot (C_3)^{\frac{\ell_1}{3}-1}$. If Player 1 uses both C_3 and C_4 multiple times, then, given that $\ell_1 \gg \ell_2$, he must almost fully pay for at least one of these cycles, thus, deviating to the run that repeats this fully-paid cycle is beneficial.

When Player 1 plays π_1, Player 2's best response is $\pi_2 = (C_4)^{\frac{\ell_2}{4}}$. In the profile $\langle \pi_1, \pi_2 \rangle$, Player 1 pays almost all the cost of C_4, so the players' costs are $(4 - \varepsilon, \varepsilon)$. This is not a NE. Indeed, since $\ell_2 \gg 0$, then by deviating to π'_1, the share of Player 1

---

[2] Best-response-dynamics (BRD) is a local-search method where in each step some player is chosen and plays his best-response strategy, given that the strategies of the other players do not change.

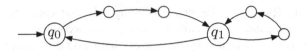

Fig. 2. A weak instance of AF games with no NE

in D_3 reduces to almost 0, and the players' costs in $\langle \pi_1', \pi_2 \rangle$, are $(3 + \varepsilon, 4 - \varepsilon)$. This profile is not a NE as Player 2's best response is $\pi_2' = D_3 \cdot (C_3)^{\frac{\ell_2}{3}-1}$. Indeed, in the profile $\langle \pi_1', \pi_2' \rangle$, the players' costs are $(4.5 - \varepsilon, 1.5 + \varepsilon)$ as they share the cost of D_3 and Player 1 pays almost all the cost of C_3. This is not a NE either, as Player 1 would deviate to the profile $\langle \pi_1, \pi_2' \rangle$, in which the players' costs are $(4 - \varepsilon, 3 + \varepsilon)$. The latter is still not a NE, as Player 2 would head back to $\langle \pi_1, \pi_2 \rangle$. We conclude that no NE exists in this game. □

The fact that a pure NE may not exist is a significant difference between standard cost-sharing games and AF games. The bad news do not end here and extend to equilibrium inefficiency. We first note that the cost of any NE is at most k times the social optimum (as otherwise, some player pays more than the cost of the SO and can benefit from migrating to his strategy in the SO). Thus, it holds that $PoS \leq PoA \leq k$. The following theorem shows that this is tight already for highly restricted instances.

Theorem 2. *The PoS in* AF *games equals the number of players. This holds already for the class of weak instances.*

Proof. We show that for every $k, \delta > 0$ there exists a simple game with k players for which the PoS is more than $k - \delta$. Given k and δ, let r be an integer such that $r > \max\{k, \frac{k-1}{\delta} - 1\}$. Consider the WFA $\mathcal{A}$ depicted in Fig. 3. Let $L = \langle \ell_1, \ell_2, \ldots, \ell_k \rangle$ for $\ell_2 = \ldots = \ell_k$ and $\ell_1 \gg \ell_2 \gg r$ denote the lengths of the objective words. Thus, Player 1 has an 'extra-long word' and the other $k - 1$ players have words of the same, long, length. Let C_r and C_{r+1} denote, respectively, the cycles of length r and $r + 1$ to the right of q_0. Let D_r denote the path of length r from q_0 to q_1, and let D_{kr} denote the 'lasso' consisting of the kr-path and the single-edge loop to the left of q_0.

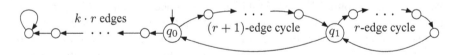

Fig. 3. A weak instance of AF games for which $PoS = k$

The social optimum of this game is to buy C_{r+1}. Its cost is $r + 1$. However, as we show, the profile P in which all players use D_{kr} is the only NE in this game. We first show that P is a NE. In this profile, Player 1 pays $r + 1 - \varepsilon$ and each other player pays $r + \varepsilon/(k-1)$. No player will deviate to a run that includes edges from the right side of $\mathcal{A}$. Next, we show that P is the only NE of this game: Every run on the right side of $\mathcal{A}$ consists of some repetitions of C_{r+1} and C_r, possibly with one traversal of D_r.

Since we consider long runs, the fact that the last cycle might be partial is ignored in the calculations below.

In the social optimum profile, Player 1 pays $r + 1 - \varepsilon$ and each of the other players pays $\varepsilon/(k - 1)$. The social optimum is not a NE as Player 1 would deviate to $D_r \cdot C_r^*$ and will reduce his cost to $r + \varepsilon'$. The other players, in turn, will also deviate to $D_r \cdot C_r^*$. In the profile in which they are all selecting a run of the form $D_r \cdot C_r^*$, Player 1 pays $r + r/k - \varepsilon > r + 1$ and prefers to return to C_{r+1}^*. The other players will join him sequentially, until the non-stable social optimum is reached. Thus, no NE that uses the right part of $\mathcal{A}$ exists. Finally, it is easy to see that no run that involves edges from both the left and right sides of $\mathcal{A}$ or includes both C_{r+1} and C_r can be part of a NE.

The cost of the NE profile is $kr+1$ and the PoS is therefore $\frac{kr+1}{r+1} = k - \frac{k-1}{r+1} > k - \delta$.

$\square$

4   Computational Complexity Issues in AF Games

In this section we study the computational complexity of two problems: finding the cost of the social optimum and finding the best-response of a player. Recall that the social optimum (SO) is a profile that minimizes the total cost the players pay. It is well-known that finding the social optimum in a network-formation game is NP-complete. We show that this hardness is carried over to simple instances of AF games. On the positive side, we identify non-trivial classes of instances, for which it is possible to compute the SO efficiently. The other issue we consider is the complexity of finding the best strategy of a single player, given the current profile, namely, the best-response of a player. In network-formation games, computing the best-response reduces to a shortest-path problem, which can be solved efficiently. We show that in AF games, the problem is NP-complete.

The proofs of the following theorems can be found in the full version. The reductions we use are from the set-cover problem, where choice of sets are related to choice of transitions.

Theorem 3. *Finding the value of the social optimum in* AF *games is NP-complete. Moreover, finding the social optimum is NP-complete already in single-worded instances that are also uniform-cost and are either single-lettered or all-accepting.*

The hardness results in Theorem 3 for single-word specification use one of two properties: either there is more than one letter, or not all states are accepting. We show that finding the SO in instances that have both properties can be done efficiently, even for specifications with arbitrary number of words.

For a language L_i over $\Sigma = \{a\}$, let $short(i) = \min_j\{a^j \in L_i\}$ denote the length of the shortest word in L_i. For a set O of languages over $\Sigma = \{a\}$, let $\ell_{max}(O) = \max_i short(i)$ denote the length of the longest shortest word in O. Clearly, any solution, in particular the social optimum, must include a run of length $\ell_{max}(O)$. Thus the cost of the social optimum is at least the cost of the cheapest run of length $\ell_{max}(O)$. Moreover, since the WFA is single-letter and all-accepting, the other players can choose runs that are prefixes of this cheapest run, and no additional transitions should be acquired. We show that finding the cheapest such run can be done efficiently.

Theorem 4. *The cost of the social optimum in a single-letter all-accepting instance* $\langle \mathcal{A}, O \rangle$ *is the cost of the cheapest run of length* $\ell_{max}(O)$. *Moreover, this cost can be found in polynomial time.*

We turn to prove the hardness of finding the best-response of a player. Our proof is valid already for a single player that needs to select a strategy on a WFA that is not used by other players (one-player game).

Theorem 5. *Finding the best-response of a player in* AF *games is NP-complete.*

5   Tractable Instances of AF Games

In the example in Theorem 1, Player 1 deviates from a run on the shortest (and cheapest) possible path to a run that uses a longer path. By doing so, most of the cost of the original path, which is a prefix of the new path and accounts to most of its cost, goes to Player 2. We consider *semi-weak* games in which the WFA is uniform-cost, all-accepting, and single-letter, but the objectives need not be a single word. We identify a property of such games that prevents this type of deviation and which guarantees that the social optimum is a NE. Thus, we identify a family of AF games in which a NE exists, finding the SO is easy, and the PoS is 1.

Definition 2. *Consider a semi-weak game* $\langle \mathcal{A}, O \rangle$. *A lasso is a path* $u \cdot v$, *where* u *is a simple path that starts from the initial state and* v *is a simple cycle. A lasso* ν *is minimal in* $\mathcal{A}$ *if* $\mathcal{A}$ *does not have shorter lassos. Note that for minimal lassos* $u \cdot v$, *we have that* $u \cap v = \emptyset$. *We say that* $\mathcal{A}$ *is* resistant *if it has no cycles or there is a minimal lasso* $\nu = u \cdot v$ *such that for every other lasso* ν' *we have* $|u \setminus \nu'| + |v| \leq |\nu' \setminus \nu|$.

Consider a resistant weak game $\langle \mathcal{A}, O \rangle$. In order to prove that the social optimum is a NE, we proceed as follows. Let ν be the lasso that is the witness for the resistance of $\mathcal{A}$. We show that the profile S^* in which all players choose runs that use only the lasso ν or a prefix of it, is a NE. The proof is technical and we go over all the possible types of deviations for a player and use the weak properties of the network along with its resistance. By Theorem 4, the cost of the profile is the SO. Hence the following. The full proof can be found in full version.

Theorem 6. *For resistent semi-weak games, the social optimum is a NE.*

A corollary of Theorem 6 is the following:

Corollary 1. *For resistant semi-weak games, we have PoS*= 1.

We note that resistance can be defined also in WFAs with non-uniform costs, with $cost(\nu)$ replacing $|\nu|$. Resistance, however, is not sufficient in the *slightly* stronger model where the WFA is single-letter and all-accepting but not uniform-cost. Indeed, given k, we show a such a game in which the PoS is kx, for a parameter x that can be arbitrarily close to 1. Consider the WFA A in Fig. 5. Note that A has a single lasso and is thus a resistant WFA. The parameter ℓ_1 is a function of x, and the players' objectives are single words of lengths $\ell_1 \gg \ell_2 \gg \ldots \gg \ell_k \gg 0$. Similar to the proof of

Theorem 2, there is only one NE in the game, which is when all players choose the left chain. The social optimum is attained when all players use the self-loop, and thus for a game in this family, $PoS = \frac{k \cdot x}{1}$. Since x tends to 1, we have $PoS = k$ for resistant all-accepting single-letter games. The proof can be found in the full version.

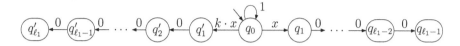

Fig. 4. A resistant all-accepting single-letter game in which the PoS tend to k

6   Surprises in Symmetric Instances

In this section we consider the class of symmetric instances, where all players share the same objective, that is, there exists a language L, such that for all $1 \leq i \leq k$, we have $L_i = L$. In such instances it is tempting to believe that the social optimum is also a NE, as all players evenly share the cost of the solution that optimizes their common objective. While this is indeed the case in all known symmetric games, we show that, surprisingly, this is not valid for AF-games, in fact already for the class of one-letter, all accepting, unit-cost and single-word instances.

We start, however, with general symmetric instances, and show that a NE need not exist.

Theorem 7. *Symmetric instances of* AF*-games need not have a pure NE.*

Proof. Consider a WFA $\mathcal{A}$ consisting of a single accepting state with two self loops, labeled $(a, 1)$ and $(b, \frac{5}{14} - \epsilon)$. Let ℓ_1 and ℓ_2 be such that $0 \ll \ell_2 \ll \ell_1$. We define $L = a^6 + ab^{\ell_1} + aab^{\ell_2} + aaab$. We denote the 4 strategies available to each of the players by A, B, C, and D, with $A = (6, 0)$ indicating 6 uses of the a transition and 0 uses of the b transition, $B = (1, \ell_1)$, $C = (2, \ell_2)$, and $D = (3, 1)$.

In order to show that there is no NE, we only have to show that the four profiles in which the players follow the same strategy are not a NE. Indeed, it is easy to see that for every other profile, one of the players would deviate to one of these profiles. Now, in profile $\langle A, A \rangle$ both players pay $\frac{1}{2}$ as they split the cost of the a-transition evenly. This is not a NE as Player 1 (or, symmetrically, Player 2) would deviate to $\langle B, A \rangle$, where he pays $\frac{1}{7}$ for the a-transition and the full price of the b-transition, which is $\frac{5}{14} - \epsilon$, thus he pays $\frac{1}{2} - \epsilon$.

In profile $\langle B, B \rangle$, both players pay $\frac{1}{2}$ for the a-transition plus $\frac{5}{2 \cdot 14} - \epsilon$ for the b-transition, which sums to $0.678 - \epsilon$. This is not a NE, as Player 1 would deviate to $\langle C, B \rangle$, where he pays $\frac{2}{3}$ for the a-transition and, as $\ell_2 \ll \ell_1$, only ϵ for the b-transition.

In profile $\langle C, C \rangle$, again both players pay $0.678 - \epsilon$. By deviating to $\langle D, C \rangle$, Player 1 reduces his payment to $\frac{3}{5} + \epsilon$. Finally, in profile $\langle D, D \rangle$, both players pay $0.678 - \epsilon$ and when deviating to $\langle A, D \rangle$, Player 1 reduces his payment to $\frac{6}{9}$.

We continue to study the PoS. Before we show that the PoS can be larger than 1, let us elaborate on the PoA. It is easy to see that in symmetric AF games, we have $PoA = k$.

This bound is achieved, as in the classic network-formation game, by a network with two parallel edges labeled by a and having costs k and 1. The players all have the same specification $L = \{a\}$. The profile in which all players select the expensive path is a NE. We show that $PoA = k$ is achieved even for weak symmetric instances.

Theorem 8. *The PoA equals the number of players, already for weak symmetric instances.*

Proof. We show a lower bound of k. The example is a generalization of the PoA in cost sharing games [2]. For k players, consider the weak instance depicted in Fig. 6, where all players have the length k. Intuitively, the social optimum is attained when all players use the loop $\langle q_0, q_0 \rangle$ and thus $OPT = 1$. The worst NE is when all players use the run $q_0 q_1 \ldots q_k$, and its cost is clearly k. Formally, there are two NEs in the game:

- The cheap NE is when all players use the loop $\langle q_0, q_0 \rangle$. This is indeed a NE because if a player deviates, he must buy at least the transition $\langle q_0, q_1 \rangle$. Thus, he pays at least 1, which is higher than $\frac{1}{k}$, which is what he pays when all players use the loop.
- The expensive NE is when all players use the run $q_0, q_1, \ldots, q_k$. This is a NE because a player has two options to deviate. Either to the run that uses only the loop, which costs 1, or to a run that uses the loop and some prefix of $q_0, q_1, \ldots, q_k$, which costs at least $1 + \frac{1}{k}$. Since he currently pays 1, he has no intention of deviating to either runs.

Since the cheap NE costs 1 and the expensive one costs k, we get $PoA = k$. □

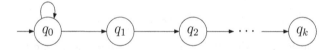

Fig. 5. The WFA $\mathcal{A}$ for which a symmetric game with $|L| = 1$ achieves $PoA = k$

We now turn to the PoS analysis. We first demonstrate the anomaly of having $PoS > 1$ with the two-player game appearing in Fig. 6. All the states in the WFA $\mathcal{A}$ are accepting, and the objectives of both players is a single long word. The social optimum is when both players traverse the loop q_0, q_1, q_0. Its cost is $2 + \epsilon$, so each player pays $1 + \frac{\epsilon}{2}$. This, however, is not a NE, as Player 1 (or, symmetrically, Player 2) prefers to deviate to the run $q_0, q_1, q_1, q_1, \ldots$, where he pays the cost of the loop q_1, q_1 and his share in the transition from q_0 to q_1. We can choose the length of the objective word and ϵ so that this share is smaller than $\frac{\epsilon}{2}$, justifying his deviation. Note that the new situation is not a NE either, as Player 2, who now pays 2, is going to join Player 1, resulting in an unfortunate NE in which both players pay 1.5.

It is not hard to extend the example from Fig. 6 to $k > 2$ players by changing the 2-valued transition to k, and adjusting ϵ and the lengths of the players accordingly. The social optimum and the only NE are as in the two-player example. Thus, the PoS in the resulting game is $1 + \frac{1}{k}$.

A higher lower bound of $1 + \frac{1}{k-1}$ is shown in the following theorem. Although both bounds tend to 1 as k grows to infinity, this bound is clearly stronger. Also, for $k = 2$, the bound $PoS = 1 + \frac{1}{k-1} = 2$ is tight. We conjecture that $\frac{k}{k-1}$ is tight for every $k > 2$.

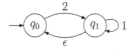

Fig. 6. The WFA $\mathcal{A}$ for which the SO in a symmetric game is not a NE

Theorem 9. *In a symmetric k-player game, the PoS is at least $\frac{k}{k-1}$.*

Proof. For $k \geq 2$, we describe a family of symmetric games for which the PoS tends to $\frac{k}{k-1}$. For $n \geq 1$, the game $G_{\epsilon,n}$ uses the WFA that is depicted in Figure 7. Note that this is a one-letter instance in which all states are accepting. The players have an identical specification, consisting of a single word w of length $\ell \gg 0$. We choose ℓ and $\epsilon = \epsilon_0 > \ldots > \epsilon_{n-1}$ as follows. Let $C_0, \ldots, C_n$ denote, respectively, the cycles with costs $(k^n + \epsilon_0), (k^{n-1} + \epsilon_1), \ldots, (k + \epsilon_{n-1}), 1$. Let $r_0, \ldots, r_n$ be lasso-runs on w that end in $C_0, \ldots, C_n$, respectively. Consider $0 \leq i \leq n - 1$ and let P_i be the profile in which all players choose the run r_i. We choose ℓ and ϵ_i so that Player 1 benefits from deviating from P_i to the run r_{i+1}, thus P_i is not a NE. Note that by deviating from r_i to r_{i+1}, Player 1 pays the same amount for the path leading to C_i. However, his share of the loop C_i decreases drastically as he uses the k^{n-i}-valued transition only once whereas the other players use it close to ℓ times. On the other hand, he now buys the loop C_{i+1} by himself. Thus, the change in his payment change is $\frac{1}{k} \cdot (k^{n-i} + \epsilon_i) - (\epsilon' + k^{n-(i+1)} + \epsilon_{i+1})$. We choose ϵ_{i+1} and ℓ so that $\frac{\epsilon_i}{k} > \epsilon' + \epsilon_{i+1}$, thus the deviation is beneficial.

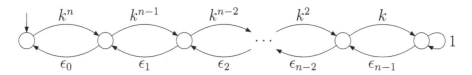

Fig. 7. The network of the identical-specification game $G_{\epsilon,n}$, in which PoS tends to $\frac{k}{k-1}$

We claim that the only NE is when all players use the run r_n. Indeed, it is not hard to see that every profile in which a player selects a run that is not from $r_0, \ldots, r_n$ cannot be a NE. Also, a profile in which two players select runs r_i and r_j, for $1 \leq i < j \leq n$, cannot be a NE as the player using r_i can decreases his payment by joining the other player in r_j. Finally, by our selection of $\epsilon_1, \ldots, \epsilon_n$, and ℓ, every profile in which all the players choose the run r_i, for $0 \leq i \leq n - 1$, is not a NE.

Clearly, the social optimum is attained when all players choose the run r_0, and its cost is $k^n + \epsilon$. Since the cost of the only NE in the game is $\sum_{0 \leq i \leq n} k^{n-i}$, the PoS in this family of games tends to $\frac{k}{k-1}$ as n grows to infinity and ϵ to 0. $\square$

Finally, we note that our hardness result in Theorem 5 implies that finding the social optimum in a symmetric AF-game is NP-complete. Indeed, since the social optimum is the cheapest run on some word in L, finding the best-response in a one-player game is equivalent to finding the social optimum in a symmetric game. This is contrast with other cost-sharing and congestion game (e.g. [21], where the social optimum in symmetric games can be computed using a reduction to max-flow).

References

1. Albers, S., Elits, S., Even-Dar, E., Mansour, Y., Roditty, L.: On Nash Equilibria for a Network Creation Game. In: Proc. 17th SODA, pp. 89–98 (2006)
2. Anshelevich, E., Dasgupta, A., Kleinberg, J., Tardos, É., Wexler, T., Roughgarden, T.: The Price of Stability for Network Design with Fair Cost Allocation. SIAM J. Comput. 38(4), 1602–1623 (2008)
3. Alur, R., Henzinger, T.A., Kupferman, O.: Alternating-time temporal logic. Journal of the ACM 49(5), 672–713 (2002)
4. Aminof, B., Kupferman, O., Lampert, R.: Reasoning about online algorithms with weighted automata. ACM Transactions on Algorithms 6(2) (2010)
5. Avni, G., Kupferman, O., Tamir, T.: Network-formation games with regular objectives. In: Muscholl, A. (ed.) FOSSACS 2014 (ETAPS). LNCS, vol. 8412, pp. 119–133. Springer, Heidelberg (2014)
6. Alpern, B., Schneider, F.B.: Recognizing safety and liveness. Distributed Computing 2, 117–126 (1987)
7. Brihaye, T., Bruyère, V., De Pril, J., Gimbert, H.: On subgame perfection in quantitative reachability games. Logical Methods in Computer Science 9(1) (2012)
8. Chatterjee, K.: Nash equilibrium for upward-closed objectives. In: Ésik, Z. (ed.) CSL 2006. LNCS, vol. 4207, pp. 271–286. Springer, Heidelberg (2006)
9. Chatterjee, K., Henzinger, T.A., Jurdzinski, M.: Games with secure equilibria. Theoretical Computer Science 365(1-2), 67–82 (2006)
10. Chatterjee, K., Henzinger, T.A., Piterman, N.: Strategy logic. In: Caires, L., Vasconcelos, V.T. (eds.) CONCUR 2007. LNCS, vol. 4703, pp. 59–73. Springer, Heidelberg (2007)
11. Chatterjee, K., Majumdar, R., Jurdziński, M.: On Nash equilibria in stochastic games. In: Marcinkowski, J., Tarlecki, A. (eds.) CSL 2004. LNCS, vol. 3210, pp. 26–40. Springer, Heidelberg (2004)
12. Chen, H., Roughgarden, T.: Network Design with Weighted Players. Theory of Computing Systems 45(2), 302–324 (2009)
13. Correa, J.R., Schulz, A.S., Stier-Moses, N.E.: Selfish Routing in Capacitated Networks. Mathematics of Operations Research 29, 961–976 (2004)
14. Daniele, M., Giunchiglia, F., Vardi, M.Y.: Improved automata generation for linear temporal logic. In: Halbwachs, N., Peled, D.A. (eds.) CAV 1999. LNCS, vol. 1633, pp. 249–260. Springer, Heidelberg (1999)
15. Droste, M., Kuich, W., Vogler, H. (eds.): Handbook of Weighted Automata. Springer (2009)
16. Dwork, C., Naor, M.: Pricing via Processing or Combatting Junk Mail. In: Brickell, E.F. (ed.) CRYPTO 1992. LNCS, vol. 740, pp. 139–147. Springer, Heidelberg (1993)
17. Fisman, D., Kupferman, O., Lustig, Y.: Rational synthesis. In: Esparza, J., Majumdar, R. (eds.) TACAS 2010. LNCS, vol. 6015, pp. 190–204. Springer, Heidelberg (2010)
18. Fabrikant, A., Luthra, A., Maneva, E., Papadimitriou, C., Shenker, S.: On a network creation game. In: Proc. 22nd PODC, pp. 347–351 (2003)
19. Feldman, M., Tamir, T.: Conflicting Congestion Effects in Resource Allocation Games. Journal of Operations Research 60(3), 529–540 (2012)
20. von Falkenhausen, P., Harks, T.: Optimal Cost Sharing Protocols for Scheduling Games. In: Proc. 12th EC, pp. 285–294 (2011)

21. Fabrikant, A., Papadimitriou, C., Talwar, K.: The Complexity of Pure Nash Equilibria. In: Proc. 36th STOC, pp. 604–612 (2004)
22. de Giacomo, G., Vardi, M.Y.: Automata-Theoretic Approach to Planning for Temporally Extended Goals. In: Biundo, S., Fox, M. (eds.) ECP 1999. LNCS, vol. 1809, pp. 226–238. Springer, Heidelberg (2000)
23. Harel, D., Pnueli, A.: On the development of reactive systems. In: LMCS. NATO Advanced Summer Institutes, vol. F-13, pp. 477–498. Springer (1985)
24. Herzog, S., Shenker, S., Estrin, D.: Sharing the "Cost" of Multicast Trees: An Axiomatic Analysis. IEEE/ACM Transactions on Networking (1997)
25. Koutsoupias, E., Papadimitriou, C.: Worst-case Equilibria. Computer Science Review 3(2), 65–69 (2009)
26. Kupferman, O., Tamir, T.: Coping with selfish on-going behaviors. Information and Computation 210, 1–12 (2012)
27. Mavronicolas, M., Milchtaich, I., Monien, B., Tiemann, K.: Congestion Games with Player-specific Constants. In: Kučera, L., Kučera, A. (eds.) MFCS 2007. LNCS, vol. 4708, pp. 633–644. Springer, Heidelberg (2007)
28. Milchtaich, I.: Weighted Congestion Games With Separable Preferences. Games and Economic Behavior 67, 750–757 (2009)
29. Mohri, M.: Finite-state transducers in language and speech processing. Computational Linguistics 23(2), 269–311 (1997)
30. Monderer, D., Shapley, L.: Potential Games. Games and Economic Behavior 14, 124–143 (1996)
31. Moulin, H., Shenker, S.: Strategyproof Sharing of Submodular Costs: Budget Balance Versus Efficiency. Journal of Economic Theory 18, 511–533 (2001)
32. Nisan, N., Roughgarden, T., Tardos, E., Vazirani, V.V.: Algorithmic Game Theory. Cambridge University Press (2007)
33. Nerode, A., Yakhnis, A., Yakhnis, V.: Concurrent programs as strategies in games. In: Proc. Logic from Computer Science, pp. 405–480 (1992)
34. Papadimitriou, C.: Algorithms, games, and the internet (Extended abstract). In: Orejas, F., Spirakis, P.G., van Leeuwen, J. (eds.) ICALP 2001. LNCS, vol. 2076, pp. 1–3. Springer, Heidelberg (2001)
35. Pnueli, A., Rosner, R.: On the synthesis of a reactive module. In: Automata, Languages and Programming, vol. 372, pp. 652–671. Springer, Heidelberg (1989)
36. Paes Leme, R., Syrgkanis, V., Tardos, E.: The curse of simultaneity. In: Innovations in Theoretical Computer Science (ITCS), pp. 60–67 (2012)
37. Reed, M.G., Syverson, P.F., Goldschlag, D.M.: Anonymous Connections and Onion Routing. IEEE J. on SAC, Issue on Copyright and Privacy Protection (1998)
38. Reif, J.H.: The complexity of two-player games of incomplete information. Journal of Computer and Systems Science 29, 274–301 (1984)
39. Rosenthal, R.W.: A Class of Games Possessing Pure-Strategy Nash Equilibria. International Journal of Game Theory 2, 65–67 (1973)
40. Tardos, E., Wexler, T.: Network Formation Games and the Potential Function Method. In: Algorithmic Game Theory. Cambridge University Press (2007)
41. Vöcking, B.: In: Nisan, N., Roughgarden, T., Tardos, E., Vazirani, V. (eds.) Algorithmic Game Theory: Selfish Load Balancing, ch. 20. Cambridge University Press (2007)

Electronic Voting: How Logic Can Help*

Véronique Cortier

LORIA - CNRS, France

Abstract. Electronic voting should offer at least the same guarantees than traditional paper-based voting systems. In order to achieve this, electronic voting protocols make use of cryptographic primitives, as in the more traditional case of authentication or key exchange protocols. All these protocols are notoriously difficult to design and flaws may be found years after their first release. Formal models, such as process algebra, Horn clauses, or constraint systems, have been successfully applied to automatically analyze traditional protocols and discover flaws. Electronic voting protocols however significantly increase the difficulty of the analysis task. Indeed, they involve for example new and sophisticated cryptographic primitives, new dedicated security properties, and new execution structures.

After an introduction to electronic voting, we describe the current techniques for e-voting protocols analysis and review the key challenges towards a fully automated verification.

1 Context

Electronic voting promises a convenient and efficient way for collecting and tallying votes, avoiding human counting errors. Several countries now use electronic voting for politically binding elections. This is for example the case of Argentina, United States, Norway, Canada, or France. However electronic voting also causes controversy. Indeed these systems have been shown to be vulnerable to attacks. For example, the Diebold machines as well as the electronic machines used in India have been attacked [44,58]. Consequently, some countries like Germany, Netherlands, or the United Kingdom have stopped electronic voting, at least momentarily [47].

Electronic voting covers two distinct families of voting systems: voting machines and Internet voting. Voting machines are computers placed at polling stations. They provide an interface for the voters to cast their vote and they process the ballots. Internet voting do not need physical polling stations: voters may simply vote using their own device (computers, smartphones, etc.) from home. In this paper we focus on Internet voting.

Internet voting raises several security challenges. Firstly, since votes need to be sent through the Internet, they obviously cannot be sent in clear. A simple solution would therefore to have all the voters encrypt their votes with the key of the voting server. At the end of the election, the server can then simply decrypt all the votes and announce the result. This is however not at all satisfactory since voters have no privacy with respect

* The research leading to these results has received funding from the European Research Council under the European Union's Seventh Framework Programme (FP7/2007-2013) / ERC grant agreement no 258865, project ProSecure.

S. Demri, D. Kapur, and C. Weidenbach (Eds.): IJCAR 2014, LNAI 8562, pp. 16–25, 2014.

to the voting authority who could easily learn the vote of everyone. Moreover, such a solution offers no transparency: voters have no way to check that the announced outcome corresponds to the votes casted by voters. Therefore, two main security properties are put forward in the context of Internet voting: *confidentiality* and *verifiability*.

- Confidentiality or vote privacy ensures that no one can learn someone else' vote. Stronger than vote privacy is *receipt-freeness*: a voter should not be able to prove how she voted, even if she is willing to. In particular, she should not be given a receipt that can prove to a third-party for who she voted. This is to prevent vote buying for example. Even stronger than receipt-freeness is *coercion resistance*: even if a voter is temporarily under the control of a coercer, she should be able to cast the vote of her choice. This is typically achieved by letting voters re-vote (without the coercer being able to notice it).
- Verifiability ensures that anyone can check that the final result corresponds to the votes. In the literature, verifiability is typically split into sub-properties: *individual verifiability* states that a voter can check that her ballot appears on the bulletin board while *universal verifiability* states that the announced result corresponds to the ballots on the bulletin board. An additional property is *eligible verifiability*: only legitimate voters can vote, at most once. Of course, all these three properties are highly desirable.

A voting system should ensure all these properties despite the presence of attackers who may intercept communications and control some of the voters. Ideally, these properties should also hold when some voting authorities are corrupted too since it is desirable that voters can trust the result without having to trust all the authorities.

There is therefore a need for rigorous foundations for formalizing and reasoning about security of voting systems. In a slightly different context, formal methods have shown their usefulness in the analysis of security protocols. This line of research has started in the late 70's with the seminal work of Dolev and Yao [42] and Even and Goldreich [43]. Since then, many decision procedures have been developed to automatically analyse security properties such as authentication or confidentiality. Current leading tools include e.g. ProVerif [20,21], Avispa [11], and Scyther [36]. They have been successfully applied to protocols of the literature as well as well-deployed protocols such as SSL and Kerberos, yielding the discoveries of flaws. A famous flaw is the "man-in-the-middle" attack found by Lowe [51] on the Needham-Schroeder asymmetric key protocol. More recently, an automated analysis [23] proved most of the secure tokens implementing the PKCS#11 standard to be broken. Similarly, a flaw was discovered using the Avispa tool on the Single-Sign-On protocol [12], used by many sites including Google.

Despite the similarities between standard security protocols and voting protocols, the current analysis techniques for standard protocols do not apply very well to voting systems. This is due to two main reasons. First, the cryptographic primitives used in e-voting are often ad-hoc and more complex than standard encryption and signatures. Second, privacy properties such as vote privacy or coercion-resistance are typically expressed as equivalence properties while the techniques developed so far mostly apply to reachability properties. We survey here the particularities of Internet voting and describe the current limitations of existing techniques.

2   Existing Systems for Internet Voting

We first start by a short overview of some existing voting systems. This list is not meant to be exhaustive. Many systems used by companies are proprietary and there is few information available. We focus here on publicly available Internet voting systems, designed for achieving both privacy and verifiability.

Helios [8] is based on a protocol proposed by Cramers *et al* [35] with a variant proposed by Benaloh [16]. It has been used at the University of Louvain-la-Neuve to elect its president (recteur) and also in student elections. The IACR (International Association for Cryptologic Research) now uses Helios to elect its board, since 2010 [1]. It makes use of homomorphic encryption and zero-knowledge proofs. Helios has been proved to offer ballot privacy [33,17,18] (provided some fix is implemented [33]) and verifiability [32]. Several variants of Helios have then been proposed to enforce more properties. For example, [37] is a variant of Helios that guarantees *everlasting privacy*, that is, vote privacy is guaranteed even if the keys of the election get broken after the election. Belenios [32] offers better verifiability, in particular even if the election server is corrupted. However, Helios is not receipt-free (nor its variants): a voter may prove how she voted. It should therefore be used in low-coercion environment only.

Civitas [29] is one of the only implemented scheme (if not the only one) that offers both verifiability and coercion-resistance. It makes use of plaintext equivalence tests, re-encryption and mixnets. Civitas is still quite complex, both in terms of usability and computational complexity. It is therefore still unclear whether it is scalable to large elections.

Norwegian Protocol [46]. Norway has conducted Internet voting trials during the parliamentary election of 2013 and 2011. For the last election in 2013, about 250 000 voters of twelve selected municipalities were offered the possibility to use Internet voting [2]. The underlying voting protocol is developed by Scytl [3,4] and is designed for both privacy and verifiability: voters are given a receipt that allow them to check that their vote has been counted, under some rather strong trust assumptions.

Several more academic voting protocols have been proposed in the literature such as the FOO protocol [45] or the Okamoto protocol [53].

3   Cryptographic Primitives

Different formal models have been designed to reason about security protocols. Most popular ones include process algebra (e.g. CSP [55], applied-pi [5], spi-calculus [7]), strand spaces [56], Horn clauses [19], or constraint systems [52,30]. They all have in common the fact that messages are represented by *terms*.

3.1   Terms

Given a signature $\mathcal{F}$, that is, a finite set of function symbols with their arity, given a set of variables $\mathcal{X}$, the set of terms $T(\mathcal{F}, \mathcal{X})$ is defined by the following grammar:

$$t, t_1, \ldots, t_n ::= x \mid f(t_1, \ldots, t_n) \qquad x \in \mathcal{X}$$

For example, a typical signature for security protocols is

$$\mathcal{F}_{enc} = \{enc, dec, pair, proj_1, proj_2\}$$

The function symbol enc represents encryption with associated decryption operator dec while pair represents concatenation with associated projectors $proj_1$ and $proj_2$. The properties of the primitives are then expressed through an *equational theory*. For (symmetric) encryption and concatenation, the usual equations are the following ones:

$$dec(enc(x, y), y) = x$$
$$proj_1(pair(x, y)) = x$$
$$proj_2(pair(x, y)) = y$$

For example, $proj_2(dec(enc(pair(a, n), k), k) = n$.

The equational theories are rather simple and belong to the class of *subterm convergent theories* [6]: they are convergent and the right member of an equational is always a subterm of the left member, or a constant. Deciding secrecy or authentication properties have been shown to be decidable both for passive [6] and active adversaries [15], for a bounded number of sessions. Some tools such as ProVerif [20] or Akiss [25] can handle arbitrary theories (with no termination guarantee of course) and typically behave well for subterm convergent theories.

3.2 Equational Theories for e-voting

Cryptographic primitives for e-voting systems are however more complex than standard primitives such as encryption or signatures. We review here some examples.

A rather standard primitive is blind signature, used for example in the FOO protocol [45]. While signatures are typically designed to be non malleable, blind signatures support some form of malleability. In FOO, voters send a blinded version of their vote to the voting authority, get it signed and then retrieve the signature of the authority on their (unblinded) vote. This property can be formalised as follows [41]:

$$unblind(sign(blind(x, z), y), z) = sign(x, y)$$

This equation means intuitively that knowing the blinding factor and the signature of a blinded message, anyone can compute the signature of the original message.

Another example comes from the Helios protocol described in Section 2. This protocol involves homomorphic encryption, that is, the combination of two encrypted votes yields the encryption of the sum of the votes. This property is at the heart of the Helios protocol since anyone can combine the votes to obtain the result (in an encrypted form). This homomorphic property can be expressed by the following equation:

$$aenc(v_1, r_1, pk) * aenc(v_2, r_2, pk) = aenc(v_1 + v_2, r_1.r_2, pk)$$

where $*$, $+$, and $.$ are associative and commutative functional symbols. Note that aenc is a ternary symbol that represents (randomized) asymmetric encryption. The second argument r represents the randomness used for encrypting. Using randomized encryption

$$\text{proj}_1(\text{pair}(x, y)) = x \tag{1}$$

$$\text{proj}_2(\text{pair}(x, y)) = y \tag{2}$$

$$\text{dec}(\text{aenc}(x_{plain}, x_{rand}, \text{pk}(x_{sk})), x_{sk}) = x_{plain} \tag{3}$$

$$\text{dec}(\text{blind}(\text{aenc}(x_{plain}, x_{rand}, \text{pk}(x_{sk})), x_{blind}), x_{sk}) = \text{blind}(x_{plain}, x_{blind}) \tag{4}$$

$$\text{aenc}(x_{pl}, x_{rand}, x_{pub}) \circ \text{aenc}(y_{pl}, y_{rand}, x_{pub}) =$$
$$\text{aenc}(x_{pl} \diamond y_{pl}, x_{rand} * y_{rand}, x_{pub}) \tag{5}$$

$$\text{renc}(\text{aenc}(x_{plain}, x_{rand}, \text{pk}(x_{sk})), y_{sk}) =$$
$$\text{aenc}(x_{plain}, x_{rand}, \text{pk}(x_{sk} + y_{sk})) \tag{6}$$

$$\text{unblind}(\text{blind}(x_{plain}, x_{blind}), x_{blind}) = x_{plain} \tag{7}$$

$$\text{Checksign}(x_{plain}, \text{vk}(x_{id}), \text{sign}(x_{plain}, x_{id})) = \text{ok} \tag{8}$$

$$\text{Checkpfk}(\text{vk}(x_{id}), \text{ball}, \text{pfk}(x_{id}, x_{rand}, x_{plain}, \text{ball})) = \text{ok}$$
$$\text{where ball} = \text{aenc}(x_{plain}, x_{rand}, x_{pub}) \tag{9}$$

$$\text{Checkpfkk}(\text{vk}(x_{id}), \text{ball}, \text{pfkk}(\text{vk}(x_{id}), x_{bk}, x_{plain}, \text{ball})) = \text{ok}$$
$$\text{where ball} = \text{renc}(x_{plain}, x_{bk}) \text{ or ball} = \text{blind}(x_{plain}, x_{bk}) \tag{10}$$

The symbols $+$, $*$, $\diamond$, and $\circ$ are assumed to be commutative and associative.

Fig. 1. Equations used in [34] to model the protocol used in Norway

is crucial in e-voting to prevent an attacker to compare encrypted votes. However, associativity and commutativity are typically not supported by existing tools for security protocols.

Other examples of primitives used in e-voting are trapdoor commitments schemes, zero-knowledge proofs, designated verifier zero-knowledge proofs, or plaintext equivalence tests. Of course, voting systems may mix several of those primitives. For the sake of illustration, we display in Figure 1 the complete equational theory used in [34] to model the protocol used in Norway. It is clearly out of reach of existing tools.

4   Security Properties

Most existing techniques developed so far for security protocols focus on reachability properties, that is, properties of the form: "for any execution trace, nothing bad happens". Confidentiality of keys or nonces as well as authentication properties are typical security properties that fall into the category of reachability properties. Ballot secrecy is however not expressed as a reachability property. Indeed, ballot privacy does not mean that the *value* of the vote remains secret. On the contrary, all the possible values of a vote (for example 0 or 1 in case of a referendum) are well-known by anyone. Therefore ballot secrecy is typically stated as an indistinguishability property [41]: an attacker should not notice any difference when Alice is voting 0 and Bob is voting 1 from the converse scenario where votes are swapped (Alice votes 1 and Bob votes 0). This can be easily expressed in process algebra calculus that has a notion of behavioral equivalence $\approx$.

$$V_{\text{Alice}}(0) \mid V_{\text{Bob}}(1) \approx V_{\text{Alice}}(1) \mid V_{\text{Bob}}(0)$$

where the process V_α represents voter α.

Coercion-resistance and receipt-freeness are also stated using equivalence properties [40].

ProVerif is one of the only tools that can check equivalence properties. It actually tries to prove a stronger property than behavioral equivalence [21] for couple of protocols that have a very similar structure. However ProVerif does not work very well on vote privacy, although it has recently improved [27]. Several recent (and preliminary) tools have been proposed to check equivalence of protocols, for a bounded number of sessions. AKiSs [25] can check (trace) equivalence for arbitrary (convergent) theories but is not guaranteed to terminate. APTE [26] checks (trace) equivalence for a large family of standard primitives (encryption, signatures, hashes, concatenation) and can handle non determinism and else branches. SPEC [38] implements a procedure for open bisimulation, a notion of equivalence stronger than the standard notion of trace equivalence.

Verifiability has not yet reached the same level of maturity than ballot privacy in terms of modeling. A first proposal has been made in [49] that provides formal definitions of both individual, universal, and eligibility verifiability. A much simpler yet probably weaker definition [48,32] states that the final outcome should contain:

- the votes of all voters that have voted and performed appropriate checks;
- a subset of the votes of the voters that did vote but did not check anything;
- at most k arbitrary valid votes where k is the number of voters under the control of the attacker.

Another approach [50] proposes a very general framework to define verifiability and also *accountability*, a notion that captures that a system should not only be verifiable but in case something wrong happened, it should be possible to blame who misbehaved. Due to its generality, the approach developed in [50] does not provide with a unique definition of verifiability. Instead one has to instantiate the framework for each voting system.

It is likely that new alternative definitions will still emerge to formally define verifiability.

5 Conclusion

Voting systems raise challenging issues to the area of formal verification of security protocols. First, tools and techniques need to shift from reachability to equivalence-based properties. The interest of equivalence-based properties is not confined to voting systems. Indeed behavioural equivalences are used more generally to formalize privacy properties such as anonymity or unlinkability in many different contexts (RFIDs [24,10], passports [28], mobile telephony systems[9]). They may also express security properties closer to game-based definitions used in cryptography. For example, learning even a single bit of a key is considered as an attack in cryptography. The fact that not even a bit of the secret shall be linked is called *strong secrecy* in symbolic models and is defined through the equivalence of two processes. More generally, game-based cryptographic definitions can be defined in symbolic models through equivalences [31,39]. New tools have been designed to automatically check equivalence of security protocols, for a bounded number of sessions. This is in particular the case of AKiSs [25], APTE [26], and SPEC [38]. For an unbounded number of sessions, the only available tool is ProVerif [20,21] which can check equivalence for pairs of protocols that have the same structure and for reasonably general equational theories.

Another major issue of e-voting systems is the complexity and variety of cryptographic primitives that include homomorphic encryption, re-encryption mixnets, zero-knowledge proofs, and trapdoor commitments. These primitives may be formalized through equational theories. However, most of them include associative and commutative symbols and are out of reach of existing tools, even for reachability properties.

Moreover, the primitives used in e-voting challenge the abstractions made in symbolic models: although the resulting equational theories are already quite complex, some equations may still be missed. In cryptography, more accurate models are used: instead of using process algebra with terms, protocols and attackers are simply any (polynomial) Turing machines. While cryptographic and symbolic models largely differ, symbolic models were shown to be *sound* with respect to cryptographic ones, that is, any protocol proved to be secure in symbolic models is deemed secure in cryptographic ones. Such a soundness result holds for most standard primitives [13,22] but very few results exist outside these standard primitives ([54] being one of the few exceptions). Some primitives like the Exclusive Or were even shown to be impossible to soundly abstract [57]. It may be therefore preferable in some cases to analyse e-voting protocols directly in cryptographic models, possibly using recently developed techniques that assist and partially automate the proof (see for example the line of research developed on EasyCrypt [14]).

To conclude, we expect e-voting to continue to foster the development of new techniques and tools in both symbolic and cryptographic approaches.

References

1. International association for cryptologic research. Elections page at,
   `http://www.iacr.org/elections/`
2. Web page of the Norwegian government on the deployment of e-voting,
   `http://www.regjeringen.no/en/dep/krd/prosjekter/`
   `e-vote-2011-project.html`
3. Documentations of the code used for the 2013 parlementary election in Norway (2013),
   `https://brukerveiledning.valg.no/Dokumentasjon/Dokumentasjon/`
   `Forms/AllItems.aspx`
4. KRD - evalg2011 platform - update for 2013 parliamentary elections (2013),
   `https://brukerveiledning.valg.no/Dokumentasjon/Dokumentasjon/`
   `Norway-2013_BulletinBoard_v1.2.pdf`
5. Abadi, M., Fournet, C.: Mobile values, new names, and secure communication. In: 28th ACM Symposium on Principles of Programming Languages, POPL 2001 (2001)
6. Abadi, M., Cortier, V.: Deciding knowledge in security protocols under equational theories. Theoretical Computer Science 367(1-2), 2–32 (2006)

7. Abadi, M., Gordon, A.D.: A Calculus for Cryptographic Protocols: The Spi Calculus. In: CCS 1997: 4th ACM Conference on Computer and Communications Security, pp. 36–47. ACM Press (1997)

8. Adida, B., de Marneffe, O., Pereira, O., Quisquater, J.-J.: Electing a university president using open-audit voting: Analysis of real-world use of Helios. In: Proceedings of the 2009 Conference on Electronic Voting Technology/Workshop on Trustworthy Elections (2009)

9. Arapinis, M., Mancini, L., Ritter, E., Ryan, M.: Privacy through pseudonymity in mobile telephony systems. In: 21st Annual Network and Distributed System Security Symposium, NDSS 2014 (2014)

10. Arapinis, M., Chothia, T., Ritter, E., Ryan, M.: Analysing Unlinkability and Anonymity Using the Applied Pi Calculus. In: CSF 2010: 23rd Computer Security Foundations Symposium, pp. 107–121. IEEE Computer Society (2010)

11. Armando, A., et al.: The AVISPA Tool for the automated validation of internet security protocols and applications. In: Etessami, K., Rajamani, S.K. (eds.) CAV 2005. LNCS, vol. 3576, pp. 281–285. Springer, Heidelberg (2005)

12. Armando, A., Carbone, R., Compagna, L., Cuellar, J., Abad, L.T.: Formal analysis of saml 2.0 web browser single sign-on: Breaking the saml-based single sign-on for google apps. In: Proceedings of the 6th ACM Workshop on Formal Methods in Security Engineering (FMSE 2008), pp. 1–10 (2008)

13. Backes, M., Pfitzmann, B.: Symmetric encryption in a simulatable Dolev-Yao style cryptographic library. In: Proc. 17th IEEE Computer Science Foundations Workshop (CSFW 2004), pp. 204–218 (2004)

14. Barthe, G., Grégoire, B., Heraud, S., Béguelin, S.Z.: Computer-aided security proofs for the working cryptographer. In: Rogaway, P. (ed.) CRYPTO 2011. LNCS, vol. 6841, pp. 71–90. Springer, Heidelberg (2011)

15. Baudet, M.: Deciding security of protocols against off-line guessing attacks. In: Proceedings of the 12th ACM Conference on Computer and Communications Security (CCS 2005), pp. 16–25. ACM Press (November 2005)

16. Benaloh, J.: Ballot casting assurance via voter-initiated poll station auditing. In: Proceedings of the Second Usenix/ACCURATE Electronic Voting Technology Workshop (2007)

17. Bernhard, D., Cortier, V., Pereira, O., Smyth, B., Warinschi, B.: Adapting Helios for provable ballot secrecy. In: Atluri, V., Diaz, C. (eds.) ESORICS 2011. LNCS, vol. 6879, pp. 335–354. Springer, Heidelberg (2011)

18. Bernhard, D., Pereira, O., Warinschi, B.: How not to prove yourself: Pitfalls of the Fiat-Shamir heuristic and applications to helios. In: Wang, X., Sako, K. (eds.) ASIACRYPT 2012. LNCS, vol. 7658, pp. 626–643. Springer, Heidelberg (2012)

19. Blanchet, B.: An efficient cryptographic protocol verifier based on prolog rules. In: Proc. of the 14th Computer Security Foundations Workshop (CSFW 2001). IEEE Computer Society Press (June 2001)

20. Blanchet, B.: An automatic security protocol verifier based on resolution theorem proving (invited tutorial). In: 20th International Conference on Automated Deduction (CADE-20) (July 2005)

21. Blanchet, B., Abadi, M., Fournet, C.: Automated verification of selected equivalences for security protocols. In: 20th IEEE Symposium on Logic in Computer Science (LICS 2005), pp. 331–340. IEEE Computer Society (June 2005)

22. Böhl, F., Cortier, V., Warinschi, B.: Deduction soundness: Prove one, get five for free. In: 20th ACM Conference on Computer and Communications Security (CCS 2013), Berlin, Germany (2013)

23. Bortolozzo, M., Centenaro, M., Focardi, R., Steel, G.: Attacking and fixing PKCS#11 security tokens. In: Proceedings of the 17th ACM Conference on Computer and Communications Security (CCS 2010), pp. 260–269. ACM Press (October 2010)

24. Brusó, M., Chatzikokolakis, K., den Hartog, J.: Formal verification of privacy for RFID systems. In: CSF 2010: 23rd Computer Security Foundations Symposium, pp. 75–88. IEEE Computer Society (2010)
25. Chadha, R., Ciobâcă, Ş., Kremer, S.: Automated verification of equivalence properties of cryptographic protocols. In: Seidl, H. (ed.) Programming Languages and Systems. LNCS, vol. 7211, pp. 108–127. Springer, Heidelberg (2012)
26. Cheval, V.: Apte: an algorithm for proving trace equivalence. In: Ábrahám, E., Havelund, K. (eds.) TACAS 2014 (ETAPS). LNCS, vol. 8413, pp. 587–592. Springer, Heidelberg (2014)
27. Cheval, V., Blanchet, B.: Proving more observational equivalences with ProVerif. In: Basin, D., Mitchell, J.C. (eds.) POST 2013 (ETAPS 2013). LNCS, vol. 7796, pp. 226–246. Springer, Heidelberg (2013)
28. Cheval, V., Cortier, V., Plet, A.: Lengths may break privacy – or how to check for equivalences with length. In: Sharygina, N., Veith, H. (eds.) CAV 2013. LNCS, vol. 8044, pp. 708–723. Springer, Heidelberg (2013)
29. Clarkson, M.R., Chong, S., Myers, A.C.: Civitas: Toward a secure voting system. In: Proc. IEEE Symposium on Security and Privacy, pp. 354–368 (2008)
30. Comon-Lundh, H., Shmatikov, V.: Intruder deductions, constraint solving and insecurity decision in presence of Exclusive Or. In: Proc. of 18th Annual IEEE Symposium on Logic in Computer Science (LICS 2003), pp. 271–280. IEEE Computer Society (2003)
31. Comon-Lundh, H., Cortier, V.: Computational soundness of observational equivalence. In: Proceedings of the 15th ACM Conference on Computer and Communications Security (CCS 2008), Alexandria, Virginia, USA, pp. 109–118. ACM Press (October 2008)
32. Cortier, V., Galindo, D., Glondu, S., Izabachene, M.: A generic construction for voting correctness at minimum cost - application to helios. Cryptology ePrint Archive, Report 2013/177 (2013)
33. Cortier, V., Smyth, B.: Attacking and fixing helios: An analysis of ballot secrecy. Journal of Computer Security 21(1), 89–148 (2013)
34. Cortier, V., Wiedling, C.: A formal analysis of the norwegian e-voting protocol. In: Degano, P., Guttman, J.D. (eds.) Principles of Security and Trust. LNCS, vol. 7215, pp. 109–128. Springer, Heidelberg (2012)
35. Cramer, R., Gennaro, R., Schoenmakers, B.: A secure and optimally efficient multi-authority election scheme. In: Fumy, W. (ed.) EUROCRYPT 1997. LNCS, vol. 1233, pp. 103–118. Springer, Heidelberg (1997)
36. Cremers, C.J.F.: The Scyther Tool: Verification, falsification, and analysis of security protocols. In: Gupta, A., Malik, S. (eds.) CAV 2008. LNCS, vol. 5123, pp. 414–418. Springer, Heidelberg (2008)
37. Cuvelier, É., Pereira, O., Peters, T.: Election verifiability or ballot privacy: Do we need to choose? In: Crampton, J., Jajodia, S., Mayes, K. (eds.) ESORICS 2013. LNCS, vol. 8134, pp. 481–498. Springer, Heidelberg (2013)
38. Dawson, J., Tiu, A.: Automating open bisimulation checking for the spi-calculus. In: Proceedings of IEEE Computer Security Foundations Symposium, CSF 2010 (2010)
39. Delaune, S., Kremer, S., Pereira, O.: Simulation based security in the applied pi calculus. In: Proceedings of the 29th Conference on Foundations of Software Technology and Theoretical Computer Science (FSTTCS 2009). Leibniz International Proceedings in Informatics, vol. 4, pp. 169–180 (December 2009)
40. Delaune, S., Kremer, S., Ryan, M.: Coercion-Resistance and Receipt-Freeness in Electronic Voting. In: CSFW 2006: 19th Computer Security Foundations Workshop, pp. 28–42. IEEE Computer Society (2006)
41. Delaune, S., Kremer, S., Ryan, M.D.: Verifying privacy-type properties of electronic voting protocols. Journal of Computer Security 17(4), 435–487 (2009)

42. Dolev, D., Yao, A.C.: On the security of public key protocols. In: Proc. of the 22nd Symp. on Foundations of Computer Science, pp. 350–357. IEEE Computer Society Press (1981)

43. Even, S., Goldreich, O.: On the security of multi-party ping-pong protocols. Technical Report. IEEE Computer Society Press (1983)

44. Feldman, A.J., Halderman, J.A., Felten, E.W.: Security analysis of the diebold accuvote-ts voting machine (2006), http://itpolicy.princeton.edu/voting/

45. Fujioka, A., Okamoto, T., Ohta, K.: A Practical Secret Voting Scheme for Large Scale Elections. In: Zheng, Y., Seberry, J. (eds.) AUSCRYPT 1992. LNCS, vol. 718, pp. 244–251. Springer, Heidelberg (1993)

46. Gjøsteen, K.: Analysis of an internet voting protocol. Cryptology ePrint Archive, Report 2010/380 (2010), http://eprint.iacr.org/

47. Esteve, J.B., Goldsmith, B., Turner, J.: International experience with e-voting. Technical report, Norwegian E-Vote Project (2012)

48. Juels, A., Catalano, D., Jakobsson, M.: Coercion-Resistant Electronic Elections. In: Chaum, D., Jakobsson, M., Rivest, R.L., Ryan, P.Y.A., Benaloh, J., Kutylowski, M., Adida, B. (eds.) Towards Trustworthy Elections. LNCS, vol. 6000, pp. 37–63. Springer, Heidelberg (2010)

49. Kremer, S., Ryan, M., Smyth, B.: Election verifiability in electronic voting protocols. In: Gritzalis, D., Preneel, B., Theoharidou, M. (eds.) ESORICS 2010. LNCS, vol. 6345, pp. 389–404. Springer, Heidelberg (2010)

50. Küsters, R., Truderung, T., Vogt, A.: Clash Attacks on the Verifiability of E-Voting Systems. In: 33rd IEEE Symposium on Security and Privacy (S&P 2012), pp. 395–409. IEEE Computer Society (2012)

51. Lowe, G.: Breaking and fixing the Needham-Schroeder public-key protocol using FDR. In: Margaria, T., Steffen, B. (eds.) TACAS 1996. LNCS, vol. 1055, pp. 147–166. Springer, Heidelberg (1996)

52. Millen, J., Shmatikov, V.: Constraint solving for bounded-process cryptographic protocol analysis. In: Proc. of the 8th ACM Conference on Computer and Communications Security, CCS 2001 (2001)

53. Okamoto, T.: Receipt-Free Electronic Voting Schemes for Large Scale Elections. In: Christianson, B., Crispo, B., Lomas, M., Roe, M. (eds.) Security Protocols 1997. LNCS, vol. 1361, pp. 25–35. Springer, Heidelberg (1998)

54. Sakurada, H.: Computational soundness of symbolic blind signatures under active attacker. In: Danger, J.-L., Debbabi, M., Marion, J.-Y., Garcia-Alfaro, J., Heywood, N.Z. (eds.) FPS 2013, vol. 8532, pp. 247–263. Springer, Heidelberg (2014)

55. Schneider, S.: Verifying authentication protocols with CSP. In: Proc. of the 10th Computer Security Foundations Workshop (CSFW 1997). IEEE Computer Society Press (1997)

56. Thayer, J., Herzog, J., Guttman, J.: Strand spaces: proving security protocols correct. IEEE Journal of Computer Security 7, 191–230 (1999)

57. Unruh, D.: The impossibility of computationally sound xor, Preprint on IACR ePrint 2010/389 (July 2010)

58. Wolchok, S., Wustrow, E., Halderman, J.A., Prasad, H.K., Kankipati, A., Sakhamuri, S.K., Yagati, V., Gonggrijp, R.: Security analysis of india's electronic voting machines. In: 17th ACM Conference on Computer and Communications Security, CCS 2010 (2010)

And-Or Tableaux for Fixpoint Logics
with Converse: LTL, CTL, PDL and CPDL

Rajeev Goré

Logic and Computation Group
Research School of Computer Science
The Australian National University,
Canberra, Australia

Abstract. Over the last forty years, computer scientists have invented
or borrowed numerous logics for reasoning about digital systems. Here,
I would like to concentrate on three of them: Linear Time Temporal
Logic (LTL), branching time Computation Tree temporal Logic (CTL),
and Propositional Dynamic Logic (PDL), with and without converse.
More specifically, I would like to present results and techniques on how
to solve the satisfiability problem in these logics, with global assumptions, using the tableau method. The issues that arise are the typical
tensions between computational complexity, practicality and scalability.
This is joint work with Linh Anh Nguyen, Pietro Abate, Linda Postniece,
Florian Widmann and Jimmy Thomson.

1 Introduction and Credits

Over the last forty years, computer scientists have invented or borrowed numerous logics for reasoning about digital systems [1]. Here, I would like to concentrate on three of them: Linear Time Temporal Logic (LTL), branching time
Computation Tree temporal Logic (CTL), and Propositional Dynamic Logic
(PDL). More specifically, I would like to present results and techniques on how
to solve the satisfiability problem in these logics, with global assumptions, using
the tableau method. The issues that arise are the typical tensions between computational time-complexity, space-complexity, practicality and scalability. This
overview is based on joint work with Linh Anh Nguyen [2, 3], Linda Postniece [4]
and Florian Widmann [5–7]. Some of the implementations have been refined by
Jimmy Thomson. The current best account with full algorithmic details and
proofs is Widmann's doctoral dissertation [8].

I have deliberately concentrated on tableaux methods, but the satisfiability
problem for some of these fixpoint logics can also be solved using resolution
methods and automata methods. These are beyond my expertise.

I assume that the reader is familiar with the syntax and semantics of propositional modal, description and fixpoint logics, the notion of global logical consequence in these logics, the associated notions of being satisfiable with respect to
a set of global assumptions (TBox) and with basic tableau methods for classical
propositional logic. I assume that all formulae are in negation normal form since

S. Demri, D. Kapur, and C. Weidenbach (Eds.): IJCAR 2014, LNAI 8562, pp. 26–45, 2014.

this reduces the number of rules. It is well-known that, in all the logics I consider, a formula can be put into negation normal form with only a polynomial increase in size, while preserving validity. I also assume that we are given a finite set $\mathcal{T}$ of "global assumptions" (TBox) and asked to solve the problem of whether ϕ is satisfiable with respect to the global assumptions $\mathcal{T}$ in the logic under consideration. Thus a formula ϕ is a global logical consequence of $\mathcal{T}$ iff the formula $\mathtt{nnf} (\neg\phi)$ is unsatisfiable with respect to $\mathcal{T}$, where $\mathtt{nnf} (.)$ is the function that returns the negation normal form of its argument.

The tableau method is a very general method for automated reasoning and has been widely applied for modal logics [9] and description logics [10]. Tableau methods usually come in two flavours as we explain shortly. Both methods build a rooted tree with some leaves duplicating ancestors, thereby giving cycles. Because the same node may be explored on multiple branches, tableau algorithms are typically suboptimal w.r.t. the known theoretical bounds for many logics. For example, the traditional tableau method for $\mathcal{ALC}$ can require double-exponential time even though the decision problem is known to be EXPTIME-complete.

For fixpoint logics like LTL, CTL and PDL, optimal tableau methods are possible if we proceed in stages with the first stage building a cyclic graph, and subsequent passes pruning nodes from the graph until no further pruning is possible or until the root node is pruned [11]. Optimality can also be obtained if we construct the set of all subsets of the Fischer-Ladner closure of the given initial formula [12]. But these methods can easily require exponential time even when it is not necessary. Indeed, the method of Fischer and Ladner will always require exponential time since it must first construct the set of all subsets of a set whose size is usually linear in the size of the given formula.

Thus a long-standing open problem in tableau methods for modal, description, and fixpoint logics has been to find a complexity-optimal and "on the fly" method for checking satisfiability which only requires exponential time when it is really necessary. We describe such tableau methods for each of the logics K, Kt (*i.e.* K with converse) and PDL. The resulting methods necessarily build graphs rather than trees. The various components can be combined non-trivially to give an on-the-fly and complexity-optimal tableau method for CPDL (*i.e.* PDL with converse) but we omit details. We also describe sub-optimal tableaux methods for these logics which build one single tree tableau and determine satisfiability in one pass by exploring this tree one branch at a time, reclaiming the space of previous branches. We describe such a method for the logic CTL, and give pointers to how to adapt such one-pass methods to LTL and PDL.

2   Traditional Modal and Description Logic Tree Tableaux

A tableau is a tree of nodes where the children of a node are created by applying a tableau rule to the parent and where each node contains a finite set of formulae. We refer to these formulae as the "contents" of a node, noting that the term "label" is also used to mean the same thing. Thus a label is *not* a name for a Kripke world as in some formulations of "labelled tableaux". The ancestors of a node are simply the nodes on the unique path from the root to that node.

$$(id)\frac{\Gamma\,;\,\neg p\,;\,p}{}\qquad(\wedge)\frac{\Gamma\,;\,\varphi\wedge\psi}{\Gamma\,;\,\varphi\,;\,\psi}\qquad(\vee)\frac{\Gamma\,;\,\varphi\vee\psi}{\Gamma\,;\,\varphi\mid\Gamma\,;\,\psi}$$

$$(\exists)\frac{\Delta\,;\,[]\,\Gamma\,;\,\langle\rangle\,\varphi_1\,;\,\cdots\,;\,\langle\rangle\,\varphi_n}{\Gamma\,;\,\varphi_1\,;\,\mathcal{T}\,\|\,\cdots\,\|\,\Gamma\,;\,\varphi_n\,;\,\mathcal{T}}$$

Δ contain only atoms and negated atoms

Fig. 1. AND/OR Tableaux Rules for Modal Logic with Global Assumptions $\mathcal{T}$

A leaf node is "closed" when it can be deemed to be unsatisfiable, usually because it contains an obvious contradiction like p and $\neg p$. A leaf is "open" when it can be deemed to be satisfiable, usually when no rule is applicable to it, but also when further rule applications are guaranteed to give an infinite (satisfiable) branch. A branch is closed/open if its leaf is closed/open. The aim of course is to use these classifications to determine whether the root node is satisfiable or unsatisfiable. But the tableau used in modal logics and those used in description logics are dual in a sense which is explained next.

Traditional modal tableaux a là Beth [13] are or-trees in that branches are caused by disjunctions only. Each "diamond" formula in a node causes the creation of a "successor world", fulfilling that formula. But such successors of a given node are created and explored one at a time, using backtracking, until one of them is closed, meaning that there is no explicit trace of previously explored "open" successors in any single tableau.

Traditional description logic tableaux are usually and-trees in that branches are caused by existential/diamond formulae only. Each disjunctive formula causes the creation of a child, one at a time, using backtracking, until one child is open, meaning that there is no explicit trace of previously explored "closed" or-children in any single tableau.

Thus, in both types of tableaux, the overall search space is really an and-or tree: traditional modal (Beth) tableaux display only the or-related branches and explore the and-related branches using backtracking while description logic tableaux do the reverse.

In all such methods, termination is obtained by "blocking" a node from expansion if the node that would be created already exists. For a detailed discussion of the various blocking methods, and the various notions of "caching" see [2].

3   And-Or Graph and Tree Tableaux for K

We unify these two views by taking a global view which considers tableaux as And-Or trees or And-Or graphs rather than as or-trees or and-trees. In particular, since the non-determinism in both traditional tableaux methods is determinised in And-Or tableaux, we need to build one and only one And-Or tableau!

Thus the And-Or tableau rules for modal logic K can be written as shown in Figure 1 where Γ and Δ are finite sets of formulae in negation normal form and $\Gamma\,;\,\varphi$ stands for the set $\Gamma\cup\{\varphi\}$.

The $(\vee)$-rule creates or-branching, indicated by "$|$" while the $\exists$-rule creates and-branching, indicated by "$||$". These are dual in the following senses:

$(\vee)$: if the set $\Gamma; \varphi \vee \psi$ is satisfiable w.r.t. $\mathcal{T}$ then the set $\Gamma; \varphi$ is satisfiable w.r.t. $\mathcal{T}$ or the set $\Gamma; \psi$ is satisfiable w.r.t. $\mathcal{T}$

$(\vee)$: if both sets $\Gamma; \varphi$ and $\Gamma; \psi$ are unsatisfiable w.r.t. $\mathcal{T}$ then so is $\Gamma; \varphi \vee \psi$

$(\exists)$: if the set $\Delta; [] \Gamma; \langle\rangle \varphi_1; \cdots ; \langle\rangle \varphi_n$ is satisfiable w.r.t. $\mathcal{T}$ then the set $\Gamma; \varphi_1; \mathcal{T}$ is satisfiable w.r.t. $\mathcal{T}$ and the set $\Gamma; \varphi_2; \mathcal{T}$ is satisfiable w.r.t. $\mathcal{T}$ and $\dots$ and the set $\Gamma; \varphi_n; \mathcal{T}$ is satisfiable w.r.t. $\mathcal{T}$.

$(\exists)$: if there is some integer $1 \leq i \leq n$, such that the set $\Gamma; \varphi_i; \mathcal{T}$ is unsatisfiable w.r.t. $\mathcal{T}$ then the set $\Delta; [] \Gamma; \langle\rangle \varphi_1; \cdots ; \langle\rangle \varphi_n$ is unsatisfiable w.r.t. $\mathcal{T}$.

We now give a non-algorithmic description of the procedure to create an and-or tableau. We have chosen this format over the more algorithmic description in [2, 14] to highlight its simplicity.

1. start with a root node and repeatedly try to apply exactly one of the rules in the order (id), $(\wedge)$, $(\vee)$, $(\exists)$ to each node but if a rule application to node x will create a copy y' of an existing node y then make y the child of x instead
2. if we apply the $(\vee)$-rule to x then x is an or-node and if we apply the rule $(\wedge)$ or $(\exists)$ to x then it is an and-node
3. whenever we apply the (id) rule to x then set the status of x to unsat, and if we cannot apply any rule to x then set its status to sat, and propagate this status through the current graph as follows:
   or-node: unsat if all its children have status unsat and sat if some child has status sat
   and-node: sat if all its children have status sat and unsat if some child has status unsat
4. when every node has been expanded in this way then set the status of all nodes with undefined status to sat and propagate as above.

A little more formally but still non-algorithmically. Given a TBox $\mathcal{T}$ and a formula ϕ, both in negation normal form, our method searches for a model which satisfies ϕ w.r.t. $\mathcal{T}$ by building an and-or graph G with root node τ containing $\mathcal{T} \cup \{\phi\}$. A node in the constructed and-or graph is a record with three attributes:

content: the set of formulae carried by the node
status: $\{\mathsf{unexpanded}, \mathsf{expanded}, \mathsf{sat}, \mathsf{unsat}\}$
kind: $\{\mathsf{and\text{-}node}, \mathsf{or\text{-}node}, \mathsf{leaf\text{-}node}\}$

The root node has initial status $\mathsf{unexpanded}$ and our method constructs the and-or graph using a traditional strategy explained shortly. But we interleave this generation strategy with a propagation phase which propagates the status of a node throughout the graph. We explain each in turn.

Our strategy for building the and-or graph applies the rules for decomposing $\wedge$ and $\vee$ repeatedly until they are no longer applicable to give a "saturated" node x, and then applies the $\exists$-rule which creates a child node for x containing $\mathcal{T} \cup \{\varphi\} \cup \{\psi \mid [] \psi \in x\}$ for each $\langle\rangle \varphi \in x$. The addition of the TBox $\mathcal{T}$ to such a

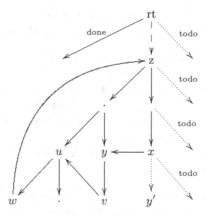

Fig. 2. Graph constructed by our algorithm for K using global caching

child is a naive way to handle global assumption (TBoxes) but suffices for our needs. We now saturate any such child to obtain a saturated node y, then apply the ∃-rule to y, and so on, until we find a contradiction, or find a repeated node, or find a saturated node which contains no ⟨⟩-formulae. For uniformity with our method for the extensions to converse and PDL, we explore/expand children in a left to right depth-first manner, although any search strategy can be used for K (or ALC) [2]. All nodes are initially given a status of **unexpanded**.

An application of (∨) to a node v causes v to be an **or-node**, while an application of (∧) or (∃) to a node v causes v to be an **and-node**. Notice that our method uses the (∨) and (∃) rules which use both or-branching and and-branching as summarised in Section 2. The crucial difference from traditional tableau methods is that we create an and-or graph rather than an and-tree or an and-tree, and we create the required child in the graph G only if it does not yet exist in the graph: this step therefore uses global caching [2]. Notice that the required child need not be an ancestor but can exist on any previous branch of the tableau. For example, as shown in Figure 2, suppose the current node is x and that the rule applied to x generates a node y' which duplicates y. The node y' is not put into G, but y becomes the child of x instead. Thus, G is really a rooted and-or tree with cross-branch edges to nodes on previously created branches like that from x to y or from v to u, or edges to ancestors like that from w to z. The problem of course is to show that this remains sound.

The propagation phase begins whenever we determine the status of a node as either **unsat** or **sat** as explained next.

A generated node that contains both p and $\neg p$ for some atomic formula p becomes a **leaf-node** with status **unsat** (i.e. unsatisfiable w.r.t. $\mathcal{T}$). A generated node to which no tableau rule is applicable becomes a **leaf-node** with status **sat** (i.e. satisfiable w.r.t. $\mathcal{T}$). Both conclusions are **irrevocable** because each relies only on classical propositional principles and not on modal principles. We therefore propagate this information to the parent node v using the kind (**or-node/and-node**) of v and the status of the children of v, treating **unsat** as irrevocably **f** and **sat** as irrevocably **t**. That is, an **or-node** gets status **sat** as soon as one of its children gets status **sat**, and gets status **unsat** when all of its children get status **unsat**. Dually for **and-nodes**. In particular, it does not matter whether the parent-child edge is a cross-branch edge or whether it is a traditional top-down edge. If these steps cannot determine the status as **sat** or **unsat**, then the rule application sets the status to **expanded** and we return to the generation phase.

The main loop ends when the status of the initial node τ becomes **sat** or **unsat**, or when no node has status **unexpanded**. In the last case, all nodes with status $\neq$ **unsat** are given status **sat** (effectively giving the status "open" to tableau branches which loop to an ancestor) and this status is propagated through the graph to obtain the status of the root node as either **unsat** or **sat**.

Theorem 1 (Soundness and Completeness). *The root node of the And-Or graph for* $\mathcal{T} \cup \{\phi\}$ *has status* **sat** *iff* ϕ *is K-satisfiable with respect to* $\mathcal{T}$.

Theorem 2 (Complexity of And-Or graph tableaux). *If the sum of the sizes of the formulae in* $\mathcal{T} \cup \{\phi\}$ *is* n *then the algorithm requires* $O(2^n)$ *space and* $O(2^n)$ *time.*

This algorithms thus uses both caching and propagation techniques and runs in EXPTIME [2].

3.1 And-Or Tree Tableaux

The method described above creates an And-Or graph as shown in Figure 2 which means that we have to keep previous branches in memory. An alternative is to only allow "loops" to ancestors. Using this strategy gives an And-Or tree which can be explored one branch at a time, and there is no need to keep previous branches in memory. We use the term And-Or tree tableaux for the resulting tableau method.

The soundness and completeness is not affected by this change, but the ability to reclaim previous branches saves memory but leads to sub-optimality.

Theorem 3 (Complexity of And-Or tree tableaux). *If the sum of the sizes of the formula in* $\mathcal{T} \cup \{\phi\}$ *is* n *then the tree-tableaux algorithm requires* $O(2^n)$ *space and* $O(2^{2^n})$ *time.*

There are n subformulae of $\mathcal{T} \cup \{\phi\}$ and hence 2^n subsets which might appear on a branch before a node repeats, hence a branch can require $O(2^n)$ space. We explore the And-Or tree one branch at a time, so we require at most this much space. An and-or tree of depth $O(2^n)$ may have $O(2^{2^n})$ or-branches. In the worse case, we have to close each branch, hence we may require $O(2^{2^n})$ time. Thus And-Or tree tableaux are sub-optimal: the satisfiability problem for K is known to be EXPTIME-complete, but our algorithm has worst-case complexity of 2EXPTIME.

The soundness shows that an ancestor loop *always* represents a "good loop" in which every node is *satisfiable*. It is this property that fails for fixpoint logics.

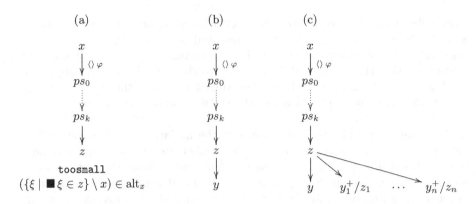

Fig. 3. The use of special node z to handle in/compatibility between states x and y. Scenario (a) occurs when x and y are incompatible. Scenario (b) occurs when x and y are compatible. Scenario (c) occurs when x and y are compatible, but y is/becomes `toosmall`.

4   And-Or Graph Tableaux for Adding Converse

Recall that the standard strategy for rule applications in tableau algorithms is to apply the rules for decomposing $\wedge$ and $\vee$ repeatedly until they are no longer applicable, giving a "saturated" node which contains only atoms, negated atoms, $[]$-formulae and $\langle\rangle$-formulae. Let us call such a "saturated" node a *state* and call the other nodes *prestates*. Thus the only rule applicable to a state x is the $\exists$-rule which creates a node containing $\mathcal{T} \cup \{\varphi\} \cup \{\psi \mid [] \psi \in x\}$ for each $\langle\rangle \varphi \in x$. The standard strategy will now saturate any such child to obtain a state y, then apply the $\exists$-rule to y, and so on, until we find a contradiction, or find a repeated node, or find a state which contains no $\exists$-formulae. Let us call x the parent state of y since all intervening nodes are not states.

When converse modalities $\blacklozenge / \blacksquare$ (inverse roles) present, we require that $\{\xi \mid \blacksquare \xi \in y\} \subseteq x$, since y is then compatible with being an R-successor of x in the putative model under construction. If some $\blacksquare \xi \in y$ has $\xi \notin x$ then x is "too small", and must be enlarged into an alternative node x^+ by adding all such ξ. If any such ξ is a complex formula then the alternative node x^+ is not "saturated", and hence not a state. So we must saturate it using the $\wedge/\vee$-rules until we reach a state. That is, a state x may conceptually be "replaced" by an alternative prestate x^+ which is an enlargement of x, and which may have to be saturated further in order to reach a state.

Our algorithm handles these "alternatives" by introducing a new type of node called a *special node*, introducing a new type of status called `toosmall`, allowing states to contain a field alt for storing these alternatives, and ensuring that a state always has a special node as its parent. When we need to replace a state x by its alternatives, the special node above x extracts these alternatives from the alt_x field and creates the required alternative nodes as explained next.

Referring to Fig. 3, suppose state x has an R-successor prestate ps_0, and further saturation of ps_0 leads to prestate ps_k, and an application of an $\wedge/\vee$-rule to p_k will give a state y. Instead of directly creating y, we create a special node z which carries the same set of formulae as would y, and make z a child of ps_k. We now check whether z is compatible with its parent state x by checking whether $\{\xi \mid \blacksquare \xi \in z\} \subseteq x$. If z is not compatible then we mark z as toosmall, and add $\{\xi \mid \blacksquare \xi \in z\} \setminus x$ to the set of alternative sets contained in alt_x, without creating y, as shown in Fig. 3(a). If z is compatible with x, we create a state y if it does not already exist, and make the new/old y a child of z, as in Fig. 3(b).

Suppose that y is compatible with x and that either y is already toosmall or becomes so later because of some descendant state w of y. In either case, the attribute alt_y then contains a number of sets $y_1, y_2, \ldots, y_n$ (say), and the toosmall status of y is propagated to the special node z. In response, z will create the alternatives $y_1^+, y_2^+, \ldots, y_n^+$ for y with $y_i^+ := y \cup y_i$. If y_i^+ is a state then our algorithm will create a special node z_i below z, and if z_i is compatible with x then y_i^+ will be created or retrieved and will become the child of z_i as in (b) else y_i^+ will not be created and z_i will be marked as toosmall as in (a). If y_i^+ is not a state then it will be created as a direct prestate child of z. Figure 3(c) captures this by using y_i^+/z_i to stand for either y_i^+ or z_i. Each of these new non-special nodes will eventually be expanded by our algorithm but now the "lapsed" special node z will be treated as a $\vee$-node.

Global State Caching. The complexities introduced by alternative nodes makes it difficult to use global caching so instead we use "global state caching": that is, the saturation phase is allowed to re-create prestates that occur on previous branches, but states cannot be duplicated so we must use cross-branch edges to their previous incarnations. The resulting algorithm runs in EXPTIME [6].

5 Traditional Tableaux Methods for Fixpoint Logics

As we have seen, modal and description logic tableaux require some form of "loop check" to guarantee termination, but fix-point logics require a further test to distinguish a "good loop" that represents a path in a model from a "bad loop" that represents an infinite branch with no hope of ever giving a model.

Most tableau-based methods for fix-point logics solve this problem using a multi-pass graph procedure [11, 15–17]. The first pass applies the tableau rules to construct a finite rooted cyclic graph. The subsequent passes prune nodes that are unsatisfiable because they contain contradictions like $\{p, \neg p\}$, and also remove nodes which give rise to "bad loops". The main practical disadvantage of such multi-pass methods is that the cyclic graph built in the first pass has a size which is *always* exponential in the size of the initial formula. So the very act of building this graph immediately causes EXPTIME behaviour even in the average case.

6   One-Pass And-Or Tree Tableaux for Fixpoint Logics

One-pass And-Or tableaux avoid this bottle-neck by building a rooted cyclic *tree* (where all cyclic edges loop back to ancestors) one branch at a time, using backtracking. The experience from one-pass tableaux for very expressive description logics [18] of similar worst-case complexity shows that their average case behaviour is often much better since the given formulae may not contain the full complexity inherent in the decision problem, particularly if the formula arises from real-world applications. Of course, there is no free lunch, since in the worst case, these one-pass methods may have significantly worse behaviour than the known optimal behaviour: 2EXPTIME than EXPTIME in the case of CTL for example. Moreover, the method for separating "good loops" from "bad loops" becomes significantly more complicated since it cannot utilise the global view offered by a graph built during a previous pass. Ideally, we want to evaluate each branch on its own during construction, or during backtracking, using only information which is "local" to this branch since this allows us to explore these branches in parallel using multiple processors.

Implemented one-pass [19, 20] and multi-pass [21] tableau provers already exist for LTL. A comparison between them [22] shows that the median running time for Janssen's highly optimised multi-pass prover for LTL is greater than the median running time for Schwendimann's not-so-optimised one-pass prover for LTL [20] *for problems which are deliberately constructed to be easy for tableau provers*, indicating that the multi-pass prover spends most of its time in the first pass building the cyclic graph. There is also a one-pass "tableau" method for propositional dynamic logic (PDL) [23] which constructs a rooted cyclic tree and uses a finite collection of automata, pre-computed from the initial formula, to distinguish "good loops" from "bad loops".

7   One-Pass And-Or Tree Tableaux for CTL

For simplicity, we ignore global assumptions (TBoxes) and concentrate on only the satisfiability problem since global assumptions can be added by a simple modification of the rules that create modal successors.

A *tableau algorithm* is a systematic search for model for a formula ϕ. The algorithm stores additional information with each node of the tableau using *histories* and *variables* [20]. A history is a mechanism for collecting extra information during proof search and passing it from parents to children. A variable is a mechanism to propagate information from children to parents.

In the following, we restrict ourselves to the tableau algorithm for CTL.

Definition 1. *A tableau node x is of the form* $(\Gamma :: \text{HCr} :: \text{mrk}, \text{uev})$ *where:*

Γ *is a set of formulae;*
HCr *is a list of the formula sets of some designated ancestors of x;*
mrk *is a Boolean valued variable indicating whether the node is marked; and*
uev *is a partial function from formulae to* $\mathbb{N}_{>0}$.

The list HCr is the only history since its value in a node is determined by the parent node, whereas mrk and uev are variables since their values in a node are determined by the children. In the following we call tableau nodes just nodes when the meaning is clear.

Informally, the value of mrk at node x is **true** if x is "closed". Since repeated nodes cause "cycles" or "loops", a node that is not "closed" is not necessarily "open" as in traditional tableaux. That is, although we have enough information to detect that further expansion of the node will cause an infinite branch, we may not yet have enough information to determine the status of the node. Informally, if a node x lies on such a "loop" in the tableau, and an "eventuality" *EU*- or *AU*-formula φ appears on this loop but remains unfulfilled, then uev of x is defined for φ by setting $\text{uev}(\varphi) = n$, where n is the height of the highest ancestor of x which is part of the loop.

We postpone the definition of a rule for a moment and proceed with the definition of a tableau.

Definition 2. *A tableau for a formula set Γ and a list of formula sets HCr is a tree of tableau nodes with root $(\Gamma :: \text{HCr} :: \text{mrk}, \text{uev})$ where the children of a node x are obtained by a single application of a rule to x (i.e. only one rule can be applied to a node). A tableau is* expanded *if no rules can be applied to any of its leaves.*

Note that mrk and uev in the definition are not given but are part of the result as they are determined by the children of the root.

Definition 3. *The partial function $\text{uev}_\perp : \text{Fml} \rightharpoonup \mathbb{N}_{>0}$ is the constant function that is undefined for all formulae (i.e. $\text{uev}_\perp(\psi) = \perp$ for all ψ).*

Note 1. In the following, we use Λ to denote a set containing only propositional variables or their negations (*i.e.* $\varphi \in \Lambda \Rightarrow \varphi = p$ or $\varphi = \neg p$ for some atom p). To focus on the "important" parts of the rule, we use "$\cdots$" for the "unimportant" parts which are passed from node to node unchanged (*e.g.* $(\Gamma :: \cdots :: \cdots)$). We define $\sim\varphi := \text{nnf}(\neg\varphi)$.

7.1   The Rules

Terminal Rule

$$(id) \ \frac{(\Gamma :: \cdots :: \text{mrk}, \text{uev})}{} \ \{p, \neg p\} \subseteq \Gamma \text{ for some atomic formula } p$$

with mrk := **true** and uev := $\text{uev}_\perp$. The intuition is that the node is "closed" so we pass this information up to the parent by putting mrk to **true**, and putting uev as undefined for all formulae.

Linear (α) Rules

$$(\wedge) \quad \frac{(\varphi \wedge \psi \,;\, \Gamma :: \cdots :: \cdots)}{(\varphi \,;\, \psi \,;\, \Gamma :: \cdots :: \cdots)} \qquad (D) \quad \frac{(AX\Delta \,;\, \Lambda :: \cdots :: \cdots)}{(EX(p_0 \vee \neg p_0) \,;\, AX\Delta \,;\, \Lambda :: \cdots :: \cdots)}$$

$$(EB) \quad \frac{(E(\varphi \, B \, \psi) \,;\, \Gamma :: \cdots :: \cdots)}{(\sim\psi \,;\, \varphi \vee EXE(\varphi \, B \, \psi) \,;\, \Gamma :: \cdots :: \cdots)}$$

$$(AB) \quad \frac{(A(\varphi \, B \, \psi) \,;\, \Gamma :: \cdots :: \cdots)}{(\sim\psi \,;\, \varphi \vee AXA(\varphi \, B \, \psi) \,;\, \Gamma :: \cdots :: \cdots)}$$

The $\wedge$-rule is standard and the D-rule captures the fact that the binary relation of a model is total by ensuring that every potential dead-end contains at least one EX-formula. The EB- and AB-rules capture the fix-point nature of the corresponding formulae according to the valid formulae $E(\varphi \, B \, \psi) \leftrightarrow \neg\psi \wedge (\varphi \vee EXE(\varphi \, B \, \psi))$ and $A(\varphi \, B \, \psi) \leftrightarrow \neg\psi \wedge (\varphi \vee AXA(\varphi \, B \, \psi))$.

These rules do not modify the histories or variables at all.

Universal Branching (β) Rules

$$(\vee) \quad \frac{(\varphi \vee \psi \,;\, \Gamma :: \cdots :: \mathrm{mrk}, \mathrm{uev})}{(\varphi \,;\, \Gamma :: \cdots :: \mathrm{mrk}_1, \mathrm{uev}_1) \mid (\psi \,;\, \Gamma :: \cdots :: \mathrm{mrk}_2, \mathrm{uev}_2)}$$

$$(EU) \quad \frac{(E(\varphi \, U \, \psi) \,;\, \Gamma :: \cdots :: \mathrm{mrk}, \mathrm{uev})}{(\psi \,;\, \Gamma :: \cdots :: \mathrm{mrk}_1, \mathrm{uev}_1) \mid (\varphi \,;\, EXE(\varphi \, U \, \psi) \,;\, \Gamma :: \cdots :: \mathrm{mrk}_2, \mathrm{uev}_2)}$$

$$(AU) \quad \frac{(A(\varphi \, U \, \psi)) \,;\, \Gamma :: \cdots :: \mathrm{mrk}, \mathrm{uev})}{(\psi \,;\, \Gamma :: \cdots :: \mathrm{mrk}_1, \mathrm{uev}_1) \mid (\varphi \,;\, AXA(\varphi \, U \, \psi) \,;\, \Gamma :: \cdots :: \mathrm{mrk}_2, \mathrm{uev}_2)}$$

with:

$$\mathrm{mrk} := \mathrm{mrk}_1 \,\&\, \mathrm{mrk}_2$$

$$\mathrm{excl}_\phi(f)(\chi) := \begin{cases} \bot & \text{if } \chi = \phi \\ f(\chi) & \text{otherwise} \end{cases}$$

$$\mathrm{uev}_1' := \begin{cases} \mathrm{uev}_1 & \text{for the } \vee\text{-rule} \\ \mathrm{excl}_{E(\varphi \, U \, \psi)}(\mathrm{uev}_1) & \text{for the } EU\text{-rule} \\ \mathrm{excl}_{A(\varphi \, U \, \psi)}(\mathrm{uev}_1) & \text{for the } AU\text{-rule} \end{cases}$$

$$\min_\bot(f,g)(\chi) := \begin{cases} \bot & \text{if } f(\chi) = \bot \text{ or } g(\chi) = \bot \\ \min(f(\chi), g(\chi)) & \text{otherwise} \end{cases}$$

$$\mathrm{uev} := \begin{cases} \mathrm{uev}_\bot & \text{if } \mathrm{mrk}_1 \,\&\, \mathrm{mrk}_2 \\ \mathrm{uev}_1' & \text{if } \mathrm{mrk}_2 \,\&\, \text{not } \mathrm{mrk}_1 \\ \mathrm{uev}_2 & \text{if } \mathrm{mrk}_1 \,\&\, \text{not } \mathrm{mrk}_2 \\ \min_\bot(\mathrm{uev}_1', \mathrm{uev}_2) & \text{otherwise} \end{cases}$$

The $\vee$-rule is standard except for the computation of uev. The EU- and AU-rules capture the fix-point nature of the EU- and AU-formulae, respectively, according to the valid formula $E(\varphi\, U\, \psi) \leftrightarrow \psi \vee (\varphi \wedge EXE(\varphi\, U\, \psi))$ and $A(\varphi\, U\, \psi) \leftrightarrow \psi \vee (\varphi \wedge AXA(\varphi\, U\, \psi))$. The intuitions of the definitions of the histories and variables are:

mrk: the value of the variable mrk is **true** if the node is "closed", so the definition of mrk just captures the "universal" nature of these rules whereby the parent node is closed if both children are closed.

excl: the definition of $\mathrm{excl}_\phi(f)(\psi)$ just ensures that $\mathrm{excl}_\phi(f)(\phi)$ is undefined.

uev$'_1$: the definition of uev$'_1$ ensures that its value is undefined for the principal formulae of the EU- and AU-rules.

min$_\perp$: the definition of min$_\perp$ ensures that we take the minimum of $f(\chi)$ and $g(\chi)$ only when both functions are defined for χ.

uev: if both children are "closed" then the parent is also closed via mrk so we ensure that uev is undefined in this case. If only the right child is closed, we take uev$'_1$, which is just uev$_1$ modified to ensure that it is undefined for the principal EU- or AU-formula. Similarly if only the left child is closed. Finally, if both children are unmarked, we define uev for all formulae that are defined in the uev of both children but map them to the minimum of their values in the children, and undefine the value for the principal formula.

Existential Branching Rule

$$(EX) \quad \frac{EX\varphi_1 \,;\, \ldots \,;\, EX\varphi_n \,;\, EX\varphi_{n+1} \,;\, \ldots \,;\, EX\varphi_{n+m} \,;\, AX\Delta \,;\, \Lambda}{\varphi_1 \,;\, \Delta \qquad \qquad \varphi_n \,;\, \Delta}$$

$$\frac{EX\varphi_1 \,;\, \ldots \,;\, EX\varphi_{n+m} \,;\, AX\Delta \,;\, \Lambda \quad :: \mathrm{HCr} :: \mathrm{mrk}, \mathrm{uev}}{\begin{array}{c} \varphi_1 \,;\, \Delta \\ :: \mathrm{HCr}_1 :: \mathrm{mrk}_1, \mathrm{uev}_1 \end{array} \;\Bigg|\, \ldots \,\Bigg|\; \begin{array}{c} \varphi_n \,;\, \Delta \\ :: \mathrm{HCr}_n :: \mathrm{mrk}_n, \mathrm{uev}_n \end{array}}$$

where:

(1) $\{p, \neg p\} \not\subseteq \Lambda$
(2) $n + m \geq 1$
(3) $\forall i \in \{1, \ldots, n\}.\, \forall j \in \{1, \ldots, \mathrm{len}(\mathrm{HCr})\}.\, \{\varphi_i\} \cup \Delta \neq \mathrm{HCr}[j]$
(4) $\forall k \in \{n+1, \ldots, n+m\}.\, \exists j \in \{1, \ldots, \mathrm{len}(\mathrm{HCr})\}.\, \{\varphi_k\} \cup \Delta = \mathrm{HCr}[j]$

with:

$$\mathrm{HCr}_i := \mathrm{HCr} \,@\, [\{\varphi_i\} \cup \Delta] \text{ for } i = 1, \ldots, n$$

$$\mathrm{mrk} := \bigvee_{i=1}^{n} \mathrm{mrk}_i \text{ or}$$
$$\exists i \in \{1, \ldots, n\}.\, \exists \psi \in \{\varphi_i\} \cup \Delta.\, \perp \neq \mathrm{uev}_i(\psi) > \mathrm{len}(\mathrm{HCr})$$

$$\mathrm{uev}_k(\cdot) := j \in \{1, \ldots, \mathit{len}(\mathrm{HCr})\} \text{ such that } \{\varphi_k\} \cup \Delta = \mathrm{HCr}[j]$$
$$\text{for } k = n+1, \ldots, n+m$$

$$\mathrm{uev}(\psi) := \begin{cases} \mathrm{uev}_j(\psi) & \text{if } \psi \in \mathrm{FmlEU} \,\&\, \psi = \varphi_j \quad (j \in \{1, \ldots, n+m\}) \\ l & \text{if } \psi \in \mathrm{FmlAU} \cap \Delta \,\& \\ & \qquad l = \max\{\mathrm{uev}_j(\psi) \neq \perp \mid j \in \{1, \ldots, n+m\}\} \\ \perp & \text{otherwise} \end{cases}$$
$$\text{(where } \max(\emptyset) := \perp)$$

Some intuitions are in order:

(1) The EX-rule is applicable if the parent node contains no α- or β-formulae and Λ, which contains propositional variables and their negations only, contains no contradictions.
(2) Both n and m can be zero, but not together.
(3) If $n > 0$, then each $EX\varphi_i$ for $1 \le i \le n$ is not "blocked" by an ancestor, and has a child containing $\varphi_i; \Delta$, thereby generating the required EX-successor;
(4) If $m > 0$, then each $EX\varphi_k$ for $n + 1 \le k \le m$ is "blocked" from creating its required child $\varphi_k; \Delta$ because some ancestor does the job;

HCr$_i$: is just the HCr of the parent but with an extra entry to extend the "history" of nodes on the path from the root down to the ith child.

mrk: captures the "existential" nature of this rule whereby the parent is marked if some child is closed or if some child contains a formula whose uev is defined and "loops" lower than the parent. Moreover, if n is zero, then mrk is set to **false** to indicate that this branch is not "closed".

uev$_k$: for $n + 1 \le k \le n + m$ the kth child is blocked by a proxy child higher in the branch. For every such k we set uev$_k$ to be the *constant* function which maps every formula to the level of this proxy child. Note that this is just a temporary function used to define uev as explained next.

uev(ψ): for an EU-formula $\psi = E(\psi_1 \, U \, \psi_2)$ such that there is a principal formula $EX\varphi_i$ with $\varphi_i = \psi$, we take uev of ψ from the child if $EX\psi$ is "unblocked", or set it to be the *level* of the proxy child higher in the branch if it is "blocked". For an AU-formula $\psi = A(\psi_1 \, U \, \psi_2) \in \Delta$, we put uev to be the maximum of the defined values from the real children and the levels of the proxy children. For all other formulae, we put uev to be undefined. The intuition is that a defined uev(ψ) tells us that there is a "loop" which starts at the parent and eventually "loops" up to some blocking node higher up on the current branch. The actual value of uev(ψ) tells us the level of the proxy because we cannot distinguish whether this "loop" is "good" or "bad" until we backtrack up to that level.

Note that the EX-rule and the id-rule are mutually exclusive since their side-conditions cannot be simultaneously true.

Proposition 1 (Termination). *Let $\phi \in$ Fml be a formula in negation normal form. Any tableau T for a node $(\{\phi\} :: \cdots :: \cdots)$ is a finite tree, hence the procedure that builds a tableau always terminates [24].*

Let $\phi \in$ Fml be a formula in negation normal form and T an expanded tableau with root $r = (\{\phi\} :: [] :: \text{mrk}, \text{uev})$: that is, the initial formula set is $\{\phi\}$ and the initial HCr is the empty list.

Theorem 4 (Soundness and Completeness). *The root r is marked iff ϕ is not satisfiable [24].*

Theorem 5 (Complexity). *The tableau algorithm runs in double exponential deterministic time and needs exponential space [24]..*

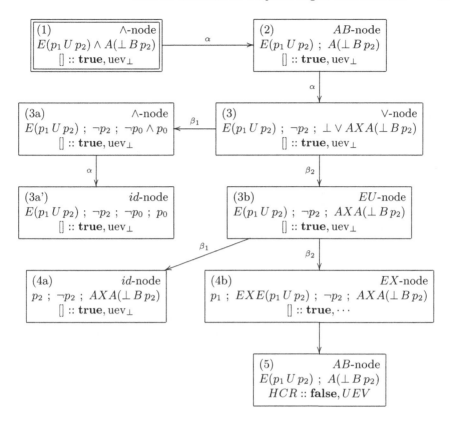

Fig. 4. An example: a tableau for $E(p_1 \, U \, p_2) \wedge A(\bot \, B \, p_2)$

7.2   A Fully Worked Example

As an example, consider the formula $E(p_1 \, U \, p_2) \wedge \neg E(\top \, U \, p_2)$ which is obviously not satisfiable. Converting the formula into negation normal form gives us $E(p_1 \, U \, p_2) \wedge A(\bot \, B \, p_2)$. Hence, any expanded tableau with root $E(p_1 \, U \, p_2) \wedge A(\bot \, B \, p_2)$ should be marked.

Figure 4 and Fig. 5 show such a tableau where the root node is node (1) in Fig. 4 and where Fig. 5 shows the sub-tableau rooted at node (5). Each node is classified as a ρ-node if rule ρ is applied to that node in the tableau. The unlabelled edges go from states to pre-states. Dotted frames indicate that the sub-tableaux at these nodes are not shown because they are very similar to sub-tableaux of other nodes: that is node (6a) behaves the same way as node (3a). Dots "$\cdots$" indicate that the corresponding values are not important because they are not needed to calculate the value of any other history or variable. The partial function UEV maps the formula $E(p_1 \, U \, p_2)$ to 1 and is undefined otherwise as explained below. The history HCR is defined as $HCR := [\{E(p_1 \, U \, p_2), A(\bot \, B \, p_2)\}]$.

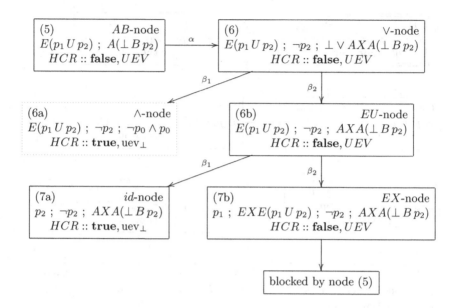

Fig. 5. An example: a tableau for $E(p_1 U p_2) \wedge A(\perp B p_2)$ (continued)

The marking of the nodes (1) to (4a) in Fig. 4 with **true** is straightforward. Note that $\perp$ is just an abbreviation for $\neg p_0 \wedge p_0$ to save some space and make things easier for the reader; the tableau procedure as described in this paper does not know about the symbol $\perp$. It is, however, not a problem to adapt the rules so that the tableau procedure can handle $\top$ and $\perp$ directly. For node (5), our procedure constructs the tableau shown in Fig. 5. The leaf (7b) is an EX-node, but it is "blocked" from creating the desired successor containing $\{E(p_1 U p_2), A(\perp B p_2)\}$ because there is a $j \in \mathbb{N}$ such that $HCr_{7b}[j] = HCR[j] = \{E(p_1 U p_2), A(\perp B p_2)\}$: namely $j = 1$. Thus the EX-rule computes $UEV(E(p_1 U p_2)) = 1$ as stated above and also puts $mrk_{7b} := $ **false**. As the nodes (7a) and (6a) are marked, the function UEV is passed on to the nodes (6b), (6), and (5) according to the corresponding β- and α-rules.

The crux of our procedure happens at node (4b) which is an EX-node with $HCr_{4b} = [\,]$ and hence $len(HCr_{4b}) = 0$. The EX-rule therefore finds a child node (5) and a formula $E(p_1 U p_2)$ in it such that $1 = UEV(E(p_1 U p_2)) = uev_5(E(p_1 U p_2)) > len(HCr_{4b}) = 0$. That is, node (4b) "sees" a child (5) that "loops lower", meaning that node (5) is the root of an "isolated" subtree which does not fulfil its eventuality $E(p_1 U p_2)$. Thus the EX-rule sets $mrk_{4b} = $ **true**, marking (4b) as "closed". The propagation of **true** to the root is then just via simple β- and α-rule applications.

7.3   One-Pass And-Or Tree Tableaux for Other Fixpoint Logics

One-pass And-Or tree tableaux were first given by Schwendimann [20] for LTL. There is a slight bug in the original formulation but a correct version can be obtained from our method for CTL by using the appropriate α/β-rules for LTL instead of CTL in our description and by changing the (EX)-rule to be linear since the premise of this rule becomes $\bigcirc\varphi; \bigcirc\Delta; \Lambda$ and the conclusion just becomes $\varphi; \Delta$. A correct implementation can be found here: http://users.cecs.anu.edu.au/~rpg/PLTLProvers/. A recent experimental comparison of it also exists [25].

One-pass And-Or tree tableaux for PDL also exist [6] and a correct implementation can be found here: http://users.cecs.anu.edu.au/~rpg/PDLProvers/

The method has been extended to the logic of common knowledge (LCK) [26].

8   On-the-Fly And-Or Graph Tableaux for PDL

The one-pass tableau given in the previous section are complexity-suboptimal: 2EXPTIME rather than EXPTIME. Next we show how to regain complexity optimality. Again, we ignore global assumptions (TBoxes) for simplicity.

Our algorithm starts at a root containing a given formula ϕ and builds an and-or tree in a depth-first and left to right manner to try to build a model for ϕ. The rules are based on the semantics of PDL and either add formulae to the current world using Smullyan's α/β rules from Table 1, or create a new world in the underlying model and add the appropriate formulae to it. For a node x, the attribute Γ_x carries this set of formulae.

The strategy for rule applications is the usual one where we "saturate" a node using the α/β-rules until they are no longer applicable, giving a "state" node s, and then, for each $\langle a \rangle \xi$ in s, we create an a-successor node containing $\{\xi\} \cup \Delta$, where $\Delta = \{\psi \mid [a]\psi \in s\}$. These successors are saturated to produce new states using the α/β-rules, and we create the successors of these new states, and so on.

Our strategy can produce infinite branches as the same node can be created repeatedly on the same branch. We therefore "block" a node from being created if this node exists already on any previous branch, thereby using global caching again, but now nodes are required to contain "focused sets of formulae" [6]. For example, in Fig. 6, if the node y' already exists in the tree, say as node y, then we create a "backward" edge from x to y (as shown) and do not create y'. If y' does not duplicate an existing node then we create y' and add a "forward" edge from x to y'. The distinction between "forward" and "backward" edges is important for the proofs. Thus our tableau is a tree of forward edges, with backward edges that either point upwards from a node to a "forward-ancestor", or point leftwards from one branch to another. Cycles can arise only via backward edges to a forward-ancestor.

Our tableau must "fulfil" every formula of the form $\langle \delta \rangle \varphi$ in a node but only eventualities, defined as those where δ contains $*$-connectives, cause problems. If $\langle \delta \rangle \varphi$ is not an eventuality, the α/β-rules reduce the size of the principal formula, ensuring fulfilment. If $\langle \delta \rangle \varphi$ is an eventuality, the main problem is the

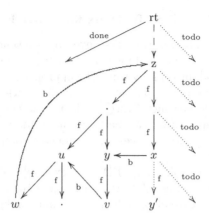

Fig. 6. Graph constructed by our algorithm using forward (f) and backward edges (b)

β-rule for formulae of the form $\langle\gamma*\rangle\varphi$. Its left child reduces $\langle\gamma*\rangle\varphi$ to a strict sub-formula φ, but the right child "reduces" it to $\langle\gamma\rangle\langle\gamma*\rangle\varphi$. If the left child is always inconsistent, this rule can "procrastinate" an eventuality $\langle\gamma*\rangle\varphi$ indefinitely and never find a world which makes φ true. This non-local property must be checked globally by tracking eventualities.

Consider Fig. 6, and suppose the current node x contains an eventuality e_x. We distinguish three cases. The first is that some path from x fulfils e_x in the existing tree. Else, the second case is that some path from x always procrastinates the fulfilment of e_x and hits a forward-ancestor of x on the current branch: e.g. the path x, y, v, u, w, z. The forward-ancestor z contains some "reduction" e_z of e_x. The path from the root to the current node x contains the only currently existing nodes which may need further expansion, and may allow z to fulfil e_z at a later stage, and hence fulfil e_x. We call the pair (z, e_z) a "potential rescuer" of e_x in Γ_x. The only remaining case is that $e_x \in \Gamma_x$ is unfulfilled, has no potential rescuers, and hence can never become fulfilled later, so x can be "closed". The machinery to distinguish these three cases and compute, if needed, all currently existing potential rescuers of every eventuality in Γ_x is described next.

A tableau node x also contains a status sts_x. The value of sts_x is the constant `closed` if the node x is closed. Otherwise, the node is "open" and sts_x contains a

Table 1. Smullyan's α- and β-notation to classify formulae

α	$\varphi \wedge \psi$	$[\gamma \cup \delta]\varphi$	$[\gamma*]\varphi$	$\langle\psi?\rangle\varphi$	$\langle\gamma;\delta\rangle\varphi$	$[\gamma;\delta]\varphi$
α_1	φ	$[\gamma]\varphi$	φ	φ	$\langle\gamma\rangle\langle\delta\rangle\varphi$	$[\gamma][\delta]\varphi$
α_2	ψ	$[\delta]\varphi$	$[\gamma][\gamma*]\varphi$	ψ		

β	$\varphi \vee \psi$	$\langle\gamma \cup \delta\rangle\varphi$	$\langle\gamma*\rangle\varphi$	$[\psi?]\varphi$
β_1	φ	$\langle\gamma\rangle\varphi$	φ	φ
β_2	ψ	$\langle\delta\rangle\varphi$	$\langle\gamma\rangle\langle\gamma*\rangle\varphi$	$\sim\psi$

function prs which maps each eventuality $e_x \in \Gamma_x$ to $\bot$ or to a set of pairs (v, e) where v is a forward-ancestor of x and e is an eventuality. The status of a node is determined from those of its children once they have all been processed. A closed child's status is propagated as usual, but the propagation of the function prs from open children is more complicated. The intuition is that we must preserve the following invariant for each eventuality $e_x \in \Gamma_x$:

> if e_x is fulfilled in the tree to the left of the path from the root to the node x then $\mathrm{prs}_x(e_x) := \bot$, else $\mathrm{prs}_x(e_x)$ is exactly the set of all potential rescuers of e_x in the current tableau.

An eventuality $e_x \in \Gamma_x$ whose $\mathrm{prs}_x(e_x)$ becomes the empty set can never become fulfilled later, so $\mathrm{sts}_x := \mathtt{closed}$, thus covering the three cases as desired.

Whenever a node n gets a status $\mathtt{closed}$, we interrupt the depth-first and left-to-right traversal and invoke a separate procedure which explicitly propagates this status transitively throughout the and-or graph rooted at n. For example, if z gets closed then so will its backward-parent w, which may also close u and so on. This propagation (update) may break the invariant for some eventuality e in this subgraph by interrupting the path from e to a node that fulfils e or to a potential rescuer of e. We must therefore ensure that the propagation (update) procedure re-establishes the invariant in these cases by changing the appropriate prs entries. At the end of the propagation (update) procedure, we resume the usual depth-first and left-to-right traversal of the tree by returning the status of n to its forward-parent. This "on-the-fly" nature guarantees that unfulfilled eventualities are detected as early as possible.

Our algorithm terminates, runs in EXPTIME, and formula ϕ is satisfiable iff the root is open [6].

9   On-the-Fly And-Or Graph Tableaux for CPDL

The methods described in the previous sections can be combined to give a complexity-optimal on-the-fly And-Or graph tableau method for CPDL but the extension is non-trivial and cannot really be described without giving an actual algorithm [7]. An implementation by Florian Widmann can be found here: http://users.cecs.anu.edu.au/~rpg/CPDLTabProver/

10   Further Work

All methods described here have been implemented (http://users.cecs.anu.edu.au/~rpg/software.html) but further work is required to add optimisations to make these methods practical on large examples and to extend them to more expressive logics like SHOIQ: but see [27, 28]. There is also the possibility to marry these methods with advances from SAT and SMT [29].

Our original aim in this endeavour was to obtain tableaux algorithms for the modal mu-calculus but we have been unable to extend our one-pass or on-the-fly methods to this logic. Similarly, we have been unable to extend our methods to handle full computation tree logic CTL*. Finally, we have been unable to find a complexity-optimal and-or graph tableaux method for CTL using our approach.

Tableaux-like methods for both CTL* and the modal mu-calculus have been given and implemented, and are an exciting avenue for future work [30–33].

References

1. Vardi, M.Y.: From philosophical to industrial logics. In: Ramanujam, R., Sarukkai, S. (eds.) Logic and Its Applications. LNCS (LNAI), vol. 5378, pp. 89–115. Springer, Heidelberg (2009)
2. Goré, R., Nguyen, L.A.: EXPTIME tableaux for ALC using sound global caching. In: Proc. of the International Workshop on Description Logics, DL 2007 (2007)
3. Goré, R.P., Nguyen, L.A.: EXPTIME tableaux with global caching for description logics with transitive roles, inverse roles and role hierarchies. In: Olivetti, N. (ed.) TABLEAUX 2007. LNCS (LNAI), vol. 4548, pp. 133–148. Springer, Heidelberg (2007)
4. Goré, R.P., Postniece, L.: An experimental evaluation of global caching for ALC (system description). In: Armando, A., Baumgartner, P., Dowek, G. (eds.) IJCAR 2008. LNCS (LNAI), vol. 5195, pp. 299–305. Springer, Heidelberg (2008)
5. Goré, R., Widmann, F.: Sound global state caching for ALC with inverse roles. In: Giese, M., Waaler, A. (eds.) TABLEAUX 2009. LNCS, vol. 5607, pp. 205–219. Springer, Heidelberg (2009)
6. Goré, R., Widmann, F.: An optimal on-the-fly tableau-based decision procedure for PDL-satisfiability. In: Schmidt, R.A. (ed.) CADE-22. LNCS, vol. 5663, pp. 437–452. Springer, Heidelberg (2009)
7. Goré, R., Widmann, F.: Optimal and cut-free tableaux for propositional dynamic logic with converse. In: Giesl, J., Hähnle, R. (eds.) IJCAR 2010. LNCS, vol. 6173, pp. 225–239. Springer, Heidelberg (2010)
8. Widmann, F.: Tableaux-based Decision Procedures for Fixpoint Logics. PhD thesis, The Australian National University, Australia (2010)
9. Goré, R.: Tableau methods for modal and temporal logics. In: D'Agostino, M., Gabbay, D.M., Hähnle, R., Posegga, J. (eds.) Handbook of Tableau Methods, pp. 297–396. Kluwer (1999)
10. Baader, F., Calvanese, D., McGuinness, D.L., Nardi, D., Patel-Schneider, P. (eds.): The Description Logic Handbook: Theory, Implementation and Applications. Cambridge University Press, Cambridge (2003)
11. Pratt, V.R.: A near-optimal method for reasoning about action. Journal of Computer and System Sciences 20(2), 231–254 (1980)
12. Fischer, M.J., Ladner, R.E.: Propositional dynamic logic of regular programs. Journal of Computer and Systems Science 18, 194–211 (1979)
13. Beth, E.: On Padoa's method in the theory of definition. Indag. Math. 15, 330–339 (1953)
14. Goré, R., Nguyen, L.: ExpTime tableaux for $\mathcal{ALC}$ using sound global caching. In: C., D., et al. (eds.) Proc. DL 2007, pp. 299–306 (2007)
15. Wolper, P.: Temporal logic can be more expressive. Information and Control 56, 72–99 (1983)
16. Ben-Ari, M., Manna, Z., Pnueli, A.: The temporal logic of branching time. In: Proceedings of Principle of Programming Langauages (1981)

17. Emerson, E.A., Halpern, J.Y.: Decision procedures and expressiveness in the temporal logic of branching time. Journal of Computer and System Sciences 30(1), 1–24 (1985)
18. Horrocks, I., Sattler, U.: A tableaux decision procedure for SHOIQ. In: IJCAI, pp. 448–453 (2005)
19. Kesten, Y., Manna, Z., McGuire, H., Pnueli, A.: A decision algorithm for full propositional temporal logic. In: Courcoubetis, C. (ed.) CAV 1993. LNCS, vol. 697, pp. 97–109. Springer, Heidelberg (1993)
20. Schwendimann, S.: A new one-pass tableau calculus for PLTL. In: de Swart, H. (ed.) TABLEAUX 1998. LNCS (LNAI), vol. 1397, pp. 277–291. Springer, Heidelberg (1998)
21. Janssen, G.: Logics for Digital Circuit Verication: Theory, Algorithms, and Applications. PhD thesis, Eindhoven University of Technology, The Netherlands (1999)
22. Hustadt, U., Konev, B.: TRP++: A temporal resolution prover. In: Baaz, M., Makowsky, J., Voronkov, A. (eds.) Collegium Logicum, pp. 65–79. Kurt Gödel Society (2004)
23. Baader, F.: Augmenting concept languages by transitive closure of roles: An alternative to terminological cycles. In: Proc. IJCAI 1991, pp. 446–451 (1991)
24. Abate, P., Goré, R., Widmann, F.: One-pass tableaux for computation tree logic. In: Dershowitz, N., Voronkov, A. (eds.) LPAR 2007. LNCS (LNAI), vol. 4790, pp. 32–46. Springer, Heidelberg (2007)
25. Schuppan, V., Darmawan, L.: Evaluating LTL satisfiability solvers. In: Bultan, T., Hsiung, P.-A. (eds.) ATVA 2011. LNCS, vol. 6996, pp. 397–413. Springer, Heidelberg (2011)
26. Abate, P., Goré, R., Widmann, F.: Cut-free single-pass tableaux for the logic of common knowledge. In: Workshop on Agents and Deduction at TABLEAUX (2007)
27. Nguyen, L.A., Golinska-Pilarek, J.: An exptime tableau method for dealing with nominals and quantified number restrictions in deciding the description logic SHOQ. In: CS&P, pp. 296–308 (2013)
28. Nguyen, L.A.: A tableau method with optimal complexity for deciding the description logic SHIQ. In: Nguyen, N.T., van Do, T., Thi, H.A. (eds.) ICCSAMA 2013. SCI, vol. 479, pp. 331–342. Springer, Heidelberg (2013)
29. Suda, M., Weidenbach, C.: A pltl-prover based on labelled superposition with partial model guidance. In: Gramlich, B., Miller, D., Sattler, U. (eds.) IJCAR 2012. LNCS (LNAI), vol. 7364, pp. 537–543. Springer, Heidelberg (2012)
30. Jungteerapanich, N.: A tableau system for the modal μ-calculus. In: Giese, M., Waaler, A. (eds.) TABLEAUX 2009. LNCS (LNAI), vol. 5607, pp. 220–234. Springer, Heidelberg (2009)
31. Friedmann, O., Latte, M., Lange, M.: A decision procedure for CTL* based on tableaux and automata. In: Giesl, J., Hähnle, R. (eds.) IJCAR 2010. LNCS, vol. 6173, pp. 331–345. Springer, Heidelberg (2010)
32. Friedmann, O., Lange, M.: A solver for modal fixpoint logics. Electr. Notes Theor. Comput. Sci. 262, 99–111 (2010)
33. Reynolds, M.: A faster tableau for CTL*. In: GandALF, pp. 50–63 (2013)

Unified Classical Logic Completeness

A Coinductive Pearl

Jasmin Christian Blanchette[1], Andrei Popescu[1,2], and Dmitriy Traytel[1]

[1] Fakultät für Informatik, Technische Universität München, Germany
[2] Institute of Mathematics Simion Stoilow of the Romanian Academy, Bucharest, Romania

Abstract. Codatatypes are absent from many programming and specification languages. We make a case for their importance by revisiting a classical result: the completeness theorem for first-order logic established through a Gentzen system. The core of the proof establishes an abstract property of possibly infinite derivation trees, independently of the concrete syntax or inference rules. This separation of concerns simplifies the presentation. The abstract proof can be instantiated for a wide range of Gentzen and tableau systems as well as various flavors of first-order logic. The corresponding Isabelle/HOL formalization demonstrates the recently introduced support for codatatypes and the Haskell code generator.

1 Introduction

Gödel's completeness theorem [12] is a major result about first-order logic (FOL). It forms the basis of results and techniques in various areas, including mathematical logic, automated deduction, and program verification. It can be stated as follows: If a set of formulas is syntactically consistent (i.e., no contradiction arises from it), then it has a model. The theorem enjoys many accounts in the literature that generalize and simplify the original proof; indeed, a textbook on mathematical logic would be incomplete without a proof of this fundamental theorem.

Formal logic has always been a battleground between semantic and syntactic methods. Generally, mathematicians belong to the semantic school, whereas computer scientists tend to take the other side of the argument. The completeness theorem, which combines syntax and semantics, is also disputed, with the result that each school has its own proof. In his review of Gallier's *Logic for Computer Science* [11], Pfenning notes the following [29]:

> All too often, proof-theoretic methods are neglected in favor of shorter, and superficially more elegant semantic arguments. [In contrast, in Gallier's book] the treatment of the proof theory of the Gentzen system is oriented towards computation with proofs. For example, a pseudo-Pascal version of a complete search procedure for first-order cut-free Gentzen proofs is presented.

In the context of completeness, the "superficially more elegant semantic arguments" are proofs that rely on Hilbert systems. These systems have several axioms but only one or two deduction rules, providing minimal support for presenting the structure of proofs or for modeling proof search. A proof of completeness based on Hilbert systems follows the *Henkin style*: It employs a heavy bureaucratic apparatus to establish facts about deduction and conservative language extensions, culminating in a nonconstructive step:

S. Demri, D. Kapur, and C. Weidenbach (Eds.): IJCAR 2014, LNAI 8562, pp. 46–60, 2014.

an application of Zorn's lemma to extend any syntactically consistent set of formulas to a maximally consistent one, from which a model is produced.

In contrast, a proof of completeness based on more elaborate Gentzen or tableau systems follows the *Beth–Hintikka style* [20]. It performs a search that builds either a finite deduction tree yielding a proof (or refutation, depending on the system) or an infinite tree from which a countermodel (or model) can be extracted. Such completeness proofs have an intuitive content that stresses the tension of the argument: The deduction system systematically tries to prove the goal; a failure yields, at the limit, a countermodel.

The intuitive appeal of the Beth–Hintikka approach comes at a price: It requires reasoning about infinite derivation trees and infinite paths. Unfortunately, convenient means to reason about infinite (or lazy) data structures are lacking in mainstream mathematics. In textbooks, at best the trees are defined rigorously (e.g., as prefix-closed sets), but the reasoning relies on the intuitive notion of trees, as Gallier does. One could argue that trees are intuitive and do not need a formal treatment, but the same holds for the syntax of formulas, which is treated very rigorously in most of the textbooks.

This paper presents a rigorous Beth–Hintikka-style proof of the completeness theorem, based on a Gentzen system. The potentially infinite trees are captured by codatatypes (also called coinductive datatypes or final coalgebras) [18]. Another novel aspect of the proof is its modularity: The core tree construction argument is isolated from the proof system and concrete formula syntax (Section 3). The abstract proof can be instantiated for a wide range of Gentzen and tableau systems as well as various flavors of FOL (Sections 4 and 5). This modularization replaces the textbook proofs by analogy. The core of the argument amounts to reasoning about a lazy functional program.

The proof is formalized in Isabelle/HOL [28] (Section 6). The tree construction makes use of a new definitional package for codatatypes [5], which automates the derivation of characteristic theorems from specifications of the constructors. Through Isabelle's code generator [14], the corecursive construction gives rise to a Haskell program that implements a semidecision procedure for validity instantiable with various proof systems, yielding verified sound and complete provers.

Conventions. Isabelle/HOL is a proof assistant based on classical higher-order logic (HOL) with Hilbert choice, the axiom of infinity, and rank-1 polymorphism. It is the logic of Gordon's original HOL system and of its many successors [13]. HOL notations are a mixture of functional programming and mathematics. We refer to Nipkow and Klein [27] for a modern introduction. In this paper, the logic is viewed as a framework for expressing mathematics, much like set theory is employed by working mathematicians. In keeping with the standard semantics of HOL, types α are identified with sets.

2 A Gentzen System for First-Order Logic

We fix a first-order language: a countably infinite set var of variables x, y, z and countable sets fsym and psym of function symbols f and predicate symbols p together with an assignment ar : fsym $\uplus$ psym $\to$ nat of numeric arities. Terms $t \in$ term are symbolic expressions built inductively from variables by application of function symbols $f \in$ fsym to tuples of arguments whose lengths respect the arities: $f(t_1, \ldots, t_{\mathsf{ar}\,f})$. Atoms $a \in$ atom are expressions of the form $p(t_1, \ldots, t_{\mathsf{ar}\,p})$, where $p \in$ psym and $t_1, \ldots, t_{\mathsf{ar}\,p} \in$ term.

Formulas φ, ψ may be atoms, negations, conjunctions, or universal quantifications. They are defined as follows:

datatype fmla = Atm atom | Neg fmla | Conj fmla fmla | All var fmla

A structure $\mathscr{S} = \left(S, (F_f)_{f \in \mathsf{fsym}}, (P_p)_{p \in \mathsf{psym}}\right)$ for the given language consists of a carrier set S, together with a function $F_f : S^n \to S$ for each n-ary $f \in \mathsf{fsym}$ and a predicate $P_p : S^n \to \mathsf{bool}$ for each n-ary $p \in \mathsf{psym}$. The notions of interpretation of a term t and satisfaction of a formula φ by a structure $\mathscr{S}$ with respect to a variable valuation $\xi : \mathsf{var} \to S$ are defined in the standard way. For terms:

$$[\![x]\!]_\xi^\mathscr{S} = \xi\, x \qquad\qquad [\![f(t_1, \ldots, t_n)]\!]_\xi^\mathscr{S} = F_f\left([\![t_1]\!]_\xi^\mathscr{S}, \ldots, [\![t_n]\!]_\xi^\mathscr{S}\right)$$

For atoms: $\mathscr{S} \models_\xi p(t_1, \ldots, t_n)$ iff $P_p\left([\![t_1]\!]_\xi^\mathscr{S}, \ldots, [\![t_n]\!]_\xi^\mathscr{S}\right)$. For formulas:

$$\begin{array}{llll}
\mathscr{S} \models_\xi \mathsf{Atm}\,a & \text{iff} & \mathscr{S} \models_\xi a & \qquad \mathscr{S} \models_\xi \mathsf{Conj}\,\varphi\,\psi \quad \text{iff} \quad \mathscr{S} \models_\xi \varphi \wedge \mathscr{S} \models_\xi \psi \\
\mathscr{S} \models_\xi \mathsf{Neg}\,\varphi & \text{iff} & \mathscr{S} \not\models_\xi \varphi & \qquad \mathscr{S} \models_\xi \mathsf{All}\,x\,\varphi \quad \text{iff} \quad \forall a \in S.\ \mathscr{S} \models_{\xi[x \leftarrow a]} \varphi
\end{array}$$

The following substitution lemma relates the notions of satisfaction and capture-avoiding substitution $\varphi[t/x]$ of a term t for a variable x in a formula φ:

Lemma 1. $\mathscr{S} \models_\xi \varphi[t/x]$ iff $\mathscr{S} \models_{\xi[x \leftarrow [\![t]\!]_\xi^\mathscr{S}]} \varphi$.

A sequent is a pair $\Gamma \rhd \Delta$ of finite formula sets. Satisfaction is extended to sequents: $\mathscr{S} \models_\xi \Gamma \rhd \Delta$ iff $(\forall \varphi \in \Gamma.\ \mathscr{S} \models_\xi \varphi) \Rightarrow (\exists \psi \in \Delta.\ \mathscr{S} \models_\xi \psi)$.

The proof system on sequents is defined inductively as follows, where the notation Γ, φ abbreviates the set $\Gamma \cup \{\varphi\}$:

$$\frac{}{\Gamma, \mathsf{Atm}\,a \rhd \Delta, \mathsf{Atm}\,a}\ \mathrm{Ax} \qquad \frac{\Gamma \rhd \Delta, \varphi}{\Gamma, \mathsf{Neg}\,\varphi \rhd \Delta}\ \mathrm{NEGL} \qquad \frac{\Gamma, \varphi \rhd \Delta}{\Gamma \rhd \Delta, \mathsf{Neg}\,\varphi}\ \mathrm{NEGR}$$

$$\frac{\Gamma, \varphi, \psi \rhd \Delta}{\Gamma, \mathsf{Conj}\,\varphi\,\psi \rhd \Delta}\ \mathrm{CONJL} \qquad \frac{\Gamma \rhd \Delta, \varphi \quad \Gamma \rhd \Delta, \psi}{\Gamma \rhd \Delta, \mathsf{Conj}\,\varphi\,\psi}\ \mathrm{CONJR}$$

$$\frac{\Gamma, \mathsf{All}\,x\,\varphi, \varphi[t/x] \rhd \Delta}{\Gamma, \mathsf{All}\,x\,\varphi \rhd \Delta}\ \mathrm{ALLL} \qquad \frac{\Gamma \rhd \Delta, \varphi[y/x]}{\Gamma \rhd \Delta, \mathsf{All}\,x\,\varphi}\ \mathrm{ALLR}\ (y\ \text{fresh})$$

The rules are applied from bottom to top. One chooses a formula from either side of the sequent, the eigenformula, and applies a rule according to the topmost connective or quantifier. For a given choice of eigenformula, at most one rule is applicable. The aim of applying the rules is to prove the sequent by building a finite derivation tree whose branches are closed by an axiom (Ax). The completeness theorem states that any sequent $\Gamma \rhd \Delta$ either is provable (denoted by $\vdash$) or has a countermodel, i.e., a structure $\mathscr{S}$ and a valuation ξ that falsify it: $\vdash \Gamma \rhd \Delta \vee (\exists \mathscr{S}, \xi.\ \mathscr{S} \not\models_\xi \Gamma \rhd \Delta)$.

3 Abstract Completeness

The proof of the completeness theorem is divided in two parts. The first part, described in this section, focuses on the core of the completeness argument in an abstract, syntax-free manner. This level captures the tension between the existence of a proof or of an abstract notion of countermodel; the latter is introduced through what we call an escape path—an infinite sequence of rule applications that "escapes" the proof attempt. The tension is distilled in a completeness result: Either there exists a finite derivation tree or there exists an infinite derivation tree with a suitable escape path. The second part maps the abstract escape path to a concrete, proof-system-specific countermodel. Section 4 performs this connection for the Gentzen system presented in Section 2.

Rule Systems. We abstract away the syntax of formulas and sequents and the specific rules of the proof system. We fix countable sets state and rule for states and rules. A state represents a formal statement in the logic. We assume that the meaning of the rules is given by an effect relation eff : rule $\rightarrow$ state $\rightarrow$ state fset $\rightarrow$ bool, where α fset denotes the set of finite subsets of α. The reading of eff r s ss is as follows: Starting from state s, applying rule r expands s into the states ss. The triple $\mathscr{R} = $ (state, rule, eff) forms a *rule system*.

Example 1. The Gentzen system from Section 2 can be presented as a rule system. The set state is the set of sequents, and rule consists of the following: a rule AX_a for each atom a; rules NEGL_φ and NEGR_φ for each formula φ; rules $\text{CONJL}_{\varphi,\psi}$ and $\text{CONJR}_{\varphi,\psi}$ for each pair of formulas φ and ψ; a rule $\text{ALLL}_{x,\varphi,t}$ for each variable x, formula φ, and term t; and a rule $\text{ALLR}_{x,\varphi}$ for each variable x and formula φ.

The eigenformula is part of the rule. Hence we have a countably infinite number of rules. The effect is defined as follows, where semicolons (;) separate set elements:

eff AX_a $(\Gamma, \text{Atm } a \triangleright \Delta, \text{Atm } a)$ $\emptyset$
eff NEGR_φ $(\Gamma \triangleright \Delta, \text{Neg } \varphi)$ $\{\Gamma, \varphi \triangleright \Delta\}$
eff NEGL_φ $(\Gamma, \text{Neg } \varphi \triangleright \Delta)$ $\{\Gamma \triangleright \Delta, \varphi\}$
eff $\text{CONJL}_{\varphi,\psi}$ $(\Gamma, \text{Conj } \varphi \psi \triangleright \Delta)$ $\{\Gamma, \varphi, \psi \triangleright \Delta\}$
eff $\text{CONJR}_{\varphi,\psi}$ $(\Gamma \triangleright \Delta, \text{Conj } \varphi \psi)$ $\{\Gamma \triangleright \Delta, \varphi; \Gamma \triangleright \Delta, \psi\}$
eff $\text{ALLL}_{x,\varphi,t}$ $(\Gamma, \text{All } x \varphi \triangleright \Delta)$ $\{\Gamma, \text{All } x \varphi, \varphi[t/x] \triangleright \Delta\}$
eff $\text{ALLR}_{x,\varphi}$ $(\Gamma \triangleright \Delta, \text{All } x \varphi)$ $\{\Gamma \triangleright \Delta, \varphi[y/x]\}$ where y is fresh for Γ and All x φ

Derivation Trees. Possibly infinite trees are represented by the following codatatype:

$$\textbf{codatatype } \alpha \text{ tree } = \text{ Node } (\text{lab}: \alpha) \text{ (sub: } (\alpha \text{ tree}) \text{ fset)}$$

This definition introduces a constructor Node : $\alpha \rightarrow (\alpha \text{ tree})$ fset $\rightarrow \alpha$ tree and two selectors lab : α tree $\rightarrow \alpha$, sub : α tree $\rightarrow (\alpha \text{ tree})$ fset. Trees have the form Node a Ts, where a is the tree's *label* and Ts is the finite set of its (immediate) *subtrees*. The **codatatype** keyword indicates that, unlike for inductive datatypes, this tree formation rule may be applied an infinite number of times.

A *step* combines the current state and the rule to be applied: step = state $\times$ rule. Derivation trees are defined as trees labeled by steps, dtree = step tree, in which the root's label (s, r) represents the proved goal s and the first (backward) applied rule r. The well-formed derivation trees are captured by the predicate wf : dtree $\rightarrow$ bool defined by the coinductive rule

$$\frac{\text{eff } r \ s \ (\text{image } (\text{fst} \circ \text{lab}) \ Ts) \quad \forall T \in Ts. \ \text{wf } T}{\text{wf } (\text{Node } (s, r) \ Ts)} \text{WF}$$

(Double lines distinguish coinductive rules from their inductive counterparts.) Thus, the predicate wf is the greatest (weakest) solution to

$$\text{wf (Node } (s,r) \ Ts) \Leftrightarrow \text{eff } r \ s \ (\text{image (fst} \circ \text{lab)} \ Ts) \wedge (\forall T \in Ts. \ \text{wf } T)$$

The term image f A denotes the image of set A through function f, and fst is the left projection operator (i.e., fst $(x, y) = x$).

The first assumption requires that the rule r from the root be applied to obtain the subtrees' labels. The second assumption requires that wellformedness hold for the immediate subtrees. The coinductive nature of the definition ensures that these properties hold for arbitrarily deep subtrees of T, even if T has infinite paths.

Proofs. The finite derivation trees—the trees that would result from an inductive data-type definition with the same constructors—can be carved out of the codatatype dtree using the predicate finite defined inductively (i.e., as a least fixpoint) by the rule

$$\frac{\forall T \in Ts. \ \text{finite } T}{\text{finite (Node } (s,r) \ Ts)} \ \text{FIN}$$

A *proof* of a state s is a finite well-formed derivation tree with the state s at its root. An infinite well-formed derivation tree represents a failed proof attempt.

Example 2. Given the instantiation of Example 1, Figure 1 shows a finite derivation tree for the sequent All x $(p(x)) \rhd$ Conj $(p(y))$ $(p(z))$ written using the familiar syntax for logical symbols. Figure 2 shows an infinite tree for the same sequent.

Escape Paths. An infinite path in a derivation tree can be regarded as a way to "escape" the proof. To represent infinite paths independently of trees, we introduce the codatatype of streams over a type α with the constructor SCons and the selectors shead and stail:

$$\textbf{codatatype } \alpha \text{ stream } = \text{SCons (shead: } \alpha) \ (\text{stail: } \alpha \text{ stream})$$

$$\cfrac{\cfrac{}{\forall x. \, p(x), \, p(y) \rhd p(y)} \, \text{AX}_{p(y)}}{\cfrac{\forall x. \, p(x) \rhd p(y)}{} \, \text{ALLL}_{x,p(x),y}} \qquad \cfrac{\cfrac{}{\forall x. \, p(x), \, p(z) \rhd p(z)} \, \text{AX}_{p(z)}}{\cfrac{\forall x. \, p(x) \rhd p(z)}{} \, \text{ALLL}_{x,p(x),z}}$$
$$\cfrac{}{\forall x. \, p(x) \rhd p(y) \wedge p(z)} \, \text{CONJR}_{p(y),p(z)}$$

Fig. 1. A proof

$$\vdots$$

$$\cfrac{\cfrac{}{\forall x. \, p(x), \, p(y) \rhd p(y)} \, \text{AX}_{p(y)}}{\cfrac{\forall x. \, p(x) \rhd p(y)}{} \, \text{ALLL}_{x,p(x),y}} \qquad \cfrac{\cfrac{\cfrac{\forall x. \, p(x), \, p(y) \rhd p(z)}{\forall x. \, p(x), \, p(y) \rhd p(z)} \, \text{ALLL}_{x,p(x),y}}{\cfrac{\forall x. \, p(x), \, p(y) \rhd p(z)}{} \, \text{ALLL}_{x,p(x),y}}}{\cfrac{\forall x. \, p(x) \rhd p(z)}{} \, \text{ALLL}_{x,p(x),y}}$$
$$\cfrac{}{\forall x. \, p(x) \rhd p(y) \wedge p(z)} \, \text{CONJR}_{p(y),p(z)}$$

Fig. 2. A failed proof attempt

The coinductive predicate ipath : dtree → step stream → bool ascertains whether a stream of steps is an infinite path in a tree:

$$\frac{T \in Ts \quad \text{ipath } T\ \sigma}{\text{ipath } (\text{Node } (s,r)\ Ts)\ (\text{SCons } (s,r)\ \sigma)} \ \text{IPATH}$$

An *escape path* is a stream of steps that can form an infinite path in a derivation tree. It is defined coinductively as the predicate epath : step stream → bool, which requires that every element in the given stream be obtained by applying an existing rule and choosing one of the resulting states:

$$\frac{\text{eff } r\ s\ ss \quad s' \in ss \quad \text{epath } (\text{SCons } (s',r')\ \sigma)}{\text{epath } (\text{SCons } (s,r)\ (\text{SCons } (s',r')\ \sigma))} \ \text{EPATH}$$

The following lemma is easy to prove by coinduction.

Lemma 2. *For any stream σ and tree T, if wf T and ipath σ T, then epath σ.*

Example 3. The stream

$$(\forall x.\ p(x) \rhd p(y) \wedge p(z)) \cdot (\forall x.\ p(x) \rhd p(z)) \cdot (\forall x.\ p(x), p(y) \rhd p(z))^{\infty}$$

where $s \cdot \sigma = \text{SCons } s\ \sigma$ and $s^{\infty} = s \cdot s \cdot \dots$ is an escape path for the tree of Figure 2.

Since the trees are finitely branching, König's lemma applies. Its proof allows us to study a first simple corecursive definition.

Lemma 3. *If the tree T is infinite, there exists an infinite path σ in T.*

Proof. By the contrapositive of FIN, if Node (s,r) Ts is infinite, there exists an infinite subtree $T \in Ts$. Let f : $\{T \in \text{dtree}.\ \neg\ \text{finite } T\} \to \{T \in \text{dtree}.\ \neg\ \text{finite } T\}$ be a function witnessing this fact—i.e., f T is an immediate infinite subtree of T. The desired infinite path p : $\{T \in \text{dtree}.\ \neg\ \text{finite } T\} \to$ step stream can be defined by primitive corecursion over the codatatype of streams: p $T = \text{SCons } (\text{lab } T)\ (\text{p } (\text{f } T))$. The predicate ipath $(\text{p } T)\ T$ holds by straightforward coinduction on the definition of ipath. □

Countermodel Paths. A countermodel path is a structure that witnesses the unprovability of a state s. Any escape path starting in s is a candidate for a countermodel path, given that it indicates a way to apply the proof rules without reaching any result. For it to be a genuine countermodel path, all possible proofs must have been attempted. More specifically, whenever a rule becomes enabled along the escape path, it is eventually applied later in the sequence. For FOL with sequents as states, such paths can be used to produce actual countermodels by interpreting as true (resp. false) all statements made along the path on the left (resp. right) of the sequents.

A rule r is *enabled* in a state s if it has an effect (i.e., $\exists ss.$ eff $r\ s\ ss$). This is written enabled $r\ s$. For any rule r, stream σ, and predicate $P : \alpha$ stream → bool:

- taken$_r$ σ iff r is taken at the start of the stream (i.e., shead $\sigma = (s,r)$ for some s);
- enabledAt$_r$ σ iff r is enabled at the beginning of the stream (i.e., if shead $\sigma = (s,r')$, then enabled $r\ s$);

$$\frac{\vdots}{\frac{\forall x.\, p(x),\, p(t_1),\, p(t_2),\, p(t_3) \rhd q(y)}{\frac{\forall x.\, p(x),\, p(t_1),\, p(t_2) \rhd q(y)}{\frac{\forall x.\, p(x),\, p(t_1) \rhd q(y)}{\forall x.\, p(x) \rhd q(y)}\, \text{ALLL}_{x,p(x),t_1}}\, \text{ALLL}_{x,p(x),t_2}}\, \text{ALLL}_{x,p(x),t_3}}\, \text{ALLL}_{x,p(x),t_4}}$$

Fig. 3. A derivation tree with a countermodel path

- ev $P\,\sigma$ ("eventually P") iff P is true for some suffix of σ;
- alw $P\,\sigma$ ("always P") iff P is true for all suffixes of σ.

A stream of steps σ is *saturated* if, at each point, any enabled rule is taken at a later point: $\forall r \in$ rule. alw $(\lambda\sigma'.\ \text{enabledAt}_r\, \sigma' \Rightarrow \text{ev taken}_r\, \sigma')\, \sigma$. A *countermodel path* for a state s is a saturated escape path σ starting at s (i.e., shead $\sigma = (s,r)$ for some r).

Example 4. The escape path given in Example 3 is not saturated, because the rule $\text{ALLL}_{x,p(x),z}$ is enabled starting from the first position but never taken.

Example 5. The escape path associated with the tree of Figure 3 is a countermodel path for $\forall x.\, p(x) \rhd q(y)$, assuming that each possible term occurs infinitely often in the sequence $t_1, t_2, \ldots$. The only enabled rules along the path are of the form $\text{ALLL}_{x,p(x),\_}$, and each is always eventually taken.

Completeness. For the proof of completeness, we assume that the set of rules satisfies the following properties:

- *Availability*: At each state, at least one rule is enabled (i.e., $\forall s.\ \exists r.\ \text{enabled } r\, s$).
- *Persistence*: At each state, if a rule is enabled but not taken, it remains enabled (i.e., $\forall s, r, r', s', ss.\ \text{enabled } r'\, s \wedge r' \neq r \wedge \text{eff } r\, s\, ss \wedge s' \in \text{set } ss \Rightarrow \text{enabled } r'\, s'$).

(We will later remove the first condition with Theorem 6.) The above conditions are local properties of the rules' effect, not global properties of the proof system. This makes them easy to verify for particular systems.

Saturation is a stronger condition than the standard properties of fairness and justice [10]. Fairness would require the rules to be continuously enabled to guarantee that they are eventually taken. The property of justice is stronger in that it would require the rules to be enabled infinitely often, but not necessarily continuously. Saturation goes further: If a rule is ever enabled, it will certainly be chosen at a later point. Saturation may seem too strong for the task at hand; however, in the presence of persistence, the notions of fairness, justice, and saturation all coincide.

Theorem 4. *Given a rule system that fulfills availability and persistence, every state admits a proof or a countermodel path.*

Proof. The proof uses the following combinators:

- stake : α stream $\to$ nat $\to$ α list maps ρ and n to the list of the first n elements of ρ;
- smap : $(\alpha \to \beta) \to \alpha$ stream $\to \beta$ stream maps f to every element of the stream;

- nats : nat stream denotes the stream of natural numbers: $0 \cdot 1 \cdot 2 \cdot 3 \cdot \ldots$;
- flat : (α list) stream $\rightarrow \alpha$ stream maps a stream of finite nonempty lists to the stream obtained by concatenating those lists;
- sdropWhile : ($\alpha \rightarrow$ bool) $\rightarrow \alpha$ stream $\rightarrow \alpha$ stream removes the maximal prefix of elements that fulfill a given predicate from a given stream (or returns an irrelevant default value if the predicate holds for all elements of the stream).

We start by constructing a stream of rules fenum in a fair fashion, so that every rule occurs infinitely often in fenum. Let enum be a stream such that its elements cover the entire set rule (which is required to be countable). Take fenum = flat (smap (stake enum) (stail nats)). Thus, if enum $= r_1 \cdot r_2 \cdot r_3 \cdot \ldots$, then fenum = $r_1 \cdot r_1 \cdot r_2 \cdot r_1 \cdot r_2 \cdot r_3 \cdot \ldots$.

Let s be a state. Using fenum, we build a derivation tree T_0 labeled with s such that all its infinite paths are saturated. Let fair be the subset of rule stream consisting of the fair streams. Clearly, any suffix of an element in fair also belongs to fair. In particular, fenum and all its suffixes belong to fair. Given $\rho \in$ fair and $s \in$ state, sdropWhile ($\lambda r. \neg$ enabled r s) ρ has the form SCons r ρ', making r the first enabled rule in ρ. Such a rule exists because, by availability, at least one rule is enabled in s and, by fairness, all the rules occur in ρ. Since enabled r s, we can pick a state set ss such that eff r s ss. We define mkTree : fair $\rightarrow$ state $\rightarrow$ dtree corecursively as mkTree ρ $s =$ Node (s, r) (image (mkTree ρ') ss).

We prove that, for all $\rho \in$ fair and s, the derivation tree mkTree ρ s is well formed and all its infinite paths are saturated. Wellformedness is obvious because at each point the continuation is built starting with the effect of a rule. For saturation, we show that if rule r is enabled at state s and ipath (mkTree ρ s) σ, then r appears along σ (i.e., there exists a state s' such that (s', r) is in σ). This follows by induction on the position of r in ρ, pos r ρ—formally, the length of the shortest list ρ_0 such that $\rho = \rho_0$ @ SCons r _, where @ denotes concatenation. Let r' be the first rule from ρ enabled at state s. If $r = r'$, then mkTree ρ s has label (s, r) already. Otherwise, ρ has the form ρ_1 @ $[r']$ @ ρ', with r not in ρ_1, hence pos r $\rho' <$ pos r ρ. From the definitions of ipath and mkTree, it follows that ipath (mkTree ρ' s') (stail σ) holds for some $s' \in ss$ such that eff r s' ss. By the induction hypothesis, r appears along stail σ, hence along σ as desired. In particular, $T_0 =$ mkTree fenum s is well formed and all its infinite paths are saturated.

Finally, if T_0 is finite, it is the desired finite derivation tree. Otherwise, by Lemma 3 (König) it has an infinite path. This path is necessarily saturated; by Lemma 2, it is the desired countermodel path.   □

Theorem 4 captures the abstract essence of arguments from the literature, although this is sometimes hard to grasp under the thick forest of syntactic details and concrete strategies for fair enumeration: A fair tree is constructed, which attempts a proof; in case of failure, the tree exhibits a saturated escape path.

If we are not interested in witnessing the proof attempt closely, Theorem 4 can be established more directly by bulding the fair path without going through an intermediate fair tree. The key observation is that if a state s has no proof and eff r s ss, there must exist a state $s' \in ss$ that has no proof. (Otherwise, we could compose the proofs of all s' into a proof of s by applying rule r.) Let pick r s ss denote such an s'. We proceed directly to the construction of a saturated escape path as a corecursive predicate mkPath : fair → {$s \in$ state. s has no proof} → step stream following the same idea as for the previous tree construction: mkPath $\rho\, s =$ SCons (s, r) (mkPath ρ' (pick $r\, s\, ss$)), where again SCons $r\, \rho' =$ sdropWhile $(\lambda r. \neg$ enabled $r\, s)\, \rho$ and ss is such that eff $r\, s\, ss$. Fairness of mkPath $\rho\, s$ follows by a similar argument as before for fairness of the tree.

Omitting the Availability Assumption. The above result assumes availability and persistence. Among these assumptions, persistence is essential: It ensures that the constructed fair path is saturated, meaning that every rule available at any point is eventually applied. Availability can be added later to the system without affecting its behavior by introducing a special "idle" rule.

Lemma 5. *A rule system $\mathscr{R} =$ (state, rule, eff) that fulfills persistence can be transformed into an equivalent rule system $\mathscr{R}_{\mathrm{idle}} =$ (state, rule$_{\mathrm{idle}}$, eff$_{\mathrm{idle}}$) that fulfills both persistence and availability, with rule$_{\mathrm{idle}} =$ rule $\cup$ {IDLE} and eff$_{\mathrm{idle}}$ behaving like eff on rule and eff$_{\mathrm{idle}}$ IDLE $s\, ss \Leftrightarrow ss = \{s\}$.*

Proof. Availability for the modified system follows from the continuous enabledness of IDLE. Persistence follows from the persistence of the original system together with the property that IDLE is continuously enabled and does not alter the state. The modified system is equivalent to the original one because IDLE does not alter the state. □

Theorem 6. *Given a rule system $\mathscr{R}$ that fulfills persistence, every state admits a proof over $\mathscr{R}$ or a countermodel path over $\mathscr{R}_{\mathrm{idle}}$.*

Proof. We first apply Theorem 4 to the system $\mathscr{R}_{\mathrm{idle}}$ to obtain that every state admits either a proof or a countermodel path, both in this system. And since $\mathscr{R}$ and $\mathscr{R}_{\mathrm{idle}}$ are equivalent, any proof of $\mathscr{R}_{\mathrm{idle}}$ yields one of $\mathscr{R}$. □

4   Concrete Completeness

The abstract completeness proof is parameterized by a rule system. This section concretizes the result for the Gentzen system from Section 2 to derive the standard completeness theorem. Example 1 recast it as a rule system; we must verify that it fulfills persistence and interpret abstract countermodel paths as actual FOL countermodels.

The Gentzen rules are persistent because they preserve the context surrounding the eigenformulas. For example, an application of AX_a (which affects only the atom a) leaves any potential enabledness of $\mathrm{ALLL}_{x,\varphi,t}$ (which affects only formulas with All at the top) unchanged; moreover, AX_a does not overlap with AX_b for $a \neq b$. A minor subtlety concerns $\mathrm{ALLR}_{x,\varphi}$, which requires the existence of a fresh y; but since the sequents are finite, we can always find a fresh variable in the infinite set var.

On the other hand, availability does not hold for the proof system; for example, the sequent $p(x) \vartriangleright q(x)$ has no enabled rule. Hence, we need Theorem 6 and its IDLE rule.

Lemma 7. *If $\Gamma \vartriangleright \Delta$ admits a countermodel path, there exist a structure $\mathscr{S}$ and a valuation $\xi :$ var $\to S$ such that $\mathscr{S} \not\models_\xi \Gamma \vartriangleright \Delta$.*

Proof. Let σ be a countermodel path for $\Gamma \vartriangleright \Delta$ (i.e., a saturated escape path with $\Gamma \vartriangleright \Delta$ as the first state). Let $\widetilde{\Gamma}$ be the union of the left-hand sides of sequents occurring in σ, and let $\widetilde{\Delta}$ be the union of the corresponding right-hand sides. Clearly, $\Gamma \subseteq \widetilde{\Gamma}$ and $\Delta \subseteq \widetilde{\Delta}$. The pair $(\widetilde{\Gamma}, \widetilde{\Delta})$ can be shown to be well behaved with respect to all the connectives and quantifiers in the following sense:

1. For all atoms a, Atm $a \notin \widetilde{\Gamma} \cap \widetilde{\Delta}$.
2. If Neg $\varphi \in \widetilde{\Gamma}$, then $\varphi \in \widetilde{\Delta}$.
3. If Neg $\varphi \in \widetilde{\Delta}$, then $\varphi \in \widetilde{\Gamma}$.
4. If Conj $\varphi \psi \in \widetilde{\Gamma}$, then $\varphi \in \widetilde{\Gamma}$ and $\psi \in \widetilde{\Gamma}$.
5. If Conj $\varphi \psi \in \widetilde{\Delta}$, then $\varphi \in \widetilde{\Delta}$ or $\psi \in \widetilde{\Delta}$.
6. If All $x\,\varphi \in \widetilde{\Gamma}$, then $\varphi[t/x] \in \widetilde{\Gamma}$ for all t.
7. If All $x\,\varphi \in \widetilde{\Delta}$, there exists a variable y such that $\varphi[y/x] \in \widetilde{\Delta}$.

These properties follow from the saturation of σ with respect to the corresponding rules. The proofs are routine. For example, if All $x\,\varphi \in \widetilde{\Gamma}$ and t is a term, $\text{ALLL}_{x,\varphi,t}$ is enabled in σ and hence eventually taken, ensuring that $\varphi[t/x] \in \widetilde{\Gamma}$.

We construct the concrete (Herbrand) countermodel $\mathscr{S} = (S, F, P)$ as follows. Let the domain S be the set term, and let ξ be the embedding of variables into terms. For each n-ary f and p and each $t_1, \ldots, t_n \in S$, we define $F_f(t_1, \ldots, t_n) = f(t_1, \ldots, t_n)$ and $P_p(t_1, \ldots, t_n) \Leftrightarrow p(t_1, \ldots, t_n) \in \widetilde{\Gamma}$.

To prove $\mathscr{S} \not\models_\xi \Gamma \rhd \Delta$, it suffices to show that $\forall \varphi \in \widetilde{\Gamma}.\ \mathscr{S} \models_\xi \varphi$ and $\forall \varphi \in \widetilde{\Delta}.\ \mathscr{S} \not\models_\xi \varphi$. These two facts follow together by induction on the depth of φ. In the base case, if Atm $a \in \widetilde{\Gamma}$, then $\mathscr{S} \models_\xi$ Atm a follows directly from the definition of $\mathscr{S}$; moreover, if Atm $a \in \widetilde{\Delta}$, then by property 1 Atm $a \notin \widetilde{\Gamma}$, hence again $\mathscr{S} \not\models_\xi$ Atm a follows from the definition of $\mathscr{S}$. The only nontrivial inductive case is All, which requires Lemma 1 (substitution). Assume All $x\,\varphi \in \widetilde{\Gamma}$. By property 6, we have $\varphi[t/x] \in \widetilde{\Gamma}$ for any t. Hence, by the induction hypothesis, $\mathscr{S} \models_\xi \varphi[t/x]$. By Lemma 1, $\mathscr{S} \models_{\xi[x \leftarrow t]} \varphi$ for all t; that is, $\mathscr{S} \models_\xi$ All $x\,\varphi$. The second fact, concerning $\widetilde{\Delta}$, follows similarly from property 7. □

Theorem 8. *For any sequent $\Gamma \rhd \Delta$, we have* $\vdash \Gamma \rhd \Delta \vee (\exists \mathscr{S}, \xi.\ \mathscr{S} \not\models_\xi \Gamma \rhd \Delta)$.

Proof. From Theorem 6 and Lemma 7. □

The rule ALLL from Section 2 stores, in the left context, a copy of the universal formula All $x\,\varphi$ when applied backward. This is crucial for concrete completeness since a fair enumeration should try all the t instances of the universally quantified variable x, which requires availability of All $x\,\varphi$ even after its use. If we labeled ALLL as $\text{ALLL}_{x,\varphi}$ instead of $\text{ALLL}_{x,\varphi,t}$, thereby delegating the choice of t to the nondeterminism of eff, the system would still be persistent as required by the abstract completeness proof, but Lemma 7 (and hence concrete completeness) would not hold—more specifically, property 6 from the lemma's proof would fail.

5 Further Concrete Instances

Theorem 6 is applicable to classical FOL Gentzen systems from the literature, in several variants: with sequent components represented as lists, multisets or sets, one-sided or two-sided, and so on. This includes the systems G', GCNF', G, and G$_=$ from Gallier [11] and the systems G1, G2, G3, GS1, GS2, and GS3 from Troelstra and Schwichtenberg [37]. Persistence is easy to check. The syntax-independent part of the argument is provided by Theorem 6, while an ad hoc step analogous to Lemma 7 is required to build a concrete countermodel.

Several FOL refutation systems based on tableaux or resolution are instances of the abstract theorem, providing that we read the abstract notion of "proof" as "refutation" and "countermodel" as "model." Nondestructive tableaux [15]—including those presented in Bell and Machover [1] and in Fitting [9]—are usually persistent when regarded as derivation systems. After an application of Theorem 6, the argument for interpreting the abstract model is similar to that for Gentzen systems (Lemma 7).

Regrettably, abstract completeness is not directly applicable beyond classical logic. It is generally not clear how to extract a specific model from a nonstandard logic from an abstract (proof-theoretic) model. Another issue is that standard sequent systems for nonclassical variations of FOL such as modal or intuitionistic logics do not satisfy persistence. A typical right rule for the modal operator $\Box$ ("must") is as follows [37]:

$$\frac{\Box\,\Gamma \rhd \Diamond\,\Delta, \varphi}{\Box\,\Gamma \rhd \Diamond\,\Delta, \Box\,\varphi} \text{ MustR}$$

To be applicable, the rule requires that all the formulas in the context surrounding the eigenformula have $\Box$ or $\Diamond$ at the top. Other rules may remove these operators, or introduce formulas that do not have them, thus disabling MustR.

Recent work targeted at simplifying completeness arguments [26] organizes modal logics as labeled transition systems, for which Kripke completeness is derived. In the proposed systems, the above rule becomes

$$\frac{\Gamma, w\,R\,w' \rhd \Delta, w' : \varphi}{\Gamma \rhd \Delta, w : \Box\,\varphi} \text{ MustR}' \quad (w' \text{ fresh})$$

The use of labels for worlds (w, w') and the bookkeeping of the accessibility relation R makes it possible to recast the rule so that only resilient facts are ever assumed about the context. The resulting proof system satisfies persistence, enabling Theorem 6. The Kripke countermodel construction is roughly as for classical FOL Gentzen systems.

6   Formalization and Implementation

The definitions, lemmas, and theorems presented in Sections 2 to 4 are formalized in the proof assistant Isabelle/HOL. The instantiation step of Section 4 is formalized for a richer version of FOL, with sorts and interpreted equality, as required by our motivating application (efficient encodings of sorts in unsorted FOL [4]). The formal development is publicly available [6, 7].

The necessary codatatypes and corecursive definitions are realized using a recently introduced definitional package [5]. The tree codatatype illustrates the support for corecursion through permutative data structures (with non-free constructors) such as finite sets, a feature that is not available in any other proof assistant.

For generating code, we make the additional assumption that the effect relation corresponds to a partial function eff′ : rule → state → (state fset) option, where the Isabelle datatype α option enriches a copy of α with a special value None.[1] From this function,

---

[1] In the proof system from Example 1, eff is not deterministic due to the rule ALLR. It can be made deterministic by refining the rule with a systematic choice of the fresh variable y.

we build the relational eff as the partial function's graph. Isabelle's code generator [14] can then produce Haskell code for the computable part of our completeness proof—the abstract prover mkTree, defined corecursively in the proof of Theorem 4:

```
data Stream a = SCons a (Stream a)
newtype FSet a = FSet [a]
data Tree a = Node a (FSet (Tree a))

fmap f (FSet xs) = FSet (map f xs)

sdropWhile p (SCons a σ) =
  if p a then sdropWhile p σ else SCons a σ

mkTree eff ρ s =
  Node (s, r) (fmap (mkTree eff ρ') (fromJust (eff r s)))
  where SCons r ρ' = sdropWhile (\r -> not (isJust (eff r s))) ρ
```

Finite sets are represented as lists. The functions isJust : α option → bool and fromJust : α option → α are the Haskell-style discriminator and selector for option. Since the Isabelle formalization is parametric over rule systems (state, rule, eff), the code for mkTree explicitly takes eff as a parameter.

Although the code generator was not designed with codatatypes in mind, it is general enough to handle them. Internally, it reduces Isabelle specifications to higher-order rewrite systems [24] and generates functional code in Haskell, OCaml, Scala, or Standard ML. Partial correctness is guaranteed regardless of the target language's evaluation strategy. However, for the guarantee to be non-vacuous for corecursive definitions, one needs a language with a lazy evaluation strategy, such as Haskell.

The verified contract of the program reads as follows: Given an available and persistent rule system (state, rule, eff), a fair rule enumeration ρ, and a state s representing the formula to prove, mkTree eff ρ s either yields a finite derivation tree of s or produces an infinite fair derivation tree whose infinite paths are all countermodel paths. These guarantees involve only partial correctness of ground term evaluation.

The generated code is a generic countermodel-producing semidecision procedure parameterized by the the proof system. Moreover, the fair rule enumeration parameter ρ can be instantiated to various choices that may perform better than the simple scheme described in Section 3.

7 Related Work

This paper joins a series of pearls aimed at reclaiming mathematical concepts and results for coinductive methods, including streams [31, 35], regular expressions [32, 34], and automata [33]. Some developments pass the ultimate test of formalization, usually in Agda and Coq, the codatatype-aware proof assistants par excellence: the sieve of Eratosthenes [3], real number basics [8], and temporal logic for red–blue trees [25].

So why write yet another formalized pearl involving coinduction? First, because we finally could—with the new codatatype package, Isabelle has caught up with its rivals in this area. Second, because, although codatatypes are a good match for the completeness theorem, there seems to be no proof in the literature that takes advantage of this.

While there are many accounts of the completeness theorem for FOL and related logics, most of them favor the more mathematical Henkin style, which obfuscates the rich structure of proof and failure. This preference has a long history. It is positively motivated by the ability to support uncountable languages. More crucially, it is negatively motivated by the lack of rigor perceived in the alternative: "geometric" reasoning about infinite trees. Negri [26] gives a revealing account in the context of modal logic, quoting reviews that were favorable to Kripke's completeness result [21] but critical of his informal argument based on infinite tableau trees.[2] Kaplan [19] remarks that "although the author extracts a great deal of information from his tableau constructions, a completely rigorous development along these lines would be extremely tedious."

A few textbooks venture in a proof-theoretic presentation of completeness, notably Gallier's [11]. Such a treatment highlights not only the structure, but also the algorithmic content of the proofs. The price is usually a lack of rigor, in particular a gap between the definition of derivation trees and its use in the completeness argument. This lack of rigor should not be taken lightly, as it may lead to serious ambiguities or errors: In the context of a tableau completeness proof development, Hähnle [15] first performs an implicit transition from finite to possibly infinite tableaux, and then claims that tableau chain suprema exist by wrongly invoking Zorn's lemma [15, Definition 3.16].[3]

The completeness theorem has been mechanized before in proof assistants. Schlöder and Koepke, in Mizar [36], formalize a Henkin-style argument for possibly uncountable languages. Building on an early insight by Krivine [22] concerning the expressibility of the completeness proof in intuitionistic second-order logic, Ilik [17] analyzes Henkin-style arguments for classical and intuitionistic logic with respect to standard and Kripke models and formalizes them in Coq (without employing codatatypes).

At least three proofs were developed using HOL-based provers. Harrison [16], in HOL Light, and Berghofer [2], in Isabelle, formalize Henkin-style arguments. Ridge and Margetson [23, 30], in Isabelle, employ proof trees constructed as graphs of nodes that carry their levels as natural numbers. This last work has the merits of analyzing the computational content of proofs in the style of Gallier [11] and discussing an OCaml implementation. Our formalization relates to this work in a similar way to which our presentation relates to Gallier's: The newly introduced support for codatatypes and corecursion in Isabelle provides suitable abstraction mechanisms for reasoning about infinite trees, avoiding boilerplate for tree manipulation based on numeric indexing. Moreover, codatatypes are mapped naturally to Haskell types, allowing Isabelle's code generator to produce certified Haskell code. Finally, our proof is more abstract and applies to several variants of FOL and beyond.

8   Conclusion

The completeness theorem is a fundamental result about classical logic. Its proof is presented in many variants in the literature. Few of these presentations emphasize the algorithmic content, and none of them uses codatatypes. Gallier's pseudo-Pascal code

---

[2] And Kripke's degree of rigor in this early paper is not far from today's state of the art in proof theory; see, e.g., Troelstra and Schwichtenberg [37].

[3] This is the only error we found in this otherwise excellent chapter on tableaux.

is inspiring, but we prefer "pseudo-Haskell," i.e., Isabelle/HOL with codatatypes. In our view, coinduction is the key to formulate an account that is both mathematically rigorous and abundant in algorithmic content. The definition of the abstract prover mkTree is stated rigorously, is accessible to functional programmers, and replaces pages of verbose descriptions.

The advantages of machine-checked metatheory are well known from programming language research, where new results are often formalized and proof assistants are used in the classroom. This paper, like its predecessor [4], reported on some steps we have taken to apply the same methods to formal logic and automated reasoning.

Acknowledgment. Tobias Nipkow made this work possible. Mark Summerfield and the anonymous reviewers suggested many textual improvements to earlier versions of this paper. Blanchette is supported by the Deutsche Forschungsgemeinschaft (DFG) project Hardening the Hammer (grant Ni 491/14-1). Popescu is supported by the DFG project Security Type Systems and Deduction (grant Ni 491/13-2) as part of the program Reliably Secure Software Systems (RS^3, priority program 1496). Traytel is supported by the DFG program Program and Model Analysis (PUMA, doctorate program 1480). The authors are listed alphabetically.

References

1. Bell, J.L., Machover, M.: A Course in Mathematical Logic. North-Holland (1977)
2. Berghofer, S.: First-order logic according to Fitting. In: Klein, G., Nipkow, T., Paulson, L. (eds.) Archive of Formal Proofs (2007),
   http://afp.sf.net/entries/FOL-Fitting.shtml
3. Bertot, Y.: Filters on coinductive streams, an application to Eratosthenes' sieve. In: Urzyczyn, P. (ed.) TLCA 2005. LNCS, vol. 3461, pp. 102–115. Springer, Heidelberg (2005)
4. Blanchette, J.C., Popescu, A.: Mechanizing the metatheory of Sledgehammer. In: Fontaine, P., Ringeissen, C., Schmidt, R.A. (eds.) FroCoS 2013. LNCS, vol. 8152, pp. 245–260. Springer, Heidelberg (2013)
5. Blanchette, J.C., Hölzl, J., Lochbihler, A., Panny, L., Popescu, A., Traytel, D.: Truly modular (co)datatypes for Isabelle/HOL. In: Klein, G., Gamboa, R. (eds.) ITP 2014. LNCS, Springer (2014)
6. Blanchette, J.C., Popescu, A., Traytel, D.: Abstract completeness. In: Klein, G., Nipkow, T., Paulson, L. (eds.) Archive of Formal Proofs (2014),
   http://afp.sf.net/entries/Abstract_Completeness.shtml
7. Blanchette, J.C., Popescu, A., Traytel, D.: Formal development associated with this paper (2014), http://www21.in.tum.de/~traytel/compl_devel.zip
8. Ciaffaglione, A., Gianantonio, P.D.: A certified, corecursive implementation of exact real numbers. Theor. Comput. Sci. 351(1), 39–51 (2006)
9. Fitting, M.: First-Order Logic and Automated Theorem Proving, 2nd edn. Graduate Texts in Computer Science. Springer (1996)
10. Francez, N.: Fairness. Texts and Monographs in Computer Science, Springer (1986)
11. Gallier, J.H.: Logic for Computer Science: Foundations of Automatic Theorem Proving. Computer Science and Technology. Harper & Row (1986)
12. Gödel, K.: Über die Vollständigkeit des Logikkalküls. Ph.D. thesis, Universität Wien (1929)
13. Gordon, M.J.C., Melham, T.F. (eds.): Introduction to HOL: A Theorem Proving Environment for Higher Order Logic. Cambridge University Press (1993)

14. Haftmann, F., Nipkow, T.: Code generation via higher-order rewrite systems. In: Blume, M., Kobayashi, N., Vidal, G. (eds.) FLOPS 2010. LNCS, vol. 6009, pp. 103–117. Springer, Heidelberg (2010)
15. Hähnle, R.: Tableaux and related methods. In: Robinson, A., Voronkov, A. (eds.) Handbook of Automated Reasoning, vol. I, pp. 100–178. Elsevier (2001)
16. Harrison, J.: Formalizing basic first order model theory. In: Grundy, J., Newey, M. (eds.) TPHOLs 1998. LNCS, vol. 1479, pp. 153–170. Springer, Heidelberg (1998)
17. Ilik, D.: Constructive Completeness Proofs and Delimited Control. Ph.D. thesis, École Polytechnique (2010)
18. Jacobs, B., Rutten, J.: A tutorial on (co)algebras and (co)induction. Bull. Eur. Assoc. Theor. Comput. Sci. 62, 222–259 (1997)
19. Kaplan, D.: Review of Kripke (1959) [21]. J. Symb. Log. 31(1966), 120–122 (1966)
20. Kleene, S.C.: Mathematical Logic. John Wiley & Sons (1967)
21. Kripke, S.: A completeness theorem in modal logic. J. Symb. Log. 24(1), 1–14 (1959)
22. Krivine, J.L.: Une preuve formelle et intuitionniste du théorème de complétude de la logique classique. Bull. Symb. Log. 2(4), 405–421 (1996)
23. Margetson, J., Ridge, T.: Completeness theorem. In: Klein, G., Nipkow, T., Paulson, L. (eds.) Archive of Formal Proofs (2004),
http://afp.sf.net/entries/Completeness.shtml
24. Mayr, R., Nipkow, T.: Higher-order rewrite systems and their confluence. Theor. Comput. Sci. 192(1), 3–29 (1998)
25. Nakata, K., Uustalu, T., Bezem, M.: A proof pearl with the fan theorem and bar induction: Walking through infinite trees with mixed induction and coinduction. In: Yang, H. (ed.) APLAS 2011. LNCS, vol. 7078, pp. 353–368. Springer, Heidelberg (2011)
26. Negri, S.: Kripke completeness revisited. In: Primiero, G., Rahman, S. (eds.) Acts of Knowledge: History, Philosophy and Logic: Essays Dedicated to Göran Sundholm, pp. 247–282. College Publications (2009)
27. Nipkow, T., Klein, G.: Concrete Semantics: A Proof Assistant Approach. Springer (to appear), http://www.in.tum.de/~nipkow/Concrete-Semantics
28. Nipkow, T., Paulson, L.C., Wenzel, M.: Isabelle/HOL. LNCS, vol. 2283. Springer, Heidelberg (2002)
29. Pfenning, F.: Review of "Jean H. Gallier: Logic for Computer Science. J. Symb. Log. 54(1), 288–289 (1989)
30. Ridge, T., Margetson, J.: A mechanically verified, sound and complete theorem prover for first order logic. In: Hurd, J., Melham, T. (eds.) TPHOLs 2005. LNCS, vol. 3603, pp. 294–309. Springer, Heidelberg (2005)
31. Roşu, G.: Equality of streams is a Π_2^0-complete problem. In: Reppy, J.H., Lawall, J.L. (eds.) ICFP 2006. ACM (2006)
32. Roşu, G.: An effective algorithm for the membership problem for extended regular expressions. In: Seidl, H. (ed.) FoSSaCS 2007. LNCS, vol. 4423, pp. 332–345. Springer, Heidelberg (2007)
33. Rutten, J.J.M.M.: Automata and coinduction (an exercise in coalgebra). In: Sangiorgi, D., de Simone, R. (eds.) CONCUR 1998. LNCS, vol. 1466, pp. 194–218. Springer, Heidelberg (1998)
34. Rutten, J.J.M.M.: Regular expressions revisited: A coinductive approach to streams, automata, and power series. In: Backhouse, R., Oliveira, J.N. (eds.) MPC 2000. LNCS, vol. 1837, pp. 100–101. Springer, Heidelberg (2000)
35. Rutten, J.J.M.M.: Elements of stream calculus (an extensive exercise in coinduction). Electr. Notes Theor. Comput. Sci. 45, 358–423 (2001)
36. Schlöder, J.J., Koepke, P.: The Gödel completeness theorem for uncountable languages. Formalized Mathematics 20(3), 199–203 (2012)
37. Troelstra, A.S., Schwichtenberg, H.: Basic Proof Theory, 2nd edn. Cambridge University Press (2000)

A Focused Sequent Calculus
for Higher-Order Logic

Fredrik Lindblad

University of Gothenburg, Chalmers University of Technology,
Gothenburg, Sweden

Abstract. We present a focused intuitionistic sequent calculus for high-er-order logic. It has primitive support for equality and mixes λ-term conversion with equality reasoning. Classical reasoning is enabled by extending the system with rules for *reductio ad absurdum* and the axiom of choice. The resulting system is proved sound with respect to Church's simple type theory. The soundness proof has been formalized in Agda. A theorem prover based on bottom-up search in the calculus has been implemented. It has been tested on the TPTP higher-order problem set with good results. The problems for which the theorem prover performs best require higher-order unification more frequently than the average higher-order TPTP problem. Being strong at higher-order unification, the system may serve as a complement to other theorem provers in the field.

1 Introduction

Benchmarking and development of automated reasoning tools for higher-order logic has been facilitated by the introduction of a new syntax and a dedicated set of higher-order problems at TPTP in 2009 [1,2]. However, higher-order unification, which is an important component of higher-order reasoning, is still a challenge for several of the established theorem proving systems.

Agda [3] is an intuitionistic higher-order logical framework based on Martin-Löf type theory. As for any interactive theorem prover, Agda users are facilitated by local automation. Agda has a plugin for this, called *Agsy* [4]. In order to be able to evaluate the performance of Agsy, an adaptation to classical higher-order logic has been developed.

This variant of Agsy, called *AgsyHOL*, turned out to be competitive in the context of TPTP. AgsyHOL performs well not least on problems involving higher-order unification. Out of the problems that AgsyHOL either solves in less than 50 milliseconds or at least 100 times faster than any other system on TPTP online (31 problems), roughly 80% of the solutions contain λ-abstractions, i.e. involve higher-order unification. For all problems that AgsyHOL solves (1722 problems) the percentage is 37.

A major idea behind Agsy and AgsyHOL is to take an inference system constructed with proof search in mind and combine it with a simple search mechanism and a small layer of search control heuristics. More precisely, the semantics

S. Demri, D. Kapur, and C. Weidenbach (Eds.): IJCAR 2014, LNAI 8562, pp. 61–75, 2014.

is represented by a proof checking algorithm defined over an explicit proof language. Each construct in the language corresponds to one of the inference rules. Proof search is achieved by applying narrowing [5], which lazily instantiates the proof candidate.

The setup results in a clear separation between logic and control. It facilitates reasoning about correctness since it is (an implementation of) the inference rules themselves which drive the search. The implementation of AgsyHOL is also relatively small, approximately 2000 lines of code excluding parsing, printing and the generic narrowing search.

Both Agsy and AgsyHOL are based on focused intuitionistic sequent calculus. Related proof search oriented inference systems include LJT [6], PTSC [7], and focused derivations [8].

In the context of higher-order logic theorem proving proof by refutation is the dominating approach, whereas AgsyHOL produces derivation proofs. Thus it can fill a role when proof readability and integration with interactive systems are of interest.

The inference system which AgsyHOL is based on includes rules for classical reasoning and has been proved correct with respect to Church's simple type theory (STT). The soundness proof has been formalized in Agda, and AgsyHOL produces derivations that can be independently checked by Agda relative to the soundness proof.

Completeness of the inference system has not been formally addressed. Although interesting on the theoretical level, we have instead focused on the performance of the implementation.

Section 2 presents the inference system which forms the basis of AgsyHOL. Section 3 discusses the heuristics which control search on top of the inference rules. Section 4 describes some more technical details of the implementation. Section 5 presents the empirical results of running AgsyHOL on the higher-order problems in TPTP. Finally, section 6 concludes and gives some pointers for future work.

2   Inference System

The system is based on sequent calculus, which is more suitable for proof search than natural deduction. In sequent calculus right and left rules, which correspond to introductions and eliminations, are both guided by formula deconstruction in a backward search. In natural deduction this is only true for introduction rules, whereas elimination rules can be arbitrarily and indefinitely applied backwards.

More precisely, the system is based on the sequent calculus for intuitionistic logic and originates from Gentzen's system LJ. The main reason for using an intuitionistic system in a classical context is that the initial motivation of the work was to evaluate the performance of Agsy, which targets intuitionistic logic.

The original system LJ suffers from significant and unnecessary nondeterminism caused by the proliferation of antecedents and the arbitrary order in which to deconstruct them. One improvement is the concept of *uniform proofs* [9].

The idea is to identify a subset of the deconstructing rules for which the order of application is unimportant. Such rules are applied in a fixed order, avoiding the corresponding branching of the search tree. However, no restrictions corresponding to uniform proofs are built in to the presented calculus. Instead, corresponding improvements were made on the search control level in the implementation, see section 3.

Another improvement over LJ is *back-chaining* or *focused derivations* [8]. This means that the calculus has two judgment forms, one for introduction and one for elimination. When switching, in the perspective of bottom-up proof construction, from introduction to elimination, one antecedent is selected. It becomes the focus of the derivation. The antecedent is then deconstructed through zero or more eliminations. The result of the elimination steps must at the end be used either to prove the succedent or to justify absurd or disjunctive elimination. This technique is a fundamental part of the presented system.

Investigations using similar calculi for proof search in higher-order logic include the presentation of a focused sequent calculus for pure type systems by Lengrand et. al. [7], called PTSC. This work is based on system LJT [6], a type theoretic variation of system LJ due to Herbelin. In contrast to these systems, the presented one has a full set of logical connectives, including equality, and is extended with rules for classical reasoning.

2.1 Syntax and Judgments

The type and formula syntax is shown in figure 1. Type subscripts in formulas will be suppressed whenever the type is clearly resolved by the context. Syntax variables are denoted by γ, τ for types and by plain capital letters for formulas. The inference system will be denoted $FSC_{=}^{ho}$.

As already mentioned, there are two interdependent judgments in intuitionistic sequent calculus with focusing. They will be called the *main judgment* and

$$
\begin{array}{llll}
\gamma, \tau ::= \imath & \text{(type of individuals)} & A, B, \ldots ::= x_\gamma & \text{(variable)} \\
\quad | \ o & \text{(type of truth values)} & \quad | \ A_{\gamma\tau} B_\tau & \text{(application)} \\
\quad | \ \gamma\tau & \text{(function type)} & \quad | \ \lambda x_\tau.\ A_\gamma & \text{(abstraction)} \\
& & \quad | \ \iota_{\gamma(o\gamma)} & \text{(choice operator)} \\
& & \quad | \ \sim A_o & \text{(negation)} \\
& & \quad | \ A_o \vee B_o & \text{(disjunction)} \\
& & \quad | \ A_o \wedge B_o & \text{(conjunction)} \\
& & \quad | \ A_o \supset B_o & \text{(implication)} \\
& & \quad | \ \forall x_\gamma.\ A_o & \text{(universal quantification)} \\
& & \quad | \ \exists x_\gamma.\ A_o & \text{(existential quantification)} \\
& & \quad | \ \top & \text{(truth)} \\
& & \quad | \ \bot & \text{(falsity)} \\
& & \quad | \ A_\gamma = B_\gamma & \text{(equality)}
\end{array}
$$

Fig. 1. Type and Formula Syntax

focusing judgment. The main and focusing judgments have the forms $\Gamma \vdash A$ and $\Gamma; B \vdash A$, where Γ represents a set of antecedent formulas, $A_1, A_2, \ldots, A_n$. An empty set of antecedents is denoted ϵ. The notation essentially follows Girard [10] and system LJT [6].

The meaning of both judgments is that the proposition represented by A is derivable from the propositions in Γ. In addition, the rules of the focusing judgment restrict the derivations in such a way that the focusing mechanism described above is imposed.

Apart from this there is a conversion judgment which determines when two formulas are equal. Instead of an axiomatic representation of equality reasoning on the level of the main judgment, this is expressed in the conversion judgment. Each step in an equality chain is justified by an antecedent. Therefore the conversion judgment, just as the other judgments, is parameterized by a set of antecedents, Γ. The conversion judgment has the form $\Gamma \vdash A \leftrightarrow_\gamma B$.

Each judgment form has some variants which express search oriented restrictions. These variants are presented along the way in the respective sub-sections.

The inference rules rely on a few standard notions, whose definitions are omitted. Replacing x by A in B is denoted $[A/x]B$. A formula A being well-formed of type γ is denoted $wf_\gamma A$. A formula A being a member of a set Γ is denoted $A \in \Gamma$, Reduction of a formula A to head normal form, A', is denoted $A \to_\beta A'$.

2.2 Main Judgment

A good principle to adhere to when constructing a calculus intended for proof search in predicate logic is to inspect the structure of a formula in a lazy manner. The rationale is that the structure of the formula may be currently unknown. By

$$\frac{A \to_\beta B \quad \Gamma \vdash_i B}{\Gamma \vdash A} \text{ intro} \qquad \frac{\Gamma \vdash \sim\sim A}{\Gamma \vdash A} \text{ RAA} \qquad \frac{B \in \Gamma \quad \Gamma; B \vdash^\uparrow A}{\Gamma \vdash A} \text{ focus}$$

$$\frac{\Gamma; \forall x_{o\gamma}.\ (\exists y_\gamma.\ x\,y) \supset x\,(\iota_{\gamma(o\gamma)}\,x) \vdash^\uparrow A}{\Gamma \vdash A} \text{ AC} \qquad \frac{\Gamma \vdash A}{\Gamma \vdash_i A \vee B} \text{ } \vee\text{-}l_l$$

$$\frac{\Gamma \vdash B}{\Gamma \vdash_i A \vee B} \text{ } \vee\text{-}l_r \qquad \frac{\Gamma \vdash A \quad \Gamma \vdash B}{\Gamma \vdash_i A \wedge B} \text{ } \wedge\text{-I} \qquad \frac{}{\Gamma \vdash_i \top} \text{ } \top\text{-I}$$

$$\frac{\Gamma, A \vdash B}{\Gamma \vdash_i A \supset B} \supset\text{-I} \qquad \frac{\Gamma, A \vdash \bot}{\Gamma \vdash_i \sim A} \sim\text{-I} \qquad \frac{\Gamma \vdash A}{\Gamma \vdash_i \forall x_\gamma.\ A} \text{ } \forall\text{-I}$$

$$\frac{wf_\gamma B \quad \Gamma \vdash [B/x]A}{\Gamma \vdash_i \exists x_\gamma.\ A} \exists\text{-I} \qquad \frac{\Gamma \vdash A \leftrightarrow_\gamma B}{\Gamma \vdash_i A_\gamma = B_\gamma} =\text{-I}$$

Fig. 2. Inference rules for $\Gamma \vdash A$

waiting as long as possible until splitting search according to the different possible constructs of the formula, as much restricting judgments as possible come into play. Therefore β-reduction of a formula is performed in direct connection to matching against its head form.

Figure 2 shows the rules of the main judgment, including the sub-form $\Gamma \vdash_i A$, which contains all introduction rules. The intro rule has this judgment as premise, but with the succedent replaced by its head normal form.

In the case of equality, $= \text{-I}$, the introduction rule simply has a conversion derivation as premise. In the introduction rule for universal quantification, $\forall\text{-I}$, x implicitly serves as the quantified variable. In the formalized soundness proof as well as in the implementation of AgsyHOL the judgments have an explicit context of universally quantified variables.

The focus rule identifies an antecedent and focuses on this in the premise. The upward arrow in the focusing judgment is explained in the next sub-section.

In order to allow a close relationship between a proof and the original problem that it solves, $FSC_{=}^{ho}$ has a rich set of logical connectives. This is also a natural consequence of the fact that the basis of the system is intuitionistic. However, the rules for negation corresponds to the standard representation, i.e. $\sim A \equiv A \supset \bot$, and could be excluded in an intuitionistic setting.

Apart from the rules mentioned so far there are two more, which enable classical reasoning, one for *reductio ad absurdum*, RAA, and one for the axiom of choice, AC. These rules could have been added as axioms except for the inability to quantify over types in the case of AC. However, both are included as rules of the system in order to facilitate search control. The AC rule, just like focus, has an focusing judgment as premise. It inserts the axiom of choice as the antecedent in focus.

2.3   Focusing Judgment

The focusing judgment has two modes, denoted $\Gamma; B \vdash^\uparrow A$ and $\Gamma; B \vdash^\downarrow A$, respectively. The first mode is used in the main judgment and the second in the conversion judgment, which is presented in next sub-section.

On top of these two modes, there is a sub-form which has the same function as the sub-form of the main judgment, but expresses elimination instead of introduction rules. It is denoted $\Gamma; B \vdash_e A$, and is used to express the restriction of β-reduction for antecedents.

Figure 3 shows the rules for the focusing judgment. The rules in which the turnstile is decorated with neither up nor down arrow apply to both modes of the judgment, with the same direction of the arrow both in the premise and conclusion. This accounts both for the focusing judgment itself and its elimination sub-form.

Derivations of the focusing judgment are sequences of elimination steps. An elimination sequence for the mode $\Gamma; B \vdash^\uparrow A$ terminates either with using the antecedent in focus to prove the succedent (the use rule), or the elimination of disjunction, absurdity or negation. A sequence for the mode $\Gamma; B \vdash^\downarrow A$ ends

$$\frac{A \to_\beta C \quad \Gamma; C \vdash_e B}{\Gamma; A \vdash B} \; \text{elim} \qquad \frac{A \to_\beta A' \quad B \to_\beta B' \quad \Gamma \vdash_s B' \leftrightarrow_o A'}{\Gamma; A \vdash^\uparrow B} \; \text{use}$$

$$\frac{\Gamma; A \vdash C}{\Gamma; A \wedge B \vdash_e C} \; \wedge\text{-E}_l \qquad \frac{\Gamma; B \vdash C}{\Gamma; A \wedge B \vdash_e C} \; \wedge\text{-E}_r$$

$$\frac{\Gamma \vdash A \quad \Gamma; B \vdash C}{\Gamma; A \supset B \vdash_e C} \; \supset\text{-E} \qquad \frac{wf_\gamma C \quad \Gamma; [C/x]A \vdash B}{\Gamma; \forall x_\gamma.\, A \vdash_e B} \; \forall\text{-E}$$

$$\frac{\Gamma; [(\iota_{\gamma(o\gamma)} (\lambda x_\gamma.\, A))/x]A \vdash B}{\Gamma; \exists x_\gamma.\, A \vdash_e B} \; \exists\text{-E} \qquad \frac{}{\Gamma; \bot \vdash^\uparrow_e A} \; \bot\text{-E}$$

$$\frac{\Gamma; (A \supset B) \wedge (B \supset A) \vdash C}{\Gamma; A_o = B_o \vdash_e C} \; =\text{-E}_{\text{bool}}$$

$$\frac{\Gamma \vdash A \supset C \quad \Gamma \vdash B \supset C}{\Gamma; A \vee B \vdash^\uparrow_e C} \; \vee\text{-E} \qquad \frac{\Gamma \vdash A}{\Gamma; \sim A \vdash^\uparrow_e B} \; \sim\text{-E}$$

$$\frac{}{\Gamma; A = B \vdash^\downarrow_e A = B} \; \text{use}_{\text{eq}} \qquad \frac{}{\Gamma; A = B \vdash^\downarrow_e B = A} \; \text{use}_{\text{eq}}\text{-symm}$$

Fig. 3. Inference rules for $\Gamma; A \vdash B$

either with use_{eq} or use_{eq}-symm, inferring an equality or its symmetric correspondence.

The existential elimination rule uses the choice operator applied to the predicate in order to represent the witness. In a intuitionistic setting a variable is introduced for this purpose, or, in the type theoretic case, the first projection of the corresponding Σ-type.

The rule $=\text{-E}_{\text{bool}}$ expresses the converse of Boolean extensionality. This rule eliminates the need to allow equality reasoning prior to head normal reduction in the introduction and elimination rules.

When using natural deduction for proof search, introduction steps must be constructed bottom up and elimination steps top down. In bottom-up construction of elimination steps, any rule is applicable indefinitely. A top-down construction on the other hand guides the search by deconstructing a concrete antecedent incrementally. Focused sequent calculus inverts the structure of eliminations such that a uniform bottom-up search effectively implements that division into an introductions part at the bottom of the derivation and an eliminations part at the top.

2.4 Conversion Judgment

The conversion judgment also has a sub-form, denoted $\Gamma \vdash_s A \leftrightarrow_\gamma B$, which restricts derivations to top-level simplification of the formulas. Figure 4 shows the rules for the conversion judgment.

$$\frac{\begin{array}{c} C \in \Gamma \quad \Gamma; C \vdash^{\downarrow} D_\gamma = E_\gamma \quad A \to_\beta A' \quad D \to_\beta D' \\ \Gamma \vdash_s A' \leftrightarrow_\gamma D' \quad \Gamma \vdash E \leftrightarrow_\gamma B \end{array}}{\Gamma \vdash A \leftrightarrow_\gamma B} \; \text{step}$$

$$\frac{A \to_\beta A' \quad B \to_\beta B' \quad \Gamma \vdash_s A' \leftrightarrow_\gamma B'}{\Gamma \vdash A \leftrightarrow_\gamma B} \; \text{simp}$$

$$\frac{\Gamma \vdash (A \supset B) \wedge (B \supset A)}{\Gamma \vdash A \leftrightarrow_o B} \; \text{ext}_{\text{bool}} \qquad \frac{\Gamma \vdash A\,x \leftrightarrow_\tau B\,x}{\Gamma \vdash A \leftrightarrow_{\tau\gamma} B} \; \text{ext}_{\text{fun}}$$

$$\frac{}{\Gamma \vdash_s x \leftrightarrow_\gamma x} \; \text{simp}_{\text{var}} \qquad \frac{\Gamma \vdash A \leftrightarrow_{\tau\gamma} C \quad \Gamma \vdash B \leftrightarrow_\gamma D}{\Gamma \vdash_s A_{\tau\gamma} B_\gamma \leftrightarrow_\tau C_{\tau\gamma} D_\gamma} \; \text{simp}_{\text{app}}$$

$$\frac{\Gamma \vdash A \leftrightarrow_\tau B}{\Gamma \vdash_s \lambda x_\gamma.\, A_\tau \leftrightarrow_{\tau\gamma} \lambda x_\gamma.\, B_\tau} \; \text{simp}_{\text{lam}}$$

$$\frac{\Gamma \vdash A \leftrightarrow_o C \quad \Gamma \vdash B \leftrightarrow_o D}{\Gamma \vdash_s A_o \vee B_o \leftrightarrow_o C_o \vee D_o} \; \text{simp}_{\text{disj}} \qquad \dots$$

Fig. 4. Inference rules for $\Gamma \vdash A \leftrightarrow_\gamma B$ (incomplete list)

In the step rule an antecedent, C, justifies, after a sequence of eliminations, rewriting the left-hand side of the conversion. The direction of rewriting is imposed by restricting the conversion of the left-hand side to simplification. In other words, the left-hand side of the succedent and one of the sides of the antecedent must have identical heads, module β-reduction. In the subsequent conversion judgment the opposite side of the antecedent appears as the left-hand side of the succedent.

The simp rule terminates an equality chain by requiring the formulas of the conversion to have identical heads. However, the premises in the simplification rules refer back to the full conversion judgment, allowing rewriting of the subformulas to take place.

As seen in the previous sub-section the rules for elimination of disjunctions, falsity and negations are not available in the inference mode of the focusing judgment, and therefore not allowed in equality reasoning. This restriction does not compromise completeness and empirical results indicate that enforcing it improves performance.

In the simplification rule for λ-abstractions, simp$_{\text{lam}}$, both sides are assumed to bind the same variable, x, and it is implicitly quantified in the premise.

The rule simp$_{\text{lam}}$ is redundant as far as completeness is concerned, due to the presence of ext$_{\text{fun}}$, which also expresses η-conversion. However, using ext$_{\text{fun}}$ when the left- and right-hand sides are both λ-abstractions introduces unnecessary substitutions in the formulas, which complicate proof search.

The simplification rules for the connectives have the expected form. The figure only includes the rule for disjunction, while the rest are omitted.

Equality reasoning is part of conversion instead of having rules for this on the level of the main judgment, which would be the case with an axiomatic representation. By interleaving simplification and rewriting, equality chains appear as locally as possible. This allows the inference system itself to guide the proof search in terms of in what parts of formula rewriting should take place. The setup also avoids the choice of in which order to do the rewriting steps in the case of independent rewriting of multiple sub-formulas.

Example 1. Let h define the axiom $x = z \wedge w = y$ and the conjecture be $f\,x\,y = f\,z\,w$. The solution reported by AgsyHOL is

```
=-I (simp-app (step <<h>> (And-El use-eq) simp-all simp-all)
              (step <<h>> (And-Er use-eq-sym) simp-all simp-all))
```

On the top level is an equality introduction, followed by a simplification of the application. For the first argument of f the left conjunct of h is used, and for the second argument the symmetrical counterpart of the right conjunct is used. The term `simp-all` is a shorthand for a conversion which contains no equality steps. Thanks to the representation of equality reasoning, the positions in the proof at which to rewrite the two sub-formulas are fixed. There is not one proof corresponding to the equality chain $f\,x\,y = f\,z\,y = f\,z\,w$ and another one corresponding to $f\,x\,y = f\,x\,w = f\,z\,w$.

2.5 Soundness

The reference logic is STT [11]. Figure 5 shows the syntax of STT and how the logical connectives of $FSC_{=}^{ho}$ are interpreted as STT formulas.

Proposition. Assume A is a well formed Boolean formula in $FSC_{=}^{ho}$. If $\epsilon \vdash A_o$ is derivable in $FSC_{=}^{ho}$ then the interpretation of a A is derivable in STT.

Proof. The proof is straightforward and conducted by induction on the structure of proof derivations. The proof has been formalized in Agda [3] and the code can be found at [12]. The proof uses de Bruijn indices and explicit quantification of variables. Consequently, the inference rule for α-conversion in STT is dropped while a new rule for adding a fresh variable is required.

$$
\begin{aligned}
&\sim A_o \equiv \mathbf{N}_{oo} A_o & A, B &::= x_\gamma & &\text{(variable)} \\
&A_o \vee B_o \equiv \mathbf{A}_{ooo} A_o B_o & &\mid \mathbf{N}_{oo} & &\text{(negation operator)} \\
&A_o \wedge B_o \equiv \;\sim (\sim A_o \vee \sim B_o) & &\mid \mathbf{A}_{ooo} & &\text{(disjunction operator)} \\
&A_o \supset B_o \equiv \;\sim A_o \vee B_o & &\mid \Pi_{o(o\gamma)} & &\text{(quantification operator)} \\
&\forall x_\gamma.\, A_o \equiv \Pi_{o(o\gamma)}(\lambda x_\gamma A_o) & &\mid \iota_{\gamma(o\gamma)} & &\text{(choice operator)} \\
&\exists x_\gamma.\, A_o \equiv \;\sim \forall x_\gamma.\;\sim A_o & &\mid A_{\gamma\tau} B_\tau & &\text{(application)} \\
&\top \equiv \forall x_o.\, (x_o \supset x_o) & &\mid \lambda x_\tau.\, A_\gamma & &\text{(abstraction)} \\
&\bot \equiv \forall x_o.\, x_o \\
&A_\gamma = B_\gamma \equiv (\lambda x_\gamma.\, \lambda y_\gamma.\, \forall z_{o\gamma}.\, (z_{o\gamma} x_\gamma \supset z_{o\gamma} y_\gamma))\, A_\gamma\, B_\gamma
\end{aligned}
$$

Fig. 5. STT terms and interpretation of logical connectives

3   Proof Search

System $FSC_{=}^{ho}$ is based on focused sequent calculus, which already facilitates proof search compared to e.g. natural deduction. It also expresses some proof search oriented details, e.g. the restrictions of where to perform β-reduction and how equality reasoning and simplification are mixed.

Another such detail is the order of B and A in the premise $\Gamma \vdash_s B \leftrightarrow_o A$ in the use rule. Empirical tests indicate that replacing the judgment with $\Gamma \vdash_s A \leftrightarrow_o B$ gives inferior performance. This outcome depends on the exact configuration of the conversion judgment rules. The conversion rules are devised in such a way that backward application constructs equality chains from left to right. The direction of unfolding equality chains can have an impact on performance whenever parts of the formulas in the conversion are unknown. In the case of the use rule the succedent is to the left and the antecedent to the right. It is reasonable to believe that if one side of an equality is more frequently uninstantiated than the other, the search is more restricted if rewriting from the more instantiated side than vice versa. Using this argument to explain the empirical difference, the succedent is presumably on average more known than the antecedent.

Although $FSC_{=}^{ho}$ itself is devised with proof search in mind, some additional restrictions and search control heuristics are needed in order to achieve an efficient theorem prover. This section presents the search control mechanisms which are part of the AgsyHOL implementation. Proof search is based on applying the inference rules of $FSC_{=}^{ho}$ backwards and constructing a complete derivation step by step. The search state consists of a derivation tree where some of the sub-derivations are incomplete. When no unknown parts of the derivation tree remain, a proof has been found and the given formula is valid.

An unknown part of a derivation may be a formula. Formula and derivation refinements are handled uniformly by defining a proof language with one construct for each inference rule. There is no separate mode of the search that performs unification. Unification effectively takes place when applying the rules for the conversion judgment backwards.

Unknown parts of formulas and proofs are represented by *meta-variables*. They only exist on the level of the narrowing algorithm and search control. Meta-variables that act as placeholders for formulas, e.g. B in the $\exists$-I rule, have the role of existential variables.

Since the set of possible partial derivations is in general infinite, search depth must be limited. In AgsyHOL search is limited by the size of the derivation tree. Instead using the maximum depth of the derivation tree as the limit is inferior according to our experience.

In section 2 it was mentioned that the concept of uniform proofs is not adopted in AgsyHOL. The reason is that even though uniform proofs preserve completeness, it does not necessarily improve the overall performance of proof search. AgsyHOL uses proof size as the notion of search depth in order to stratify the search. Finding the first solution at a small depth is often more important in practice than keeping the number of solutions at a minimum. Most notably, the adoption of a uniform application of the right rule for implication is not preferable in our experience. The rule extends the set of antecedents which increases the number of choices for elimination, so it should be invoked conservatively.

The following sub-sections present the important aspects of search control in AgsyHOL. The spectrum of configurations for the parameters involved has not been explored sufficiently in order for the exact numbers to be relevant. Therefore the presentation is not very specific, but intended to give a rough view of the amount of and nature of search control we found necessary to include.

3.1 Customized Order of Refining Sub-derivations

When choosing which incomplete part of the derivation to refine, a system of priorities is used. Different priorities are assigned to the judgments of the calculus. At each step in the search an incomplete sub-derivation or formula, which has an unresolved judgment with the highest priority, is chosen. The principle which has been followed is to assign a higher priority to judgments which typically lead to a smaller degree of branching. The judgments of $FSC_=^{ho}$ are listed below in decreasing order of priority in AgsyHOL.

- Simplification judgment, $\Gamma \vdash_s A \leftrightarrow_\gamma B$, flexible-rigid situation
- Conversion judgment, $\Gamma \vdash A \leftrightarrow_\gamma B$
- Focusing judgment, $\Gamma; B \vdash A$
- Main judgment, $\Gamma \vdash A$
- Simplification judgment, flexible-flexible situation

The list shows that resolving unification has top priority. Moreover, deciding whether to proceed with or end an elimination sequence has higher priority than the main judgment, so that how to end a sequence is explored before attacking any of its sub-problems.

3.2 Weighted Refinement of the Derivation Tree

The proof constructs are associated with individual weights. The size of a proof is taken to be the sum of the weights of the constructs it contains. The weights have been chosen by trial and error aiming to maximize performance. The following list summarizes the weights used in AgsyHOL.

- Introduction rules have a small weight, except for implication which has a larger weight.
- The use of a hypothesis has a small weight if it has a low degree of generality and large weight when it is highly general. The generality of a hypothesis is measured by the portion of the statement which is constituted by variables. This is an ad hoc mechanism but improves success rate on problems which have a large amount of hypotheses.
- The rules RAA and AC have a large weight.
- The rules ext_{bool}, ext_{fun} have a large weight.

- The rule step has a large weight.
- Simplification rules have a small weight.
- Refining a formula in the presence of a conversion constraint without pending substitutions has a small weight. In such situations unification becomes first-order and there is no real search taking place. When there are postponed substitutions, refining the formula has a slightly larger weight.

The reader may refer to the source code [12] for the exact relation between weights.

3.3 Unification

Applying the rules of conversion in $FSC^{ho}_{\doteq}$ effectively implements full higher-order unification. General higher-order unification is undecidable. However, in practice this is often, including for AgsyHOL, not very problematic. Therefore restricting unification to some decidable fragment, like pattern unification, has not been investigated.

Formulas which are unknown are constructed incrementally during search, just as the derivation tree constructs. Since formulas are higher-order, substitutions must be postponed at meta-variables and performed once they are instantiated. This is handled by using explicit substitutions in the implementation.

How to deal with occurs checks in higher-order logic is not straightforward. One observation is that a meta-variable may occur in a pending substitution and that it is not in general clear whether such an occurrence will cause circularity. A second observation is that since the formulas are constructed step by step, a circularity check the way it is done in first-order unification will not catch circularity caused by the interplay between several equality constraints.

AgsyHOL has the following way to deal with occurs check. For conversion judgments the systems keeps track of which formula heads surround the current sub-formulas. Instantiating one of the formulas to a head which is already among the surrounding heads is discouraged by associating it with a large weight. The following simple example should clarify this mechanism.

Assume $X_1, X_2, \ldots$ are meta-variables. Given the equality

$$X_1 = f\ (g\ X_1)$$

X_1 will first be refined to $f\ X_2$. The equality is then simplified to

$$X_2 = g\ (f\ X_2)$$

keeping in mind that the head f has been traversed. Then X_2 is refined to $g\ X_3$ and simplification yields

$$X_3 = f\ (g\ X_3)$$

where both f and g now have been traversed. Refining X_3 to $f\ X_4$ in this situation is assigned a large weight since f is already among the traversed heads.

3.4 β-Reduction

Since the search state may include partially instantiated formulas β-reduction may take place in formulas which are not completely type checked. There is therefore a risk of non-termination. There are various ways to overcome this problem. Some are theoretically appealing, such as the one based on linearity/non-linearity discrimination, presented in e.g. Lengrand [13].

However, AgsyHOL has a very straightforward mechanism to deal with non-termination of reduction. It simply limits the number of consecutive reduction steps and lets this limit increase with the search depth.

4 Implementation

The source code of AgsyHOL is available at [12] and the system can be invoked on Systems on TPTP. It is written in Haskell, and is, as already mentioned, composed of three parts, namely a generic search algorithm, semantics and search control. The search algorithm is an extension of lazy narrowing [5].

Lazy narrowing is a search procedure which, given a function, f, and a fixed output, y, exploits non-strictness in f in order to efficiently find inputs, X, such that $f(X) = y$. Lazy narrowing refines the input step-wise and has meta-variables which represent yet unknown parts of it.

The semantics of $FSC_{=}^{ho}$ is represented by a proof checking algorithm unaware of meta-variables. The proof checker takes a boolean formula and a proof, and decides whether the proof is a valid derivation of the formula. Proof search is achieved by applying lazy narrowing on the proof checking function with a given formula and an unknown proof, i.e. the proof given to the proof checker is initially a fully uninstantiated meta-variable.

The narrowing algorithm is first-order. In order to be able use it to construct higher-order formulas, the formulas have a first-order representation. In AgsyHOL de Bruijn-indices represented by integers are used. The search control added on top of the narrowing algorithm keeps track of the size of the variable scope.

The extensions of lazy narrowing used in AgsyHOL are concurrent conjunction [14], and customizable depth weights and priorities for instantiation. Concurrent conjunction enables inference rules with multiple premises to result in a collection of simultaneously active constraints. This improves the pruning accomplished by the narrowing algorithm. It is also a prerequisite for the mechanism of priorities, since otherwise there would always be only one blocking meta-variable.

The search control part of the theorem prover consists of annotations in the proof checker which specify the weights and priorities associated with each branching point at which execution may stop because of the presence of a meta-variable.

The lazy narrowing mechanism is implemented as an embedded library in Haskell. It uses depth-first search with back-tracking and iterated deepening. There are functional logic programming languages, such as Curry [15], which natively implement lazy narrowing. These languages however do not fit the needs regarding the extensions for expressing priorities and costs used in AgsyHOL.

The proof checker in AgsyHOL uses a sequent calculus style, not only of eliminations, but also applications in formulas. This means that instead of binary application, applications are represented by a head, which is a variable, and a list of arguments. Having this representation of application simplifies the interplay between conversion and head normal reduction in the way that only the top level construct has to be inspected in order to determine whether a formula is head normal.

AgsyHOL reads the THF0 syntax [2], the basic higher-order syntax in TPTP. When translating from THF0 to the internal representation of formulas definitions are inlined. In STT there is only one set of individuals. THF permits multiple sets and this is reflected in the implementation of $FSC_{=}^{ho}$ in AgsyHOL. The representation of applications in AgsyHOL syntactically disallows β-redexes, which are therefore reduced away in the translation.

The use of RAA is problematic since it is universally applicable and difficult to know when needed. The weight of this rule is therefore high in AgsyHOL. This has the effect that proofs that require classical reasoning can be time consuming to find. In order to improve this, each problem is transformed using double negation elimination and the de Morgan laws. The transformation done in Agsy-HOL aims to limit the need for classical reasoning by minimizing the number of negations. This makes it solve around 10% more problems in TPTP (within the standard 300 seconds time limit) compared to having no transformation. On the other hand, bringing the formulas to negation normal form does not improve the performance on the problem set at all. Another alternative would be to use double-negation transformation in order to avoid the need of the RAA rule. This however substantially lowers the success rate of AgsyHOL.

5   Empirical Result

AgsyHOL has been tested on the higher-order problems in the TPTP version 6.0.0, January 2014. There are 3025 problems in total. Figure 6 shows a comparison between AgsyHOL and the other higher-order theorem provers available

system	solutions	unique
Satallax—2.7	2080	73
Isabelle—2013	1883	3
Isabelle-HOT—2013	1876	2
AgsyHOL—1.0	**1722**	**17**
LEO-II–1.6.0	1700	20
TPS—3.120601S1b	1469	25
cocATP—0.1.8	577	0

Fig. 6. AgsyHOL compared to other theorem provers in TPTP

on Systems on TPTP. All theorem provers had 300 seconds to solve each problem on a standard desktop computer. The table shows the number of confirmed theorems after timeout and how many unique solutions each system has.

AgsyHOL positions itself below the two top performing systems, Satallax and Isabelle, and gets a similar result as LEO-II. There is some distance to the best performing system. We however consider the performance of AgsyHOL to be surprisingly good, given its straightforward approach of searching for derivation proofs.

Example 2. The TPTP problem NUM636^2 declares one and succ, intended to represent the natural numbers. It then defines the axioms one_is_first, succ_injective and induction, and conjectures that for all x, succ $x \neq x$. The solution produced by AgsyHOL is the following:

```
1 Forall-I (\#0.(elim <<induction>>
2 (Forall-E (\#1:$i.(Not (Eq $i (<<succ>> (#1)) (#1))))
3 (Implies-E (And-I
4 (Not-I (\#1.(elim <<one_is_first>> (Forall-E (<<one>>)
5 (Not-E (=-I (step #1 use-eq simp-all simp-all)))))))
6 (Forall-I (\#1.(Implies-I (\#2.(Not-I (\#3.
7 (elim #2 (Not-E (=-I (step <<succ_injective>>
8 (Forall-E (<<succ>> (#1)) (Forall-E (#1)
9 (Implies-E (=-I (step #3 use-eq simp-all simp-all)) use-eq)))
10 simp-all simp-all))))))))))
11 ) (Forall-E (#0) (use simp-all))))))
```

Although reading the details of the solution requires knowledge about the exact formulation of the axioms and conjecture, the top-level structure of the proof can readily be identified. On line 1 the proof starts by using induction. Line 2 contains the generated λ-term which states the induction hypothesis. Lines 4 and 5 prove the base case using one_is_first. Lines 6–10 prove the induction step using the induction hypothesis, #2, and succ_injective.

6   Conclusions and Future Work

One weakness of AgsyHOL is how it deals with classical reasoning, the RAA and AC rules. These rules are axiomatic and universally applicable. AgsyHOL tends to be weak on problems that require classical reasoning. One direction for future work would be to improve this situation, possibly by switching to classical sequent calculus.

The empirical result shows that a calculus like the presented one, which defines a language of derivation proofs, can be the basis of a competitive theorem prover. The approach can appear naive and it is interesting that it works so well in practice.

As stated in the introduction, AgsyHOL is strong on problems involving higher-order unification. By dealing with derivation steps and formulas uniformly, unification becomes a fully integrated part of the proof search. Being good at dealing with higher-order unification the technique could have an impact on the field of automated higher-order theorem proving.

Producing derivation proofs, it could also be valuable in the area of automation of interactive theorem provers.

References

1. Sutcliffe, G.: The TPTP problem library and associated infrastructure. J. Autom. Reason. 43(4), 337–362 (2009)
2. Sutcliffe, G., Benzmüller, C.: Automated reasoning in higher-order logic using the TPTP THF infrastructure. Journal of Formalized Reasoning 3(1), 1–27 (2010)
3. Bove, A., Dybjer, P., Norell, U.: A brief overview of Agda — a functional language with dependent types. In: Berghofer, S., Nipkow, T., Urban, C., Wenzel, M. (eds.) TPHOLs 2009. LNCS, vol. 5674, pp. 73–78. Springer, Heidelberg (2009)
4. Lindblad, F.: Higher-order proof construction based on first-order narrowing. Electron. Notes Theor. Comput. Sci. 196, 69–84 (2008)
5. Middeldorp, A., Okui, S., Ida, T.: Lazy narrowing: Strong completeness and eager variable elimination. Theoretical Computer Science 167, 95–130 (1995)
6. Herbelin, H.: A lambda-calculus structure isomorphic to Gentzen-style sequent calculus structure. In: Pacholski, L., Tiuryn, J. (eds.) CSL 1994. LNCS, vol. 933, pp. 61–75. Springer, Heidelberg (1995)
7. Lengrand, S., Dyckhoff, R., McKinna, J.: A focused sequent calculus framework for proof search in pure type systems. Logical Methods in Computer Science 7(1) (2011)
8. Andreoli, J.: Logic programming with focusing proofs in linear logic. Journal of Logic and Computation 2, 297–347 (1992)
9. Miller, D., Nadathur, G., Pfenning, F., Scedrov, A.: Uniform proofs as a foundation for logic programming. Annals of Pure and Applied Logic 51(12), 125–157 (1991)
10. Girard, J.Y.: A new constructive logic: Classical logic. Mathematical Structures in Computer Science 1(3), 255–296 (1991)
11. Henkin, L.: Completeness in the theory of types. J. Symb. Log. 15(2), 81–91 (1950)
12. Lindblad, F.: AgsyHOL source code and Agda formalization (2012), https://github.com/frelindb/agsyHOL
13. Lengrand, S.: Normalisation & Equivalence in Proof Theory & Type Theory. PhD thesis, Université Paris 7 & University of St Andrews (2006)
14. Lindblad, F.: Property directed generation of first-order test data. In: Trends in Functional Programming. Intellect, vol. 8, pp. 105–123 (2008)
15. Hanus, M.: Curry: An integrated functional logic language. Language report (March 2006), http://www.informatik.uni-kiel.de/~curry/report.html

SAT-Based Decision Procedure for Analytic Pure Sequent Calculi*

Ori Lahav and Yoni Zohar

School of Computer Science, Tel Aviv University, Israel
orilahav@post.tau.ac.il, yoni.zohar@cs.tau.ac.il

Abstract. We identify a wide family of analytic sequent calculi for propositional non-classical logics whose derivability problem can be uniformly reduced to SAT. The proposed reduction is based on interpreting these calculi using non-deterministic semantics. Its time complexity is polynomial, and, in fact, linear for a useful subfamily. We further study an extension of such calculi with *Next* operators, and show that this extension preserves analyticity and is subject to a similar reduction to SAT. A particular interesting instance of these results is a HORNSAT-based linear-time decision procedure for Gurevich and Neeman's primal infon logic and several natural extensions of it.

1 Introduction

Sequent calculi provide a flexible well-behaved proof-theoretic framework for a huge variety of different logics. Usually, they allow us to perform proof-search for the corresponding logic. The fundamental property of cut-elimination is traditionally proven, as it often guarantees the adequacy of a given sequent calculus for this task. Nevertheless, a great deal of ingenuity is required for developing an efficient proof-search algorithms for cut-free sequent calculi (see, e.g., [12]).

In this work we identify a general case in which it is possible to replace proof-search by SAT solving. While SAT is NP-complete, it is considered "easy" when it comes to real-world applications. Indeed, there are many off-the-shelf SAT solvers, that, despite an exponential worst-case time complexity, are considered extremely efficient (see, e.g., [14]).

We focus on a general family of relatively simple sequent calculi, called *pure sequent calculi*. Roughly speaking, these are propositional fully-structural calculi (calculi that include the structural rules: exchange, contraction and weakening), whose derivation rules do not enforce any limitations on the context formulas (following [1], the adjective "pure" stands for this requirement). We do not assume that the calculi enjoy cut-elimination. Instead, we formulate an analyticity property, that generalizes the usual subformula property, and show that the derivability problem in each *analytic* pure calculus can be reduced to (the complement of) SAT. This result applies to a wide range of sequent calculi for

---

* This research was supported by The Israel Science Foundation (grant no. 280-10).

S. Demri, D. Kapur, and C. Weidenbach (Eds.): IJCAR 2014, LNAI 8562, pp. 76–90, 2014.

different non-classical logics, including important three and four valued logics and various paraconsistent logics.

To achieve this result we utilize an alternative semantic view of pure sequent calculi. For that, we have extended the correspondence between sequent calculi and their *bivaluation* semantics from [7], so the semantics is tied to the set of formulas allowed to be used in derivations. The derivability problem in a given analytic sequent calculus is then replaced by small countermodel search, which can be translated into a SAT instance. In turn, one can construct a countermodel from a satisfying assignment given by the SAT solver in the form of a bivaluation (or a functional Kripke model when *Next* operators are involved, see below).

The efficiency of the proposed SAT-based decision procedure obviously depends on the time complexity of the reduction. This complexity, as we show, is $O(n^k)$, where n is the size of the input sequent and k is determined according to the structure of the particular calculus. For a variety of useful calculi, we obtain a *linear time* reduction. This paves the way to efficient uniform decision procedures for all logics that can be covered in this framework. In particular, we identify a subfamily of calculi for which the generated SAT instances consist of *Horn clauses*. In these calculi the derivability problem can be decided in linear time by applying the reduction and using a linear time HORNSAT solver [13].

In Section 6 we extend this method to analytic pure calculi augmented with a finite set of *Next* operators. These are often employed in temporal logics. Moreover, in primal infon logic [11] *Next* operators, as we show, play the role of quotations, which are indispensable in the application of this logic for the access control language DKAL. We show that all analytic pure calculi, satisfying a certain natural requirement, can be augmented with *Next* operators, while retaining their analyticity. In turn, the general reduction to SAT is extended to analytic calculi with *Next* operators, based on a (possibly non-deterministic) Kripke-style semantic characterization. A HORNSAT-based decision procedure for primal infon logic with quotations is then obtained as a particular instance. In addition, in this general framework we are able to formulate several extensions of primal infon logic with additional natural rules, making it somewhat "closer" to classical logic, and still decidable in linear time.

Related Works. Our method generalizes the reduction given in [6] of quotations-free primal infon logic to classical logic. For the case of primal infon logic with quotations the proposed reduction produces practically equivalent outputs to the reduction in [8] from this logic to Datalog. A general methodology for translating derivability questions in Hilbertian deductive systems to Datalog was introduced in [9]. However, this method may produce infinitely many Datalog premises, and then it is difficult to use for computational purposes. In contrast, the reduction proposed in this paper always produces finite SAT instances. This is possible due to our focus on *analytic* calculi. Since Hilbertian systems are rarely analytic, we handle Gentzen-type calculi.

Due to lack of space, some proofs are omitted, and will appear in an extended version.

2   Preliminaries

A *propositional language* $\mathcal{L}$ consists of a countably infinite set of atomic variables $At = \{p_1, p_2, \ldots\}$ and a finite set $\Diamond_{\mathcal{L}}$ of propositional connectives. The set of all n-ary connectives of $\mathcal{L}$ is denoted by $\Diamond_{\mathcal{L}}^n$. We identify $\mathcal{L}$ with its set of well-formed formulas (e.g. when writing $\psi \in \mathcal{L}$ or $\mathcal{F} \subseteq \mathcal{L}$). A *sequent* is a pair $\langle \Gamma, \Delta \rangle$ (denoted by $\Gamma \Rightarrow \Delta$) where Γ and Δ are finite sets of formulas. We employ the standard sequent notations, e.g. when writing expressions like $\Gamma, \psi \Rightarrow \Delta$ or $\Rightarrow \psi$. The union of sequents is defined by $(\Gamma_1 \Rightarrow \Delta_1) \cup (\Gamma_2 \Rightarrow \Delta_2) = (\Gamma_1 \cup \Gamma_2 \Rightarrow \Delta_1 \cup \Delta_2)$. For a sequent $\Gamma \Rightarrow \Delta$, $frm(\Gamma \Rightarrow \Delta) = \Gamma \cup \Delta$. This notation is naturally extended to sets of sequents. Given $\mathcal{F} \subseteq \mathcal{L}$, we say that a formula φ is an $\mathcal{F}$-*formula* if $\varphi \in \mathcal{F}$ and that a sequent s is an $\mathcal{F}$-*sequent* if $frm(s) \subseteq \mathcal{F}$. A *substitution* is a function from At to some propositional language. A substitution σ is naturally extended to any propositional language by $\sigma(\Diamond(\psi_1, \ldots, \psi_n)) = \Diamond(\sigma(\psi_1), \ldots, \sigma(\psi_n))$ for every compound formula $\Diamond(\psi_1, \ldots, \psi_n)$. Substitutions are also naturally extended to sets of formulas, sequents and sets of sequents. In what follows, $\mathcal{L}$ denotes an arbitrary propositional language.

3   Pure Sequent Calculi

In this section we define the family of pure sequent calculi, and provide some examples for known calculi that fall in this family.

Definition 1. A *pure rule* is a pair $\langle S, s \rangle$ (denoted by S / s) where S is a finite set of sequents and s is a sequent. The elements of S are called the *premises* of the rule and s is called the *conclusion* of the rule. An *application* of a pure rule $\{s_1, \ldots s_n\} / s$ is any inference step of the form

$$\frac{\sigma(s_1) \cup c \quad \ldots \quad \sigma(s_n) \cup c}{\sigma(s) \cup c}$$

where σ is a substitution and c is a sequent (called a *context sequent*). The sequents $\sigma(s_i) \cup c$ are called the *premises* of the application and $\sigma(s) \cup c$ is called the *conclusion* of the application. The set S of premises of a pure rule is usually written without set braces, and its elements are separated by ";".

Note that we differentiate between rules and their applications, and use different notations for them.

Example 1. The following are pure rules:

$$p_1 \Rightarrow p_2 \,/ \Rightarrow p_1 \supset p_2 \qquad \Rightarrow p_1; p_2 \Rightarrow \,/\, p_1 \supset p_2 \Rightarrow \qquad /\Rightarrow p_1 \supset p_1$$

Applications of these rules have respectively the forms:

$$\frac{\Gamma, \psi_1 \Rightarrow \psi_2, \Delta}{\Gamma \Rightarrow \psi_1 \supset \psi_2, \Delta} \qquad \frac{\Gamma \Rightarrow \psi_1, \Delta \quad \Gamma, \psi_2 \Rightarrow \Delta}{\Gamma, \psi_1 \supset \psi_2 \Rightarrow \Delta} \qquad \frac{}{\Gamma \Rightarrow \psi \supset \psi, \Delta}$$

Note that the usual rule for introducing implication on the right-hand side in intuitionistic logic is not a pure rule, since it allows only *left* context formulas.

In turn, pure sequent calculi are finite sets of pure rules. To make them fully-structural (in addition to defining sequents as pairs of *sets*), the weakening rule, the identity axiom and the cut rule are allowed to be used in derivations.

Definition 2. A *pure calculus* is a finite set of pure rules. A (standard) *proof* in a pure calculus $\mathbf{G}$ is defined as usual, where in addition to applications of the pure rules of $\mathbf{G}$, the following standard application schemes may be used:

$$(weak)\frac{\Gamma \Rightarrow \Delta}{\Gamma',\Gamma \Rightarrow \Delta, \Delta'} \qquad (id)\ \overline{\Gamma, \psi \Rightarrow \psi, \Delta} \qquad (cut)\frac{\Gamma \Rightarrow \psi, \Delta \quad \Gamma, \psi \Rightarrow \Delta}{\Gamma \Rightarrow \Delta}$$

Henceforth, we consider only pure rules and pure calculi, and may refer to them simply as *rules* and *calculi*. By an $\mathcal{L}$-rule ($\mathcal{L}$-calculus) we mean a rule (calculus) that includes only connectives from $\mathcal{L}$.

Notation 1. For an $\mathcal{L}$-calculus $\mathbf{G}$, a set $\mathcal{F} \subseteq \mathcal{L}$ of formulas, and an $\mathcal{F}$-sequent s, we write $\vdash_{\mathbf{G}}^{\mathcal{F}} s$ if there is a proof of s in $\mathbf{G}$ consisting only of $\mathcal{F}$-sequents. For $\mathcal{F} = \mathcal{L}$, we write $\vdash_{\mathbf{G}} s$.

Example 2. The propositional fragment of Gentzen's fundamental sequent calculus for classical logic can be directly presented as a pure calculus, denoted henceforth by $\mathbf{LK}$. It consists of the following rules:

$(\neg \Rightarrow) \qquad \Rightarrow p_1 \ / \ \neg p_1 \Rightarrow \qquad\qquad (\Rightarrow \neg) \qquad p_1 \Rightarrow \ / \ \Rightarrow \neg p_1$

$(\wedge \Rightarrow) \quad p_1, p_2 \Rightarrow \ / \ p_1 \wedge p_2 \Rightarrow \qquad (\Rightarrow \wedge) \Rightarrow p_1; \Rightarrow p_2 \ / \ \Rightarrow p_1 \wedge p_2$

$(\vee \Rightarrow) \ p_1 \Rightarrow; p_2 \Rightarrow \ / \ p_1 \vee p_2 \Rightarrow \qquad (\Rightarrow \vee) \quad \Rightarrow p_1, p_2 \ / \ \Rightarrow p_1 \vee p_2$

$(\supset\Rightarrow) \Rightarrow p_1; p_2 \Rightarrow \ / \ p_1 \supset p_2 \Rightarrow \qquad (\Rightarrow\supset) \quad p_1 \Rightarrow p_2 \ / \ \Rightarrow p_1 \supset p_2$

Besides $\mathbf{LK}$ there are many sequent calculi for non-classical logics (admitting cut-elimination) that fall in this framework. These include calculi for well-known three and four-valued logics, various calculi for paraconsistent logics, and all canonical and quasi-canonical sequent systems [3,4,5,7].

Example 3. The calculus for (quotations free) primal infon logic from [11], can be directly presented as a pure calculus, that we call $\mathbf{P}$. It consists of the rules $(\wedge \Rightarrow)$, $(\Rightarrow \wedge)$, $(\Rightarrow \vee)$ and $(\supset\Rightarrow)$ of $\mathbf{LK}$, together with the two rules $\Rightarrow p_2 \ / \ \Rightarrow p_1 \supset p_2$ and $\emptyset \ / \ \Rightarrow \top$.

Example 4. The calculus from [3] for da Costa's historical paraconsistent logic C_1 can be directly presented as a pure calculus, that we call $\mathbf{G}_{C_1}$. It consists of the rules of $\mathbf{LK}$ except for $(\neg \Rightarrow)$ that is replaced by the following rules:

$p_1 \Rightarrow \ / \ \neg\neg p_1 \Rightarrow$

$\Rightarrow p_1; \Rightarrow \neg p_1 \ / \ \neg(p_1 \wedge \neg p_1) \Rightarrow \qquad\qquad \neg p_1 \Rightarrow; \neg p_2 \Rightarrow \ / \ \neg(p_1 \wedge p_2) \Rightarrow$

$\neg p_1 \Rightarrow; p_2, \neg p_2 \Rightarrow \ / \ \neg(p_1 \vee p_2) \Rightarrow \qquad p_1, \neg p_1 \Rightarrow; \neg p_2 \Rightarrow \ / \ \neg(p_1 \vee p_2) \Rightarrow$

$p_1 \Rightarrow; p_2, \neg p_2 \Rightarrow \ / \ \neg(p_1 \supset p_2) \Rightarrow \qquad p_1, \neg p_1 \Rightarrow; \neg p_2 \Rightarrow \ / \ \neg(p_1 \supset p_2) \Rightarrow$

3.1 Analyticity

Our goal in this paper is to provide a general effective tool to solve the *derivability problem* for a given pure calculus.

Definition 3. The *derivability problem* for an $\mathcal{L}$-calculus **G** is given by:
Input: An $\mathcal{L}$-sequent s.   **Question:** Does $\vdash_{\mathbf{G}} s$?

Obviously, one cannot expect to have decision procedures for the derivability problem for all pure calculi.[1] Thus we require our calculi to admit a *generalized analyticity property*. Analyticity is a crucial property of proof systems. In the case of fully-structural propositional sequent calculi it usually implies their decidability and consistency (the fact that the empty sequent is not derivable). Roughly speaking, a calculus is analytic if whenever a sequent s is provable in it, s can be proven using only the "syntactic material available inside s". This "material" is usually taken to consist of all subformulas occurring in s, and then analyticity amounts to the (global) subformula property. However, weaker restrictions on the formulas that are allowed to appear in proofs of a given sequent may suffice for decidability. Next we introduce a generalized analyticity property based on an extended notion of a subformula. In what follows, $\odot$ denotes an arbitrary set of unary connectives (assumed to be a subset of $\diamondsuit_{\mathcal{L}}^{1}$).

Definition 4. A formula φ is a $\odot$-*subformula* of a formula ψ if either φ is a subformula of ψ or $\varphi = \circ\psi'$ for some $\circ \in \odot$ and proper subformula ψ' of ψ.

Note that the $\odot$-subformula relation is transitive.

Notation 2. $sub^{\odot}(\psi)$ denotes the set of $\odot$-subformulas of ψ. This notation is extended to sets of formulas and sequents in the obvious way.

Example 5. $sub^{\{\neg\}}(\neg(p_1 \supset p_2)) = \{p_1, p_2, \neg p_1, \neg p_2, p_1 \supset p_2, \neg(p_1 \supset p_2)\}$.

Definition 5. An $\mathcal{L}$-calculus **G** is called $\odot$-*analytic* if $\vdash_{\mathbf{G}} s$ implies $\vdash_{\mathbf{G}}^{sub^{\odot}(s)} s$ for every $\mathcal{L}$-sequent s.

Note that $sub^{\emptyset}(\varphi)$ is the set of usual subformulas of φ, and so $\emptyset$-analyticity is the usual subformula property.

Example 6. The calculi **LK**, **P** and $\mathbf{G}_{C_1}$ (presented in previous examples) admit cut-elimination. This, combined with the structure of their rules, directly entails that **LK** and **P** are $\emptyset$-analytic, while $\mathbf{G}_{C_1}$ is $\{\neg\}$-analytic. Example 10 below shows an extension of **P** that does not admit cut-elimination, but is still $\emptyset$-analytic.

Example 7. A cut-free sequent calculus for Łukasiewicz three-valued logic was presented in [2]. This calculus, that we call $\mathbf{G}_3$, can be directly presented as a pure calculus. For example, the rules involving implication are the following:
$$\neg p_1 \Rightarrow; p_2 \Rightarrow; \Rightarrow p_1, \neg p_2 / p_1 \supset p_2 \Rightarrow \qquad p_1 \Rightarrow p_2; \neg p_2 \Rightarrow \neg p_1 / \Rightarrow p_1 \supset p_2$$
$$p_1, \neg p_2 \Rightarrow / \neg(p_1 \supset p_2) \Rightarrow \qquad\qquad \Rightarrow p_1; \Rightarrow \neg p_2 / \Rightarrow \neg(p_1 \supset p_2)$$
The structure of its rules, together with the fact that this calculus admits cut-elimination, directly entail that $\mathbf{G}_3$ is $\{\neg\}$-analytic.

---

[1] Any Hilbert calculus H (without side conditions on rule applications) can be translated to a pure sequent calculus $\mathbf{G}_H$, by taking a rule of the form $\Rightarrow \psi_1; \ldots; \Rightarrow \psi_n / \Rightarrow \psi$ for each Hilbert-style derivation rule $\psi_1, \ldots, \psi_n / \psi$ (where $n = 0$ for axioms). It is easy to show that ψ is derivable from Γ in H iff $\vdash_{\mathbf{G}_H} \Gamma \Rightarrow \psi$.

To end this section, we point out a useful property of pure calculi. We call a rule *axiomatic* if it has an empty set of premises. In turn, a calculus is *axiomatic* if it consists solely of axiomatic rules. We show that every calculus is equivalent (in the sense defined below) to an axiomatic calculus, obtained by "multiplying out" the rules, and "moving" the formulas in the premises to the opposite side of the conclusion.

Definition 6. A *component* of a sequent $\Gamma \Rightarrow \Delta$ is any sequent of the form $\psi \Rightarrow$ where $\psi \in \Gamma$ or $\Rightarrow \psi$ where $\psi \in \Delta$. A sequent s is called a *combination* of a set S of sequents if there are distinct sequents $s_1, \ldots, s_n$ and respective components $s'_1, \ldots, s'_n$ such that $S = \{s_1, \ldots, s_n\}$ and $s = s'_1 \cup \ldots \cup s'_n$.

Definition 7. Let $r = S / \Gamma \Rightarrow \Delta$ be a rule. The set $Ax(r)$ consists of all axiomatic rules of the form $\emptyset / \Gamma, \Delta' \Rightarrow \Gamma', \Delta$ where $\Gamma' \Rightarrow \Delta'$ is a combination of S. In turn, given a calculus $\mathbf{G}$, $Ax(\mathbf{G})$ denotes the calculus obtained from $\mathbf{G}$ by replacing each non-axiomatic rule r of $\mathbf{G}$ by $Ax(r)$.

Example 8. For $r = \neg p_1 \Rightarrow; p_2 \Rightarrow; \Rightarrow p_1, \neg p_2 / p_1 \supset p_2 \Rightarrow$, $Ax(r)$ consists of the axiomatic rules $\emptyset / p_1, p_1 \supset p_2 \Rightarrow \neg p_1, p_2$ and $\emptyset / \neg p_2, p_1 \supset p_2 \Rightarrow \neg p_1, p_2$.

Proposition 1. *Let $\mathbf{G}$ be an $\mathcal{L}$-calculus. For every set $\mathcal{F} \subseteq \mathcal{L}$ and $\mathcal{F}$-sequent s, if $\vdash_{\mathbf{G}}^{\mathcal{F}} s$ then $\vdash_{Ax(\mathbf{G})}^{\mathcal{F}} s$. For $\mathcal{F} = \mathcal{L}$ the converse holds as well. Moreover, if $\mathbf{G}$ is $\odot$-analytic then so is $Ax(\mathbf{G})$.*

As happens for **LK**, it is likely that $Ax(\mathbf{G})$ does not admit cut-elimination even when $\mathbf{G}$ does.

4   Semantics for Pure Sequent Calculi

In this section we present a semantic view of pure calculi, that plays a major role in the reduction of their derivability problem to SAT. For that matter, we follow [7] and use *bivaluations* – functions assigning a binary truth value to each formula. Pure rules are naturally translated into conditions on bivaluations. In order to have finite models, we strengthen the correspondence in [7] and consider *partial* bivaluations. These correspond exactly to derivations that are confined to a certain set of formulas.

Definition 8. A *bivaluation* is a function v from some set $dom(v)$ of formulas in some propositional language to $\{0, 1\}$. A bivaluation v is extended to $dom(v)$-sequents by: $v(\Gamma \Rightarrow \Delta) = 1$ iff $v(\varphi) = 0$ for some $\varphi \in \Gamma$ or $v(\varphi) = 1$ for some $\varphi \in \Delta$. v is extended to sets of $dom(v)$-sequents by: $v(S) = min \{v(s) \mid s \in S\}$, where $min \ \emptyset = 1$. Given a set $\mathcal{F}$ of formulas, by an $\mathcal{F}$-*bivaluation* we refer to a bivaluation v with $dom(v) = \mathcal{F}$.

Definition 9. A bivaluation v *respects* a rule S / s if $v(\sigma(S)) \leq v(\sigma(s))$ for every substitution σ such that $\sigma(frm(S / s)) \subseteq dom(v).[2]$ v is called $\mathbf{G}$-*legal* for a calculus $\mathbf{G}$ if it respects all rules of $\mathbf{G}$.

---

[2] frm is extended to pure rules in the obvious way, i.e. $frm(S / s) = frm(S) \cup frm(s)$.

Example 9. A $\{p_1, \neg\neg p_1\}$-bivaluation v respects the rule $p_1 \Rightarrow / \neg\neg p_1 \Rightarrow$ iff either $v(p_1) = v(\neg\neg p_1) = 0$ or $v(p_1) = 1$. Note that **LK**-legal bivaluations are exactly usual classical valuation functions.

Theorem 1 (Soundness and Completeness). *Let* **G** *be an* $\mathcal{L}$-*calculus,* $\mathcal{F}$ *be a set of* $\mathcal{L}$-*formulas, and* s *be an* $\mathcal{F}$-*sequent. Then,* $\vdash_{\mathbf{G}}^{\mathcal{F}} s$ *iff* $v(s) = 1$ *for every* **G**-*legal* $\mathcal{F}$-*bivaluation* v.

Using Theorem 1, we are able to formulate a semantic property that corresponds exactly to $\odot$-analyticity:

Definition 10. An $\mathcal{L}$-calculus **G** is called *semantically* $\odot$-*analytic* if every **G**-legal bivaluation v can be extended to a **G**-legal $\mathcal{L}$-bivaluation, provided that $dom(v)$ is a finite subset of $\mathcal{L}$ closed under $\odot$-subformulas.

Theorem 2. *An* $\mathcal{L}$-*calculus* **G** *is* $\odot$-*analytic iff it is semantically* $\odot$-*analytic.*

Proof. Suppose that there is an $\mathcal{L}$-sequent s such that $\vdash_{\mathbf{G}} s$ and $\nvdash_{\mathbf{G}}^{sub^{\odot}(s)} s$. According to Theorem 1, there exists a **G**-legal $sub^{\odot}(s)$-bivaluation v such that $v(s) = 0$, but $u(s) = 1$ for every **G**-legal $\mathcal{L}$-bivaluation u. Therefore, v cannot be extended to a **G**-legal $\mathcal{L}$-bivaluation. In addition, $dom(v) = sub^{\odot}(s)$ is finite and closed under $\odot$-subformulas.

For the converse, suppose that v is a **G**-legal bivaluation, $dom(v)$ is finite and closed under $\odot$-subformulas, and v cannot be extended to a **G**-legal $\mathcal{L}$-bivaluation. Let $\Gamma = \{\psi \in dom(v) \mid v(\psi) = 1\}$, $\Delta = \{\psi \in dom(v) \mid v(\psi) = 0\}$, and $s = \Gamma \Rightarrow \Delta$. Then $dom(v) = sub^{\odot}(s)$ and $v(s) = 0$. We show that $u(s) = 1$ for every **G**-legal $\mathcal{L}$-bivaluation u. Indeed, every such u does not extend v, and so $u(\psi) \neq v(\psi)$ for some $\psi \in dom(v)$. Then, $u(\psi) = 0$ if $\psi \in \Gamma$, and $u(\psi) = 1$ if $\psi \in \Delta$. In either case, $u(s) = 1$. By Theorem 1, $\nvdash_{\mathbf{G}}^{sub^{\odot}(s)} s$ and $\vdash_{\mathbf{G}} s$. □

The left-to-right direction of Theorem 2 is used to prove the correctness of the reduction in the next section. The converse provides a semantic method to prove $\odot$-analyticity, that can be used alternatively to deriving analyticity as a consequence of cut-elimination.

Example 10. An extension of primal infon logic, that we call **EP**, extends the calculus **P** (see Example 3) with the following classically valid axiomatic rules:

$$\emptyset / \Rightarrow \bot \supset p_1 \qquad \emptyset / p_1 \vee p_1 \Rightarrow p_1 \qquad \emptyset / \Rightarrow p_1 \supset p_1$$
$$\emptyset / \bot \Rightarrow \qquad \emptyset / p_1 \vee p_2 \Rightarrow p_2 \vee p_1 \qquad \emptyset / \Rightarrow (p_1 \wedge p_2) \supset p_1$$
$$\emptyset / \bot \vee p_1 \Rightarrow p_1 \qquad \emptyset / p_1 \vee (p_1 \wedge p_2) \Rightarrow p_1 \qquad \emptyset / \Rightarrow (p_1 \wedge p_2) \supset p_2$$
$$\emptyset / p_1 \vee \bot \Rightarrow p_1 \qquad \emptyset / (p_1 \wedge p_2) \vee p_1 \Rightarrow p_1 \qquad \emptyset / \Rightarrow p_2 \supset (p_1 \supset p_2)$$

Note that none of these rules is derivable in **P**. It is possible to prove that **EP** is $\emptyset$-analytic by showing that it is semantically $\emptyset$-analytic and applying Theorem 2.

5   Reduction to Classical Satisfiability

In this section we present a reduction from the derivability problem for a given $\odot$-analytic pure calculus to the complement of SAT. SAT instances are taken to be CNFs represented as sets of clauses, where clauses are sets of literals (that is, atomic variables and their negations, denoted by overlines). The set $\{x_\psi \mid \psi \in \mathcal{L}\}$ is used as the set of atomic variables in the SAT instances. The translation of sequents to SAT instances is naturally given by:

Definition 11. For a sequent $\Gamma \Rightarrow \Delta$:

$$\text{SAT}^+(\Gamma \Rightarrow \Delta) := \{\{\overline{x_\psi} \mid \psi \in \Gamma\} \cup \{x_\psi \mid \psi \in \Delta\}\}\,.$$

$$\text{SAT}^-(\Gamma \Rightarrow \Delta) := \{\{x_\psi\} \mid \psi \in \Gamma\} \cup \{\{\overline{x_\psi}\} \mid \psi \in \Delta\}\,.$$

This translation captures the semantic interpretation of sequents. Indeed, given an $\mathcal{L}$-bivaluation v and a classical assignment u that assigns true to x_ψ iff $v(\psi) = 1$, we have that for every $\mathcal{L}$-sequent s: $v(s) = 1$ iff u satisfies $\text{SAT}^+(s)$, and $v(s) = 0$ iff u satisfies $\text{SAT}^-(s)$. Now, in order for a bivaluation to be **G**-legal for some calculus **G**, it should satisfy the semantic restrictions arising from the rules of **G**. These restrictions can be directly encoded as SAT instances (as done, e.g., in [17] for the particular case of the classical truth tables). For this purpose, the use of $Ax(\mathbf{G})$ (see Definition 7) instead of **G** is technically convenient.

Definition 12. The SAT instance associated with a given $\mathcal{L}$-calculus **G**, an $\mathcal{L}$-sequent s and a set $\odot \subseteq \Diamond_{\mathcal{L}}^1$ is given by:

$$\text{SAT}^\odot(\mathbf{G}, s) := \bigcup \left\{\text{SAT}^+(\sigma(s')) \mid \emptyset / s' \in Ax(\mathbf{G}), \sigma(frm(s')) \subseteq sub^\odot(s)\right\}\,.$$

Example 11. Consider the $\{\neg\}$-analytic calculus $\mathbf{G}_3$ for Łukasiewicz three-valued logic from Example 7. Following Example 8, $Ax(\mathbf{G}_3)$ contains the axiomatic rules $\emptyset / p_1, p_1 \supset p_2 \Rightarrow \neg p_1, p_2$ and $\emptyset / \neg p_2, p_1 \supset p_2 \Rightarrow \neg p_1, p_2$. Given a sequent s, $\text{SAT}^{\{\neg\}}(\mathbf{G}_3, s)$ includes the clause $\{\overline{x_{\psi_1}}, \overline{x_{\psi_1 \supset \psi_2}}, x_{\neg\psi_1}, x_{\psi_2}\}$ and the clause $\{\overline{x_{\neg\psi_2}}, \overline{x_{\psi_1 \supset \psi_2}}, x_{\neg\psi_1}, x_{\psi_2}\}$ for every formula of the form $\psi_1 \supset \psi_2$ in $sub^{\{\neg\}}(s)$.

Theorem 3. *Let* **G** *be a* $\odot$-*analytic* $\mathcal{L}$-*calculus and* s *be an* $\mathcal{L}$-*sequent. Then* $\vdash_{\mathbf{G}} s$ *iff* $SAT^\odot(\mathbf{G}, s) \cup SAT^-(s)$ *is unsatisfiable.*

Proof. Suppose that $\nvdash_{\mathbf{G}} s$. By Proposition 1, $\nvdash_{Ax(\mathbf{G})} s$. By Theorem 1, there exists an $Ax(\mathbf{G})$-legal $\mathcal{L}$-bivaluation v such that $v(s) = 0$. The classical assignment u that assigns true to a variable x_ψ iff $v(\psi) = 1$ satisfies $\text{SAT}^\odot(\mathbf{G}, s) \cup \text{SAT}^-(s)$.

For the converse, let u be a classical assignment satisfying the SAT instance $\text{SAT}^\odot(\mathbf{G}, s) \cup \text{SAT}^-(s)$. Consider the $sub^\odot(s)$-bivaluation v defined by $v(\psi) = 1$ iff u assigns true to x_ψ. It is easy to see that since u satisfies $\text{SAT}^\odot(\mathbf{G}, s)$, v is $Ax(\mathbf{G})$-legal. u also satisfies $\text{SAT}^-(s)$, and hence $v(s) = 0$. Since **G** is $\odot$-analytic, so is $Ax(\mathbf{G})$ (by Proposition 1). By Theorem 2, $Ax(\mathbf{G})$ is semantically $\odot$-analytic, and so v can be extended to an $Ax(\mathbf{G})$-legal $\mathcal{L}$-bivaluation. Theorem 1 entails that $\nvdash_{Ax(\mathbf{G})} s$. By Proposition 1, it follows that $\nvdash_{\mathbf{G}} s$.   □

Now, we show that the above reduction is computable in polynomial time.

Definition 13. A rule S / s is called k-$\odot$-*closed* if there are $\varphi_1, \ldots, \varphi_k \in frm(s)$ (called *main formulas*) such that $frm(S / s)$ consists only of $\odot$-subformulas of the φ_i's. A calculus is k-$\odot$-*closed* if each of its rules is k'-$\odot$-closed for some $k' \leq k$.

Example 12. **LK** and **P** (see Examples 2 and 3) are 1-$\emptyset$-closed. $\mathbf{G}_{C_1}$ (see Example 4) is 1-$\{\neg\}$-closed. **EP** (see Example 10) is 2-$\emptyset$-closed, because of the rule $\emptyset \,/\, p_1 \vee p_2 \Rightarrow p_2 \vee p_1$.

Remark 1. Every axiomatic calculus is k-$\circledcirc$-closed for some k (e.g., the maximal number of formulas in its rules). As seen in Proposition 1, every calculus **G** is equivalent to the axiomatic calculus $Ax(\mathbf{G})$. Moreover, if **G** is k-$\circledcirc$-closed, then so is $Ax(\mathbf{G})$.

Theorem 4. *Let* **G** *be a* k-$\circledcirc$-*closed* $\mathcal{L}$-*calculus. Given an* $\mathcal{L}$-*sequent* s, *the SAT instance* $SAT^{\circledcirc}(\mathbf{G}, s) \cup SAT^{-}(s)$ *is computable in* $O(n^k)$ *time, where* n *is the length of the string representing* s.

Proof (sketch). The following algorithm computes $SAT^{\circledcirc}(\mathbf{G}, s) \cup SAT^{-}(s)$:

1. Build a parse tree for the input using standard techniques. As usual, every node represents an occurrence of some subformula in s.
2. Using, e.g., the linear-time algorithm from [10], compress the parse tree into an ordered dag by maximally unifying identical subtrees. After the compression, the nodes of the dag represent subformulas of s, rather than occurrences. Hence we may identify nodes with their corresponding formulas.
3. Traverse the dag. For every $\circ \in \circledcirc$ and node v that has a parent, add a new parent labeled with $\circ$, if such a parent does not exist. To check this it is possible to maintain in each node v a constant-size list of all unary connectives in $\circledcirc$ that label the parents of v. Note that after these additions, the nodes of the dag one-to-one correspond to $sub^{\circledcirc}(s)$.
4. $SAT^{-}(s)$ is obtained by traversing the dag and generating $\{x_\psi\}$ for every ψ on the left-hand side of s and $\{\overline{x_\psi}\}$ for every ψ on the right-hand side of s.
5. $SAT^{\circledcirc}(\mathbf{G}, s)$ is generated by looping over all rules in $Ax(\mathbf{G})$. For each rule $\emptyset \,/\, s'$ with main formulas $\varphi_1, \ldots, \varphi_{k'}$ ($k' \leq k$), go over all k'-tuples of nodes in the dag. For each k' nodes $v_1, \ldots, v_{k'}$ check whether $v_1, \ldots, v_{k'}$ match the pattern given by $\varphi_1, \ldots, \varphi_{k'}$, and if so, construct a mapping h from the formulas in $sub^{\circledcirc}(s')$ to their matching nodes. Then construct a clause consisting of a literal $\overline{x_{h(\varphi)}}$ for every φ on the left-hand side of s', and a literal $x_{h(\varphi)}$ for every φ on the right-hand side of s'. Note that only a constant depth of the sub-dags rooted at $v_1, \ldots, v_{k'}$ is considered - that is the complexity of $\varphi_1, \ldots, \varphi_{k'}$, in addition to parents labeled with elements from $\circledcirc$. These are independent of the input sequent s. To see that we generate exactly all required clauses, note that a substitution σ satisfies $\sigma(\mathit{frm}(s')) \subseteq sub^{\circledcirc}(s)$ iff $\sigma(\{\varphi_1, \ldots, \varphi_{k'}\}) \subseteq sub^{\circledcirc}(s)$. Thus a substitution σ satisfies $\sigma(\mathit{frm}(s')) \subseteq sub^{\circledcirc}(s)$ iff there are k' nodes matching the patterns given by $\varphi_1, \ldots, \varphi_{k'}$.

Steps 1,2,3,4 require linear time. Each pattern matching in step 5 is done in constant time, and so handling a k'-$\circledcirc$-closed rule takes $O(n^{k'})$ time. Thus step 5 requires $O(n^k)$ time. $\square$

Remark 2. We employ the same standard computation model of analysis of algorithms used in [11]. A linear time implementation of this algorithm cannot afford the variables x_ψ to literally include a full string representation of ψ. Thus we assume that each node has a key that can be printed and manipulated in constant time (e.g., its memory address).

Corollary 1. *For any $\odot$-analytic calculus* **G**, *the derivability problem for* **G** *is in co-NP.*

The reduction runs in linear time for 1-$\odot$-closed calculi. In such cases, it is natural to identify calculi whose SAT instances can be decided in linear time. This is the case, for example, for instances consisting of *Horn clauses* [13].

Definition 14. A rule r is called a *Horn rule* if $\#_L(r) + \#_R(r) \le 1$, where $\#_L(r)$ is the number of premises of r whose left-hand side is not empty, and $\#_R(r)$ is the number of formulas on the right-hand side of the conclusion of r. A calculus is called a *Horn calculus* if each of its rules is a Horn rule.

Proposition 2. *Let* **G** *be a Horn $\mathcal{L}$-calculus and s be an $\mathcal{L}$-sequent. Then $SAT^{\odot}(\mathbf{G}, s)$ consists solely of Horn clauses.*

Corollary 2. *Let* **G** *be a $\odot$-analytic, 1-$\odot$-closed Horn $\mathcal{L}$-calculus. The derivability problem for* **G** *can be decided in linear time using a HORNSAT solver.*

Example 13. The derivability problem for **EP** (see Example 10) is decidable in quadratic time, as **EP** is a $\emptyset$-analytic, 2-$\emptyset$-closed Horn calculus. Excluding the rule $\emptyset \, / \, p_1 \vee p_2 \Rightarrow p_2 \vee p_1$ results in a 1-$\emptyset$-closed Horn calculus, whose derivability problem can be decided in linear time. The linear time algorithm for **P** from [6] is also an instance of this method.

6 Next Operators

In this section we extend the framework to accommodate *Next* operators, that are often employed in temporal logics. In primal infon logic [11], they play the role of quotations (see Example 14 below). In what follows, $\circledast$ denotes an arbitrary finite set of unary connectives (*Next* operators), and $\mathcal{L}_\circledast$ denotes the propositional language obtained by augmenting $\mathcal{L}$ with $\circledast$ (we assume that $\Diamond_{\mathcal{L}} \cap \circledast = \emptyset$). A sequence $\bar{*} = *_1 \ldots *_m$ ($m \ge 0$) of elements of $\circledast$ is called a $\circledast$-*prefix*. Given a set $\mathcal{F} \subseteq \mathcal{L}_\circledast$ and a $\circledast$-prefix $\bar{*}$, we denote the set $\{\bar{*}\psi \mid \psi \in \mathcal{F}\}$ by $\bar{*}\mathcal{F}$. This notation is extended to sequents and sets of sequents in the obvious way. We now extend pure calculi with new rules for *Next* operators.

Definition 15. A $\circledast$-*proof* in a calculus **G** is defined similarly to a standard proof (see Definition 2), where in addition to (*weak*), (*id*) and (*cut*), the following scheme may be used for any $* \in \circledast$:

$$(*i) \qquad \frac{\Gamma \Rightarrow \Delta}{*\Gamma \Rightarrow *\Delta}$$

For an $\mathcal{L}$-calculus **G**, a set $\mathcal{F} \subseteq \mathcal{L}_\circledast$, and an $\mathcal{L}_\circledast$-sequent s, we write $\vdash_{\mathbf{G}_\circledast}^{\mathcal{F}} s$ (or $\vdash_{\mathbf{G}_\circledast} s$ if $\mathcal{F} = \mathcal{L}_\circledast$) if there is a $\circledast$-proof of s in **G** consisting only of $\mathcal{F}$-sequents.

($*i$) is a usual rule for *Next* in the temporal logic LTL (i.e., for $\circledast = \{X\}$, we have $\vdash_{\mathbf{LK}_\circledast} \Rightarrow \psi$ iff ψ is valid in the *Next*-only fragment of LTL; see, e.g., [16]). It is also used for $\square$ (and $\Diamond$) in the modal logic $KD!$ of functional Kripke frames.

Remark 3. Applications of $\mathcal{L}$-rules in $\circledast$-proofs may include $\mathcal{L}_\circledast$-formulas. For example, using the rule $\Rightarrow p_2 / \Rightarrow p_1 \supset p_2$, it is possible to derive the sequent $*p_3 \Rightarrow *p_1 \supset p_2$ from $*p_3 \Rightarrow p_2$.

Example 14. The quotations employed in primal infon logic [11] are unary connectives of the form q `said`, where q ranges over a finite set of principals. If we take $\circledast$ to include these connectives, we have that $\vdash_{\mathbf{P}_\circledast} \Gamma \Rightarrow \psi$ (see Example 3) iff ψ is derivable from Γ in the Hilbert system for primal infon logic given in [11]. This can be shown by induction on the lengths of the proofs.

Next, we define Kripke-style semantics for calculi with *Next* operators.

Definition 16. A *biframe* for $\circledast$ is a tuple $\mathcal{W} = \langle W, \mathcal{R}, \mathcal{V} \rangle$ where:

1. W is a set of elements called *worlds*. Henceforth, we may identify $\mathcal{W}$ with this set (e.g., when writing $w \in \mathcal{W}$ instead of $w \in W$).
2. $\mathcal{R}$ is a function assigning a binary relation on W to every $* \in \circledast$. We write $\mathcal{R}_*$ instead of $\mathcal{R}(*)$, and $\mathcal{R}_*[w]$ denotes the set $\{w' \in W \mid w\mathcal{R}_*w'\}$.
3. $\mathcal{V}$ is a function assigning a bivaluation to every $w \in \mathcal{W}$, such that for every $w \in \mathcal{W}$, $* \in \circledast$ and formula ψ: if $*\psi \in dom(\mathcal{V}(w))$ and $\psi \in dom(\mathcal{V}(w'))$ for every $w' \in \mathcal{R}_*[w]$, then $\mathcal{V}(w)(*\psi) = \min\{\mathcal{V}(w')(\psi) \mid w' \in \mathcal{R}_*[w]\}$. Henceforth, we write $\mathcal{V}_w$ instead of $\mathcal{V}(w)$.

Furthermore, if $dom(\mathcal{V}_w) = \mathcal{F}$ for every $w \in \mathcal{W}$, we refer to $\mathcal{W}$ as an $\mathcal{F}$-*biframe*.

Definition 17. A biframe $\langle W, \mathcal{R}, \mathcal{V} \rangle$ for $\circledast$ is called *functional* if $\mathcal{R}_*$ is a functional relation (that is, a total function from W to W) for every $* \in \circledast$. In this case we write $\mathcal{R}_*(w)$ to denote the unique element $w' \in W$ satisfying $w\mathcal{R}_*w'$.

Definition 18. A biframe $\langle W, \mathcal{R}, \mathcal{V} \rangle$ for $\circledast$ is called **G**-*legal* for an $\mathcal{L}$-calculus **G** if $\mathcal{V}_w$ is **G**-legal for every $w \in W$ (see Definition 9).

Theorem 5 (Soundness and Completeness). *Let* **G** *be an $\mathcal{L}$-calculus, $\mathcal{F}$ be a set of $\mathcal{L}_\circledast$-formulas, and s be an $\mathcal{F}$-sequent. Then, $\vdash_{\mathbf{G}_\circledast}^{\mathcal{F}} s$ iff $\mathcal{V}_w(s) = 1$ for every* **G**-*legal functional $\mathcal{F}$-biframe $\langle W, \mathcal{R}, \mathcal{V} \rangle$ for $\circledast$ and $w \in W$.*

Generally speaking, soundness is proved by induction on the length of the $\circledast$-proof. The fact that the biframes are functional is essential for the soundness of ($*i$). Completeness is proved using a canonical countermodel construction.

Remark 4. Note that similar results hold for usual rules for introducing $\square$. For example, if we take the usual rule used in the system for the modal logic K (which, unlike ($*i$), allows only one formula on the right-hand side), we can prove soundness and completeness as above with respect to *all* **G**-legal $\mathcal{F}$-biframes. Similarly, for other known sequent rules for $\square$ (as those of the systems for $K4$,

KB, $S4$, and $S5$, see [18]) it is possible to show a similar general soundness and completeness with respect to **G**-legal $\mathcal{F}$-biframes satisfying the corresponding condition (transitivity, symmetry, etc.). Nevertheless, the reduction to SAT proposed below applies only for $(*i)$.

Next we extend the reduction from Section 5 to analytic calculi with *Next* operators. This is done for a large family of calculi that we call *standard*.

Definition 19. An atomic variable $p \in At$ is called *lonely* in some rule r if $p \in frm(r)$, but p is not a proper subformula of any formula in $frm(r)$. A calculus is called *standard* if none of its rules has lonely atomic variables.

As before, we use $\{x_\psi \mid \psi \in \mathcal{L}_\circledcirc\}$ as the set of atomic variables in the SAT instances. Nevertheless, while the reduction above was based on $\odot$-subformulas, the current reduction is based on $\odot$-*local formulas*. This notion generalizes the *local formulas relation* from [15].

Definition 20. $loc^\odot(\psi)$, the set of formulas that are $\odot$-*local* to an $\mathcal{L}_\circledcirc$-formula ψ, is inductively defined as follows: *1)* $loc^\odot(p) = \{p\}$ for every atomic variable $p \in At$; *2)* $loc^\odot(\diamond(\psi_1, \ldots, \psi_n)) = \{\diamond(\psi_1, \ldots, \psi_n)\} \cup \{\circ\psi_i \mid \circ \in \odot, 1 \le i \le n\} \cup \bigcup_{i=1}^n loc^\odot(\psi_i)$ for every $\diamond \in \diamondsuit_\mathcal{L}^n$ and formulas $\psi_1, \ldots, \psi_n$; *3)* $loc^\odot(*\psi) = *loc^\odot(\psi)$ for every $* \in \circledast$ and formula ψ. This definition is extended to sequents in the obvious way, i.e. $loc^\odot(s) = \bigcup\{loc^\odot(\varphi) \mid \varphi \in frm(s)\}$.

Note that for $\circledast = \emptyset$, we have $loc^\odot(\psi) = sub^\odot(\psi)$ for every formula ψ.

Example 15. For $\circledast = \{\flat, \sharp\}$,
$loc^{\{\neg\}}(\flat(\sharp p_1 \supset p_2)) = \{\flat\sharp p_1, \flat\neg\sharp p_1, \flat p_2, \flat\neg p_2, \flat(\sharp p_1 \supset p_2)\}$.

Definition 21. The SAT instance associated with an $\mathcal{L}$-calculus **G**, an $\mathcal{L}_\circledcirc$-sequent s and a set $\odot \subseteq \diamondsuit_\mathcal{L}^1$ is given by:

$$\mathrm{SAT}_\circledast^\odot(\mathbf{G}, s) := \bigcup \left\{ \mathrm{SAT}^+(\bar{*}\sigma(s')) \mid \emptyset / s' \in Ax(\mathbf{G}), \bar{*}\sigma(frm(s')) \subseteq loc^\odot(s) \right\}.$$

Theorem 6. *Let* **G** *be a standard $\odot$-analytic $\mathcal{L}$-calculus and s be an $\mathcal{L}_\circledcirc$-sequent. Then* $\vdash_{\mathbf{G}_\circledcirc} s$ *iff* $SAT_\circledast^\odot(\mathbf{G}, s) \cup SAT^-(s)$ *is unsatisfiable.*

Generally speaking, the main difficulty in the proof of this theorem is to construct a countermodel for s (in the form of a **G**-legal functional $\mathcal{L}_\circledcirc$-biframe for $\circledast$) out of a satisfying assignment u of $\mathrm{SAT}_\circledast^\odot(\mathbf{G}, s) \cup SAT^-(s)$. Thus if $\nvdash_{\mathbf{G}_\circledcirc} s$, the full proof of Theorem 6 actually provides a way to translate the classical assignment that satisfies $\mathrm{SAT}_\circledast^\odot(\mathbf{G}, s) \cup SAT^-(s)$ into a countermodel of s. This is done in two steps. First, we translate u into a $\odot$-*closed* **G**-legal functional biframe $\mathcal{W}$ which is not a model of s:

Definition 22. A set of $\mathcal{L}_\circledcirc$-formulas is called $\odot$-*closed* if whenever it contains a formula of the form $\diamond(\varphi_1, \ldots, \varphi_n)$ (for some $\diamond \in \diamondsuit_\mathcal{L}$) it also contains φ_i and $\circ\varphi_i$ for every $1 \le i \le n$ and $\circ \in \odot$. A biframe $\langle W, \mathcal{R}, \mathcal{V} \rangle$ for $\circledast$ is called $\odot$-*closed* if the following hold for every $w \in W$: $dom(\mathcal{V}_w)$ is $\odot$-closed and finite; and for every $* \in \circledast$, if $*\psi \in dom(\mathcal{V}_w)$, then $\psi \in dom(\mathcal{V}_{w'})$ for every $w' \in \mathcal{R}_*[w]$.

Given an assignment u that satisfies $\text{SAT}^\circledcirc_\circledast(\mathbf{G}, s) \cup \text{SAT}^-(s)$, a $\circledcirc$-*closed* $\mathbf{G}$-legal functional biframe $\mathcal{W} = \langle W, \mathcal{R}, \mathcal{V} \rangle$ is constructed as follows:

1. W is the set of all $\circledast$-prefixes.
2. For every $* \in \circledast$ and $\bar{*} \in W$, $\mathcal{R}_*(\bar{*}) = \bar{*}*$.
3. $\mathcal{V}_{\bar{*}}$ is defined by induction on the length of $\bar{*}$: $dom(\mathcal{V}_\epsilon) = loc^\circledcirc(s)$ and $\mathcal{V}_\epsilon(\psi) = 1$ iff u satisfies x_ψ;[3] $dom(\mathcal{V}_{*_1 \ldots *_n}) = \{\varphi \mid *_n \varphi \in dom(\mathcal{V}_{*_1 \ldots *_{n-1}})\}$ and $\mathcal{V}_{*_1 \ldots *_n}(\psi) = \mathcal{V}_{*_1 \ldots *_{n-1}}(*_n \psi)$.

Then, the following theorem is used to extend $\mathcal{W}$ to a full $\mathbf{G}$-legal $\mathcal{L}_\circledast$-biframe.

Definition 23. A biframe $\langle W, \mathcal{R}, \mathcal{V} \rangle$ for $\circledast$ *extends* a biframe $\langle W', \mathcal{R}', \mathcal{V}' \rangle$ for $\circledast$ if $W = W'$, $\mathcal{R} = \mathcal{R}'$, and $\mathcal{V}_w$ extends $\mathcal{V}'_w$ for every $w \in W$.

Theorem 7. *Let $\mathbf{G}$ be a standard semantically $\circledcirc$-analytic $\mathcal{L}$-calculus, and $\mathcal{W}$ be a $\mathbf{G}$-legal $\circledcirc$-closed biframe for $\circledast$ with $dom(\mathcal{V}_w) \subseteq \mathcal{L}_\circledcirc$ for every $w \in W$. Then $\mathcal{W}$ can be extended to a $\mathbf{G}$-legal $\mathcal{L}_\circledast$-biframe for $\circledast$.*

For the case that $\circledcirc = \emptyset$, the polynomial time algorithm from Section 5 can be modified to accommodate *Next* operators.

Theorem 8. *Let $\mathbf{G}$ be a k-$\emptyset$-closed $\mathcal{L}$-calculus. Given an $\mathcal{L}_\circledast$-sequent s, it is possible to compute $\text{SAT}^\emptyset_\circledast(\mathbf{G}, s) \cup \text{SAT}^-(s)$ in $O(n^k)$ time, where n is the length of the string representing s.*

Proof (sketch). The algorithm from the proof of Theorem 4 is reused with several modifications. As in [11], an auxiliary trie (an ordered tree data structure commonly used for string processing) for $\circledast$-prefixes is constructed in linear time, and every node in the input parse tree has a pointer to a node in this trie. Now each node in the parse tree corresponds to an occurrence of a formula that is $\emptyset$-local to s. The tree is then compressed to a dag as in the proof of Theorem 4. The nodes of the dag one-to-one correspond to the $\emptyset$-local formulas of s. The rest of the algorithm is exactly as in the proof of Theorem 4 with $\circledcirc = \emptyset$. □

For a Horn calculus $\mathbf{G}$, $\text{SAT}^\circledcirc_\circledast(\mathbf{G}, s) \cup \text{SAT}^-(s)$ consists of Horn clauses for every sequent s. When $\mathbf{G}$ is 1-$\emptyset$-closed and $\emptyset$-analytic, a linear time decision procedure for the derivability problem for $\mathbf{G}$ with *Next* operators is obtained by applying a HORNSAT solver on $\text{SAT}^\emptyset_\circledast(\mathbf{G}, s) \cup \text{SAT}^-(s)$.

Example 16. Example 13 works as is for the extension of $\mathbf{P}$ or $\mathbf{EP}$ with any finite set of *Next* operators.

Example 17. The linear time fragment of dual-Horn clauses can be utilized as well. For example, consider the ($\emptyset$-analytic) calculus $\mathbf{P}_d$ that consists of the rules $(\vee \Rightarrow)$, $(\Rightarrow \vee)$, $(\wedge \Rightarrow)$ of $\mathbf{LK}$ and the following ones for "dual primal implication":
$$(\prec \Rightarrow) \, p_1 \Rightarrow \, / \, p_1 \prec p_2 \Rightarrow \qquad (\Rightarrow \prec) \Rightarrow p_1; p_2 \Rightarrow \, / \Rightarrow p_1 \prec p_2$$
For any sequent s, $\text{SAT}^\circledcirc_\circledast(\mathbf{P}_d, s) \cup \text{SAT}^-(s)$ consists of dual-Horn clauses. Thus the derivability problem for $\mathbf{P}_d$ with *Next* operators can be decided in linear time.

---

[3] ϵ denotes the empty $\circledast$-prefix.

6.1 On Analyticity of Pure Calculi with *Next* Operators

At this point, a natural question arises: does the extension of a calculus with *Next* operators preserve the $\odot$-analyticity of the calculus? In this final section we provide a positive answer to this question, based on Theorem 7 above that was used to prove the correctness of the reduction.

Definition 24. An $\mathcal{L}$-calculus $\mathbf{G}$ is called $\odot$-*analytic with* $\circledast$ if $\vdash_{\mathbf{G}_\circledast} s$ implies $\vdash_{\mathbf{G}_\circledast}^{sub^\odot(s)} s$ for every $\mathcal{L}_\circledast$-sequent s.

Theorem 9. *A* standard $\mathcal{L}$-calculus $\mathbf{G}$ *is* $\odot$-*analytic iff it is* $\odot$-*analytic with* $\circledast$.

Proof. Suppose that $\mathbf{G}$ is $\odot$-analytic. By Theorem 2 it is also semantically $\odot$-analytic. Let s be an $\mathcal{L}_\circledast$-sequent such that $\nvdash_{\mathbf{G}_\circledast}^{sub^\odot(s)} s$. By Theorem 5, there exists a $\mathbf{G}$-legal functional $sub^\odot(s)$-biframe $\mathcal{W} = \langle W, \mathcal{R}, \mathcal{V} \rangle$ and $w \in W$ such that $\mathcal{V}_w(s) = 0$. $\mathcal{W}$ is $\odot$-closed, and by Theorem 7, it can be extended to a $\mathbf{G}$-legal functional $\mathcal{L}_\circledast$-biframe $\mathcal{W}' = \langle W, \mathcal{R}, \mathcal{V}' \rangle$ for $\circledast$. After this extension, we still have $\mathcal{V}'_w(s) = 0$. Theorem 5 implies that $\nvdash_{\mathbf{G}_\circledast} s$. For the converse, suppose that $\mathbf{G}$ is $\odot$-analytic with $\circledast$. Assume that $\vdash_{\mathbf{G}} s$ for some $\mathcal{L}$-sequent s. Hence, $\vdash_{\mathbf{G}_\circledast} s$. Consequently, there is a $\circledast$-proof of s in $\mathbf{G}$ that consists only of $sub^\odot(s)$-formulas. This proof cannot contain applications of $(*i)$, and therefore, $\vdash_{\mathbf{G}}^{sub^\odot(s)} s$. $\square$

Example 18. Since $\mathbf{P}$ and $\mathbf{EP}$ are $\emptyset$-analytic and standard, they are also $\emptyset$-analytic with $\circledast$. In contrast, the Hilbert system for primal infon logic in [11] admits a similar property that involves local formulas rather than subformulas.

Remark 5. Further to Remark 4, it can be similarly shown that the extension of a standard pure calculus with any usual rule for $\square$ preserves analyticity. In particular, we did not assume in Theorem 7 that the biframes are functional.

7 Conclusions and Further Research

We have identified a wide family of calculi for which the derivability problem can be solved using off-the-shelf SAT solvers. Our method was presented for pure calculi, and later extended to accommodate *Next* operators. The produced SAT instances do not encode derivations, whose lengths might not be polynomially bounded. Instead, they represent the (non-) existence of polynomially bounded countermodels in the form of partial bivaluations or Kripke frames.

The proposed reduction is limited to analytic pure calculi, as it relies on their straightforward bivaluation semantic presentation. Nevertheless, some of the theoretic developments presented in this paper can be extended to different families of calculi. For example, following Remark 5, the fact that analyticity is preserved when pure calculi are augmented with *Next* operators, holds also for other introduction rules for modalities. Such extensions, as well as studying multi-ary modalities in this context, are left for future work. In addition, we plan to extend the methods of this paper to analytic many-sided sequent calculi, that are more expressive than ordinary two-sided calculi. Finally, it is interesting to study possible applications of logics (besides primal logic) that can be reduced to efficient fragments of SAT (e.g., 2SAT).

References

1. Avron, A.: Simple consequence relations. Inf. Comput. 92(1), 105–139 (1991)
2. Avron, A.: Classical gentzen-type methods in propositional many-valued logics. In: Fitting, M., Orłowska, E. (eds.) Beyond Two: Theory and Applications of Multiple-Valued Logic. Studies in Fuzziness and Soft Computing, vol. 114, pp. 117–155. Physica-Verlag HD (2003)
3. Avron, A., Konikowska, B., Zamansky, A.: Modular construction of cut-free sequent calculi for paraconsistent logics. In: 2012 27th Annual IEEE Symposium on Logic in Computer Science (LICS), pp. 85–94 (2012)
4. Avron, A., Konikowska, B., Zamansky, A.: Cut-free sequent calculi for c-systems with generalized finite-valued semantics. Journal of Logic and Computation 23(3), 517–540 (2013)
5. Avron, A., Lev, I.: Non-deterministic multiple-valued structures. Journal of Logic and Computation 15(3), 241–261 (2005)
6. Beklemishev, L., Gurevich, Y.: Propositional primal logic with disjunction. Journal of Logic and Computation (2012)
7. Béziau, J.-Y.: Sequents and bivaluations. Logique et Analyse 44(176), 373–394 (2001)
8. Bjørner, N., de Caso, G., Gurevich, Y.: From primal infon logic with individual variables to datalog. In: Erdem, E., Lee, J., Lierler, Y., Pearce, D. (eds.) Correct Reasoning. LNCS, vol. 7265, pp. 72–86. Springer, Heidelberg (2012)
9. Blass, A., Gurevich, Y.: Abstract hilbertian deductive systems, infon logic, and datalog. Information and Computation 231, 21–37 (2013)
10. Cai, J., Paige, R.: Using multiset discrimination to solve language processing problems without hashing. Theoretical Computer Science 145(12), 189–228 (1995)
11. Cotrini, C., Gurevich, Y.: Basic primal infon logic. Journal of Logic and Computation (2013)
12. Degtyarev, A., Voronkov, A.: The inverse method. In: Handbook of Automated Reasoning, vol. 1, pp. 179–272 (2001)
13. Dowling, W.F., Gallier, J.H.: Linear-time algorithms for testing the satisfiability of propositional horn formulae. The Journal of Logic Programming 1(3), 267–284 (1984)
14. Gomes, C.P., Kautz, H., Sabharwal, A., Selman, B.: Satisfiability solvers. In: Handbook of Knowledge Representation. Foundations of Artificial Intelligence, vol. 3, pp. 89–134. Elsevier (2008)
15. Gurevich, Y., Neeman, I.: Logic of infons: The propositional case. ACM Trans. Comput. Logic 9, 1–9 (2011)
16. Kawai, H.: Sequential calculus for a first order infinitary temporal logic. Mathematical Logic Quarterly 33(5), 423–432 (1987)
17. Kowalski, R.: Logic for Problem-solving. North-Holland Publishing Co., Amsterdam (1986)
18. Wansing, H.: Sequent systems for modal logics. In: Gabbay, D.M., Guenthner, F. (eds.) Handbook of Philosophical Logic, 2nd edn., vol. 8, pp. 61–145. Springer (2002)

A Unified Proof System for QBF Preprocessing[*]

Marijn J.H. Heule[1], Martina Seidl[2], and Armin Biere[2]

[1] Department of Computer Science, The University of Texas at Austin, USA
marijn@cs.utexas.edu
[2] Institute for Formal Models and Verification, JKU Linz, Austria
{martina.seidl,biere}@jku.at

Abstract. For quantified Boolean formulas (QBFs), preprocessing is essential to solve many real-world formulas. The application of a preprocessor, however, prevented the extraction of proofs for the original formula. Such proofs are required to independently validate correctness of the preprocessor's rewritings and the solver's result. Especially for universal expansion proof checking was not possible so far. In this paper, we introduce a unified proof system based on three simple and elegant *quantified resolution asymmetric tautology* (QRAT) rules. In combination with an extended version of universal reduction, they are sufficient to efficiently express *all* preprocessing techniques used in state-of-the-art preprocessors including universal expansion. Moreover, these rules give rise to new preprocessing techniques. We equip our preprocessor bloqqer with QRAT proof logging and provide a proof checker for QRAT proofs.

1   Introduction

Effectively checking the result returned by a QBF solver has been an open challenge for a long time [1,2,3,4,5,6,7]. The current state-of-the-art is to simply dump Q-resolution proofs and to validate their structure. This approach has two major drawbacks. On the one hand the proofs might get extremely large and cannot be produced due to technical limitations. On the other hand, there are solving and preprocessing techniques for which it is not known if and how they translate to Q-resolution.

Due to the diversity of techniques in state-of-the-art preprocessors [8,9], it is not straightforward to provide a checker which verifies the output of the preprocessor. In fact, it would be preferable to translate the different preprocessing techniques to a canonical representation which then can be checked easily. Some efforts go in this direction by using Q-resolution. If a resolution proof is available, then checking is polynomial w.r.t. the proof size. However, the proof itself might become exponentially large and already writing down the proof might be costly.

---

[*] This work was supported by the Austrian Science Fund (FWF) through the national research network RiSE (S11408-N23), Vienna Science and Technology Fund (WWTF) under grant ICT10-018, DARPA contract number N66001-10-2-4087, and the National Science Foundation under grant number CCF-1153558.

S. Demri, D. Kapur, and C. Weidenbach (Eds.): IJCAR 2014, LNAI 8562, pp. 91–106, 2014.
© Springer International Publishing Switzerland 2014

Furthermore, it is not known for all preprocessing techniques how to express them in terms of resolution, what is the case for universal expansion [10].

In propositional logic, the RUP proof checking format [11] is extremely successful because it simply logs the learnt clauses and provides an easy checking criterion. For optimization purposes, recently, the DRUP extension has been presented [12] which provides elimination criteria for redundant clauses. It has been recognized that RUP and DRUP can be characterized with the *resolution asymmetric tautology property* (RAT) which has been originally developed in the context of propositional preprocessing for characterizing and comparing the strength of the various techniques. In this paper, we extend RAT [13] to QRAT, the *quantified resolution asymmetric tautology property*, and introduce novel clause addition and elimination techniques using QRAT. On this basis, we capture the state-of-the-art preprocessing techniques in a uniform manner what allows us to develop a checker verifying the correctness of a QBF preprocessor. Moreover, checking QRAT proofs is polynomial in the proof size. We integrated QRAT-based tracing in our preprocessor bloqqer [8] and implemented an efficient checker for QRAT proofs.

2   Preliminaries

We consider QBFs in prenex conjunctive normal form (PCNF). A QBF in PCNF has the structure $\Pi.\psi$ where the prefix Π has the form $Q_1 X_1 Q_2 X_2 \ldots Q_n X_n$ with disjoint variable sets X_i and $Q_i \in \{\forall, \exists\}$. The formula ψ is a propositional formula in conjunctive normal form, i.e., a conjunction of clauses. A clause is a disjunction of literals and a literal is either a variable (positive literal) or a negated variable (negative literal). The variable of a literal is denoted by $\mathsf{var}(l)$ where $\mathsf{var}(l) = x$ if $l = x$ or $l = \bar{x}$. The negation of a literal l is denoted by $\bar{l}$. The quantifier $\mathsf{Q}(\Pi, l)$ of a literal l is Q_i if $\mathsf{var}(l) \in X_i$. Let $\mathsf{Q}(\Pi, l) = Q_i$ and $\mathsf{Q}(\Pi, k) = Q_j$, then $l \leq_\Pi k$ if $i \leq j$. We sometimes write formulas in CNF as sets of clauses and clauses as sets of literals. We consider only closed QBFs, so ψ contains only variables which occur in the prefix. The variables occurring in the prefix of ϕ are given by $\mathsf{vars}(\phi)$. The subformula ψ_l consisting of all clauses of matrix ψ containing literal l is defined by $\psi_l = \{C \mid l \in C, C \in \psi\}$. By $\top$ and $\bot$ we denote the truth constants true and false. QBFs are interpreted as follows: a QBF $\forall x \Pi.\psi$ is false iff $\Pi.\psi[x/\top]$ or $\Pi.\psi[x/\bot]$ is false where $\Pi.\psi[x/t]$ is the QBF obtained by replacing all occurrences of variable x by t. Respectively, a QBF $\exists x \Pi.\psi$ is false iff both $\Pi.\psi[x/\top]$ and $\Pi.\psi[x/\bot]$ are false. If the matrix ψ of a QBF ϕ contains the empty clause after eliminating the truth constants according to standard rules, then ϕ is false. Accordingly, if the matrix ψ of QBF ϕ is empty, then ϕ is true. Two QBFs are *satisfiability equivalent* iff they have the same truth value.

Models and countermodels of QBFs can either be described intensionally in form of Herbrand and Skolem functions [1] or extensionally in form of subtrees of assignment trees. An *assignment tree* of a QBF ϕ is a complete binary tree of depth $|\mathsf{vars}(\phi) + 1|$ where the non-leaf nodes of each level are *associated* with a variable of ϕ. The order of the associated variables in the tree respects the order of the variables in the prefix of ϕ. A non-leaf node associated with variable x has one outgoing edge *labelled* with x and one outgoing edge *labelled* with $\bar{x}$. Each path starting from the root of the tree represents a (partial) variable assignment. We also write a path as a sequence of literals. A path τ from the root node to a leaf is a complete assignment and the leaf is labelled with the value of the QBF under τ. Nodes associated with existential variables act as OR-nodes, while universal nodes act as AND-nodes. Respectively, a node is labelled either with $\top$ or with $\bot$. A QBF is true (satisfiable) iff its root is labelled with $\top$. A QBF is false (unsatisfiable) iff its root is labelled with $\bot$. By τ^x and τ_x we denote the partial assignments obtained from the complete assignment τ with $\tau = \tau^x l \tau_x$ where $\mathsf{var}(l) = x$. A QBF ϕ with $\mathsf{vars}(\phi) = \{x_1, \dots, x_n\}$ under (partial) assignment τ is the QBF $\phi[x_1/t_1, \dots, x_n/t_n]$ where $t_i = \top$ if $x_i \in \tau$, $t_i = \bot$ if $\bar{x}_i \in \tau$, and $t_i = x_i$ otherwise.

Example 1. Consider the QBF $\exists a \forall b \exists c \forall d \exists e.(a \vee b \vee \bar{c} \vee \bar{d} \vee e)$ and the path from the root $\tau = a\bar{b}\bar{c}d\bar{e}$. Then we have partial assignments $\tau^c = a\bar{b}$ and $\tau_c = d\bar{e}$.

A *pre-model* M of QBF ϕ is a subtree of the assignment tree of ϕ such that (1) for each universal node in M, both children are in M; (2) for each existential node in M, exactly one of the children is in M; and (3) the root of the assignment tree is in M. A pre-model M of QBF ϕ is a *model* of ϕ if in addition each node in M is labelled with $\top$. Obviously, only a true QBF can have a model. A false QBF has at least one countermodel, which is defined dually as follows. In a *pre-countermodel* M existential nodes have two children, whereas universal nodes have only one and the root of the assignment tree is in M. A pre-countermodel M is a *countermodel* if each node is labelled with $\bot$. Two QBFs are *logically equivalent* iff they have the same set of (counter) models modulo variable names.

3   QRAT: Quantified Resolution Asymmetric Tautologies

The QRAT proof system, introduced below, provides the basis for satisfiability equivalence preserving clause addition, clause deletion, and clause modification techniques. To this end, we first have to recapitulate the notion of *QBF resolvents* (resolvents for short) and introduce the concept of *asymmetric literal addition*.

Definition 1 (Resolvent). *Given two non-tautological clauses C and D with $x \in C$ and $\bar{x} \in D$, the resolvent over pivot variable x is $(C \backslash \{x\}) \cup (D \backslash \{\bar{x}\})$.*

Note that we do not restrict the pivot element to existential variables as it is usually done in the literature. Furthermore, for the moment, we do not consider universal reduction rule necessary for the completeness of Q-resolution.

Definition 2 (Asymmetric Literal Addition). *Given a QBF $\Pi.\psi$ and a clause C. The clause $\mathsf{ALA}(\psi, C)$ is the unique clause obtained by repeatedly applying the extension rule $C := C \cup \{\bar{l}\}$ if $\exists l_1, \dots, l_k \in C$ and $(l_1 \vee \dots \vee l_k \vee l) \in \psi$ called asymmetric literal addition to C until fixpoint.*

Asymmetric literal addition is well understood for propositional logic [14]. For QBF, a variant called *hidden literal addition* has been described in [8] where it is (unnecessarily) required that the l_i occur to the left of l in the prefix.

The new definition for QBF used in this paper is the same in the propositional case. Thus $\phi[C/C']$ with $C' = \mathsf{ALA}(\psi, C)$ has exactly the same (propositional) models as ϕ, which lifts to QBF equivalence, since the values the leaves of assignment trees do not change. As consequence we have the following lemma.

Lemma 1. *Let* $\phi = \Pi.\psi \cup \{C\}$ *be a QBF and* $C' = \mathsf{ALA}(\psi, C)$ *be obtained from* C *by asymmetric literal addition. Further, let* $\phi' = \phi[C/C']$. *Then* ϕ *and* ϕ' *are logically equivalent.*

A clause C is called an *asymmetric tautology* (AT) w.r.t. ψ if $\mathsf{ALA}(\psi, C)$ is a tautology. ALA, AT, and resolution as introduced above are sufficient to define the RAT proof system for propositional logic. For QBFs, we must additionally consider quantifier dependencies which we capture by the notion of *outer clauses* and *outer resolvents*.

Definition 3 (Outer Clause). *Let* C *be a clause occurring in QBF* $\Pi.\psi$. *The outer clause of* C *on literal* $l \in C$, *denoted by* $\mathcal{O}(\Pi, C, l)$, *is given by the clause* $\{k \mid k \in C, k \leq_\Pi l, k \neq l\}$.

Definition 4 (Outer Resolvent). *Let* C *be a clause with* $l \in C$ *and and* D *a clause occurring in QBF* $\Pi.\psi$ *with* $\bar{l} \in D$. *The outer resolvent of* C *with* D *on literal* l *w.r.t.* Π, *denoted by* $\mathcal{R}(\Pi, C, D, l)$, *is given by the clause* $O \cup (C \backslash \{l\})$ *if* $Q(\Pi, l) = \forall$ *and by* $O \cup C$ *if* $Q(\Pi, l) = \exists$ *assuming* $O = \mathcal{O}(\Pi, D, \bar{l})$.

Definition 5 (Quantified Resolution Asymmetric Tautology (QRAT)). *Given a QBF* $\Pi.\psi$ *and a clause* C. *Then* C *has* QRAT *on literal* $l \in C$ *with respect to* $\Pi.\psi$ *iff it holds for all* $D \in \psi_{\bar{l}}$ *that* $\mathsf{ALA}(\psi, R)$ *is a tautology for the outer resolvent* $R = \mathcal{R}(\Pi, C, D, l)$.

The intuition behind these definitions is almost identical to the propositional case [15]: consider potential resolvents of a clause on a certain literal with resolution candidates containing the negation of the picked literal. If all of them are "redundant", or more precisely asymmetric tautologies in the context of this paper, then this clause is redundant too and can be added or removed.

The important difference to the propositional case is that inner variables, w.r.t. the pivot variable resolved upon, might have different values for different choices of universal literals, and thus one can not simply apply resolution blindly before checking for redundancy of the resolvent. Inner literals in the resolution candidates should be ignored. This is the same restriction as for quantified blocked clauses [8]. As it turns out, for existential pivots, it is possible to have a slightly more general version, i.e., the pivot literal can be included in the outer resolvent, while in previous work this was not the case, and for universal pivots, it is not allowed. The QRAT proof system uses this observation to establish syntactical redundancy detection criteria to safely add, remove, and modify clauses.

Lemma 2. *Given a clause C which has* QRAT *w.r.t. a QBF $\Pi.\psi$ on an existential literal $l \in C$ with* var$(l) = x$. *If there is an assignment $\sigma = \tau^x \bar{l} \tau_x$ that falsifies C, but satisfies ψ then the assignment τ^x satisfies all $D \in \psi$ with $\bar{l} \in D$.*

Proof. Let $D \in \psi$ be a clause with $\bar{l} \in D$, $\sigma(C) = \bot$, and $O = \mathcal{O}(\Pi, D, \bar{l})$. In order to show $\tau^x(D) = \top$ by contradiction we assume that $\tau^x(O) = \bot$. This leads to $\sigma(R) = \bot$ for the outer resolvent $R = O \cup C$ too (note that we do not remove l from C). By induction on the order of literals added to R in computing $\mathsf{ALA}(\psi, R)$ we show that $\sigma(l') = \bot$ for all literals l' in $\mathsf{ALA}(\psi, R)$. This is clear for all $l' \in R$. Assume $l_1, \ldots, l_{k-1}$ are from R or have been added through ALA extensions and further assume there is a clause $E = \{l_1, \ldots, l_{k-1}, l_k\} \in \psi$ which is used to add $\neg l_k$ next. Observe that $\sigma(E) = \sigma(\psi) = \top$ and by the induction hypothesis we have $\sigma(l_1) = \cdots = \sigma(l_{k-1}) = \bot$, which leads to $\sigma(l_k) = \top$. This concludes the induction proof resulting in $\sigma(\mathsf{ALA}(R)) = \bot$, which is impossible for the tautology $\mathsf{ALA}(R)$. The assumption is invalid and thus $\tau^x(O) = \tau^x(D) = \top$. $\square$

Theorem 1. *Given a QBF $\phi = \Pi.\psi$ and a clause $C \in \psi$ with* QRAT *on an existential literal $l \in C$ with respect to QBF $\phi' = \Pi'.\psi'$ where $\psi' = \psi \setminus \{C\}$ and Π' is Π without the variables of C not occurring in ψ'. Then ϕ and ϕ' are satisfiability equivalent.*

Proof. If ϕ is satisfiable then ϕ' is also satisfiable, since all models of ϕ are also models of ϕ'. In the following, we show that if ϕ' is satisfiable then ϕ is also satisfiable. Let M be a model for ϕ', which is not a model for ϕ. Then for every assignment $\tau^x \bar{l} \tau_x$ in M which satisfies $\psi' = \psi \setminus \{C\}$ and falsifies C we replace in M' all assignments $\tau^x \bar{l} \rho_x$ by $\tau^x l \rho_x$. Now we need to show that all these $\tau^x l \rho_x$ still satisfy ψ'. Since $\tau^x \bar{l} \rho_x$ satisfies all clauses in ψ', the only clauses in ψ' that can be falsified by $\tau^x l \rho_x$ must contain literal $\bar{l}$. Lemma 2 shows, however, that all clauses $D \in \psi$ with $\bar{l} \in D$ are satisfied by τ^x and hence by $\tau^x l \rho_x$. Thus the resulting pre-model M' turns out to be a model of ϕ. $\square$

In order to remove or to add a clause which has QRAT on literal l requires l to be existential. The following example illustrates that it would not be sound either to allow for universal variables or to ignore the variable dependency restrictions.

Example 2. Consider the false QBF $\exists x \forall y.(x \vee y) \wedge (\bar{x} \vee \bar{y})$. Clause $(x \vee y)$ has QRAT on y w.r.t. $(\bar{x} \vee \bar{y})$, but eliminating $(x \vee y)$ does not preserve unsatisfiability. Hence, one cannot remove clauses based on QRAT on a universal literal. If we would drop the variable dependency restriction, then $(x \vee y)$ would have QRAT on x w.r.t. $(\bar{x} \vee \bar{y})$. Again, removing $(x \vee y)$ does not preserve unsatisfiability.

The elimination of a clause which has QRAT or AT w.r.t. a QBF ϕ is called QRATE. We write QRATE also as $\Pi.\psi \cup \{C\} \xrightarrow{\text{QRATE}} \Pi.\psi$. Analogously, QRAT allows the introduction of clauses. The addition of a clause which has QRAT or AT w.r.t. a QBF ϕ is called QRATA. We write QRATA also as $\Pi.\psi \xrightarrow{\text{QRATA}} \Pi'.\psi \cup \{C\}$. Note that the added clause may contain variables which do not occur in the original QBF. Then the prefix has to be extended by these variables for getting a closed QBF again. These variables may be quantified arbitrarily and put at any position within the prefix.

Example 3. Consider the true QBF $\Pi.\psi = \forall a\, \exists b, c.(a \vee b) \wedge (\bar{a} \vee c) \wedge (b \vee \bar{c})$. Clause $(a \vee c)$ has QRAT on c w.r.t. $\Pi.\psi$: the only clause that contains literal $\bar{c}$ is $(b \vee \bar{c})$, which produces the outer resolvent $(a \vee b \vee c)$. $\mathsf{ALA}(\psi, (a \vee b \vee c)) = (a \vee \bar{a} \vee b \vee \bar{b} \vee c \vee \bar{c})$ is a tautology. Therefore, QRATA can add $(a \vee c)$ to ψ. Now consider a new existential variable d in the innermost quantifier block. The clause $(\bar{b} \vee c \vee d)$ has QRAT on c (and d) w.r.t. ψ. Adding $(\bar{b} \vee c \vee d)$ to ψ will result in the true QBF $\forall a\, \exists b, c, d.(a \vee \bar{b}) \wedge (\bar{a} \vee c) \wedge (b \vee \bar{c}) \wedge (\bar{b} \vee c \vee d)$.

However, as we will show below, one can remove universal literals if they have QRAT. This is similar to the pure literal elimination rule (see next section) which is a clause elimination technique if the pure literal is existentially quantified and which is a literal elimination technique if the pure literal is universally quantified.

For the proof of the following theorem we need the concept of "dual assignment". The dual of an assignment σ in a model for a universal literal is the unique τ in the same model obtained from flipping this literal in σ but keeping all literals before l and all universal literals after l untouched, or more formally:

Definition 6. *Given a model M of a QBF and $\sigma = \sigma^x l \sigma_x \in M$ and a literal l with $\sigma(l) = \top$ then $\tau = \sigma^x \bar{l} \tau_x \in M$ is the dual of σ w.r.t. l iff all universal literals in σ_x are the same in τ_x, e.g. for all universal literals k we have $\sigma_x(k) = \tau_x(k)$.*

Note, that existential literals in σ_x and τ_x might have opposite signs.

Theorem 2. *Given QBF $\phi_0 = \Pi.\psi$ and $\phi = \Pi.\psi \cup \{C\}$ where C has QRAT on a universal literal $l \in C$ with respect to ϕ_0. Further, let $\phi' = \Pi.\psi \cup \{C'\}$ with $C' = C \setminus \{l\}$. Then ϕ and ϕ' are satisfiability equivalent.*

Proof. We need to show that if ϕ is satisfiable, then ϕ' is satisfiable. The reverse is trivial. Let M be a model of ϕ. We are going to define a model M' for ϕ' from M as follows. All assignments σ in M with $\sigma(C') = \top$ are kept in M'. If $\sigma(C') = \bot$ and $\tau = \sigma^x \bar{l} \tau_x$ is the dual assignment of σ w.r.t. l in M, then we replace σ by $\sigma' = \sigma^x l \tau_x$, which is the same as the dual τ of σ w.r.t. l but with l flipped. It is apparent that the set of assignments M' defined this way actually forms a tree and thus a pre-model. Further, note, that C' is satisfied on all paths in M', since either $\sigma(C') = \top$ or otherwise (if $\sigma(C') = \bot$) we have $\sigma'(C') = \tau(C') = \tau(C) = \top$. As in the proof of Theorem 1 we assume that the pre-model M' defined above is not a model. Then we have $\sigma' \in M'$, a clause $D \in \psi$ with $\sigma'(D) = \bot$, and σ' was obtained from σ, by replacing σ by the dual τ of σ w.r.t. l with l flipped and $\sigma(C') = \bot$. Since $\tau(D) = \top$ and σ' differs from τ only for l, we know that $\bar{l}$ single satisfies D in σ', thus $\bar{l} \in D$. The outer clause $O = \mathcal{O}(\Pi, D, \bar{l}) \subset D$ has the property $\sigma'(O) = \bot$ and since $\sigma^x = \sigma'^x$ we derive that $\sigma(R) = \bot$. Observe that $\sigma(R) = \bot$ for the outer resolvent $R = \mathcal{R}(\Pi, C, D, l) = O \cup C'$ (note that l is removed, e.g., $l \notin C'$). Using similar arguments as in the proof of Lemma 2 we can show that all the literals added to R are false under σ. This is in contradiction to the assumption that $\mathsf{ALA}(\psi, R)$ is a tautology. As a consequence M' is a model of ϕ' and ϕ' is satisfiable too. □

In principle, a universal literal on which a clause C has QRAT w.r.t. to a QBF ϕ may be safely removed from C or vice versa added. In the following, we only need the elimination of universal literals. The elimination of a universal literal l from a clause C which has QRAT on l w.r.t. a QBF ϕ is called QRATU. We write QRATU also as $\Pi.\psi \cup \{C\} \xrightarrow{\text{QRATU}} \Pi.\psi \cup \{C \setminus \{l\}\}$.

The definition of outer resolvent depends on quantification. QRATU is not sound if we allow the existential variant of outer resolvent for a universal literal.

Example 4. Consider the true QBF $\forall x \exists y, z.(\bar{x} \vee \bar{y}) \wedge (\bar{x} \vee \bar{z}) \wedge (x \vee y) \wedge (x \vee y \vee z)$. Both resolution candidates for resolving $(x \vee y \vee z)$ on x lead to an empty outer clause. Incorrectly using the existential variant would keep x in the outer resolvent which is identical to the original clause $(x \vee y \vee z)$, which in turn is subsumed. However, removing x from the original clause $(x \vee y \vee z)$ makes the QBF false and thus it is incorrect to keep a universally quantified pivot in the outer resolvent before checking for asymmetric tautology.

4 Preprocessing for QBFs

For successfully solving quantified Boolean formulas (QBF), the introduction of an additional preprocessing step has been shown to be extremely beneficial to focus the search of many solvers. Frequently, preprocessing is crucial to solve a QBF formula. In general, the preprocessed formula is not logically equivalent, but satisfiability equivalent. Below, we introduce the most prominent techniques for preprocessing used in state-of-the-art tools.

We can distinguish three types of rules: (1) clause elimination rules; (2) clause modification rules; and (3) clause addition rules. Table 1 summarizes the preprocessing techniques and their necessary *preconditions*. We omit showing their soundness as this is extensively discussed in the referenced literature.

Clause Elimination Rules remove clauses while preserving unsatisfiability. *Tautology elimination* (E1) removes clauses containing a positive and negative occurrence of a variable. *Subsumption* (E2) removes clauses that are a superset of other clauses. *Existential pure literal elimination* (E3) removes all clauses with an existential literal that occurs only positive or only negative in the formula. *Quantified blocked clause elimination* (E4) removes clauses which contain a variable producing only tautological resolvents when used as pivot.

Clause Modification Rules add, remove, and rename literals. The *universal reduction rule* (M1) removes a universal literal if it is the innermost literal in a clause. The *strengthening rule* (M2) relies on clauses produced by resolution which subsume one of its antecedents. If an existential literal l occurs in a clause of size one, then *unit literal elimination* (M3) allows to remove clauses containing l and literal occurrences $\bar{l}$. *Universal pure literal elimination* (M4) removes a universal literal if it occurs only in one polarity in the whole formula. Covered literal addition (M5) extends a clause with literals that occur in all non-tautological resolvents. Finally, the *equivalence replacement rule* (M6) substitutes the occurrence of a literal l (and $\bar{l}$) by a literal k (and $\bar{k}$) if clauses of the form $(l \vee \bar{k})$ and $(\bar{l} \vee k)$ are in the formula. Literal l must be existentially quantified and $l \geq_\Pi k$.

Clause Addition Rules extend the formula with new clauses, while modifying and removing old ones. The *variable elimination rule* (A1), also known as DP resolution, replaces the clauses in which a certain existential variable occurs by all non-tautological resolvents on that variable. The *universal expansion rule* (A2) removes an innermost universal variable by duplicating and modifying all clauses that contain one or more innermost existential variables.

Table 1. Preprocessing Rules

	name	rewriting rule	precondition	
E1.	tautology elimination	$\Pi.\psi, C \vee l \vee \bar{l} \xrightarrow{\text{Taut}} \Pi.\psi$	none	clause elimination
E2.	subsumption	$\Pi.\psi, C, D \xrightarrow{\text{Subs}} \Pi.\psi, C$	$C \subseteq D$	
E3.	exist. pure literal elim.	$\Pi.\psi, C_1 \vee l, \ldots, C_n \vee l \xrightarrow{\text{Pure}_\exists} \Pi.\psi$	$Q(\Pi, l) = \exists,$ $\bar{l} \notin \psi \wedge C_1 \wedge \ldots \wedge C_n$	
E4.	blocked clause elimination	$\Pi.\psi, C \xrightarrow{\text{QBCE}} \Pi.\psi$	$\exists y \in C \text{ with } Q(\Pi, y) = \exists,$ $\forall D \in \psi \text{ with } \bar{y} \in D:$ $l, \bar{l} \in C \otimes_y D \text{ with } l \leq_\Pi y$	
M1.	universal reduction	$\Pi.\psi, C \vee l \xrightarrow{\text{URed}} \Pi.\psi, C$	$Q(\Pi, l) = \forall,$ $\not\exists k \in C \text{ with } l <_\Pi k$	clause modification
M2.	strengthening	$\Pi.\psi, l \vee C, \bar{l} \vee D \xrightarrow{\text{Str}} \Pi.\psi, C, \bar{l} \vee D$	$C \subseteq D$	
M3.	unit literal elimination	$\Pi.\psi, l, C_1 \vee \bar{l}, \ldots, C_n \vee \bar{l},$ $D_1 \vee l, \ldots, D_m \vee l$ $\xrightarrow{\text{Unit}} \Pi.\psi, C_1, \ldots, C_n$	$Q(\Pi, l) = \exists$	
M4.	univ. pure literal elim.	$\Pi.\psi, C_1 \vee l, \ldots, C_n \vee l$ $\xrightarrow{\text{Pure}_\forall} \Pi.\psi, C_1, \ldots, C_n$	$Q(\Pi, l) = \forall,$ $\bar{l} \notin \psi \wedge C_1 \wedge \ldots \wedge C_n$	
M5.	covered literal addition	$\Pi.\psi, C \xrightarrow{\text{QCLA}} \Pi.\psi, C \vee l$	$\exists y \in C \text{ with } Q(\Pi, y) = \exists,$ $\forall D \in \psi \text{ with } \bar{y} \in D:$ $l \in D \text{ or } k, \bar{k} \in C \otimes_y D$ $\text{with } k, l \leq_\Pi y$	
M6.	equivalence replacement	$\Pi.\psi, \bar{l} \vee k, l \vee \bar{k} \xrightarrow{\text{Equiv}} \Pi.\psi[l/k]$	$Q(\Pi, l) = \exists, k \leq_\Pi l$	
A1.	variable elimination	$\Pi \exists y.\psi, C_1 \vee \bar{y}, \ldots, C_n \vee \bar{y},$ $D_1 \vee y, \ldots, D_m \vee y$ $\xrightarrow{\text{VElim}} \Pi.\psi \bigwedge_{\substack{1 \leq i \leq n \\ 1 \leq j \leq m}} (C_i \cup D_j)$	$Q(\Pi, y) = \exists,$ $y \notin \text{vars}(\psi)$	clause addition
A2.	universal expansion	$\Pi \forall x \exists Y.\psi, C_1 \vee \bar{x}, \ldots, C_n \vee \bar{x},$ $D_1 \vee x, \ldots, D_m \vee x, E_1, \ldots, E_p$ $\xrightarrow{\text{UExp}}$ $\Pi \exists Y Y'.\psi, C_1, \ldots, C_n, D'_1, \ldots, D'_m,$ $E_1, \ldots, E_p, E'_1, \ldots, E'_p$	$Q(\Pi, x) = \forall,$ $\exists y_i \in \text{vars}(E_j), y_i \notin \text{vars}(\psi)$ $x \notin \text{vars}(\psi \wedge C_i \wedge D_j \wedge E_k),$ $D'_i = D_i[y_1/y'_1, \ldots, y_n/y'_n],$ $E'_i = E_i[y_1/y'_1, \ldots, y_n/y'_n]$	

5   Representing Preprocessing Techniques with **QRAT**

The QRAT proof system as presented above provides clause elimination and addition rules as in propositional logic when the pivot variable is existentially quantified. Further, QRAT allows for the removal/addition of variables in the case of universal pivots. This is almost sufficient to express the preprocessing rules introduced in the previous section. The only missing element is universal reduction, which also marks the difference between propositional resolution and resolution for QBF. To this end, we introduce the concept of *extended universal resolution* what is based on Theorem 4.9 of Van Gelder's work on resolution path dependency schemes [16]. In the following, we do not introduce the concept of resolution path dependencies, but we describe the universal literal elimination criterion according to the terminology used in the rest of the paper.

Definition 7 (Inner Clause). *Let C be a clause occurring in QBF $\Pi.\psi$. The inner clause of C on literal $l \in C$, denoted by $\mathcal{I}(\Pi, C, l)$, is given by the clause $\{k \mid k \in C, k = \bar{l} \text{ or } k >_{\Pi} l\}$.*

Lemma 3. *Given a QBF formula $\Pi.\psi$, let $\mathcal{E}(\Pi, C, l)$ be the unique clause obtained by repeatedly applying the extension rule*

$$C := C \cup \mathcal{I}(\Pi, D, l) \text{ if exists } k \in C, D \in \psi \text{ with } \bar{k} \in D, Q(k) = \exists, \text{ and } k >_{\Pi} l$$

until fixpoint. Given a QBF $\Pi.\psi \wedge \{E\}$ with a universal literal $l \in E$ such that $\bar{l} \notin \mathcal{E}(\Pi, E, l)$. Then, the removal of l from E is satisfiability preserving.

Lemma 3 is a generalization of the universal reduction rule which we call *extended universal reduction* in the following. For the application of extended universal reduction we write $\Pi.\psi \cup \{C\} \xrightarrow{\text{EUR}} \Pi.\psi \cup \{C \setminus \{l\}\}$.

Now we are able to express the preprocessing techniques shown in Table 1 with only four rules: QRATE, QRATA, QRATU, and EUR. Table 2 shows the translations for the clause elimination techniques, Table 3 for the clause modification techniques, and Table 4 for the clause addition techniques. We refer to Table 1 for the preconditions for the application of the preprocessing rules.

Tautologies, subsumed clauses as well as blocked clauses have QRAT, so only one application of QRATE is necessary for their removal. If an existential literal is pure than all clauses in which it occurs are blocked w.r.t. this literal and therefore can be omitted by multiple applications of QRATE.

For strengthening a clause $C \vee l$, we first add the resolvent with $D \vee \bar{l}$ which is C. Now, $C \vee l$ is subsumed and can, as we have discussed before, be removed by QRATE. To express unit literal elimination, we first add clauses C_i, i.e., the resolvents of $C_i \vee \bar{l}$ and l. Then $C_i \vee \bar{l}$ become QRAT and can be removed. Now the literal l occurs only in one polarity and hence, the clauses containing l can be removed by QRATE (cf., existential pure literal elimination). Universal pure literal elimination simply maps to multiple applications of QRATU such that l does not occur in the formula anymore. If a universal literal l is removed from a clause C, this can naturally be expressed by extended universal resolution.

Table 2. Clause Elimination Rules

preprocessing rule	rewriting
$\Pi.\psi, C \vee l \vee \bar{l} \xrightarrow{\text{Taut}} \Pi.\psi$	$\Pi.\psi, C \vee l \vee \bar{l} \xrightarrow{\text{QRATE}} \Pi.\psi$
$\Pi.\psi, C, D \xrightarrow{\text{Subs}} \Pi.\psi, C$	$\Pi.\psi, C, D \xrightarrow{\text{QRATE}} \Pi.\psi, C$
$\Pi.\psi, C_1 \vee l, \ldots, C_n \vee l \xrightarrow{\text{Pure}_\exists} \Pi.\psi$	$\Pi.\psi, C_1 \vee l, \ldots, C_n \vee l \xrightarrow{\text{QRATE*}} \Pi.\psi$
$\Pi.\psi, C \xrightarrow{\text{QBCE}} \Pi.\psi$	$\Pi.\psi, C \xrightarrow{\text{QRATE}} \Pi.\psi$

Table 3. Clause Modification Rules

preprocessing rule	rewriting
$\Pi.\psi, C \vee l, D \vee \bar{l}$ $\xrightarrow{\text{Str}} \Pi.\psi, C, D \vee \bar{l}$	$\Pi.\psi, C \vee l, D \vee \bar{l}$ $\xrightarrow{\text{QRATA}} \Pi.\psi, C, C \vee l, D \vee \bar{l} \xrightarrow{\text{QRATE}} \Pi.\psi, C, D \vee \bar{l}$
$\Pi.\psi, C_1 \vee \bar{l}, \ldots, C_n \vee \bar{l},$ $l, D_1 \vee l, \ldots, D_m \vee l$ $\xrightarrow{\text{Unit}} \Pi.\psi, C_1, \ldots, C_n$	$\Pi.\psi, C_1 \vee \bar{l}, \ldots, C_n \vee \bar{l}, l, D_1 \vee l, \ldots, D_m \vee l$ $\xrightarrow{\text{QRATA*}} \Pi.\psi, C_1 \vee \bar{l}, \ldots, C_n \vee \bar{l},$ $l, D_1 \vee l, \ldots, D_m \vee l, C_1, \ldots, C_n$ $\xrightarrow{\text{QRATE*}} \Pi.\psi, l, C_1, \ldots, C_n \xrightarrow{\text{QRATE}} \Pi.\psi, C_1, \ldots, C_n$
$\Pi.\psi, C_1 \vee l, \ldots, C_n \vee l$ $\xrightarrow{\text{Pure}_\vee} \Pi.\psi, C_1, \ldots, C_n$	$\Pi.\psi, C_1 \vee l, \ldots, C_n \vee l$ $\xrightarrow{\text{QRATU*}} \Pi.\psi, C_1, \ldots, C_n$
$\Pi.\psi, C \vee l$ $\xrightarrow{\text{URed}} \Pi.\psi, C$	$\Pi.\psi, C \vee l$ $\xrightarrow{\text{EUR}} \Pi.\psi, C$
$\Pi.\psi, \bar{l} \vee k, l \vee \bar{k}$ $\xrightarrow{\text{Equiv}} \Pi.\psi[l/k]$	$\Pi.\psi, C_1 \vee l, \ldots, C_n \vee l, D_1 \vee \bar{l}, \ldots, D_m \vee \bar{l}, \bar{l} \vee k, l \vee \bar{k}$ $\xrightarrow{\text{QRATA*}} \Pi.\psi, C_1 \vee l, \ldots, C_n \vee l, D_1 \vee \bar{l}, \ldots, D_m \vee \bar{l},$ $\bar{l} \vee k, l \vee \bar{k}, C_1 \vee k, \ldots, C_n \vee k, D_1 \vee \bar{k}, \ldots, D_m \vee \bar{k}$ $\xrightarrow{\text{QRATE*}}$ $\Pi.\psi, \bar{l} \vee k, l \vee \bar{k}, C_1 \vee k, \ldots, C_n \vee k, D_1 \vee \bar{k}, \ldots, D_m \vee \bar{k}$ $\xrightarrow{\text{QRATE*}} \Pi.\psi, C_1 \vee k, \ldots, C_n \vee k, D_1 \vee \bar{k}, \ldots, D_m \vee \bar{k}$
$\Pi.\psi, C$ $\xrightarrow{\text{QCLA}} \Pi.\psi, C \vee l$	$\Pi.\psi, C$ $\xrightarrow{\text{QRATA}} \Pi.\psi, C, C \vee l \xrightarrow{\text{QRATE}} \Pi.\psi, C \vee l$

If l is a covered literal for C w.r.t. $\Pi.\psi$, then $C \vee l$ has QRAT w.r.t. $\Pi.\psi$. After adding $C \vee l$ using QRATA, C gets QRAT and can be removed using QRATE. Quantified covered clause elimination [8] is a clause elimination procedure that extends clauses with covered literals until clauses become blocked. To represent this procedure, we add an intermediate clause for each covered literal addition. When the clause is blocked, it can be eliminated using QRATE.

If a literal l shall be substituted by a literal k due to equivalence replacement, the formula has to contain the binary clauses $(l \vee \bar{k})$ and $(\bar{l} \vee k)$. Then first the clauses $C_i \vee k$ and $D_j \vee \bar{k}$ are added by resolution, i.e., by QRATA. All clauses containing l and $\bar{l}$ are asymmetric tautologies, because $k, \bar{k} \in \mathsf{ALA}(\psi, C_i \vee l)$ and can therefore be removed by QRATE.

Variable elimination is rewritten as follows. First all possible non-tautological resolvents on elimination variable y are added with QRATA. Then all clauses containing y or $\bar{y}$ become QRAT and can be eliminated by QRATE.

Finally, we describe universal expansion using redundancy elimination and addition rules. Consider the QBF $\Pi\forall x\exists Y.\psi$ from which we want to eliminate the innermost universal variable x. Let $E = \{E_i \mid E_i \in \psi_Y, x \notin E_i, \bar{x} \notin E_i\}$. In the first step, we add clauses $E_i \vee \bar{x}$ (which are subsumed by E_i) using QRATA. This is necessary, because we later need to eliminate E_i. We introduce conditional equivalences represented by the clauses $x \vee y_j \vee \bar{y}'_j$ and $x \vee \bar{y}_j \vee y'_j$ for all $y_j \in Y$ and append $\exists Y'$ to the prefix. Now we copy all original clauses with literal y_j, but without $\bar{x}$ and add literal x in case it is not already present. The conditional equivalences allow to treat original and primed copies of clauses with x as alternative. One version can be exchanged for the other as long the equivalence clauses are there. We add the primed copies and afterwards remove the original ones. Now all clauses E_i are asymmetric tautologies and can be removed. Next, we remove the conditional equivalences $x \vee y_j \vee \bar{y}'_j$ and $x \vee \bar{y}_j \vee y'_j$ which have QRAT on the y_j after removal of the E_i clauses. At this point, clauses containing variables from Y do not contain x and clauses with variables from Y' do not contain $\bar{x}$. So extended universal reduction can remove the literals x and $\bar{x}$.

Table 4. Clause Addition Rules. Clauses are added / removed in order of appearance.

preprocessing rule	rewriting
$\Pi\exists y.\psi, C_1 \vee \bar{y}, \ldots, C_n \vee \bar{y},$ $D_1 \vee y, \ldots, D_m \vee y$ $\overset{\mathsf{VElim}}{\Longrightarrow} \Pi.\psi \bigwedge\limits_{\substack{1\le i\le n \\ 1\le j\le m}} (C_i \cup D_j)$	$\Pi\exists y.\psi, C_1 \vee y, \ldots, C_n \vee y, D_1 \vee \bar{y}, \ldots, D_m \vee \bar{y}$ $\xrightarrow{\mathsf{QRATA*}} \Pi\exists y.\psi, C_1 \vee y, \ldots, C_n \vee y,$ $D_1 \vee \bar{y}, \ldots, D_m \vee \bar{y}, C_1 \cup D_1, \ldots, C_n \cup D_m$ $\xrightarrow{\mathsf{QRATE*}} \Pi.\psi, C_1 \cup D_1, \ldots, C_n \cup D_m$
$\Pi\forall x\exists Y.\psi,$ $C_1 \vee \bar{x}, \ldots, C_n \vee \bar{x},$ $D_1 \vee x, \ldots, D_m \vee x,$ $E_1, \ldots, E_p$ $\overset{\mathsf{UExp}}{\Longrightarrow} \Pi\exists YY'.\psi,$ $C_1, \ldots, C_n, E_1, \ldots, E_p,$ $D'_1, \ldots, D'_m, E'_1, \ldots, E'_p$	$\Pi\forall x\exists Y.\psi, C_1 \vee \bar{x}, \ldots, C_n \vee \bar{x},$ $D_1, \vee x, \ldots, D_m \vee x, E_1, \ldots, E_p$ $\xrightarrow{\mathsf{QRATA*}} \Pi\forall x\exists YY'.\psi, C_1 \vee \bar{x}, \ldots, C_n \vee \bar{x},$ $D_1 \vee x, \ldots, D_m \vee x, E_1, \ldots, E_p, E_1 \vee \bar{x}, \ldots, E_p \vee x,$ $x \vee y_1 \vee \bar{y}'_1, \ldots, x \vee y_{|Y|} \vee \bar{y}'_{|Y|},$ $x \vee \bar{y}_1 \vee y'_1, \ldots, x \vee \bar{y}_{|Y|} \vee y'_{|Y|},$ $D'_1 \vee x, \ldots, D'_m \vee x, E'_1 \vee x, \ldots, E'_p \vee x$ $\xrightarrow{\mathsf{QRATE*}} \Pi\forall x\exists YY'.\psi, C_1 \vee \bar{x}, \ldots, C_n \vee \bar{x},$ $E_1 \vee \bar{x}, \ldots, E_p \vee \bar{x}, D'_1 \vee x, \ldots, D'_m \vee x,$ $E'_1 \vee x, \ldots, E'_p \vee x$ $\xrightarrow{\mathsf{EUR*}} \Pi\exists YY'.\psi, C_1, \ldots, C_n, E_1, \ldots, E_p,$ $D'_1, \ldots, D'_m, E'_1, \ldots, E'_p$

6   QRAT Proofs

This section describes our new proof format for QBF formulas, how to check it and an experimental evaluation. The syntax of the proof format is very similar to the DRUP proof format [12] for CNF formulas. We extend the DRUP syntax to express elimination of universal literals. Furthermore, the redundancy check is different than proofs in DRUP because we deal with QBF formulas.

6.1   The QRAT Proof Format

Proofs are sequences of clause additions, deletions, and modifications. They are build using three kind of lines: addition (QRATA), deletion (QRATE), and universal elimination (QRATU and EUR). Addition lines have no prefix and are unconstrained in the sense that one can add any clause at any point in the proof. Clause deletion lines, with prefix "d", and universal elimination lines, with prefix "u", are restricted. The clause after a "d" or "u" prefix must be either present in the original formula or as a clause added earlier in the proof.

Let $\Pi.\psi$ be a QBF formula and P be a QRAT proof for $\Pi.\psi$. We denote the number of lines in a proof P by $|P|$. For each $i \in \{0, \ldots, |P|\}$, we define a CNF formula ψ_P^i below. C_i refers to the clause on line i of P and l_i refers to the first literal on line i of P.

$$\psi_P^i := \begin{cases} \psi & \text{if } i = 0; \\ \psi_P^{i-1} \setminus \{C_i\} & \text{if the prefix of } C_i \text{ is "d"}; \\ \psi_P^{i-1} \setminus \{C_i\} \cup \{C_i \setminus \{l_i\}\} & \text{if the prefix of } C_i \text{ is "u"}; \\ \psi_P^{i-1} \cup \{C_i\} & \text{otherwise.} \end{cases}$$

A proof P is called a *satisfaction proof* for QBF formula $\Pi.\psi$ if the following two properties hold. First, for all $i \in \{1, \ldots, |P|\}$, if clause C_i has prefix "d", then it must have QRAT on l_i with respect to ψ_P^i. In case l_i is universally quantified, we check whether $\mathsf{ALA}(\psi_P^i, C_i)$ is a tautology. Second, $\psi_P^{|P|}$ must be empty.

A proof P is called a *refutation proof* for QBF formula $\Pi.\psi$ if the following three properties hold. First, for all $i \in \{1, \ldots, |P|\}$, if clause C_i has no prefix, then it must have QRAT on l_i with respect to ψ_P^{i-1}. In case l_i is universally quantified, we check whether $\mathsf{ALA}(\psi_P^{i-1}, C_i)$ is a tautology. Second, for all $i \in \{1, \ldots, |P|\}$, if clause C_i has has prefix "u", then l_i must be universally quantified. Additionally, C_i must have either QRAT on l_i with respect to ψ_P^{i-1}, or l_i can be removed using EUR. Third, $C_{|P|}$ must be the empty clause (without a prefix). Fig. 1 shows a true and a false QBF and a QRAT proof for both.

A universal elimination line in satisfaction proofs can be replaced by a clause addition and deletion line to obtain another satisfaction proof. Simply add the clause without its first literal, and afterwards delete the subsumed clause. For example, consider the line "u 1 2 3 0" in a satisfaction proof. This line can be replaced by "2 3 0" followed by "d 1 2 3 0". Consequently, any satisfaction proof can be converted such that it contains only addition and deletion lines.

true QBF formula	satisfaction proof	false QBF formula	refutation proof
p cnf 3 3		p cnf 3 3	
a 1 0	-1 -2 0	a 1 0	-2 0
e 2 3 0	d 3 -1 0	e 2 3 0	d -2 -3 0
1 2 0	d -3 -2 0	1 2 0	1 0
-1 3 0	d -2 -1 0	1 3 0	u 1 0
-2 -3 0	d 2 1 0	-2 -3 0	0

Fig. 1. Two QBFs formulas and QRAT proofs. On the left a true QBF with a satis-faction proof next to it. On the right a false QBF with a refutation next to it. The formulas and proofs are spaced to improve readability. Proofs consist of three kind of lines: addition (no prefix), deletion ("d " prefix) and universal elimination ("u " prefix).

Recall that QRATA can add clauses that contain new variables. The QRAT proof format does not support describing the quantifier block for new variables. For all known preprocessing techniques, newly introduced variables are placed in the innermost active existential quantifier block. Consequently, the QRAT format assumes this convention for all new variables.

6.2 Checking QRAT Proofs

Although the syntax for QRAT proof is identical for true and false QBFs, vali-dating a proof is different. For true QBFs only the clause deletion lines (the ones with a "d " prefix) have to be checked, while for false QBFs, all the lines except the clause deletion lines have to be checked.

The easiest, but rather expensive, method to validate proofs checks the re-dundancy of each clause: for true QBFs all deletion lines and for false QBFs all addition and universal elimination lines. However, one can check proofs more efficiently my marking involved clauses during each redundancy check. That way the checker can be restricted to validate marked clauses only. The marking pro-cedure is a bit tricky. In short, it marks all involved clauses that were required to compute the last unique implication point from each conflict.

Checking only marked clauses was proposed to check clausal proofs of CNF formulas efficiently [11]. For false QBFs, the checking is similar to the SAT case: during initialization the empty clause is marked. Refutation proofs should be validated in reverse order, starting with the marked empty clause. For true QBFs the procedure is different: initially all original clauses are marked and satisfaction proofs are checked in chronological order. When a clause is deleted that was not marked by any redundancy check, the clause can be skipped.

Example 5. Consider the true QBF $\Pi.\psi = \forall a\, \exists b, c.(a \lor b) \land (\bar{a} \lor c) \land (\bar{b} \lor \bar{c})$. This is the same QBF as in Fig. 1 (left). Fig. 1 also shows the satisfaction proof $P := (\bar{a} \lor \bar{b}), \mathsf{d}(c \lor \bar{a}), \mathsf{d}(\bar{c} \lor \bar{b}), \mathsf{d}(\bar{b} \lor \bar{a}), \mathsf{d}(b \lor a)$. Satisfaction proofs are checked in chronological order. So, first, $(\bar{a} \lor \bar{b})$ is added, afterwards $(c \lor \bar{a})$ is removed, until all original clauses and all added clauses have been deleted.

6.3   Implementation

We equipped our preprocessor bloqqer [8] with QRAT-based tracing as described in Section 5. In contrast to previous extensions of bloqqer [7,17] we hardly had to modify its internal behavior. Hence, with QRAT-based tracing, we have the first QBF preprocessor fully supporting proof generation for true and false formulas.

We implemented an efficient QRAT checker QRATtrim, which is based on DRUPtrim [12], a clausal proof checking tool for CNF. It uses the optimizations of Section 6.2, such as validating marked clauses only and checking satisfaction and refutation proofs in chronological and reverse order, respectively.

Evaluations on the benchmark sets of the QBF evaluations 2010 and 2012 indicate that the power of the preprocessor is hardly reduced by enabling QRAT-based tracing. The benchmark set of 2010 (resp. 2012) contains 64 (resp. 32) true instances and 86 (resp. 36) false instances which can be solved by using only bloqqer. These formulas turn out to be extremely hard for conflict/solution-driven clause/cube learning solvers like DepQBF [18], which can only solve 26 (resp. 2) true formulas and 57 (resp. 14) false formulas. For the other formulas DepQBF timed-out, given a time limit of 900 seconds. The resolution-proof producing version of bloqqer which was presented in [7] is able to evaluate 28 (resp. 22) true formulas and 57 (resp. 22) false formulas. Please note that the resolution-proof producing version of bloqqer did not time out for the unsolved formulas. Less formulas are solved because the techniques for which no translation to resolution is presented in [7] are simply turned off. Our new QRAT-based proof producing version of bloqqer solves 63 (resp. 32) true formulas and 86 (resp. 36) false formulas, i.e., only one formula less is solved. We could verify all but two QRAT proofs. For solving these formulas, miniscoping is necessary – which is not yet supported by our checker, but can be realized by taking dependencies into account while computing outer clauses. If we turn off miniscoping and increase the bounds for variable elimination, these formulas can be solved and checked as well. For satisfiable formulas, solving is twice as fast as checking on average, in particular we have 1.4s (2.2s) for solving and 3.2s (5.9s) for checking. For unsatisfiable formulas, checking is considerably faster: we have 18.7s (30.1s) for solving and 5.2s (8.6s) for checking. Our bloqqer extension, the proof checking tools as well as the details on our experiments are available at http://fmv.jku.at/bloqqer.

7   Conclusion

We presented a proof system which captures recent preprocessing and solving techniques for QBF in a uniform manner. Based on *asymmetric tautologies*, the proof system consists only of four simple rules. We showed how state-of-the-art preprocessing techniques can be represented within this proof system. Our rules QRATE, QRATU, and QRATA may be applied as preprocessing rules themselves similar as QBCE and we plan to integrate them in our preprocessor. We deal with all the challenges regarding certificates and preprocessing for QBF recently listed in [7], namely: can we (1) produce polynomially-verifiable certificates for true QBFs in the context of preprocessing, (2) narrow the performance gap between solving with and without certificate generation; and (3) develop methods to deal with universal expansion and other techniques. First, the size of our certificates for true QBFs is polynomial in the solving time and certificate checking can be done in polynomial time. Second, the overhead of emitting certificates is small and all existing preprocessing techniques are supported. Third, our proof system can simulate universal expansion and other existing techniques. Future work will focus on rewriting search based QBF solver techniques [18] to the **QRAT** proof system and extracting Skolem functions [4] from **QRAT** proofs.

References

1. Benedetti, M.: Extracting Certificates from Quantified Boolean Formulas. In: IJ-CAI, pp. 47–53. Professional Book Center (2005)
2. Kleine Büning, H., Subramani, K., Zhao, X.: Boolean Functions as Models for Quantified Boolean Formulas. J. Autom. Reasoning 39(1), 49–75 (2007)
3. Jussila, T., Biere, A., Sinz, C., Kroning, D., Wintersteiger, C.M.: A first step towards a unified proof checker for QBF. In: Marques-Silva, J., Sakallah, K.A. (eds.) SAT 2007. LNCS, vol. 4501, pp. 201–214. Springer, Heidelberg (2007)
4. Niemetz, A., Preiner, M., Lonsing, F., Seidl, M., Biere, A.: Resolution-Based Certificate Extraction for QBF. In: Cimatti, A., Sebastiani, R. (eds.) SAT 2012. LNCS, vol. 7317, pp. 430–435. Springer, Heidelberg (2012)
5. Janota, M., Grigore, R., Marques-Silva, J.: On Checking of Skolem-based Models of QBF. In: RCRA 2012 (2012)
6. Van Gelder, A.: Certificate Extraction from Variable-Elimination QBF Preprocessors. In: QBF, pp. 35–39 (2013),
   http://fmv.jku.at/qbf2013/reportQBFWS13.pdf
7. Janota, M., Marques-Silva, J.: On QBF Proofs and Preprocessing. In: McMillan, K., Middeldorp, A., Voronkov, A. (eds.) LPAR-19 2013. LNCS, vol. 8312, pp. 473–489. Springer, Heidelberg (2013)
8. Biere, A., Lonsing, F., Seidl, M.: Blocked clause elimination for QBF. In: Bjørner, N., Sofronie-Stokkermans, V. (eds.) CADE 2011. LNCS, vol. 6803, pp. 101–115. Springer, Heidelberg (2011)
9. Giunchiglia, E., Marin, P., Narizzano, M.: sQueezeBF: An Effective Preprocessor for QBFs Based on Equivalence Reasoning. In: Strichman, O., Szeider, S. (eds.) SAT 2010. LNCS, vol. 6175, pp. 85–98. Springer, Heidelberg (2010)
10. Biere, A.: Resolve and expand. In: Hoos, H.H., Mitchell, D.G. (eds.) SAT 2004. LNCS, vol. 3542, pp. 59–70. Springer, Heidelberg (2005)
11. Goldberg, E.I., Novikov, Y.: Verification of proofs of unsatisfiability for CNF formulas. In: DATE, pp. 10886–10891 (2003)
12. Heule, M.J.H., Hunt Jr., W.A., Wetzler, N.: Trimming while checking clausal proofs. In: FMCAD, pp. 181–188. IEEE (2013)
13. Heule, M.J.H., Hunt Jr., W.A., Wetzler, N.: Verifying refutations with extended resolution. In: Bonacina, M.P. (ed.) CADE 2013. LNCS, vol. 7898, pp. 345–359. Springer, Heidelberg (2013)
14. Heule, M.J.H., Järvisalo, M., Biere, A.: Clause elimination procedures for CNF formulas. In: Fermüller, C.G., Voronkov, A. (eds.) LPAR-17. LNCS, vol. 6397, pp. 357–371. Springer, Heidelberg (2010)

15. Järvisalo, M., Heule, M.J.H., Biere, A.: Inprocessing rules. In: Gramlich, B., Miller, D., Sattler, U. (eds.) IJCAR 2012. LNCS, vol. 7364, pp. 355–370. Springer, Heidelberg (2012)
16. Van Gelder, A.: Variable Independence and Resolution Paths for Quantified Boolean Formulas. In: Lee, J. (ed.) CP 2011. LNCS, vol. 6876, pp. 789–803. Springer, Heidelberg (2011)
17. Könighofer, R., Seidl, M.: Partial witnesses from preprocessed quantified boolean formulas. Accepted for DATE 2014 (2014)
18. Lonsing, F., Biere, A.: DepQBF: A Dependency-Aware QBF Solver. JSAT 7(2-3), 71–76 (2010)

The Fractal Dimension of SAT Formulas*

Carlos Ansótegui[1], Maria Luisa Bonet[2], Jesús Giráldez-Cru[3], and Jordi Levy[3]

[1] DIEI, Univ. de Lleida
carlos@diei.udl.cat
[2] LSI, UPC
bonet@lsi.upc.edu
[3] IIIA-CSIC
{jgiraldez,levy}@iiia.csic.es

Abstract. Modern SAT solvers have experienced a remarkable progress on solving industrial instances. Most of the techniques have been developed after an intensive experimental process. It is believed that these techniques exploit the underlying structure of industrial instances. However, there is not a precise definition of the notion of structure.

Recently, there have been some attempts to analyze this structure in terms of complex networks, with the long-term aim of explaining the success of SAT solving techniques, and possibly improving them.

We study the fractal dimension of SAT instances with the aim of complementing the model that describes the structure of industrial instances. We show that many industrial families of formulas are self-similar, with a small fractal dimension. We also show how this dimension is affected by the addition of learnt clauses during the execution of SAT solvers.

1 Introduction

The SAT community has been able to come up with successful SAT solvers for industrial applications. However, nowadays we can hardly explain why these solvers are so efficient working on industrial SAT instances with hundreds of thousands of variables and not on random instances with hundreds of variables. The common wisdom is that the success of modern SAT/CSP solvers is correlated to their ability to exploit the hidden structure of real-world instances [13]. Unfortunately, there is no precise definition of the notion of structure.

Parallelly, the community of complex networks has produced tools for describing and analyzing the structure of social, biological and communication networks [1] which can explain some interactions in the real-world. *Preferential attachment* (where the probability that a new edge is attached to a node is proportional to its degree) has been proposed as the responsible of scalefree structure in real-world graphs [5]. Thus, in the web, the probability of a web page to get new connections is proportional to its popularity (the number of connections it already has). In cite [9], it is proposed *similarity* (where nodes

---

* This research has been partially founded by the MINECO research project TASSAT (TIN2010-20967).

S. Demri, D. Kapur, and C. Weidenbach (Eds.): IJCAR 2014, LNAI 8562, pp. 107–121, 2014.

tend to get connected to similar nodes, according to some topological distance) as a mechanism that, together with preferential attachment or *popularity*, explains the structure of some real-world graphs. This explains the self-similarity property observed in many real-world graphs [11].

Representing SAT instances as graphs, we can use some of the techniques from complex networks to characterize the structure of SAT instances. Recently, some progress has been made in this direction. It is known that many industrial instances have the *small-world* property [12], exhibit high *modularity* [4], and have a *scale-free structure* [2]. In this later work, it is shown that in many formulas the number of occurrences of a variable (i.e. the degree of graph nodes) follows a powerlaw distribution with *hub* variables having a huge number of occurrences. A method to generate scale-free random instances is proposed in [3]. They show that SAT solvers specialized on industrial formulas perform better than random-specialized solvers on these scale-free random instances. In [7], the *eigenvector centrality* of variables in industrial instances is analyzed. They show that it is correlated with some aspects of SAT solvers. For instance, decision variables selected by the SAT solvers are usually the most central variables in the formula. However, how these analyses may help to improve the performance of SAT solvers is not known at this stage.

The contribution of this paper is to analyze the existence of self-similarity in industrial SAT instances. The existence of a self-similar structure would mean that after *rescaling* (replacing groups of nodes by a single node, for example), we would observe the same kind of structure. It would also mean that the diameter d^{max} of the graph grows as $d^{max} \sim n^{1/d}$, where d is the fractal dimension of the graph, and not as $d^{max} \sim \log n$, as in random graphs or small-world graphs. Therefore, actions in some part of the graph (like variable instantiation) may not *propagate* to other parts as fast as in random graphs. Our analysis shows that many industrial formulas are self-similar. We think that the self-similarity, as well as the scale-free structure, is already present in many of the problems encoded as SAT instances. Thus, for instance, hardware-verification instances may have this structure because the circuits they encode already have this structure.

Studying graph properties of formulas has several direct applications. One of them, is the generation of industrial-like random SAT instances. Understanding the structure of industrial instances is a first step towards the development of random instance generators, reproducing the features of industrial instances. This would allow us to generate industrial-like random instances of a predefined size and structure to support the testing of industrial SAT solvers under development. Related work in this direction can be found in [3].

Another potential application is to improve portfolio approaches [14,6] which are solutions to the algorithm selection problem. State-of-the-art SAT Portfolios compute a set of features of SAT instances in order to select the best solver from a predefined set to be run on a particular SAT instance. It is reasonable to think that more informative structural features of SAT instances can help to improve portfolios.

Our experimental investigation shows that most industrial instances are self-similar, and their dimension ranges between 2 and 4. In the case of crafted instances, they also exhibit a clear self-similar behaviour, but their fractal dimensions are bigger in some cases. On the other hand, random instances are clearly not self-similar. We also show that using a very reduced set of complex networks properties we are able to classify industrial instances into families quite accurately.

Finally, we have investigated how the addition of learnt clauses during the execution of a SAT solver affects the dimension of the working instance. The addition of learnt clauses increases the fractal dimension, as expected. However, we show that modern SAT solvers produce a smooth increase, that suggests that SAT solvers tend to work locally. In contrast, the substitution of the learnt clauses by random clauses of the same size, produces a much bigger increase in the dimension.

The paper proceeds as follows. We introduce the fractal dimension of graphs in Section 2. In Section 3, we define the notion of fractal dimension of a SAT formula and compare it with the notion of diameter of a SAT formula. Then, we analyze whether SAT instances represented as graphs have a fractal dimension in Section 4. In Section 5, we study the effect of learnt clauses on the fractal dimension. Section 6 contains the conclusions. All the software used in the paper is available at http://www.iiia.csic.es/~jgiraldez.

2   Fractal Dimension of a Graph

We can define a notion of fractal dimension of a graph following the principle of self-similarity. We will use the definition of box covering by Hausdorff [8].

Definition 1. *The **distance** between two nodes is the minimum number of edges we need to follow to go from one node to the other.*
*The **diameter** d^{max} of a graph is the maximal distance between any two nodes of the graph.*
*Given a graph G, a **box** B of size l is a subset of nodes such that the distance between any pair of them is strictly smaller than l.*
We say that a set of boxes covers a graph, if every node of the graph is in some box. Let $N(l)$ be the minimum number of boxes of size l required to cover the graph.
*We say that a graph has the **self-similarity** property if the function $N(l)$ decreases polynomially, i.e. $N(l) \sim l^{-d}$, for some value d. In this case, we call d the **dimension** of the graph.*

Notice that $N(1)$ is equal to the number of nodes of G, and $N(d^{max}+1)$ is the number of connected components of the graph.

Lemma 1. *Computing the function $N(l)$ is NP-hard.[1]*

---
[1] In [10] the same result is stated, but there, they prove the wrong reduction. They reduce the computation of $N(2)$ to the graph coloring problem.

PROOF: We prove that computing $N(2)$ is already NP-hard by reducing the graph coloring problem to the computation of $N(2)$. Given a graph G, let $\overline{G}$, the complement of G, be a graph with the same nodes, and where any pair of distinct nodes are connected in $\overline{G}$ iff they are not connected in G. Boxes of size 2 in $\overline{G}$ are cliques, thus they are sets of nodes of G without an edge between them. Therefore, the minimal number of colors needed to color G is equal to the minimal number of cliques needed to cover $\overline{G}$, i.e. $N(2)$. ∎

There are several efficient algorithms that approximate $N(l)$. They compute upper bounds of $N(l)$. They are called *burning* algorithms (see [10]). Following a greedy strategy, at every step they try to select the box that covers (burns) the maximal number of uncovered (unburned) nodes. Although they are polynomial algorithms, we still need to do some further approximations to make the algorithms of practical use in very large graphs.

First, instead of boxes, we will use *circles*.

Definition 2. *A **circle** of radius r and center c is a subset of nodes of G such that the distance between any of them and the node c is strictly smaller that r.*

Let $N(r)$ be the minimum number of circles of radius r required to cover a graph.

Notice that any circle of radius r is inside of a box of size $2r - 1$ (the opposite is in general false) and any box of size l is inside a circle of radius l (it does not matter what node of the box we use as center). Notice also that every radius r and center c characterizes a unique circle.

According to Hausdorff's dimension definition, $N(r) \sim r^{-d}$ also characterizes self-similar graphs of dimension d. We can approximate this fractal dimension using the *Maximum-Excluded-Mass-Burning (MEMB)* algorithm [10], which works as follows: Consider a graph G and a radius r. We compute an upper bound of the number of circles with radius r necessary to cover the graph $N(r)$. We start with all nodes set to unburned. At every step, for every possible node c, we compute the number of unburned nodes covered by the circle of center c and radius r, then select the node c that maximizes this number, and burn the new covered nodes.

The MEMB algorithm is still too costly for our purposes. We apply the following strategy to make the algorithm more efficient. We order the nodes according to their degree: $\langle c_1, \ldots, c_n \rangle$ such that $degree(c_i) \geq degree(c_j)$, when $i > j$. Now, for $i = 1$ to n, if c_i is not burned, then select the circle of center c_i and radius r (even if it does not maximizes the number of unburned covered nodes), and burn all its unburned nodes. We call this algorithm *Burning by Node Degree (BND)*, and describe it in Alg. 1. After we give the definition of fractal dimension of a SAT instance, we will compare the accuracy and efficiency of algorithms MEMB and BND in subsection 4.1 to justify the use of algorithm BND in our experimentation.

Algorithm 1: Burning by Node Degree (**BND**)

Input: Graph $G = (V, E)$
Output: vector[int] N

1   $N[1] := |V|$;
2   int $i := 2$;
3   **while** $N[i-1] > connectedComponents(G)$ **do**
4   vector[bool] $burned(|V|)$;
5      $N[i] := 0$;
6      $burned := \{$**false**$, \ldots,$ **false**$\}$;
7   **while** $existsUnburnedNode(burned)$ **do**
8         $c := highestDegreeUnburnedNode(G, burned)$;
9         $S := circle(c, i)$; // circle with center c and radius i;
10   **foreach** $x \in S$ **do**
11           $burned[x] :=$ **true**;
12        $N[i] + +$;
13   i := i+1;

---

3   The Fractal Dimension of SAT Instances

Given a SAT instance, we can build a graph from it. Here, we propose two models.

Definition 3. *Given a SAT formula, the* **Clause-Variable Incidence Graph** **(CVIG)** *associated to it is a* bipartite *graph whose nodes are the set of variables and the set of clauses, and its edges connect a variable and a clause whenever that variable occurs in the clause.*

The **Variable Incidence Graph** **(VIG)** *associated to a formula is a graph whose nodes represent the set of variables, and an edge between two nodes indicates the existence of a clause containing both variables.*

In this paper we analyze the function $N(r)$ for the graphs obtained from a SAT instance following the VIG and CVIG models. These two functions are denoted $N(r)$ and $N^b(r)$, respectively, and they relate to each other as follows.

Lemma 2. *If $N(r) \sim r^{-d}$ then $N^b(r) \sim r^{-d}$.*
If $N(r) \sim e^{-\beta r}$ then $N^b(r) \sim e^{-\frac{\beta}{2} r}$.

PROOF: Notice that, for any formula, given a circle of radius r in the VIG model, using the same center and radius $2r - 1$ we can cover the same variable nodes in the CVIG model. With radius $2r$ we can also cover all clauses adjacent to some covered variable. Hence $N^b(2r) \leq N(r)$.

Conversely, given a circle of radius $2r$ in the CVIG model, we consider two possibilities. If the center is a variable node, we cover the same variables in the VIG model using a circle of radius r and the same center. If the center is a clause c, to cover the same variables in the VIG model, we need a circle of radius $r + 1$ centered in a variable node adjacent to c. Hence $N(r + 1) \leq N^b(2r)$.

Therefore $N(r+1) \leq N^b(2\,r) \leq N(r)$, and $N(r) \sim N^b(2\,r)$. From this asymptotic relation, we can derive the two implications stated in the lemma. ∎

Previous lemma states that if a SAT formula is (fully) self-similar, then in both models, VIG and CVIG, the fractal dimension is the same. In such case, if we plot $N(r)$ as a function of r in double-logarithmic axes, we obtain a line with slope $-d$. If $N(r)$ decays exponentially (as in random SAT formulas), then the decay factor in the CVIG model is half of the decay factor in the VIG model. In such case, if we plot $N(r)$ in semi-logarithmic axes, we obtain a line with slope $-\beta$. We will always plot $N(r)$ in double-logarithmic axes. Thus, when $N(r)$ decays exponentially, we will observe a concave curve.

3.1 Fractal Dimension versus Diameter

The function $N(r)$ determines the *maximal radius* r^{max} of a connected graph, defined as the minimum radius of a circle covering the whole graph minus one: $N(r^{max}+1) = 1$. The maximal radius and the *diameter* d^{max} of a graph are also related, because $r^{max} \leq d^{max} \leq 2\,r^{max}$. From these relations we can conclude the following.

Lemma 3. *For self-similar graphs or SAT formulas (where $N(r) \sim r^{-d}$), the diameter is $d^{max} \approx n^{1/d}$, where d is the fractal dimension.*
In graphs or SAT formulas where $N(r) \sim e^{-\beta\,r}$, the diameter is $d^{max} \approx \frac{\log n}{\beta}$.

PROOF: The diameter of a graph and the maximal radius are related as $r^{max} \leq d^{max} \leq 2\,r^{max}$. Notice that, by definition of the function $N(r)$, we have $N(1) = n$, where n is the number of nodes, and $N(r^{max}+1) = 1$.

Assuming $N(r) = C\,r^{-d}$ and replacing r by 1 we get $C = n$. Then, replacing r by $r^{max}+1$, we get $1 = N(r^{max}+1) = n\,(r^{max}+1)^{-d}$. Hence, $r^{max} = n^{1/d} - 1$.

Assuming $N(r) = C\,e^{-\beta\,r}$ and replacing r by 1 we get $C = n\,e^{\beta}$. Then, replacing r by $r^{max}+1$, we get $1 = N(r^{max}+1) = n\,e^{-\beta\,(r^{max})}$. Hence, $r^{max} = \frac{\log n}{\beta}$. ∎

The diameter, as well as the *typical distance*[2] L of a graph, have been widely used in the characterization of graphs. For instance, *small world graphs* [12] are characterized as those graphs with a small typical distance $L \sim \log n$ and a large clustering coefficient. This definition works well for *families* of graphs because then we can quantify the typical distance as a function on the number of nodes. But it is quite imprecise in the case of individual graphs, because it is difficult to decide what is a "small" distance and a "large" clustering coefficient, for a concrete graph. Moreover, the diameter and the typical distance of a graph are measures quite expensive to compute in practice (for huge graphs, as the ones representing many industrial SAT formulas), even though there is a quadratic algorithm. In fact, our approximation to the fractal dimension can be computed more efficiently than the diameter.

---

[2] The typical distance of a graph is the average of the distances between any two nodes.

Since we are interested in characterizing the structure of formulas, the fractal dimension is a better measure because it is independent of the size. Thus, formulas of the same family (and similar structure), but very distinct size, will have similar dimension and $N(r)$ function shape.

4   Experimental Evaluation

We have conducted an exhaustive analysis of the 300 industrial SAT instances and the 300 crafted instances of the SAT Competition 2013[3], and 90 random 3CNF formulas of 10^5 variables at different clause/variable ratios. We will see that most industrial and crafted instances are self-similar and have a small fractal dimension, i.e. $N(r) \sim r^{-d}$, for small d. In random instances $N(r)$ decays exponentially, i.e. $N(r) \sim e^{-\beta r}$.

Before presenting the results of this evaluation, let us justify the use of the BND algorithm to calculate the fractal dimension, instead of the MEMB algorithm.

4.1   The Accuracy of the BND Algorithm

In order to evaluate how accurate the algorithm BND is, we compare it to the MEMB algorithm presented in [10].

We run both algorithms for the set of 300 industrial instances of the SAT Competition 2013 with a timeout of 30 minutes. While the BND algorithm finishes for all the 300 instances, MEMB is only able to approximate $N^b(r)$ in 17 instances. Moreover, while the average run-time of BND for these instances is 0.11 seconds, MEMB takes an average of 10 minutes and 7.2 seconds to compute them. On the other hand, the approximations of $N^b(r)$ computed by MEMB and BND are very similar (see Fig. 1).

Since the MEMB algorithm is more accurate than the BND algorithm, the upper bounds of $N^b(r)$ that MEMB calculates are below the ones calculated by BND. The real values of $N^b(r)$ are probably even lower in the final points (where the approximation is less accurate).

4.2   Random Formulas

Random 2SAT formulas in the VIG model correspond to Endös-Renyi graphs. It is known that these formulas have a phase transition point at $m/n = 1$ where formulas pass from satisfiable to unsatisfiable with probability one. It is also known that at $m/n = 0.5$ there is a percolation threshold. Formulas below this point have an non-connected associated VIG graph, and above this threshold there is a major connected component. In the percolation point the formula is self-similar with a fractal dimension $d = 2$. Above this point $N(r)$ decays exponentially. To the best of our knowledge, a result of this kind is not known for random 3CNF formulas.

---

[3] http://satcompetition.org/2013/

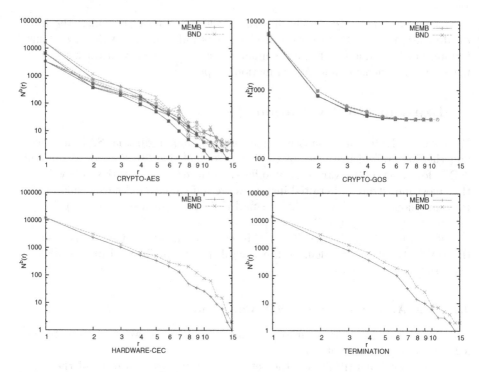

Fig. 1. Upper bounds for $N^b(r)$ obtained with MEMB and BND algorithms, for the 17 industrial instances that MEMB is able to compute in 30 minutes, grouped by families

Experimentally, we observe that the function $N(r)$ only depends on the clause/variable ratio m/n, and not on the number of variables (this is not shown in figures). In the phase transition point $m/n = 4.25$, the function has the form $N(r) \sim e^{-2.3\,r}$, i.e. it decays exponentially with $\beta = 2.3$ (see Fig. 2). Hence, $r^{max} = \frac{\log n}{2.3} + 1$. For instance, for $n = 10^5$ variables, random formulas have a radius $r^{max} \approx 6$. For bigger values of m/n, the decay β is bigger. In the CVIG model, we observe the same behavior. However, in this case, in the phase transition point, $N(r)$ decays exponentially with $\beta = 1.16 \approx 2.3/2$. Hence, the decay is just half of the decay of the VIG model, as we expected by Lemma 2.

For random 3CNF formulas, we have experimentally found a percolation threshold at $m/n \approx 0.17$. At this point the principal connected component also exhibits a fractal dimension $d = 2$.

4.3   Industrial Instances

Analyzing industrial instances, we observe that most of them are self-similar, and most dimensions ranges between 2 and 4. In the SAT Competition 2013, instances are grouped into families. In many of these families, all instances have the same fractal dimension, being this dimension a characteristic of the family. See, for instance, families *crypto-sha* or *diagnosis* in Fig. 3. Notice that the size

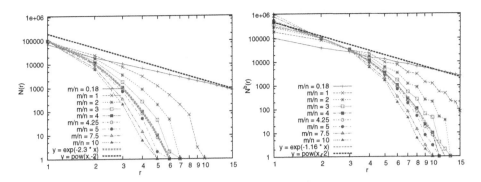

Fig. 2. Functions $N(r)$ for VIG (left), and $N^b(r)$ for CVIG (right), for 3CNF random formulas with distinct values of m/n. Formulas are generated using $n = 10^5$ variables and taking the major connected component, except for $m/n = 0.18$, where $n = 10^6$.

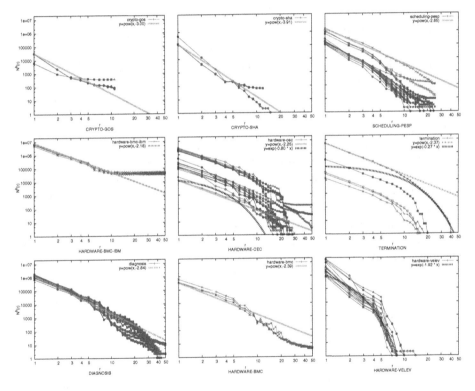

Fig. 3. Function $N^b(r)$ for some industrial SAT formulas grouped by families

of the formulas does not affect the value of the dimension (in the representation the function can be higher or lower, but with the same slope).

In general, the polynomial decay is clearer for small values of r. Moreover, in this area, the slope is the same for all instances of the same family of formulas.

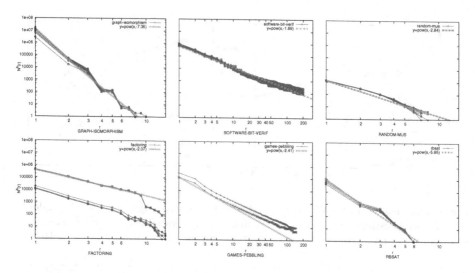

Fig. 4. Function $N^b(r)$ for some crafted SAT formulas grouped by families

For big values of r, we must make some considerations. First, the upper bound on $N^b(r)$ that we calculate can be a bad approximation. Second, there are two phenomena that we can identify. In some cases there is an abrupt decay, but the whole function can not be approximated by an exponential function (see some *hardware-cec* or *termination* instances, for instance). This decay in the number of required tiles can be due to a small number of edges connecting distant areas of the graph. These edges have no effect for small values of r, but may drop down the number of tiles for big values of r. In some other cases (see *hardware-bmc-ibm*, for instance), there is a long tail. In this case, it is due to the existence of (small) unconnected components in the graph. If we compute $N(r)$ only for the major component, this tail disappears.[4]

Finally, all instances of the *hardware-velev* family have a $N(r)$ function with exponential decay, i.e. are not self-similar.

4.4   Crafted Instances

Studying crafted instances, we see that most of them are self-similar. However, their fractal dimension have bigger values than the ones of the industrial formulas (some values are even bigger than 7).

The crafted instances of the SAT Competition 2013, as well as the industrial instances of this competition, are grouped into families. In general, we find that many families exhibit an homogeneus curve of $N(r)$ in all their instances. Moreover, in many of these families, $N(r)$ has a polynomial decay (i.e., all instances have the same fractal dimension). The fractal dimension of crafted formulas ranges from 1.5 to 7.5. In Fig. 4 we represent some crafted families.

---

[4]   In the figures, we can subtract from $N(r)$ the number of unconnected components, as an approximation, since most are covered with a few tiles.

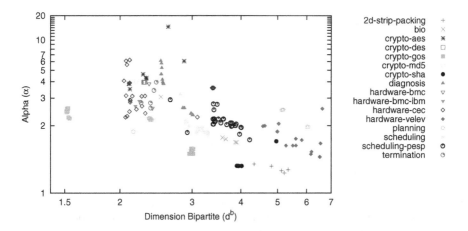

Fig. 5. Distribution of families according to the exponent α of the powerlaw distribution of node degrees, and fractal dimensions d^b at a fine-grained scale. Heterogeneous families (*software-bit-verif* and *software-bmc*) are not plotted.

4.5 Fractal Dimension at Fine-Grained Scale

If a graph is self-similar, then it has the same structure at all scales. We could replace groups of nodes tiled by a box by a single node, obtaining another graph with the same structure. In our experiments, we observe that this is the case for small values of r (for small values of r, function $N(r) \approx C\,r^{-d}$). However, this is more arguable for big values of r. Perhaps this is because the graph is not self-similar at large scale (coarse-grained), or because our approximation of $N(r)$ is not precise enough. If the formula has a small refutation, this will be visualized in our VIG or CVIG graphs as a small cycle. This means that what is really relevant is the fractal dimension looking at the graph at small scale (fine-grained dimension). In other words, we think that, more than whether there exists a self-similar structure, what is important, is the value of the fractal dimension at fine-grained, i.e. the slope of the function $N(r)$ for small values of r.

In our next experiment, we try to classify industrial instances according to their fractal dimension at fine-grained, and the exponent α of the powerlaw distribution of node degrees (see [2] for a description of how to compute exponent α). We will also note these fine-grained dimensions as d and d^b for the VIG and CVIG, respectively. We compute them as the interpolation, by linear regression, of $\log N(r)$ vs. $\log r$. We use the values of $N(r)$ and $N^b(r)$, for $r = 1, \ldots, 6$. Experimentally, we see that these approximations are accurate enough. As we can see in Fig. 5, just with the fractal dimension d^b and the powerlaw exponent α, we are able to determine which family an instance belongs to.

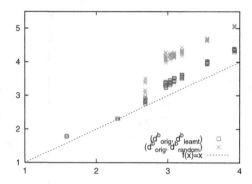

Fig. 6. Relation between the original fractal dimension d^b_{orig}, and the dimension d^b_{learnt} after adding learned clauses, or after adding random clauses d^b_{rand}, in random 3CNF formulas. Learnt clauses are computed after 10^3 conflicts.

5   The Effect of Learning

State-of-the-art SAT solvers add learnt clauses from conflicts during their execution. When a learnt clause is unitary, it can be propagated simplifying the original formula. Given a unitary clause x, clauses with literal x are completely removed, and literals $\neg x$ are removed from the formula. Learnt clauses of bigger length create new relations between variables, i.e., new edges in the VIG model.

Both, the addition of learnt clauses, and the simplification of formulas, due to unitary learnt clauses, may affect the dimension of the formula. The addition of edges in a graph (preserving the nodes) always increases its dimension, because tiles may cover more nodes, and the number $N(r)$ of tiles required to cover the graph decreases, whereas $N(1)$ is preserved. This contributes to increase the slope of function $N(r)$, hence the dimension. The effect of simplifications due to unitary learnt clauses is more difficult to predict, since we remove satisfied clauses (edges in the VIG model), but also nodes (decreasing $N(1)$).

We have conducted some experiments to analyze how the fractal dimension evolves during the execution of the SAT solver. First we show the effect of introducing learnt clauses in random 3CNF instances with 10^5 variables and distinct clause/variable ratios. In these instances almost all learnt clauses are not unitary, hence we do not remove variable nodes. In the VIG model, the addition of these learnt clauses introduces edges, and increases the dimension. In the CVIG model dimension also increases due to the same reason. In Fig. 6, we plot the dimension d^b_{learnt} after adding learnt clauses w.r.t. the original dimension d^b_{orig}. We observe that the addition of learnt clauses increases the dimension of the formula. This increase is bigger for formulas with higher clause/variable ratio. In order to *quantify* the increase in the dimension, we repeat the same experiment replacing learnt clauses by random clauses of the same size, and computing the

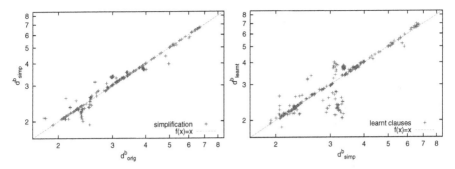

Fig. 7. Relation between the original fractal dimension d^b_{orig} and the fractal dimension d^b_{simp} after simplifying the formula with the unitary learnt clauses (left), and relation between the fractal dimension d^b_{simp} and the fractal dimension d^b_{learnt} after simplification and adding learnt clauses (right), for all industrial formulas. Learnt clauses are the result of 10^3 conflicts.

new dimension d^b_{random} (results are also shown in Fig. 6). We observe that in this second experiment the increase in the dimension is bigger than adding learnt clauses: $d^b_{random} \geq d^b_{learnt} \geq d^b_{orig}$. This means that learnt clauses, even in these random formulas, tend to connect variables that were already *close* in the graph. Therefore, their effect in the dimension is not as important as adding random clauses. In industrial instances some of the learnt clauses are unitary. We have analyzed separately the effect of simplifying the formula using these unitary clauses, and the effect of adding non-unitary learnt clauses. In the first case, when we learn x, and remove satisfied clauses containing x, we may remove edges connecting pairs of variables of those clauses. This contributes to decrease the dimension. However, we also remove the variable node x and the clauses nodes satisfied by x ($N(1)$ decreases). The effect of this second transformation on the graph cannot be predicted. Experimentally, we observe that simplifying the formula using unitary learnt clauses tends to decrease the dimension of the VIG and CVIG graph (see Fig. 7). The only exceptions are the *crypto-sha* and the *crypto-gos* families where a great number of variable nodes are removed.

In Fig. 8 we show the change in the dimension after 10^3, 10^4 and 10^5 conflicts. We observe that at the beginning dimensions may increase or decrease slightly. However, after 10^5 conflicts, the dimension clearly increases in most of the cases. Finally, in Fig. 9 we *quantify* the variation of the dimension due to the addition of learnt clauses, compared with the addition of the same number of random clauses with the same sizes. The effect of random clauses is much more significant, i.e., most of learnt clauses do not contribute to make tiles bigger (i.e. to reduce the number of needed tiles). They mainly connect nodes inside the tiles, i.e. nodes that where already close. Therefore, learning acts quite locally in the formula.

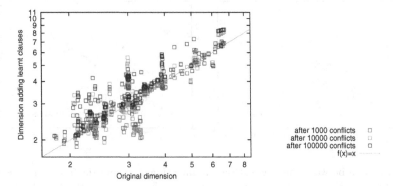

Fig. 8. Relation between the original fractal dimension and the fractal dimensions after learning clauses, in industrial formulas

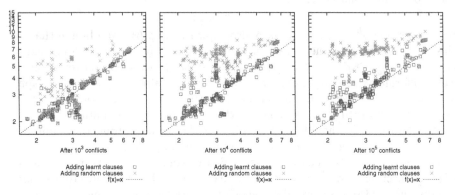

Fig. 9. Relation between the original fractal dimension and the fractal dimensions after adding learnt clauses, or after adding random clauses, in industrial formulas

6   Conclusions

We conclude that many industrial instances are self-similar, with most fractal dimensions ranging between 2 and 4. Fractal dimension, typical distances and graph diameter are related (small dimension implies big distance and diameter). Hence, industrial SAT instances have a big diameter (intuitively, we need long chains of implications to propagate a variable instantiation to others). We observe the same behaviour in crafted instances, although the fractal dimension is bigger in some cases. On the other hand, random instances are not self-similar.

We have also observed that fractal dimension increases due to learnt clauses. Moreover, the increase is specially abrupt in instances that show exponential decays (for instance, in the family *hardware-velev* or random formulas). This increase is bigger, if we substitute learnt clauses by random clauses of the same size. Therefore, learning *does not* contribute very much to connect distant parts of the formula, as one could think.

We have proved that we can determine the family an industrial instance belongs according to their fractal dimension at fine-grained, and the exponent α of the powerlaw distribution of node degrees. This is of interest for the development of portfolio solvers.

As future work, we plan to investigate how to develop industrial-like random instance generators to produce instances whose structural graph features such as the fractal dimension, the α exponent or the modularity are similar to the ones of industrial instances.

References

1. Albert, R., Jeong, H., Barabási, A.-L.: The diameter of the WWW. Nature 401, 130–131 (1999)
2. Ansótegui, C., Bonet, M.L., Levy, J.: On the structure of industrial SAT instances. In: Gent, I.P. (ed.) CP 2009. LNCS, vol. 5732, pp. 127–141. Springer, Heidelberg (2009)
3. Ansótegui, C., Bonet, M.L., Levy, J.: Towards industrial-like random SAT instances. In: IJCAI 2009, pp. 387–392 (2009)
4. Ansótegui, C., Giráldez-Cru, J., Levy, J.: The community structure of SAT formulas. In: Cimatti, A., Sebastiani, R. (eds.) SAT 2012. LNCS, vol. 7317, pp. 410–423. Springer, Heidelberg (2012)
5. Barabasi, A.L., Albert, R.: Emergence of scaling in random networks. Science 286, 509–512 (1999)
6. Kadioglu, S., Malitsky, Y., Sabharwal, A., Samulowitz, H., Sellmann, M.: Algorithm selection and scheduling. In: Lee, J. (ed.) CP 2011. LNCS, vol. 6876, pp. 454–469. Springer, Heidelberg (2011)
7. Katsirelos, G., Simon, L.: Eigenvector centrality in industrial SAT instances. In: Milano, M. (ed.) CP 2012. LNCS, vol. 7514, pp. 348–356. Springer, Heidelberg (2012)
8. Mandelbrot, B.B.: The fractal geometry of nature. Macmillan (1983)
9. Papadopoulos, F., Kitsak, M., Serrano, M., Bogu, M., Krioukov, D.: Popularity versus similarity in growing networks. Nature 489, 537–540 (2012)
10. Song, C., Gallos, L.K., Havlin, S., Makse, H.A.: How to calculate the fractal dimension of a complex network: the box covering algorithm. Journal of Statistical Mechanics: Theory and Experiment 2007(03), P03006 (2007)
11. Song, C., Havlin, S., Makse, H.A.: Self-similarity of complex networks. Nature 433, 392–395 (2005)
12. Walsh, T.: Search in a small world. In: IJCAI 1999, pp. 1172–1177 (1999)
13. Williams, R., Gomes, C.P., Selman, B.: Backdoors to typical case complexity. In: IJCAI 2003, pp. 1173–1178 (2003)
14. Xu, L., Hutter, F., Hoos, H.H., Leyton-Brown, K.: SATzilla: portfolio-based algorithm selection for SAT. J. Artif. Int. Res. 32(1), 565–606 (2008)

A Gentle Non-disjoint Combination
of Satisfiability Procedures*

Paula Chocron[1,3], Pascal Fontaine[2], and Christophe Ringeissen[3]

[1] Universidad de Buenos Aires, Argentina
[2] INRIA, Université de Lorraine & LORIA, Nancy, France
[3] INRIA & LORIA, Nancy, France

Abstract. A satisfiability problem is often expressed in a combination
of theories, and a natural approach consists in solving the problem by
combining the satisfiability procedures available for the component theories. This is the purpose of the combination method introduced by Nelson
and Oppen. However, in its initial presentation, the Nelson-Oppen combination method requires the theories to be signature-disjoint and stably
infinite (to guarantee the existence of an infinite model). The notion of
gentle theory has been introduced in the last few years as one solution
to go beyond the restriction of stable infiniteness, but in the case of disjoint theories. In this paper, we adapt the notion of gentle theory to the
non-disjoint combination of theories sharing only unary predicates (plus
constants and the equality). Like in the disjoint case, combining two theories, one of them being gentle, requires some minor assumptions on the
other one. We show that major classes of theories, i.e. Löwenheim and
Bernays-Schönfinkel-Ramsey, satisfy the appropriate notion of gentleness
introduced for this particular non-disjoint combination framework.

1 Introduction

The design of satisfiability procedures has attracted a lot of interest in the last
decade due to their ubiquity in SMT (Satisfiability Modulo Theories [4]) solvers
and automated reasoners. A satisfiability problem is very often expressed in a
combination of theories, and a very natural approach consists in solving the problem by combining the satisfiability procedures available for each of them. This is
the purpose of the combination method introduced by Nelson and Oppen [15].
In its initial presentation, the Nelson-Oppen combination method requires the
theories in the combination to be (1) signature-disjoint and (2) stably infinite
(to guarantee the existence of an infinite model). These are strong limitations,
and many recent advances aim to go beyond disjointness and stable infiniteness.
Both corresponding research directions should not be opposed. In both cases,
the problems are similar, i.e. building a model of $\mathcal{T}_1 \cup \mathcal{T}_2$ from a model of $\mathcal{T}_1$ and

* This work has been partially supported by the project ANR-13-IS02-0001-01 of the
Agence Nationale de la Recherche, by the European Union Seventh Framework Pro-
gramme under grant agreement no. 295261 (MEALS), and by the STIC AmSud
MISMT

S. Demri, D. Kapur, and C. Weidenbach (Eds.): IJCAR 2014, LNAI 8562, pp. 122–136, 2014.
© Springer International Publishing Switzerland 2014

a model of $\mathcal{T}_2$. This is possible if and only if there exists an isomorphism between the restrictions of the two models to the shared signature [24]. The issue is to define a framework to enforce the existence of this isomorphism. In the particular case of disjoint theories, the isomorphism can be obtained if the domains of the models have the same cardinality, for instance infinite; several classes of *kind* theories (shiny [25], polite [19], gentle [9]) have been introduced to enforce a (same) domain cardinality on both sides of the combination. For extensions of Nelson-Oppen to non-disjoint cases, e.g. in [24,27], cardinality constraints also arise. In this paper, we focus on non-disjoint combinations for which the isomorphism can be simply constructed by satisfying some cardinality constraints. More precisely, we extend the notion of gentle theory to the non-disjoint combination of theories sharing only unary predicates (plus constants and the equality). Some major classes of theories fit in our non-disjoint combination framework.

Contributions. The first contribution is to introduce a class of $\mathcal{P}$-*gentle* theories, to combine theories sharing a finite set of unary predicates symbols $\mathcal{P}$. The notion of $\mathcal{P}$-gentle theory extends the one introduced for the disjoint case [9]. Roughly speaking, a $\mathcal{P}$-gentle theory has nice cardinality properties not only for domains of models but also more locally for all Venn regions of shared unary predicates. We present a combination method for unions of $\mathcal{P}$-gentle theories sharing $\mathcal{P}$. The proposed method can also be used to combine a $\mathcal{P}$-gentle theory with another arbitrary theory for which we assume the decidability of satisfiability problems with cardinality constraints. This is a natural extension of previous works on combining non-stably infinite theories, in the straight line of combination methods à la Nelson-Oppen. Two major classes of theories are $\mathcal{P}$-gentle, namely the Löwenheim and Bernays-Schönfinkel-Ramsey (BSR) classes.

We characterize precisely the cardinality properties satisfied by Löwenheim theories. As a side contribution, bounds on cardinalities given in [8] have been improved, and we prove that our bounds are optimal. Our new result establishes that Löwenheim theories are $\mathcal{P}$-gentle.

We prove that BSR theories are also $\mathcal{P}$-gentle. This result relies on a non-trivial extension of Ramsey's Theorem on hypergraphs. This extension should be considered as another original contribution, since it may be helpful as a general technique to construct a model preserving the regions.

Related Work. Our combination framework is a way to combine theories with sets. The relation between (monadic) logic and sets is as old as logic itself, and this relation is particularly clear for instance considering Aristotle Syllogisms. It is however useful to again study monadic logic, and more particularly the Löwenheim class, and in view of the recent advances in combinations with non-disjoint and non-stably infinite theories.

In [26], the authors focus on the satisfiability problem of unions of theories sharing set operations. The basic idea is to reduce the combination problem into a satisfiability problem in a fragment of arithmetic called $BAPA$ (Boolean Algebra and Presburger Arithmetic). Löwenheim and BSR classes are also considered, but infinite cardinalities were somehow defined out of their reduction scheme, whilst infinite cardinalities are smoothly taken into account in our combination framework. In [26], BSR was shown to be reducible to Presburger. We here give a detailed proof. We believe such a proof is useful since it is more complicated that it may appear. In particular, our proof is based on an original (up to our knowledge) extension of Ramsey's Theorem to accommodate a domain partitioned into (Venn) regions. Finally, the notion of $\mathcal{P}$-gentleness defined and used here is stronger than semi-linearity of Venn-cardinality, and allows non-disjoint combination with more theories, e.g. the guarded fragment.

In [21,22], a locality property is used to properly instantiate axioms connecting two disjoint theories. Hence, the locality is a way to reduce (via instantiation) a non-disjoint combination problem to a disjoint one. In that context, cardinality constraints occur when considering bridging functions over a data structure with some cardinality constraints on the underlying theory of elements [28,21,23].

In [12], Ghilardi proposed a very general model-theoretic combination framework to obtain a combination method à la Nelson-Oppen when $\mathcal{T}_1$ and $\mathcal{T}_2$ are two *compatible* extensions of the same shared theory (satisfying some properties). This framework relies on an application of the Robinson Joint Consistency Theorem (roughly speaking, the union of theories is consistent if the intersection is complete). Using this framework, several shared fragments of arithmetic have been successfully considered [12,16,17]. Due to its generality, Ghilardi's approach is free of cardinality constraints.

It is also possible to consider a general semi-decision procedure for the unsatisfiability problem modulo $\mathcal{T}_1 \cup \mathcal{T}_2$, e.g. a superposition calculus. With the rewrite-based approach initiated in [3], the problem reduces to proving the termination of this calculus. General criteria have been proposed to get modular termination results for superposition, when $\mathcal{T}_1$ and $\mathcal{T}_2$ are either disjoint [2] or non-disjoint [20]. Notice that the superposition calculus can also be used as a deductive engine to entail some cardinality constraints, as shown in [5].

Structure of the paper. Section 2 introduces some classical notations and definitions. In Section 3, we introduce the notion of $\mathcal{P}$-gentle theory and we present the related combination method for unions of theories sharing a (non-empty finite) set $\mathcal{P}$ of unary predicate symbols. All the theories in the Löwenheim class and in the BSR class are $\mathcal{P}$-gentle, as shown respectively in Section 4 and in Section 5. A simple example is given in Section 6. The conclusion (Section 7) discusses the current limitations of our approach and mentions some possible directions to investigate. Our extension of Ramsey's Theorem can be found in Appendix A.

2   Notation and Basic Definitions

A first-order language is a tuple $\mathcal{L} = \langle \mathcal{V}, \mathcal{F}, \mathcal{P} \rangle$ such that $\mathcal{V}$ is an enumerable set of variables, while $\mathcal{F}$ and $\mathcal{P}$ are sets of function and predicate symbols. Every function and predicate symbol is assigned an arity. Nullary predicate symbols are called proposition symbols, and nullary function symbols are called constant symbols. A first-order language is called relational if it only contains function symbols of arity zero. A relational formula is a formula in a relational language. Terms, atomic formulas and first-order formulas over the language $\mathcal{L}$ are defined in the usual way. In particular an atomic formula is either an equality, or a predicate symbol applied to the right number of terms. Formulas are built from atomic formulas, Boolean connectives ($\neg$, $\wedge$, $\vee$, $\Rightarrow$, $\equiv$), and quantifiers ($\forall$, $\exists$). A literal is an atomic formula or the negation of an atomic formula. Free variables are defined in the usual way. A formula with no free variables is closed, and a formula without variables is ground. A universal formula is a closed formula $\forall x_1 \ldots \forall x_n.\varphi$ where φ is quantifier-free. A (finite) theory is a (finite) set of closed formulas. Two theories are disjoint if no predicate symbol in P or function symbol in F appears in both theories, except constants and equality.

An interpretation $\mathcal{I}$ for a first-order language $\mathcal{L}$ provides a non empty domain D, a total function $\mathcal{I}[f] : D^r \to D$ for every function symbol f of arity r, a predicate $\mathcal{I}[p] \subseteq D^r$ for every predicate symbol p of arity r, and an element $\mathcal{I}[x] \in D$ for every variable x. The cardinality of an interpretation is the cardinality of its domain. The notation $\mathcal{I}_{x_1/d_1,\ldots,x_n/d_n}$ for $x_1, \ldots, x_n$ different variables stands for the interpretation that agrees with $\mathcal{I}$, except that it associates $d_i \in D$ to the variable x_i, $1 \leq i \leq n$. By extension, an interpretation defines a value in D for every term, and a truth value for every formula. We may write $\mathcal{I} \models \varphi$ whenever $\mathcal{I}[\varphi] = \top$. Given an interpretation $\mathcal{I}$ on domain D, the *restriction* $\mathcal{I}'$ of $\mathcal{I}$ on $D' \subseteq D$ is the unique interpretation on D' such that $\mathcal{I}$ and $\mathcal{I}'$ interpret predicates, functions and variables the same way on D'. An *extension* $\mathcal{I}'$ of $\mathcal{I}$ is an interpretation on a domain D' including D such that $\mathcal{I}'$ restricted to D is $\mathcal{I}$.

A model of a formula (theory) is an interpretation that evaluates the formula (resp. all formulas in the theory) to true. A formula or theory is satisfiable if it has a model; it is unsatisfiable otherwise. A formula G is $\mathcal{T}$-satisfiable if it is satisfiable in the theory $\mathcal{T}$, that is, if $\mathcal{T} \cup \{G\}$ is satisfiable. A $\mathcal{T}$-model of G is a model of $\mathcal{T} \cup \{G\}$. A formula G is $\mathcal{T}$-unsatisfiable if it has no $\mathcal{T}$-models. In our context, a theory $\mathcal{T}$ is *decidable* if the $\mathcal{T}$-satisfiability problem for sets of (ground) literals is decidable in the language of $\mathcal{T}$ (extended with fresh constants).

Consider an interpretation $\mathcal{I}$ on a language with unary predicates $p_1, \ldots, p_n$ and some elements D in the domain of this interpretation. Every element $d \in D$ belongs to a *Venn region* $v(d) = v_1 \ldots v_n \in \{\top, \bot\}^n$ where $v_i = \mathcal{I}[p_i](d)$. We denote by $D_v \subseteq D$ the set of elements of D in the Venn region v. Notice also that, for a language with n unary predicates, there are 2^n Venn regions. Given an interpretation $\mathcal{I}$, D^c denotes the subset of elements in D associated to constants by $\mathcal{I}$. Naturally, D^c_v denotes the set of elements associated to constants that are in the Venn region v.

3   Gentle Theories Sharing Unary Predicates

From now on, we assume that $\mathcal{P}$ is a non-empty finite set of unary predicates. A $\mathcal{P}$-*union* of two theories $\mathcal{T}_1$ and $\mathcal{T}_2$ is a union sharing only $\mathcal{P}$, a set of constants and the equality.

Definition 1. *An arrangement $\mathcal{A}$ for finite sets of constant symbols S and unary predicates $\mathcal{P}$ is a maximal satisfiable set of equalities and inequalities $a = b$ or $a \neq b$ and literals $p(a)$ or $\neg p(a)$, with $a, b \in S$, $p \in \mathcal{P}$.*

There are only a finite number of arrangements for given sets S and $\mathcal{P}$.

Given a theory $\mathcal{T}$ whose signature includes $\mathcal{P}$ and a model $\mathcal{M}$ of $\mathcal{T}$ on domain D, the $\mathcal{P}$-*cardinality* κ is the tuple of cardinalities of all Venn regions of $\mathcal{P}$ in $\mathcal{M}$ (κ_v will denote the cardinality of the Venn region v). The following theorem (specialization of general combination lemmas in e.g. [24,25]) states the completeness of the combination procedure for $\mathcal{P}$-unions of theories:

Theorem 1. *Consider a $\mathcal{P}$-union of theories $\mathcal{T}_1$ and $\mathcal{T}_2$ whose respective languages $\mathcal{L}_1$ and $\mathcal{L}_2$ share a finite set S of constants, and let L_1 and L_2 be sets of literals, respectively in $\mathcal{L}_1$ and $\mathcal{L}_2$. Then $L_1 \cup L_2$ is $\mathcal{T}_1 \cup \mathcal{T}_2$-satisfiable if and only if there exist an arrangement $\mathcal{A}$ for S and $\mathcal{P}$, and a $\mathcal{T}_i$-model $\mathcal{M}_i$ of $\mathcal{A} \cup L_i$ with the same $\mathcal{P}$-cardinality for $i = 1, 2$.*

The *spectrum* of a theory $\mathcal{T}$ is the set of $\mathcal{P}$-cardinalities of its models. The above theorem can thus be restated as:

Corollary 1. *The $\mathcal{T}_1 \cup \mathcal{T}_2$-satisfiability problem for sets of literals is decidable if, for any sets of literals $\mathcal{A} \cup L_1$ and $\mathcal{A} \cup L_2$ it is possible to decide if the intersection of the spectrums of $\mathcal{T}_1 \cup \mathcal{A} \cup L_1$ and of $\mathcal{T}_2 \cup \mathcal{A} \cup L_2$ is non-empty.*

To characterize the spectrum of the decidable classes considered in this paper, we introduce the notion of *cardinality constraint*. A *finite* cardinality constraint is simply a $\mathcal{P}$-cardinality with only finite cardinalities. An *infinite* cardinality constraint is given by a $\mathcal{P}$-cardinality κ with only finite cardinalities and a non-empty set of Venn regions V, and stands for all the $\mathcal{P}$-cardinalities κ' such that $\kappa'_v \geq \kappa_v$ if $v \in V$, and $\kappa'_v = \kappa_v$ otherwise. The *spectrum* of a finite set of cardinality constraints is the union of all $\mathcal{P}$-cardinalities represented by each cardinality constraint. It is now easy to define the class of theories we are interested in:

Definition 2. *A theory $\mathcal{T}$ is $\mathcal{P}$-gentle if, for every set L of literals in the language of $\mathcal{T}$, the spectrum of $\mathcal{T} \cup L$ is the spectrum of a computable finite set of cardinality constraints.*

Notice that a $\mathcal{P}$-gentle theory is (by definition) decidable. To relate the above notion with the gentleness in the disjoint case [9], observe that if p is a unary predicate symbol not occurring in the signature of the theory $\mathcal{T}$, then $\mathcal{T} \cup \{\forall x.p(x)\}$ is $\{p\}$-gentle if and only if $\mathcal{T}$ is gentle.

If a theory is $\mathcal{P}$-gentle, then it is $\mathcal{P}'$-gentle for any non-empty subset $\mathcal{P}'$ of $\mathcal{P}$. It is thus interesting to have $\mathcal{P}$-gentleness for the largest possible $\mathcal{P}$. Hence, when $\mathcal{P}$ is not explicitly given for a theory, we assume that $\mathcal{P}$ denotes the set of unary predicates symbols occurring in its signature. In the following sections we show that the Löwenheim theories and the BSR theories are $\mathcal{P}$-gentle.

The union of two $\mathcal{P}$-gentle theories is decidable, as a corollary of the following modularity result:

Theorem 2. *The class of $\mathcal{P}$-gentle theories is closed under $\mathcal{P}$-union.*

Proof. If we consider the $\mathcal{P}$-union of two $\mathcal{P}$-gentle theories with respective spectrums $\mathcal{S}_1$ and $\mathcal{S}_2$, then we can build some finite set of cardinality constraints whose spectrum is $\mathcal{S}_1 \cap \mathcal{S}_2$. □

Some very useful theories are not $\mathcal{P}$-gentle, but in practical cases they can be combined with $\mathcal{P}$-gentle theories. To define more precisely the class of theories $\mathcal{T}'$ that can be combined with a $\mathcal{P}$-gentle one, let us introduce the $\mathcal{T}'$-*satisfiability problem with cardinality constraints*: given a formula and a finite set of cardinality constraints, the problem amounts to check whether the formula is satisfiable in a model of $\mathcal{T}$ whose $\mathcal{P}$-cardinality is in the spectrum of the cardinality constraints. As a direct consequence of Corollary 1:

Theorem 3. $\mathcal{T} \cup \mathcal{T}'$-*satisfiability is decidable if the theory $\mathcal{T}$ is $\mathcal{P}$-gentle and $\mathcal{T}'$-satisfiability with cardinality constraints is decidable.*

Notice that $\mathcal{T}$-satisfiability with cardinality constraints is decidable for most common theories, e.g. the theories handled in SMT solvers. This gives the theoretical ground to add to the SMT solvers any number of $\mathcal{P}$-gentle theories sharing unary predicates.

From the results in the rest of the paper, it will also follow that the non-disjoint union (sharing unary predicates) of BSR and Löwenheim theories with one decidable theory accepting further constraints of the form $\forall x \,. ((\neg)p_1(x) \wedge \ldots (\neg)p_n(x)) \Rightarrow (x = a_1 \vee \ldots x = a_m)$ is decidable. For instance, the guarded fragment with equality accepts such further constraints and the superposition calculus provides a decision procedure [11]. Thus any theory in the guarded fragment can be combined with Löwenheim and BSR theories sharing unary predicates.

In the disjoint case, any decidable theory expressed as a finite set of first-order axioms can be combined with a gentle theory [9]. Here this is not the case anymore. Indeed, consider the theory $\psi = \varphi \vee \exists x \, p(x)$ where p does not occur in φ; any set of literals is satisfiable in the theory ψ if and only if it is satisfiable in the theory of equality. If the satisfiability problem of literals in the theory φ is undecidable, the $\mathcal{P}$-union of ψ and the Löwenheim theory $\forall x \, \neg p(x)$ will also be undecidable.

4   The Löwenheim Class

We first review some classical results about this class and refer to [6] for more details. A Löwenheim theory is a finite set of closed formulas in a relational language containing only unary predicates (and no functions except constants). This class is also known as first-order relational monadic logic. Usually one distinguishes the Löwenheim class with and without equality. The Löwenheim class has the finite model property (and is thus decidable) even with equality. Full monadic logic *without equality*, i.e. the class of finite theories over a language containing symbols (predicates and functions) of arity at most 1, also has the finite model property. Considering monadic logic with equality, the class of finite theories over a language containing only unary predicates and just two unary functions is already undecidable. With only one unary function, however, the class remains decidable [6], but does not have the finite model property anymore. Since the spectrum for this last class is significantly more complicated [13] than for the Löwenheim class we will here only focus on the Löwenheim class with equality (only classes with equality are relevant in our context), that is, without functions. More can be found about monadic first-order logic in [6,8]. In particular, a weaker version of Corollary 2 (given below) can be found in [8].

Previously [9,1], combining theories with non-stably infinite theories took advantage of "pumping" lemmas, allowing — for many decidable fragments — to build models of arbitrary large cardinalities. The following theorem is such a pumping lemma, but it considers the cardinalities of the Venn regions and not only the global cardinality.

Lemma 1. *Assume $\mathcal{T}$ is a Löwenheim theory with equality. Let q be the number of variables in $\mathcal{T}$. If there exists a model $\mathcal{M}$ on domain D with $|D_v \setminus D^c| \geq q$, then, for each cardinality $q' \geq q$, there is a model extension or restriction $\mathcal{M}'$ of $\mathcal{M}$ on domain D' such that $|D'_v \setminus D^c| = q'$ and $D'_{v'} = D_{v'}$ for all $v' \neq v$.*

Proof. Two interpretations $\mathcal{I}$ (on domain D) and $\mathcal{I}'$ (on domain D') for a formula ψ are *similar* if

- $|(D_v \cap D'_v) \setminus D^c| \geq q$;
- $D_{v'} = D'_{v'}$ for each Venn region v' distinct from v;
- $\mathcal{I}[a] = \mathcal{I}'[a]$ for each constant in ψ;
- $\mathcal{I}[x] = \mathcal{I}'[x]$ for each variable free in ψ.

Considering $\mathcal{M}$ as above, we can build a model $\mathcal{M}'$ as stated in the theorem, such that $\mathcal{M}$ and $\mathcal{M}'$ are similar. Indeed similarity perfectly defines a model with respect to another, given the cardinalities of the Venn regions.

We now prove that, given a Löwenheim formula ψ (or a set of formulas), two similar interpretations for ψ give the same truth value to ψ and to each sub-formula of ψ.

The proof is by induction on the structure of the (sub-)formula ψ. It is obvious if ψ is atomic, since similar interpretations assign the same value to variables and constants. If ψ is $\neg\varphi_1$, $\varphi_1 \vee \varphi_2$, $\varphi_1 \wedge \varphi_2$ or $\varphi_1 \Rightarrow \varphi_2$, the result holds if it also holds for φ_1 and φ_2.

Assume $\mathcal{I}$ makes true the formula $\psi = \exists x\,\varphi(x)$. Then there exists some $d \in D$ such that $\mathcal{I}_{x/d}$ is a model of $\varphi(x)$. If $d \in D'$, then $\mathcal{I}'_{x/d}$ is similar to $\mathcal{I}_{x/d}$ and, by the induction hypothesis, it is a model of $\varphi(x)$; $\mathcal{I}'$ is thus a model of ψ. If $d \notin D'$, then $d \in D_v$ and $|(D_v \cap D'_v) \setminus D^c| \geq q$. Furthermore, since the whole formula contains at most q variables, $\varphi(x)$ contains at most $q - 1$ free variables besides x. Let $x_1, \ldots, x_m$ be those variables. There exists some $d' \in (D_v \cap D'_v) \setminus D^c$ such that $d' \neq \mathcal{I}[x_i]$ for all $i \in \{1, \ldots, m\}$. By structural induction, it is easy to show that $\mathcal{I}_{x/d}$ and $\mathcal{I}_{x/d'}$ give the same truth value to $\varphi(x)$. Furthermore $\mathcal{I}_{x/d'}$ and $\mathcal{I}'_{x/d'}$ are similar. $\mathcal{I}'$ is thus a model of ψ. To summarize, if $\mathcal{I}$ is a model of ψ, $\mathcal{I}'$ is also a model of ψ. By symmetry, if $\mathcal{I}'$ is a model of ψ, $\mathcal{I}$ is also a model of ψ. The proof for formulas of the form $\forall x\,\varphi(x)$ is dual. □

Lemma 1 has the following consequence on the acceptable cardinalities for the models of a Löwenheim theory:

Corollary 2. *Assume $\mathcal{T}$ is a Löwenheim theory with equality with n distinct unary predicates. Let r and q be respectively the number of constants and variables in $\mathcal{T}$. If $\mathcal{T}$ has a model of some cardinality κ strictly larger than $r + 2^n \max(0, q - 1)$, then $\mathcal{T}$ has models of each cardinality equal or larger than $\min(\kappa, r + q\, 2^n)$.*

Proof. If a model with such a cardinality exists, then there are Venn regions v such that $|D_v \setminus D^c| \geq q$. Then the number of elements in these Venn regions can be increased to any arbitrary larger cardinality, thanks to Lemma 1. If $\kappa > r + q\, 2^n$, it means some Venn regions v are such that $|D_v \setminus D^c| > q$, and by eliminating elements in such Venn regions (using again Lemma 1), it is possible to obtain a model of cardinality $r + q\, 2^n$.   □

In [8], the limit is $q\, 2^n$, q being the number of constants plus the maximum number of nested quantifiers. Now q is more precisely set to the number of variables, and the constants are counted separately. Moreover, $\max(0, q - 1)$ replaces the factor q.

The case where q and r are both 0 corresponds to pure propositional logic (Löwenheim theories without variables and constants), where the size of the domain is not relevant. With $q = 1$ (one variable), there is no way to compare two elements (besides the ones associated to constants) and enforce them to be equal. It is still possible to constrain the domain to be of size at most r, using constraints like $\forall x\,.\, x = c_1 \vee \ldots \vee x = c_r$, but any model with one element not associated to a constant can be extended to a model of arbitrary cardinality (by somehow duplicating any number of time this element). Notice also that it is possible to set a lower bound on the size of the domain that can be $r + 2^n$. Consider for instance a set of sentences of the form $\exists x.(\neg)p_1(x) \vee \ldots (\neg)p_n(x)$; there are 2^n such formulas, each enforcing one Venn region to be non-empty.

Using several variables, a Löwenheim formula can enforce upper bounds larger than r on cardinalities. For $q = 2$, it is indeed easy to build a formula that has only models of cardinality at most $(q - 1)\, 2^n = 2^n$:

$$\forall x \forall y\,.\, \Big[\ \bigwedge_{0 < i < j \leq n} p_i(x) = p_j(y)\ \Big] \Rightarrow x = y.$$

With a larger number of variables, the following formula ($q \geq 2$)

$$\forall x_1 \ldots \forall x_q\,.\, \Big[\ \bigwedge_{\substack{0 < i < j \leq n \\ 0 < i' < j' \leq q}} p_i(x_{i'}) = p_j(x_{j'})\ \Big] \Rightarrow \bigvee_{0 < i' < j' \leq q} x_{i'} = x_{j'}$$

enforces the cardinality of the domain to be at most $(q - 1)\, 2^n$. To obtain a formula with constants that accepts only models of cardinality up to $r + 2^n \max(0, q - 1)$, it suffices to add as a guard in the above formula the conjunctive sets of atoms expressing that the variables are disjoint from the r constants. So the above condition in Corollary 2 is the strongest one.

Besides the finite model property and the decidability of Löwenheim theories, Corollary 2 also directly entails the $\mathcal{P}$-gentleness:

Theorem 4. *Löwenheim theories on a language with unary predicates in $\mathcal{P}$ are $\mathcal{P}$-gentle.*

5   The Bernays-Schönfinkel-Ramsey Class

A Bernays-Schönfinkel-Ramsey (BSR for short) theory is a finite set of formulas of the form $\exists^*\forall^*\varphi$, where φ is a first-order formula which is function-free (but constants are allowed) and quantifier-free. Bernays and Schönfinkel first proved the decidability of this class without equality; Ramsey later proved that it remains decidable with equality. More can be found about BSR theories in [6]. Ramsey also gave some (less known) results about the spectrum of BSR theories [18]. We here give a proof that BSR theories are $\mathcal{P}$-gentle.

For simplicity, we will assume that existential quantifiers are Skolemized. In the following, a BSR theory is thus a finite set of universal function-free closed first-order formulas.

Lemma 2. *Let $\mathcal{T}$ be a BSR theory, and $\mathcal{M}$ be a model of $\mathcal{T}$ on domain D. Then any restriction $\mathcal{M}'$ of $\mathcal{M}$ on domain D' with $D^c \subseteq D' \subseteq D$ is a model of $\mathcal{T}$.*

Proof. Consider $\mathcal{M}$ and $\mathcal{M}'$ as above. Since $\mathcal{M}$ is a model of $\mathcal{T}$, for each closed formula $\forall x_1 \ldots x_n . \varphi$ in $\mathcal{T}$ (where φ is function-free and quantifier-free), and for all $d_1, \ldots, d_n \in D' \subseteq D$, $\mathcal{M}_{x_1/d_1,\ldots,x_n/d_n}$ is a model of φ. This also means that, for all $d_1, \ldots, d_n \in D'$, $\mathcal{M}'_{x_1/d_1,\ldots,x_n/d_n}$ is a model of φ, and finally that $\mathcal{M}'$ is a model of $\forall x_1 \ldots x_n . \varphi$.   □

Intuitively, this states that the elements not assigned to ground terms (i.e. the constants) can be eliminated from a model of a BSR theory. It is known [18,9] that for any BSR theory $\mathcal{T}$ there is a computable finite number k such that if $\mathcal{T}$ has a model of cardinality greater or equal to k, then it has a model of any cardinality larger than k. Later in this section, we prove that the same occurs locally for each Venn region.

The notion of n-repetitive models, which we now define, is instrumental for this. Informally, a model is n-repetitive if it is symmetric for those elements of its domain that are not assigned to constants in the theory.

Definition 3. *An interpretation $\mathcal{I}$ on domain D for a BSR theory $\mathcal{T}$ is n-repetitive for a set V of Venn regions if, for each $v \in V$, $|D_v \setminus D^c| \geq n$ and there exists a total order $\prec$ on elements in $D_v \setminus D^c$ such that*

- *for every r-ary predicate symbol p in $\mathcal{T}$*
- *for all $d_1, \ldots, d_r \in D$, and $d'_1, \ldots, d'_r \in D$ with*
    - $|\{d_1, \ldots, d_r\} \setminus D^c| \leq n$
    - $d'_i = d_i$ *if d_i or $d'_i \in D^c \cup \bigcup_{v \notin V} D_v$*
    - $v(d'_i) = v(d_i)$

- $d_i' \prec d_j'$ iff $d_i \prec d_j$, if for some $v' \in V$, $d_i, d_j \in D_{v'} \setminus D^c$

we have $\mathcal{I}[p](d_1, \dots, d_r) = \mathcal{I}[p](d_1', \dots, d_r')$.

Notice that a same interpretation can be n-repetitive for several Venn regions at the same time. Also, the above definition allows $D_v \setminus D^c$ to be empty for every $v \notin V$. Previously [9] (without distinguishing regions) we showed that one can decide if a BSR theory $\mathcal{T}$ is n-repetitive by building another BSR theory that is satisfiable if and only if $\mathcal{T}$ is n-repetitive. The same occurs to n-repetitiveness for Venn regions.

Theorem 5. *Consider a BSR theory $\mathcal{T}$ with n variables and a model $\mathcal{M}$ on domain D. If $\mathcal{M}$ is n-repetitive for the Venn regions V then, for any (finite or infinite) cardinalities $\kappa_v \geq |D_v|$ ($v \in V$), $\mathcal{T}$ has a model $\mathcal{M}'$ extension of $\mathcal{M}$ on domain D' such that $|D_v'| = \kappa_v$ if $v \in V$ and $D_{v'}' = D_{v'}$ for all $v' \notin V$.*

Proof. Assume that $\prec$ are the total orders mentioned in Definition 3. We first build an extension $\mathcal{M}'$ of $\mathcal{M}$ as specified in the theorem, and later prove it is a model of $\mathcal{T}$.

Let E be the set of new elements $E = D' \setminus D$, and fix arbitrary total orders (again denoted by $\prec$) on $D_v' \setminus D^c$ for all $v \in V$ that extend the given orders on $D_v \setminus D^c$. Since $\mathcal{M}'$ is an extension of $\mathcal{M}$, the interpretation of the predicate symbols is already defined when all arguments belong to D. When some arguments belong to E, the truth value of an r-ary predicate p is defined as follows:

- $(d_1', \dots, d_r') \notin \mathcal{M}'[p]$ for $|\{d_1', \dots, d_r'\} \setminus D^c| > n$: the interpretation of p over tuples with more than n elements outside D^c is fixed arbitrarily. Indeed, such tuples are irrelevant for the evaluation of the formulas of $\mathcal{T}$: terms occurring as arguments of a predicate are either variables or constants, and no more than n variables occur in any formula of $\mathcal{T}$.
- otherwise, to determine $\mathcal{M}'[p](d_1', \dots, d_r')$, first choose $d_1, \dots d_r \in D$ such that $d_1', \dots, d_r'$ and $d_1, \dots d_r$ are related to each other just like in Definition 3. This is possible since, for every Venn region v for which the interpretation is repetitive, there are at least n elements in $D_v \setminus D^c$. Then $(d_1', \dots, d_r') \in \mathcal{M}'[p]$ iff $(d_1, \dots, d_r) \in \mathcal{M}[p]$. Observe that all possible choices of $d_1, \dots, d_n$ lead to the same definition because $\mathcal{M}$ is n-repetitive.

The construction is such that $\mathcal{M}'$ is also n-repetitive for the same regions. It is also a model of $\mathcal{T}$: all formulas in $\mathcal{T}$ are of the form $\forall x_1 \dots x_m . \varphi(x_1, \dots, x_m)$, with $m \leq n$. For all $d_1' \dots, d_m' \in D'$, if $\{d_1', \dots, d_m'\} \subseteq D$ then

$$\mathcal{M}'_{x_1/d_1', \dots, x_m/d_m'}[\varphi(x_1, \dots, x_m)] = \mathcal{M}_{x_1/d_1', \dots, x_m/d_m'}[\varphi(x_1, \dots, x_m)]$$

since $\mathcal{M}'$ is an extension of $\mathcal{M}$. Otherwise, let $d_1, \dots, d_m \in D$ be some elements related to $d_1', \dots, d_m'$ like in Definition 3. Since $\mathcal{M}'$ is n-repetitive,

$$\mathcal{M}'_{x_1/d_1', \dots, x_m/d_m'}[\varphi(x_1, \dots, x_m)] = \mathcal{M}'_{x_1/d_1, \dots, x_m/d_m}[\varphi(x_1, \dots, x_m)]$$
$$= \mathcal{M}_{x_1/d_1, \dots, x_m/d_m}[\varphi(x_1, \dots, x_m)].$$

In both cases, $\mathcal{M}'_{x_1/d_1', \dots, x_m/d_m'}[\varphi(x_1, \dots, x_m)]$ evaluates to true, and therefore $\mathcal{M}'$ is a model of $\forall x_1 \dots x_n . \varphi(x_1, \dots, x_m)$. $\qquad \square$

Now it is possible to state that the full spectrum of a BSR theory only depends on (a finite set of) $\mathcal{P}$-cardinalities κ such that, for all Venn region v, $\kappa_v \leq k$ for some finite cardinality k only depending on the theory. The proof requires an extension of Ramsey's Theorem which can be found in the appendix A.

Theorem 6. *Given a BSR theory $\mathcal{T}$ with n variables, there exists a number k computable from the theory, such that, if $\mathcal{T}$ has a model $\mathcal{M}$ on domain D such that $|D_v \setminus D^c| \geq k$ for Venn regions $v \in V$, then it has a model which is n-repetitive for Venn regions V.*

Proof. Using Lemma 2, we can assume that $\mathcal{T}$ has a (sufficiently large) finite model $\mathcal{M}$ on domain D. We can assume without loss of generality that $\mathcal{M}$ is such that, for every predicate p of the language, $(d_1, \ldots d_r) \notin \mathcal{M}[p]$ whenever there are more than n elements in $\{d_1, \ldots d_r\} \setminus D^c$; indeed, these interpretations play no role in the truth value of a formula with n variables.

Let $\prec$ be an order on $D \setminus D^c$. Given two ordered (with respect to $\prec$) sequences $e_1, \ldots, e_n$ and $e'_1, \ldots, e'_n$ of elements in $D \setminus D^c$ such that $v(e_i) = v(e'_i)$ ($1 \leq i \leq n$), we say that the configurations for $e_1, \ldots, e_n$ and $e'_1, \ldots, e'_n$ agree if for every r-ary predicate p, and for every $d_1, \ldots, d_r \in D^c \cup \{e_1, \ldots, e_n\}$, $(d_1, \ldots, d_r) \in \mathcal{M}[p]$ iff $(d'_1, \ldots, d'_r) \in \mathcal{M}[p]$, with $d'_i = e'_j$ if $d_i = e_j$ for some j, and $d'_i = d_i$ otherwise. Notice that there are only a finite number of disagreeing configurations for n elements in $D \setminus D^c$: more precisely a configuration is determined by at most $b = \sum_p (n + |D^c|)^{\text{arity}(p)}$ Boolean values, where the sum ranges over all predicates in the theory. Thus the number of disagreeing configurations is bounded by $C = 2^b$.

Interpreting configurations as colors, one can use the extension of Ramsey's Theorem given in Appendix A: according to Theorem 7, there is a computable function f such that, for any $N \in \mathbb{N}$, if $|D \setminus D^c|_V \geq f(n, N, C)$, then there exists a model on $D' \subseteq D$ with $|D' \setminus D^c|_V \geq N$ for which configurations agree if they have the same number of elements in each Venn region of V. Taking $N = n$, this is actually building a n-repetitive restriction of $\mathcal{M}$. $\square$

The BSR class obviously has the finite model property, and is decidable. Lemma 2 and Theorems 5 and 6 above also prove that BSR theories are (gentle and) $\mathcal{P}$-gentle:

Corollary 3. *BSR theories on a language including unary predicates in $\mathcal{P}$ are $\mathcal{P}$-gentle.*

A simple constructive proof of this corollary would consider the finite number of all $\mathcal{P}$-cardinalities κ such that $\kappa_v \leq k$ (where k comes from Theorem 6). All such $\mathcal{P}$-cardinalities can be understood as cardinality constraints, the extendable Venn regions being the ones for which $\kappa_v > k$. Of course this construction is highly impractical, since it uses some kind of Ramsey numbers, known to be extremely large. In practice, we believe there are much better constructions: the important elements of the domain are basically only the ones associated to constants, and theoretical upper bounds are not met in non-artificial cases.

6   Example: Non-Disjoint Combination of Order and Sets

To illustrate the kind of theories that can be handled in our framework, consider a simple yet informative example with a BSR theory defining an ordering $<$ and augmented with clauses connecting the ordering $<$ and the sets p and q (we do not distinguish sets and their related predicates):

$$\mathcal{T}_1 = \begin{cases} \forall x.\ \neg(x < x) \\ \forall x, y, z.\ (x < y \wedge y < z) \Rightarrow x < z \\ \forall x, y.\ (p(x) \wedge \neg p(y)) \Rightarrow x < y \\ \forall x, y.\ (q(x) \wedge \neg q(y)) \Rightarrow x < y \end{cases}$$

and a Löwenheim theory

$$\mathcal{T}_2 = \begin{cases} \exists y \forall x.\ (p(x) \wedge q(x)) \equiv x = y \\ \forall x \exists y.\ p(x) \Rightarrow (x \neq y \wedge q(y)) \end{cases}$$

Notice that $\mathcal{T}_2$ is not a BSR theory due to the $\forall\exists$ quantification of its second axiom, but both theories $\mathcal{T}_1$ and $\mathcal{T}_2$ are actually $\mathcal{P}$-gentle. The theory $\mathcal{T}_1$ imposes either $p \cap \overline{q}$ or $\overline{p} \cap q$ to be empty (we will assume that the domain is non-empty and simplify the cardinality constraints accordingly). The theory $\mathcal{T}_2$ imposes the cardinality of $p \cap q$ to be exactly 1, and the cardinality of $\overline{p} \cap q$ to be at least 1. The following table collects the cardinality constraints:

	$\mathcal{T}_1$		$\mathcal{T}_2$
$\overline{p} \cap \overline{q}$	≥ 0	≥ 0	≥ 0
$\overline{p} \cap q$	0	≥ 0	≥ 1
$p \cap \overline{q}$	≥ 0	0	≥ 0
$p \cap q$	≥ 0	≥ 0	1

The theory $\mathcal{T}_1 \cup \mathcal{T}_2$ imposes $p \cap \overline{q}$ to be empty, in other words $p \subseteq q$. Moreover, the cardinality of $p \cap q$ is 1, and so it implies that the cardinality of p is 1. Hence, the set

$$\mathcal{T}_1 \cup \mathcal{T}_2 \cup \{p(a), p(b), a \neq b\}$$

is unsatisfiable. As a final comment, there could be theories using directly the Venn cardinalities as integer variables. For instance, imagine a constraint stating $|p| > 1$ in a theory including linear arithmetic on integers. This would of course be unsatisfiable with $\mathcal{T}_1 \cup \mathcal{T}_2$.

7   Conclusion

The notion of gentleness was initially presented as a tool to combine non-stably infinite disjoint theories. In this paper, we have introduced a notion of $\mathcal{P}$-gentleness which is well-suited for combining theories sharing (besides constants and the equality) only unary predicates in a set $\mathcal{P}$. The major contributions of this paper are that the Löwenheim theories and BSR theories are $\mathcal{P}$-gentle.

A corollary is that the non-disjoint union (sharing unary predicates) of Löwenheim theories, BSR theories, and decidable theories accepting further constraints of the form $\forall x.\,((\neg)p_1(x) \wedge \ldots (\neg)p_n(x)) \Rightarrow (x = a_1 \vee \ldots x = a_m)$ is decidable.

Our combination method is limited to shared unary predicates. Unfortunately, the theoretical limitations are strong for a framework sharing predicates with larger arities: for instance even the guarded fragment with two variables and transitivity constraints is undecidable [10], although the guarded fragment (or first-order logic with two variables) is decidable, and transitivity constraints can be expressed in BSR. The problem of combining theories with only a shared dense order has however been successfully solved [12,14]. In that specific case, there is again an implicit infiniteness argument that could be possibly expressed as a form of extended gentleness, to reduce the isomorphism construction problem into solving some appropriate extension of cardinality constraints. A clearly challenging problem is to identify an appropriate extended notion of gentleness for some particular binary predicates.

Also in future works, the reduction approach (Löwenheim and BSR theories can be simplified to a subset of Löwenheim) may be useful as a simplification procedure for sets of formulas that can be seen as non-disjoint (sharing unary predicates only) combinations of BSR, Löwenheim theories and an arbitrary first-order theory: this would of course not provide a decision procedure, but refutational completeness can be preserved. More generally we also plan to study how superposition-based satisfiability procedures could benefit from a non-disjoint (sharing unary predicates) combination point of view. In particular, superposition-based satisfiability procedures could be used as deductive engines with the capability to exchange constraints à la Nelson-Oppen.

The results here are certainly too combinatorially expensive to be directly applicable. However, this paper paves the theoretical grounds for mandatory further works that would make such combinations practical. There are important incentives since the BSR and Löwenheim fragments are quite expressive: for instance, it is possible to extend the language of SMT solvers with sets and cardinalities. Many formal methods are based on logic languages with sets. Expressive decision procedures (even if they are not efficient) including e.g. sets and cardinalities will help proving the often small but many verification conditions stemming from these applications.

References

1. Areces, C., Fontaine, P.: Combining theories: The Ackerman and Guarded fragments. In: Tinelli, C., Sofronie-Stokkermans, V. (eds.) FroCoS 2011. LNCS (LNAI), vol. 6989, pp. 40–54. Springer, Heidelberg (2011)
2. Armando, A., Bonacina, M.P., Ranise, S., Schulz, S.: New results on rewrite-based satisfiability procedures. ACM Trans. Comput. Log. 10(1) (2009)
3. Armando, A., Ranise, S., Rusinowitch, M.: A rewriting approach to satisfiability procedures. Inf. Comput. 183(2), 140–164 (2003)

4. Barrett, C., Sebastiani, R., Seshia, S.A., Tinelli, C.: Satisfiability modulo theories. In: Biere, A., Heule, M.J.H., van Maaren, H., Walsh, T. (eds.) Handbook of Satisfiability. Frontiers in Artificial Intelligence and Applications, vol. 185, ch. 26, pp. 825–885. IOS Press (February 2009)

5. Bonacina, M.P., Ghilardi, S., Nicolini, E., Ranise, S., Zucchelli, D.: Decidability and undecidability results for Nelson-Oppen and rewrite-based decision procedures. In: Furbach, U., Shankar, N. (eds.) IJCAR 2006. LNCS (LNAI), vol. 4130, pp. 513–527. Springer, Heidelberg (2006)

6. Börger, E., Grädel, E., Gurevich, Y.: The Classical Decision Problem. In: Perspectives in Mathematical Logic. Springer, Berlin (1997)

7. Chocron, P., Fontaine, P., Ringeissen, C.: A Gentle Non-Disjoint Combination of Satisfiability Procedures (Extended Version). Research Report 8529, Inria (2014), http://hal.inria.fr/hal-00985135

8. Dreben, B., Goldfarb, W.D.: The Decision Problem: Solvable Classes of Quantificational Formulas. Addison-Wesley, Reading (1979)

9. Fontaine, P.: Combinations of theories for decidable fragments of first-order logic. In: Ghilardi, S., Sebastiani, R. (eds.) FroCoS 2009. LNCS (LNAI), vol. 5749, pp. 263–278. Springer, Heidelberg (2009)

10. Ganzinger, H., Meyer, C., Veanes, M.: The two-variable guarded fragment with transitive relations. In: Logic In Computer Science (LICS), pp. 24–34. IEEE Computer Society (1999)

11. Ganzinger, H., Nivelle, H.D.: A superposition decision procedure for the guarded fragment with equality. In: Logic In Computer Science (LICS), pp. 295–303. IEEE Computer Society Press (1999)

12. Ghilardi, S.: Model-theoretic methods in combined constraint satisfiability. Journal of Automated Reasoning 33(3-4), 221–249 (2004)

13. Gurevich, Y., Shelah, S.: Spectra of monadic second-order formulas with one unary function. In: Logic In Computer Science (LICS), pp. 291–300. IEEE Computer Society, Washington, DC (2003)

14. Manna, Z., Zarba, C.G.: Combining decision procedures. In: Aichernig, B.K., Maibaum, T. (eds.) Formal Methods at the Crossroads. From Panacea to Foundational Support. LNCS, vol. 2757, pp. 381–422. Springer, Heidelberg (2003)

15. Nelson, G., Oppen, D.C.: Simplification by cooperating decision procedures. ACM Trans. on Programming Languages and Systems 1(2), 245–257 (1979)

16. Nicolini, E., Ringeissen, C., Rusinowitch, M.: Combinable extensions of Abelian groups. In: Schmidt, R.A. (ed.) CADE 2009. LNCS (LNAI), vol. 5663, pp. 51–66. Springer, Heidelberg (2009)

17. Nicolini, E., Ringeissen, C., Rusinowitch, M.: Combining satisfiability procedures for unions of theories with a shared counting operator. Fundam. Inform. 105(1-2), 163–187 (2010)

18. Ramsey, F.P.: On a Problem of Formal Logic. Proceedings of the London Mathematical Society 30, 264–286 (1930)

19. Ranise, S., Ringeissen, C., Zarba, C.G.: Combining data structures with nonstably infinite theories using many-sorted logic. In: Gramlich, B. (ed.) FroCos 2005. LNCS (LNAI), vol. 3717, pp. 48–64. Springer, Heidelberg (2005)

20. Ringeissen, C., Senni, V.: Modular termination and combinability for superposition modulo counter arithmetic. In: Tinelli, C., Sofronie-Stokkermans, V. (eds.) FroCoS 2011. LNCS (LNAI), vol. 6989, pp. 211–226. Springer, Heidelberg (2011)

21. Sofronie-Stokkermans, V.: Locality results for certain extensions of theories with bridging functions. In: Schmidt, R.A. (ed.) CADE 2009. LNCS (LNAI), vol. 5663, pp. 67–83. Springer, Heidelberg (2009)

22. Sofronie-Stokkermans, V.: On combinations of local theory extensions. In: Voronkov, A., Weidenbach, C. (eds.) Ganzinger Festschrift. LNCS, vol. 7797, pp. 392–413. Springer, Heidelberg (2013)

23. Suter, P., Dotta, M., Kuncak, V.: Decision procedures for algebraic data types with abstractions. In: Hermenegildo, M.V., Palsberg, J. (eds.) Principles of Programming Languages (POPL), pp. 199–210. ACM (2010)

24. Tinelli, C., Ringeissen, C.: Unions of non-disjoint theories and combinations of satisfiability procedures. Theoretical Computer Science 290(1), 291–353 (2003)

25. Tinelli, C., Zarba, C.G.: Combining non-stably infinite theories. Journal of Automated Reasoning 34(3), 209–238 (2005)

26. Wies, T., Piskac, R., Kuncak, V.: Combining theories with shared set operations. In: Ghilardi, S., Sebastiani, R. (eds.) FroCoS 2009. LNCS (LNAI), vol. 5749, pp. 366–382. Springer, Heidelberg (2009)

27. Zarba, C.G.: Combining sets with cardinals. J. Autom. Reasoning 34(1), 1–29 (2005)

28. Zhang, T., Sipma, H.B., Manna, Z.: Decision procedures for term algebras with integer constraints. Inf. Comput. 204(10), 1526–1574 (2006)

A   An Extension of Ramsey's Theorem

We define an *n-subset* of S to be a subset of n elements of S. An *n-hypergraph* of S is a set of n-subsets of S. In particular, a 2-hypergraph is an (undirected) graph. The *complete n-hypergraph* of S is the set of all n-subsets of S, and its *size* is the cardinality of S. An n-hypergraph G is *colored* with c colors if there is a coloring function that assigns one color to every n-subset in G. In particular, a colored 2-hypergraph (that is, a colored graph), is a graph where all edges are assigned a color. Consider a set S of elements partitioned into disjoint regions $R = \{R_1, \ldots R_m\}$. We say that a set $S' \subseteq S$ has *region size* larger than x and note $|S'|_R \geq x$ if $|S' \cap R_i| \geq x$ for all $i \in \{1, \ldots, m\}$. We also say that an n-hypergraph is *region-monochromatic* if the color of each hyperedge only depends on the number of elements belonging to each region. Two hyperedges are said of the *same kind* if they have the same number of elements in each region; all hyperedges of the same kind of a region-monochromatic hypergraph thus have the same color. The following extension[1] of Ramsey's Theorem holds:

Theorem 7. *There exists a computable function f such that,*

- *for every number of colors c*
- *for every $n, N \in \mathbb{N}$*
- *for every complete n-hypergraph G on S colored with c colors*

if $|S|_R \geq f(n, N, c)$, then there exists a complete region-monochromatic n-sub-hypergraph of G on some $S' \subseteq S$ with $|S'|_R \geq N$.

Proof. The full proof can be found in [7].   □

---

[1] The classical Ramsey's Theorem is the case with only one region.

A Rewriting Strategy to Generate Prime Implicates in Equational Logic

Mnacho Echenim[1,2], Nicolas Peltier[1,4], and Sophie Tourret[1,3]

[1] Grenoble Informatics Laboratory
[2] Grenoble INP - Ensimag
[3] Université Grenoble 1
[4] CNRS

Abstract. Generating the prime implicates of a formula consists in finding its most general consequences. This has many fields of application in automated reasoning, like planning and diagnosis, and although the subject has been extensively studied (and still is) in propositional logic, very few have approached the problem in more expressive logics because of its intrinsic complexity. This paper presents one such approach for flat ground equational logic. Aiming at efficiency, it intertwines an existing method to generate all prime implicates of a formula with a rewriting technique that uses atomic equations to simplify the problem by removing constants during the search. The soundness, completeness and termination of the algorithm are proven. The algorithm has been implemented and an experimental analysis is provided.

1 Introduction

The automated generation of the prime implicates of a formula (i.e. its most general consequences, shortened as p.i. from this point on) has been a topic of interest in automated reasoning because of its various applications (e.g. program analysis, knowledge representation...). Generating the p.i. of a formula allows one to extract relevant information, and is useful for instance to remove redundant variables, to simplify the formula, to identify sufficient conditions, etc. The notion of p.i. and their duals, prime implicants, were first introduced for propositional logic in 1955 [21] and from that point on, a lot of algorithms were designed for their computation. Efficient algorithms for computing p.i. in propositional logic use either variants of the resolution rule [5,12,13,24] or decomposition-based approaches in the spirit of the DPLL method [3,4,10,11,15,17,22,23]. However, most applications of automated reasoning (e.g. in program verification) require the handling of properties and theories that cannot be expressed in propositional logic, hence the need to extend tools such as p.i. generators to more expressive logics (such as quantifier-free equational logic or first-order logic). One of the strong points of the decomposition methods is that they can be applied to all kinds of formulæ, while resolution-based methods can only deal with formulæ in clausal normal form. On the other hand, decomposition methods are designed to handle only finitely many propositional variables, which greatly impairs the

S. Demri, D. Kapur, and C. Weidenbach (Eds.): IJCAR 2014, LNAI 8562, pp. 137–151, 2014.

possibility of extending such algorithms to more expressive logics. However, extending resolution-based methods is not a trivial task either, since crucial characteristics such as completeness or termination of propositional algorithms are lost in more expressive logics, rendering them useless. Thus, such an extension must be designed very carefully, which explains why a domain that was extensively studied over the past sixty years contains so few results outside of propositional logic. Nevertheless, methods have been devised for generating p.i. in first-order logic, based mostly either on first-order resolution [16] or the sequent calculus [18]. These methods can handle equational reasoning, the domain targeted in this paper, by adding equality axioms. Other techniques also exist, such as [14] which proposes a built-in method for handling equational reasoning based on an analysis of unification failures. These approaches are very general but not well-suited for real-world applications since termination is not ensured (even for standard decidable classes); furthermore they include no technique for removing equational redundancies, which is a major source of inefficiency. [27] uses the superposition calculus [1] to generate *positive* and *unit* p.i. for specific theories. However, as shown in [9,14], the superposition calculus is not complete for non-positive or non-unit p.i. Some work has also been done on domains near first-order logic: [2] focuses on a modal logic and presents an extensive study of p.i. in this context. [7] devises a technique for computing some specific prime implicants called *minimum satisfying assignments* in several theories, provided there exists a decision procedure for testing the satisfiability of first-order formulæ in the considered theory (e.g. Presburger arithmetic). This technique is applied in [6] to perform a semi-automated bug detection. [25,26] propose an approach to synthesize p.i.-like constraints ensuring that a system satisfies some invariant or safety properties. This approach relies on external provers to check satisfiability of first-order formulæ in some base theories. It is very generic and modular, however no automated method is presented to simplify the obtained constraints.

This paper focuses on the generation of p.i. in function-free equational logic: the considered formulæ are boolean combinations of equations between constants. This research stems from the design of a method for abductive reasoning in first-order logic [8], in which a superposition-based calculus is devised to generate function-free consequences of first-order formulæ. This superposition procedure is sound and deduction-complete (for ground function-free implicates; the absence of function symbols is not really restrictive, since functions can be reduced to equalities by adding substitutivity axioms [20]) but the obtained formulæ contain many redundancies which make them hard to analyze. The automated generation of their p.i. allows for the elimination of these redundancies and the computation of a minimal representation of the given formulæ. Note that in [25] a similar lack of parsimony is also identified as one of the main issues. In [9], we designed a tool that is capable of efficiently finding the most general consequences of a quantifier-free equational formula with no function symbols. The proposed algorithm is somewhat similar to the resolution-based p.i. generation method for propositional logic of [5] in its structure, but uses built-in techniques to handle the properties of the equality predicate. This affects both the representation of clauses, i.e. the way they are stored and tested for redundancy, and their generation: instead of using the resolution method, new inference rules that can be viewed as a form of relaxed paramodulation are defined. An implementation of this algorithm, named KPARAM, was compared to state-of-the-art propositional p.i. generation tools by respectively feeding in ground equational formulæ and equivalent propositional abstractions. The KPARAM tool outperforms the propositional one in most cases, but it performs badly on some problems. A careful analysis of the experimental results has shown that this is due, for a large part, to the lack of an efficient technique for handling equational simplifications. Obviously, a most natural and efficient way of handling an equation $a \simeq b$ is to uniformly replace one of the terms, say a, by the other, b, thus yielding a simpler problem. It is clear that this operation preserves satisfiability, because a formula $F \wedge a \simeq b$ is satisfiable iff $F[b/a]$ is. The application of such a strategy in the context of p.i. generation raises two important and related issues. First, how to reconstruct the set of implicates of the original problem $F \wedge a \simeq b$, from that of $F[b/a]$? This is not obvious, since, although rewriting preserves satisfiability, it does not in general preserve the set of implicates, as shown later. Second, how to intertwine the systematic application of the rewriting operation with the overall algorithm used to handle the clauses incrementally? This last point is important because the equational simplifications are not transparent to the overall process of p.i. generation. Adjustments are needed, that obviously should not counterbalance the gain of handling equations. In this paper, we investigate both issues and provide solutions for each of them, yielding a much more efficient algorithm for generating implicates of ground equational formulæ. This algorithm is proved to be sound, terminating and complete, thus generating all implicates of the input up to redundancy in a finite time. Experimental comparisons show that equational propagation improves the performances of the algorithm by several orders of magnitude.

Structure of the Paper. In Sect. 2 the original strategy from [9] is introduced along with the notations necessary to follow the technical part of the article. Section 3 is a presentation of the rewriting algorithm used in the first step of the strategy and the theoretical properties (completeness and termination) of the global algorithms are provided in Sect. 4. An experimental comparison is conducted in Sect. 5 and the final section contains a summary of the obtained results, along with some lines of future work. Due to space restriction, the proofs are omitted (a technical report containing the proofs is available on the authors web pages).

2   On Equational Logic and Prime Implicate Generation

This section contains the necessary definitions about equational logic along with a simplified presentation of the starting point of our work, namely the p.i. generation algorithm of [9].

2.1 Equational Logic

Let Σ be a finite set of *constants* denoted by $a,b,c...$ We assume a total order $\prec$ on Σ. We also write $a \succ b$ if $b \prec a$. A *literal* l is either an *atom* (or *equation*) $a \simeq b$ (where $a, b \in \Sigma$ and $\simeq$ is the symbol for semantic equality), or the negation of an atom (or *disequation*) $a \not\simeq b$. A literal written $a \bowtie b$ denotes either $a \simeq b$ or $a \not\simeq b$ and by commutativity $a \bowtie b$ and $b \bowtie a$ are considered equivalent. The literal l^c denotes the complement of l, i.e. $a \not\simeq b$ if $l = a \simeq b$ and $a \simeq b$ if $l = a \not\simeq b$. A *clause* C is a disjunction (or multiset) of literals. C^+ is the clause composed of the atoms in C and C^- is composed of the disequations in C. A clause is *positive* if $C^- = \emptyset$. The empty clause is denoted by $\square$ and $|C|$ is the number of literals in C. An *atomic clause* is a positive unit clause. A *formula* S is a set of clauses. For every clause C, $\neg C$ denotes the formula $\{\{l^c\} \mid l \in C\}$.

An *equational interpretation* $\mathcal{I}$ is a partition of Σ into equivalence classes. Given two constant symbols a and b, we write $a =_{\mathcal{I}} b$ if a and b belong to the same equivalence class in $\mathcal{I}$, and in this case we say that $a \simeq b$ is *true* in $\mathcal{I}$ (respectively, if $a \neq_{\mathcal{I}} b$ then $a \not\simeq b$ is true in $\mathcal{I}$). This notation is extended to literals: $a \bowtie b =_{\mathcal{I}} c \bowtie d$ means that both literals have the same sign and that either $a =_{\mathcal{I}} c$ and $b =_{\mathcal{I}} d$ or $a =_{\mathcal{I}} d$ and $b =_{\mathcal{I}} c$ (this implies that both literals have the same truth value in $\mathcal{I}$, but the converse does not hold). A clause C is true in $\mathcal{I}$ if C contains at least one literal that is true in $\mathcal{I}$ and a formula S is true in $\mathcal{I}$ if all clauses in S are true in $\mathcal{I}$. Let E represent either a literal, a clause or a formula, then $\mathcal{I} \models E$ means that E is true in $\mathcal{I}$ and in this case $\mathcal{I}$ is called a *model* of E. The notation $E \models E'$ means that all the models of E are also models of E'. If $E \models E'$ and $E' \models E$ then we write $E \equiv E'$. A *tautology* is a clause that is true in all equational interpretations. Unless stated otherwise, only non-tautological clauses will be considered. A *contradiction*, e.g. $\square$ or $a \not\simeq a$, is a clause with no model. To each clause C we associate a special interpretation $\mathcal{I}_C$ such that $a =_{\mathcal{I}_C} b$ iff $\neg C \models a \simeq b$. To lighten notations we write $a =_C b$ instead of $a =_{\mathcal{I}_C} b$. Note that $\mathcal{I}_C \models C$ iff C is a tautology. The following related notations are also used: $[a]_C \overset{\text{def}}{=} \{b \in \Sigma \mid a =_C b\}$ is the equivalence class of a in $\mathcal{I}_C$ and $a_{|C} \overset{\text{def}}{=} \min_{\succ} [a]_C$ is the *representative* of the class $[a]_C$.

In the original method of p.i. generation from [9], a critical point is redundancy detection. To deal with the constraints induced by the equality axioms, we define a redundancy criterion named *eq-subsumption*, essentially equivalent to semantic entailment.

Definition 1. *Let C, D be two clauses. The clause D eq-subsumes C (written $D \leq_{eq} C$) iff the two following conditions hold:*

- *for all $a, b \in \Sigma$, if $\neg D \models a \simeq b$ then $\neg C \models a \simeq b$;*
- *for every positive literal $l \in D$, there exists a literal $l' \in C$ such that $l =_C l'$.*

$D <_{eq} C$ *means that $D \leq_{eq} C$ and $C \not\leq_{eq} D$. If S, S' are formulæ, we write $S \leq_{eq} C$ if $\exists D \in S$, $D \leq_{eq} C$ and $S \leq_{eq} S'$ if $\forall C \in S'$, $S \leq_{eq} C$. A clause C is redundant in S if either C is a tautology or there exists a clause $D \in S$ such that $D <_{eq} C$. A clause set S is* subsumption-minimal *if it contains no redundant clause, i.e. $\forall C \in S$, C is not redundant in S.*

Example 2. Let $C = a \not\simeq b \vee b \not\simeq c \vee a \simeq d$ and $D = a \not\simeq c \vee b \simeq d$. Then $\mathcal{I}_C = \{\{a, b, c\}, \{d\}\}$ and $\mathcal{I}_D = \{\{a, c\}, \{b\}, \{d\}\}$, thus $D \leq_{eq} C$ because $\{a, c\} \subseteq \{a, b, c\}$ and $a \simeq d =_C c \simeq d$. On the other hand $a \not\simeq e \vee a \simeq d \not\leq_{eq} C$ and $a \not\simeq b \vee b \simeq e \not\leq_{eq} C$, respectively because $\{a, e\} \not\subseteq \{a, b, c\}$ and because $b \simeq e \neq_C a \simeq d$.

This criterion offers a syntactic method to detect equational entailment.

Theorem 3. (Th. 8 of [9]) *Let C and D be two clauses. If C is not a tautology then $D \models C$ iff $D \leq_{eq} C$.*

To reduce the work of the redundancy detection algorithms, a normal form is defined for clauses which projects all equivalent clauses onto a single one, thus drastically reducing the number of clauses to be considered.

Definition 4. *The normal form of a non-tautological clause C is:*

$$C_\downarrow \stackrel{def}{=} \left(\bigvee_{a \in \Sigma, a \neq a_{\downarrow C}} a \not\simeq a_{\downarrow C} \right) \vee \left(\bigvee_{a \simeq b \in C} a_{\downarrow C} \simeq b_{\downarrow C} \right)$$

and all the literals in $C_\downarrow$ occur only once. A formula S is in normal form (denoted by $S_\downarrow$) iff all its non-tautological clauses are in normal form. The normal form of a set of atomic clauses $U = \{a_i \simeq b_i\}_{i \in \{1...n\}}$ is the set of atomic clauses $U_\downarrow \stackrel{def}{=} \{a'_j \simeq b'_j\}_{j \in \{1...m\}}$ such that $(\bigvee_{i=1}^{n} a_i \not\simeq b_i)_\downarrow = \bigvee_{j=1}^{m} a'_j \not\simeq b'_j$.

Proposition 5. (Prop. 4 of [9]) *For every non-tautological clause C, $C_\downarrow$ is equivalent to C. Furthermore, if D and C are equivalent and non-tautological then $C_\downarrow = D_\downarrow$.*

Example 6. Let $C = a \not\simeq b \vee b \not\simeq c \vee a \not\simeq c \vee b \simeq d$. If $a \succ b \succ c \succ d$ then the normal form of C is $C_\downarrow = b \not\simeq c \vee a \not\simeq c \vee c \simeq d$.

In all algorithms, we assume that the formulæ are always subsumption-minimal. If a formula S is described by set operations, the redundant clauses are automatically removed. The process is straightforward for all operations except the difference operation, which is defined as follows: for two formulæ S_1 and S_2, $S_1 \backslash S_2 \stackrel{def}{=} \{C \in S_1 \mid \forall D \in S_2, D \not\models C\}$. Note that if $S_2 \subseteq S_1$ then, since S_1 is subsumption-minimal, only clauses actually belonging to S_2 are removed from S_1. The subject of clause manipulations is not developed further in this article since it is not essential for the understanding of the present paper and there is no significant change in their use w.r.t. the algorithm described in [9].

2.2   Implicate Generation

Definition 7. *A clause C is an* implicate *of a formula S if $S \models C$. An implicate C is a* prime implicate *of S if C is not a tautology, and for every clause D such that $S \models D$, either $D \not\models C$ or $C \models D$. Given a formula S, $PI(S)$ is the set of all the p.i. of S.*

Intuitively, a p.i. of a formula is a consequence that is as general as possible. If any information is removed from it, it is no longer a consequence of the original formula.

Example 8. Let $S = \{a \simeq b \vee d \simeq e, a \simeq c, d \not\simeq e\}$. Both $c \simeq b$ and $c \simeq b \vee d \simeq e$ are implicates of S, but only $c \simeq b$ is a p.i., because $c \simeq b <_{eq} c \simeq b \vee d \simeq e$.

A standard method for testing the satisfiability of equational clause sets is the superposition calculus. This calculus is refutationally complete, meaning that it generates a contradiction from every unsatisfiable set. It is however not complete for deduction, since we may have $S \models C$ even if C cannot be generated from S.

Example 9. Consider $S = \{a \simeq b, c \not\simeq d \vee a \not\simeq d\}$ with $a \succ b \succ c \succ d$. By superposition, the only new implicate that can be generated is $c \not\simeq d \vee b \not\simeq d$ but there are other implicates of S such as $b \not\simeq c \vee a \not\simeq d$.

The implicates of an equational formula are generated using a relaxed paramodulation calculus that permits the replacement of arbitrary constants (instead of identical ones), by adding equality conditions in the resulting clause. For example, the paramodulation rule usually applies between a clause $C[a]$ (a clause C containing the constant a) and $a \simeq b \vee D$, yielding the clause $C[b] \vee D$. In our setting, the clauses $C[a']$ where $a \neq a'$ and $a \simeq b \vee D$ generate $a \not\simeq a' \vee C[b] \vee D$ which can be understood as "if $a \simeq a'$ holds then so does $C[b] \vee D$". Formally, the following rules define the so-called $\mathcal{K}$-paramodulation calculus.

$$\text{Paramodulation (P):} \quad \frac{a \simeq b \vee C \qquad a' \simeq c \vee D}{a \not\simeq a' \vee b \simeq c \vee C \vee D}$$

$$\text{Factorization (F):} \quad \frac{a \simeq b \vee a' \simeq b' \vee C}{a \simeq b \vee a \not\simeq a' \vee b \not\simeq b' \vee C}$$

$$\text{Negative Multi-Paramodulation (M):} \quad \frac{\bigvee_{i=1}^{n}(a_i \not\simeq b_i) \vee P_1 \qquad c \simeq d \vee P_2}{\bigvee_{i=1}^{n}(a_i \not\simeq c \vee d \not\simeq b_i) \vee P_1 \vee P_2}$$

The rule M can be applied to one or several disequations at once.

Example 10. Let $C = a \not\simeq d \vee c \not\simeq e \vee d \simeq e$ and $D = a \simeq b$. Using M, it is possible to generate the clause $(a \not\simeq a \vee)b \not\simeq d \vee c \not\simeq e \vee d \simeq e$ by selecting only $a \not\simeq d$ in C and the clause $(a \not\simeq a \vee)b \not\simeq d \vee a \not\simeq c \vee b \not\simeq e \vee d \simeq e$ by selecting both $a \not\simeq d$ and $c \not\simeq e$.

Theorem 11. (Th. 13 of [9]) *The $\mathcal{K}$-paramodulation calculus is complete for deduction, i.e. it generates all the implicates of any formula up to redundancy.*

A formula S is *saturated up to redundancy* iff all clauses that can be inferred from premises in S using the rules P, F or M are either in S or redundant w.r.t. S. Note that unlike the superposition calculus, no ordering restrictions are imposed on the premises. This is needed for completeness, as shown in the following example.

Example 12. Let $C = b \simeq c$ and $D = a \simeq c$, with $a \succ b \succ c$. Usual ordering restrictions prevent the generation of the clause $a \simeq b$ because c, being the smallest constant, cannot be replaced by b.

In the algorithms, the following additional notation related to the $\mathcal{K}$-paramodulation calculus is needed.

Definition 13. *Let S be a formula, and C be a clause. $S_{\vdash i, C}$ is the set of all clauses obtained from $S \cup \{C\}$ by exactly i steps of $\mathcal{K}$-paramodulation such that at least one parent of each $\mathcal{K}$-paramodulation step is C. Similarly, we denote by $S_{\vdash C}$ the set of all clauses (up to redundancy) generated by any number of $\mathcal{K}$-paramodulation steps from $S \cup \{C\}$ where C is always one of the parents.*

Algorithm 1. KPARAM(S)

$T := \emptyset$
$S_1 := S$
while $S_1 \neq \emptyset$ **do**
    Choose a clause $C \in S_1$
    if $T \not\leq_{eq} C$ **then**
        $T := T \cup \{C\} \backslash \{D \in T \mid C \leq_{eq} D\}$
        $S_1 := (S_1 \cup T_{\vdash 1, C}) \backslash \{C\}$
    else
        $S_1 := S_1 \backslash \{C\}$
    end if
end while
return T

To generate only the p.i. of a formula, the redundant implicates must be deleted as soon as possible to avoid using them to generate other redundant implicates. For this purpose, the algorithm KPARAM (Algorithm 1, originally proposed in [9]) selects implicates one at a time from a *waiting set* S_1 and uses the selected clause in $\mathcal{K}$-paramodulation inferences with previously selected clauses to generate new implicates. The newly generated clauses are stored in the waiting set and a new implicate can then be selected. Redundant clauses found during the process are removed. The non-redundant used clauses are stored in the *processed set* T[1]. This procedure was proved to be sound, complete and terminating.

3   Atomic Rewriting

To improve the performance of the algorithm described in Sect. 2, we incorporate a rewriting strategy, atomic rewriting (otherwise known as equational simplification), to the process of implicate generation. It simplifies the problem by reducing the number of constants it contains. The underlying principle is

---

[1] KPARAM is an instance of the *given clause* algorithm in the *Otter* variant [19].

simple: assume that an atomic clause $a \simeq b$ is an implicate of a formula S. It is clear that for every model $\mathcal{M}$ of S, necessarily $a =_{\mathcal{M}} b$. Since a and b are always equal, they can be substituted with each other and it is actually possible to entirely replace one of these constants with the other in the formula, storing the atom $a \simeq b$ apart to avoid any loss of information. Note that the removal of an atom can lead to the generation of new ones as shown in the following example.

Example 14. Consider the formula $S = \{a \simeq b, a \not\simeq b \vee c \not\simeq d, a \not\simeq c \vee a \simeq e, b \not\simeq c \vee b \simeq e, c \simeq a \vee c \simeq b\}$. Using the atom $a \simeq b$, the formula S can be rewritten into $S' = \{b \simeq b, b \not\simeq b \vee c \not\simeq d, b \not\simeq c \vee b \simeq e, b \not\simeq c \vee b \simeq e, c \simeq b \vee c \simeq b\}$ and further simplified into $S'' = \{c \not\simeq d, b \not\simeq c \vee b \simeq e, c \simeq b\}$. Since S'' contains the atom $c \simeq b$, it can in turn be rewritten in $S^{(3)} = \{c \not\simeq d, b \not\simeq b \vee b \simeq e\}$, etc., until only $c \not\simeq d$ remains in the formula.

We introduce the following notations:

Notation 15. *Let S be a formula and U be a set of atomic clauses:*

- *for a and b constants with $a \succ b$, $S[b/a]$ is the formula S where every occurrence of a is replaced by b,*
- *$S[U]$ is the set S where every constant a is replaced by $\min\{a' \mid U \models a \simeq a'\}$. For example, if U contains $a \simeq b$ and $a \simeq c$ with $a \succ b \succ c$ then both a and b are replaced by c in $S[U]$.*

In what follows, we will invoke a procedure ATOMREWRITE that recursively removes the atomic clauses appearing in a formula and rewrites all the remaining clauses according to the clauses extracted, until no atom remains. ATOMREWRITE(S) returns the pair $\langle S', U \rangle$ made of the rewritten formula S' and a set U of extracted atomic clauses such that $S \models U$, and $S' = S[U]$ where S' contains no atomic clause. Note that U does not necessarily contain all the atomic clauses that are logical consequences of S. However, it necessarily contains all those occurring in S and those generated by atomic rewriting.

Example 16. Assume that $S = \{a \simeq b, a \not\simeq b \vee c \not\simeq d, c \simeq d \vee e \simeq f\}$. Then invoking ATOMREWRITE(S) returns $S' = \{c \not\simeq d, c \simeq d \vee e \simeq f\}$ and $U = \{a \simeq b\}$, even though it is simple to verify that $S \models e \simeq f$.

New "hidden" atomic implicates like the one in the previous example can be generated at any iteration of the strategy, hence the atomic rewriting should be applicable not only on the initial clause set but also on the newly generated clauses. However, in order to preserve completeness, some clauses occurring in the processed set must then be transferred back to the waiting set (i.e. resp. T_1 and S_1 in Algorithms 1 and 2), otherwise some inferences involving the rewritten clauses of the processed set can never occur. A straightforward way to ensure completeness would be to transfer all clauses back to the waiting set, but this yields a very inefficient algorithm. The following definition introduces a refined criterion that strongly reduces the number of clauses that must be reprocessed.

Definition 17. *A clause D is $\langle a, b\rangle$-neutral if $D^+[b/a] = (D[b/a]_\downarrow)^+$. The function* NEUTRAL$(D, a, b)$ *returns true iff D is $\langle a, b\rangle$-neutral.*

This property means that the replacement of a by b does not affect the representatives of the equivalence classes occurring in the positive part of a clause, even if it contains both a and b.

Example 18. Consider the clauses $C = a \not\simeq c \vee b \not\simeq d \vee c \simeq e$ and $D = a \not\simeq c \vee b \not\simeq c \vee c \simeq d$, with $a \succ b \succ c \succ d$. C is not $\langle a, b\rangle$-neutral since $C^+[b/a] = c \simeq e$ while $(C[b/a]_\downarrow)^+ = d \simeq e$. On the other hand D is $\langle a, b\rangle$-neutral because $D^+[b/a] = (D[b/a]_\downarrow)^+ = c \simeq d$.

In the procedure SPLITATOMREWRITE (Algorithm 2) we therefore assume that every non-$\langle a, b\rangle$-neutral clause occurring in T_1 is transferred back to S_1 after rewriting. This procedure takes as an input a pair (T, S) of processed set/waiting set, and returns the new sets after rewriting every atomic clause they contain. $\langle a, b\rangle$-neutrality is the key ensuring the completeness of the algorithm. The informal and intuitive justification is that, if C is $\langle a, b\rangle$-neutral, then all inferences that can be performed on the clause $C[b/a]$ can be "simulated" by inferences with descendants of C. Hence the clause $C[b/a]$ does not need to be considered again.

Algorithm 2. SPLITATOMREWRITE(T, S)

$U_1 := \{a \simeq b \in T \cup S\}_\downarrow$
$T_1 := T$ // T_1 *is the processed set*
$S_1 := S$ // S_1 *is the waiting set*
$U := \emptyset$
while $U_1 \neq \emptyset$ **do**
     extract a clause $a \simeq b$ from U_1 and put it into U
     $T_2 := \{D \mid \exists D' \in T_1, D = (D'[b/a])_\downarrow \wedge$ NEUTRAL$(D', a, b)\}$
     $S_2 := (S_1[b/a])_\downarrow \cup \left\{(T_1[b/a])_\downarrow \setminus T_2\right\}$
     $U_1 := (U_1 \cup \{u \simeq v \in T_2 \cup S_2\})_\downarrow$
     $T_1 := T_2$
     $S_1 := S_2$
end while
return $\langle T_1, S_1, U\rangle$

When initializing U_1, taking atoms directly from S is possible because every unit implicate of a non-contradictory formula is one of its p.i., since no clause other than $\square$ and itself subsumes it. Note that the replacement of a by b implicitly deletes the clause $a \simeq b$ from the sets T_2 and S_2.

Lemma 19. SPLITATOMREWRITE *terminates.*

4   Prime Implicate Generation: A New Algorithm

The new algorithm combines $\mathcal{K}$-paramodulation with atomic rewriting to simplify the p.i. computation on the fly. This process is presented in Subsection 4.1 and results in the generation of the set of non-atomic p.i. of the simplified problem together with the set of atomic clauses collected during the search. The recovery of the p.i. of the original formula is described in Subsection 4.2. From this point on, any clause appearing in an algorithm is assumed to be in normal form.

4.1   Integration of the Atomic Rewriting

Algorithm 3. SATURATERW(S)

$\langle S_1, U_1 \rangle := $ ATOMREWRITE(S)
$T_1 := \emptyset$
while $S_1 \neq \emptyset$ **do**
   Choose a clause $C \in S_1$
   $S_2 := S_1 \setminus \{C\}$
   if $T_1 \not\leq_{eq} C$ **then**
      $T_2 := T_1 \cup \{C\}$
      $R_1 := (T_2)_{\vdash 1, C}$
      $\langle T_3, S_3, U_2 \rangle := $ SPLITATOMREWRITE$(T_2, (S_2 \cup R_1))$
      $U_1 := U_1 \cup U_2$
      $T_1 := T_3$
      $S_1 := S_3$
   else
      $S_1 := S_2$
   end if
end while
return $\langle T_1, U_1 \rangle$

As can be seen in Algorithm 3, atomic rewritings are added to the original procedure both during the initialization phase, where a call to ATOMREWRITE removes the atomic clauses occurring in the original formula, and at each iteration of the main loop, where SPLITATOMREWRITE is used. SATURATERW(S) returns the pair $\langle T, U \rangle$ where T is the set of clauses eventually obtained by saturation and U is the set of atomic clauses collected during proof search (and deleted from the search space by ATOMREWRITE or SPLITATOMREWRITE).

Lemma 20. *The algorithm* SATURATERW *terminates.*

Theorem 21. *Let S be a formula. If $\langle T, U \rangle = $ SATURATERW(S), then T is saturated up to redundancy and contains no positive unit clauses while U contains only positive unit clauses. Additionally $S \models U$ and $T \equiv S[U]$.*

By Theorem 11, we deduce that T contains all its own p.i. These p.i. are also implicates of S (since $S \models T$), but it is clear that T does not in general contain all the p.i. of S. For instance this set also includes U and all clauses that can be inferred from T and U. Reconstructing the set of p.i. of S is the subject of the next section.

Algorithm 4. COMPUTEPI(T, U)

$T_1 := T$
for all $C \in U$ **do**   // U is in normal form
    $T_1 := T_1 \cup \{C\}$
    $R := (T_{1 \vdash 1, C}) \backslash T_1$ // R contains only newly generated clauses
    while $R \neq \emptyset$ **do**
        $T_1 := T_1 \cup R$
        $R := (R_{\vdash 1, C}) \backslash T_1$
    end while
end for
return T_1

4.2   Recovery of the Main Solution

The invocation of SATURATERW on an initial clause set S generates a saturated set of non-atomic clauses T and a normalized set of atomic clauses U. To recover the set of p.i. of S from T and U the principle of COMPUTEPI is to apply the $\mathcal{K}$-paramodulation calculus between the p.i. of T and all atomic clauses in U. In this way, for each atom extracted from S by SATURATERW, COMPUTEPI generates the missing implicates, i.e. those containing the constants that had been previously removed. The essential point (which ensures the efficiency of the approach) is that it is not necessary to apply any inference between the newly generated clauses: only the inferences involving U need to be considered. Formally, what renders COMPUTEPI efficient is the fact that all the implicates of a set of clauses $S \cup \{a \simeq b\}$ (with $a \succ b$) are eq-subsumed by clauses recursively obtained by $\mathcal{K}$-paramodulation between $a \simeq b$ and the p.i. of $S[b/a]$ as stated in Lemma 22.

Lemma 22. *Let S be a formula, $a \simeq b$ be a literal such that $a \succ b$ and $S' = (PI(S[b/a]))_{\vdash a \simeq b}$. Let D be a clause such that $S \cup \{a \simeq b\} \models D$, then $S' \leq_{eq} D$. Thus $S' \equiv S \cup \{a \simeq b\}$ and S' is saturated up to redundancy.*

Lemma 23. COMPUTEPI *terminates.*

The following theorem states that the proposed algorithm, composed of successive calls to SATURATERW and COMPUTEPI, is complete, i.e., that it computes all the p.i. of the input formula.

Theorem 24. *Let S be a formula, $\langle T, U \rangle = $ SATURATERW(S) and $S' = $ COMPUTEPI$(T, U_\downarrow)$. Then S' is the set of p.i. of S.*

5   Experimental Results

Both KPARAM and KPARAMRW have been implemented in Ocaml[2]. Below is an experimental comparison of both tools. The benchmark is made of a thousand

---

[2] See http://membres-lig.imag.fr/tourret/documents/kparam.tgz for the source code.

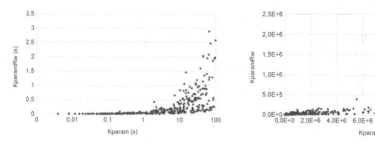

Fig. 1. Execution Time   **Fig. 2.** Generated Implicates

ground flat equational formulæ of a reasonable size[3] that were randomly generated. All tests were conducted on a machine equipped with an Intel core i5-3470
CPU and 4x2 GB of RAM, with a timeout of 100 seconds when not explicitly
said otherwise.

A first result worth mentioning is that in KPARAMRW the execution time of
COMPUTEPI is quasi-negligible no matter what the total execution time is: the
maximum is less than one second and the mean is 0.09 seconds. In general, it
always represents less that 1 percent of the total execution time. Another interesting indicator of the relative superiority of KPARAMRW compared to KPARAM
is the fact that while 15% of the benchmark reaches timeout before terminating
with KPARAM, only 9% does so with KPARAMRW. An additional 45% of the
formulæ have no atomic p.i. and are thus of little interest to us since KPARAM
and KPARAMRW merely coincide on such problems. Results concerning the remaining 46% of the benchmark are presented on Fig. 1 & 2. On Fig. 1 the
gain of going from KPARAM to KPARAMRW with regards to the execution time
can be observed. A logarithmic scale is used for the X axis to highlight that
this graph empirically indicates an exponential gain for our benchmark. The
results on Fig. 2 were obtained with a timeout of 5 minutes and compare the
number of implicates generated by KPARAM and KPARAMRW (for readability
issues the scales of the X and Y axis differ). There are two kinds of dots represented on the graph: filled diamonds and X's, the latter representing tests
for which KPARAM reaches the 5 minutes timeout before terminating. It shows
that some problems with atomic p.i. that KPARAM cannot solve by computing
more than a million implicates can be solved by KPARAMRW with less than two
hundred thousand implicates generated. We also compared our algorithms with
Zres [24][4], a state-of-the-art tool for p.i. generation in propositional logic that
uses a resolution-based algorithm together with ZBDDs for storing clause sets.
This system was chosen because it outperforms all other available propositional
systems on all our examples. To the best of our knowledge, besides KPARAM no

---

[3] Each test is made of 6 clauses with a maximum of 5 literals, using 8 constants. Although the size of the initial formula is small, hundreds of thousand or even millions
of implicates are often generated, leading to hundreds of them being eventually kept
as prime.

[4] Many thanks to Laurent Simon for providing the executable.

Table 1. Percentage of Tests Executed Twice Faster than `Zres`

Number of Generated Atoms	0	1	> 1	> 0	Total
KPARAM	64%	26%	23%	25%	45%
KPARAMRW	64%	83%	80%	82%	73%

complete p.i. computation tool is available for equational logic[5]. To make the comparison possible, the equational formulæ of the benchmark were translated into equivalent propositional formulæ by abstracting literals away and adding suitable instances of the equality axioms. This straightforward translation is obviously not the most efficient existing method, but it has the advantage of being simple. It still gives a rough execution time reference with which to compare the new algorithm, keeping into account that the time needed for translating the result back to equational logic and removing the redundancies was omitted, so as to underestimate this time. As shown in Table 1, this comparison proved useful by giving an insight of where to improve the original algorithm. The main observation on the line corresponding to KPARAM is that `Zres` is a lot more efficient than this algorithm as soon as atomic implicates appear in the formulæ (only 25% of the tests are faster than `Zres`, while 64% are faster when there are no atomic implicates), which was the motivation for designing KPARAMRW in the first place. As can be seen in the second line of the table, KPARAMRW is a good answer to this problem since an additional 57% of the problems with atomic implicates turn out faster than `Zres` with KPARAMRW, for a total of 82% of these tests being at least twice faster than the state-of-the-art tool. The results also distinguish between formulæ with a single atomic implicate (72%) and several ones (28%). A slight improvement of the performances is noted for the latter, but not as significant as the gap between none and one atomic implicate.

6   Conclusion

In this paper, a new algorithm for the generation of p.i. in ground flat equational logic was presented. It is based on a previous version introduced in [9]. The main idea of this algorithm is to isolate atomic equations to reduce the number of constants handled by the p.i. generator. Although in some applications it may be possible to directly use the simplified results along with the extracted equations, we also devised a way to recover the p.i. of the original input in a efficient way. This new algorithm is terminating, sound and complete and outperforms the previous one when atomic implicates are present. According to our experimental results, the gain is empirically exponential in time. This system can be used in connection with the calculus presented in [8] to efficiently generate ground implicates of first-order theories.

---

[5] To our knowledge, there exists only one tool, integrated in the Mistral solver [7], that is seemingly similar to KPARAM. However, in contrast to it, the Mistral tool is not complete (it does not compute all the p.i.) hence no comparison is possible.

An idea to improve atomic rewriting is to find a faster way to generate all the atomic equations entailed by the input formula instead of waiting for them to appear during the inference steps. To do so, the $\mathcal{K}$-paramodulation calculus could be replaced with a more efficient calculus specifically tailored to directly generate all atomic implicates, so that, after a unique rewriting step, the $\mathcal{K}$-paramodulation calculus can be used to generate all remaining non-atomic implicates. To extend further the atomic rewriting strategy, it should be possible to apply it to any equation appearing in the formula in a "divide and conquer" way. Any clause of the form $a \simeq b \vee C$ would then lead to two recursive calls of the strategy, one where $a \simeq b$ is true where the rewriting applies and the other where only C remains. It is still unclear whether this idea is efficient because of two problems: merging the results of the two recursive calls is by no means a simple task, and the fact that there are two calls on formulæ that differ only by one clause may generate a lot of redundant computation steps, thus slowing down the whole process. These questions need a thorough investigation and are one of our objectives for future work.

Up to now, our system has been mainly tested on randomly computed instances. We now plan to apply it, in conjunction with an implementation of the calculus described in [8], to more concrete problems in system verification, particularly for checking properties of algorithms operating on arrays or pointer-based data-structures.

References

1. Bachmair, L., Ganzinger, H.: Rewrite-based Equational Theorem Proving with Selection and Simplification. Journal of Logic and Computation 3(4), 217–247 (1994)
2. Bienvenu, M.: Prime implicates and prime implicants in modal logic. In: Proceedings of the National Conference on Artificial Intelligence, p. 379. AAAI Press, MIT Press, Menlo Park, Cambridge (1999, 2007)
3. Bittencourt, G.: Combining syntax and semantics through prime form representation. Journal of Logic and Computation 18(1), 13–33 (2008)
4. Coudert, O., Madre, J.C.: A new method to compute prime and essential prime implicants of boolean functions. In: Knight, Savage (eds.) Advanced Research in VLSI and Parallel Systems, pp. 113–128 (1992)
5. De Kleer, J.: An improved incremental algorithm for generating prime implicates. In: Proceedings of the tenth National Conference on Artificial Intelligence, pp. 780–785. AAAI Press, Menlo Park (1992)
6. Dillig, I., Dillig, T., Aiken, A.: Automated error diagnosis using abductive inference. In: ACM SIGPLAN Notices, pp. 181–192. ACM (2012)
7. Dillig, I., Dillig, T., McMillan, K.L., Aiken, A.: Minimum satisfying assignments for SMT. In: Madhusudan, P., Seshia, S.A. (eds.) CAV 2012. LNCS, vol. 7358, pp. 394–409. Springer, Heidelberg (2012)
8. Echenim, M., Peltier, N.: A calculus for generating ground explanations. In: Gramlich, B., Miller, D., Sattler, U. (eds.) IJCAR 2012. LNCS (LNAI), vol. 7364, pp. 194–209. Springer, Heidelberg (2012)
9. Echenim, M., Peltier, N., Tourret, S.: An approach to abduction in equational logic. In: Proceeding of the 23d International Joint Conference on Artificial Intelligence, pp. 531–538. AAAI Press (2013)

10. Errico, B., Pirri, F., Pizzuti, C.: Finding prime implicants by minimizing integer programming problems. In: AI-CONFERENCE-, pp. 355–362. World Scientific Publishing (1995)
11. Jackson, P., Pais, J.: Computing Prime Implicants. In: Stickel, M.E. (ed.) CADE 1990. LNCS, vol. 449, pp. 543–557. Springer, Heidelberg (1990)
12. Jackson, P.: Computing prime implicates incrementally. In: Kapur, D. (ed.) CADE 1992. LNCS, vol. 607, pp. 253–267. Springer, Heidelberg (1992)
13. Kean, A., Tsiknis, G.: An incremental method for generating prime implicants/implicates. Journal of Symbolic Computation 9(2), 185–206 (1990)
14. Knill, E., Cox, P.T., Pietrzykowski, T.: Equality and abductive residua for Horn clauses. Theoretical Computer Science 120(1), 1–44 (1993)
15. Manquinho, V.M., Oliveira, A.L., Marques-Silva, J.: Models and algorithms for computing minimum-size prime implicants. In: Proceedings of the International Workshop on Boolean Problems (1998)
16. Marquis, P.: Extending abduction from propositional to first-order logic. In: Jorrand, P., Kelemen, J. (eds.) FAIR 1991. LNCS, vol. 535, pp. 141–155. Springer, Heidelberg (1991)
17. Matusiewicz, A., Murray, N.V., Rosenthal, E.: Tri-based set operations and selective computation of prime implicates. In: Kryszkiewicz, M., Rybinski, H., Skowron, A., Raś, Z.W. (eds.) ISMIS 2011. LNCS (LNAI), vol. 6804, pp. 203–213. Springer, Heidelberg (2011)
18. Mayer, M.C., Pirri, F.: First order abduction via tableau and sequent calculi. Logic Journal of IGPL 1(1), 99–117 (1993)
19. McCune, W., Wos, L.: Otter-the CADE-13 competition incarnations. Journal of Automated Reasoning 18(2), 211–220 (1997)
20. Meir, O., Strichman, O.: Yet another decision procedure for equality logic. In: Etessami, K., Rajamani, S.K. (eds.) CAV 2005. LNCS, vol. 3576, pp. 307–320. Springer, Heidelberg (2005)
21. Quine, W.V.: A way to simplify truth functions. The American Mathematical Monthly 62(9), 627–631 (1955)
22. Ramesh, A., Becker, G., Murray, N.V.: CNF and DNF considered harmful for computing prime implicants/implicates. Journal of Automated Reasoning 18(3), 337–356 (1997)
23. Rymon, R.: An se-tree-based prime implicant generation algorithm. Annals of Mathematics and Artificial Intelligence 11(1-4), 351–365 (1994)
24. Simon, L., Del Val, A.: Efficient consequence finding. In: International Joint Conference on Artificial Intelligence, pp. 359–365. Lawrence Erlbaum Associates ltd. (2001)
25. Sofronie-Stokkermans, V.: Hierarchical reasoning for the verification of parametric systems. In: Giesl, J., Hähnle, R. (eds.) IJCAR 2010. LNCS (LNAI), vol. 6173, pp. 171–187. Springer, Heidelberg (2010)
26. Sofronie-Stokkermans, V.: Hierarchical reasoning and model generation for the verification of parametric hybrid systems. In: Bonacina, M.P. (ed.) CADE 2013. LNCS (LNAI), vol. 7898, pp. 360–376. Springer, Heidelberg (2013)
27. Tran, D., Ringeissen, C., Ranise, S.: al.: Combination of convex theories: Modularity, deduction completeness, and explanation. Journal of Symbolic Computation 45(2), 261–286 (2010)

Finite Quantification in Hierarchic Theorem Proving

Peter Baumgartner[1], Joshua Bax[1], and Uwe Waldmann[2]

[1] NICTA* and Australian National University, Canberra, Australia
[2] MPI für Informatik, Saarbrücken, Germany

Abstract. Many applications of automated deduction require reasoning in first-order logic modulo background theories, in particular some form of integer arithmetic. A major unsolved research challenge is to design theorem provers that are "reasonably complete" even in the presence of free function symbols ranging into a background theory sort. In this paper we consider the case when all variables occurring below such function symbols are quantified over a finite subset of their domains. We present a non-naive decision procedure for background theories extended this way on top of black-box decision procedures for the EA-fragment of the background theory. In its core, it employs a *model-guided* instantiation strategy for obtaining pure background formulas that are equi-satisfiable with the original formula. Unlike traditional finite model finders, it avoids exhaustive instantiation and, hence, is expected to scale better with the size of the domains. Our main results in this paper are a correctness proof and first experimental results.

1 Introduction

Many applications of automated deduction require reasoning in first-order logic modulo background theories, in particular some form of integer arithmetic. A major unsolved research challenge is to design theorem provers that are "reasonably complete" for quantified formulas, in particular in presence of free function symbols ranging into a background theory sort ("free BG-sorted operators", for short). Such formulas arise frequently when reasoning on data structures with specific properties, e.g., *symmetric* arrays over integers and *sorted* lists over integers. Modelling such data structures is easy when full quantification and free integer-sorted function symbols are available to axiomatize the array access function and the list head function respectively.

Unfortunately, (refutationally) complete theorem proving in the presence of free BG-sorted operators is intractable in general. For instance, just adding one free predicate symbol to linear integer arithmetic results in a Π_1^1-hard validity problem [12]. Theorem proving approaches hence have to circumvent this problem in one way or the other. On the one hand, SMT-solvers [18] generally use instantiation heuristics [10,16] for reducing the input problem to a quantifier-free one, and these are complete only in rather restricted cases [11]. On the other hand, approaches rooted in first-order theorem proving either are incomplete; do not accept free BG-sorted operators at all [13,21,9,5] or, are complete only for certain fragments or under certain conditions [3,1,14,6,7].

---

* NICTA is funded by the Australian Government through the Department of Communications and the Australian Research Council through the ICT Centre of Excellence Program.

S. Demri, D. Kapur, and C. Weidenbach (Eds.): IJCAR 2014, LNAI 8562, pp. 152–167, 2014.

In practice, lack of completeness is a major concern in, e.g., software verification applications, which frequently require disproving non-valid proof obligations. In such cases, incomplete theorem provers run out of resources or report "unknown" instead of detecting non-validity. We address this problem by working with quantification over *finite* segments of the background sorts, e.g., the integers. Our underlying methodology assumes that from a user's point of view, data structures over the integers can often be supplanted by data structures over reasonably large finite segments of the integers, say, from −Maxint to +Maxint, as good-enough approximations. As no other restrictions apply, our method should be widely applicable in practice. Our method is also refutationally sound wrt. the standard semantics. That is, if our algorithm determines unsatisfiability wrt. finite domains, the given clause set is also unsatisfiable wrt. unbounded domains. Because of that, our approach can be seen as an extension of current quantifier instantiation heuristics by being able to determine satisfiability wrt. finite domains.

If all quantifiers range over finite domains, decidability can be recovered in a trivial way by exhaustive instantiation and calling a suitable SMT-solver afterwards. Of course, this naive approach does not scale with the domain size and cannot be expected to work well in practice. This problem has often been observed in the context of finite-model finding [22,23,15,8,4,20,19]. While our method is also based on instantiation, it is (often) far less prolific than the naive method.

More precisely, our method accepts as input a set of *finitely quantified clauses*. A clause is finitely quantified if every variable occurring below a free BG-sorted operator is quantified over a finite segment of its domain. The core idea is to give the free BG-sorted operators a *default* interpretation that is then stepwise refined. This default interpretation maps every free BG-sorted operator to a constant function, and refinements are done by finding exceptions to that in a conflict-driven way. After each refinement, the given clause set is transformed into a certain form whose satisfiability can be decided by existing reasoners in a black-box fashion. Suitable reasoners are, e.g., theorem provers implementing hierarchic superposition [3,7] and, with one more simple transformation step, SMT-solvers for the EA-fragment of the background theory. The procedure stops after finitely many (hopefully few) refinement steps, either with a representation of a model or a set of ground instances obtained from exceptions which demonstrates the unsatisfiability of the given clause set.

We preview our method with an example. Let N be the following clause set:

(1) read(write($a, i, x), i$) $\approx x$ (4) $1 \leq$ m $\wedge$ m < 1000

(2) read(write($a, i, x), j$) $\approx$ read(a, j) $\vee i \approx j$ (5) read(a, m) $<$ read(a, m + 1)

(3) read(a, i) $\leq$ read(a, j) $\vee \neg(i < j) \vee i \notin [1..1000^i] \vee j \notin [1..1000^j]$

where $t \in [l..h]$ abbreviates the formula $l \leq t \wedge t \leq h$ for any integer-sorted terms t, l and h. Variables are typeset in italics, e.g, x, and operators in sans-serif, e.g., read, a and m. The axioms (1) and (2) are the standard axioms for integer-sorted arrays with integer indices. The axiom (3) states that the array a is sorted within the domain [1..1000] for i and j. Annotating the upper bounds as 1000^i and 1000^j facilitates replacing them with different values for a given variable, see below. The clauses (4) constrains the integer constant m to the stated range. The task is to confirm that N is satisfiable.

In order to check satisfiability with hierarchic superposition, the input clause set has to be *sufficiently complete* (cf. Section 2). In the example, sufficient completeness means that in every model of (1)-(5) wrt. pure first-order logic every ground read-term must be equal to some background term. With the axioms (1) and (2) every write-term inside of a read-term can be eliminated, and so the only critical terms are applications of read to the array constant a. The clauses (3) and (5) constrain the interpretation of terms of the form $\text{read}(a, t)$ but do not enforce sufficient completeness. Achieving sufficient completeness for *ground* clauses like (5) is easy, one just needs to add "definitions" like (5b) $\text{read}(a, m) \approx n_0$ and (5c) $\text{read}(a, m + 1) \approx n_1$ where n_0 and n_1 are fresh integer-sorted parameters (symbolic constants) and replace the clause (5) by (5a) $n_0 < n_1$. Indeed, our transformation does all that (and so does our earlier calculus in [7]).

The more difficult part concerns the non-ground clause (3). Our procedure generalizes the above mechanism of introducing definitions and applying them to the non-ground case (see Section 3). For that, it uses a candidate model which initially is the *default interpretation* that maps all read-terms of a particular shape to the *same* arbitrary symbolic constant. This results in the following transformation of clause (3):

(3a) $n_3 \leq n_4 \lor \neg(i < j) \lor i \notin [1..1000^i] \lor j \notin [1..1000^j]$
(3b) $\text{read}(a, i) \approx n_3 \lor i \notin [1..1000^i]$ (3c) $\text{read}(a, j) \approx n_4 \lor j \notin [1..1000^j]$

Clauses (3b) and (3c) are the definitions for the default interpretation, one per occurrence of a read-term in (3), and clause (3a) is clause (3) after applying these definitions.

The new clause set $N_1 = \{(1), (2), (3a)–(3c), (4), (5a)–(5c)\}$ now needs to be checked for satisfiability. Because the clause set N_1 is sufficiently complete and hierarchic superposition decides the underlying fragment, we get a definite result.

The clause set N_1 is in fact unsatisfiable. Because this only means that N is not satisfied using the current model candidate, the search for a model needs to continue. This is done by refining the default interpretation at a critical point that is responsible for unsatisfiability. Our algorithm determines that point as an adjacent one to a maximal sub-domain that results in satisfiability. In the example, this is the sub-domain $[1..999^i]$ for the variable i and the point is 1000. That is, the set N_2 obtained from N_1 by replacing everywhere 999^i by 1000^i is satisfiable, while adding back 1000 to $[1..999^i]$ makes it unsatisfiable again. The refinement then is done by excluding the point 1000 from the default interpretation and providing a separate definition for it. The corresponding transformation of clause (3) hence looks as follows :

(3a1) $n_{31} \leq n_4 \lor \neg(i < j) \lor i \notin [1..1000^i] \setminus \{1000\} \lor j \notin [1..1000^j]$
(3a2) $n_{32} \leq n_4 \lor \neg(1000 < j) \lor j \notin [1..1000^j]$
(3b1) $\text{read}(a, i) \approx n_{31} \lor i \notin [1..1000^i] \setminus \{1000\}$ (3c) $\text{read}(a, j) \approx n_4 \lor j \notin [1..1000^j]$
(3b2) $\text{read}(a, 1000) \approx n_{32}$

Clauses (3b1) and (3b2) provide the modified definitions and clauses (3a1) and (3a2) are the correspondingly rewritten versions of (3). Let $N_3 = \{(1), (2), (3a1) - (3c), (4), (5a) - (5c)\}$ be the result of the current transformation step.

The clause set N_3 is still unsatisfiable. In the next round, the new upper bounds required for the clauses in N_3 to have satisfiability are 999^j and 1000^i. Transforming clause (3) wrt. the points 1000 for j and 1000 for i from the previous step gives:

$(3a1)$ $n_{31} \leq n_{41} \vee \neg(i < j) \vee i \notin [1..1000^i] \setminus \{1000\} \vee j \notin [1..1000^j] \setminus \{1000\}$

$(3a2)$ $n_{32} \leq n_{41} \vee \neg(1000 < j) \vee j \notin [1..1000^j] \setminus \{1000\}$

$(3a3)$ $n_{31} \leq n_{42} \vee \neg(i < 1000) \vee i \notin [1..1000^j] \setminus \{1000\}$

$(3a4)$ $n_{32} \leq n_{42} \vee \neg(1000 < 1000)$

$(3b1)$ $\mathsf{read}(\mathsf{a}, i) \approx n_{31} \vee i \notin [1..1000^i] \setminus \{1000\}$ $(3b2)$ $\mathsf{read}(\mathsf{a}, 1000) \approx n_{32}$

$(3c1)$ $\mathsf{read}(\mathsf{a}, j) \approx n_{41} \vee j \notin [1..1000^j] \setminus \{1000\}$ $(3c2)$ $\mathsf{read}(\mathsf{a}, 1000) \approx n_{42}$

Let $N_4 = \{(1), (2), (3a1) - (3c2), (4), (5a) - (5c)\}$ be the result of the current transformation step. This time, N_4 is satisfiable, and so is N, with the same models. If I is any such model we have $I(\mathsf{m}) = 999$, $I(\mathsf{read}(\mathsf{a}, i)) = k$, for some integer k and all $i = 1..999$, and $I(\mathsf{read}(\mathsf{a}, 1000)) = l$ for some integer $l > k$. (We present the general model finding procedure and its correctness results in Section 4.)

The example is solved after two iterations of transformation steps. In general, each transformation step needs $O(m \cdot \log(n))$ prover calls to determine the sub-intervals and the next point as explained above, where m is the number of variables in the given clause set after making clauses variable-disjoint and n is the size of the largest domain. In total, with $m = 2$ and $n = 1000$ this accounts for $2 \cdot (m \cdot \log(n)) \leq 40$ theorem prover calls, however each one rather simple. By contrast, the full ground instantiation of the clauses (3)-(5) has a size of $n^m = 10^6$ which, in general, grows too quickly for current theorem provers or SMT solvers. In the worst case, though, our method also requires full ground instantiation (but is not worse). This happens when the default interpretation is unsuitable for the whole domain, so that separate definitions are needed for all points to establish (un)satisfiability. In Section 5 we report on first experimental results.

Related Work

Related work comes from several directions. Procedures for computing models of first-order logic formulas *without background theories* have a long tradition in automated reasoning. MACE-style model finding [8] utilizes translation into propositional SAT or into EPR [4] for deciding satisfiability wrt. a given candidate domain size k; SEM-style model finding [22,23,15] utilizes constraint solving techniques, again wrt. k. The main problem is scalability wrt. both the domain size k and the number of variables in the input clause set, which severely limits the applicability of both styles in practice. Recently, Reynolds, Tinelli, Goel, Krstić, Deters and Barrett proposed a finite model finding procedure in the SMT framework that addresses this problem by on-demand instantiation techniques [20,19]. This way, their work is conceptually related to ours, but, unlike ours, they allow quantification only over variables ranging into the free sort. An extension for quantifying variables over background domains such as the integers does not seem straightforward and is left as future work in [20].

Heuristic instantiation is the state of the art technique for handling quantified formulas in SMT-solvers [10,16]. These heuristics perform impressively well in practice, but in general are incomplete even for pure first-order logic. Ge and deMoura [11] propose a technique where the ground terms used for instantiation come from solving certain set constraints. They obtain completeness results for the fragment where every variable occurs only as an argument of a free function or predicate symbol. Interestingly, they also use the notion of a default interpretation in a similar way as we do. However, even with certain extensions their approach remains incomparable to ours. For example, terms like $f(x + y)$ are disallowed, but are acceptable in our approach when x and y are finitely quantified.

Regarding related work in first-order theorem proving, the problem we are considering has been tackled in the framework of the hierarchic superposition calculus [3]. Weidenbach and Kruglow [14] consider the case when all background-sorted terms are ground, similarly to our calculus in [7]. In [6] we have identified a certain syntactic fragment that enables complete reasoning.

2   Hierarchic Theorem Proving

Hierarchic superposition [3,7] is a calculus for automated reasoning in a hierarchic combination of first-order logic and some background theory, for instance some form of arithmetic. We consider the following scenario:[1]

We assume that we have a *background ("BG")* prover that accepts as input a set of clauses over a *BG signature* $\Sigma_B = (\Xi_B, \Omega_B)$, where Ξ_B is a set of *BG sorts* and Ω_B is a set of *BG operators*. Terms/clauses over Σ_B and BG-sorted variables are called *BG terms/clauses*. The BG prover decides the satisfiability of Σ_B-clause sets w. r. t. a *BG specification*, that is, a class of term-generated Σ_B-interpretations (called *BG models*) that is closed under isomorphisms. We assume that Ω_B contains a set of distinguished constant symbols $\Omega_B^D \subseteq \Omega_B$ that has the property that any two distinct $d_1, d_2 \in \Omega_B^D$ are interpreted by different elements in every BG model. We refer to these constant symbols as *(BG) domain elements*. We also assume that Σ_B contains infinitely many *parameters*, that is, additional constants that may be interpreted freely by arbitrary elements of the appropriate domain. In examples we use $\{0, 1, 2, \ldots\}$ to denote BG domain elements, $\{+, -, <, \leq\}$ to denote (non-parameter) BG operators, and the possibly subscripted letters $\{x, y\}$ and $\{\alpha, \beta\}$ to denote variables and parameters, respectively. We assume that the BG specification is the class of all models of linear integer arithmetic (LIA).

The *foreground ("FG")* theorem prover accepts as input clauses over a signature $\Sigma = (\Xi, \Omega)$, where $\Xi_B \subseteq \Xi$ and $\Omega_B \subseteq \Omega$. The sorts in $\Xi_F = \Xi \setminus \Xi_B$ and the operator symbols in $\Omega_F = \Omega \setminus \Omega_B$ are called *FG sorts* and *FG operators*. The intended semantics is that of *conservative extensions of the BG specification*, i. e., Σ-interpretations whose restriction to Σ_B is a model of the BG specification. Below we refer to satisfiability in this sense as *$\mathcal{B}$-satisfiability*.

We use $\{a, b, c, f, g\}$ to denote FG operators. A Σ-term is an *FG term* if it is not a BG term, that is, if it contains at least one FG operator or FG variable (and analogously for equations, literals, or clauses). We emphasize that for an FG operator $f : \xi_1 \ldots \xi_n \to \xi_0$ in Ω_F any of the ξ_i may be a BG sort. Consequently, FG terms may have BG sorts. Every FG operator f with a BG range sort $\xi_0 \in \Xi_B$ is called a *free BG-sorted (FG) operator*.

After abstracting out BG terms other than BG domain elements and variables that occur as subterms of FG terms,[2] the FG prover saturates the set of Σ-clauses using the

---

[1] Due to a lack of space, we can only give a brief overview of the calculus and of the semantics of hierarchic specifications. We refer to [7] for the details.

[2] *Abstracting out* a term t that occurs in a clause $C[t]$ means replacing $C[t]$ by $x \not\approx t \vee C[x]$ for a new variable x.

inference rules of hierarchic superposition, such as, e. g.,

Negative superposition
$$\frac{l \approx r \vee C \qquad s[u] \not\approx t \vee D}{\mathrm{abstr}((s[r] \not\approx t \vee C \vee D)\sigma)}$$

if (i) neither l nor u is a BG term, (ii) u is not a variable, (iii) σ is an mgu of l and u, (iv) σ maps all BG variables to BG terms, (v) $r\sigma \not\succeq l\sigma$, (vi) $(l \approx r)\sigma$ is strictly maximal in $(l \approx r \vee C)\sigma$, (vii) the first premise does not have selected literals, (viii) $t\sigma \not\succeq s\sigma$, and (ix) if the second premise has selected literals, then $s \not\approx t$ is selected in the second premise, otherwise $(s \not\approx t)\sigma$ is maximal in $(s \not\approx t \vee D)\sigma$.

These differ from the standard superposition inference rules [2] mainly in that only the FG parts of clauses are overlapped and that any BG clauses derived during the saturation are instead passed to the BG prover. The BG prover implements an inference rule

Close
$$\frac{C_1 \quad \cdots \quad C_n}{\square} \qquad \text{if } C_1, \ldots, C_n \text{ are BG clauses and } \{C_1, \ldots, C_n\} \text{ is unsatisfiable w. r. t. the BG specification.}$$

As soon as one of the two provers detects a contradiction, the input clause set has been shown to be $\mathcal{B}$-unsatisfiable.

There are two requirements for the refutational completeness of hierarchic superposition. The first one is *sufficient completeness*: We must be able to prove that every ground BG-sorted FG term is equal to some BG term. Sufficient completeness of a set of Σ-clauses is a property that is not even recursively enumerable. For certain classes of Σ-clause sets, however, it is possible to establish a variant of sufficient completeness automatically [14,7]: If all BG-sorted FG terms in the input are ground, it suffices to show that each BG-sorted FG term *in the input* is equal to some BG term. This can be achieved by adding a *definition* $\alpha_t \approx t$ for every BG-sorted FG term t occurring in a clause $C[t]$, where α_t is a new parameter (BG constant); afterwards $C[t]$ can be replaced by $C[\alpha_t]$.

Since we can only pass *finite* clause sets to a BG prover, there is a second requirement for refutational completeness, namely the compactness of the BG specification. A specification is called *compact*, if every set of formulas that is unsatisfiable w. r. t. the specification has a finite unsatisfiable subset.

3   Finite Domain Transformation

We are interested in refutationally complete hierarchic theorem proving in the presence of free BG-sorted FG operators. Unfortunately, just adding one free predicate symbol to linear integer arithmetic results in a Π_1^1-hard validity problem. To circumvent this problem, we work with a modified semantics and introduce a concept of finite quantification of BG variables. This allows us to remove all free BG-sorted FG operators by a *finite domain transformation*, introduced next, and use existing reasoning methods as decision procedures on the result.

Let $\xi \in \mathcal{E}_B$ be a BG sort. By a *finite ξ-domain* Δ we mean any possibly empty finite set $\{d_1, \ldots, d_n\} \subseteq \Omega_B^D$ of ξ-sorted domain elements d_i. Set membership in Δ can be expressed by a BG formula $\mathcal{F}_\Delta[x]$ in one free ξ-sorted variable x whose extension is exactly the set Δ, in every $\mathcal{B}$-interpretation. One can always take $\mathcal{F}_\Delta[x] = x \approx d_1 \vee \cdots \vee x \approx d_n$, but if supported by the BG logic, as in the case of integer arithmetic, it may be advantageous to use "compact" representations like $\mathcal{F}_\Delta[x] = 1 \leq x \wedge x \leq 20$ instead.

We use set-theoretic expressions for finite ξ-domains, in particular of the form $\Delta \setminus \Gamma$, where Γ is a finite set of domain elements of the proper sort. In the previous example, e.g., $\mathcal{F}_{\Delta\setminus\{3,5\}}[x] = 1 \leq x \wedge x \leq 20 \wedge x \not\approx 3 \wedge x \not\approx 5$. Instead of $\mathcal{F}_\Delta[x]$ and $\mathcal{F}_{\Delta\setminus\Gamma}[x]$ we generally write $x \in \Delta$ and $x \in \Delta \setminus \Gamma$, respectively, and $x \notin \Delta$ and $x \notin \Delta \setminus \Gamma$ for their negations. We call these expressions *domain predicates* and treat them as literals in clauses instead of expanding them.

Definition 3.1. A finitely quantified clause *is a Σ-clause of the form $D \vee x_1 \notin \Delta_{x_1} \vee \cdots \vee x_n \notin \Delta_{x_n}$ such that D does not contain domain predicates, $n \geq 0$, $x_i \neq x_j$ for $1 \leq i < j \leq n$, and every variable occurring below a free BG-sorted operator in D is among $x_1, \ldots, x_n$.*

For example, $f(x + 1) > \alpha + y \vee y > 0 \vee x \notin [1..1000]$ is finitely quantified.

Example 3.2. Let N consist of the following two finitely quantified clauses:

(C_1) $f(x_1) > x_1 \vee x_1 \notin [1..1000]$
(C_2) $f(x_2 + 3) < 10 \vee \neg(x_2 > 2) \vee x_2 \notin [1..1000]$

We formally have $\Delta_{x_1} = \Delta_{x_2} = [1..1000]$, and in C_1 the pseudo-literal $x_1 \notin [1..1000]$ is short for $\neg(1 \leq x_1 \leq 1000)$.   □

Where $\mathbf{x} = (x_1, \ldots, x_n)$, let $\Delta_{\mathbf{x}}$ denote the $\mathbf{x}$-indexed list $(\Delta_{x_1}, \ldots, \Delta_{x_n})$ of sets of domain elements. We extend usual set operations pointwise to $\mathbf{x}$-indexed lists $\Pi_{\mathbf{x}}$ and $\Delta_{\mathbf{x}}$ of sets of domain elements. For instance $\Pi_{\mathbf{x}} \subseteq \Delta_{\mathbf{x}}$ iff $\Pi_x \subseteq \Delta_x$, for each $x \in \mathbf{x}$.

We are going to define the earlier mentioned finite domain transformation for evaluating finitely quantified clauses under a given interpretation. It takes as input a finitely quantified clause $C[\Delta_{\mathbf{x}}]$ and sets of points $\Pi_{\mathbf{x}}$ that provide possible exceptions to interpreting the free BG-sorted operators as the constant function on the domains $\Delta_{\mathbf{x}}$ as specified by the default interpretation.

Definition 3.3 (Finite Domain Transformation). *Let $C[\Delta_{\mathbf{x}}] = D \vee x_1 \notin \Delta_{x_1} \vee \cdots \vee x_n \notin \Delta_{x_n}$ be a finitely quantified clause and $\Pi_{\mathbf{x}} \subseteq \Delta_{\mathbf{x}}$ a list of sets of domain elements.*

Let $\mathrm{Cls}_C := \emptyset$ and $\mathrm{Def}_C := \emptyset$ be initially empty sets of Σ-clauses. For every partition $\{y_1, \ldots, y_k\} \uplus \{z_1, \ldots, z_l\}$ of $\{x_1, \ldots, x_n\}$ do the following:

For all substitutions $\gamma = [z_1 \mapsto d_1, \ldots z_l \mapsto d_l]$ such that $d_m \in \Pi_{z_m}$:

1. *Let $E := D\gamma$*
2. *While E has the form $E[t]$ where t is a minimal term with a free BG-sorted operator at the top-level do the following:*
   (a) *Let α be a fresh parameter*
   (b) *Add to Def_C the clause $t \approx \alpha \vee y_1 \notin \Delta_{y_1} \setminus \Pi_{y_1} \vee \cdots \vee y_k \notin \Delta_{y_k} \setminus \Pi_{y_k}$*
   (c) *Set $E := E[\alpha]$*
3. *Add to Cls_C the clause $E \vee y_1 \notin \Delta_{y_1} \setminus \Pi_{y_1} \vee \cdots \vee y_k \notin \Delta_{y_k} \setminus \Pi_{y_k}$*

The result is the pair $FD(C, \Pi_{\mathbf{x}}) = (\text{Cls}_C, \text{Def}_C)$, *the* finite domain transformation of C.

By the minimality of t in (2) we mean that no proper subterm of t is built with a free BG-sorted operator. The finite domain transformation removes from the given finitely quantified clause C every occurrence of a term t built with some free BG-sorted symbol. Recall from Definition 3.1 that all variables in t are among $\mathbf{x} = (x_1, \ldots, x_n)$. The removal of t distinguishes whether x_i is interpreted as an element of $\Delta_i \setminus \Pi_i$ or as an element $d_i \in \Pi_i$. This is done in all possible ways by exhaustive partitioning of the variables $\mathbf{x}$ and exhausting the substitution γ for all possible assignments for x_i. The set $\Delta_i \setminus \Pi_i$ specifies those domain elements for which the interpretation of t is undistinguished, and the set Π_i specifies those domain elements for which the interpretation of t is distinguished, by taking different parameters α per substitution γ. In step (b) corresponding definitions for t are put into Def_C. Step (c) applies these definitions to the current clause E.

On complexity: the result $FD(C, \Pi_{\mathbf{x}})$ contains $O(|\mathbf{x}|^{|\Pi_{\mathbf{x}}|+1})$ clauses. This is, because for every $x_i \in \mathbf{x}$ a choice is made for either instantiating x_i exhaustively with all elements from $\Pi_{\mathbf{x}}$ if $x_i \in \{z_1, \ldots, z_n\}$, or otherwise not doing so, which explains the " $+ 1$". (Extracting out subterms does not affect the complexity.) In the worst case $\Pi_{\mathbf{x}} = \Delta_{\mathbf{x}}$ and all clauses stemming from the latter case are tautological. The complexity in this case is $O(|\mathbf{x}|^{|\Pi_{\mathbf{x}}|})$, which is the same as with ground-instantiation based MACE-style model finders.

Example 3.2 (continued). Let $\Pi_{(x_1)} = (\{9\})$. Then $FD(C_1, \Pi_{(x_1)})$ consists of the clauses

(C_{11}) $\alpha_1 > x_1 \vee x_1 \notin [1..1000] \setminus \{9\}$ (C_{13}) $\alpha_2 > 9$
(C_{12}) $f(x_1) \approx \alpha_1 \vee x_1 \notin [1..1000] \setminus \{9\}$ (C_{14}) $f(9) \approx \alpha_2$

where $\text{Cls}_{C_1} = \{C_{11}, C_{13}\}$ and $\text{Def}_{C_1} = \{C_{12}, C_{14}\}$. The left clauses stem from partitioning $\{x_1\}$ as $\{y_1\} \uplus \emptyset$, and the right clauses from $\emptyset \uplus \{z_1\}$. There are two occurrences of $\Delta_{x_1} = [1..1000]$. $\qquad\square$

There are no restrictions on nesting free BG-sorted operators, although none of our examples shows that. For example, a literal like $f(x + g(y, \beta)) > f(y) + y$ is perfectly acceptable. The possible nesting of free BG-sorted operators necessitates the while-loop in step (2) in Definition 3.3; removing all of them in a single step is not possible.

The sets of domain elements Δ_x occurring in clauses in $FD(C, \Pi_{\mathbf{x}})$ are all within pseudo-literals of the form $x \notin \Delta_x \setminus \Pi_x$. Hence, both Cls_C and Def_C are of the form $\text{Cls}_C[\Delta_{\mathbf{x}}]$ and $\text{Def}_C[\Delta_{\mathbf{x}}]$. Moreover, in $FD(C, \Pi_{\mathbf{x}})$, every free BG-sorted operator f occurs only in a clause of the form $f(t_1, \ldots, t_n) \approx \alpha \vee D$ in Def_C where no t_i and no literal in D contains any free BG-sorted operator.

The finite domain transformation is generalized to clause sets by taking the union of the finite domain transformations applied to its members. More precisely, let $N = \{C_1[\Delta_{\mathbf{x}_1}], \ldots, C_m[\Delta_{\mathbf{x}_m}]\}$ be a finite set of finitely quantified clauses. Let us assume that the clauses in N have been renamed apart, so that the lists of variables $\mathbf{x}_i$ are pairwise disjoint, for all $i = 1..m$. By definition, each $\mathbf{x}_i$ consists of pairwise different variables, too. This allows us to take $\mathbf{x}$ as the concatenation of all $\mathbf{x}_i$'s and to write $\Delta_{\mathbf{x}}$ for the concatenation of all $\Delta_{\mathbf{x}_m}$'s. The clause set N hence is of the form $N[\Delta_{\mathbf{x}}]$. Now let $(\text{Cls}_{C_i}, \text{Def}_{C_i}) = \text{FD}(C_i, \Pi_{\mathbf{x}})$ and define $\text{FD}(N, \Pi_{\mathbf{x}}) = (\text{Cls}_N, \text{Def}_N)$ where $\text{Cls}_N = \bigcup_{i=1..m} \text{Cls}_{C_i}$ and $\text{Def}_N = \bigcup_{i=1..m} \text{Def}_{C_i}$.

Below, we usually denote $\text{FD}(N, \Pi_{\mathbf{x}})$ as a single clause set $M[\Delta_{\mathbf{x}}] = \text{Cls}_N \cup \text{Def}_N$. The following result follows immediately:

Proposition 3.5. *Let $N[\Delta_{\mathbf{x}}]$ be a set of finitely quantified clauses and $\Pi_{\mathbf{x}} \subseteq \Delta_{\mathbf{x}}$. Then $\text{FD}(N, \Pi_{\mathbf{x}})$ is sufficiently complete.*

Proposition 3.5 is one of the ingredients that allows us to argue for hierarchic superposition [7] as a decision procedure for $\mathcal{B}$-satisfiability of the clause sets $\text{FD}(C, \Pi_{\mathbf{x}})$. We also need a termination argument for derivations (compactness, cf. Section 2, is unproblematic then). This is easy, for instance, in the absence of non-ground FG-sorted operators only finitely many superposition steps exist and all of these are between the clauses in Def_C, and then only at the top-level – that is, between the literals $f(t_1, \ldots, t_n) \approx \alpha$. Alternatively one can use SMT-solvers after removing all free BG-operators by exhaustive application of a superposition-like inference rule that from premises $f(t_1, \ldots, t_n) \approx \alpha \vee D$ and $f(s_1, \ldots, s_n) \approx \beta \vee E$ derives the clause $s_1 \not\approx t_1 \vee \cdots \vee s_n \not\approx t_n \vee \alpha \approx \beta \vee D \vee E$. In general, hierarchic superposition can be used if it is guaranteed to terminate on $\text{FD}(C, \Pi_{\mathbf{x}})$. This applies, e.g., to the example in the introduction.

The notation $M[\Delta_{\mathbf{x}}]$ makes it easy to modify the sets Δ_x in pseudo-literals in clauses in M. More precisely, if $\Delta_{\mathbf{x}} = (\ldots, \Delta_x, \ldots)$ for some $x \in \mathbf{x}$ and Γ is a set of domain elements with the same sort as x, we denote by $\Delta_{\mathbf{x}}[x \mapsto \Gamma]$ the update of $\Delta_{\mathbf{x}}$ at index x by Γ, i.e., the list $(\ldots, \Gamma_x, \ldots)$. Correspondingly, $C[\Delta_{\mathbf{x}}[x \mapsto \Gamma]]$ is the clause that is obtained from $C[\Delta_{\mathbf{x}}]$ be replacing Δ_x by Γ_x everywhere. For clause sets $N[\Delta_{\mathbf{x}}]$ we define $N[\Delta_{\mathbf{x}}[x \mapsto \Gamma]]$ analogously.

Example 3.2 (continued). The clause set N is of the form $N[\Delta_{\mathbf{x}}]$ where $\mathbf{x} = (x_1, x_2)$ and $\Delta_{x_1} = \Delta_{x_2} = [1..1000]$. Now let $\Pi_{\mathbf{x}} = (\{9\}, \{6\})$. Then $M[\Pi_{\mathbf{x}}] = \text{FD}(N, \Pi_{\mathbf{x}}) = (\text{Cls}_{C_1} \cup \text{Cls}_{C_2}) \cup (\text{Def}_{C_1} \cup \text{Def}_{C_2})$ where $\text{Cls}_{C_2} = \{C_{21}, C_{23}\}$, $\text{Def}_{C_2} = \{C_{22}, C_{24}\}$ and

(C_{21}) $\alpha_3 < 10 \vee \neg(x_2 > 2) \vee x_2 \notin [1..1000] \setminus \{6\}$	(C_{23}) $\alpha_4 < 10 \vee \neg(6 > 2)$
(C_{22}) $f(x_2 + 3) \approx \alpha_3 \vee x_2 \notin [1..1000] \setminus \{6\}$	(C_{24}) $f(6 + 3) \approx \alpha_4$

The clause set $M[\Delta_{\mathbf{x}}[x_2 \mapsto \emptyset]] = M[(\{9\}, \emptyset)]$ is obtained by replacing the two occurrences of $\Delta_{x_2} = [1..1000]$ in C_{21} and C_{22} by the empty interval []. □

We conclude this section with some lemmas that will be needed in the proof of the main correctness result, Theorem 4.2 below. In each of them, $N[\Delta_{\mathbf{x}}]$ is a set of finitely quantified clauses, $\Pi_{\mathbf{x}} \subseteq \Delta_{\mathbf{x}}$, $(\text{Cls}_N, \text{Def}_N) = \text{FD}(N, \Pi_{\mathbf{x}})$, and $M = \text{Cls}_N \cup \text{Def}_N$.

Lemma 3.7. $\text{Cls}_N \cup \text{Def}_N$ *is $\mathcal{B}$-satisfiable iff $N \cup \text{Def}_N$ is $\mathcal{B}$-satisfiable.*

Proof. For the if-direction assume that $N \cup \mathrm{Def}_N$ is $\mathcal{B}$-satisfiable. It suffices to show that $N \cup \mathrm{Cls}_N \cup \mathrm{Def}_N$ is $\mathcal{B}$-satisfiable. Observe that all clauses in Cls_N can be seen to be obtained by paramodulation inferences from clauses in $N \cup \mathrm{Def}_N$, which are all logical consequences of $N \cup \mathrm{Def}_N$.

For the only-if direction assume that $\mathrm{Cls}_N \cup \mathrm{Def}_N$ is $\mathcal{B}$-satisfiable. The definitions in Def_N are exhaustive in the sense that any instance C of a finitely quantified clause in N obtained by ground instantiation with domain elements is congruent with some clause in Cls_N obtained by paramodulation with clauses in Def_N. This entails that $N \cup \mathrm{Cls}_N \cup \mathrm{Def}_N$ is $\mathcal{B}$-satisfiable, and hence so is $N \cup \mathrm{Def}_N$.   □

Lemma 3.8. *If $M[\emptyset_\mathbf{x}]$ is $\mathcal{B}$-unsatisfiable then N and N' are $\mathcal{B}$-unsatisfiable, where N' is obtained from N by removing from all clauses all domain predicates.*

Proof. Assume that $M[\emptyset_\mathbf{x}]$ is $\mathcal{B}$-unsatisfiable. Every clause in $M[\Delta_\mathbf{x}]$ that contains a pseudo-literal of the form $x \notin \Delta_x \setminus \Pi_x$, for some $x \in \mathbf{x}$, becomes a tautology in $M[\emptyset_\mathbf{x}]$ after replacing $x \notin \Delta_x \setminus \Pi_x$ by $x \notin \emptyset \setminus \Pi_x$. Deleting all these tautologies leaves us with a ($\mathcal{B}$-unsatisfiable) set $M' \subseteq M[\emptyset_\mathbf{x}]$. All clauses in M' are either ground definitions in Def_N of the form $t \approx \alpha$ (cf. Definition 3.3), or clauses in Cls_N that are obtained by (re-)peated) paramodulation of the sub-clause D of a clause $C \in N$ (cf. again Definition 3.3) such that all instantiated domain predicates in the instance $C\gamma$ are satisfied. Clearly, adding such definitions to N preserves $\mathcal{B}$-satisfiability. The $\mathcal{B}$-unsatisfiability of both N and N' then follows from the soundness of paramodulation.   □

Lemma 3.9. *Let $\Gamma_\mathbf{x}$ be a vector of sets of domain elements of the proper sorts. For every $x \in \mathbf{x}$ and $d \in \Pi_x$, if $M[\Gamma_\mathbf{x}]$ is $\mathcal{B}$-satisfiable then $M[\Gamma_\mathbf{x}[x \mapsto \Gamma_x \cup \{d\}]]$ is $\mathcal{B}$-satisfiable.*

Proof. All occurrences of Γ_x in clauses in $M[\Gamma_\mathbf{x}]$ are within pseudo-literals of the form $x \notin \Gamma_x \setminus \Pi_x$. We are given $d \in \Pi_x$. It follows trivially that $\Gamma_x \setminus \Pi_x$ and $(\Gamma_x \cup \{d\}) \setminus \Pi_x$ are the same sets, which immediately entails the claim.   □

Example 3.2 (continued). Let $M[\Delta_{(x_1)}] = \mathrm{FD}(C_1, \Pi_{(x_1)})$ from above. Let $\Gamma_{(x_1)} = ([5..500])$ and $d = 9$. Then $M[\Gamma_{(x_1)}[x_1 \mapsto \Gamma_{x_1} \cup \{d\}]]$ consists of the clauses

(C'_{11}) $\alpha_1 > x_1 \vee x_1 \notin ([5..500] \cup \{9\}) \setminus \{9\}$ (C_{13}) $\alpha_2 > 9$
(C'_{12}) $\mathsf{f}(x_1) \approx \alpha_1 \vee x_1 \notin ([5..500] \cup \{9\}) \setminus \{9\}$ (C_{14}) $\mathsf{f}(9) \approx \alpha_2$

Lemma 3.9 requires $d \in \Pi_x$. Adding d to Γ_x does not change anything, as d is again removed from $\Gamma_x \cup \{d\}$: the sets $([5..500] \cup \{9\}) \setminus \{9\}$ and $[5..500] \setminus \{9\}$ are the same.   □

4   Checking Satisfiability

Next we define a procedure checkSAT for checking the $\mathcal{B}$-satisfiability of sets of finitely quantified clauses. It repeatedly applies the finite domain transformation wrt. growing sets of exception points. It stops if a transformed set has been found that is either $\mathcal{B}$-satisfiable or serves to demonstrate $\mathcal{B}$-unsatisfiability.

```
1   algorithm checkSAT(N[Δx])
2   // returns "B-satisfiable" or "B-unsatisfiable"
3   var Πx := ∅x // The current set of exceptions
4   while true {
5     let M = FD(N, Πx)
6     if M is B-satisfiable return "B-satisfiable" // justified by Lemma 3.7
7     if M[∅x] is B-unsatisfiable return "B-unsatisfiable" // justified by Lemma 3.8
8     let (x, d) = find(M)
9     Πx := Πx[x ↦ Πx ∪ {d}]
10  }
```

```
1   algorithm find(M[Δx])
2   // returns a pair (x, d) such that x ∈ x and d ∈ Δx \ Πx
3   let (x1, ..., xn) = x
4   for i = 1 to n {
5     if M[∅(x1,...,xi) · Δ(xi+1,...,xn)] is B-satisfiable {
6       let Γ ⊆ Δxi and d ∈ Γ such that
7         M[∅(x1,...,xi-1) · Γxi · Δ(xi+1,...,xn)] is B-unsatisfiable and
8         M[∅(x1,...,xi-1) · (Γ \ {d})xi · Δ(xi+1,...,xn)] is B-satisfiable // see text
9       return (xi, d) // from Lemma 3.9 it follows d ∈ Δx \ Πx as claimed
10    }
11  }
```

We tacitly assume that the B-satisfiability tests in checkSAT and find are effective. This is always the case, e.g., if there are no FG operators other than free BG-sorted operators and the EA-fragment of the background theory is decidable.

Let us go through the run of checkSAT(N), where $N = \{C_1, C_2\}$ from Example 3.2. Let $\Pi_x^1 = (\emptyset, \emptyset)$ be the initially empty set of exceptions set in line 3. For $M^1 = FD(N, \Pi_x^1)$ in line 5 none of the termination cases applies, hence find is called. The condition in the for-loop in find is satisfied for $i = 1$. In line 6, a suitable set Γ is the interval $[1..9]$ and $d = 9$, as $M^1[([1..9], \Delta_{x_2})]$ is B-unsatisfiable and $M^1[([1..9] \setminus \{9\}, \Delta_{x_2})]$ is B-unsatisfiable. The call of find(M^1) hence returns the pair $(x_1, 9)$. (In the proof of Lemma 4.1 below we show how Γ and $d \in \Gamma$ can be found efficiently by binary search in the case of (linear) integer arithmetic.)

The updated set Π_x^2 in checkSAT now is $(\{9\}, \emptyset)$ and we get $M^2[\Delta_x] = FD(N, \Pi_x^2)$ in the next iteration. Again, the termination tests do not apply and find(M^2) is called. This time $M^2[(\emptyset, \Delta_{x_2})]$ is B-unsatisfiable and the result of find(M^2) is $(x_2, 6)$.

The updated set Π_x^3 hence is $(\{9\}, \{6\})$ and $M^3[\Delta_x] = FD(N, \Pi_x^3)$ consists of the clauses C_{11}–C_{14} and C_{21}–C_{24} already shown above. In the next iteration, the set $M^3[\emptyset_x]$ is built, which is obtained by replacing the sets $\Delta_{x_1} = \Delta_{x_2} = [1..1000]$ everywhere by the empty interval $[]$:

(C'_{11}) $\alpha_1 > x_1 \lor x_1 \notin [] \setminus \{9\}$	(C_{13}) $\alpha_2 > 9$
(C'_{12}) $f(x_1) \approx \alpha_1 \lor x_1 \notin [] \setminus \{9\}$	(C_{14}) $f(9) \approx \alpha_2$
(C'_{21}) $\alpha_3 < 10 \lor \neg(x_2 > 2) \lor x_2 \notin [] \setminus \{6\}$	(C_{23}) $\alpha_4 < 10 \lor \neg(6 > 2)$
(C'_{22}) $f(x_2 + 3) \approx \alpha_3 \lor x_2 \notin [] \setminus \{6\}$	(C_{24}) $f(6 + 3) \approx \alpha_4$

By construction, all clauses affected by the replacement are tautological. Yet, the set $M^3[\emptyset_{\mathbf{x}}]$ is $\mathcal{B}$-unsatisfiable, which can be seen easily from the clauses in the right column. The algorithm returns "$\mathcal{B}$-unsatisfiable". This is indeed correct, as, by construction, the remaining non-tautological clauses contain and use definitions for *ground* instances of the f-terms only. Because of that, our method is sound wrt. $\mathcal{B}$-unsatisfiability even for non-finitely quantified clause sets as expressed in Lemma 3.8 above.

Notice that find searches for the set Γ wrt. the whole set $M = \mathrm{FD}(N, \Pi_{\mathbf{x}}) = \mathrm{Cls}_N \cup \mathrm{Def}_N$. It would be tempting to fix Def_N and search only wrt. Cls_N (or vice versa) but this would be unsound. An example for that is the clause set $N = \{f(x) \geq 0 \vee x \notin \Delta, f(3) \approx 3, f(4) \approx 4\}$, where $\Delta = [0..1000]$. Using the default interpretation we get $\mathrm{Cls}_N = \{\alpha_1 \geq 0 \vee x \notin \Delta, \alpha_2 \approx 3, \alpha_3 \approx 4\}$ and $\mathrm{Def}_N = \{f(x) \approx \alpha_1 \vee x \notin \Delta, f(3) \approx \alpha_2, f(4) \approx \alpha_3\}$. While $\mathrm{Cls}_N[\emptyset] \cup \mathrm{Def}_N$ is $\mathcal{B}$-unsatisfiable, N is $\mathcal{B}$-satisfiable. Hence the procedure in that form would be unsound.

Lemma 4.1. *Whenever* find *is called from* checkSAT *on line 8 then the if-clause in the for-loop in* find *is executed for some i, and* find *returns a pair* (x_i, d) *such that* $x_i \in \mathbf{x}$ *and* $d \in \Delta_{x_i} \setminus \Pi_{x_i}$.

Proof. Assume find$(M[\Delta_{\mathbf{x}}])$ is executed and that $\mathbf{x}$ is of the form $(x_1, \ldots, x_n)$. Because the test in line 7 in checkSAT has not applied it follows that the condition in line 5 in find is satisfied for some i in $1, \ldots, n$. Among all these values, the if-clause is executed for the least one. That is, $M[\emptyset_{(x_1,\ldots,x_i)} \cdot \Delta_{(x_{i+1},\ldots,x_n)}]$ is $\mathcal{B}$-satisfiable and $M[\emptyset_{(x_1,\ldots,x_j)} \cdot \Delta_{(x_{j+1},\ldots,x_n)}]$ is $\mathcal{B}$-unsatisfiable, for all j with $1 \leq j < i \leq n$. Because $i \geq 1$ we can rewrite the former and obtain that $M[\emptyset_{(x_1,\ldots,x_{i-1})} \cdot \emptyset_{x_i} \cdot \Delta_{(x_{i+1},\ldots,x_n)}]$ is $\mathcal{B}$-satisfiable. Furthermore, $M[\emptyset_{(x_1,\ldots,x_{i-1})} \cdot \Delta_{x_i} \cdot \Delta_{(x_{i+1},\ldots,x_n)}]$ is $\mathcal{B}$-unsatisfiable: if $i = 1$ this follows from the fact that the test in line 6 in checkSAT has not applied, and if $i > 1$ this follows from the minimality of i. This shows that Γ and a $d \in \Gamma$ exists as claimed in lines 7 and 8.

In our main application of integer arithmetic the set Γ and $d \in \Gamma$ can be determined efficiently, as follows: We assume the set Δ_{x_i} is an interval of the form $[l..u]$ for some numbers l and u with $l < u$. From the above it follows there is a maximal number u' with $l < u' \leq u$ such that $\Gamma := [l..u']$ is as claimed. The number u' can be determined by binary search in the interval $[l + 1..u]$. By maximality, u' is the desired element d. $\square$

For termination of checkSAT, instead of determining the pair (x, d) in line 11 by the call to find, one could choose any (x, d) such that the current set $\Pi_{\mathbf{x}}$ grows. An advantage of using find, however, is that the relevant ground instances of the clauses $C_1[x_1]$ and $C_2[x_2]$, which are $C_1[9]$ and $C_2[6]$, have been found through semantic guidance by refining the default interpretation in only two steps.

In general terms, checkSAT/find realizes a *heuristic* that tries to search for a model by deviating from the current interpretation only when a conflict arises. The conflict is identified by the point d for the variable x_i in Line 8 of find. The next round of checkSAT continues with the correspondingly updated current interpretation by adding d to Π_{x_i}, which may stop now with "satisfiable", "unsatisfiable" or continue the search. We summarize the essential properties of checkSAT in our main result as follows.

Theorem 4.2 (Correctness of checkSAT). *For any set N of finitely quantified clauses,* checkSAT(N) *terminates with the correct result "$\mathcal{B}$-satisfiable" or "$\mathcal{B}$-unsatisfiable"* for N. Moreover, in case of "$\mathcal{B}$-unsatisfiable" the non-domain restricted version of N is $\mathcal{B}$-unsatisfiable, which is obtained from N by removing from all clauses all domain predicates.

Proof. Termination follows from the fact that find always returns some pair (x, d) such that $x \in \mathbf{x}$ and $d \in \Delta_x \setminus \Pi_x$, as shown in Lemma 4.1. Hence, the set Π_x grows monotonically in line 12 in checkSAT and there are only finitely many elements in Δ_x available for that. Correctness follows from the lemmas in Section 3 as referenced in the comments in checkSAT. □

5 Experimental Results

We have implemented the checkSAT/find algorithm on top of the hierarchic superposition prover Beagle [7].[3] The implementation is prototypical and currently serves only to try out the ideas in the paper. Table 1 summarizes the experiments we carried out.

Table 1. Experimental results. Problem 4 is $\{f(x) \not\approx x \vee x \notin \Delta,\ f(5) \approx 8,\ f(8) \approx 5\}$

#	Problem	$|\Delta|$	#Iter	#TP	Time
1	$f(x) > 1 + y \vee y < 0 \vee x \notin \Delta$	any	1	1	<1
2	$g(x) \approx x \vee g(x) \approx x + 1 \vee \neg(x \geq 0)$	10	9	32	5.5
	$g(x) \approx -x \vee \neg(x < 0)$	20	20	86	55
	$f(x) < g(x) \vee x \notin \Delta$				
3	$f(x_1, x_2, x_3, x_4) > x_1 + x_2 + x_3 + x_4 \vee$	any	1	1	<1
	$x_1 \notin \Delta \vee x_2 \notin \Delta \vee x_3 \notin \Delta \vee x_4 \notin \Delta$				

	4- see caption			5- see Section 1			6- see Example 3.2			6alt- see text		
$|\Delta|$	#Iter	#TP	Time	#Iter	#TP	Time	#Iter	#TP	Time	#Iter	#TP	Time
10	2	5	<1	3	15	2.3	3	12	<1	5	25	1.5
20	2	6	<1	3	17	2.6	3	14	<1	15	87	4.4
50	2	8	<1	3	19	2.8	3	19	1.1	34	239	23
100	2	9	<1	3	21	2.8	3	21	1.1	59	456	181
200	2	10	<1	3	23	2.8	3	23	1.2			
500	2	11	<1	3	25	2.9	3	24	1.2			
1000	2	12	<1	3	27	3.0	3	26	1.3			
2000	2	13	<1	3	29	3.0	3	28	1.4			
5000	2	15	<1	3	33	3.5	3	32	1.5			

We have tried six problems, some of them with varying domain sizes. The problems (1) and (6) are $\mathcal{B}$-unsatisfiable, the others $\mathcal{B}$-satisfiable. The "Problem" column contains the individual clause sets. The column "$|\Delta|$" gives the size of the finite domains uniformly used in the problem clauses, e.g., $|\Delta| = 50$ means the range $[1..50]$. The column "#Iter" is the number of while-loop iterations in checkSAT needed to solve the problem for the given Δ. The column "#TP" is the number of theorem prover calls (Beagle calls)

---

[3] http://users.cecs.anu.edu.au/~baumgart/systems/beagle/

stemming from the various $\mathcal{B}$-satisfiability checks in checkSAT/find. Finally, "Time" is the total CPU time needed to solve the problem. All experiments were carried out on a Linux desktop with a quad-core Intel i7 cpu running at 2.8 GHz. For comparison, we have also run Microsoft's SMT-solver Z3 [17], version 4.1, on our examples, using the obvious formula representation of the domains Δ.

Some comments on the individual problems. Problem (1) is trivially solved, for any Δ. In fact, the default interpretation is sufficient for that. Notice that the variable y is not finitely quantified (and does not need to be). Z3 reports "unknown" on problem (1), but, surprisingly it solves the essentially same problem $f(x) > y \lor y < 0$ quickly. Problem (2) is meant to showcase our algorithm in conjunction with Beagle's theorem proving capabilities. The function symbol g is "sufficiently complete" defined by the first two clauses, and only the third clause containing the function symbol f needs finite quantification. Z3 could not solve this problem within three minutes. We devised problem (3) to get some insight into Z3's capabilities on the problems we are interested in. While it is trivial for our approach, Z3 seems to instantiate the clause in problem (3). Clearly, there is a scalability issue here, as for about $|\Delta| > 60$ the problem becomes unsolvable in reasonable time.

As a side note, we found Z3s performance impressive, and it could solve problems (4)–(6) in very short time. Indeed, we plan to integrate Z3 in our approach and expect much better performance on many problems (Beagle's theory reasoning component is a rather slow implementation of Cooper's quantifier elimination algorithm.)

Problem (4) is a simple test of the default interpretation/exception mechanism. Problem (5) is the one in the Introduction, and problem (6) is our running example.

The problems (4), (5) and (6) scale very well, as expected. The first two are proven satisfiable using the default interpretation and a fixed number of exception points. In problem (4) these are easily discovered from the problem and in (5) the exceptions are quickly discovered by the search. Similarly, in problem (6) the definition for $f(9)$ is found quickly, which is the only one needed to establish unsatisfiability. However, this requires to search first the domain of x_1, then x_2 (cf. Example 3.2). With the other way round we obtain much worse scaling behavior, cf. the entry "6alt" in Table 1.

6   Conclusions

We have presented a method for deciding hierarchic satisfiability, or satisfiability modulo theories, of first-order clause sets where all variables are quantified over finite subsets of background domains. The method tries to construct a model by stepwise amending a default interpretation in a conflict-driven way by utilizing a decision procedure for the EA-fragment of the background theory. It may also terminate with a set of ground instances witnessing that no model exists. For space reasons and for clarity we have focused in this paper on the basic principles and leave extensions for future work. Here are some ideas.

Richer input language: One important extension concerns foreground-sorted variables and operators, like the array-sorted variable a and the write-operator in clauses (1) and (2) in the introduction. In the example we got away without further modifications because the axioms (1) and (2) do not pose problems for sufficient completeness and for termination of hierarchic superposition. The question is under which conditions this is possible in general. One could also try to enumerate finite segments of the foreground domains in a Herbrand fashion, similarly as with background domains.

Our method can also be applied to certain richer syntactic fragments that require a full-fledged theorem prover for hierarchic specifications instead of a decision procedure for the background theory. However, this would "reverse" the common architecture by invoking that foreground reasoner from within an outer loop. This is problematic, however, because the foreground reasoner might not terminate or be incomplete. To fix that, it should be possible, under certain conditions, to instead integrate the checkSAT as an inference rule into, say, hierarchic superposition and apply it only to finitely quantified clauses as defined above. (This would directly generalize the Define-rule in [7].)

Alternative default interpretation: Taking the constant function as the default interpretation for free BG-sorted operators is not always a good choice. For example, for the clause $f(x) \approx x \vee x \notin [1..1000]$ our method needs to amend the default interpretation at every point. Fortunately, any interpretation can be used as a default, and the identity function as the default interpretation for f leads immediately to a model. (On the other hand, in this example f is already sufficiently defined and could possibly be excepted from the transformation in the first place.)

Bernays-Schönfinkel fragment: The hierarchic superposition calculus can immediately be instantiated with, say, an instance-based method for deciding background theories that are given as a set of EPR-clauses. Our method, or the extensions above, could possibly be used to integrate arithmetic reasoners, instance-based methods and superposition in a beneficial way.

References

1. Althaus, E., Kruglov, E., Weidenbach, C.: Superposition modulo linear arithmetic SUP(LA). In: Ghilardi, S., Sebastiani, R. (eds.) FroCoS 2009. LNCS (LNAI), vol. 5749, pp. 84–99. Springer, Heidelberg (2009)
2. Bachmair, L., Ganzinger, H.: Rewrite-based equational theorem proving with selection and simplification. Journal of Logic and Computation 4(3), 217–247 (1994)
3. Bachmair, L., Ganzinger, H., Waldmann, U.: Refutational theorem proving for hierarchic first-order theories. Appl. Algebra Eng. Commun. Comput 5, 193–212 (1994)
4. Baumgartner, P., Fuchs, A., de Nivelle, H., Tinelli, C.: Computing finite models by reduction to function-free clause logic. Journal of Applied Logic 7(1), 58–74 (2009)
5. Baumgartner, P., Tinelli, C.: Model evolution with equality modulo built-in theories. In: Bjørner, N., Sofronie-Stokkermans, V. (eds.) CADE 2011. LNCS (LNAI), vol. 6803, pp. 85–100. Springer, Heidelberg (2011)
6. Baumgartner, P., Waldmann, U.: Hierarchic superposition: Completeness without compactness. In: Kosta, M., Sturm, T. (eds.) MACIS (2013)
7. Baumgartner, P., Waldmann, U.: Hierarchic superposition with weak abstraction. In: Bonacina, M.P. (ed.) CADE 2013. LNCS (LNAI), vol. 7898, pp. 39–57. Springer, Heidelberg (2013)
8. Claessen, K., Sörensson, N.: New techniques that improve MACE-style finite model building. In: Baumgartner, P., Fermüller, C.G. (eds.) CADE-19 Workshop: Model Computation – Principles, Algorithms, Applications (2003)

9. Ganzinger, H., Korovin, K.: Theory instantiation. In: Hermann, M., Voronkov, A. (eds.) LPAR 2006. LNCS (LNAI), vol. 4246, pp. 497–511. Springer, Heidelberg (2006)

10. Ge, Y., Barrett, C.W., Tinelli, C.: Solving quantified verification conditions using satisfiability modulo theories. In: Pfenning, F. (ed.) CADE 2007. LNCS (LNAI), vol. 4603, pp. 167–182. Springer, Heidelberg (2007)

11. Ge, Y., de Moura, L.: Complete instantiation for quantified formulas in satisfiabiliby modulo theories. In: Bouajjani, A., Maler, O. (eds.) CAV 2009. LNCS, vol. 5643, pp. 306–320. Springer, Heidelberg (2009)

12. Halpern, J.: Presburger Arithmetic With Unary Predicates is Π_1^1-Complete. Journal of Symbolic Logic 56(2), 637–642 (1991)

13. Korovin, K., Voronkov, A.: Integrating linear arithmetic into superposition calculus. In: Duparc, J., Henzinger, T.A. (eds.) CSL 2007. LNCS, vol. 4646, pp. 223–237. Springer, Heidelberg (2007)

14. Kruglov, E., Weidenbach, C.: Superposition decides the first-order logic fragment over ground theories. In: Mathematics in Computer Science, pp. 1–30 (2012)

15. McCune, W.: Mace4 reference manual and guide. Technical Report ANL/MCS-TM-264, Argonne National Laboratory (2003)

16. de Moura, L., Bjørner, N.S.: Efficient E-matching for SMT solvers. In: Pfenning, F. (ed.) CADE 2007. LNCS (LNAI), vol. 4603, pp. 183–198. Springer, Heidelberg (2007)

17. de Moura, L., Bjørner, N.S.: Z3: An efficient SMT solver. In: Ramakrishnan, C.R., Rehof, J. (eds.) TACAS 2008. LNCS, vol. 4963, pp. 337–340. Springer, Heidelberg (2008)

18. Nieuwenhuis, R., Oliveras, A., Tinelli, C.: Solving SAT and SAT modulo theories: from an abstract Davis-Putnam-Logemann-Loveland Procedure to DPLL(T). Journal of the ACM 53(6), 937–977 (2006)

19. Reynolds, A., Tinelli, C., Goel, A., Krstić, S.: Finite model finding in SMT. In: Sharygina, N., Veith, H. (eds.) CAV 2013. LNCS, vol. 8044, pp. 640–655. Springer, Heidelberg (2013)

20. Reynolds, A., Tinelli, C., Goel, A., Krstić, S., Deters, M., Barrett, C.: Quantifier instantiation techniques for finite model finding in SMT. In: Bonacina, M.P. (ed.) CADE 2013. LNCS (LNAI), vol. 7898, pp. 377–391. Springer, Heidelberg (2013)

21. Rümmer, P.: A constraint sequent calculus for first-order logic with linear integer arithmetic. In: Cervesato, I., Veith, H., Voronkov, A. (eds.) LPAR 2008. LNCS (LNAI), vol. 5330, pp. 274–289. Springer, Heidelberg (2008)

22. Slaney, J.: Finder (finite domain enumerator): Notes and guide. Technical Report TR-ARP-1/92, Australian National University, Automated Reasoning Project, Canberra (1992)

23. Zhang, J., Zhang, H.: SEM: a system for enumerating models. In: Mellish, C. (ed.) IJCAI 1995. Morgan Kaufmann (1995)

Computing All Implied Equalities
via SMT-Based Partition Refinement

Josh Berdine and Nikolaj Bjørner

Microsoft Research
{jjb,nbjorner}@microsoft.com

Abstract. Consequence finding is used in many applications of deduction. This paper develops and evaluates a suite of optimized SMT-based algorithms for computing equality consequences over arbitrary formulas and theories supported by SMT solvers. It is inspired by an application in the SLAYER analyzer, where our new algorithms are commonly 10–$100x$ faster than simpler algorithms. The main idea is to incrementally refine an initially coarse partition using models extracted from a solver. Our approach requires only $O(N)$ solver calls for N terms, but in the worst case creates $O(N^2)$ fresh subformulas. Simpler algorithms, in contrast, require $O(N^2)$ solver calls. We also describe an asymptotically superior algorithm that requires $O(N)$ solver calls and only $O(N \log N)$ fresh subformulas. We evaluate algorithms which reduce the number of fresh formulas required either by using specialized data structures or by relying on subformula sharing.

Keywords: Implied Equalities, Consequence Finding, Satisfiability Modulo Theories, Decision Procedures, Congruence Closure, Software Verification.

1 Introduction

We define and evaluate optimized algorithms for computing all equalities between terms of a fixed set that are implied by a fixed constraint. As a example, consider the formula

$$\Phi : (a \simeq b \wedge b[i] \simeq c) \vee (a[i]-4 \simeq d \wedge f(d) \simeq d+3 \wedge f(d)+1 \simeq c)$$

where _[_] denotes array selection. Φ implies the equality $a[i] \simeq c$, but not $a \simeq b$, $a[i] \simeq b[i]$, nor $d \simeq c$. On the other hand, the formula

$$\Phi' : \Phi \vee (b[i] \simeq c \wedge c \simeq d)$$

does not imply any equality between $a, b, c, d, a[i]$ and $b[i]$. We describe and evaluate algorithms that require a number of SMT-solver calls that is at worst linear in the number of terms N, although they may potentially create quadratically-many fresh literals. Naïve algorithms for computing all equalities implied by a formula require $O(N^2)$ SMT-solver calls. The simplest, called *Basic Partition Merging* (BPM), is 10–$1000x$ slower than a model-based variant, *Model-based Partition Merging* (MPM). Starting with *Basic* (BPR) and *Incremental* (IPR) *Partition Refinement* algorithms, we examine several variants that are all significantly (1–$100x$) faster than MPM. In a quest of asymptotically better solutions, we also outline an algorithm that requires only $O(N \log N)$ fresh subformulas.

S. Demri, D. Kapur, and C. Weidenbach (Eds.): IJCAR 2014, LNAI 8562, pp. 168–183, 2014.
© Springer International Publishing Switzerland 2014

Application and Evaluation of Equality Inference.   The experimental evaluation of the algorithms uses problem instances that the SLAYER [2] program analyzer encounters while attempting to verify memory safety properties of C programs. SLAYER relies on learning implied equalities, and previously used simpler algorithms that would sometimes exhibit catastrophic performance. This prompted the development of more sophisticated algorithms. We evaluated both the folklore and the new algorithms in the context of the SLAYER tool and found that the simpler algorithms are impractical because they either require a quadratic number of solver calls, or are non-incremental: they effectively reset the solver state between several calls. The practical evaluation demonstrates the advantages of the new algorithms.

Consequence finding is a central component of abstract interpreters. The set of reachable states can be approximated by the least fixed point of a predicate transformer, and analyses based on abstract interpretation commonly develop special classes and representations of logical formulas where approximations of the least fixed point can be computed. For example, analyses using the octagon abstract domain [10] can be thought of as computing consequences as a conjunction of constraints using unit coefficients and two variables per inequality. The TVLA system [9,15] is distinguished as it produces shape formulas as a result of bottom-up evaluation of Horn clauses.

The SLAYER tool synthesizes separation logic formulas in an approximation of the least fixed-point semantics of programs. This approach serves as a foundation for a verification tool for analyzing heap properties of C programs. In order to compute precise abstractions, SLAYER relies on learning all implied equalities from a formula. An example symbolic state is the formula $\exists x, y.\ \Phi * \Psi$, where Φ is defined above and

$$\Psi : \mathsf{list}(p, x) * x \mapsto c * \mathsf{list}(q, y) * y \mapsto a[i] \ .$$

SLAYER weakens the formula to $\mathsf{list}(p, c) * \mathsf{list}(q, c)$. The first step of this abstraction is to replace $\mathsf{list}(p, x) * x \mapsto c$ with $\mathsf{list}(p, c)$, forgetting that x is on the list from p to c. This rewrite would not be performed if x could begin a shared tail between two lists. Checking this requires learning that neither $x \simeq q$, $x \simeq y$ nor $x \simeq a[i]$ are implied, any of which would make x begin a tail shared between p and q. Finally, $\mathsf{list}(q, y) * y \mapsto c$ is rewritten to $\mathsf{list}(q, c)$. This inference requires learning $a[i] \simeq c$, as previously discussed.

Note that the core of checking the shared-tail condition is to compare the sets of predecessors in the transitive closure of the equivalence closure of the union of the $\mapsto$ and list relations. The use of equivalence closure necessitates an eager computation of the equality consequences of formulas.

Related Work.   Classical congruence closure [6] infers equalities from conjunctions of equalities with uninterpreted functions. In contrast, the problem addressed here is to infer congruences modulo arbitrary formulas (e.g., clauses instead of conjunctions), over theories supported by SMT solvers. Satisfiability Modulo Theories solvers that are based on the DPLL(T) architecture [12] use congruence closure to check satisfiability of a conjunction of assumed equalities with respect to assumed disequalities. Saturation based theorem provers for classical first-order logic use superposition inference rules to deduce new equalities from old ones. The algorithmic problems of unification and congruence closure have received significant attention and enjoyed several celebrated results, such as linear time algorithms for unification [13,14] and efficient congruence closure algorithms [6]. Saturation procedures rely on term indexing and unification algorithms and higher performance saturation engines strike a trade-off between indexing data structures and the unification algorithms that are used in practice [8]. Similarly, SMT solvers use congruence closure algorithms that have shown to perform well in their context of use [5,11]. For instance, Z3's congruence closure algorithm uses a variant of union-find with eager path compression.

Equality inference of Boolean functions is a deeply studied subject in the context of circuit verification [3,4] because when checking equality between two circuits it can be a significant advantage to know that two sub-circuits compute the same Boolean function. Some of the main techniques use binary decision diagrams, BDDs, for identifying sub-circuit equivalence. Note that equivalence of Boolean functions is a special case of the problem we consider here: we are determining equivalence of not only Boolean functions, but functions with any signature (e.g., functions over reals and integers). Additionally, for the sub-circuit equivalence problem, it suffices to find enough equalities to speed up the equivalence check, while we consider the problem of computing the complete set. Detecting and using equivalences can also have profound effects on SAT solving [7]. The use of SAT sweeping is critical for reducing the set of candidate equivalences. SAT sweeping and possible generalizations to SMT provides orthogonal value to the algorithms developed here.

Organization. Section 2 first recalls a few technical preliminaries. Section 3 discusses the simpler algorithms and then presents the Basic and Incremental Partition Refinement algorithms and variants. The practical context where the algorithms are used is discussed in Section 4. Section 5 closes with a thorough evaluation.

2   Preliminaries

We assume some basic familiarity with SMT solving and, to a greater degree, congruence closure.

SMT, Models and Formulas.   We use standard notions of sorts, terms, formulas and interpretations. Formulas are terms of Boolean sort. Terms are built using a first-order signature where functions, such as $+$ and $\_[\_]$ can be interpreted. We assume that interpretations, denoted $\mathcal{M}$ can be used to evaluate terms to values. For example, if $\mathcal{M} \models x+2 \simeq 1$, then $\mathcal{M}(x) = -1$.

Union-Find.   Our algorithms maintain partitions and use the well-known union and find routines [16]. Given a domain E of nodes and a domain $\mathcal{P}$ of partitions of E, a partition P of $\{1, \ldots, N\}$ is a set of disjoint subsets, called classes, that cover the set. We assume the following routines:

- *find* : $\mathcal{P} \times E \to E$, such that $find(P, i)$ returns a unique representative from the equivalence class of i. Thus, $find(P, find(P, i)) = find(P, i)$.
- *union* : $\mathcal{P} \times E \times E \to \mathcal{P}$, such that the equivalence classes of i and j are merged in the result of $union(P, i, j)$. Thus $find(union(P, i, j), i) = find(union(P, i, j), j)$.

We furthermore use a nonstandard routine to split equivalence classes by removing an element from a class and creating a new singleton class:

- *remove* : $\mathcal{P} \times E \rightarrow \mathcal{P}$, such that $find(remove(P, i), j) = i$ if and only if $i = j$.

We only need to call $remove(P, i)$ in the case when i is not an equivalence class representative. So if we implement *union* using eager path compression, then *remove* is realized by detaching the removed node from a doubly-linked list.

3   Algorithms for Implied Equalities

The problem of computing all implied equalities can be stated as: given a formula Φ and a set of terms $t_1, \ldots, t_N$, find a partition P of $\{1, \ldots, N\}$, such that for every $p, q \in P$ and $s \in p, t \in q$: $(\Phi \rightarrow s \simeq t)$ is valid if and only if $p = q$. To set the stage for our partition refinement algorithms, we begin by discussing some simpler algorithms.

3.1   Basic Partition Merging (BPM)

The most straightforward approach for finding all implied equalities is to check whether an equality is implied for each pair of terms. In more detail, create a partition P of the indices $\{1, \ldots, N\}$ by checking whether $\Phi \rightarrow t_i \simeq t_j$ for each pair $1 \leq i < j \leq N$. One can save a redundant check if j is already merged with another index $k < j$. The union-find data structure can be used to maintain the partition as it is built. Asymptotically, this straightforward algorithm requires $O(N^2)$ solver calls in the worst case.

Example 1 Consider again the formula Φ : $(a \simeq b \,\wedge\, b[i] \simeq c) \,\vee\, (a[i]{-}4 \simeq d \,\wedge\, f(d) \simeq d{+}3 \,\wedge\, f(d){+}1 \simeq c)$. We wish to partition the terms $\{a, b, c, d, a[i], b[i]\}$ in the context of Φ. Only the formula $\Phi \rightarrow a[i] \simeq c$ is valid, so the resulting partition is $\{\{a\}, \{b\}, \{c, a[i]\}, \{d\}, \{b[i]\}\}$. Other validity checks fail. Once $a[i]$ and c are found to be equal, it is redundant to check both $\Phi \rightarrow a[i] \simeq d$ and $\Phi \rightarrow c \simeq d$.   ⌐

3.2   Model-Based Partition Merging (MPM)

The previous algorithm uses very little information between solver calls. More useful information is available when the SMT solver produces *models*. Whenever checking that $\Phi \rightarrow t_i \simeq t_j$ is valid, we do in fact check dually whether $\Phi \wedge t_i \not\simeq t_j$ is unsatisfiable. If it is satisfiable, then a model can be extracted that satisfies Φ and the disequality $t_i \not\simeq t_j$. The model may also distinguish other terms that are not tested for equality. So the set of *potential* equalities that have to be tested can be reduced by inspecting the model after each satisfiability check, exploiting the property that if two terms evaluate to distinct values in a model, their equality cannot be implied. While this algorithm still requires $O(N^2)$ solver calls, all but one of them will be satisfiable, in contrast to BPM where each merge is the result of an unsatisfiable query. This is relevant since solvers often can solve satisfiable queries faster than unsatisfiable ones.

Example 2 Continuing with Φ, suppose that it has a model where $i = 0$, $a[i] = b[i] = c = 1$, $d = a[1] = 2$, $b[1] = 3$. We then know that it only makes sense to check equalities among $\{a[i], b[i], c\}$ instead of the full set including $\{d, a, b\}$.   ⌐

3.3 Basic Partition Refinement (BPR)

We saw that models play a role dual to validity checks: they indicate what equalities are *not* implied. We can take this idea to its fullest extent and develop an algorithm that splits partitions based on models instead of merging partitions based on validity checks. If we ensure that every satisfiability check splits at least one class, then this approach requires at worst only a linear number of solver calls. To start with, we can check satisfiability of $\Phi \wedge \bigvee_{i>1} t_1 \not\simeq t_i$. Any model of the formula must satisfy at least one disequality. The model also produces a more refined partition: only terms that are equal in the current model have to be compared for disequality. So suppose P is the current partition, then we define the formula

$$SomeDiff: \bigvee_{p \in P,\, |p|>1,\, i \in p} \;\; \bigvee_{j \in p,\, i \neq j} t_i \not\simeq t_j$$

such that $\Phi \wedge SomeDiff$ is satisfiable if and only if some class can be split.

Example 3 Continuing the previous example, we create the predicate $SomeDiff : c \not\simeq a[i] \vee c \not\simeq b[i] \vee a \not\simeq b$ corresponding to the partition $\{\{c, a[i], b[i]\}, \{d\}, \{a, b\}\}$. The formula $\Phi \wedge SomeDiff$ is satisfiable where $c \not\simeq b[i]$ is true, $c \not\simeq a[i]$ is false (so $c \simeq a[i]$), and $a \not\simeq b$ is true. The next formula $\Phi \wedge c \not\simeq a[i]$ is unsatisfiable, so we learn that $c \simeq a[i]$ is implied.   ⌐

This method only requires a linear number of solver calls. Yet, it is either non-incremental (the *SomeDiff* constraint is retracted between each solver call), or it requires quadratic space: it has to create fresh disequality literals between subsequent calls and has to create new clauses at every call. That is, it is unclear whether the complexity has simply been shifted around, or in other words, if $O(N)$ solver calls which each receive $O(N^2)$ new constraints is an improvement over $O(N^2)$ solver calls which each receive $O(N)$ new constraints. We will now develop algorithms that address time and space deficiencies of BPR.

3.4 Incremental Partition Refinement (IPR)

The IPR algorithm refines BPR by allowing reuse of literals and clauses between iterations. The algorithm creates a binary heap of propositional variables. The heap has a leaf for each term, where the proposition is constrained to hold if and only if the term is not equal to its representative. The proposition of each internal node holds if and only if one of its children does. The root of this heap is therefore equivalent to *SomeDiff*.

Example 4 Consider computing the equality partition of terms $\{a[i]-4, a, b, c, d, a[i], b[i]\}$ implied, again, by the formula $\Phi : (a \simeq b \wedge b[i] \simeq c) \vee (a[i]-4 \simeq d \wedge f(d) \simeq d+3 \wedge f(d)+1 \simeq c)$. The algorithm starts with the initial partition $P = \{\{1, \ldots, 7\}\}$ and initializes a proposition heap ϕ of size 12, adding the leaf and internal node constraints. The tree structure induced, as well as the associations between propositions ϕ_i and nodes, and between leaves and terms, is shown in the figure.   ⌐

The heap is used in several ways. The first use we describe is to use the truth values of the propositions in a model $\mathcal{M}$ to determine what partitions to refine.

Example 5 Continued: Φ conjoined with the proposition heap constraints is satisfiable, and suppose the extracted model $\mathcal{M}$ satisfies $a[i]-4 \simeq d$, $c \simeq a[i] \simeq b[i]$, $a \simeq b$, $c \not\simeq d$, $a \not\simeq d$ and $a \not\simeq c$. Therefore (assuming for the sake of presentation, that the *find* routine returns the element t_i of a class with minimal i) $\mathcal{M}$ will assign ϕ_6 and ϕ_{10} to *false* and the other propositions to *true*. This is because the term associated with ϕ_6, $t_1 = a[i]-4$ was the representative of the single class, and ϕ_{10} is associated to $t_5 = d$, which is still equal to its representative $a[i]-4$ in $\mathcal{M}$. All other terms are no longer equal to their existing representative. ⌐

The heap is only updated along paths where a model $\mathcal{M}$ established that candidate equalities were found to be not implied.

Example 6 Continued: The constraints for the new partition are constructed by updating the proposition heap using a depth-first traversal starting from the root 0, excluding sub-trees whose propositions do not hold. Since $\mathcal{M}(\phi_i) = \text{true}$ for all i except 6, 10, traversal proceeds from the root 0 to the leaf 7. To update the heap at 7, first the partition is refined to $remove(P, 2)$ by removing the associated term, $t_2 = a$, from its current class. Then ϕ_7 is overwritten with a fresh proposition, and the updated partition is used to conjoin $\phi_7 \leftrightarrow t_2 \not\simeq t_{find(P,2)}$ for the fresh ϕ_7 to Φ, reestablishing the leaf constraint. Next, 8, associated with $t_3 = b$, is visited. This proceeds similarly to the update for 7, except that now, since $\mathcal{M}$ satisfies $a \simeq b$, the partition is updated to $union(remove(P, 3), 3, 2)$, removing b from its current class and merging it into the class of a. The updated partition is again used to extend Φ to reestablish the leaf constraint for a fresh ϕ_8. Next, to update the internal node 3, ϕ_3 is overwritten with a fresh proposition and Φ is extended to reestablish the internal constraint $\phi_3 \leftrightarrow (\phi_7 \vee \phi_8)$ for the new propositions. Updating the heap proceeds similarly until reaching 10. Since $\mathcal{M}(\phi_{10}) = \text{false}$, ϕ_{10} and its existing constraints are reused. Updating then proceeds similarly, refining the partition to $P = \{\{1, 5\}, \{2, 3\}, \{4, 6, 7\}\}$. ⌐

Note how the leaf constraints, where each term is disequal to its representative, mirror an eagerly path-compressed union-find representation of the current partition. This design is significant since it ensures that leaf propositions for representatives are always inconsistent (representatives are their own representatives), and hence an equivalence class representative will never be chosen for removal from its current class. Without this guarantee, the simple implementation of the nonstandard *remove* routine described in Section 2 would not suffice.

Algorithm 1 distills the examples as the IPR algorithm. The binary heap of propositional variables is represented as an array, which is initialized with fresh propositional variables on line 3. Line 4 then adds the heap constraints described above to the input formula. Following the approach illustrated by the preceding examples, the *update* procedure uses the values of the ϕ_i to traverse the part of the binary heap containing satisfied disequalities, constructing new disequalities for the leaves where the partition changes, and rebuilding the binary heap of propositions reusing as many subformulas as possible, so that ϕ_0 is again equivalent to the disjunction of the disequalities between each term and its current representative. Incrementally updating the partition in

---

Algorithm 1: Incremental Partition Refinement (IPR)

Input: formula Φ and set of terms $\{t_1, \ldots, t_N\}$
Output: equality partition P of the set $\{1, \ldots, N\}$

1   $P \leftarrow \{\{1, \ldots, N\}\}$
2   **foreach** $i = 0 \ldots 2N - 2$ **do**
3      $\phi_i \leftarrow$ fresh propositional variable

4   $\Phi \leftarrow \Phi \wedge \phi_0 \wedge \bigwedge\limits_{i=0}^{N-2} (\phi_i \leftrightarrow (\phi_{2i+1} \vee \phi_{2i+2})) \wedge \bigwedge\limits_{i=1}^{N} (\phi_{i+N-2} \leftrightarrow t_i \not\simeq t_{find(P,i)})$

5   **while** Φ *is satisfiable* **do**
6      $\mathcal{M} \leftarrow$ interpretation satisfying Φ
7      $Q \leftarrow \emptyset$
8   **Procedure** $update(i)$ **is**
9   **if** $\mathcal{M}(\phi_i) = true$ **then**
10            $\phi_i \leftarrow$ fresh propositional variable
11   **if** $i < N - 1$ **then**   *// i is internal*
12               $update(2i + 1)$
13               $update(2i + 2)$
14               $\Phi \leftarrow \Phi \wedge (\phi_i \leftrightarrow (\phi_{2i+1} \vee \phi_{2i+2}))$
15   **else**   *// i is a leaf*
16   **let** $j = i - (N - 2)$ *// leaf i is associated with term t_j*
17   **let** $k = find(P, j)$
18   **assert** $k \neq j$
19               $P \leftarrow remove(P, j)$
20   **if** $\langle k, \mathcal{M}(t_j) \rangle \notin \mathbf{dom}\, Q$ **then**
21                  $Q[\langle k, \mathcal{M}(t_j) \rangle] \leftarrow j$
22   **else**
23   **let** $h = Q[\langle k, \mathcal{M}(t_j) \rangle]$
24                  $P \leftarrow union(P, j, h)$
25               $\Phi \leftarrow \Phi \wedge (\phi_i \leftrightarrow t_j \not\simeq t_{find(P,j)})$

26      $update(0)$
27      $\Phi \leftarrow \Phi \wedge \phi_0$

28 **return** P

---

lines 18–24 uses a temporary map Q from pairs of (representatives of) classes of the previous partition and values to classes of the updated partition. This map is used to merge terms in the updated partition that are both in the same class of the previous partition and given the same value by $\mathcal{M}$. This is necessary to avoid the refined P breaking more equalities than $\mathcal{M}$ refuted. Lines 10–25 are executed for each j that should change class, that is, such that $\mathcal{M}$ satisfies $t_j \not\simeq t_{find(P,j)}$. First j is removed from its existing class. (Note that since $\mathcal{M}$ satisfies $t_j \not\simeq t_{find(P,j)}$, $j \neq find(P,j)$, so the *remove* operation is not problematic.) Then if there is not already another class for terms of j's previous class, k, and current value, $\mathcal{M}(t_j)$, then record j as such a class. Otherwise, merge j into the existing class h. Therefore, the effect of lines 26–27 is to strictly refine

P while preserving all equalities that are true in $\mathcal{M}$, and to extend Φ to admit only models which violate the new P.

Example 7 Continued: The map Q is initially empty, so when node 7, associated with $t_2 = a$, of the heap is updated, Q is updated with a mapping from $\langle 1, \mathcal{M}(a) \rangle$ to 2 to record that 2 is the new representative for terms currently in the class $\{\{1, \ldots, 7\}\}$ that are also equal to a in $\mathcal{M}$. Then, when 8, associated with $t_3 = b$, is updated, since $\mathcal{M}$ satisfies $a \simeq b$, Q contains a mapping for $\langle 1, \mathcal{M}(b) \rangle$. So P is updated to merge the classes of 3 and 2.

Continuing with the second iteration, $P = \{\{1, 5\}, \{2, 3\}, \{4, 6, 7\}\}$, and Φ is still satisfiable. Suppose the new model $\mathcal{M}$ still satisfies $a[i] \simeq c$, and now satisfies $a \not\simeq b$, $a[i] - 4 \not\simeq d$ and $c \not\simeq b[i]$, as well as $d \simeq b[i]$. Therefore $\mathcal{M}$ will satisfy ϕ_8, ϕ_{10}, ϕ_{12}, as well as their ancestors, but no others. The updates for 8 and 10 will proceed similarly to the first iteration described above. The update for 12 is similar, but illustrates the necessity of keying the map Q on pairs of the *current class and* value. Note that $\mathcal{M}$ equates $t_5 = d$ and $t_7 = b[i]$, but t_5 and t_7 are in distinct classes of P. Therefore when updating 12, the lookup at line 20 does not find an existing entry, and so does not merge the classes of 7 and 5. Keying Q on only the term values would result in merging classes that have already been split. This would violate the property that the partition is monotonically refined, which is crucial to making only a linear number of solver calls.

Finally at the end of the second iteration, $P = \{\{1\}, \{2\}, \{3\}, \{4, 6\}, \{5\}, \{7\}\}$, and Φ is unsatisfiable.   ⌐

Compared to the BPR algorithm, Algorithm 1 constructs fewer, but still $O(N^2)$, new disequalities. The problem is that terms may change representative many times. Consider a case where no equalities are implied, and an execution where at iteration $i \in 1 \ldots N$ the interpretation satisfies all disequalities $t_j \not\simeq t_{find(P,j)}$ for $i < j \leq N$. Therefore at iteration i, $N - i$ new disequalities will be created, and overall $\sum_{i=1}^{N} N - i = \frac{1}{2} N(N - 1)$ disequalities are created. We here use problem instances encountered in practice to justify that this worst-case scenario is unlikely.

Assumption-Based IPR (ABIPR).   We also experimented with a slight variation of IPR that uses assumptions to control the contents of leaf nodes. The variant uses fresh propositional variables a_i and assertions of the form $a_i \rightarrow (\phi_i \leftrightarrow t_i \not\simeq t_{find(P,i)})$. With these constraints conjoined to Φ, the satisfiability check is replaced with a satisfiability check subject to also assuming the a_i for each leaf. This avoids accumulating assertions and is potentially more incremental. Indeed our experimental evaluation did show improvements in performance for this alternative over IPR, yet the improvements were not major for our evaluation suite.

Incrementality via Term Sharing (HIPR).   Some solvers ensure maximal sharing of terms, that is, if a term that occurs in an existing constraint is constructed, the existing term is reused for the newly constructed one. For such solvers, asserting a constraint that is, e.g., a disjunction of N disequations where most already occur in existing constraints is not significantly more expensive than asserting only the new disequations. Native support for n-ary disjunction is also beneficial in this situation.

We experimented with a hybrid incremental partition refinement (HIPR) algorithm that is a hybrid between BPR and IPR. Like BPR, it asserts a disjunction of N disequations at each refinement iteration. The particular disequations are those at the leaves of IPR's proposition heap, thereby ensuring that many disequations will be shared with those from the previous constraints. In this way, much of the benefit of IPR's incrementality may be realized without the overhead of manipulating the proposition heap. Indeed, our experimental evaluation indicates that, for our evaluation suite, the overhead of manipulating the propositions sometimes significantly outweighs the overhead of repeatedly reconstructing existing disequations.

3.5   Space-Optimized Partition Refinement

We can do even better asymptotically. In the following we outline, omitting lower-level details, a partition refinement based algorithm that takes $O(N)$ iterations and has an overhead of $O(N \log N)$ fresh sub-terms.

The idea is as follows: Similar to Algorithm 1 we will maintain a binary tree rooted in ϕ_0 that covers the current disjunction of disequalities. In contrast to that algorithm, however, we represent each class of a partition as a disjunction *chain* of the form

$$t_1 \not\simeq t_2 \vee t_2 \not\simeq t_3 \vee \cdots \vee t_{N-1} \not\simeq t_N \tag{1}$$

instead of a disjunction *star* of the form

$$t_1 \not\simeq t_j \vee t_2 \not\simeq t_j \vee \cdots \vee t_N \not\simeq t_j$$

where $1 \leq j \leq N$ is the equivalence class representative.

We maintain a two level binary tree where internal nodes are labeled by literals ϕ_i and constraints of the form $\phi_i \leftrightarrow (\phi_{2i+1} \vee \phi_{2i+2})$.

The lower level summarizes the disjunction of asserted disequalities within a current class. The upper level summarizes the disjunction of disequalities for classes of size at least 2. Let us consider the case where one of the classes is refined. So suppose that Φ is satisfiable with model $\mathcal{M}$ and that K out of the $N-1$ disequalities are satisfied by $\mathcal{M}$. The sub-tree that covers the previous class is refined into a tree that covers the new classes. We claim that we require at most $2 \cdot (\log N + K)$ fresh literals for the sub-tree. To see this, consider the sequence (1) where K out of the $N-1$ literals are true and the rest are false. If there are more than two contiguous literals that are false, then two neighbors must be summarized by an internal literal (a literal ϕ_i). We will reuse this summary when rebuilding the tree for the new constraints. The number of literals it takes to cover a sequence of false literals is therefore at most $\log N$. So we can build a tree of size $\log N + K$ above these together with the K fresh literals in the leaves. The total number of fresh literals required to cover the new equivalence classes is therefore $2 \cdot (\log N + K)$ (half for the leaves and half for the internal nodes). We also have to ensure that the upper and lower-level trees are balanced so that we can update at most $O(\log N)$ literals from the root to the leaves when updating the partitions. So consider a set of balanced trees of different heights. We extract an *almost balanced* tree by eagerly joining any two trees of the same height until all trees have different heights. Then create an almost balanced tree out of the remaining trees. 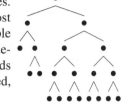 The maximal depth of the resulting two-level tree is then at most $\log N + 1$. By *almost balanced* we mean a binary tree (example on the right) that satisfies the following predicate $AB(n)$, defined recursively over binary trees for nodes n: $AB(n)$ holds if n is a leaf, or n is balanced, or if $n.right$ is balanced, $height(n.left) < height(n.right)$ and $AB(n.left)$.

The construction ensures that the number of fresh literals introduced is at most $O(N \log N)$. To justify this, let N_0 be the original number of terms. During one iteration we introduce M_j new classes. The new classes have sizes $a_1 N, \ldots, a_{M_j} N$, where $\sum_{i=1}^{M_j} a_i = 1$ and $\frac{N-1}{N} \geq a_1 \geq a_2 \geq \cdots \geq a_{M_j} \geq \frac{1}{N}$, so that a_1 is the fraction in the largest class. Thus, K is proportional to $(1 - a_1)N$; the contribution $\log N$ from the false literals is dominated by N. The cost of partitioning a class with N ($N \leq N_0$) terms is bounded by

$$T(N, N_0) \leq O(M_j \log N_0) + O((1 - a_1)N) + \sum_{i=1}^{M_j} T(a_i N, N_0), \quad N > 1$$

where $T(1, N_0) = O(1)$, $O(M_j \log N_0)$ is the cost of updating the upper layer tree, $O((1 - a_1)N)$ is the number of literals introduced for updating the lower level tree for the current class, and $T(a_i N, N_0)$ is the number of fresh literals used for further refining the new classes. Since there can be at most N classes, we have $\sum_j M_j \leq N$. We claim that the recurrence $T(N, N_0)$ is bounded by $O(N \log N_0)$, so $T(N_0, N_0)$ is bounded by $O(N \log N)$. To verify this, first separate the contribution $O(M_j \log N_0)$ from T. The contribution expands into a sum $\sum_j O(M_j \log N_0)$, which by the bound on $\sum_j M_j$, is $O(N \log N_0)$. Thus,

$$T(N, N_0) \leq O(N \log N_0) + T'(N), \text{ where } T'(N) \leq O((1-a_1)N) + \sum_{i=1}^{M_j} T'(a_i N).$$

We can over-approximate the cost $\sum_{i=2}^{M_j} T'(a_i N)$ by $T'((1 - a_1)N)$ (as the cost of the latter includes splitting a class of size $(1 - a_1)N$ into smaller classes), so the bound to analyze is: $T'(N) \leq O((1-a_1)N) + T'(a_1 N) + T'((1 - a_1)N)$. The upper bound for the size contribution a_1 of the largest class decreases in each unfolding because $\frac{M-1}{M} < \frac{N-1}{N}$ for $M < N$, so we will just assume that it remains fixed at a_1. Thus, the depth of unfolding T' is bounded by $- \log N / \log a_1$, and the contribution of $O((1 - a_1)N)$ in each level adds up to $(1 - a_1)$, so the overall cost is the product $O(N \log N)$.

Space-Optimized Partition Refinement via Term Sharing (HSOPR). Similar to the hybrid algorithm between BPR and IPR, we experimented with an algorithm that asserts a disjunction of N disequations at each refinement iteration, but uses chains rather than stars of disequations following (1). This yields an algorithm with $O(N^2)$ space complexity as there are $O(N)$ iterations, each of which creates a disjunction of size N. However, at most $O(N \log N)$ of the disequalities are fresh, resulting in a high degree of incrementality.

4   Practicalities

There are several important details that are significant in an implementation of these algorithms. We will describe the most prolific ones here that we encountered in the context of Z3. We believe these are generic issues.

Canonicity.   The model-based algorithms all rely on a solver providing models with the ability to evaluate terms. The requirement is that if two terms t_1, t_2 are the same under an interpretation, then the evaluation under a model $\mathcal{M}$ is the same: $\mathcal{M}(t_1) = \mathcal{M}(t_2)$. We say that the interpretations are *canonizing* for a sort. Z3 produces canonizing interpretations for sorts Booleans, bit-vectors, integers, reals (for linear constraints), and algebraic data types that use sorts with canonizing interpretations. An example algebraic data type that is canonizing is the sort of finite lists of integers. Another example is finite lists of finite lists of integers. Terms of sort finite lists over arrays are on the other hand not canonizing. The implementation falls back to a version of basic partition merging for terms of non-canonizing sorts.

Array Values.   Z3 does not produce canonizing interpretations for arrays. So if we are given terms t_1, t_2, t_3 whose sorts are (one-dimensional) arrays, Z3's evaluation of these terms under $\mathcal{M}$ does not produce canonical values. There is a simple trick, however, that takes care of arrays in many cases: Due to extensional equality of arrays, the equality partition for $\{t_1, t_2, t_3\}$ under Φ is the same as the equality partition for $\{t_1[i], t_2[i], t_3[i]\}$ under Φ, where i is a fresh index variable.

Pre-partitioning Based on Sorts.   SLAYER queries for partitions of several sorts of terms at the same time. We found that the merging-based algorithms benefited significantly from pre-partitioning the terms by sort. It was particularly important to distinguish terms in the image of the translation of array terms above from those that genuinely have the same sort as the range of the array. In such cases, not distinguishing by sorts leads to logically more difficult problems. The implementations of the refinement algorithms include an optimization where the initial partition is not taken to be the coarsest one, but is computed from the model generated by the first satisfiability check. This first model will yield an initial partition which distinguishes all terms of distinct sorts, except those produced by the array translation, which the implementation explicitly separates from the others.

Knowing the Terms.   Z3 can provide an evaluator given a model $\mathcal{M} \models \Phi$ that evaluates subterms in Φ. To force all terms $t_1, \ldots, t_N$ to be in Φ, we initialize Φ to $\Phi \wedge K(t_1) \wedge \cdots \wedge K(t_N)$, where K is a fresh predicate (*Known*).

Diversity.   All algorithms that rely on models require fewer iterations if the models are as *diverse* as possible. For example, if Φ is consistent with all $t_1, \ldots, t_N$ evaluating to different values, we are done in a single iteration. But Z3 is not required to produce diverse models. In the case of algebraic data types Z3 searches for models by building small instances. So for lists, Z3 always attempts to set a term to *nil* (the empty list). For arithmetic, Z3 supports a configuration, `arith.random_initial_value=true`, for shaking up initial values. Otherwise values of variables default to 0.

5 Empirical Evaluation

We used SLAYER running on device driver benchmarks to evaluate the algorithms. Statistics were gathered as SLAYER was running, to accurately reflect the actual mode of usage, which is through Z3's incremental programmatic interface. No individual implied equalities queries exhausted time or memory resources, although there are fewer queries for some algorithms in cases where the client analyzer exhausted resources. SMT-LIB2 benchmark files for most queries were generated during separate runs and are available online.[1] The companion technical report [1] contains many more details.

Figs. 1–6 each compare two algorithms. Results are reported only for the instances where the formula is satisfiable, all the algorithms behave equivalently with inconsistent formulas and so those points only add clutter. The right y-axis and top x-axis are the run times for the two algorithms. Times are reported in seconds, where query times measured below 50ms have been reported as 50ms since such instances are uninterestingly easy and accurately measuring such short times is problematic. Instances that are quickly solved by only one algorithm appear on an axis. A solid $y = x$ line is shown, as well as dotted lines indicating speedup and slowdown factors of $10x$, $100x$, and so on. Each plot includes a solid trend line that has been fit to the data, for what it is worth given the very high degree of variation and delicacy of nonlinear fitting. Each plot also includes two lines from upper-left to lower-right, associated with the left y-axis and bottom x-axis. The solid line indicates the number of instances where the algorithm on the right y-axis was faster than the algorithm on the top x-axis by at least the left y-coordinate seconds, and vice versa for the dashed line. The key reports the area under these curves, representing the cumulative speedups.

Fig. 1 compares the run times of the naïve Basic Partition Merging and semi-naïve Model-based Partition Merging algorithms. The conclusion is extremely clear-cut: despite the fact that the algorithms have the same theoretical complexity, in virtually all cases the model-based algorithm shows 10–$5000x$ speedups.

Fig. 2 compares the run times of the MPM and new Incremental Partition Refinement algorithms. Here the results are still very clear-cut, though not as dramatic as with the comparison to the most basic algorithm. There is a scattering of, generally easier, instances where MPM outperforms IPR, but the bulk of the harder instances see between 10–$100x$ speedups with IPR.

The assumption-based algorithm, ABIPR, does not offer dramatic benefits over IPR. Fig. 3 shows that while ABIPR is slightly faster overall, and is trending to scale slightly better on the harder instances, there are many instances on which IPR is faster. Fig. 4 compares BPR to ABIPR, showing that for our evaluation suite manipulating the proposition heap results in a significant overhead. ABIPR is faster on the easier instances, but BPR has larger speedups. So while on our benchmark suite the overall time spent by BPR is slightly higher than ABIPR, there is a slight trend toward BPR scaling better.

Fig. 5 compares HIPR versus BPR, showing that the hybrid incremental algorithm is overall somewhat faster than the basic version.

---

[1] http://research.microsoft.com/en-us/um/cambridge/projects/
slayer/gie_benchmarks.tgz

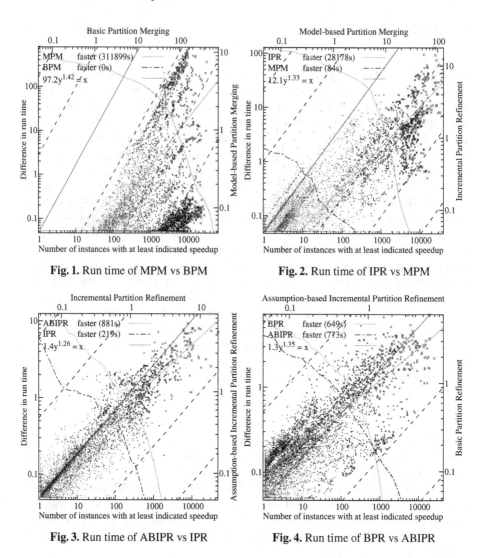

Fig. 1. Run time of MPM vs BPM   **Fig. 2.** Run time of IPR vs MPM

Fig. 3. Run time of ABIPR vs IPR   **Fig. 4.** Run time of BPR vs ABIPR

Fig. 6 shows that the HSOPR algorithm based on the $O(N \log N)$-space algorithm has some overhead relative to HIPR leading to slower performance on the easier instances, but scales better on the harder instances.

To provide an overall picture of all the algorithms discussed, Fig. 7 shows the number of instances solved within a given run time. From this we see that:

– BPM is much slower than the others.
– MPM is much faster than BPM but still significantly slower than the others.
– HIPR is fastest on easy instances, but is overtaken by HSOPR as it scales better.
– BPR and HSOPR are slower on easier instances, but scale better than IPR and ABIPR, eventually overtaking them.

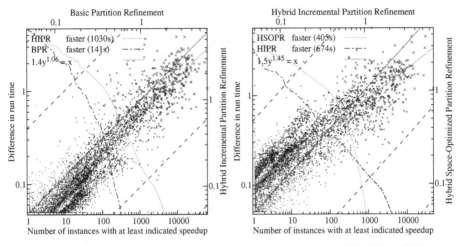

Fig. 5. Run time of HIPR vs BPR

Fig. 6. Run time of HSOPR vs HIPR

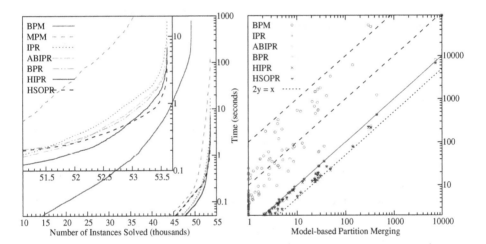

Fig. 7. Run time vs No. instances solved

Fig. 8. Analysis time relative to MPM

The implementations of BPR and HSOPR are similar. They require a little more work outside of the SMT solver than the IPR, ABIPR and especially HIPR.

Fig. 8 compares the overall SLAYER analysis run times using each algorithm relative to using MPM. The results show that, while computing implied equalities is only one sub-algorithm, improving it still yields considerable speedups of 10–$100x$ or more over BPM and an additional $2x$ over MPM. SLAYER hardly works with BPM, computing implied equalities is *the* bottleneck. With MPM, computing implied equalities is no longer the only bottleneck, but a significant speedup is still achieved by the refinement-based algorithms. The differences between the various refinement-based algorithms are not as apparent on full analysis runs, though HIPR is most often fastest, and ABIPR and IPR have some notable wins on hard instances.

In summary, the model-based algorithms dramatically outperform BPM. Among the model-based algorithms, partition refinement is clearly superior to the partition merging done by MPM. Between the partition refinement algorithms, relying on sub-formula sharing to achieve incrementality is at least as effective as using the proposition heap. And finally, the additional reuse enabled by following the $O(N \log N)$-space algorithm results in noticeably-better scaling.

6   Conclusions

Prompted by benchmarks from an application in program analysis, we developed efficient algorithms for inferring implied equalities. It generalizes congruence closure along two dimensions: from conjunctions of equations to general Boolean formulas, and from the free theory of uninterpreted functions to the variety of theories the employed solver supports. To our knowledge, only initial basic algorithms had been previously proposed. The overall result is a drastic reduction in solver calls over simple algorithms. An empirical evaluation using non-synthetic, but single-source and generally short-running, benchmarks shows speedups exceeding 10–$100x$ over the implementation of Model-based Partition Merging previously available in Z3, which is already almost never less than $10x$, and up to $5000x$, faster than the basic algorithm.

References

1. Berdine, J., Bjørner, N.: Computing all implied equalities via SMT-based partition refinement. Tech. Rep. MSR-TR-2014-57, Microsoft Research (2014)
2. Berdine, J., Cook, B., Ishtiaq, S.: SLAyer: Memory safety for systems-level code. In: Gopalakrishnan, G., Qadeer, S. (eds.) CAV 2011. LNCS, vol. 6806, pp. 178–183. Springer, Heidelberg (2011)
3. Berman, C., Trevillyan, L.: Functional comparison of logic designs for VLSI circuits. In: Kannan, L.N. (ed.) ICCAD, pp. 456–459. IEEE Computer Society (1989)
4. Brand, D.: Verification of large synthesized designs. In: Lightner, M.R., Jess, J.A.G. (eds.) ICCAD, pp. 534–537. IEEE Computer Society (1993)
5. Detlefs, D., Nelson, G., Saxe, J.B.: Simplify: A theorem prover for program checking. J. ACM 52(3), 365–473 (2005)
6. Downey, P.J., Sethi, R., Tarjan, R.E.: Variations on the common subexpression problem. J. ACM 27(4), 758–771 (1980)
7. Heule, M., Biere, A.: Blocked clause decomposition. In: McMillan, K., Middeldorp, A., Voronkov, A. (eds.) LPAR-19 2013. LNCS, vol. 8312, pp. 423–438. Springer, Heidelberg (2013)
8. Hoder, K., Voronkov, A.: Comparing unification algorithms in first-order theorem proving. In: Mertsching, B., Hund, M., Aziz, Z. (eds.) KI 2009. LNCS (LNAI), vol. 5803, pp. 435–443. Springer, Heidelberg (2009)
9. Lev-Ami, T., Sagiv, M.: TVLA: A system for implementing static analyses. In: SAS 2000. LNCS, vol. 1824, pp. 280–302. Springer, Heidelberg (2000)
10. Miné, A.: The octagon abstract domain. Higher-Order and Symbolic Computation 19(1), 31–100 (2006)

11. Nieuwenhuis, R., Oliveras, A.: Fast congruence closure and extensions. Inf. Comput. 205(4), 557–580 (2007)
12. Nieuwenhuis, R., Oliveras, A., Tinelli, C.: Solving SAT and SAT Modulo Theories: From an abstract Davis–Putnam–Logemann–Loveland procedure to DPLL(T). J. ACM 53(6), 937–977 (2006)
13. Paterson, M., Wegman, M.N.: Linear unification. J. Comput. Syst. Sci. 16(2), 158–167 (1978)
14. Robinson, J.A.: Computational logic: The unification computation. In: Meltzer, B., Michie, D. (eds.) Machine Intelligence 6, pp. 63–72. Edinburgh University Press (1971)
15. Sagiv, S., Reps, T.W., Wilhelm, R.: Parametric shape analysis via 3-valued logic. ACM Trans. Program. Lang. Syst. 24(3), 217–298 (2002)
16. Tarjan, R.E.: Efficiency of a good but not linear set union algorithm. J. ACM 22(2), 215–225 (1975)

Proving Termination of Programs Automatically with AProVE[*]

Jürgen Giesl[1], Marc Brockschmidt[2], Fabian Emmes[1], Florian Frohn[1],
Carsten Fuhs[3], Carsten Otto[6], Martin Plücker[1], Peter Schneider-Kamp[4],
Thomas Ströder[1], Stephanie Swiderski[7], and René Thiemann[5]

[1] RWTH Aachen University, Germany
[2] Microsoft Research Cambridge, UK
[3] University College London, UK
[4] University of Southern Denmark, Denmark
[5] University of Innsbruck, Austria
[6] andrena objects AG, Germany
[7] Interactive Pioneers GmbH, Germany

Abstract. AProVE is a system for automatic termination and complexity proofs of Java, C, Haskell, Prolog, and term rewrite systems (TRSs). To analyze programs in high-level languages, AProVE automatically converts them to TRSs. Then, a wide range of techniques is employed to prove termination and to infer complexity bounds for the resulting TRSs. The generated proofs can be exported to check their correctness using automatic certifiers. For use in software construction, we present an AProVE plug-in for the popular Eclipse software development environment.

1 Introduction

AProVE (Automated Program Verification Environment) is a tool for automatic termination and complexity analysis. While previous versions (described in [19, 20]) only analyzed termination of term rewriting, the new version of AProVE also analyzes termination of Java, C, Haskell, and Prolog programs. Moreover, it also features techniques for automatic complexity analysis and permits the certification of automatically generated termination proofs. To analyze programs, AProVE uses an approach based on symbolic execution and abstraction [11] to transform the input program into a *symbolic execution graph*[1] that represents all possible computations of the input program. Language-specific features (such as sharing effects of heap operations in Java, pointer arithmetic and memory safety in C, higher-order functions and lazy evaluation in Haskell, or extra-logical predicates in Prolog) are handled when generating this graph. Thus, the exact definition of the graph depends on the considered programming language. For termination or complexity analysis, the graph is transformed into a TRS. The success of AProVE

---

[*] Supported by the DFG grant GI 274/6-1 and the FWF grant P22767. Most of the research was done while the authors except R. Thiemann were at RWTH Aachen.
[1] In earlier papers, this was often called a *termination graph*.

S. Demri, D. Kapur, and C. Weidenbach (Eds.): IJCAR 2014, LNAI 8562, pp. 184–191, 2014.
© Springer International Publishing Switzerland 2014

at the annual international *Termination Competition* demonstrates that our rewriting-based approach is well suited for termination analysis of real-world programming languages.[2] A graphical overview of our approach is displayed on the side.[3]

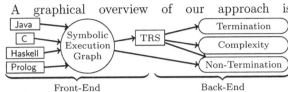

Technical details on the techniques for transform- ing programs to TRSs and for analyzing TRSs can be found in [5–9, 15–18, 21–23, 27, 28, 30]. In the current paper, we focus on their implementation in AProVE, which we now made available as a plug-in for the popular Eclipse software development environment [13]. In this way, AProVE can already be applied during program construction (e.g., by analyzing termination of single Java methods for user-specified classes of inputs). In addition to the full version of AProVE, we also made AProVE's front-ends for the different programming languages available as separate programs. Thus, they can be coupled with other external tools that operate on TRSs, integer transition systems, or symbolic execution graphs. These external tools can then be used as alternative back-ends. Finally, AProVE can also be accessed directly via a web interface [2].

We describe the use of AProVE for the different programming languages and TRSs in Sect. 2. To increase the reliability of the generated proofs, AProVE supports their certification, cf. Sect. 3. We end with a short conclusion in Sect. 4.

2   AProVE and Its Graphical User Interface in Eclipse

AProVE and its graphical user interface are available as an Eclipse plug-in at [2] under "Download". After the initial installation, "Check for Updates" in the "Help" menu of Eclipse also checks for updates of AProVE. As Eclipse and AProVE are written in Java, they can be used on most operating systems.

2.1   Analyzing Programming Languages

The screenshot on the next page shows the main features of our AProVE plug- in. Here, AProVE is applied on a Java (resp. Java Bytecode (JBC)) program in the file List.jar and tries to prove termination of the main method of the class List, which in turn calls the method contains. (The source code is shown in the editor window (**B**).) Files in an Eclipse project can be analyzed by right-clicking on the file in Eclipse's Project Explorer (**A**) and selecting "Launch AProVE".[4]

When AProVE is launched, the proof (progress) can be inspected in the Proof Tree View (**C**). Here, problems (e.g., programs, symbolic execution graphs, TRSs, ...) alternate with proof steps that modify problems, where "⇐" indicates sound

---

[2] See http://www.termination-portal.org/wiki/Termination_Competition

[3] While termination can be analyzed for Java, C, Haskell, Prolog, and TRSs, the current version of AProVE analyzes complexity only for Prolog and TRSs.

[4] An initial "ExampleProject" with several examples in different programming lan- guages can be created by clicking on the "AProVE" entry in Eclipse's menu bar.

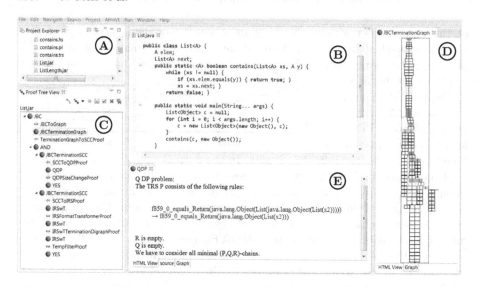

and "⇔" indicates sound and complete steps. This information is used to propagate information from child nodes to the parent node. A green (resp. red) bullet in front of a problem means that termination of the problem is proved (resp. disproved) and a yellow bullet denotes an unsuccessful (or unfinished) proof. Since the root of the proof tree is always the input problem, the color of its bullet indicates whether AProVE could show its termination resp. non-termination.

To handle Java-specific features, AProVE first constructs a symbolic execution graph **(D)** from the program [5–7, 28]. From the cycles of this graph, TRSs are created whose termination implies termination of the original program.[5] Double-clicking on a problem or proof step in the proof tree shows detailed information about them. For example, the symbolic execution graph can be inspected by double-clicking on the node JBCTerminationGraph and selecting the **Graph** tab in the **Problem View (D)**. This graph can be navigated with the mouse, allowing to zoom in on specific nodes or edges. Similarly, one of the generated TRSs is shown in the **Problem View (E)**. For *non*-termination proofs [6], witness executions are provided in the **Problem View**. In contrast to termination proofs, these analyses are performed directly on the symbolic execution graph.

The buttons in the upper right part of the **Proof Tree View (C)** interact with AProVE (e.g., ▮ aborts the analysis). When AProVE is launched, the termination proof is attempted with a time-out of 60 seconds. If it is aborted, one can right-click on a node in the proof tree and by selecting "Run", one can continue the proof at this node (here, one may also specify a new time-out).

For Java programs, there are two options to specify which parts of the program are analyzed. AProVE can be launched on a jar (Java archive) file, and then tries

---

[5] These TRSs are represented as *dependency pair problems* [21] ("QDP" in **(C)**).

to prove termination of the `main` method of the archive's "main class".[6] Alternatively, to use AProVE during software development, single Java methods can be analyzed. Eclipse's Outline View (reachable via "Window" and "Show View") shows the methods of a class opened by a double-click in Eclipse's Project Explorer. An initial "JavaProject" with a class List can be created via the "AProVE" entry in Eclipse's menu bar. Right-clicking on a method in the Outline View and choosing "Launch AProVE" leads to the configuration dialog on the side. It can be used to specify the sharing and shape of the method's input values. Each argument can be tree-shaped, DAG-shaped, or arbitrary (i.e., possibly cyclic) [7]. Furthermore, one can specify which arguments may be sharing. Similarly, one can provide assumptions about the contents of static fields. There are also two short-cut buttons which lead to the best- and the worst-case assumption. Moreover, under "AProVE options", one can adjust the desired time-out for the termination proof and under "Problem selection", one has the option to replace AProVE's default strategy with alternative user-defined strategies (a general change of AProVE's strategy is possible via the "AProVE" entry in Eclipse's main menu).

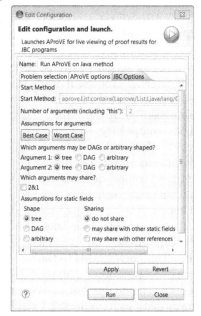

C [30], Haskell [22], and Prolog [23] are handled similarly. The function, start terms, or queries to be analyzed can be specified in the input file (as in the *Termination Competition*). Otherwise the user is prompted when the analysis starts. For Prolog, AProVE can also infer asymptotic upper bounds on the number of evaluation steps (i.e., unification attempts) and prove determinacy (i.e., that there is at most one solution).

All our programming language front-ends first construct symbolic execution graphs, which are then used to extract the information relevant for termination as a TRS. Thus, analyzing implementations of the same algorithm in different languages leads to very similar TRSs, as AProVE identifies that the reason for termination is always the same. For example, implementations of a `contains` algorithm in different languages all terminate for the same reason on (finite acyclic) lists, since the length of the list decreases in each recursive call or iteration.

---

[6] See `http://www.termination-portal.org/wiki/Java_Bytecode` for the conventions of the *Termination Competition*, which also specify certain restrictions on the Java programs. In particular, similar to many other termination provers, AProVE treats built-in data types like `int` in Java as unbounded integers $\mathbb{Z}$. Thus, a termination proof is only valid under the assumption that no overflows occur.

2.2 Analyzing Term Rewrite Systems

To prove termination of TRSs, AProVE implements a combination of numerous techniques within the dependency pair framework [21]. To deal with the pre-defined type of integers in programming languages, AProVE also handles TRSs with built-in integers, using extensions of the dependency pair framework proposed in [16, 18]. To solve the arising search problems (e.g., for well-founded orders), AProVE relies on SAT- and SMT-based techniques like [1, 9, 17, 29]. As SAT solvers, AProVE uses SAT4J [24] and MiniSAT [14]. Like AProVE, SAT4J is implemented in Java and hence, AProVE calls it for small SAT instances, where it is very efficient. MiniSAT is used on larger SAT instances, but as it is invoked as an external process, it leads to a small overhead. As SMT solvers, AProVE uses Yices [12] and Z3 [25]. Non-termination of TRSs is detected by suitable adaptions of narrowing [15].

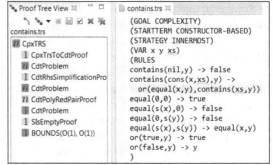

For complexity analysis, AProVE infers runtime complexity of innermost rewriting. Runtime complexity means that one only considers initial terms $f(t_1, \ldots, t_m)$ where $t_1, \ldots, t_m$ represent data (thus, they are already in normal form). This corresponds to the setting in program analysis. Similarly, the analysis of innermost rewriting is motivated by the fact that the transformations from Sect. 2.1 yield TRSs where it suffices to consider innermost rewriting in the back-end. (Polynomial) upper bounds on the runtime complexity are inferred by an adaption of dependency pairs for complexity analysis [27]. To solve the resulting search problems, AProVE re-uses the techniques from termination analysis to generate suitable well-founded orders. As shown in the screenshot, AProVE easily infers that the above TRS has linear asymptotic complexity. More precisely, the ∏ at the root node of the proof tree means that initial terms $f(t_1, \ldots, t_m)$ of size n only have evaluations of length $\mathcal{O}(n)$.[7]

3 Partial Certification of Generated Proofs

Like any large software product, AProVE had (and very likely still has) bugs. To allow verification of its results, it can export generated termination proofs as machine-readable CPF (Certification Problem Format)[8] files by clicking on the button 🖫 of the Proof Tree View. Independent certifiers can then check the validity of all proof steps. Examples for such certifiers are CeTA [31], CiME/Coccinelle

---

[7] Moreover, proof steps also result in complexities (e.g., ▯ or ∏). More precisely, in each proof step, a problem P is transformed into a new problem P' and a complexity c. Then the complexity of P is bounded by the maximum of P''s complexity and of c.

[8] See http://cl-informatik.uibk.ac.at/software/cpf/

[10], and CoLoR/Rainbow [4]. Their correctness has been formally proved using Isabelle/HOL [26] or Coq [3]. To certify a proof in AProVE's GUI, one can also call CeTA directly using the button ☑ of the Proof Tree View.

Some proof techniques (like the transformation of programming languages to TRSs in AProVE) are not yet formalized in CPF. Until now, proofs with such steps could not be certified at all. As a solution, we extended CPF by an additional element unknownProof for proof steps which are not supported by CPF. In the certification, unknownProof is treated as an axiom of the form $P_0 \longleftarrow P_1 \wedge \ldots \wedge P_n$. This allows to prove $P_1, \ldots, P_n$ instead of the desired property P_0. Each P_i can be an arbitrary property such as (non-)termination of some TRS, and P_i's subproof can be checked by the certifier again. In this way, it is possible to certify large parts of *every* termination proof generated by AProVE. For example, now 90% of AProVE's proof steps for termination analysis of the 4367 TRSs in the *termination problem data base (TPDB)*[9] can be certified by CeTA.

Moreover, we added a new CPF element unknownInput for properties that cannot be expressed in CPF, like termination of a Java program. The only applicable proof step to such a property is unknownProof. Using unknownInput, CPF files for *every* proof can be generated. Now the program transformations in AProVE's front-end correspond to unknown proof steps on unknown inputs, but the reasoning in AProVE's back-end can still be checked by a certifier (i.e., proof steps can transform unknownInput into objects that are expressible in CPF).

Due to this new *partial* certification, three bugs of AProVE have been revealed (and fixed) which could be exploited to prove termination of a non-terminating TRS. These bugs had not been discovered before by certification, as the errors occurred when analyzing TRSs resulting from logic programs. If one is only interested in completely certified proofs, the "AProVE" entry in Eclipse's main menu allows to change AProVE's default strategy to a "certifiable" strategy which tries to use proof techniques that can be exported to CPF whenever possible.

4   Conclusion

We presented a new version of AProVE to analyze termination of TRSs and programs for four languages from prevailing programming paradigms. Moreover, AProVE analyzes the runtime complexity of Prolog programs and TRSs. We are currently working on extending AProVE's complexity analysis to Java as well [8].

AProVE's power is demonstrated by its performance in the annual *Termination Competition*, where it won almost all categories related to termination of Java, Haskell, Prolog, and to termination or innermost runtime complexity of TRSs. Moreover, AProVE participated very successfully in the *SV-COMP* competition[10] at *TACAS* which featured a category for termination of C programs for the first time in 2014. AProVE's automatically generated termination proofs can be exported to (partially) check them by automatic certifiers. Our tool is available as a plug-in of the well-known Eclipse software development environment.

---

[9] The *TPDB* is the collection of examples used in the annual *Termination Competition*.
[10] See http://sv-comp.sosy-lab.org/2014/

Moreover, the front-ends of AProVE for the different programming languages are also available separately in order to couple them with alternative back-ends. To download AProVE or to access it via a web interface, we refer to [2].

References

1. Alias, C., Darte, A., Feautrier, P., Gonnord, L.: Multi-dimensional rankings, program termination, and complexity bounds of flowchart programs. In: Cousot, R., Martel, M. (eds.) SAS 2010. LNCS, vol. 6337, pp. 117–133. Springer, Heidelberg (2010)
2. AProVE, http://aprove.informatik.rwth-aachen.de/
3. Bertot, Y., Castéran, P.: Coq'Art. Springer (2004)
4. Blanqui, F., Koprowski, A.: CoLoR: A Coq library on well-founded rewrite relations and its application to the automated verification of termination certificates. Mathematical Structures in Computer Science 4, 827–859 (2011)
5. Brockschmidt, M., Otto, C., Giesl, J.: Modular termination proofs of recursive Java Bytecode programs by term rewriting. In: Schmidt-Schauß, M. (ed.) RTA 2011. LIPIcs, vol. 10, pp. 155–170. Dagstuhl Publishing (2011)
6. Brockschmidt, M., Ströder, T., Otto, C., Giesl, J.: Automated detection of non-termination and NullPointerExceptions for Java Bytecode. In: Beckert, B., Damiani, F., Gurov, D. (eds.) FoVeOOS 2011. LNCS, vol. 7421, pp. 123–141. Springer, Heidelberg (2012)
7. Brockschmidt, M., Musiol, R., Otto, C., Giesl, J.: Automated termination proofs for Java programs with cyclic data. In: Madhusudan, P., Seshia, S.A. (eds.) CAV 2012. LNCS, vol. 7358, pp. 105–122. Springer, Heidelberg (2012)
8. Brockschmidt, M., Emmes, F., Falke, S., Fuhs, C., Giesl, J.: Alternating runtime and size complexity analysis of integer programs. In: Ábrahám, E., Havelund, K. (eds.) TACAS 2014 (ETAPS). LNCS, vol. 8413, pp. 140–155. Springer, Heidelberg (2014)
9. Codish, M., Giesl, J., Schneider-Kamp, P., Thiemann, R.: SAT solving for termination proofs with recursive path orders and DPs. JAR 49(1), 53–93 (2012)
10. Contejean, E., Courtieu, P., Forest, J., Pons, O., Urbain, X.: Automated certified proofs with CiME3. In: Schmidt-Schauß, M. (ed.) RTA 2011. LIPIcs, vol. 10, pp. 21–30. Dagstuhl Publishing (2011)
11. Cousot, P., Cousot, R.: Abstract interpretation: A unified lattice model for static analysis of programs by construction or approximation of fixpoints. In: Graham, R.M., Harrison, M.A., Sethi, R. (eds.) POPL 1977, pp. 238–252. ACM Press (1977)
12. Dutertre, B., de Moura, L.M.: The Yices SMT solver (2006), tool paper at http://yices.csl.sri.com/tool-paper.pdf
13. Eclipse, http://www.eclipse.org/
14. Eén, N., Sörensson, N.: An extensible SAT-solver. In: Giunchiglia, E., Tacchella, A. (eds.) SAT 2003. LNCS, vol. 2919, pp. 502–518. Springer, Heidelberg (2004)
15. Emmes, F., Enger, T., Giesl, J.: Proving non-looping non-termination automatically. In: Gramlich, B., Miller, D., Sattler, U. (eds.) IJCAR 2012. LNCS (LNAI), vol. 7364, pp. 225–240. Springer, Heidelberg (2012)
16. Falke, S., Kapur, D., Sinz, C.: Termination analysis of C programs using compiler intermediate languages. In: Schmidt-Schauß, M. (ed.) RTA 2011. LIPIcs, vol. 10, pp. 41–50. Dagstuhl Publishing (2011)

17. Fuhs, C., Giesl, J., Middeldorp, A., Schneider-Kamp, P., Thiemann, R., Zankl, H.: SAT solving for termination analysis with polynomial interpretations. In: Marques-Silva, J., Sakallah, K.A. (eds.) SAT 2007. LNCS, vol. 4501, pp. 340–354. Springer, Heidelberg (2007)

18. Fuhs, C., Giesl, J., Plücker, M., Schneider-Kamp, P., Falke, S.: Proving termination of integer term rewriting. In: Treinen, R. (ed.) RTA 2009. LNCS, vol. 5595, pp. 32–47. Springer, Heidelberg (2009)

19. Giesl, J., Thiemann, R., Schneider-Kamp, P., Falke, S.: Automated termination proofs with AProVE. In: van Oostrom, V. (ed.) RTA 2004. LNCS, vol. 3091, pp. 210–220. Springer, Heidelberg (2004)

20. Giesl, J., Schneider-Kamp, P., Thiemann, R.: AProVE 1.2: Automatic termination proofs in the dependency pair framework. In: Furbach, U., Shankar, N. (eds.) IJCAR 2006. LNCS (LNAI), vol. 4130, pp. 281–286. Springer, Heidelberg (2006)

21. Giesl, J., Thiemann, R., Schneider-Kamp, P., Falke, S.: Mechanizing and improving dependency pairs. JAR 37(3), 155–203 (2006)

22. Giesl, J., Raffelsieper, M., Schneider-Kamp, P., Swiderski, S., Thiemann, R.: Automated termination proofs for Haskell by term rewriting. TOPLAS 33(2), 7:1–7:39 (2011)

23. Giesl, J., Ströder, T., Schneider-Kamp, P., Emmes, F., Fuhs, C.: Symbolic evaluation graphs and term rewriting — A general methodology for analyzing logic programs. In: De Schreye, D., Janssens, G., King, A. (eds.) PPDP 2012, pp. 1–12. ACM Press (2012)

24. Le Berre, D., Parrain, A.: The SAT4J library, release 2.2. JSAT 7, 59–64 (2010)

25. de Moura, L., Bjørner, N.S.: Z3: An efficient SMT solver. In: Ramakrishnan, C.R., Rehof, J. (eds.) TACAS 2008. LNCS, vol. 4963, pp. 337–340. Springer, Heidelberg (2008)

26. Nipkow, T., Paulson, L.C., Wenzel, M.T.: Isabelle/HOL. LNCS, vol. 2283. Springer, Heidelberg (2002)

27. Noschinski, L., Emmes, F., Giesl, J.: Analyzing innermost runtime complexity of term rewriting by dependency pairs. JAR 51(1), 27–56 (2013)

28. Otto, C., Brockschmidt, M., von Essen, C., Giesl, J.: Automated termination analysis of Java Bytecode by term rewriting. In: Lynch, C. (ed.) RTA 2010. LIPIcs, vol. 6, pp. 259–276. Dagstuhl Publishing (2010)

29. Podelski, A., Rybalchenko, A.: A complete method for the synthesis of linear ranking functions. In: Steffen, B., Levi, G. (eds.) VMCAI 2004. LNCS, vol. 2937, pp. 239–251. Springer, Heidelberg (2004)

30. Ströder, T., Giesl, J., Brockschmidt, M., Frohn, F., Fuhs, C., Hensel, J., Schneider-Kamp, P.: Proving termination and memory safety for programs with pointer arithmetic. In: Demri, S., Kapur, D., Weidenbach, C. (eds.) IJCAR 2014. LNCS (LNAI), vol. 8562, pp. 204–218. Springer, Heidelberg (2014)

31. Thiemann, R., Sternagel, C.: Certification of termination proofs using CeTA. In: Berghofer, S., Nipkow, T., Urban, C., Wenzel, M. (eds.) TPHOLs 2009. LNCS, vol. 5674, pp. 452–468. Springer, Heidelberg (2009)

Locality Transfer: From Constrained Axiomatizations to Reachability Predicates

Matthias Horbach and Viorica Sofronie-Stokkermans

University Koblenz-Landau, Koblenz, Germany, and
Max-Planck-Institut für Informatik Saarbrücken, Germany

Abstract. In this paper, we build upon our previous work in which we used constrained clauses in order to finitely represent infinite sets of clauses and proved that constrained axiomatizations are local if they are saturated under a version of resolution. We extend this result by identifying situations in which locality of saturated axiomatizations is maintained if we enrich the base theory by introducing new predicates (often reachability predicates) instead of using constraints for these properties.

1 Introduction

Many problems in computer science (e.g. in verification) can be reduced to checking the satisfiability of conjunctions of literals with respect to a theory. For efficient and accurate reasoning it is essential to reduce the search space without losing completeness and to make modular or hierarchical reasoning possible. In [13], we introduced the class of *local theory extensions* for which (i) complete instantiation schemes exist, and (ii) hierarchical and modular reasoning is possible. Locality is a property of an *axiomatization of a theory*; therefore it is very important to *recognize locality* of a set of clauses, and *to obtain local axiomatizations* by transforming non-local sets of clauses into local ones. In [2, 3], a link between (order)-locality and saturation under ordered (hyper)resolution is presented; this allows us to obtain, by saturation, local axiomatizations for a theory from non-local ones. Sometimes, however, the saturation process may not terminate. In [6] we showed that in order to obtain finite representations of possibly infinite sets of clauses we can use *constrained clauses*; we used a sound and complete ordered resolution and superposition calculus for constrained clauses and established a link between saturation in our calculus and order locality.

In spite of their advantages, axiomatizations using constrained clauses are difficult to export to standard first-order provers. In this paper we show that we can avoid using constrained clauses if we introduce new relations which encode the constraints. Our main contributions can be described as follows:

- We analyze possibilities of changing local sets of constrained clauses into local sets of clauses, by encoding the constraints using additional predicates.
- We identify situations in which the properties of the new predicates (which include reachability) can be encoded using first-order logic.
- We then identify situations in which locality results for the constrained clause set can be transferred to the new presentation defined this way.

S. Demri, D. Kapur, and C. Weidenbach (Eds.): IJCAR 2014, LNAI 8562, pp. 192–207, 2014.
© Springer International Publishing Switzerland 2014

Structure of the paper. The paper is structured as follows: In Sect. 2 we introduce the terminology we use and present the relevant results on local theory extensions. In Sect. 3 we identify a class of constrained clauses for which instantiation can be effectively computed. In Sect. 4 we analyze possibilities of changing local sets of constrained clauses into local sets of clauses, by encoding the constraints using additional predicates. In Sect. 5 we identify extensions of theories of absolutely free constructors for which locality is preserved after this transformation. Sect. 6 contains an overview of possible applications and further work.

2   Preliminaries

In this section we introduce the terminology and main results used in the paper.

2.1   General Definitions

We build on the notions of [1, 7] and shortly recall here the most important concepts concerning terms and orderings and the specific extensions (concerning constrained clauses) needed in this article. For simplicity, we restrict to single-sorted signatures (the many-sorted case works similarly).

Terms and Clauses. Let $\Pi = (\Sigma, \mathsf{Pred})$ be a *signature* consisting of a set Σ of function symbols of fixed arity and a set Pred of predicate symbols of fixed arity, and let X be a countably infinite set of variables such that X and Σ are disjoint. Terms s, t, equations $s \approx t$, atoms $P(\vec{t})$, atoms, literals, clauses and Horn clauses are defined as usual. We denote by $\mathsf{T}_\Sigma(X)$ the set of all terms over Σ and X and by T_Σ the set of all ground terms over Σ. To improve readability, term tuples $(t_1, \ldots, t_n)$ will often be denoted by $\vec{t}$.

Substitution Expressions. A *substitution* σ is a map from a finite set $X' \subseteq X$ of variables to $\mathsf{T}_\Sigma(X)$. The application of σ to a term t or a term tuple $\vec{t}$ is denoted by $t\sigma$ or $\vec{t}\sigma$, respectively. *Substitution expressions* are built over substitutions and constructors $\circ$ (composition), $|$ (disjunction), and $*$ (loop) of arity 2, 2 and 1, respectively. Substitution expressions are denoted as $\bar{\sigma}, \bar{\tau}$. We will often write $\bar{\sigma} \circ \bar{\tau}$ as $\bar{\sigma}\bar{\tau}$ and $\bar{\sigma}\bar{\sigma}^*$ as $\bar{\sigma}^+$. The *domain* $\mathrm{dom}(\bar{\sigma})$ and the *variable range* $\mathrm{VRan}(\bar{\sigma})$ of a substitution expression are defined as follows: For a substitution $\sigma : \{x_1, \ldots, x_n\} \to \mathsf{T}_\Sigma(X)$, we define $\mathrm{dom}(\sigma) = \{x_1, \ldots, x_n\}$ and $\mathrm{VRan}(\sigma) = \mathrm{vars}(x_1\sigma, \ldots, x_n\sigma)$. For complex substitution expressions, we have

$$\mathrm{dom}(\bar{\sigma} \circ \bar{\tau}) = \mathrm{dom}(\bar{\sigma}) \qquad\qquad \mathrm{VRan}(\bar{\sigma} \circ \bar{\tau}) = \mathrm{VRan}(\bar{\tau})$$
$$\mathrm{dom}(\bar{\sigma}_1 | \bar{\sigma}_2) = \mathrm{dom}(\bar{\sigma}_1) \cup \mathrm{dom}(\bar{\sigma}_2) \qquad \mathrm{VRan}(\bar{\sigma}_1 | \bar{\sigma}_2) = \mathrm{VRan}(\bar{\sigma}_1) \cap \mathrm{VRan}(\bar{\sigma}_2)$$
$$\mathrm{dom}(\bar{\sigma}^*) = \mathrm{dom}(\bar{\sigma}) \qquad\qquad \mathrm{VRan}(\bar{\sigma}^*) = \mathrm{dom}(\bar{\sigma}) \cup \mathrm{VRan}(\bar{\sigma})$$

These notions are only intuitive for well-formed expressions: A substitution expression $\bar{\sigma}$ is *well-formed*, if (i) for each subexpression $\bar{\tau}_1 \circ \bar{\tau}_2$ of $\bar{\sigma}$, $\mathrm{VRan}(\bar{\tau}_1) = \mathrm{dom}(\bar{\tau}_2)$, (ii) for each subexpression $\bar{\tau}_1 | \bar{\tau}_2$, $\mathrm{dom}(\bar{\tau}_1) = \mathrm{dom}(\bar{\tau}_2)$ and $\mathrm{VRan}(\bar{\tau}_1) = \mathrm{VRan}(\bar{\tau}_2)$, and (iii) for each subexpression $\bar{\tau}^*$, $\mathrm{VRan}(\bar{\tau}) = \mathrm{dom}(\bar{\tau})$.

Constrained Clauses. A *constrained clause* $\alpha \,\|\, C$ consists of a clause C and a *regular constraint* α of the form $(x_1 \approx y_1, \ldots, x_n \approx y_n)\bar{\sigma}$, also written as $(\vec{x} \approx \vec{y})\bar{\sigma}$, such that x_i, y_i are variables and $\bar{\sigma}$ is a well-formed substitution expression with domain $\{x_1, y_1, \ldots, x_n, y_n\}$. If a regular constraint α does not contain any equations, we call $\alpha \parallel C$ *unconstrained* and identify it with its clausal part C. If $\alpha = (\vec{x} \approx \vec{y})\bar{\tau}$ is a regular constraint, then $\alpha\sigma$ is defined as $(\vec{x} \approx \vec{y})\bar{\tau}\sigma'$, where $\sigma' : \mathrm{VRan}(\bar{\tau}) \rightarrow \mathsf{T}_\Sigma(X)$ maps z to $z\sigma$ if $z \in \mathrm{dom}(\sigma)$ and to z otherwise. The application $(\alpha \parallel C)\sigma$ of a substitution to a constrained clause is then defined as $\alpha\sigma \parallel C\sigma$. The set of *ground instances* of a constrained clause $\alpha \parallel C$ consists of all ground clauses $C\sigma$ for which $\alpha\sigma$ is a satisfiable ground constraint. This means that regular constraints are interpreted syntactically. To improve readability of constraints, we sometimes use straightforward expansions of the substitution expressions. For example, $(x \approx y)\{x \mapsto x, y \mapsto s(y)\}^*$ might be written as $x \approx s^*(y)$.

Denotations and Models of Constrained Clauses. We define the *denotation* $[\![\bar{\sigma}]\!]$ of a substitution expression $\bar{\sigma}$ inductively as follows:

$$[\![\sigma]\!] = \{\sigma\} \qquad\qquad [\![\bar{\sigma}\bar{\tau}]\!] = \{\sigma\tau \mid \sigma \in [\![\bar{\sigma}]\!], \tau \in [\![\bar{\tau}]\!]\}$$
$$[\![\bar{\sigma}_1 | \bar{\sigma}_2]\!] = [\![\bar{\sigma}_1]\!] \cup [\![\bar{\sigma}_2]\!] \qquad\qquad [\![\bar{\sigma}^*]\!] = \bigcup_{n \geq 0}[\![\bar{\sigma}^n]\!]$$

Here $\bar{\sigma}^0$ denotes the identity substitution on $\mathrm{dom}\,\bar{\sigma}$ and $\bar{\sigma}^{n+1} = \bar{\sigma} \circ \bar{\sigma}^n$.

The semantics of the application of substitution expressions to terms and clauses and the semantics of constrained clause sets are defined just as one would expect by identifying a substitution expression with its denotation and by identifying a constrained clause $(\vec{x} \approx \vec{y})\bar{\sigma} \parallel C$ with the (potentially infinite) clause set $\{\vec{x}\sigma \approx \vec{y}\sigma \rightarrow C \mid \sigma \in [\![\bar{\sigma}]\!]\}$. Models and satisfiability of constrained clause sets are then defined straightforwardly using this set (cf. [7] for details).

Orderings. A (strict partial) *ordering* $\prec$ on a set S is a transitive and irreflexive binary relation on S. It is *total* if $s \prec t$ or $t \prec s$ whenever $s \neq t$. It is *well-founded* if there is no infinite descending chain $s_1 \succ s_2 \succ \ldots$ of elements of S. A well-founded ordering $\prec$ on $\mathsf{T}_\Sigma(X)$ is a *reduction ordering* if $t \prec t'$ implies $u[t\sigma] \prec u[t'\sigma]$ for all $t, t', u \in \mathsf{T}_\Sigma(X)$ and all substitutions σ.

Let $\prec_\mathsf{T}$ be an ordering on $\mathsf{T}_\Sigma(X)$ and let $\prec$ be an ordering on atoms over $\mathsf{T}_\Sigma(X)$. Then $\prec$ is *compatible* with $\prec_\mathsf{T}$ if $A_1 \prec A_2$ whenever every term in A_1 is strictly bounded by a term in A_2, i.e. if for each term t_1 in A_1 there is a term t_2 in A_2 such that $t_1 \prec_\mathsf{T} t_2$. Any ordering $\prec$ on atoms can be extended to clauses in a standard way (by setting $\neg A \succ A$ for all atoms A and taking the multiset extension of $\prec$). Let $\prec_\mathsf{T}$ be a term ordering and $\prec$ be an atom ordering. A ground clause is *reductive (w.r.t. $\prec_\mathsf{T}$ and $\prec$)* if all of its $\prec_\mathsf{T}$-maximal terms appear in the $\prec$-maximal atoms. A constrained clause is *reductive (w.r.t. $\prec_\mathsf{T}$ and $\prec$)* if all its ground instances are reductive (cf. [3]).

Inferences. A (clausal) *inference rule* is a relation on (constrained) clauses. Its elements are called *inferences* and written as

$$\frac{C_1 \ldots C_k}{C} \qquad \text{resp.} \qquad \frac{\alpha_1 \parallel C_1 \ldots \alpha_k \parallel C_k}{\alpha \parallel C} .$$

The clauses $C_1, \ldots, C_k$ (resp. the constrained clauses $\alpha_1 \parallel C_1, \ldots, \alpha_k \parallel C_k$) are called the *premises* and C ($\alpha \parallel C$) the *conclusion* of the inference. An *inference system* is a set of inference rules. In what follows we will use the standard inference rules for ordered resolution and hyperresolution (with selection) (cf. e.g. [1]), as well extensions to constrained clauses (cf. [7]).

Redundancy. Let N be a set of clauses. A ground clause C is *redundant* w.r.t. N (and $\prec$) if it is entailed by ground instances of N which are smaller than C w.r.t. $\prec$. A constrained clause is *redundant* w.r.t. a constrained clause set N (and $\prec$) if all of its ground instances are redundant. An inference is *redundant* w.r.t. N if its conclusion is redundant w.r.t. N or if a premise C is redundant w.r.t. $N \setminus \{C\}$. A constrained clause set N is *saturated* (w.r.t. a given inference system) if each inference with premises in N is redundant w.r.t. N.

2.2 Local Theories and Theory Extensions

Local Theories. The notion of local set of Horn clauses (or local Horn theory) was introduced by Givan and McAllester in [5] A *local set of Horn clauses* is a set of Horn clauses $\mathcal{K}$ such that, for any ground Horn clause C, $\mathcal{K} \models C$ only if already $\mathcal{K}[C] \models C$ (where $\mathcal{K}[C]$ is the set of instances of $\mathcal{K}$ in which all terms are subterms of ground terms in either $\mathcal{K}$ or C). Since the size of $\mathcal{K}[G]$ is polynomial in the size of G for a fixed $\mathcal{K}$ and satisfiability of sets of ground Horn clauses can be checked in linear time, the validity of ground Horn clauses w.r.t. local Horn theories can be checked in polynomial time. In [2, 3], Basin and Ganzinger defined *order locality*. Given a term ordering $\prec$, we say that a set $\mathcal{K}$ of clauses entails a ground clause C bounded by $\prec$ (notation: $\mathcal{K} \models_{\prec} C$), iff there is a proof of $\mathcal{K} \models C$ from those ground instances of clauses in $\mathcal{K}$ in which (under $\preceq$) each term is smaller than or equal to some term in C. A set of clauses $\mathcal{K}$ is *local with respect to* $\prec$ if whenever $\mathcal{K} \models C$ for a ground clause C, then $\mathcal{K} \models_{\prec} C$.

Local Theory Extensions. In [13] the notion of locality for Horn clauses is extended to the notion of *local extension* of a base theory.

Let $\mathcal{T}_0$ be an arbitrary theory with signature $\Pi_0 = (\Sigma_0, \mathsf{Pred})$, where the set of function symbols is Σ_0. Let $\Pi = (\Sigma_0 \cup \Sigma, \mathsf{Pred}) \supseteq \Pi_0$ be an extension by a non-empty set Σ of new function symbols and let $\mathcal{K}$ be a set of (implicitly universally closed) clauses in the extended signature. We will denote by Π^c the extension of Π with a fixed countable set of fresh constants. In what follows, we assume that all ground clauses we refer to contain symbols in Π^c. We say that an extension $\mathcal{T}_0 \cup \mathcal{K}$ of $\mathcal{T}_0$ is *local* if it satisfies the following condition[1]:

(Loc)   For every set G of ground clauses in Π^c it holds that
$\mathcal{T}_0 \cup \mathcal{K} \cup G \models \bot$ if and only if $\mathcal{T}_0 \cup \mathcal{K}[G] \cup G \models \bot$

where $\mathcal{K}[G]$ consists of those instances of $\mathcal{K}$ in which the terms starting with *extension functions* are in the set $\mathsf{est}(\mathcal{K}, G)$ of extension ground terms (i.e. terms starting with a function in Σ) which already occur in G or $\mathcal{K}$. (Note that the variables in clauses in $\mathcal{K}$ which do not occur below extension functions will not be instantiated in $\mathcal{K}[G]$.) The notion of local theory extension generalizes the notion of *local theories*. In [9, 10] we generalized condition (Loc) by considering

---

[1] It is easy to check that the formulation we give here and that in [13] are equivalent.

operators on sets of ground terms. This allows us to be more flexible w.r.t. the instances needed. Let Ψ be a map associating with every set T of ground terms a set $\Psi(T)$ of ground terms. For any set G of (augmented) ground Π^c-clauses we write $\mathcal{K}[\Psi_{\mathcal{K}}(G)]$ for $\mathcal{K}[\Psi(\text{est}(\mathcal{K}, G))]$ (the set consists of those instances of $\mathcal{K}$ in which the terms starting with *extension functions* are in $\Psi(\text{est}(\mathcal{K}, G))$). We can define a version of locality (Loc^Ψ) in which the set of terms used in the instances of the axioms is described using the map Ψ.

(Loc^Ψ)   For every set G of ground clauses in Π^c it holds that
$$\mathcal{T}_0 \cup \mathcal{K} \cup G \models \bot \text{ if and only if } \mathcal{T}_0 \cup \mathcal{K}[\Psi_{\mathcal{K}}(G)] \cup G \models \bot.$$

Extensions satisfying condition (Loc^Ψ) are called Ψ-local. Local theory extensions are Ψ-local, where Ψ is the identity operator. The order-local theories introduced in [3] satisfy a $\Psi^{\preceq}$-locality condition, where for every set T of ground clauses $\Psi^{\preceq}(T) = \{s \mid s \text{ ground term and } s \preceq t \text{ for some } t \in T\}$, where $\prec$ is the order on terms considered in [3].

Hierarchical Reasoning. Let $\mathcal{T}_0 \subseteq \mathcal{T} = \mathcal{T}_0 \cup \mathcal{K}$ be a theory extension satisfying (Loc^Ψ). To check the satisfiability w.r.t. $\mathcal{T}$ of a formula G, where G is a set of ground Π^c-clauses, we proceed as follows:

Step 1: By locality, $\mathcal{T} \cup G \models \bot$ iff $\mathcal{T}_0 \cup \mathcal{K}[\Psi_{\mathcal{K}}(G)] \cup G \models \bot$.

Step 2: Purification. We purify $\mathcal{K}[\Psi_{\mathcal{K}}(G)] \cup G$ (by introducing, in a bottom-up manner, new constants c_t for subterms $t = f(g_1, \ldots, g_n)$ with $f \in \Sigma$, g_i ground Π_0^c-terms, and corresponding definitions $c_t \approx t$) and obtain the set of formulae $\mathcal{K}_0 \cup G_0 \cup D$, where D consists of definitions $c_t \approx f(g_1, \ldots, g_n)$, where $f \in \Sigma$, c_t is a constant, $g_1, \ldots, g_n$ are ground Π_0^c-terms, and $\mathcal{K}_0, G_0$ are Π_0^c-formulae.

Step 3: Reduction to testing satisfiability in $\mathcal{T}_0$. We reduce the problem to testing satisfiability in $\mathcal{T}_0$ by replacing D with the following set of clauses:
$$\mathsf{Con}_0 = \{\bigwedge_{i=1}^{n} c_i \approx d_i \rightarrow c \approx d \mid f(c_1, \ldots, c_n) \approx c, f(d_1, \ldots, d_n) \approx d \in D\}.$$
This yields a sound and complete hierarchical reduction to a satisfiability problem in the base theory $\mathcal{T}_0$:

Theorem 1 ([9]). *Let $\mathcal{K}$ and G be as specified above. Assume that $\mathcal{T}_0 \subseteq \mathcal{T}_0 \cup \mathcal{K}$ satisfies condition (Loc^Ψ). Let $\mathcal{K}_0 \cup G_0 \cup \mathsf{Con}_0$ be obtained from $\mathcal{K}[\Psi_{\mathcal{K}}(G)] \cup G$ by purification (cf. Step 2). Then $\mathcal{T}_0 \cup \mathcal{K} \cup G \models \bot$ if and only if $\mathcal{T}_0 \cup \mathcal{K}_0 \cup G_0 \cup \mathsf{Con}_0 \models \bot$.*

Thus, satisfiability of ground clauses G as above w.r.t. $\mathcal{T}$ is decidable provided that $\Psi_{\mathcal{K}}(G)$ is effective, $\mathcal{K}[\Psi_{\mathcal{K}}(G)]$ is finite and $\mathcal{K}_0 \cup G_0 \cup \mathsf{Con}_0$ belongs to a decidable fragment of $\mathcal{T}_0$.

Locality and Embeddability. The (Ψ-)locality of an extension can be recognized by proving embeddability of partial models into total models assuming that the extension clauses are flat and linear; for details see [13, 9, 10]. The locality proofs given there also explain how to construct models of satisfiable ground (extended) clauses starting from models of their instances. We used the link between locality and embeddability for identifying various classes of local theory extensions. Some examples are given below:

Theories of absolutely free data structures. Let $\mathsf{AbsFree}_{\Sigma_0} = (\bigcup_{c \in \Sigma_0}(\mathsf{Injective}_c) \cup (\mathsf{Acyclic}_c)) \cup \bigcup_{c,d \in \Sigma, c \neq d} \mathsf{Disjoint}_{c,d}$, where:

$\mathsf{Injective}_c \quad c(x_1,\ldots,x_n) \approx c(y_1,\ldots,y_n) \rightarrow x_j \approx y_j \quad$ for all $j \in \{1,\ldots,n\}$,

$\mathsf{Disjoint}_{c,d} \quad c(x_1,\ldots,x_n) \not\approx d(y_1,\ldots,y_m)$

$\mathsf{Acyclic}_c \quad c(t_1,\ldots,t_n) \not\approx x$ if x occurs in some $t_i \quad$ (an axiom schema)

Theorem 2 ([14]). *The following theories are local:*

(a) The theory $\mathsf{AbsFree}_{\Sigma_0}$ of absolutely free constructors in Σ_0.

(b) $\mathcal{T}_{cs} = \mathsf{AbsFree}_{\Sigma_0} \cup \mathsf{Sel}(\Sigma_0)$, where $\mathsf{Sel}(\Sigma_0) = \bigcup_{c \in \Sigma_0} \bigcup_{i=1}^{a(c)} \mathsf{Sel}(sel_i^c, c)$ axiomatizes a family of selectors $sel_1^c,\ldots,sel_n^c$, where $a(c)$ is the arity of c, corresponding to constructors $c \in \Sigma_0$, where:

$\mathsf{Sel}(s_i^c, c) \quad \forall x, x_1,\ldots,x_{a(c)} \quad x \approx c(x_1,\ldots,x_{a(c)}) \rightarrow sel_i^c(x) \approx x_i$

The arguments in [14] can be used to prove locality if the axioms for selectors contain in addition to $\mathsf{Sel}(s_i^c, c)$ also the axiom:

$\forall x, x_1,\ldots,x_{a(d)} \quad x \approx d(x_1,\ldots,x_{a(d)}) \rightarrow sel_i^c(x) \approx x$ if $d \neq c$.

Locality and Saturation. In [2, 3], Ganzinger and Basin established a link between saturation and order locality, and used these results for automated complexity analysis, and for obtaining local axiomatizations from non-local ones.

Theorem 3 ([2, 3]). *Let $\prec_T$ be a well-founded (possibly partial) term ordering and $\prec$ a compatible and total atom ordering. Let $\mathcal{K}$ be a set of clauses without equality which is reductive w.r.t. $\prec_T$ and $\prec$. If $\mathcal{K}$ is saturated w.r.t. $\prec$-ordered resolution, then $\mathcal{K}$ is order local w.r.t. $\prec$.*

Example 4. *Theorem 3 is used in [4] for proving (by saturation) the locality of the presentation* Int *of the set of integers with successor and predecessor:*

(1) $\quad p(x) \approx y \rightarrow s(y) \approx x \qquad$ (3) $\quad p(x) \approx p(y) \rightarrow y \approx x$

(2) $\quad s(x) \approx y \rightarrow p(y) \approx x \qquad$ (4) $\quad s(x) \approx s(y) \rightarrow y \approx x$

together with an explicit axiomatization of the predicate $\approx$ by congruence axioms.

However, using saturation for detecting locality or for generating local presentations from non-local ones has the following drawbacks: (i) Equality cannot be used as a built-in predicate: If the clauses contain the equality predicate then the congruence axioms have to be added explicitly. (ii) The size of the saturated sets of clauses can be very large. Often, in fact, infinitely many clauses are generated.

Example 5 ([6]). *We illustrate the last problem by two examples:*

(1) From the clause set $\{x \leq x, x \leq y \wedge y \leq z \rightarrow x \leq z, \ f(x) \leq f(s(x))\}$ by saturation under ordered resolution we obtain an infinite set containing all clauses of the form $f(x) \leq f(s^n(x))$, where $n \geq 0$.

(2) From the clause set $\{nat(0), \ nat(x), y \approx s(x) \rightarrow nat(y)\}$ by saturation under ordered resolution and equality factoring we obtain an infinite set containing all clauses of the form $nat(s^n(0))$.

In such cases, a usual resolution-based theorem prover will not be able to detect saturation, since the set of clauses which are generated is infinite.

To reduce the size of a representation, we can use constrained clauses. In [6] we showed that Thm. 3 has a counterpart for constrained clauses with equality:

Theorem 6 ([6]). *Let $\prec_T$ be a reduction ordering and $\prec$ a compatible and total atom ordering. Let N be a set of constrained clauses that is reductive w.r.t. $\prec_T$ and saturated under the superposition calculus w.r.t. $\prec$ for constrained clauses described in [7]. Assume that each constrained clause in N with positive equational atoms contains either a unique positive equation which is also maximal, or a negative equation which is maximal. Let C be a ground clause whose antecedent does not contain any equations. Then $N \models C$ iff $N \models_\prec C$.*

The restrictions on N guarantee that constraint equations can be regarded syntactically and that all clauses that are derived in the constraint superposition calculus [7] from $N \cup \neg C$ contain equational atoms only negatively and no inferences into constraints are necessary.

The advantage of using constrained clauses is that in many cases it allows us to obtain a finite symbolic representation for possibly infinite sets of clauses.

Example 7 ([6]). *The constrained resolution calculus in [7] allows us to obtain a finite representation for the theory in Example 5(1) by using constrained clauses of the form $y \approx s^*(x) \parallel f(x) \leq f(y)$. For the theory described in Example 5(2), we obtain the saturated set of constrained clauses $\{x \approx s^*(0) \parallel nat(x)\}$.*

3   Enumerating Ground Instances of Constrained Clauses

Locality of theories and theory extensions guarantees that in order to disprove a set G of ground clauses we only need to look at certain ground instances of the local axiomatization (without loss of completeness). Note that the constraints are used as a notation for encoding a family of (syntactic) instances of a clause. The same holds when we compute given instances of such constrained clauses. We have $(\alpha \parallel C)[\Psi(G)] \subseteq C[\Psi(G)]$. A ground instance $C\tau$ from $C[\Psi(G)]$ is in $(\alpha \parallel C)[\Psi(G)]$ if the constraint $\alpha\tau$ holds. Unfortunately, it is in general not possible to effectively enumerate all such instances. This is due to the fact that satisfiability of constraints is not decidable (cf. [7]).

We show that enumerating ground instances is possible for an important class of constraints; we introduce a notation for them below:

Notation 8 *The expression $\vec{x} \approx \vec{y}\bar{\sigma}$ denotes a constraint $(\vec{x} \approx \vec{y})\bar{\sigma}$ where (i) $\bar{\sigma}$ is the identity on $\vec{x}$ and (ii) no new occurrences of variables in $\vec{x}$ are introduced by $\bar{\sigma}$, i.e. $\vec{x} \cap \text{vars}(\vec{y}\bar{\sigma}) = \emptyset$.*

In our experience, such constraints appear very frequently. Examples are the local axiomatization of monotonicity in Example 7, as well as the following axiomatization of the theory of absolutely free constructors $c_1, \ldots, c_m$ with arities $n_1, \ldots, n_m$ (which satisfies the requirements of Thm. 6):

- Injectivity: $c_i(x_1, \ldots, x_n) \approx c_i(y_1, \ldots, y_n) \rightarrow x_j \approx y_j$ for all $j \in \{1, \ldots, n\}$;
- Disjointness: $c_i(x_1, \ldots, x_{n_i}) \not\approx c_j(y_1, \ldots, y_{n_j})$ for $i < j$;
- Acyclicity: $x \approx y(\sigma_{11} | \ldots | \sigma_{mn_m})^+ \parallel x \not\approx y$, where $\sigma_{ij} = \{y \mapsto c_i(\ldots, y, \ldots)\}$
  with y at position j and fresh variables at all other positions.

Theorem 9. *For every set N of constrained clauses with only constraints of the form $\vec{x} \approx \vec{y}\bar{\sigma}$ and every set T of ground terms we can effectively enumerate $N[T]$.*

Proof: A ground instance $C\tau$ of $C[T]$ is in $(\alpha \parallel C)[T]$ if after applying the substitution τ the constraint holds. Whether this is the case can be decided for contraints of the form $\vec{x} \approx \vec{y}\bar{\sigma}$ by solving the matching problem $\vec{x}\tau = \vec{y}\bar{\sigma}\tau$, where $\vec{x}\tau$ consists of ground terms only. This is done by a reduction in the style of the standard unification algorithm by Martelli and Montanari, recursively unfolding the substitution expression:

$$\begin{array}{ll}
(\vec{s}_1, f(\vec{s}_2)) \approx (\vec{t}_1, g(\vec{t}_2))\bar{\sigma} \rightsquigarrow \bot & \vec{s} \approx \vec{y}\bar{\sigma}\bar{\tau} \rightsquigarrow \vec{s} \approx \vec{t}\bar{\tau} \text{ if } \vec{y}\bar{\sigma} = \vec{t} \\
(\vec{s}_1, f(\vec{s}_2)) \approx (\vec{t}_1, f(\vec{t}_2))\bar{\sigma} \rightsquigarrow (\vec{s}_1, \vec{s}_2) \approx (\vec{t}_1, \vec{t}_2)\bar{\sigma} & \vec{s} \approx \vec{y}\bar{\sigma}^* \bar{\tau} \rightsquigarrow \vec{s} \approx \vec{y}(\bar{\tau} | \bar{\sigma}\bar{\sigma}^* \bar{\tau}) \\
\vec{s} \approx \vec{y}(\bar{\sigma}_1 | \bar{\sigma}_2)\bar{\tau} \rightsquigarrow \vec{s} \approx \vec{y}\bar{\sigma}_1\bar{\tau} \mid \vec{s} \approx \vec{y}\bar{\sigma}_2\bar{\tau} & (s, \vec{t}) \approx (x, \vec{y}) \rightsquigarrow \vec{t} \approx \vec{y}\{x \mapsto s\}
\end{array}$$

We split the derivation into two branches for disjunctions. If at least one of the branches terminates without yielding $\bot$, the constraint is satisfiable.   □

Example 10. *Consider the constrained clause $(x \approx y)\sigma^* \parallel f(y) \leq f(x)$ with $\sigma = \{x \mapsto x, \ y \mapsto s(y)\}$, which axiomatizes monotonicity of a function f over the integers. Let $G = (d \approx s^3(c) \wedge f(c) > f(d))$. Then $\Psi(G) = \{d, c, s(c), s^2(c), s^3(c)\}$ and $((x \approx y)\sigma^* \parallel f(y) \leq f(x))[\Psi(G)]$ consists of the clauses:*

$$f(c) \leq f(s^i(c)) \text{ for } i \in \{0, 1, 2, 3\} \quad \text{and} \quad f(d) \leq f(d).$$

4 Enriching the Language

Axioms consisting of constrained clauses are difficult to export to standard first-order provers. In addition, there are limitations in using them if we are interested in checking satisfiability in Herbrand models of the axioms, in which all Skolem constants of G are interpreted as ground terms.

Example 11. *Let $N = \{x \approx s^*(0) \parallel nat(x)\}$. If a is a constant, we cannot directly express the positive fact that there is an n such that $a = s^n(0)$, so we cannot prove $nat(x) \rightarrow nat(s(x))$ from N: For the ground clause set $G = \{nat(a), \neg nat(s(a))\}$, we cannot derive a contradiction using the axiomatization N. (In fact, $N \cup G$ has a model with the universe $\{s^n(0) \mid n \in \mathbb{N}\} \cup \{s^n(a) \mid n \in \mathbb{N}\}$.)*

To address these limitations, and to obtain more natural axiomatizations, we *enrich the language with additional predicates* encoding the constraints and analyze situations in which locality results can be obtained for these new presentations.

Let N be a set of clauses with constraints of the form $\vec{x} \approx \vec{y}\bar{\sigma}$ satisfying the conditions from Thm. 6. Assume that N is saturated (hence local). Our goal is to use the form of the constraints to (conservatively) extend the language – e.g.

by defining new predicates – in order to obtain local presentations expressible as sets of clauses without constraints. For this we extend the signature by adding a new predicate symbol $R_{\bar{\sigma}}$ for every constraint $\alpha = \vec{x} \approx \vec{y}\bar{\sigma}$ and rewriting every constrained clause of the form $\alpha \| C$ into the clause $R_{\bar{\sigma}}(\vec{x}, \vec{y}) \rightarrow C$. We can encode the properties of these new relations (i) by giving a complete description for all ground terms, or (ii) by providing axiomatizations.

4.1   Concrete Description of the New Relations

In a first step, we assume that we give a complete description of the newly introduced relations for all ground terms; we will have to ensure that saturation is preserved when changing the clauses in N and adding the encoding. For this it is necessary that, whenever we eliminate the constraint from a constrained clause $\alpha \| C$, the ground literal $\neg R(\vec{x}, \vec{y})$ in the resulting clause $R(\vec{x}, \vec{y}) \rightarrow C$ does not influence the size of that clause too much. This can be achieved by ignoring constraint equations $x \approx y$ where x does not appear in the clausal part.

Example 12. *The clause $x \approx s(y) \| P(y)$ is not translated to $R(x, y) \rightarrow P(y)$, where the R literal dominates all ground instances of the form $R(s(t), t) \rightarrow P(t)$, but to $R() \rightarrow P(y)$, where R is the predicate corresponding to the empty constraint, or equivalently to $\rightarrow P(y)$.*

We show that if N is saturated under superposition for constraint clauses, this holds for the clause sets where constraints have been replaced by new predicates.

Theorem 13. *Let N be a consistent set of constrained clauses over $\Pi = (\Sigma, \mathsf{Pred})$ with constraints of the form $\vec{x} \approx \vec{y}\bar{\sigma}$. Let N' be obtained from N by replacing the constraints by fresh literals as explained above. Let $\mathcal{R}$ be the set of ground instances of these literals, such that $R_{\bar{\sigma}}(\vec{x}, \vec{y})\tau \in \mathcal{R}$ if $(\vec{x} \approx \vec{y})\bar{\sigma}\tau$ is valid, and $\neg R_{\bar{\sigma}}(\vec{x}, \vec{y})\tau \in \mathcal{R}$ otherwise. Let $\prec_T$ be a reduction ordering and $\prec$ a compatible and total atom ordering which ensures that R-literals are not maximal in any clause in N'. If N is (i) saturated under ordered resolution w.r.t. $\prec$, or (ii) saturated under constrained superposition w.r.t. $\prec$, or (iii) Horn and peak saturated under ordered hyperresolution w.r.t. $\prec$, then the same holds for $N' \cup \mathcal{R}$.*

It can be seen that for every set of ground clauses G in the signature of N, $N \cup G$ is satisfiable iff $N' \cup \mathcal{R} \cup G$ is satisfiable. By Thm. 13, if N is a set of constrained clauses in first-order logic without equality which is saturated under ordered resolution, or it satisfies the assumptions in Thm. 6 and is saturated under constrained superposition, so $N' \cup \mathcal{R} \cup G \models \bot$ iff $(N' \cup \mathcal{R})[\Psi^{\preceq}(G)] \cup G \models \bot$.

4.2   Axiomatization for the Newly Introduced Predicates

There are some problems with the approach described above. Assume that G contains constants which are not in Σ. Then, instances of the constraints $\alpha = \vec{x} \approx \vec{y}\bar{\sigma}$ in N containing such constants might be true for some interpretations of the constants and false for other interpretations. In such a situation we cannot add the corresponding instances of the R_σ-literals to $\mathcal{R}$. If σ is a substitution expression we need to express the properties of R_σ in terms of the substitutions used in σ. For this, we introduce for every constraint of the form $\vec{x} \approx \vec{y}\sigma$ a set of first-order formulae $\mathsf{Def}_{\bar{\sigma}}$ describing the relation $R_{\bar{\sigma}}(\vec{x}, \vec{y})$ by induction on the structure of α as follows:

Definition 14 (Axiomatizing the properties of the new relations)

(1) For atomic substitutions σ: $\mathsf{Def}_\sigma := R_\sigma(\vec{x}, \vec{y}) \leftrightarrow (x_1 \approx \sigma(y_1) \wedge \cdots \wedge x_n \approx \sigma(y_n))$
(2) For compositions: $\mathsf{Def}_{\bar{\sigma}\bar{\tau}} := R_{\bar{\sigma}\bar{\tau}}(\vec{x}, \vec{y}) \leftrightarrow \exists \vec{z}\, (R_{\bar{\sigma}}(\vec{x}, \vec{z}) \wedge R_{\bar{\tau}}(\vec{z}, \vec{y}))$
(3) For disjunctions: $\mathsf{Def}_{\bar{\sigma}_1 | \bar{\sigma}_2} := R_{\bar{\sigma}_1 | \bar{\sigma}_2}(\vec{x}, \vec{y}) \leftrightarrow (R_{\bar{\sigma}_1}(\vec{x}, \vec{y}) \vee R_{\bar{\sigma}_2}(\vec{x}, \vec{y}))$
(4) For iterations: $\mathsf{Def}_{\bar{\sigma}^*}$, axiom expressing the fact that $R_{\bar{\sigma}^*}$ is the reflexive and transitive closure of $R_{\bar{\sigma}}$.

Example 15. Let σ be a substitution with $\sigma(y) = s(y)$ and $\sigma(x) = x$. Let τ be the substitution with $\tau(y) = s(s(y))$ and $\tau(x) = x$. Let $\alpha = x \approx y\sigma$ (which corresponds to $x \approx s(y)$), and $\beta = x \approx y\tau$ (which corresponds to $x \approx s(s(y))$). Then Def_σ: $R_\sigma(x, y) \leftrightarrow (x \approx s(y))$; Def_τ: $R_\tau(x, y) \leftrightarrow (x \approx s(s(y)))$; $\mathsf{Def}_{\sigma\tau}$: $R_{\sigma\tau}(x, y) \leftrightarrow \exists z(R_\sigma(x, z) \wedge R_\tau(z, y))$; $\mathsf{Def}_{\sigma | \tau}$: $R_{\sigma | \tau}(x, y) \leftrightarrow (R_\sigma(x, y) \vee R_\tau(x, y))$.

It can be shown that if N is a set of constrained clauses and G is a set of ground clauses, $N \cup G$ is satisfiable iff $N' \cup G$ has a model satisfying definitions (1)-(3) in Def. 14 and such that for every substitution expression $\bar{\sigma}$, $R_{\bar{\sigma}^*}$ is the reflexive and transitive closure of $R_{\bar{\sigma}}$. The definitions of relations $R_{\bar{\sigma}}$ where $\bar{\sigma}$ is a composition or a disjunction of substitutions can be expressed using first-order formulae over relations associated with the component substitutions of $\bar{\sigma}$. The problematic part is finding a set of axioms which expresses the fact that $R_{\bar{\sigma}^*} = (R_{\bar{\sigma}})^*$. It is known that transitive closure can be encoded in first-order logic (in an extension of the original signature) if we are interested only in finite models. An axiomatization for reflexive transitive closure which is sound and complete for finite domains was proposed in [12]; an axiomatization which completely describes transitive closure over finite, acyclic graphs was given in [11]:

$$T_1(R) : \forall x, y(R^*(x, y) \leftrightarrow (x \approx y \vee \exists z(R(x, z) \wedge R^*(z, y)))).$$

In [11] it is shown that for checking validity of formulae which contain predicates of the form R^* only positively, only the $\leftarrow$ implication $T_1^\leftarrow(R)$ of $T_1(R)$ is needed. Having a first order axiomatization does not mean that we have decidability (see the discussions on this in [11]). Consider for instance the axiom $T_2(R)$:

$$T_2(R) : \forall x, y(R^*(x, y) \leftrightarrow (x \approx y \vee \exists z(R^*(x, z) \wedge R(z, y)))).$$

Neither $T_2(R)$ nor $T_1(R)$ can be proved from the other without induction; similarly for the transitivity of the congruence closure [11].

5   Extensions of Theories of Absolutely Free Constructors

To identify situations in which axiomatizations with good properties exist, we make the following assumptions:

(A1) We consider sets $N = N_0 \cup N_1$ of constrained clauses where:
- N_0 is a set of axioms for $\mathcal{T}_{cs}$, e.g. $\mathsf{AbsFree}_{\Sigma_0} \cup \mathsf{Sel}(\Sigma_0)$ (cf. Thm. 2, [14]);
- the extension $N_0 \subseteq N = N_0 \cup N_1$ is a local theory extension with the property that in N_1 every variable occurs below an extension function and every weak partial model can be extended to a total model with the same universes for the base sorts. (This happens for instance in the case of local extensions with function symbols whose arguments are of base sorts, and where the codomain is of a new sort.)

(A2) The clauses in N contain only constraints of the form $\vec{x} \approx \vec{y}\sigma$; where σ used only constructors, and $\prec$ is a simplification ordering.

We analyze the special situation in which the properties of the new relation symbols can be described by a set of formulae Def (of the form (1)-(4) in Def. 14) which can be expressed as a set of clauses without additional Skolem functions. This is the case, for instance, when the following conditions hold:

(C1) Only composition of basic substitutions is allowed; iteration is only allowed for substitutions with $\sigma(x) \succeq x$ for every variable x.

(C2) The set G of clauses for which we want to check satisfiability contains only negative atoms starting with a relation of form R_{σ^*}.

We first analyze satisfiability w.r.t. absolutely free models, possibly with additional generators (5.1) and then satisfiability in the absolutely free (Herbrand) model over Σ_0, with no additional generators (5.2).

5.1   Satisfiability w.r.t. Absolutely Free Models

We first show that under the assumptions above locality is preserved if we replace the constraints in the clauses using new predicate symbols and we specify their properties.

Theorem 16. *Let $N_1, \mathcal{R}, N_1'$ be sets of clauses as in Theorem 13. Assume that conditions (A1), (A2) and (C1) hold for $N = N_0 \cup N_1$. Let G be a set of ground clauses satisfying condition (C2). Then the following are equivalent:*

(1) $\mathcal{R} \cup N_0 \cup N_1' \cup G$ has a model A in which the interpretations of the new relations satisfy definitions (1)–(3), and such that $R_{\sigma^*}^A = (R_\sigma^A)^*$.

(2) $(\mathsf{Def} \cup N_0 \cup N_1')[\Psi^{\preceq}(G)] \cup G$ has a finite partial model (having as elements the finite set of interpretations of terms in $\Psi^{\preceq}(G)$), where Def consists of all definitions of the form (1)-(4) in Def. 14 (which are in this case clauses without additional Skolem functions), and where the definitions of type (4) are axiomatized by the set of clauses consisting for every $R_{\bar\sigma^*}$ of $T_1^{\leftarrow}(R_{\bar\sigma})$.

Example 17. *Let $N_f = \{x \approx s^*(y) \,\|\, f(y) \leq f(x)\} \cup \mathsf{Pre}$ be the (local) constrained axiomatization of the theory of monotonicity for the function f, where Pre contains the reflexivity and transitivity axioms for $\leq$. Let N_0 be the axiomatization of the absolutely free constructor s. Note that $x \approx s^*(y)$ is a shorthand for $x \approx y\sigma^*$,* *where $\sigma(x) = x$ and $\sigma(y) = s(y)$. The set of clauses obtained from N_f by replacing the constraints with new literals is $N'_f = \{R_{\sigma^*}(x,y) \rightarrow f(y) \leq f(x)\} \cup \mathsf{Pre}$. For defining R_σ we use the axiom $\mathsf{Def}_\sigma := \{R_\sigma(x,y) \leftrightarrow x \approx s(y)\}$. Let Def_{σ^*} be an axiomatization for R_{σ^*} as R_σ^* (e.g. as in [11]).*

Let $G_1 := a \approx s(b) \wedge \neg R_{\sigma^*}(a,b) \wedge \neg f(b) \leq f(a)$ (containing only negative R_{σ^*}-literals). The set of clauses $N_0 \cup N_f \cup G_1$ is satisfiable if and only if $N_0 \cup N'_f \cup \mathcal{R} \cup G_1$ is satisfiable. By Thm. 16 this happens if and only if $(\mathsf{Def} \wedge N'_f)[\Psi^{\preceq}(G_1)] \cup G_1$ is satisfiable, where $\mathsf{Def} = \mathsf{Def}_\sigma \wedge T_1^{\leftarrow}(R_\sigma)$. We check the satisfiability of this set of ground clauses using the hierarchical reduction in Theorem 1.

D	$G_1^c \cup N_0[\Psi^{\preceq}(G)]_0$	$G_1^R \cup \mathsf{Def}[\Psi^{\preceq}(G)]_0$	G_1^e	$N'_f[\Psi^{\preceq}(G)]_0$	
$a_1 \approx f(a)$	$a \approx s(b)$	$\neg R_{\sigma^*}(a,b)$	$\neg a_1 \leq b_1$	$a_1 \leq a_1$	$R_{\sigma^*}(a,b) \rightarrow a_1 \leq b_1$
$b_1 \approx f(b)$	$\neg a \approx s(a)$	$\mathsf{Def}_\sigma[\Psi^{\preceq}(G)]$		$b_1 \leq b_1$	$R_{\sigma^*}(b,a) \rightarrow b_1 \leq a_1$
	$\neg b \approx s(b)$	$T_1^{\leftarrow}(R_\sigma)[\Psi^{\preceq}(G)]$			$\ldots$

Let $G_2 := (b \approx s(a) \wedge R_{\sigma^*}(c,b) \wedge d \approx s(c) \wedge \neg f(a) \leq f(d))$. G_2 contains positive R_{σ^*}-literals, so Theorem 16 cannot be used. We present a proof of unsatisfiability of $\mathsf{Def}_\sigma \cup T_1(R_\sigma) \cup T_2(R_\sigma) \cup N'_f \cup G_2$: From $b \approx s(a)$ and $d \approx s(c)$ and from Def_σ it follows that $R_\sigma(b,a)$ and $R_\sigma(d,c)$. From $R_\sigma(d,c), R_{\sigma^*}(c,b)$ and $T_1(R_\sigma)$ it follows that $R_{\sigma^*}(d,b)$. From $R_{\sigma^*}(d,b), R_\sigma(b,a)$ and $T_2(R_\sigma)$ it follows that $R_{\sigma^*}(d,a)$. From this and $N_f[\Psi^{\preceq}(G)]_0$ we derive $a_1 \leq d_1$, which together with $\neg a_1 \leq d_1$ leads to a contradiction.

5.2 Satisfiability w.r.t. Herbrand Models over Σ_0

Let $\mathcal{T}_{cs} := \mathsf{AbsFree}_{\Sigma_0} \cup \mathsf{Sel}(\Sigma_0)$ be the theory of absolutely free constructors in a finite set Σ_0 with corresponding selectors mentioned in Theorem 2. We analyze the problem of checking satisfiability of ground clauses w.r.t. such axiomatizations *in Herbrand models having as universe the set* T_{Σ_0} *of ground terms over the signature of* $\mathcal{T}_{cs}$ *(and no additional generators)*. Since the set T_{Σ_0} is infinite, we cannot use the results on first-order definability for transitive closure mentioned before. We will overcome the problem by noting the following fact:

Lemma 18 *For every element t in the absolutely free algebra T_{Σ_0}, the number of terms in the set $T = \bigcup_{i \in \mathbb{N}} T_i$, where $T_0 = \{t\}$ and $T_{i+1} = \{sel_j^c(t') \mid sel_j^c$ selector and $t' \in T_i\}$ is finite.*

Example 19. *(i) For the theory of one unary constructor s with selector p and one constant constructor 0, T_{Σ_0} is the set of natural numbers; for every natural number m the set of elements of the form $p^n(m)$ with $n \in \mathbb{N}$ is finite. (ii) For the theory of one binary constructor cons with selector tail and one constant constructor nil (modeling the theory of finite, acyclic lists), from every node in a linked, finite acyclic list, nil can be reached in finitely many steps using tail.*

It is known that $T_{\Sigma_0} \models t \approx c(t_1, \ldots, t_n) \leftrightarrow \bigwedge_{i=1}^n sel_i^c(t) \approx t_i \wedge \bigwedge_{i=1}^n t \not\approx t_i$. We can transform all clauses and all constraints using this equivalence, such that all constructors are eliminated. (The constrained clause $x \approx s^*(y) \parallel f(y) \leq f(x)$ can be rewritten to $y \approx p^*(x) \parallel f(y) \leq f(x)$.)

For the sake of simplicity we restrict here to signatures which contain one constructor with arity n and selectors $sel_1, \ldots, sel_n$ and m constructors $c_1, \ldots, c_m$ with arity 0. Let $\mathcal{T}_{sel}$ be axiomatized by the following axioms:

$$(S1) \quad \bigwedge_{i=1}^{n} sel_i(x) \approx sel_i(y) \rightarrow x \approx y \vee (\bigvee_{i=1}^{m} x \approx c_i \vee \bigvee_{i=1}^{m} y \approx c_i)$$
$$(S2) \quad sel_i(x) \approx x \rightarrow \bigvee_{i=1}^{n} x \approx c_i$$

In order to ensure that we check ground satisfiability over T_{Σ_0}, we will add the following axiom (after extending the signature to a signature $\Sigma_{\mathcal{R}}$ by adding the relations of the form $R_{\bar{\sigma}}$):

$$(\textsf{Fin}) \quad R_{(sel_1|\ldots|sel_k)^*}(c_1^0, x) \vee \cdots \vee R_{(sel_1|\ldots|sel_k)^*}(c_m^0, x)$$

where $sel_1, \ldots, sel_k$ are all selectors and $c_1^0, \cdots c_m^0$ are all 0-ary constructors.[2]

Theorem 20. *Let A be a model of an extension of $\mathcal{T}_{cs}$ with additional relations $\{R_{\bar{\sigma}} \mid \bar{\sigma}$ substitution expression occurring in $\mathcal{K}\}$, where $\mathcal{K}$ is a family of constrained clauses satisfying conditions (1)-(4). Assume that $\textsf{Fin}$ holds in A. Then A has as universe the set T_{Σ_0}.*

Having first order axiomatizations for relations of the form $R_{\bar{\sigma}^*}$ and transitive closure of $R_{\bar{\sigma}}$ does not necessarily imply that we have already methods for efficiently reasoning in such theories. We show that under assumptions (A1), (A2), (C1) and (C2), when restricting to models with universe T_{Σ_0}, checking satisfiability of sets of ground clauses G w.r.t. a set of saturated constrained clauses $N_0 \cup N_1$, where N_0 is a set of axioms for $\mathcal{T}_{cs}$, can be reduced to checking satisfiability of $N_0 \cup N_1' \cup \textsf{Def} \cup G$ (where N_1' is obtained from N_1 by replacing every clause $\vec{x} \approx \vec{y}\bar{\sigma}\|C$ in N with $R_\sigma(\vec{x}, \vec{y}) \rightarrow C$ and $\textsf{Def}$ is a set of clauses specifying definitions for the newly introduced relations) w.r.t. certain finite structures with universe T_G, defined by:

$$T_G = \bigcup_{i \in \mathbb{N}} T_i, \text{ where } T_0 = \textsf{st}(G) \text{ and } T_{i+1} = \{sel_i(t) \mid t \in T_i\}.$$

Theorem 21. *Let N_1 be a set of constrained clauses. Assume that conditions (A1), (A2) hold for $N = N_0 \cup N_1$. Assume that the definitions of the newly introduced predicate symbols can be expressed as a set $\textsf{Def}$ of clauses without additional Skolem functions. For every set G of ground clauses satisfying condition (C2), the following are equivalent:*

(1) *$N_0 \cup N_1 \cup G$ has a model A with support T_{Σ_0} in which all relations satisfy conditions (1)-(3) in Def. 14 and for every σ, $R_{\sigma^*}^A = (R_\sigma^A)^*$.*

(2) *$(\mathcal{T}_{sel} \cup \textsf{Def} \cup N_1')[T_G] \cup G$ has a finite model P having as universe a subset of T_Σ containing all 0-ary constructors and with the property that all terms in T_G are defined in P.*

---

[2] This is different from the approach used in [7] in which satisfiability is checked in the minimal Herbrand model; to ensure that all constants can be expressed using the initial signature, existential variables are added to the constraints.

Sometimes the assumptions can be relaxed. One of the simpler cases studied in the literature are theories with relations defined by $R(x,y) \leftrightarrow y \approx f(x)$, where f is a unary function symbol. We will therefore restrict here to analyzing the situation in which we have one unary constructor and several 0-ary constructors. For this case we use an adaptation of the axiomatization of finite, acyclic lists with reachability proposed in [16]. For the theory of one binary constructor, the results in [15] can be adapted; we think that the results in [15] can be used, with small changes, also for the theory of one n-ary constructor.

Example 22. *Let R_{p^*} be the transitive closure of R_p, where $R_p(x,y) \leftrightarrow x \approx p(y)$. We adapt the axiomatization used in [16] to obtain an axiomatization Def_{p^*} for R_{p^*} consisting in the formulae (1)–(9). We included also the clause Fin : $(R_{p^*}(0,x))$, expressing the fact that from every element we can reach 0 in finitely many steps; this guarantees that we check satisfiability in the initial model:*

(1) $R_{p^*}(0,x)$ (2) $p(x) \approx p(y) \to x \approx y \vee x \approx 0 \vee y \approx 0$

(3) $R_{p^*}(x,x)$ (4) $R_p(x,y) \to R_{p^*}(x,y)$

(5) $p(x) \approx x \to x \approx 0$ (6) $R_{p^*}(x,y) \to x \approx y \vee R_{p^*}(x,p(y))$

(7) $R_{p^*}(x,y) \wedge R_{p^*}(y,x) \to x \approx y$ (8) $R_{p^*}(x,y) \wedge R_{p^*}(y,z) \to R_{p^*}(x,z)$

(9) $R_{p^*}(x,y) \wedge R_{p^*}(x,z) \to R_{p^*}(y,z) \vee R_{p^*}(z,y)$

The locality proof in [16] uses the link between locality and embeddability; it can be adapted to prove that $\mathcal{T}_{sel} \cup \mathsf{Def}_{p^*}$ is a local theory, or alternatively that the extension $N_1 \subseteq \mathcal{T}_{sel} \cup \mathsf{Def}_{p^*}$ is local, where N_1 consists of the axioms (1), (3), (8) and (9) in Def_{p^*}.

Example 23. *Consider the saturated axiomatization $N = \{x \approx s^*(0) \,\|\, nat(x)\}$ of the natural numbers. We transform N by using the selector instead and associating new predicate symbols with the constraints as follows: Since $x \approx s^*(0)$ iff $p^*(x) \approx 0$. we introduce new predicates R_σ and R_{σ^*} axiomatized by Def_σ: $R_\sigma(x,y) \leftrightarrow p(x) \approx y$; the axiomatization for R_{σ^*} is the one in Example 22. Then $N' := R_{\sigma^*}(x,0) \to nat(x)$. Let Fin be $R_{\sigma^*}(x,0)$.*

Let $G = \{nat(a), \neg nat(s(a))\}$. By Thm. 21 we know that if $\mathcal{T}_s$ is the theory of one unary constructor s and one 0-ary constructor 0, then there is a model of $N' \cup G \cup \mathsf{Fin}$ (which has as support $\mathcal{T}_s$, i.e. the set of natural numbers) if and only if $(\mathcal{T}_{sel} \cup N' \cup \mathsf{Fin})[\mathcal{T}_G] \cup G$ has a model. Note that although T_G is finite, we cannot estimate its size. We will start by instantiating the clauses in $\mathcal{T}_{sel} \cup N' \cup \mathsf{Fin}$ using terms occurring in G. From $\neg nat(s(a))$ we derive $\neg R_{\sigma^*}(s(a),0)$, which together with Fin leads to a contradiction. We do not need other instances.

6   Conclusions and Future Work

This paper continues our previous work [6], where we used constrained clauses to finitely represent possibly infinite sets of clauses and showed that constrained axiomatizations are local if they are saturated under a version of resolution. Here we showed that we can encode the constraints by extending the signature with new predicates – which often bear a particularly useful semantics – in such a way that locality is guaranteed. We focused on a special type of constrained clauses containing only constraints of the form $\vec{x} \approx \vec{y}\bar{\sigma}$, where $\vec{x} \cap \text{vars}(\vec{y}\bar{\sigma}) = \emptyset$ and identified extensions of theories of absolutely free constructors for which locality is preserved after this transformation.

Beyond what we presented in this paper, there are other ways of using such language extensions in more general settings, which we would like to explore in future work. Of particular interest are theories with fixpoints. Consider for instance the theory of a monotone function f over a theory with an underlying complete $\wedge$-semilattice structure with top element 1. We can use the fact that greatest fixpoints exist to replace constraints of the form $x \approx f^*(1) \parallel c \leq x$ with the unconstrained clause $c \leq \text{gfp}(f)$ In future work we plan to apply this type of reasoning to the description logics $\mathcal{EL}$ or $\mathcal{ALC}$ with fixpoints (e.g. for computing uniform interpolants). Another promising research direction is inductive theorem proving: Originally, constrained clauses of this flavor were introduced in [8] to reason about entailment in Herbrand models, and in particular in minimal models. The link between saturation and locality established in [6] and in the current paper opens a completely new avenue to explore the connection between locality and efficient reasoning in minimal models.

Acknowledgments. We thank the reviewers for their helpful comments. This work was partly supported by the German Research Council (DFG) as part of the Transregional Collaborative Research Center "Automatic Verification and Analysis of Complex Systems" (SFB/TR 14 AVACS, see www.avacs.org).

References

[1] Bachmair, L., Ganzinger, H.: Rewrite-based equational theorem proving with selection and simplification. J. of Logic and Computation 4(3), 217–247 (1994)

[2] Basin, D., Ganzinger, H.: Complexity analysis based on ordered resolution. In: Proc. LICS 1996, pp. 456–465. IEEE Computer Society Press (1996)

[3] Basin, D., Ganzinger, H.: Automated complexity analysis based on ordered resolution. Journal of the ACM 48(1), 70–109 (2001)

[4] Ganzinger, H.: Relating semantic and proof-theoretic concepts for polynomial time decidability of uniform word problems. In: Proc. LICS 2001, pp. 81–92. IEEE Computer Society Press (2001)

[5] Givan, R., McAllester, D.: New results on local inference relations. In: Principles of Knowledge Representation and Reasoning: Proceedings of the Third International Conference (KR 1992), pp. 403–412. Morgan Kaufmann Press (1992)

[6] Horbach, M., Sofronie-Stokkermans, V.: Obtaining finite local theory axiomatizations via saturation. In: Fontaine, P., Ringeissen, C., Schmidt, R.A. (eds.) FroCoS 2013. LNCS (LNAI), vol. 8152, pp. 198–213. Springer, Heidelberg (2013)

[7] Horbach, M., Weidenbach, C.: Deciding the inductive validity of $\forall \exists^*$ queries. In: Grädel, E., Kahle, R. (eds.) CSL 2009. LNCS, vol. 5771, pp. 332–347. Springer, Heidelberg (2009)

[8] Horbach, M., Weidenbach, C.: Superposition for Fixed Domains. ACM Transactions on Computational Logic 11(4), 27:1–27:35 (2010)

[9] Ihlemann, C., Jacobs, S., Sofronie-Stokkermans, V.: On local reasoning in verification. In: Ramakrishnan, C.R., Rehof, J. (eds.) TACAS 2008. LNCS, vol. 4963, pp. 265–281. Springer, Heidelberg (2008)

[10] Ihlemann, C., Sofronie-Stokkermans, V.: On hierarchical reasoning in combinations of theories. In: Giesl, J., Hähnle, R. (eds.) IJCAR 2010. LNCS, vol. 6173, pp. 30–45. Springer, Heidelberg (2010)

[11] Lev-Ami, T., Immerman, N., Reps, T.W., Sagiv, M., Srivastava, S., Yorsh, G.: Simulating reachability using first-order logic with applications to verification of linked data structures. Logical Methods in Computer Science 5(2) (2009)

[12] Klaessen, K.: Expressing transitive closure for finite domains in pure first-order logic. Unpublished manuscript

[13] Sofronie-Stokkermans, V.: Hierarchic reasoning in local theory extensions. In: Nieuwenhuis, R. (ed.) CADE 2005. LNCS (LNAI), vol. 3632, pp. 219–234. Springer, Heidelberg (2005)

[14] Sofronie-Stokkermans, V.: Locality results for certain extensions of theories with bridging functions. In: Schmidt, R.A. (ed.) CADE-22. LNCS (LNAI), vol. 5663, pp. 67–83. Springer, Heidelberg (2009)

[15] Wies, T., Muñiz, M., Kuncak, V.: An efficient decision procedure for imperative tree data structures. In: Bjørner, N., Sofronie-Stokkermans, V. (eds.) CADE 2011. LNCS (LNAI), vol. 6803, pp. 476–491. Springer, Heidelberg (2011)

[16] Wies, T., Muñiz, M., Kuncak, V.: Deciding functional lists with sublist sets. In: Joshi, R., Müller, P., Podelski, A. (eds.) VSTTE 2012. LNCS, vol. 7152, pp. 66–81. Springer, Heidelberg (2012)

Proving Termination and Memory Safety for Programs with Pointer Arithmetic*

Thomas Ströder[1], Jürgen Giesl[1], Marc Brockschmidt[2], Florian Frohn[1],
Carsten Fuhs[3], Jera Hensel[1], and Peter Schneider-Kamp[4]

[1] LuFG Informatik 2, RWTH Aachen University, Germany
[2] Microsoft Research Cambridge, UK
[3] Dept. of Computer Science, University College London, UK
[4] IMADA, University of Southern Denmark, Denmark

Abstract. Proving termination automatically for programs with explicit pointer arithmetic is still an open problem. To close this gap, we introduce a novel abstract domain that can track allocated memory in detail. We use it to automatically construct a *symbolic execution graph* that represents all possible runs of the program and that can be used to prove memory safety. This graph is then transformed into an *integer transition system*, whose termination can be proved by standard techniques. We implemented this approach in the automated termination prover AProVE and demonstrate its capability of analyzing C programs with pointer arithmetic that existing tools cannot handle.

1 Introduction

Consider the following standard C implementation of `strlen` [23,30], computing the length of the string at pointer `str`. In C, strings are usually represented as a pointer `str` to the heap, where all following memory cells up to the first one that contains the value 0 are allocated memory and form the value of the string.

```
int strlen(char* str) {char* s = str; while(*s) s++; return s-str;}
```

To analyze algorithms on such data, one has to handle the interplay between addresses and the values they point to. In C, a violation of *memory safety* (e.g., dereferencing `NULL`, accessing an array outside its bounds, etc.) leads to undefined behavior, which may also include non-termination. Thus, to prove termination of C programs with low-level memory access, one must also ensure memory safety. The `strlen` algorithm is memory safe and terminates because there is some address `end` $\geq$ `str` (an *integer property* of `end` and `str`) such that `*end` is 0 (a *pointer property* of `end`) and all addresses `str` $\leq$ `s` $\leq$ `end` are allocated. Other typical programs with pointer arithmetic operate on arrays (which are just sequences of memory cells in C). In this paper, we present a novel approach to prove memory safety and termination of algorithms on integers and pointers automatically. To avoid handling the intricacies of C, we analyze programs in the platform-independent intermediate representation (IR) of the LLVM compilation framework [17]. Our approach works in three steps: First, a *symbolic execution graph* is created

---

* Supported by DFG grant GI 274/6-1 and Research Training Group 1298 (*AlgoSyn*).

S. Demri, D. Kapur, and C. Weidenbach (Eds.): IJCAR 2014, LNAI 8562, pp. 208–223, 2014.
© Springer International Publishing Switzerland 2014

that represents an over-approximation of all possible program runs. We present our abstract domain based on *separation logic* [22] and the automated construction of such graphs in Sect. 2. In this step, we handle all issues related to memory, and in particular prove memory safety of our input program. In Sect. 3, we describe the second step of our approach, in which we generate an *integer transition system* (ITS) from the symbolic execution graph, encoding the essential information needed to show termination. In the last step, existing techniques for integer programs are used to prove termination of the resulting ITS. In Sect. 4, we compare our approach with related work and show that our implementation in the termination prover AProVE proves memory safety and termination of typical pointer algorithms that could not be handled by other tools before.

2   From **LLVM** to Symbolic Execution Graphs

In Sect. 2.1, we introduce concrete LLVM states and *abstract* states that represent *sets* of concrete states, cf. [9]. Based on this, Sect. 2.2 shows how to construct symbolic execution graphs automatically. Sect. 2.3 presents our algorithm to *generalize* states, needed to always obtain *finite* symbolic execution graphs.

To simplify the presentation, we restrict ourselves to a single LLVM function without function calls and to types of the form in (for n-bit integers), in* (for pointers to values of type in), in**, in***, etc. Like many other approaches to termination analysis, we disregard integer overflows and assume that variables are only instantiated with signed integers appropriate for their type. Moreover, we assume a 1 byte data alignment (i.e., values may be stored at any address).

2.1   Abstract Domain

Consider the strlen function from Sect. 1. In the corresponding LLVM code,[1] str has the type i8*, since it is a pointer to the string's first character (of type i8). The program is split into the *basic blocks* entry, loop, and done. We will explain this LLVM code in detail when constructing the symbolic execution graph in Sect. 2.2.

```
define i32 @strlen(i8* str) {

entry: 0: c0 = load i8* str
       1: c0zero = icmp eq i8 c0, 0
       2: br i1 c0zero, label done, label loop

loop:  0: olds = phi i8* [str,entry],[s,loop]
       1: s = getelementptr i8* olds, i32 1
       2: c = load i8* s
       3: czero = icmp eq i8 c, 0
       4: br i1 czero, label done, label loop

done:  0: sfin = phi i8* [str,entry],[s,loop]
       1: sfinint = ptrtoint i8* sfin to i32
       2: strint = ptrtoint i8* str to i32
       3: size = sub i32 sfinint, strint
       4: ret i32 size }
```

Concrete LLVM states consist of the program counter, the values of local variables, and the state of the memory. The program counter is a 3-tuple (b_{prev}, b, i), where b is the name of the current basic block, b_{prev} is the previously executed

---

[1] This LLVM program corresponds to the code obtained from strlen with the Clang compiler [8]. To ease readability, we wrote variables without "%" in front (i.e., we wrote "str" instead of "%str" as in proper LLVM) and added line numbers.

block,[2] and i is the index of the next instruction. So if *Blks* is the set of all basic blocks, then the set of code positions is $Pos = (Blks \cup \{\varepsilon\}) \times Blks \times \mathbb{N}$. We represent assignments to the local program variables $\mathcal{V_P}$ (e.g., $\mathcal{V_P} = \{\texttt{str}, \texttt{c0}, \ldots\}$) as functions $s : \mathcal{V_P} \to \mathbb{Z}$. The state of the memory is represented by a partial function $m : \mathbb{N}_{>0} \to \mathbb{Z}$ with finite domain that maps addresses to integer values. So a concrete LLVM state is a 3-tuple $(p, s, m) \in Pos \times (\mathcal{V_P} \to \mathbb{Z}) \times (\mathbb{N}_{>0} \to \mathbb{Z})$.

To model violations of memory safety, we introduce a special state *ERR* to be reached when accessing non-allocated memory. So (p, s, m) denotes only memory safe states where all addresses in m's domain are allocated. Let $\to_{\mathsf{LLVM}}$ be LLVM's evaluation relation on concrete states, i.e., $(p, s, m) \to_{\mathsf{LLVM}} (\overline{p}, \overline{s}, \overline{m})$ holds iff (p, s, m) evaluates to $(\overline{p}, \overline{s}, \overline{m})$ by executing one LLVM instruction. Similarly, $(p, s, m) \to_{\mathsf{LLVM}} ERR$ means that the instruction at position p accesses an address where m is undefined. An LLVM program is *memory safe* for (p, s, m) iff there is no evaluation $(p, s, m) \to^+_{\mathsf{LLVM}} ERR$, where $\to^+_{\mathsf{LLVM}}$ is the transitive closure of $\to_{\mathsf{LLVM}}$.

To formalize *abstract* states that stand for sets of concrete states, we use a fragment of *separation logic* [22]. Here, an infinite set of symbolic variables $\mathcal{V}_{sym}$ with $\mathcal{V}_{sym} \cap \mathcal{V_P} = \varnothing$ can be used in place of concrete integers. We represent abstract states as tuples (p, LV, KB, AL, PT). Again, $p \in Pos$ is the program counter. The function $LV : \mathcal{V_P} \to \mathcal{V}_{sym}$ maps every local variable to a symbolic variable. To ease the generalization of states in Sect. 2.3, we require injectivity of LV. The *knowledge base* $KB \subseteq QF\_IA(\mathcal{V}_{sym})$ is a set of pure quantifier-free first-order formulas that express integer arithmetic properties of $\mathcal{V}_{sym}$.

The *allocation list* AL contains expressions of the form $alloc(v_1, v_2)$ for $v_1, v_2 \in \mathcal{V}_{sym}$, which indicate that $v_1 \leq v_2$ and that all addresses between v_1 and v_2 are allocated. Finally, PT is a set of "points-to" atoms $v_1 \hookrightarrow_{\mathtt{ty}} v_2$ where $v_1, v_2 \in \mathcal{V}_{sym}$ and $\mathtt{ty}$ is an LLVM type. This means that the value v_2 of type $\mathtt{ty}$ is stored at the address v_1. Let $size(\mathtt{ty})$ be the number of bytes required for values of type $\mathtt{ty}$ (e.g., $size(\mathtt{i8}) = 1$ and $size(\mathtt{i32}) = 4$). As each memory cell stores one byte, $v_1 \hookrightarrow_{\mathtt{i32}} v_2$ means that v_2 is stored in the four cells at the addresses $v_1, \ldots, v_1 + 3$.

Definition 1 (Abstract States). Abstract states *have the form* (p, LV, KB, AL, PT) *where* $p \in Pos$, $LV : \mathcal{V_P} \to \mathcal{V}_{sym}$ *is injective,* $KB \subseteq QF\_IA(\mathcal{V}_{sym})$, $AL \subseteq \{alloc(v_1, v_2) \mid v_1, v_2 \in \mathcal{V}_{sym}\}$, *and* $PT \subseteq \{(v_1 \hookrightarrow_{\mathtt{ty}} v_2) \mid v_1, v_2 \in \mathcal{V}_{sym}, \mathtt{ty}$ *is an* LLVM *type*$\}$. *Additionally, there is a state ERR for violations of memory safety.*

We often identify LV with the set of equations $\{\mathtt{x} = LV(\mathtt{x}) \mid \mathtt{x} \in \mathcal{V_P}\}$ and extend LV to a function from $\mathcal{V_P} \uplus \mathbb{Z}$ to $\mathcal{V}_{sym} \uplus \mathbb{Z}$ by defining $LV(z) = z$ for all $z \in \mathbb{Z}$. As an example, consider the following abstract state for our $\texttt{strlen}$ program:

$$((\varepsilon, \mathbf{entry}, 0), \quad \{\mathtt{str} = u_{\mathtt{str}}, \ldots, \mathtt{size} = u_{\mathtt{size}}\}, \quad \{z = 0\}, \qquad (\dagger)$$
$$\{alloc(u_{\mathtt{str}}, v_{end})\}, \qquad \qquad \{v_{end} \hookrightarrow_{\mathtt{i8}} z\}).$$

It represents states at the beginning of the **entry** block, where $LV(\mathtt{x}) = u_{\mathtt{x}}$ for all $\mathtt{x} \in \mathcal{V_P}$, the memory cells between $LV(\mathtt{str}) = u_{\mathtt{str}}$ and v_{end} are allocated, and the value at the address v_{end} is z (where the knowledge base implies $z = 0$).

To define the semantics of abstract states a, we introduce the formulas $\langle a \rangle_{SL}$ and $\langle a \rangle_{FO}$. The separation logic formula $\langle a \rangle_{SL}$ defines which concrete states are

---

[2] $\mathbf{b}_{prev}$ is needed for $\mathbf{phi}$ instructions (cf. Sect. 2.2). In the beginning, we set $\mathbf{b}_{prev} = \varepsilon$.

represented by a. The first-order formula $\langle a \rangle_{FO}$ is used to construct symbolic execution graphs, allowing us to use standard SMT solving for all reasoning in our approach. Moreover, we also use $\langle a \rangle_{FO}$ for the subsequent generation of integer transition systems from the symbolic execution graphs. In addition to KB, $\langle a \rangle_{FO}$ states that the expressions $alloc(v_1, v_2) \in AL$ represent disjoint intervals and that two addresses must be different if they point to different values in PT.

In $\langle a \rangle_{SL}$, we combine the elements of AL with the separating conjunction "$*$" to ensure that different allocated memory blocks are disjoint. Here, as usual $\varphi_1 * \varphi_2$ means that φ_1 and φ_2 hold for disjoint parts of the memory. In contrast, the elements of PT are combined by the ordinary conjunction "$\wedge$". So $v_1 \hookrightarrow_{ty} v_2 \in PT$ does not imply that v_1 is different from other addresses occurring in PT. Similarly, we also combine the two formulas resulting from AL and PT by "$\wedge$", as both express different properties of memory addresses.

Definition 2 (Representing States by Formulas). *For $v_1, v_2 \in \mathcal{V}_{sym}$, let $\langle alloc(v_1, v_2) \rangle_{SL} = v_1 \leq v_2 \wedge (\forall x. \exists y.\, (v_1 \leq x \leq v_2) \Rightarrow (x \hookrightarrow y))$. Due to the two's complement representation, for any LLVM type ty, we define $\langle v_1 \hookrightarrow_{ty} v_2 \rangle_{SL} =$*

$$\langle v_1 \hookrightarrow_{size(ty)} v_3 \rangle_{SL} \wedge (v_2 \geq 0 \Rightarrow v_3 = v_2) \wedge (v_2 < 0 \Rightarrow v_3 = v_2 + 2^{8 \cdot size(ty)}),$$

where $v_3 \in \mathcal{V}_{sym}$ is fresh. Here,[3] $\langle v_1 \hookrightarrow_0 v_3 \rangle_{SL} = true$ and $\langle v_1 \hookrightarrow_{n+1} v_3 \rangle_{SL} = v_1 \hookrightarrow (v_3 \bmod 256) \wedge \langle (v_1 + 1) \hookrightarrow_n (v_3 \operatorname{div} 256) \rangle_{SL}$. Then $a = (p, LV, KB, AL, PT)$ is represented by[4] $\langle a \rangle_{SL} = LV \wedge KB \wedge (*_{\varphi \in AL} \langle \varphi \rangle_{SL}) \wedge (\bigwedge_{\varphi \in PT} \langle \varphi \rangle_{SL})$.

Moreover, the following first-order information on $\mathcal{V}_{sym}$ is deduced from an abstract state $a = (p, LV, KB, AL, PT)$. Let $\langle a \rangle_{FO}$ be the smallest set with

$$\begin{aligned}
\langle a \rangle_{FO} = \; & KB \; \cup \; \{v_1 \leq v_2 \mid alloc(v_1, v_2) \in AL\} \; \cup \\
& \{v_2 < w_1 \vee w_2 < v_1 \mid alloc(v_1, v_2), alloc(w_1, w_2) \in AL, (v_1, v_2) \neq (w_1, w_2)\} \; \cup \\
& \{v_1 \neq w_1 \mid (v_1 \hookrightarrow_{ty} v_2), (w_1 \hookrightarrow_{ty} w_2) \in PT \; and \; \models \langle a \rangle_{FO} \Rightarrow v_2 \neq w_2\}.
\end{aligned}$$

Let $\mathcal{T}(\mathcal{V}_{sym})$ be the set of all arithmetic terms containing only variables from $\mathcal{V}_{sym}$. Any function $\sigma : \mathcal{V}_{sym} \to \mathcal{T}(\mathcal{V}_{sym})$ is called an *instantiation*. Thus, σ does not instantiate $\mathcal{V}_{\mathcal{P}}$. Instantiations are extended to formulas in the usual way, i.e., $\sigma(\varphi)$ instantiates every $v \in \mathcal{V}_{sym}$ that occurs free in φ by $\sigma(v)$. An instantiation is called *concrete* iff $\sigma(v) \in \mathbb{Z}$ for all $v \in \mathcal{V}_{sym}$. Then an abstract state a at position p represents those concrete states (p, s, m) where (s, m) is a *model* of $\sigma(\langle a \rangle_{SL})$ for a concrete instantiation σ of the symbolic variables. So for example, the abstract state (†) on the previous page represents all concrete states $((\varepsilon, \texttt{entry}, 0), s, m)$ where m is a memory that stores a string at the address $s(\texttt{str})$.[5]

---

[3] We assume a little-endian data layout (where least significant bytes are stored in the lowest address). A corresponding representation could also be defined for big-endian layout. This layout information is necessary to decide which concrete states are represented by abstract states, but it is not used when constructing symbolic execution graphs (i.e., our remaining approach is independent of such layout information).

[4] We identify *sets* of first-order formulas $\{\varphi_1, ..., \varphi_n\}$ with their conjunction $\varphi_1 \wedge ... \wedge \varphi_n$.

[5] The reason is that then there is an address $end \geq s(\texttt{str})$ such that $m(end) = 0$ and m is defined for all numbers between $s(\texttt{str})$ and end. Hence, $(s, m) \models \sigma(\langle a \rangle_{SL})$ holds for an instantiation with $\sigma(u_x) = s(x)$ for all $x \in \mathcal{V}_{\mathcal{P}}$, $\sigma(v_{end}) = end$, and $\sigma(z) = 0$.

It remains to define when (s, m) is a *model* of a formula from our fragment of separation logic. For $s : \mathcal{V_P} \to \mathbb{Z}$ and any formula φ, let $s(\varphi)$ result from replacing all $\mathtt{x} \in \mathcal{V_P}$ in φ by $s(\mathtt{x})$. Note that by construction, local variables $\mathtt{x}$ are never quantified in our formulas. Then we define $(s, m) \models \varphi$ iff $m \models s(\varphi)$.

We now define $m \models \psi$ for formulas ψ that may still contain symbolic variables from $\mathcal{V}_{sym}$ (this is needed for Sect. 2.2). As usual, all free variables $v_1, \ldots, v_n$ in ψ are implicitly universally quantified, i.e., $m \models \psi$ iff $m \models \forall v_1, \ldots v_n. \psi$. The semantics of arithmetic operations and relations and of first-order connectives and quantifiers is as usual. In particular, we define $m \models \forall v. \psi$ iff $m \models \sigma(\psi)$ holds for all instantiations σ where $\sigma(v) \in \mathbb{Z}$ and $\sigma(w) = w$ for all $w \in \mathcal{V}_{sym} \setminus \{v\}$.

We still have to define the semantics of $\hookrightarrow$ and $*$ for variable-free formulas. For $z_1, z_2 \in \mathbb{Z}$, let $m \models z_1 \hookrightarrow z_2$ hold iff $m(z_1) = z_2$.[6] The semantics of $*$ is defined as usual in separation logic: For two partial functions $m_1, m_2 : \mathbb{N}_{>0} \to \mathbb{Z}$, we write $m_1 \perp m_2$ to indicate that the domains of m_1 and m_2 are disjoint and $m_1 \cdot m_2$ denotes the union of m_1 and m_2. Then $m \models \varphi_1 * \varphi_2$ iff there exist $m_1 \perp m_2$ such that $m = m_1 \cdot m_2$ where $m_1 \models \varphi_1$ and $m_2 \models \varphi_2$.

As usual, "$\models \varphi$" means that φ is a tautology, i.e., that $(s, m) \models \varphi$ holds for any $s : \mathcal{V_P} \to \mathbb{Z}$ and $m : \mathbb{N}_{>0} \to \mathbb{Z}$. Clearly, $\models \langle a \rangle_{SL} \Rightarrow \langle a \rangle_{FO}$, i.e., $\langle a \rangle_{FO}$ contains first-order information that holds in every concrete state represented by a.

2.2 Constructing Symbolic Execution Graphs

We now show how to automatically generate a *symbolic execution graph* that over-approximates all possible executions of a given program. For this, we present symbolic execution rules for some of the most important LLVM instructions. Other instructions can be handled in a similar way, cf. [26]. Note that in contrast to other formalizations of LLVM's operational semantics [31], our rules operate on *abstract* instead of concrete states to allow a *symbolic* execution of LLVM. In particular, we also have rules for refining and generalizing abstract states.

Our analysis starts with the set of initial states that one wants to analyze for termination, e.g., all states where $\mathtt{str}$ points to a *string*. So in our example, we start with the abstract state (†). Fig. 1 depicts the symbolic execution graph for $\mathtt{strlen}$. Here, we omitted the component $AL = \{alloc(u_{\mathtt{str}}, v_{end})\}$, which stays the same in all states in this example. We also abbreviated parts of LV, KB, PT by "...". Instead of $v_{end} \hookrightarrow_{i8} z$ and $z = 0$, we directly wrote $v_{end} \hookrightarrow 0$, etc.

The function $\mathtt{strlen}$ starts with loading the character at address $\mathtt{str}$ to $\mathtt{c0}$. Let $p : ins$ denote that ins is the instruction at position p. Our first rule handles the case $p :$ "$\mathtt{x = load \ ty* \ ad}$", i.e., the value of type $\mathtt{ty}$ at the address $\mathtt{ad}$ is assigned to the variable $\mathtt{x}$. In our rules, let a always denote the abstract state *before* the execution step (i.e., above the horizontal line of the rule). Moreover, we write $\langle a \rangle$ instead of $\langle a \rangle_{FO}$. As each memory cell stores one byte, in the $\mathtt{load}$-rule we first have to check whether the addresses $\mathtt{ad}, \ldots, \mathtt{ad} + size(\mathtt{ty}) - 1$ are allocated, i.e., if there is an $alloc(v_1, v_2) \in AL$ such that $\langle a \rangle \Rightarrow (v_1 \leq LV(\mathtt{ad}) \wedge$

---

[6] We use "$\hookrightarrow$" instead of "$\mapsto$" in separation logic, since $m \models z_1 \mapsto z_2$ would imply that $m(z)$ is undefined for all $z \neq z_1$. This would be inconvenient in our formalization, since PT usually only contains information about a *part* of the allocated memory.

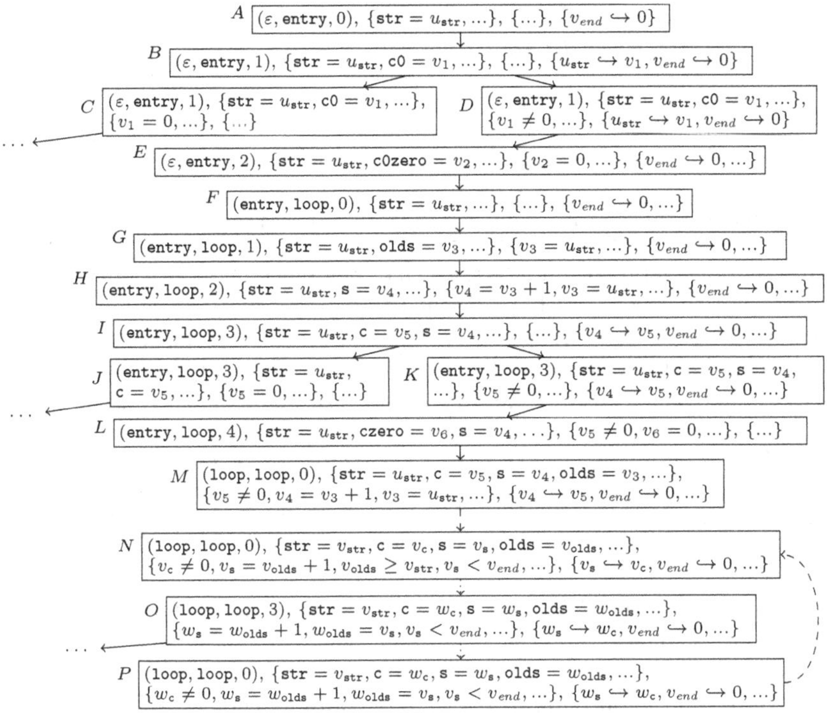

A $(\varepsilon, \mathtt{entry}, 0), \{\mathtt{str} = u_{\mathtt{str}}, ...\}, \{...\}, \{v_{end} \hookrightarrow 0\}$

B $(\varepsilon, \mathtt{entry}, 1), \{\mathtt{str} = u_{\mathtt{str}}, \mathtt{c0} = v_1, ...\}, \{...\}, \{u_{\mathtt{str}} \hookrightarrow v_1, v_{end} \hookrightarrow 0\}$

C $(\varepsilon, \mathtt{entry}, 1), \{\mathtt{str} = u_{\mathtt{str}}, \mathtt{c0} = v_1, ...\}, \{v_1 = 0, ...\}, \{...\}$

D $(\varepsilon, \mathtt{entry}, 1), \{\mathtt{str} = u_{\mathtt{str}}, \mathtt{c0} = v_1, ...\}, \{v_1 \neq 0, ...\}, \{u_{\mathtt{str}} \hookrightarrow v_1, v_{end} \hookrightarrow 0\}$

E $(\varepsilon, \mathtt{entry}, 2), \{\mathtt{str} = u_{\mathtt{str}}, \mathtt{c0zero} = v_2, ...\}, \{v_2 = 0, ...\}, \{v_{end} \hookrightarrow 0, ...\}$

F $(\mathtt{entry}, \mathtt{loop}, 0), \{\mathtt{str} = u_{\mathtt{str}}, ...\}, \{...\}, \{v_{end} \hookrightarrow 0, ...\}$

G $(\mathtt{entry}, \mathtt{loop}, 1), \{\mathtt{str} = u_{\mathtt{str}}, \mathtt{olds} = v_3, ...\}, \{v_3 = u_{\mathtt{str}}, ...\}, \{v_{end} \hookrightarrow 0, ...\}$

H $(\mathtt{entry}, \mathtt{loop}, 2), \{\mathtt{str} = u_{\mathtt{str}}, \mathtt{s} = v_4, ...\}, \{v_4 = v_3 + 1, v_3 = u_{\mathtt{str}}, ...\}, \{v_{end} \hookrightarrow 0, ...\}$

I $(\mathtt{entry}, \mathtt{loop}, 3), \{\mathtt{str} = u_{\mathtt{str}}, \mathtt{c} = v_5, \mathtt{s} = v_4, ...\}, \{...\}, \{v_4 \hookrightarrow v_5, v_{end} \hookrightarrow 0, ...\}$

J $(\mathtt{entry}, \mathtt{loop}, 3), \{\mathtt{str} = u_{\mathtt{str}}, \mathtt{c} = v_5, ...\}, \{v_5 = 0, ...\}, \{...\}$

K $(\mathtt{entry}, \mathtt{loop}, 3), \{\mathtt{str} = u_{\mathtt{str}}, \mathtt{c} = v_5, \mathtt{s} = v_4, ...\}, \{v_5 \neq 0, ...\}, \{v_4 \hookrightarrow v_5, v_{end} \hookrightarrow 0, ...\}$

L $(\mathtt{entry}, \mathtt{loop}, 4), \{\mathtt{str} = u_{\mathtt{str}}, \mathtt{czero} = v_6, \mathtt{s} = v_4, ...\}, \{v_5 \neq 0, v_6 = 0, ...\}, \{...\}$

M $(\mathtt{loop}, \mathtt{loop}, 0), \{\mathtt{str} = u_{\mathtt{str}}, \mathtt{c} = v_5, \mathtt{s} = v_4, \mathtt{olds} = v_3, ...\}, \{v_5 \neq 0, v_4 = v_3 + 1, v_3 = u_{\mathtt{str}}, ...\}, \{v_4 \hookrightarrow v_5, v_{end} \hookrightarrow 0, ...\}$

N $(\mathtt{loop}, \mathtt{loop}, 0), \{\mathtt{str} = v_{\mathtt{str}}, \mathtt{c} = v_c, \mathtt{s} = v_{\mathtt{s}}, \mathtt{olds} = v_{\mathtt{olds}}, ...\}, \{v_c \neq 0, v_{\mathtt{s}} = v_{\mathtt{olds}} + 1, v_{\mathtt{olds}} \geq v_{\mathtt{str}}, v_{\mathtt{s}} < v_{end}, ...\}, \{v_{\mathtt{s}} \hookrightarrow v_c, v_{end} \hookrightarrow 0, ...\}$

O $(\mathtt{loop}, \mathtt{loop}, 3), \{\mathtt{str} = v_{\mathtt{str}}, \mathtt{c} = w_c, \mathtt{s} = w_{\mathtt{s}}, \mathtt{olds} = w_{\mathtt{olds}}, ...\}, \{w_{\mathtt{s}} = w_{\mathtt{olds}} + 1, w_{\mathtt{olds}} = v_{\mathtt{s}}, v_{\mathtt{s}} < v_{end}, ...\}, \{w_{\mathtt{s}} \hookrightarrow w_c, v_{end} \hookrightarrow 0, ...\}$

P $(\mathtt{loop}, \mathtt{loop}, 0), \{\mathtt{str} = v_{\mathtt{str}}, \mathtt{c} = w_c, \mathtt{s} = w_{\mathtt{s}}, \mathtt{olds} = w_{\mathtt{olds}}, ...\}, \{w_c \neq 0, w_{\mathtt{s}} = w_{\mathtt{olds}} + 1, w_{\mathtt{olds}} = v_{\mathtt{s}}, v_{\mathtt{s}} < v_{end}, ...\}, \{w_{\mathtt{s}} \hookrightarrow w_c, v_{end} \hookrightarrow 0, ...\}$

Fig. 1. Symbolic execution graph for `strlen`

$LV(\mathtt{ad}) + size(\mathtt{ty}) - 1 \leq v_2)$ is valid. Then, we reach a new abstract state where the previous position $p = (\mathtt{b}_{prev}, \mathtt{b}, i)$ is updated to the position $p^+ = (\mathtt{b}_{prev}, \mathtt{b}, i+1)$ of the next instruction in the same basic block, and we set $LV(\mathtt{x}) = w$ for a fresh $w \in \mathcal{V}_{sym}$. If we already know the value at the address $\mathtt{ad}$ (i.e., if there are $w_1, w_2 \in \mathcal{V}_{sym}$ with $\models \langle a \rangle \Rightarrow (LV(\mathtt{ad}) = w_1)$ and $w_1 \hookrightarrow_{\mathtt{ty}} w_2 \in PT$) then we add $w = w_2$ to KB. Otherwise, we add $LV(\mathtt{ad}) \hookrightarrow_{\mathtt{ty}} w$ to PT. We used this rule to obtain B from A in Fig. 1. In a similar way, one can also formulate a rule for `store` instructions that store a value at some address in the memory (cf. [26]).

---

load from allocated memory ($p :$ "`x = load ty* ad`" with $\mathtt{x}, \mathtt{ad} \in \mathcal{V}_\mathcal{P}$)

$$\frac{(p, \ LV \uplus \{\mathtt{x} = v\}, \ KB, \ AL, \ PT)}{(p^+, \ LV \uplus \{\mathtt{x} = w\}, \ KB \cup \{w = w_2\}, \ AL, \ PT)} \quad \text{if}$$

- there is $alloc(v_1, v_2) \in AL$ with $\models \langle a \rangle \Rightarrow (v_1 \leq LV(\mathtt{ad}) \wedge LV(\mathtt{ad}) + size(\mathtt{ty}) - 1 \leq v_2)$,
- there are $w_1, w_2 \in \mathcal{V}_{sym}$ with $\models \langle a \rangle \Rightarrow (LV(\mathtt{ad}) = w_1)$ and $w_1 \hookrightarrow_{\mathtt{ty}} w_2 \in PT$,
- $w \in \mathcal{V}_{sym}$ is fresh

$$\frac{(p, \ LV \uplus \{\mathtt{x} = v\}, \ KB, \ AL, \ PT)}{(p^+, \ LV \uplus \{\mathtt{x} = w\}, \ KB, \ AL, \ PT \cup \{LV(\mathtt{ad}) \hookrightarrow_{\mathtt{ty}} w\})} \quad \text{if}$$

- there is $alloc(v_1, v_2) \in AL$ with $\models \langle a \rangle \Rightarrow (v_1 \leq LV(\mathtt{ad}) \wedge LV(\mathtt{ad}) + size(\mathtt{ty}) - 1 \leq v_2)$,
- there are no $w_1, w_2 \in \mathcal{V}_{sym}$ with $\models \langle a \rangle \Rightarrow (LV(\mathtt{ad}) = w_1)$ and $w_1 \hookrightarrow_{\mathtt{ty}} w_2 \in PT$,
- $w \in \mathcal{V}_{sym}$ is fresh

If `load` accesses an address that was not allocated, then memory safety is violated and we reach the *ERR* state.

---

load from unallocated memory (p: "x = load ty* ad" with $x, ad \in V_P$)

$$\frac{(p, LV, KB, AL, PT)}{ERR} \quad \text{if}$$

there is no $alloc(v_1, v_2) \in AL$ with $\models \langle a \rangle \Rightarrow (v_1 \leq LV(ad) \wedge LV(ad) + size(ty) - 1 \leq v_2)$

The instructions `icmp` and `br` in **strlen's** `entry` block check if the first character c0 is 0. In that case, we have reached the end of the string and jump to the block **done**. So for "x = icmp eq ty t_1, t_2", we check if the state contains enough information to decide whether the values t_1 and t_2 of type ty are equal. In that case, the value 1 resp. 0 (i.e., *true* resp. *false*) is assigned to x.[7]

---

icmp (p: "x = icmp eq ty t_1, t_2" with $x \in V_P$ and $t_1, t_2 \in V_P \cup \mathbb{Z}$)

$$\frac{(p, LV \uplus \{x = v\}, KB, AL, PT)}{(p^+, LV \uplus \{x = w\}, KB \cup \{w = 1\}, AL, PT)} \quad \begin{array}{l} \text{if} \models \langle a \rangle \Rightarrow (LV(t_1) = LV(t_2)) \\ \text{and } w \in V_{sym} \text{ is fresh} \end{array}$$

$$\frac{(p, LV \uplus \{x = v\}, KB, AL, PT)}{(p^+, LV \uplus \{x = w\}, KB \cup \{w = 0\}, AL, PT)} \quad \begin{array}{l} \text{if} \models \langle a \rangle \Rightarrow (LV(t_1) \neq LV(t_2)) \\ \text{and } w \in V_{sym} \text{ is fresh} \end{array}$$

The previous rule is only applicable if *KB* contains enough information to evaluate the condition. Otherwise, a case analysis needs to be performed, i.e., one has to *refine* the abstract state by extending its knowledge base. This is done by the following rule which transforms an abstract state into *two* new ones.[8]

---

refining abstract states (p: "x = icmp eq ty t_1, t_2", $x \in V_P$, $t_1, t_2 \in V_P \cup \mathbb{Z}$)

$$\frac{(p, LV, KB, AL, PT)}{(p, LV, KB \cup \{\varphi\}, AL, PT) \mid (p, LV, KB \cup \{\neg\varphi\}, AL, PT)}$$

if φ is $LV(t_1) = LV(t_2)$ and both $\not\models \langle a \rangle \Rightarrow \varphi$ and $\not\models \langle a \rangle \Rightarrow \neg\varphi$

For example, in state B of Fig. 1, we evaluate "c0zero = icmp eq i8 c0, 0", i.e., we check whether the first character c0 of the string str is 0. Since this cannot be inferred from B's knowledge base, we refine B to the successor states C and D and call the edges from B to C and D *refinement edges*. In D, we have c0 = v_1 and $v_1 \neq 0$. Thus, the `icmp`-rule yields E where c0zero = v_2 and $v_2 = 0$. We do not display the successors of C that lead to a program end.

The conditional branching instruction `br` is very similar to `icmp`. To evaluate "br i1 t, label b_1, label b_2", one has to check whether the current state contains enough information to conclude that t is 1 (i.e., *true*) or 0 (i.e., *false*). Then the evaluation continues with block b_1 resp. b_2. This rule allows us to create the successor F of E, where we jump to the block **loop**.

---

[7] Other integer comparisons (for $<, \leq, \dots$) are handled analogously.

[8] Analogous refinement rules can also be used for other conditional LLVM instructions.

---

br (p: "**br i1** t, **label** b_1, **label** b_2" with $t \in \mathcal{V}_\mathcal{P} \cup \{0,1\}$ and $b_1, b_2 \in \mathit{Blks}$)

$$\frac{(p, LV, KB, AL, PT)}{((\mathbf{b}, \mathbf{b}_1, 0), LV, KB, AL, PT)} \quad \text{if } p = (\mathbf{b}_{prev}, \mathbf{b}, i) \text{ and } \models \langle a \rangle \Rightarrow (LV(t) = 1)$$

$$\frac{(p, LV, KB, AL, PT)}{((\mathbf{b}, \mathbf{b}_2, 0), LV, KB, AL, PT)} \quad \text{if } p = (\mathbf{b}_{prev}, \mathbf{b}, i) \text{ and } \models \langle a \rangle \Rightarrow (LV(t) = 0)$$

Next, we have to evaluate a **phi** instruction. These instructions are needed due to the static single assignment form of LLVM. Here, "**x = phi ty** $[t_1, \mathbf{b}_1]$, ..., $[t_n, \mathbf{b}_n]$" means that if the previous block was $\mathbf{b}_j$, then the value t_j is assigned to **x**. All $t_1, \ldots, t_n$ must have type **ty**. Since we reached state F in Fig. 1 after evaluating the **entry** block, we obtain the state G with **olds** $= v_3$ and $v_3 = u_{str}$.

---

phi (p: "**x = phi ty** $[t_1, \mathbf{b}_1]$, ..., $[t_n, \mathbf{b}_n]$" with $\mathbf{x} \in \mathcal{V}_\mathcal{P}$, $t_i \in \mathcal{V}_\mathcal{P} \cup \mathbb{Z}$, $\mathbf{b}_i \in \mathit{Blks}$)

$$\frac{(p, \ LV \uplus \{\mathbf{x} = v\}, \ KB, \ AL, \ PT)}{(p^+, \ LV \uplus \{\mathbf{x} = w\}, \ KB \cup \{w = LV(t_j)\}, \ AL, \ PT)} \quad \begin{array}{l} \text{if } p = (\mathbf{b}_j, \mathbf{b}, k) \text{ and} \\ w \in \mathcal{V}_{sym} \text{ is fresh} \end{array}$$

The **strlen** function traverses the string using a pointer **s** that is increased in each iteration. The **loop** terminates, since eventually **s** reaches the last memory cell of the string (containing 0). Then one jumps to **done**, converts the pointers **s** and **str** to integers, and returns their difference. To perform the required pointer arithmetic, "**ad$_2$ = getelementptr ty* ad$_1$, in** t" increases **ad$_1$** by the size of t elements of type **ty** (i.e., by $size(\mathbf{ty}) \cdot t$) and assigns this address to **ad$_2$**.[9]

---

getelementptr (p: "**ad$_2$ = getelementptr ty* ad$_1$, in** t", **ad$_1$**, **ad$_2$** $\in \mathcal{V}_\mathcal{P}$, $t \in \mathcal{V}_\mathcal{P} \cup \mathbb{Z}$)

$$\frac{(p, \ LV \uplus \{\mathbf{ad}_2 = v\}, \ KB, \ AL, \ PT)}{(p^+, LV \uplus \{\mathbf{ad}_2 = w\}, KB \cup \{w = LV(\mathbf{ad}_1) + size(\mathbf{ty}) \cdot LV(t)\}, AL, PT)} \quad w \in \mathcal{V}_{sym} \text{ fresh}$$

In Fig. 1, this rule is used for the step from G to H, where LV and KB now imply **s** = **str** + 1. In the step to I, the character at address **s** is loaded to **c**. To ensure memory safety, the **load**-rule checks that **s** is in an allocated part of the memory (i.e., that $u_{str} \leq u_{str} + 1 \leq v_{end}$). This holds because $\langle H \rangle$ implies $u_{str} \leq v_{end}$ and $u_{str} \neq v_{end}$ (as $u_{str} \hookrightarrow v_1, v_{end} \hookrightarrow 0 \in PT$ and $v_1 \neq 0 \in KB$). Finally, we check whether **c** is 0. We again perform a refinement which yields the states J and K. State K corresponds to the case **c** $\neq 0$ and thus, we obtain **czero** = 0 in L and branch back to instruction 0 of the **loop** block in state M.

2.3 Generalizing Abstract States

After reaching M, one unfolds the loop once more until one reaches a state $\widetilde{M}$ at position (**loop**, **loop**, 0) again, analogous to the first iteration. To obtain *finite* symbolic execution graphs, we *generalize* our states whenever an evaluation visits a program position twice. Thus, we have to find a state that is more general than $M = (p, LV_M, KB_M, AL, PT_M)$ and $\widetilde{M} = (p, LV_{\widetilde{M}}, KB_{\widetilde{M}}, AL, PT_{\widetilde{M}})$. For readability, we again write "$\hookrightarrow$" instead of "$\hookrightarrow_{i8}$". Then $p = (\mathbf{loop}, \mathbf{loop}, 0)$ and

---

[9] Since we do not consider the handling of data structures in this paper, we do not regard **getelementptr** instructions with more than two parameters.

$$AL = \{alloc(u_{\mathtt{str}}, v_{end})\}$$

$$LV_M = \{\mathtt{str} = u_{\mathtt{str}}, \mathtt{c} = v_5, \mathtt{s} = v_4, \mathtt{olds} = v_3, \ldots\}$$

$$LV_{\widetilde{M}} = \{\mathtt{str} = u_{\mathtt{str}}, \mathtt{c} = \tilde{v}_5, \mathtt{s} = \tilde{v}_4, \mathtt{olds} = \tilde{v}_3, \ldots\}$$

$$PT_M = \{u_{\mathtt{str}} \hookrightarrow v_1, v_4 \hookrightarrow v_5, v_{end} \hookrightarrow z\}$$

$$PT_{\widetilde{M}} = \{u_{\mathtt{str}} \hookrightarrow v_1, v_4 \hookrightarrow v_5, \tilde{v}_4 \hookrightarrow \tilde{v}_5, v_{end} \hookrightarrow z\}$$

$$KB_M = \{v_5 \neq 0, v_4 = v_3 + 1, v_3 = u_{\mathtt{str}}, v_1 \neq 0, z = 0, \ldots\}$$

$$KB_{\widetilde{M}} = \{\tilde{v}_5 \neq 0, \tilde{v}_4 = \tilde{v}_3 + 1, \tilde{v}_3 = v_4, v_4 = v_3 + 1, v_3 = u_{\mathtt{str}}, v_1 \neq 0, z = 0, \ldots\}.$$

Our aim is to construct a new state N that is more general than M and $\widetilde{M}$, but contains enough information for the remaining proof. We now present our heuristic for *merging* states that is used as the basis for our implementation.

To merge M and $\widetilde{M}$, we keep those constraints of M that also hold in $\widetilde{M}$. To this end, we proceed in two steps. First, we create a new state $N = (p, LV_N, KB_N, AL_N, PT_N)$ using fresh symbolic variables $v_{\mathtt{x}}$ for all $\mathtt{x} \in \mathcal{V}_P$ and define

$$LV_N = \{\mathtt{str} = v_{\mathtt{str}}, \mathtt{c} = v_{\mathtt{c}}, \mathtt{s} = v_{\mathtt{s}}, \mathtt{olds} = v_{\mathtt{olds}}, \ldots\}.$$

Matching N's fresh variables to the variables in M and $\widetilde{M}$ yields mappings with $\mu_M(v_{\mathtt{str}}) = u_{\mathtt{str}}$, $\mu_M(v_{\mathtt{c}}) = v_5$, $\mu_M(v_{\mathtt{s}}) = v_4$, $\mu_M(v_{\mathtt{olds}}) = v_3$, and $\mu_{\widetilde{M}}(v_{\mathtt{str}}) = u_{\mathtt{str}}$, $\mu_{\widetilde{M}}(v_{\mathtt{c}}) = \tilde{v}_5$, $\mu_{\widetilde{M}}(v_{\mathtt{s}}) = \tilde{v}_4$, $\mu_{\widetilde{M}}(v_{\mathtt{olds}}) = \tilde{v}_3$. By injectivity of LV_M, we can also define a pseudo-inverse of μ_M that maps M's variables to N by setting $\mu_M^{-1}(LV_M(\mathtt{x})) = v_{\mathtt{x}}$ for $\mathtt{x} \in \mathcal{V}_P$ and $\mu_M^{-1}(v) = v$ for all other $v \in \mathcal{V}_{sym}$ ($\mu_{\widetilde{M}}^{-1}$ works analogously).

In a second step, we use these mappings to check which constraints of M also hold in $\widetilde{M}$. So we set $AL_N = \mu_M^{-1}(AL) \cap \mu_{\widetilde{M}}^{-1}(AL) = \{alloc(v_{\mathtt{str}}, v_{end})\}$ and

$$PT_N = \mu_M^{-1}(PT_M) \cap \mu_{\widetilde{M}}^{-1}(PT_{\widetilde{M}})$$
$$= \{v_{\mathtt{str}} \hookrightarrow v_1, v_{\mathtt{s}} \hookrightarrow v_{\mathtt{c}}, v_{end} \hookrightarrow z\} \cap \{v_{\mathtt{str}} \hookrightarrow v_1, v_4 \hookrightarrow v_5, v_{\mathtt{s}} \hookrightarrow v_{\mathtt{c}}, v_{end} \hookrightarrow z\}$$
$$= \{v_{\mathtt{str}} \hookrightarrow v_1, v_{\mathtt{s}} \hookrightarrow v_{\mathtt{c}}, v_{end} \hookrightarrow z\}.$$

It remains to construct KB_N. We have $v_3 = u_{\mathtt{str}}$ ("olds = str") in $\langle M \rangle$, but $\tilde{v}_3 = v_4$, $v_4 = v_3 + 1$, $v_3 = u_{\mathtt{str}}$ ("olds = str + 1") in $\langle \widetilde{M} \rangle$. To keep as much information as possible in such cases, we rewrite equations to inequations before performing the generalization. For this, let $\langle\!\langle M \rangle\!\rangle$ result from extending $\langle M \rangle$ by $t_1 \geq t_2$ and $t_1 \leq t_2$ for any equation $t_1 = t_2 \in \langle M \rangle$. So in our example, we obtain $v_3 \geq u_{\mathtt{str}} \in \langle\!\langle M \rangle\!\rangle$ ("olds $\geq$ str"). Moreover, for any $t_1 \neq t_2 \in \langle M \rangle$, we check whether $\langle M \rangle$ implies $t_1 > t_2$ or $t_1 < t_2$, and add the respective inequation to $\langle\!\langle M \rangle\!\rangle$. In this way, one can express sequences of inequations $t_1 \neq t_2$, $t_1 + 1 \neq t_2$, $\ldots$, $t_1 + n \neq t_2$ (where $t_1 \leq t_2$) by a single inequation $t_1 + n < t_2$, which is needed for suitable generalizations afterwards. We use this to derive $v_4 < v_{end} \in \langle\!\langle M \rangle\!\rangle$ ("s $< v_{end}$") from $v_4 = v_3 + 1$, $v_3 = u_{\mathtt{str}}$, $u_{\mathtt{str}} \leq v_{end}$, $u_{\mathtt{str}} \neq v_{end}$, $v_4 \neq v_{end} \in \langle M \rangle$.

We then let KB_N consist of all formulas φ from $\langle\!\langle M \rangle\!\rangle$ that are also implied by $\langle \widetilde{M} \rangle$, again translating variable names using μ_M^{-1} and $\mu_{\widetilde{M}}^{-1}$. Thus, we have

$$\langle\!\langle M \rangle\!\rangle = \{v_5 \neq 0, v_4 = v_3 + 1, v_3 = u_{\mathrm{str}}, v_3 \geq u_{\mathrm{str}}, v_4 < v_{end}, \ldots\}$$

$$\mu_M^{-1}(\langle\!\langle M \rangle\!\rangle) = \{v_c \neq 0, v_s = v_{\mathrm{olds}} + 1, v_{\mathrm{olds}} = v_{\mathrm{str}}, v_{\mathrm{olds}} \geq v_{\mathrm{str}}, v_s < v_{end}, \ldots\}$$

$$\mu_{\widetilde{M}}^{-1}(\langle \widetilde{M} \rangle) = \{v_c \neq 0, v_s = v_{\mathrm{olds}} + 1, v_{\mathrm{olds}} = v_4, v_4 = v_3 + 1, v_3 = v_{\mathrm{str}}, v_s < v_{end}, \ldots\}$$

$$KB_N = \{v_c \neq 0, v_s = v_{\mathrm{olds}} + 1, v_{\mathrm{olds}} \geq v_{\mathrm{str}}, v_s < v_{end}, \ldots\}.$$

Definition 3 (Merging States). Let $a = (p, LV_a, KB_a, AL_a, PT_a)$ and $b = (p, LV_b, KB_b, AL_b, PT_b)$ be abstract states. Then $c = (p, LV_c, KB_c, AL_c, PT_c)$ results from merging the states a and b if

- $LV_c = \{\mathbf{x} = v_\mathbf{x} \mid \mathbf{x} \in \mathcal{V}_\mathcal{P}\}$ for fresh pairwise different symbolic variables $v_\mathbf{x}$. Moreover, we define $\mu_a(v_\mathbf{x}) = LV_a(\mathbf{x})$ and $\mu_b(v_\mathbf{x}) = LV_b(\mathbf{x})$ for all $\mathbf{x} \in \mathcal{V}_\mathcal{P}$ and let μ_a and μ_b be the identity on all remaining variables from $\mathcal{V}_{sym}$.

- $AL_c = \mu_a^{-1}(AL_a) \cap \mu_b^{-1}(AL_b)$ and $PT_c = \mu_a^{-1}(PT_a) \cap \mu_b^{-1}(PT_b)$. Here, the "inverse" of the instantiation μ_a is defined as $\mu_a^{-1}(v) = v_\mathbf{x}$ if $v = LV_a(\mathbf{x})$ and $\mu_a^{-1}(v) = v$ for all other $v \in \mathcal{V}_{sym}$ (μ_b^{-1} is defined analogously).

- $KB_C = \{ \varphi \in \mu_a^{-1}(\langle\!\langle a \rangle\!\rangle) \mid \models \mu_b^{-1}(\langle\!\langle b \rangle\!\rangle) \Rightarrow \varphi \}$, where

$$\langle\!\langle a \rangle\!\rangle = \langle a \rangle \cup \{ t_1 \geq t_2, t_1 \leq t_2 \mid t_1 = t_2 \in \langle a \rangle \}$$
$$\cup \{ t_1 > t_2 \mid t_1 \neq t_2 \in \langle a \rangle, \models \langle a \rangle \Rightarrow t_1 > t_2 \}$$
$$\cup \{ t_1 < t_2 \mid t_1 \neq t_2 \in \langle a \rangle, \models \langle a \rangle \Rightarrow t_1 < t_2 \}.$$

In Fig. 1, we do not show the second loop unfolding from M to $\widetilde{M}$, and directly draw a *generalization edge* from M to N, depicted by a dashed arrow. Such an edge expresses that all concrete states represented by M are also represented by the more general state N. Semantically, a state $\bar{a}$ is a generalization of a state a iff $\models \langle a \rangle_{SL} \Rightarrow \mu(\langle \bar{a} \rangle_{SL})$ for some instantiation μ. To automate our procedure, we define a weaker relationship between a and $\bar{a}$. We say that $\bar{a} = (p, \overline{LV}, \overline{KB}, \overline{AL}, \overline{PT})$ is a *generalization* of $a = (p, LV, KB, AL, PT)$ with the instantiation μ whenever the conditions (b)-(e) of the following rule are satisfied.

---

generalization with μ $\dfrac{(p, LV, KB, AL, PT)}{(p, \overline{LV}, \overline{KB}, \overline{AL}, \overline{PT})}$ if

(a) a has an incoming evaluation edge,[10]
(b) $LV(\mathbf{x}) = \mu(\overline{LV}(\mathbf{x}))$ for all $\mathbf{x} \in \mathcal{V}_\mathcal{P}$,
(c) $\models \langle a \rangle \Rightarrow \mu(\overline{KB})$,
(d) if $alloc(v_1, v_2) \in \overline{AL}$, then $alloc(\mu(v_1), \mu(v_2)) \in AL$,
(e) if $(v_1 \hookrightarrow_{\mathsf{ty}} v_2) \in \overline{PT}$, then $(\mu(v_1) \hookrightarrow_{\mathsf{ty}} \mu(v_2)) \in PT$

Clearly, then we indeed have $\models \langle a \rangle_{SL} \Rightarrow \mu(\langle \bar{a} \rangle_{SL})$. Condition (a) is needed to avoid cycles of refinement and generalization steps in the symbolic execution graph, which would not correspond to any computation.

---

[10] *Evaluation* edges are edges that are not refinement or generalization edges.

Of course, many approaches are possible to compute such generalizations (or "widenings"). Thm. 4 shows that the merging heuristic from Def. 3 satisfies the conditions of the generalization rule. Thus, since N results from merging M and $\widetilde{M}$, it is indeed a *generalization* of M. Thm. 4 also shows that if one uses the merging heuristic to compute generalizations, then the construction of symbolic execution graphs always terminates when applying the following strategy:

- If there is a path from a state a to a state b, where a and b are at the same program position, where b has an incoming evaluation edge, and where a has no incoming refinement edge, then we check whether a is a generalization of b (i.e., whether the corresponding conditions of the generalization rule are satisfied). In that case, we draw a generalization edge from b to a.

- Otherwise, remove a's children, and add a generalization edge from a to the merging c of a and b. If a already had an incoming generalization edge from some state q, then remove a and add a generalization edge from q to c instead.

Theorem 4 (Soundness and Termination of Merging). *Let c result from merging the states a and b as in Def. 3. Then c is a generalization of a and b with the instantiations μ_a and μ_b, respectively. Moreover, if a is not already a generalization of b, then $|\langle\!\langle c\rangle\!\rangle| + |AL_c| + |PT_c| < |\langle\!\langle a\rangle\!\rangle| + |AL_a| + |PT_a|$. Here, for any conjunction φ, let $|\varphi|$ denote the number of its conjuncts. Thus, the above strategy to construct symbolic execution graphs always terminates.*[11]

In our example, we continue symbolic execution in state N. Similar to the execution from F to M, after 6 steps another state P at position $(\text{loop}, \text{loop}, 0)$ is reached. In Fig. 1, dotted arrows abbreviate several evaluation steps. As N is again a generalization of P using an instantiation μ with $\mu(v_c) = w_c$, $\mu(v_s) = w_s$, and $\mu(v_{\text{olds}}) = w_{\text{olds}}$, we draw a generalization edge from P to N. The construction of the symbolic execution graph is finished as soon as all its leaves correspond to $\texttt{ret}$ instructions (for "return").

Based on this construction, we now connect the symbolic execution graph to memory safety of the input program. We say that a concrete LLVM state (p, s, m) is *represented* by the symbolic execution graph iff the graph contains an abstract state a at position p where $(s, m) \models \sigma(\langle a\rangle_{SL})$ for some concrete instantiation σ.

Theorem 5 (Memory Safety of LLVM Programs). *Let $\mathcal{P}$ be an LLVM program with a symbolic execution graph $\mathcal{G}$. If $\mathcal{G}$ does not contain the abstract state ERR, then $\mathcal{P}$ is memory safe for all LLVM states represented by $\mathcal{G}$.*

3   From Symbolic Execution Graphs to Integer Systems

To prove termination of the input program, we extract an *integer transition system* (ITS) from the symbolic execution graph and then use existing tools to prove its termination. The extraction step essentially restricts the information

---

[11] The proofs for all theorems can be found in [26].

in abstract states to the integer constraints on symbolic variables. This conversion of memory-based arguments into integer arguments often suffices for the termination proof. The reason for considering only $\mathcal{V}_{sym}$ instead of $\mathcal{V}_{\mathcal{P}}$ is that the conditions in the abstract states only concern the symbolic variables and therefore, these are usually the essential variables for proving termination.

For example, termination of `strlen` is proved by showing that the pointer `s` is increased as long as it is smaller than v_{end}, the symbolic end of the input string. In Fig. 1, this is explicit since $v_s < v_{end}$ is an invariant that holds in all states represented by N. Each iteration of the cycle increases the value of v_s.

Formally, *ITSs* are graphs whose nodes are abstract states and whose edges are *transitions*. For any abstract state a, let $\mathcal{V}(a)$ denote the symbolic variables occurring in a. Let $\mathcal{V} \subseteq \mathcal{V}_{sym}$ be the finite set of all symbolic variables occurring in states of the symbolic execution graph. A *transition* is a tuple $(a, CON, \bar{a})$ where $a, \bar{a}$ are abstract states and the *condition* $CON \subseteq QF\_IA(\mathcal{V} \uplus \mathcal{V}')$ is a set of pure quantifier-free formulas over the variables $\mathcal{V} \uplus \mathcal{V}'$. Here, $\mathcal{V}' = \{v' \mid v \in \mathcal{V}\}$ represents the values of the variables *after* the transition. An *ITS state* (a, σ) consists of an abstract state a and a concrete instantiation $\sigma : \mathcal{V} \to \mathbb{Z}$. For any such σ, let $\sigma' : \mathcal{V}' \to \mathbb{Z}$ with $\sigma'(v') = \sigma(v)$. Given an ITS $\mathcal{I}$, (a, σ) *evaluates* to $(\bar{a}, \bar{\sigma})$ (denoted "$(a, \sigma) \to_{\mathcal{I}} (\bar{a}, \bar{\sigma})$") iff $\mathcal{I}$ has a transition $(a, CON, \bar{a})$ with $\models (\sigma \cup \bar{\sigma}')(CON)$. Here, we have $(\sigma \cup \bar{\sigma}')(v) = \sigma(v)$ and $(\sigma \cup \bar{\sigma}')(v') = \bar{\sigma}'(v') = \bar{\sigma}(v)$ for all $v \in \mathcal{V}$. An ITS $\mathcal{I}$ is *terminating* iff $\to_{\mathcal{I}}$ is well founded.[12]

We convert symbolic execution graphs to ITSs by transforming every edge into a transition. If there is a generalization edge from a to $\bar{a}$ with an instantiation μ, then the new value of any $v \in \mathcal{V}(\bar{a})$ in $\bar{a}$ is $\mu(v)$. Hence, we create the transition $(a, \langle a \rangle \cup \{v' = \mu(v) \mid v \in \mathcal{V}(\bar{a})\}, \bar{a})$.[13] So for the edge from P to N in Fig. 1, we obtain the condition $\{w_s = w_{olds} + 1, w_{olds} = v_s, v_s < v_{end}, v'_{str} = v_{str}, v'_{end} = v_{end}, v'_c = w_c, v'_s = w_s, \ldots\}$. This can be simplified to $\{v_s < v_{end}, v'_{end} = v_{end}, v'_s = v_s + 1, \ldots\}$.

An evaluation or refinement edge from a to $\bar{a}$ does not change the variables of $\mathcal{V}(a)$. Thus, we construct the transition $(a, \langle a \rangle \cup \{v' = v \mid v \in \mathcal{V}(a)\}, \bar{a})$.

So in the ITS resulting from Fig. 1, the condition of the transition from A to B contains $\{v'_{end} = v_{end}\} \cup \{u'_x = u_x \mid x \in \mathcal{V}_{\mathcal{P}}\}$. The condition for the transition from B to D is the same, but extended by $v'_1 = v_1$. Hence, in the transition from A to B, the value of v_1 can change arbitrarily (since $v_1 \notin \mathcal{V}(A)$), but in the transition from B to D, the value of v_1 must remain the same.

Definition 6 (ITS from Symbolic Execution Graph). *Let $\mathcal{G}$ be a symbolic execution graph. Then the corresponding integer transition system $\mathcal{I}_{\mathcal{G}}$ has one transition for each edge in $\mathcal{G}$:*

- *If the edge from a to $\bar{a}$ is not a generalization edge, then $\mathcal{I}_{\mathcal{G}}$ has a transition from a to $\bar{a}$ with the condition $\langle a \rangle \cup \{v' = v \mid v \in \mathcal{V}(a)\}$.*

---

[12] For programs starting in states represented by an abstract state a_0, it would suffice to prove termination of all $\to_{\mathcal{I}}$-evaluations starting in ITS states of the form (a_0, σ).

[13] In the transition, we do not impose the additional constraints of $\langle \bar{a} \rangle$ on the post-variables $\mathcal{V}'$, since they are checked anyway in the next transition which starts in $\bar{a}$.

- *If there is a generalization edge from a to $\bar{a}$ with the instantiation μ, then $\mathcal{I}_{\mathcal{G}}$ has a transition from a to $\bar{a}$ with the condition $\langle a \rangle \cup \{v' = \mu(v) \mid v \in \mathcal{V}(\bar{a})\}$.*

From the non-generalization edges on the path from N to P in Fig. 1, we obtain transitions whose conditions contain $v'_{end} = v_{end}$ and $v'_{s} = v_{s}$. So v_{s} is increased by 1 in the transition from P to N and it remains the same in all other transitions of the graph's only cycle. Since the transition from P to N is only executed as long as $v_{s} < v_{end}$ holds (where v_{end} is not changed by any transition), termination of the resulting ITS can easily be proved automatically.

The following theorem shows the soundness of our approach.

Theorem 7 (Termination of LLVM Programs). *Let $\mathcal{P}$ be an LLVM program with a symbolic execution graph $\mathcal{G}$ that does not contain the state ERR. If $\mathcal{I}_{\mathcal{G}}$ is terminating, then $\mathcal{P}$ is also terminating for all LLVM states represented by $\mathcal{G}$.*

4   Related Work, Experiments, and Conclusion

We developed a new approach to prove memory safety and termination of C (resp. LLVM) programs with explicit pointer arithmetic and memory access. It relies on a representation of abstract program states which allows an easy automation of the rules for symbolic execution (by standard SMT solving). Moreover, this representation is suitable for generalizing abstract states and for generating integer transition systems. In this way, LLVM programs are translated fully automatically into ITSs amenable to automated termination analysis.

Previous methods and tools for termination analysis of imperative programs (e.g., AProVE [4,5], ARMC [24], COSTA [1], Cyclist [7], FuncTion [29], Julia [25], KITTeL [12], LoopFrog [28], TAN [16], TRex [14], T2 [6], Ultimate [15], ...) either do not handle the heap at all, or support dynamic data structures by an abstraction to integers (e.g., to represent sizes or lengths) or to terms (representing finite unravelings). However, most tools fail when the control flow depends on explicit pointer arithmetic and on detailed information about the contents of addresses. While the general methodology of our approach was inspired by our previous work on termination of Java [4,5], in the current paper we lift such techniques to prove termination and memory safety of programs with explicit pointer arithmetic. This requires a fundamentally new approach, since pointer arithmetic and memory allocation cannot be expressed in the Java-based techniques of [4,5].

We implemented our technique in the termination prover AProVE using the SMT solvers Yices [11] and Z3 [20] in the back-end. A preliminary version of our implementation participated very successfully in the *International Competition on Software Verification (SV-COMP)* [27] at *TACAS*, which featured a category for termination of C programs for the first time in 2014. To evaluate AProVE's power, we performed experiments on a collection of 208 C programs from several sources, including the *SV-COMP 2014* termination category and standard string algorithms from [30] and the OpenBSD C library [23]. Of these 208 programs, 129 use pointers and 79 only operate on integers.

To prove termination of low-level C programs, one also has to ensure their memory safety. While there exist several tools to prove memory safety of C programs, many of them do not handle explicit byte-accurate pointer arithmetic (e.g., Thor [19] or SLAyer [3]) or require the user to provide the needed loop invariants (as in the Jessie plug-in of Frama-C [21]). In contrast, our approach can prove memory safety of such algorithms fully automatically. Although our approach is targeted toward termination and only analyzes memory safety as a prerequisite for termination, it turned out that on our collection, AProVE is more powerful than the leading publicly available tools for proving memory safety. To this end, we compared AProVE with the tools CPAchecker [18] and Predator [10] which reached the first and the third place in the category for *memory safety* at *SV-COMP 2014*.[14] For the 129 pointer programs in our collection, AProVE can show memory safety for 102 examples, whereas CPAchecker resp. Predator prove memory safety for 77 resp. 79 examples (see [2] for details).

To evaluate the power of our approach for proving termination, we compared AProVE to the other tools from the termination category of *SV-COMP 2014*. In addition, we included the termination analyzer KITTeL [12] in our evaluation, which operates on LLVM as well. On the side, we show the performance of the tools on integer and pointer programs when using a time limit of 300 seconds for

	79 integer programs					129 pointer programs				
	T	N	F	TO	RT	T	N	F	TO	RT
AProVE	67	0	11	1	19.6	91	0	19	19	58.6
FuncTion	11	0	66	2	23.1	-	-	-	-	-
KITTeL	58	0	12	9	0.2	9	0	1	119	0.2
T2	55	0	23	1	1.8	6	0	123	0	3.6
TAN	31	0	37	11	2.4	3	0	124	2	10.6
Ultimate	57	4	12	6	3.2	-	-	-	-	-

each example. Here, we used an Intel Core i7-950 processor and 6 GB of memory. "**T**" gives the number of examples where termination could be proved, "**N**" is the number of examples where non-termination could be shown, "**F**" states how often the tool failed in less than 300 seconds, "**TO**" gives the number of time-outs (i.e., examples for which the tool took longer than 300 seconds), and "**RT**" is the average run time in seconds for those examples where the tool proved termination or non-termination. For pointer programs, we omitted the results for those tools that were not able to prove termination of any examples.

Most other termination provers ignore the problem of memory safety and just prove termination under the *assumption* that the program is memory safe. So they may also return "Yes" for memory unsafe programs and may treat read accesses to the heap as non-deterministic input. Since AProVE constructs symbolic execution graphs to prove memory safety and to infer suitable invariants needed for termination proofs, its runtime is often higher than that of other tools. On the other hand, the table shows that our approach is slightly more powerful than the other tools for integer programs (i.e., our graph-based technique is also suitable for programs on integers) and it is clearly the most powerful one for

---

[14] The second place in this category was reached by the bounded model checker LLBMC [13]. However, in general such tools only disprove, but cannot verify memory safety.

pointer programs. The reason is due to our novel representation of the memory which handles pointer arithmetic and keeps information about the contents of addresses. For details on our experiments and to access our implementation in AProVE via a web interface, we refer to [2]. In future work, we plan to extend our approach to recursive programs and to inductive data structures defined via struct (e.g., by integrating existing shape analyses based on separation logic).

Acknowledgments. We are grateful to the developers of the other tools for termination or memory safety [6,10,12,15,16,18,29] for their help with the experiments.

References

1. Albert, E., Arenas, P., Codish, M., Genaim, S., Puebla, G., Zanardini, D.: Termination analysis of Java Bytecode. In: Barthe, G., de Boer, F.S. (eds.) FMOODS 2008. LNCS, vol. 5051, pp. 2–18. Springer, Heidelberg (2008)
2. AProVE, http://aprove.informatik.rwth-aachen.de/eval/Pointer/
3. Berdine, J., Cook, B., Ishtiaq, S.: SLAyer: Memory safety for systems-level code. In: Gopalakrishnan, G., Qadeer, S. (eds.) CAV 2011. LNCS, vol. 6806, pp. 178–183. Springer, Heidelberg (2011)
4. Brockschmidt, M., Ströder, T., Otto, C., Giesl, J.: Automated detection of non-termination and NullPointerExceptions for JBC. In: Beckert, B., Damiani, F., Gurov, D. (eds.) FoVeOOS 2011. LNCS, vol. 7421, pp. 123–141. Springer, Heidelberg (2012)
5. Brockschmidt, M., Musiol, R., Otto, C., Giesl, J.: Automated termination proofs for Java programs with cyclic data. In: Madhusudan, P., Seshia, S.A. (eds.) CAV 2012. LNCS, vol. 7358, pp. 105–122. Springer, Heidelberg (2012)
6. Brockschmidt, M., Cook, B., Fuhs, C.: Better termination proving through cooperation. In: Sharygina, N., Veith, H. (eds.) CAV 2013. LNCS, vol. 8044, pp. 413–429. Springer, Heidelberg (2013)
7. Brotherston, J., Gorogiannis, N., Petersen, R.L.: A generic cyclic theorem prover. In: Jhala, R., Igarashi, A. (eds.) APLAS 2012. LNCS, vol. 7705, pp. 350–367. Springer, Heidelberg (2012)
8. Clang compiler, http://clang.llvm.org
9. Cousot, P., Cousot, R.: Abstract interpretation: a unified lattice model for static analysis of programs by construction or approximation of fixpoints. In: Graham, R.M., Harrison, M.A., Sethi, R. (eds.) POPL 1977, pp. 238–252. ACM Press (1977)
10. Dudka, K., Peringer, P., Vojnar, T.: Predator: A shape analyzer based on symbolic memory graphs (competition contribution). In: Ábrahám, E., Havelund, K. (eds.) TACAS 2014 (ETAPS). LNCS, vol. 8413, pp. 412–414. Springer, Heidelberg (2014)
11. Dutertre, B., de Moura, L.M.: The Yices SMT solver (2006), tool paper at http://yices.csl.sri.com/tool-paper.pdf
12. Falke, S., Kapur, D., Sinz, C.: Termination analysis of C programs using compiler intermediate languages. In: Schmidt-Schauß, M. (ed.) RTA 2011. LIPIcs, vol. 10, pp. 41–50. Dagstuhl Publishing (2011)
13. Falke, S., Merz, F., Sinz, C.: LLBMC: Improved bounded model checking of C using LLVM (competition contribution). In: Piterman, N., Smolka, S.A. (eds.) TACAS 2013 (ETAPS 2013). LNCS, vol. 7795, pp. 623–626. Springer, Heidelberg (2013)
14. Harris, W.R., Lal, A., Nori, A.V., Rajamani, S.K.: Alternation for termination. In: Cousot, R., Martel, M. (eds.) SAS 2010. LNCS, vol. 6337, pp. 304–319. Springer, Heidelberg (2010)

15. Heizmann, M., Hoenicke, J., Leike, J., Podelski, A.: Linear ranking for linear lasso programs. In: Van Hung, D., Ogawa, M. (eds.) ATVA 2013. LNCS, vol. 8172, pp. 365–380. Springer, Heidelberg (2013)
16. Kroening, D., Sharygina, N., Tsitovich, A., Wintersteiger, C.M.: Termination analysis with compositional transition invariants. In: Touili, T., Cook, B., Jackson, P. (eds.) CAV 2010. LNCS, vol. 6174, pp. 89–103. Springer, Heidelberg (2010)
17. Lattner, C., Adve, V.S.: LLVM: A compilation framework for lifelong program analysis & transformation. In: CGO 2004, pp. 75–88. IEEE (2004)
18. Löwe, S., Mandrykin, M., Wendler, P.: CPAchecker with sequential combination of explicit-value analyses and predicate analyses (competition contribution). In: Ábrahám, E., Havelund, K. (eds.) TACAS 2014 (ETAPS). LNCS, vol. 8413, pp. 392–394. Springer, Heidelberg (2014)
19. Magill, S., Tsai, M.H., Lee, P., Tsay, Y.K.: Automatic numeric abstractions for heap-manipulating programs. In: Hermenegildo, M.V., Palsberg, J. (eds.) POPL 2010, pp. 211–222. ACM Press (2010)
20. de Moura, L., Bjørner, N.S.: Z3: An efficient SMT solver. In: Ramakrishnan, C.R., Rehof, J. (eds.) TACAS 2008. LNCS, vol. 4963, pp. 337–340. Springer, Heidelberg (2008)
21. Moy, Y., Marché, C.: Modular inference of subprogram contracts for safety checking. J. Symb. Comput. 45(11), 1184–1211 (2010)
22. O'Hearn, P.W., Reynolds, J.C., Yang, H.: Local reasoning about programs that alter data structures. In: Fribourg, L. (ed.) CSL 2001. LNCS, vol. 2142, pp. 1–19. Springer, Heidelberg (2001)
23. http://fxr.watson.org/fxr/source/lib/libsa/strlen.c?v=OPENBSD
24. Podelski, A., Rybalchenko, A.: ARMC: The logical choice for software model checking with abstraction refinement. In: Hanus, M. (ed.) PADL 2007. LNCS, vol. 4354, pp. 245–259. Springer, Heidelberg (2007)
25. Spoto, F., Mesnard, F., Payet, É.: A termination analyser for Java Bytecode based on path-length. ACM TOPLAS 32(3) (2010)
26. Ströder, T., Giesl, J., Brockschmidt, M., Frohn, F., Fuhs, C., Hensel, J., Schneider-Kamp, P.: Automated termination analysis for programs with pointer arithmetic. Tech. Rep. AIB 2014-05 available from [2] and from http://aib.informatik.rwth-aachen.de
27. SV-COMP at TACAS 2014, http://sv-comp.sosy-lab.org/2014/
28. Tsitovich, A., Sharygina, N., Wintersteiger, C.M., Kroening, D.: Loop summarization and termination analysis. In: Abdulla, P.A., Leino, K.R.M. (eds.) TACAS 2011. LNCS, vol. 6605, pp. 81–95. Springer, Heidelberg (2011)
29. Urban, C.: The abstract domain of segmented ranking functions. In: Logozzo, F., Fähndrich, M. (eds.) SAS 2013. LNCS, vol. 7935, pp. 43–62. Springer, Heidelberg (2013)
30. Wikibooks C Programming, http://en.wikibooks.org/wiki/C_Programming/
31. Zhao, J., Nagarakatte, S., Martin, M.M.K., Zdancewic, S.: Formalizing the LLVM IR for verified program transformations. In: Field, J., Hicks, M. (eds.) POPL 2012, pp. 427–440. ACM Press (2012)

QBF Encoding of Temporal Properties and QBF-Based Verification*

Wenhui Zhang

State Key Laboratory of Computer Science
Institute of Software, Chinese Academy of Sciences
P.O. Box 8718, Beijing 100190, China
zwh@ios.ac.cn

Abstract. SAT and QBF solving techniques have applications in various areas. One area of the applications of SAT-solving is formal verification of temporal properties of transition system models. Because of the restriction on the structure of formulas, complicated verification problems cannot be naturally represented with SAT-formulas succinctly. This paper investigates QBF-applications in this area, aiming at the verification of branching-time temporal logic properties of transition system models. The focus of this paper is on temporal logic properties specified by the extended computation tree logic that allows some sort of fairness, and the main contribution of this paper is a bounded semantics for the extended computation tree logic. A QBF encoding of the temporal logic is then developed from the definition of the bounded semantics, and an implementation of QBF-based verification follows from the QBF encoding. Experimental evaluation of the feasibility and the computational properties of such a QBF-based verification algorithm is reported.

1 Introduction

SAT and QBF solving techniques have applications in various areas [10,13,9]. One area of the applications of SAT-solving is formal verification of temporal properties of transition system models [1,14,2,15,19,12,6]. In various situations, it can be used to quickly determine whether a property is violated and is considered as a complementary approach to the standard BDD-based verification approaches [3,16,5]. Therefore a large number of research works has been devoted into this direction. However, because of the restriction on the structure of formulas, complicated verification problems cannot be naturally represented with SAT-formulas succinctly. This paper investigates QBF-applications in this area, aiming at the verification of branching-time temporal logic properties of transition system models. Branching-time temporal logic properties involve operators that may require the existence of certain kinds of paths starting at different states of a system model. This requires the use of quantifiers in the encoding of such properties.

* Supported by the National Natural Science Foundation of China under Grant No. 61272135 and the 973 Program of China under Grant No. 2014CB340701.

S. Demri, D. Kapur, and C. Weidenbach (Eds.): IJCAR 2014, LNAI 8562, pp. 224–239, 2014.

This paper focuses on QBF encoding and QBF-based verification of temporal properties specified by the extended computation tree logic that allows some sort of fairness [8]. The main contribution of this paper is a bounded semantics for the extended computation tree logic. A QBF encoding of the temporal logic is then developed from the definition of the bounded semantics, and an implementation of QBF-based verification follows from the QBF encoding. One of the particular aspects of this implementation is that it can handle properties (branching properties combined with fairness) that are not handled by well known model checking tools such as Spin [11] and NuSMV [4]. Finally experimental evaluation of the feasibility of the QBF-based verification relative to BDD-based verification is reported.

2   Preliminaries

We recall the definition of transition system models and that of the extended computation tree logic.

2.1   Transition System Models

Let AP be a set of propositional symbols. A finite state system may be represented by a Kripke structure which is a quadruple $M = \langle S, T, I, L \rangle$ where S is a set of states, $T \subseteq S \times S$ is a transition relation which is total, $I \subseteq S$ is a set of initial states and $L : S \to 2^{AP}$ is a labeling function that maps each state to a subset of propositions of AP. A Kripke structure is also called a model.

Transitions A transition from a state s to another state s' is denoted $s \to s'$. $s \to s'$ iff $(s, s') \in T$.

Paths. An infinite path is an infinite sequence of states $\pi = \pi_0 \pi_1 \cdots$ such that $\pi_i \to \pi_{i+1}$ for all $i \geq 0$.

Computations. A computation of M is an infinite path such that the initial state of the path is in I.

Notations. Let $\pi = \pi_0 \pi_1 \cdots$ be a path. We use $\pi(s)$ to denote a path π with $\pi_0 = s$. Then $\exists \pi(s).\varphi$ means that there is a path π with $\pi_0 = s$ such that φ holds, and $\forall \pi(s).\varphi$ means that for every path π with $\pi_0 = s$, φ holds.

2.2   Extended Computation Tree Logic (eCTL)

Properties of a transition system model may be specified by temporal logic formulas. Extended computation tree logic [8] is a propositional branching-time temporal logic that extends the computation tree logic (CTL) introduced by Emerson and Clarke [7] with possibility to express simple fairness constraints. For brevity, the extended computation tree logic is hereafter denoted eCTL.

Syntax Let p range over AP. The set of eCTL formulas Φ over AP is defined as follows:

$$\Phi ::= p \mid \neg\Phi \mid \Phi \wedge \Phi \mid \Phi \vee \Phi \mid A\Psi \mid E\Psi$$

$$\Psi ::= X\Phi \mid F\Phi \mid G\Phi \mid \overset{\infty}{F}\Phi \mid \overset{\infty}{G}\Phi \mid (\Phi\,U\,\Phi) \mid (\Phi\,R\,\Phi)$$

The formulas of Φ are eCTL formulas, and the formulas of Ψ are auxiliary path formulas. The property of a finite state system may be specified by an eCTL formula, and conversely, the truth of such a formula may be evaluated in a finite state system.

Definition 1. *(Semantics of eCTL)* *Let p denote a propositional symbol, and $\varphi, \varphi_0, \varphi_1$ denote eCTL formulas, ψ, ψ_0 denote path formulas. Let s be a state and π be a path of M. Let $M, s \models \varphi$ denote the relation that φ holds on s of M, and $M, \pi \models \psi$ denote that ψ holds on π of M. The relation $M, s \models \varphi$ and $M, \pi \models \psi$ are defined as follows.*

$M, s \models p$ iff $p \in L(s)$
$M, s \models \neg\varphi_0$ iff $M, s \not\models \varphi_0$
$M, s \models \varphi_0 \wedge \varphi_1$ iff $M, s \models \varphi_0$ and $M, s \models \varphi_1$
$M, s \models \varphi_0 \vee \varphi_1$ iff $M, s \models \varphi_0$ or $M, s \models \varphi_1$
$M, s \models A\psi_0$ iff $\forall\pi(s).(M, \pi \models \psi_0)$
$M, s \models E\psi_0$ iff $\exists\pi(s).(M, \pi \models \psi_0)$
$M, \pi \models X\varphi_0$ iff $M, \pi_1 \models \varphi_0$
$M, \pi \models F\varphi_0$ iff $\exists k \geq 0.M, \pi_k \models \varphi_0$
$M, \pi \models G\varphi_0$ iff $\forall k \geq 0.M, \pi_k \models \varphi_0$
$M, \pi \models \overset{\infty}{F}\varphi_0$ iff $\forall i \geq 0.\exists k \geq i.M, \pi_k \models \varphi_0$
$M, \pi \models \overset{\infty}{G}\varphi_0$ iff $\exists i \geq 0.\forall k \geq i.M, \pi_k \models \varphi_0$
$M, \pi \models \varphi_0 U \varphi_1$ iff $\exists k \geq 0.(M, \pi_k \models \varphi_1 \wedge \forall 0 \leq j < k.(M, \pi_j \models \varphi_0))$
$M, \pi \models \varphi_0 R \varphi_1$ iff $\forall k \geq 0.(M, \pi_k \models \varphi_1 \vee \exists 0 \leq j < k.(M, \pi_j \models \varphi_0))$

Definition 2. $M \models \varphi$ iff $M, s \models \varphi$ for all $s \in I$.

Negation Normal Form. An eCTL formula is in the negation normal form (NNF), if the negation $\neg$ is applied only to propositional symbols. Every eCTL formula can be transformed into an equivalent formula in NNF by using the following equivalences.

$\neg(\varphi \wedge \psi)$	$\equiv \neg\varphi \vee \neg\psi$	$\neg(\varphi \vee \psi)$	$\equiv \neg\varphi \wedge \neg\psi$
$\neg AX\,\varphi$	$\equiv EX\neg\varphi$	$\neg EX\,\varphi$	$\equiv AX\neg\varphi$
$\neg AF\,\varphi$	$\equiv EG\neg\varphi$	$\neg EF\,\varphi$	$\equiv AG\neg\varphi$
$\neg AG\,\varphi$	$\equiv EF\neg\varphi$	$\neg EG\,\varphi$	$\equiv AF\neg\varphi$
$\neg \overset{\infty}{AF}\,\varphi$	$\equiv \overset{\infty}{EG}\neg\varphi$	$\neg \overset{\infty}{EF}\,\varphi$	$\equiv \overset{\infty}{AG}\neg\varphi$
$\neg \overset{\infty}{AG}\,\varphi$	$\equiv \overset{\infty}{EF}\neg\varphi$	$\neg \overset{\infty}{EG}\,\varphi$	$\equiv \overset{\infty}{AF}\neg\varphi$
$\neg A(\varphi\,U\,\psi)$	$\equiv E(\neg\varphi\,R\,\neg\psi)$	$\neg E(\varphi\,U\,\psi)$	$\equiv A(\neg\varphi\,R\,\neg\psi)$
$\neg A(\varphi\,R\,\psi)$	$\equiv E(\neg\varphi\,U\,\neg\psi)$	$\neg E(\varphi\,R\,\psi)$	$\equiv A(\neg\varphi\,U\,\neg\psi)$

Without loss of generality, we only consider formulas in NNF. Formulas not in NNF are considered as an abbreviation of the equivalent ones in NNF.

3   Bounded Semantics

Before presenting the QBF encoding of temporal properties, we develop a bounded semantics for eCTL. This bounded semantics extends that of the previous works for that of the existential fragment of CTL [14] and that of CTL [19]. For convenience, we fix the model under consideration to be $M = \langle S, T, I, L \rangle$ in the rest of this paper.

Finite Paths and k-Paths. A finite path is a finite prefix of an infinite path. Let $k \geq 0$. A k-path of M is a finite path of M with length $k + 1$. π is a k-path, if $\pi = \pi_0 \cdots \pi_k$ such that $\pi_i \in S$ for $i = 0, ..., k$ and $(\pi_i, \pi_{i+1}) \in T$ for $i = 0, ..., k - 1$.

Bounded Models. The k-model of M is a quadruple $M_k = \langle S, Ph_k, I, L \rangle$ where Ph_k is the set of all k-paths of M. M_k can be considered as an approximation of M.

Paths with Repeating States (rs-paths). An rs-path is a finite path that contains repeating states (i.e., at least two states are the same). Let $rs(\pi)$ denote that π is an rs-path. An important property of such a path is that if π is a prefix of π', then $rs(\pi) \rightarrow rs(\pi')$. For the ideas of k-paths, k-models, and rs-paths, the reader is referred to [1,14,19].

Definition 3. *(Bounded Semantics of eCTL) Let p denote a propositional symbol, and $\varphi, \varphi_0, \varphi_1$ denote eCTL formulas, ψ, ψ_0 denote path formulas. Let s be a state of M and π be a k-path of Ph_k. Let $M_k, s \models \varphi$ denote the relation that φ holds on s of M_k, and $M_k, \pi \models \psi$ denote that ψ holds on π of M_k. The relation $M_k, s \models \varphi$ and $M_k, \pi \models \psi$ are defined as follows.*

$M_k, s \models p$ iff $p \in L(s)$
$M_k, s \models \neg p$ iff $p \notin L(s)$
$M_k, s \models \varphi_0 \wedge \varphi_1$ iff $(M_k, s \models \varphi_0)$ and $(M_k, s \models \varphi_1)$
$M_k, s \models \varphi_0 \vee \varphi_1$ iff $(M_k, s \models \varphi_0)$ or $(M_k, s \models \varphi_1)$
$M_k, s \models A\psi$ iff $\forall \pi(s).(M_k, \pi \models \psi)$
$M_k, s \models E\psi$ iff $\exists \pi(s).(M_k, \pi \models \psi)$
$M_k, \pi \models X\varphi_0$ iff $k \geq 1 \wedge (M_k, \pi_1 \models \varphi_0)$
$M_k, \pi \models F\varphi_0$ iff $\exists i \leq k.(M_k, \pi_i \models \varphi_0)$
$M_k, \pi \models G\varphi_0$ iff $rs(\pi) \wedge (\forall i \leq k.(M_k, \pi_i \models \varphi_0))$
$M_k, \pi \models \overset{\infty}{F}\varphi_0$ iff $rs(\pi) \wedge \forall i < l \leq k.(\pi_i = \pi_l \rightarrow \exists i < j \leq l.(M_k, \pi_j \models \varphi_0))$
$M_k, \pi \models \overset{\infty}{G}\varphi_0$ iff $rs(\pi) \wedge \forall i < l \leq k.(\pi_i = \pi_l \rightarrow \forall i < j \leq l.(M_k, \pi_j \models \varphi_0))$
$M_k, \pi \models \varphi_0 U \varphi_1$ iff $\exists i \leq k.(M_k, \pi_i \models \varphi_1 \wedge \forall j < i.(M_k, \pi_j \models \varphi_0))$
$M_k, \pi \models \varphi_0 R \varphi_1$ iff $\forall i \leq k.(M_k, \pi_i \models \varphi_1 \vee \exists j < i.(M_k, \pi_j \models \varphi_0)) \wedge (\exists j \leq k.(M_k, \pi_j \models \varphi_0) \vee rs(\pi))$

Definition 4. $M_k \models \varphi$ iff $M_k, s \models \varphi$ for all $s \in I$.

Let $|M|$ denote the number of states of M.

Lemma 1. If $M, s \models \varphi$, then there is a $k \geq 0$ such that $M_k, s \models \varphi$.

Proof: The proof is done by structural induction. For brevity (due to the page limit), we prove the two cases where φ is respectively $A\overset{\infty}{G}\varphi_0$ and $A\overset{\infty}{F}\varphi_0$, and omit the rest of the cases. Let $k = |M|$.

- Suppose that $M, s \models A\overset{\infty}{G}\varphi_0$ holds.
  Since $k = |M|$ and the transition relation is total, the length of every k-path is greater than $|M|$. Then for every k-path π, $rs(\pi)$ holds. We only need to show for every k-path π starting at s the following holds.

$$\forall i < l \leq k.(\pi_i = \pi_l \to \forall i < j \leq l.(M_k, \pi_j \models \varphi_0)).$$

  Assume $\pi_i = \pi_l$ for a k-path π. Then $\pi' = \pi_0 \cdots \pi_i(\pi_{i+1} \cdots \pi_l)^\omega$ is an infinite path starting at $s = \pi_0$. Since π' satisfies $\overset{\infty}{G}\varphi_0$, we have $\forall i < j \leq l.(M, \pi_j \models \varphi_0)$. Then according to the induction hypothesis, $\forall i < j \leq l.(M_k, \pi_j \models \varphi_0)$.
- Suppose that $M, s \models A\overset{\infty}{F}\varphi_0$ holds.
  Since $k = |M|$, for every k-path π, $rs(\pi)$ holds. We only need to show for every k-path π starting at s the following holds.

$$\forall i < l \leq k.(\pi_i = \pi_l \to \exists i < j \leq l.(M_k, \pi_j \models \varphi_0)).$$

  Assume $\pi_i = \pi_l$ for a k-path π. Then $\pi' = \pi_0 \cdots \pi_i(\pi_{i+1} \cdots \pi_l)^\omega$ is an infinite path starting at $s = \pi_0$. Since π' satisfies $\overset{\infty}{F}\varphi_0$, we have $\exists i < j \leq l.(M, \pi_j \models \varphi_0)$. Then according to the induction hypothesis, $\exists i < j \leq l.(M_k, \pi_j \models \varphi_0)$.

Lemma 2. If $M_k, s \models \varphi$ for $k \geq |M|$, then $M, s \models \varphi$.

Proof: The proof is done by structural induction. For brevity, we prove the two cases where φ is respectively $A\overset{\infty}{G}\varphi_0$ and $A\overset{\infty}{F}\varphi_0$, and omit the rest of the cases.

- Suppose that $M_k, s \models A\overset{\infty}{G}\varphi_0$ holds for k.
  Then for every k-path π' starting at s, we have $M_k, \pi' \models \overset{\infty}{G}\varphi_0$, i.e.,

$$rs(\pi') \wedge \forall i < l \leq k.(\pi'_i = \pi'_l \to \forall i < j \leq l.(M_k, \pi'_j \models \varphi_0)).$$

  Assume that $M, s \models A\overset{\infty}{G}\varphi_0$ does not hold. We show that this is a contradiction. According to this assumption, there is an infinite path starting at s such that $M, \pi \models \overset{\infty}{G}\varphi_0$ does not hold. Then we can construct an infinite path $\pi' = \pi'_0 \cdots \pi'_i(\pi'_{i+1} \cdots \pi'_l)^\omega$ starting at s such that $l \leq |M|$, $\pi'_x \neq \pi'_y$ for all $x < y < l$, $\pi'_i = \pi'_l$, and $M, \pi'_j \not\models \varphi_0$ for some $i < j \leq l$. Let π'' be a k-path with $\pi'_0 \cdots \pi'_i \pi'_{i+1} \cdots \pi'_l$ as its prefix (this is possible, since $l \leq |M| \leq k$).

Then according to the premise of the lemma, $M_k, \pi'' \models \overset{\infty}{G}\varphi_0$ holds. Then we have $M_k, \pi'_j \models \varphi_0$, and by the induction hypothesis, $M, \pi'_j \models \varphi_0$. This is a contradiction, which proves the lemma.

– Suppose that $M_k, s \models \overset{\infty}{AF}\varphi_0$ holds for k.

Then for every k-path π' starting at s, we have $M_k, \pi' \models \overset{\infty}{F}\varphi_0$, i.e.,

$$rs(\pi') \wedge \forall i < l \leq k.(\pi'_i = \pi'_l \rightarrow \exists i < j \leq l.(M_k, \pi'_j \models \varphi_0)).$$

Assume that $M, s \models \overset{\infty}{AF}\varphi_0$ does not hold. We show that this is a contradiction. According to this assumption, there is an infinite path starting at s such that $M, \pi \models \overset{\infty}{F}\varphi_0$ does not hold. Then we can construct an infinite path $\pi' = \pi'_0 \cdots \pi'_i (\pi'_{i+1} \cdots \pi'_l)^\omega$ starting at s such that $l \leq |M|$, $\pi'_x \neq \pi'_y$ for all $x < y < l$, $\pi'_i = \pi'_l$, and $M, \pi'_j \not\models \varphi_0$ for all $i < j \leq l$. Let π'' be a k-path with $\pi'_0 \cdots \pi'_i \pi'_{i+1} \cdots \pi'_l$ as its prefix (this is possible, since $l \leq |M| \leq k$). Then according to the premise of the lemma, $M_k, \pi'' \models \overset{\infty}{F}\varphi_0$ holds. Then we have $\exists i < j \leq l.(M_k, \pi'_j \models \varphi_0)$, and by the induction hypothesis, $\exists i < j \leq l.(M, \pi'_j \models \varphi_0)$. This is a contradiction, which proves the lemma.

Lemma 3. *If $M_k, s \models \varphi$, then $M_{k+1}, s \models \varphi$.*

Proof: The proof is done by structural induction. For brevity, we prove the case where φ is $\overset{\infty}{AG}\varphi_0$, and omit the rest of the cases. Suppose that $M_k, s \models \overset{\infty}{AG}\varphi_0$ holds for k.

Then for every k-path ζ starting at s, we have $M_k, \zeta \models \overset{\infty}{G}\varphi_0$, i.e.,

$$rs(\zeta) \wedge \forall i < l \leq k.(\zeta_i = \zeta_l \rightarrow \forall i < j \leq l.(M_k, \zeta_j \models \varphi_0)).$$

Then according to the induction hypothesis, we have

$$rs(\zeta) \wedge \forall i < l \leq k.(\zeta_i = \zeta_l \rightarrow \forall i < j \leq l.(M_{k+1}, \zeta_j \models \varphi_0)).$$

The goal is to prove that $M_{k+1}, s \models \overset{\infty}{AG}\varphi_0$ holds, i.e., for every $(k+1)$-path π of M_{k+1} starting at s, the following two properties hold.

(1) $rs(\pi)$
(2) $\forall i < l \leq k+1.(\pi_i = \pi_l \rightarrow \forall i < j \leq l.(M_{k+1}, \pi_j \models \varphi_0))$

Let $\pi = \pi_0 \cdots \pi_k \pi_{k+1}$ be a $(k+1)$-path starting at s. Since $\pi' = \pi_0 \cdots \pi_k$ is a k-path, we have the following fact.

$$rs(\pi') \wedge \forall i < l \leq k.(\pi_i = \pi_l \rightarrow \forall i < j \leq l.(M_{k+1}, \pi_j \models \varphi_0))$$

Since $rs(\pi')$ implies property (1), and the cases of property (2) where $l \leq k$ are covered by the fact, we only need to show

$$\forall i < k+1.(\pi_i = \pi_{k+1} \rightarrow \forall i < j \leq k+1.(M_{k+1}, \pi_j \models \varphi_0)).$$

Since $rs(\pi')$ holds, there is $x < y \leq k$ such that $\pi_x = \pi_y$.
Assume $\pi_i = \pi_{k+1}$. We divide the rest of the proof into two cases:

- $x < i < y$:

  Let π'' be the concatenation of $\pi_0 \cdots \pi_x$, $\pi_{y+1} \cdots \pi_{k+1}$, $\pi_{i+1} \cdots \pi_y$.

  Then π'' is a prefix of a k-path starting at s and, since $\pi_x = \pi_y$, every state s' between π_x and π_y satisfies φ_0 ($M_k, s' \models \varphi_0$ according to the premise and $M_{k+1}, s' \models \varphi_0$ according to the induction hypothesis).

  Therefore $\forall i < j \leq k + 1.(M_{k+1}, \pi_j \models \varphi_0)$.

- $i \leq x$ or $y \leq i$:

  Let π'' be the path obtained by removing $\pi_{x+1} \cdots \pi_y$ from π.

  Then π'' is a prefix of a k-path starting at s and, therefore every state between π_i and π_{k+1} (of π'') satisfies φ_0, according to the premise and the induction hypothesis.

  In addition, every state in the partial-path $\pi_{x+1} \cdots \pi_y$ also satisfies φ_0 (this is needed in case $i \leq x$). Therefore we have $\forall i < j \leq k + 1.(M_{k+1}, \pi_j \models \varphi_0)$.

Theorem 1 (Soundness and Completeness). $M, s \models \varphi$ iff $M_k, s \models \varphi$ for some $k \geq 0$.

The soundness and completeness of the bounded semantics follows from Lemma 1, Lemma 2 and Lemma 3.

Corollary 1. $M \models \varphi$ iff $M_k \models \varphi$ for some $k \geq 0$.

4   QBF Encoding and QBF-Based Verification

From the bounded semantics, a QBF-based characterization of eCTL formulas, extending that of CTL formulas [20], can be developed as follow. Let $k \geq 0$. Let $u_0, ..., u_k$ be a finite sequence of state variables. The sequence $u_0, ..., u_k$ (denoted by $\vec{u}$) is intended to be used as a representation of a path of M_k. This is captured by the following definition of $P_k(\vec{u})$.

Definition 5

$$P_k(\vec{u}) := \bigwedge_{j=0}^{k-1} T(u_j, u_{j+1})$$

Every assignment to the set of state variables $\{u_0, ..., u_k\}$ satisfying $P_k(\vec{u})$ represents a valid k-path of M. Let $rs_k(\vec{u})$ denote that the k-path represented by $\vec{u}$ is an rs-path. Formally, we have the following definition of $rs_k(\vec{u})$.

Definition 6

$$rs_k(\vec{u}) := \bigvee_{x=0}^{k-1} \bigvee_{y=x+1}^{k} u_x = u_y.$$

Let $p \in AP$ be a proposition symbol and $p(v)$ be the propositional formula such that $p(v)$ is true whenever v is assigned the truth value representing a state s in which p holds.

Definition 7 (Transformation of eCTL Formulas). *Let $k \geq 0$. Let v be a state variable and φ be an eCTL formula. The encoding $[[\varphi, v]]_k$ is defined as follows.*

$$
\begin{aligned}
[[p, v]]_k &= p(v) \\
[[\neg p, v]]_k &= \neg p(v) \\
[[\varphi \vee \psi, v]]_k &= [[\varphi, v]]_k \vee [[\psi, v]]_k \\
[[\varphi \wedge \psi, v]]_k &= [[\varphi, v]]_k \wedge [[\psi, v]]_k \\[4pt]
[[A\varphi, v]]_k &= \forall \vec{u}.(P(\vec{u}) \wedge v = u_0 \rightarrow [[\varphi, \vec{u}]]_k) \\
[[E\varphi, v]]_k &= \exists \vec{u}.(P(\vec{u}) \wedge v = u_0 \wedge [[\varphi, \vec{u}]]_k) \\[4pt]
[[X\varphi, \vec{u}]]_k &= k \geq 1 \wedge [[\varphi, u_1]]_k \\
[[F\psi, \vec{u}]]_k &= \bigvee_{j=0}^{k}[[\psi, u_j]]_k \\
[[G\psi, \vec{u}]]_k &= \bigwedge_{j=0}^{k}[[\psi, u_j]]_k \wedge rs_k(\vec{u})) \\
[[\overset{\infty}{F}\psi, \vec{u}]]_k &= rs_k(\vec{u}) \wedge \bigwedge_{i=0}^{k}(\bigwedge_{l=i+1}^{k}(u_i = u_l \rightarrow \bigvee_{j=i+1}^{l}[[\psi, u_j]]_k)) \\
[[\overset{\infty}{G}\psi, \vec{u}]]_k &= rs_k(\vec{u}) \wedge \bigwedge_{i=0}^{k}(\bigwedge_{l=i+1}^{k}(u_i = u_l \rightarrow \bigwedge_{j=i+1}^{l}[[\psi, u_j]]_k)) \\
[[\varphi U\psi, \vec{u}]]_k &= \bigvee_{j=0}^{k}([[\psi, u_j]]_k \wedge \bigwedge_{t=0}^{j-1}[[\varphi, u_t]]_k) \\
[[\varphi R\psi, \vec{u}]]_k &= \bigwedge_{j=0}^{k}([[\psi, u_j]]_k \vee \bigvee_{t=0}^{j-1}[[\varphi, u_t]]_k) \wedge (\bigvee_{t=0}^{k}[[\varphi, u_t]]_k \vee rs_k(\vec{u}))
\end{aligned}
$$

Note that the transition relation of M is total, and therefore every finite path either can be extended to a k-path or has a k-path as its prefix. Let $v(s)$ denote that the state variable v has been assigned a value corresponding to the state s. The following theorem follows from the transformation scheme.

Theorem 2. *Let φ be an eCTL formula. $M_k, s \models \varphi$ iff $[[\varphi, v(s)]]_k$ holds.*

Let $I(v)$ denote the propositional formula that restricts potential values of v to the initial states of M.

Corollary 2. *Let φ be an eCTL formula. $M \models \varphi$ iff there is a $k \geq 0$ such that $\forall v.(I(v) \rightarrow [[\varphi, v]]_k)$, and $M \not\models \varphi$ iff there is a $k \geq 0$ such that $\exists v.(I(v) \wedge [[\neg\varphi, v]]_k)$.*

Following from Theorem 1, we have $M \models \varphi$ iff there is a $k \geq 0$ such that $M_k \models \varphi$. According to Theorem 2, we have $M \models \varphi$ iff there is a $k \geq 0$ such that $\forall v.(I(v) \rightarrow [[\varphi, v]]_k)$. The second part of the corollary is shown as follows.

- Suppose that $M \not\models \varphi$.
  Then $(\exists s \in I, M, s \not\models \varphi)$, and therefore $(\exists s \in I, M, s \models \neg\varphi)$.
  According to Theorem 1, $(\exists s \in I, \exists k \geq 0, M_k, s \models \neg\varphi)$.
  Therefore there is a $k \geq 0$ such that $\exists s \in I$, $M_k, s \models \neg\varphi$ holds, and then there is a $k \geq 0$ such that $\exists v.(I(v) \wedge [[\neg\varphi, v]]_k)$, according to Theorem 2.

- On the other hand, suppose that $M \models \varphi$.
  Then $\forall s \in I, M, s \models \varphi$, and therefore $\neg(\exists s \in I, M, s \models \neg\varphi)$.
  According to Theorem 1, $\neg(\exists s \in I, \exists k \geq 0, M_k, s \models \neg\varphi)$.
  Therefore $\neg(\exists k \geq 0, \exists s \in I, M_k, s \models \neg\varphi)$.
  Therefore $\neg(\exists k \geq 0, \exists v.(I(v) \wedge [[\neg\varphi, v]]_k))$, according to Theorem 2.

Bounded Correctness Checking Let φ be an eCTL formula. Following from Corollary 2, we can formulate a bounded correctness checking algorithm for $M \models \varphi$, as follows.

Init $k = 0$;
If $\forall v.(I(v) \rightarrow [[\varphi, v]]_k)$ holds, report that φ holds;
If $\exists v.(I(v) \wedge [[\neg\varphi, v]]_k)$ holds, report that φ does not hold;
Increase k, go to the first "if"-test;

The correctness and the termination are guaranteed by Corollary 2. The algorithm is a combination of checking whether φ holds directly by the bounded semantics, and on the other hand checking whether φ does not hold also by the bounded semantics. The latter part is in accordance with the traditional bounded model checking approach [1].

5   Implementation and Experimental Evaluation

The implementation of the bounded correctness checking algorithm involves the following functionalities:

- For the finite state program, convert the program into a Boolean program;
- Produce a Boolean formula for the initial states (i.e., $I(v)$);
- Produce a Boolean formula for the transition relation (i.e., $T(v, v')$);
- For the property specified by an eCTL formula with a given k, produce a QBF-formula according to the transformation scheme;
- Combine the QBF-formula with the Boolean formula representing the initial states;
- Apply a QBF-solving algorithm to check the truth of the combined formula.

The proposed QBF-based approach has been implemented in a verification tool, denoted VERDS[1], and an experimental evaluation has been carried out. We first present an example to show the application of the approach, and then the experimental evaluation is reported.

5.1   An Illustrative Example

The example is a concurrent program representing a formulation of Peterson's mutual exclusion algorithm [18] as a first order transition system [17]. Let a, b be variables of enumeration type which have respectively the domain $\{s_0, ..., s_3\}$ and $\{t_0, ..., t_3\}$. Let x, y, t be variables of Boolean type. The program consists of two processes: A and B with the following specification:

---
[1] http://lcs.ios.ac.cn/~zwh/verds/

Process A:	Process B:
$a = s_0 \longrightarrow (y, t, a) := (1, 1, s_1)$	$b = t_0 \longrightarrow (x, t, b) := (1, 0, t_1)$
$a = s_1 \wedge (x = 0 \vee t = 0) \longrightarrow (a) := (s_2)$	$b = t_1 \wedge (y = 0 \vee t = 1) \longrightarrow (b) := (t_2)$
$a = s_2 \longrightarrow (y, a) := (0, s_3)$	$b = t_2 \longrightarrow (x, b) := (0, t_3)$
$a = s_2 \longrightarrow (a) := (s_2)$	$b = t_2 \longrightarrow (b) := (t_2)$
$a = s_3 \longrightarrow (y, t, a) := (1, 1, s_1)$	$b = t_3 \longrightarrow (x, t, b) := (1, 0, t_1)$

Let the formula specifying the set of the initial states be $a = s_0 \wedge b = t_0 \wedge x = y = 0$. The value of t is arbitrary at the initial state. The following explains the meaning of some of the constants.

$a = s_i$: process A is waiting for entering the critical region when $i = 1$, is in the critical region when $i = 2$, has left the critical region when $i = 3$.
$b = t_i$: process B is waiting for entering the critical region when $i = 1$, is in the critical region when $i = 2$, has left the critical region when $i = 3$.

Let the following be the properties of the program we want to verify.

p_1: $AF(a = s_2 \vee b = t_2)$
p_2: $AG(\neg(a = s_2 \wedge b = t_2))$
p_3: $\overset{\infty}{AG}(((a = s_1) \rightarrow AF(a = s_2)) \wedge ((b = t_1) \rightarrow AF(b = t_2)))$
p_4: $\overset{\infty}{AG}(((a = s_1) \rightarrow EF(a = s_2)) \wedge ((b = t_1) \rightarrow EF(b = t_2)))$

The first 2 properties are simple CTL properties for mutual exclusion algorithms, and the last 2 properties are particular eCTL properties.

Verification The input to the verification tool VERDS must be written in the language specified in [21]. Let the input be as follows, in which the temporal operator $\overset{\infty}{AG}$ is written as AFG.

```
VVM
VAR    x:0..1; y:0..1; t:0..1; a:{s0,s1,s2,s3}; b:{t0,t1,t2,t3};
INIT   x=0; y=0; a=s0; b=t0;
TRANS a=s0:              (y,t,a):=(1,1,s1);
       a=s1&(x=0|t=0):  (a):=(s2);
       a=s2:             (y,a):=(0,s3);
       a=s2:             (a):=(s2);
       a=s3:             (y,t,a):=(1,1,s1);
       b=t0:             (x,t,b):=(1,0,t1);
       b=t1&(y=0|t=1):  (b):=(t2);
       b=t2:             (x,b):=(0,t3);
       b=t2:             (b):=(t2);
       b=t3:             (x,t,b):=(1,0,t1);
SPEC   AF((a=s2|b=t2));
       AG(!(a=s2&b=t2));
       AFG((!a=s1|AF(a=s2))&(!b=t1|AF(b=t2)));
       AFG((!a=s1|EF(a=s2))&(!b=t1|EF(b=t2)));
```

Suppose that the input is contained in the file "tn1mutex.vvm". For checking the i-th property, we use the following command, where i is to be replaced by a given number.

```
verds -QBF -ck i tn1mutex.vvm
```

The verification result for the third property (with $i = 3$) is shown as follows.

```
VERSION:       verds 1.45 - JAN 2014
FILE:          tn1mutex.vvm
PROPERTY:      AFG((!(a = 1)|AF(a = 2)&(!(b = 1)|AF(b = 2))))
INFO:          applying an internal QBF-solver
bound =        0
.

.
bound =        4
CONCLUSION: FALSE
```

The verification process for the other properties are similar. A summary of the verification results is as follows, where the first row specifies the properties, and 2nd and 3rd row show respectively the satisfiability of the formula and the least k for certifying the satisfiability.

	p_1	p_2	p_3	p_4
T/F	T	T	F	T
k	3	10	4	10

The above example shows that, for some problem instances, the satisfiability or unsatisfiability may be determined when k is relatively small. In such cases, the QBF-based verification may have advantage over the traditional symbolic model checking approach. In the following, we present a comparison of such an approach with the traditional model checking approach.

5.2   Experimental Evaluation

This subsection contains a summary of an experimental evaluation of QBF-based verification implemented in VERDS (to be referred to as VERDS-QBF in the rest of the paper). The experimental evaluation compares this QBF-based verification with BDD-based verification implemented in NuSMV [4] version 2.5.0[2]. The comparison is based on the use of two types of random Boolean programs and 24 properties. A description of the programs and the properties is as follows.

Remarks. The comparison is not meant to draw a conclusion on which verification approach is better. Rather, it will show that there is a large number of cases on which one approach is better than the other, and vice versa, and in this sense the two approaches may be considered complementary.

---

[2] http://nusmv.irst.itc.it/

Programs with Concurrent Processes. The parameters of the first set of random Boolean programs are as follows:

a: number of processes
b: number of all variables
c: number of share variables
d: number of local variables in a process

The shared variables are initially set to a random value in $\{0, 1\}$, and the local variables are initially set to 0. For each process, the shared variables and the local variables are assigned the negation of a variable randomly chosen from these variables.

Programs with Concurrent Sequential Processes. The parameters of the second set of random Boolean programs are as follows, in addition to a, b, c, d specified above.

t: number of transitions in a process
p: number of parallel assignments in each transition

For each concurrent sequential process, besides the b Boolean variables, there is a local variable representing program locations, with e possible values. The shared variables are initially set to a random value in $\{0, 1\}$, and the local variables are initially set to 0. For each transition of a process, p pairs of shared variables and local variables are randomly chosen among the shared variables and the local variables, such that the first element of such a pair is assigned the negation of the second element of the pair. Transitions are numbered from 0 to $t - 1$, and are executed consecutively, and when the end of the sequence of the transitions is reached, it loops back to the execution of the transition numbered 0.

Types of Properties. The properties are specified by a subset of 24 eCTL formulas (which are actually all CTL properties, since BDD-based CTL model checking in NuSMV is used as the reference in the evaluation). These properties involve AG, AF properties, and more complicated ones specified with different combinations of operators with one or two levels of nesting (with two levels of nesting when AX or EX is involved). Properties p_{01} to p_{12} are shown below, where v_i are global variables.

$p_{01}:$	$AG(\bigvee_{i=1}^{c} v_i)$	$p_{07}:$	$A(v_1 \, U \, A(v_2 \, U \, \bigvee_{i=3}^{c} v_i)$
$p_{02}:$	$AF(\bigvee_{i=1}^{c} v_i)$	$p_{08}:$	$A(v_1 \, U \, E(v_2 \, U \, \bigvee_{i=3}^{c} v_i)$
$p_{03}:$	$AG(v_1 \to AF(v_2 \wedge \bigvee_{i=3}^{c} v_i))$	$p_{09}:$	$A(v_1 \, U \, A(v_2 \, R \, \bigvee_{i=3}^{c} v_i)$
$p_{04}:$	$AG(v_1 \to EF(v_2 \wedge \bigvee_{i=3}^{c} v_i))$	$p_{10}:$	$A(v_1 \, U \, E(v_2 \, R \, \bigvee_{i=3}^{c} v_i)$
$p_{05}:$	$EG(v_1 \to AF(v_2 \wedge \bigvee_{i=3}^{c} v_i))$	$p_{11}:$	$A(AXv_1 \, R \, AX \, A(v_2 \, U \, \bigvee_{i=3}^{c} v_i)$
$p_{06}:$	$EG(v_1 \to EF(v_2 \wedge \bigvee_{i=3}^{c} v_i))$	$p_{12}:$	$A(EXv_1 \, R \, EX \, E(v_2 \, U \, \bigvee_{i=3}^{c} v_i)$

Properties p_{13} to p_{24} are similar to p_{01} to p_{12} where the difference is that $\wedge$ and $\bigvee$ are replaced by respectively $\vee$ and $\bigwedge$.

Experimental Setup. The comparison of advantage and disadvantage is based on the time used for the verification problem instances. The experimental data were obtained by running the tools on a Linux platform. For QBF-based verification, the following command is used for running VERDS-QBF.

```
verds -QBF filename
```

For BDD-based verification, we run NuSMV (without counter-example generation) by the following command.

```
NuSMV -dcx filename
```

The option -dcx is for avoiding the generation of counter-examples. This option is used, since the corresponding use of VERDS does not generate counter-examples.

Experimental Data for Programs with Concurrent Processes. For this type of programs, we test different sizes of the programs with 3 processes ($a = 3$), and let b vary over the set of values $\{12, 24, 36\}$, then set $c = b/2$, $d = c/a$. Each of the 24 properties is tested on 20 test cases for each value of b. For brevity, for each type of properties, a summary of the experimental data is presented in the left part of Fig. 1, where N is the number of test cases, T is the number of test cases in which the property is true, F is the number of test cases in which the property is false, adv is the number of cases in which VERDS-QBF has an advantage with respect to the usage of time. In this part of the evaluation, VERDS-QBF has advantage in 1190 of 1440 test cases. On the relative advantage of verification and falsification, VERDS-QBF has better advantage in the case of falsification.

Experimental Data for Programs with Concurrent Sequential Processes. For this type of programs, we test different sizes of the programs with 2 processes ($a = 2$), and let b vary over the set of values $\{12, 16, 20\}$, and then set $c = b/2$, $d = c/a$, $t = c$, and $p = 4$. Similarly, each property is tested on 20 test cases for each value of b, and a summary of the experimental data is presented in the right part of Fig. 1. In this part of the evaluation, VERDS-QBF has advantage in 739 of 1440 test cases. On the relative advantage of verification and falsification, VERDS-QBF has also better advantage in falsification in this part of the evaluation.

Summary. Based on the total of 2880 test cases[3], the experimental evaluation[4] shows that the QBF-based verification does not have advantage in verifying any of the properties that start with *AG*. On the other hand, the QBF-based verification may have advantages in parts (ranging from a few percent to a large percent) of the test cases of other types of verification and falsification problems (including falsification of AG properties). On the relative advantage of verification and falsification, VERDS-QBF has better advantage in falsification in both

---

[3] Available at http://lcs.ios.ac.cn/~zwh/tr/verds130ee.rar

[4] Details available at http://lcs.ios.ac.cn/~zwh/tr/verds130eeq.pdf

Data for Concurrent Processes:

property	adv/T	adv/F	adv/N
p_{01}	-	60/60	60/60
p_{02}	60/60	-	60/60
p_{03}	0/3	43/57	43/60
p_{04}	0/60	-	0/60
p_{05}	46/53	1/7	47/60
p_{06}	53/60	-	53/60
p_{07}	60/60	-	60/60
p_{08}	60/60	-	60/60
p_{09}	50/52	5/8	55/60
p_{10}	60/60	-	60/60
p_{11}	13/13	45/47	58/60
p_{12}	60/60	-	60/60
p_{13}	-	60/60	60/60
p_{14}	3/3	55/57	58/60
p_{15}	0/8	38/52	38/60
p_{16}	0/60	-	0/60
p_{17}	46/56	1/4	47/60
p_{18}	53/60	-	53/60
p_{19}	5/5	54/55	59/60
p_{20}	16/21	30/39	46/60
p_{21}	3/3	56/57	59/60
p_{22}	3/3	56/57	59/60
p_{23}	-	60/60	60/60
p_{24}	24/31	11/29	35/60
sum	615/791	575/649	1190/1440

Data for Concurrent Seq. Processes:

property	adv/T	adv/F	adv/N
p_{01}	0/53	2/7	2/60
p_{02}	60/60	-	60/60
p_{03}	0/10	1/50	1/60
p_{04}	0/60	-	0/60
p_{05}	0/46	0/14	0/60
p_{06}	0/60	-	0/60
p_{07}	60/60	-	60/60
p_{08}	60/60	-	60/60
p_{09}	36/54	0/6	36/60
p_{10}	53/60	-	53/60
p_{11}	33/47	0/13	33/60
p_{12}	52/60	-	52/60
p_{13}	-	60/60	60/60
p_{14}	4/4	4/56	8/60
p_{15}	0/10	1/50	1/60
p_{16}	0/60	-	0/60
p_{17}	0/48	0/12	0/60
p_{18}	0/60	-	0/60
p_{19}	8/8	52/52	60/60
p_{20}	13/18	27/42	40/60
p_{21}	4/4	54/56	58/60
p_{22}	4/4	55/56	59/60
p_{23}	-	59/60	59/60
p_{24}	8/16	29/44	37/60
sum	395/862	344/578	739/1440

Fig. 1. Experimental Data for the two Types of Programs

of the types of programs. In summary, QBF-based verification has advantage in more than 50 percent of the test cases, which are well distributed among verification and falsification of universal properties.

6   Concluding Remarks

Bounded semantics of eCTL and QBF-based characterization of eCTL based on such a semantics have been presented. A verification algorithm of eCTL properties based on solving QBF-formulas has then been established.

The traditional application area of SAT-based verification has mainly been on the error detection of various universal properties such as LTL and the universal fragments of CTL* [1,14,15]. QBF-based verification presented in this paper applies to the set of eCTL properties (that may be specified with both universal and existential path quantifiers), and can handle verification and falsification problems with bounded models. Furthermore, one of the particular aspects of this implementation is that it can handle properties, for instance, of the form

$A\overset{\infty}{G}(p \to EFq)$, that are not handled by well known model checking tools such as Spin [11] and NuSMV [4].

Experimental evaluation of such an approach has been presented. The test cases have shown that QBF-based verification and BDD-based verification have their own advantages and may be considered complementary in the verification of different problem instances.

The efficiency of QBF-based verification depends very much on the QBF-solving techniques. External QBF-solvers may be used to increase the efficiency of the verification. Improving the efficiency by optimizing the QBF-based encoding and by enhancing QBF-solving techniques remains as future works.

References

1. Biere, A., Cimatti, A., Clarke, E., Zhu, Y.: Symbolic Model Checking without BDDs. In: Cleaveland, W.R. (ed.) TACAS/ETAPS 1999. LNCS, vol. 1579, pp. 193–207. Springer, Heidelberg (1999)
2. Biere, A., Cimmatti, A., Clarke, E., Strichman, O., Zhu, Y.: Bounded Model Checking. Advances in Computers, vol. 58. Academic Press (2003)
3. Burch, J.R., Clarke, E.M., McMillan, K.L., Dill, D.L., Hwang, J.: Symbolic model checking: 10^{20} states and beyond. LICS, pp. 428–439 (1990)
4. Cimatti, A., Clarke, E.M., Giunchiglia, F., Roveri, M.: NUSMV: A New Symbolic Model Verifier. In: Halbwachs, N., Peled, D.A. (eds.) CAV 1999. LNCS, vol. 1633, pp. 495–499. Springer, Heidelberg (1999)
5. Clarke, E.M., Grumberg, O., Peled, D.: Model Checking. The MIT Press (1999)
6. Duan, Z., Tian, C., Yang, M., He, J.: Bounded Model Checking for Propositional Projection Temporal Logic. In: Du, D.-Z., Zhang, G. (eds.) COCOON 2013. LNCS, vol. 7936, pp. 591–602. Springer, Heidelberg (2013)
7. Emerson, E.A., Clarke, E.M.: Using Branching-time Temporal Logics to Synthesize Synchronization Skeletons. Sci. of Comp. Prog. 2(3), 241–266 (1982)
8. Emerson, E.A., Halpern, J.Y.: "Sometimes" and "Not Never" revisited: on branching versus linear time temporal logic. J. ACM 33(1), 151–178 (1986)
9. Goultiaeva, A., Van Gelder, A., Bacchus, F.: A Uniform Approach for Generating Proofs and Strategies for Both True and False QBF Formulas. In: IJCAI 2011, pp. 546–553 (2011)
10. Hoffmann, J., Gomes, C.P., Selman, B., Kautz, H.A.: SAT Encodings of State-Space Reachability Problems in Numeric Domains. In: IJCAI 2007, pp. 1918–1923 (2007)
11. Holzmann, G.J.: The model checker Spin. IEEE Transactions on Software Engineering 23(5), 279–295 (1997)
12. Kemper, S.: SAT-based verification for timed component connectors. Sci. Comput. Program. 77(7-8), 779–798 (2012)
13. Kontchakov, R., Pulina, L., Sattler, U., Schneider, T., Selmer, P., Wolter, F., Zakharyaschev, M.: Minimal Module Extraction from DL-Lite Ontologies Using QBF Solvers. In: IJCAI 2009, pp. 836–841 (2009)
14. Penczek, W., Wozna, B., Zbrzezny, A.: Bounded Model Checking for the Universal Fragment of CTL. Fundamenta Informaticae 51, 135–156 (2002)
15. Wozna, B.: ATCL* properties and Bounded Model Checking. Fundam. Inform. 63(1), 65–87 (2004)

16. McMillan, K.L.: Symbolic Model Checking. Kluwer Academic Publisher (1993)
17. Peled, D.A.: Software Reliability Methods. Springer (2001)
18. Peterson, G.L.: Myths About the Mutual Exclusion Problem. Information Processing Letters 12(3), 115–116 (1981)
19. Zhang, W.: Bounded Semantics of CTL and SAT-based Verification. In: Breitman, K., Cavalcanti, A. (eds.) ICFEM 2009. LNCS, vol. 5885, pp. 286–305. Springer, Heidelberg (2009)
20. Zhang, W.: Bounded Semantics of CTL. Institute of Software, Chinese Academy of Sciences. Technical Report ISCAS-LCS-10-16 (2010)
21. Zhang, W.: VERDS modeling language, http://lcs.ios.ac.cn/~zwh/verds/

Introducing Quantified Cuts
in Logic with Equality

Stefan Hetzl[1], Alexander Leitsch[2], Giselle Reis[2],
Janos Tapolczai[1], and Daniel Weller[1]

[1] Institut für Diskrete Mathematik und Geometrie, Technische Universität Wien
[2] Institut für Computersprachen, Technische Universität Wien

Abstract. Cut-introduction is a technique for structuring and compressing formal proofs. In this paper we generalize our cut-introduction
method for the introduction of quantified lemmas of the form $\forall x.A$
(for quantifier-free A) to a method generating lemmas of the form
$\forall x_1 \ldots \forall x_n.A$. Moreover, we extend the original method to predicate
logic with equality. The new method was implemented and applied to
the TSTP proof database. It is shown that the extension of the method
to handle equality and quantifier-blocks leads to a substantial improvement of the old algorithm.

1 Introduction

Computer-generated proofs are typically analytic, i.e., they only contain logical material that also appears in the statement of the theorem. This is due to
the fact that analytic proof systems have a considerably smaller search space
which makes proof-search practically feasible. In the case of sequent calculus,
proof-search procedures typically work on the cut-free fragment. But also resolution is essentially analytic as resolution proofs satisfy the subformula property
of first-order logic. One interesting property of non-analytic proofs is their considerably smaller length. The exact difference depends on the logic (or theory)
under consideration, but it is typically enormous. In (classical and intuitionistic)
first-order logic there are proofs with cut of length n whose theorems have only
cut-free proofs of length 2_n (where $2_0 = 1$ and $2_{n+1} = 2^{2_n}$) (see [16] and [12]).
The length of a proof plays an important role in many situations such as human
readability, space requirements and time requirements for proof checking. For
most of these situations general-purpose data compression methods cannot be
used as the compressed representation is not a proof in a standard calculus anymore and hence does not allow fast processing, e.g. linear time proof checking. It
is therefore of high practical interest to develop methods of proof transformation
which produce non-analytic and hence potentially much shorter proofs.

Work on cut-introduction can be found at a number of different places in
the literature. Closest to our work are other approaches which aim to abbreviate or structure a given input proof. In [20] an algorithm for the introduction
of atomic cuts that is capable of exponential proof compression is presented.

S. Demri, D. Kapur, and C. Weidenbach (Eds.): IJCAR 2014, LNAI 8562, pp. 240–254, 2014.

The method [5] for propositional logic is shown to never increase the size of proofs more than polynomially. Another approach to the compression of first-order proofs by introduction of definitions for abbreviating terms is [19]. There is a large body of work on the generation of non-analytic formulas carried out by numerous researchers in various communities. Methods for lemma generation are of crucial importance in inductive theorem proving which frequently requires generalization [1], see e.g. [10] for a method in the context of rippling [2] which is based on failed proof attempts. In automated theory formation [3,4], an eager approach to lemma generation is adopted. This work has, for example, led to automated classification results of isomorphism classes [14] and isotopy classes [15] in finite algebra. See also [11] for an approach to inductive theory formation.

Our method of *algorithmic cut-introduction*, based on the inversion of Gentzen's cut-elimination method, has been defined in [8] and [7]. The method in [8] works on a cut-free **LK**-proof φ of a prenex skolemized end-sequent S and consists of the following steps: (1) extraction of a set of terms T from φ, (2) computation of a compressed representation of T, (3) construction of the cut formula, (4) improvement of the solution by computation of smaller cut-formulas, and (5) construction of an **LK**-proof with the universal cut formula obtained in (4) and instantiation of the quantifiers with the terms obtained in (2). It has been shown in [8] that the method is capable of compressing cut-free proofs quadratically. The paper [7] generalized the method to the introduction of arbitrarily many universal cut formulas, where the steps defined above are roughly the same, though the improvement of the solution (step 4) and the final construction of the proof with cuts (step 5) are much more difficult. The method of introducing arbitrarily many universal cuts in [7] leads even to an exponential compression of proof length. Still the methods described above were mainly designed for a theoretical analysis of the cut-introduction problem rather than for practical applications. In particular, they lacked efficient handling of equality (as they were defined for predicate logic without equality) and the introduction of several universal quantifiers in cut formulas (all cut formulas constructed in [7] are of the form $\forall x.A$ for a single variable x and a quantifier-free formula A).

In this paper we generalize our cut-introduction method to predicate logic with equality and to the construction of a (single) quantified cut containing blocks of universal quantifiers. The efficient compression of the terms (step 2) and the improvement of the solution (step 4) require new and non-trivial techniques. Moreover, we applied the new method in large-scale experiments to proofs generated by prover9 on the TPTP library. This empirical evaluation demonstrates the feasibility of our method on realistic examples.

2   Proofs and Herbrand Sequents

Throughout this paper we consider predicate logic with equality. For practical reasons equality will not be axiomatized but handled via substitution rules. We extend the sequent calculus **LK** to the calculus **LK$_=$** by allowing sequents of the form $\rightarrow t = t$ as initial sequents and adding the following rules:

$$\frac{\Gamma \to \Delta, s = t \quad A[s], \Pi \to \Lambda}{A[t], \Gamma, \Pi \to \Delta, \Lambda} \text{ El1} \qquad \frac{\Gamma \to \Delta, t = s \quad A[s], \Pi \to \Lambda}{A[t], \Gamma, \Pi \to \Delta, \Lambda} \text{ El2}$$

$$\frac{\Gamma \to \Delta, s = t \quad \Pi \to A[s], \Lambda}{\Gamma, \Pi \to A[t], \Delta, \Lambda} \text{ Er1} \qquad \frac{\Gamma \to \Delta, t = s \quad \Pi \to A[s], \Lambda}{\Gamma, \Pi \to A[t], \Delta, \Lambda} \text{ Er2}$$

LK$_=$ is sound and complete for predicate logic with equality.

For convenience we write a substitution $[x_1 \backslash t_1, \ldots, x_n \backslash t_n]$ in the form $[\bar{x} \backslash \bar{t}]$ for $\bar{x} = (x_1, \ldots, x_n)$ and $\bar{t} = (t_1, \ldots, t_n)$. A *strong quantifier* is a $\forall$ ($\exists$) quantifier with positive (negative) polarity. We restrict our investigations to end-sequents in prenex form without strong quantifiers.

Definition 1. *A Σ_1-sequent is a sequent of the form*

$$\forall x_1 \cdots \forall x_{k_1} F_1, \ldots, \forall x_1 \cdots \forall x_{k_p} F_p \to \exists x_1 \cdots \exists x_{k_{p+1}} F_{p+1}, \ldots, \exists x_1 \cdots \exists x_{k_q} F_q.$$

for quantifier free F_i.

Note that the restriction to Σ_1-sequents does not constitute a substantial restriction as one can transform every sequent into a validity-equivalent Σ_1-sequent by skolemisation and prenexing.

Definition 2. *A sequent S is called* E-valid *if it is valid in predicate logic with equality; S is called a* quasi-tautology *[13] if S is quantifier-free and E-valid.*

Definition 3. *The length of a proof φ, denoted by $|\varphi|$, is defined as the number of inferences in φ. The quantifier-complexity of φ, written as $|\varphi|_q$, is the number of weak quantifier-block introductions in φ.*

2.1   Extraction of Terms

Herbrand sequents of a sequent S are sequents consisting of instantiations of S which are quasi-tautologies. The formal definition is:

Definition 4. *Let S be a Σ_1-sequent as in Definition 1 and let H_i be a finite set of k_i-vectors of terms for every $i \in \{1, \ldots, q\}$. We define $\mathcal{F}_i = \{F_i[\bar{x}_i \backslash \bar{t}] \mid \bar{t} \in H_i\}$ if $H_i \neq \emptyset$ and $\mathcal{F}_i = \{F_i\}$ if $H_i = \emptyset$. Let*

$$S^* : \quad \mathcal{F}_1 \cup \cdots \cup \mathcal{F}_p \to \mathcal{F}_{p+1} \cup \ldots \cup \mathcal{F}_q.$$

If S^* is a quasi-tautology then it is called a Herbrand sequent *of S and $H:(H_1, \ldots, H_q)$ is called a* Herbrand structure *of S. We define the* instantiation complexity *of S^* as $|S^*|_i = \sum_{i=1}^{q} |H_i|$.*

Note that, in the instantiation complexity of a Herbrand sequent, we only count the formulas obtained by instantiation.

Example 1. Consider the language containing a constant symbol a, unary function symbols f, s, a binary predicate symbol P, and the sequent S defined below. We write f^n, s^n for n-fold iterations of f and s and omit parentheses around the argument of a unary symbol when convenient. Let

$$S : P(f^4a, a), \forall x. fx = s^2x, \forall xy(P(sx, y) \supset P(x, sy)) \rightarrow P(a, f^4a)$$

and $H = (H_1, H_2, H_3, H_4)$ for

$H_1 = \emptyset, \ H_4 = \emptyset, \ H_2 = \{a, fa, f^2a, f^3a\},$
$H_3 = \{(s^3f^2a, a), (s^2f^2a, sa), (sf^2a, s^2a), (f^2a, s^3a), (s^3a, f^2a), (s^2a, sf^2a), (sa, s^2f^2a), (a, s^3f^2a)\}.$

Then

$$\mathcal{F}_1 = \{P(f^4a, a)\}, \ \mathcal{F}_4 = \{P(a, f^4a)\},$$
$$\mathcal{F}_2 = \{fa = s^2a, \ f^2a = s^2fa, \ f^3a = s^2f^2a, \ f^4a = s^2f^3a\}$$
$$\mathcal{F}_3 = \{P(s^4f^2a, a) \supset P(s^3f^2a, sa), P(s^3f^2a, sa) \supset P(s^2f^2a, s^2a),$$
$$P(s^2f^2a, s^2a) \supset P(sf^2a, s^3a), P(sf^2a, s^3a) \supset P(f^2a, s^4a),$$
$$P(s^4a, f^2a) \supset P(s^3a, sf^2a), P(s^3a, sf^2a) \supset P(s^2a, s^2f^2a),$$
$$P(s^2a, s^2f^2a) \supset P(sa, s^3f^2a), P(sa, s^3f^2a) \supset P(a, s^4f^2a)\}.$$

A Herbrand-sequent S^* corresponding to H is then $\mathcal{F}_1 \cup \mathcal{F}_2 \cup \mathcal{F}_3 \rightarrow \mathcal{F}_4$. Note that $fa = s^2a, \ f^2a = s^2fa, \ f^3a = s^2f^2a, \ f^4a = s^2f^3a \models f^4a = s^8a$.
The instantiation complexity of S^* is 12. S^* is a quasi-tautology but not a tautology.

Theorem 1 (mid-sequent theorem). *Let S be a Σ_1-sequent and π a cut-free proof of S. Then there is a Herbrand-sequent S^* of S s.t. $|S^*|_i \leq |\pi|_q$.*

Proof. This result is proven in [6] (section IV, theorem 2.1) for **LK**, but the proof for **LK$_=$** is basically the same. By permuting the inference rules, one obtains a proof π' from π which has an upper part containing only propositional inferences and the equality rules (which can be shifted upwards until they are applied to atoms only) and a lower part containing only quantifier inferences. The sequent between these parts is called *mid-sequent* and has the desired properties.

S^* can be obtained by tracing the introduction of quantifier-blocks in the proof, which for every formula $Q\bar{x}_i.F_i$ in the sequent (where $Q \in \{\forall, \exists\}$) yields a set of term tuples H_i, and then computing the sets of formulas $\mathcal{F}_i$.

The algorithm for introducing cuts described here relies on computing a compressed representation of the Herbrand structure, which is explained in Section 3. Note, though, that the Herbrand structure $(H_1, \ldots, H_q)$ is a list of sets of term tuples (i.e. each H_i is a set of tuples $\bar{t}$ used to instantiate the formula F_i). In order to facilitate computation and representation, we will add to the language fresh function symbols $f_1, \ldots, f_q$. Each f_i will be applied to the tuples of the set H_i, therefore transforming a list of sets of tuples into a set of terms. In this new set, each term will have an f_k as its head symbol, that indicates to which formula the arguments of f_k belong.

Example 2. Using this new notation, the Herbrand structure H of the previous example is now represented as the set of terms:

$$T : \{f_2(a), f_2(fa), f_2(f^2a), f_2(f^3a), f_3(s^3f^2a, a), f_3(s^2f^2a, sa), \ldots, f_3(a, s^3f^2a)\}.$$

Henceforth we will refer to the transformed Herbrand structure as the *term set* of a proof.

3   Computing a Decomposition

We shall now describe an algorithm for computing a compressed representation of a term set T. Term sets will be represented by decompositions which are defined as follows:

Definition 5. *Let $T = \{t_1, \ldots, t_n\}$ be a set of ground terms. A decomposition D of T is a pair, written as $U \circ_{\bar{\alpha}} W$, where U is a set of terms containing the variables $\alpha_1, \ldots, \alpha_m$, and*

$$W = \left\{ \bar{w}_1 = \begin{pmatrix} w_{1,1} \\ \vdots \\ w_{1,m} \end{pmatrix}, \ldots, \bar{w}_q = \begin{pmatrix} w_{q,1} \\ \vdots \\ w_{q,m} \end{pmatrix} \right\} \text{ is a set}$$

of vectors of ground terms s.t. $T = U \circ_{\bar{\alpha}} W = \{u[\bar{\alpha} \backslash \bar{w}] \mid u \in U, \bar{w} \in W\}$. The size of a decomposition $U \circ_{\bar{\alpha}} W$ is $|U| + |W|$. When it is clear that the variables in question are $\alpha_1, \ldots, \alpha_m$, we just write $U \circ W$.

In [7], we gave an algorithm that treated the special case where $m = 1$. Here, we will extend that approach with a *generalized Δ-vector* Δ_G, which, together with a so-called Δ-table, can compute decompositions with an arbitrary m. Δ_G, given in Algorithm 1, computes a *simple decomposition*, i.e. a decomposition with only one term in U. The Δ-table stores such decompositions and builds more complex ones out of them. Due to space reasons, the algorithm can only be sketched here, for details the interested reader is referred to the technical report [18].

Definition 6. *Let T be a set of ground terms. The Δ-table for T is a set of key/value-entries, where each entry is of the form $W \Rightarrow U'$, where U' is a set $\{(u_1, T_1), \ldots, (u_q, T_q)\}$, W is a set of ground term vectors, u_i is a term containing variables, q is some a priori unknown bound different for each line, and T_i is a subset of T s.t. the following two conditions are satisfied:*

1. *For every entry $W \Rightarrow \{(u_1, T_1), \ldots, (u_q, T_q)\}$, $\{u_i\} \circ W$ is a decomposition of T_i (for $1 \leq i \leq q$).*
2. *For every $T' \subseteq T$, there is a pair $W \Rightarrow U'$ in the Δ-table s.t. $(u, T') \in U'$.*

In practice, condition 2 is relaxed in order to improve performance. Whenever a subset T' of T has only a *trivial* decomposition, i.e. $\Delta_G(T') = (\alpha_i, T')$, its information is not added to the Δ-table. Moreover, no superset of T' is considered from this point on, since we know that these will also have trivial decompositions.

Algorithm 1. Generalized Δ-vector Δ_G

---

function $\Delta_G(t_1, \ldots, t_n$: a list of terms)
   return transposeW($\Delta_G'(t_1, \ldots, t_n)$)
end function
function $\Delta_G'(t_1, \ldots, t_n$: a list of terms)
   if $t_1 = t_2 = \ldots = t_n \wedge n > 0$ **then**   ▷ case 1: all terms identical
      return $(t_1, ())$
   else if $t_i = f(t_1^i, \ldots, t_m^i)$ for $1 \leq i \leq n$ **then**   ▷ case 2: recurse
      $(\bar{w}_1, \ldots, \bar{w}_q) \leftarrow \bigsqcup_{1 \leq j \leq m} \pi_2(\Delta_G(t_j^1, \ldots, t_j^n))$ ▷ $\bigsqcup \equiv$ concatenation
      $u_j \leftarrow \pi_1(\Delta_G(t_j^1, \ldots, t_j^n))$ for all $j \in \{1, \ldots, m\}$
      return merge($f(u_1, \ldots, u_m), (\bar{w}_1, \ldots, \bar{w}_q)$) ▷ merge all α_i, α_j where $\bar{w}_i = \bar{w}_j$
   else   ▷ case 3: introduce new α
      return $(\alpha_{\text{FRESH}}, (t_1, \ldots, t_n))$
   end if
end function

Note that this implies that $\Delta_G(T)$ is almost never computed. An actual decomposition for T is found by iterating over the Δ-table and, for each entry, trying to find a set of pairs $\{(u_{i_1}, T_{i_1}), \ldots, (u_{i_q}, T_{i_q})\} \subseteq U'$ such that $\{u_{i_1}, \ldots, u_{i_n}\} \circ W$ generates T.

Theorem 2 (Soundness). *Let T be a term set. If $U \circ W$ is extracted from iterating over the Δ-table, then $U \circ W$ is a decomposition of T.*

Proof. See [18].

In fact, a stronger result holds: for every decomposition, there exists a unique normal form, and iterating over the Δ-table will only return decompositions in such a normal form. For details, see [18]. To illustrate the algorithm, we compute a decomposition of the term set of Example 2. We remark that, for our cut-introduction method, we are interested in a *decomposition* $(U_1, \ldots, U_q) \circ W$ of a Herbrand structure $H = (H_1, \ldots, H_q)$ which has the property that $H_j = \{u[\bar{a} \backslash \bar{w}] \mid u \in U_j, \bar{w} \in W\}$. This is trivially obtained from a decomposition $U \circ W$ of the term set of a Herbrand structure by setting $U_j = \{u \mid f_j(u) \in U\}$.

Example 3. Let $T = T_2 \cup T_3$ with

$$T_2 = \{t_1 : f_2(a), \ t_2 : f_2(fa), \ t_3 : f_2(f^2a), \ t_4 : f_2(f^3a)\}$$
$$T_3 = \{t_5 : f_3(s^3f^2a, a), \ t_6 : f_3(s^2f^2a, sa), \ \ldots, \ t_{12} : f_3(a, s^3f^2a)\}$$

be a term set corresponding to the Herbrand structure $H = (H_1, H_2, H_3, H_4)$:

$$H_1 = \emptyset, \ H_4 = \emptyset, H_2 = \{a, fa, f^2a, f^3a\},$$
$$H_3 = \{(s^3f^2a, a), (s^2f^2a, sa), \ldots, (a, s^3f^2a)\}$$

We now compute Δ_G for every subset of T — consider for instance the subset $T' = \{f_3(s^3f^2a, a), f_3(s^2f^2a, sa)\} \subseteq T$:

$$\Delta_G(f_3(s^3f^2a, a), f_3(s^2f^2a, sa)) = (f_3(s^2\alpha_1, \alpha_2), \left\{ \begin{pmatrix} sf^2a \\ a \end{pmatrix}, \begin{pmatrix} f^2a \\ sa \end{pmatrix} \right\}) = (u, W).$$

If the Δ-table already has an entry $W \Rightarrow U'$, we add (u, T') to U'. If not, we insert a new entry $W \Rightarrow \{(u, T')\}$. After Δ_G has been computed for all subsets, we iterate through it, looking for simple decompositions that can be composed into a decomposition of T. We find the entry

$$
\begin{aligned}
W &\Rightarrow U_1' \cup U_2' \cup U_3' \cup U_4' \\
U_1' &= \{\} \\
U_2' &= \{(f_2(\alpha_1), \{t_1, t_3\}), \ (f_2(f\alpha_1), \{t_2, t_4\}), \ (f_2(\alpha_2), \{t_1, t_3\}) \ (f_2(f\alpha_2), \{t_2, t_4\})\} \\
U_3' &= \{(f_3(s^3\alpha_1, \alpha_2), \{t_5, t_9\}), \ (f_3(s^2\alpha_1, s\alpha_2), \{t_6, t_{10}\}), \\
&\qquad (f_3(s\alpha_1, s^2\alpha_2), \{t_7, t_{11}\}), \ (f_3(\alpha_1, s^3\alpha_2), \{t_8, t_{12}\})\} \\
U_4' &= \{\} \\
W &= \left\{ \begin{pmatrix} f^2 a \\ a \end{pmatrix}, \begin{pmatrix} a \\ f^2 a \end{pmatrix} \right\}
\end{aligned}
$$

and can see that $U_i' \circ W = T_i$ for $1 \leq i \leq 4$. Therefore, $(U_1' \cup U_2' \cup U_3' \cup U_4') \circ W$ is a decomposition of T. We then translate this decomposition T back into a decomposition of H by removing the function symbols f_2 and f_3 from U_2' & U_3' (the empty sets U_1' and U_4' can be disregarded):

$$
\begin{aligned}
U &= (U_2, U_3), \\
U_2 &= \{\alpha_1, \ f\alpha_1, \ \alpha_2, \ f\alpha_2\}, \\
U_3 &= \{(s^3\alpha_1, \alpha_2), \ (s^2\alpha_1, s\alpha_2), \ (s\alpha_1, s^2\alpha_2), \ (\alpha_1, s^3\alpha_2)\}, \\
W &= \left\{ \begin{pmatrix} f^2 a \\ a \end{pmatrix}, \begin{pmatrix} a \\ f^2 a \end{pmatrix} \right\}.
\end{aligned}
$$

4   Computing a Cut-Formula

After having computed a decomposition as described in Section 3, the next step consists in computing a cut-formula based on that decomposition. A decomposition D specifies the instances of quantifier blocks in a proof with a $\forall$-cut, but does not contain information about the propositional structure of the cut formula to be constructed. The problem to find the appropriate propositional structure is reflected in the following definition.

Definition 7. *Let S be a Σ_1-sequent and F_i, k_i as in Definition 1, H be a Herbrand structure for S, and $D: U \circ W$ a decomposition of H with $V(D) = \{\alpha_1, \ldots, \alpha_n\}$. Let $U = (U_1, \ldots, U_q)$ and $W = \{\bar{w}_1, \ldots, \bar{w}_k\}$, where the $\bar{w}_j$ are n-vectors of terms not containing variables in $V(D)$, and $\mathcal{F}_i' = \{F_i[\bar{x}_i \backslash \bar{t}] \mid \bar{t} \in U_i\}$ for $U_i \neq \emptyset$ and $\mathcal{F}_i' = \{F_i\}$ for $U_i = \emptyset$. Furthermore let X be an n-place predicate variable. Then the sequent*

$$
S^{\sim}: \quad X\bar{\alpha} \supset \bigwedge_{i=1}^{k} X\bar{w}_i, \mathcal{F}_1', \ldots, \mathcal{F}_p' \to \mathcal{F}_{p+1}', \ldots, \mathcal{F}_q'.
$$

is called a schematic extended Herbrand sequent of S w.r.t. D. The instantiation complexity of $S^{\sim}$, denoted by $|S^{\sim}|_i$, is defined as $k + \sum_{i=1}^{q} |U_i|$.

Definition 8. *Let $S^\sim$ be a schematic extended Herbrand sequent of S w.r.t. a decomposition D as in Definition 7 and A be a formula with $V(A) \subseteq \{\alpha_1, \ldots, \alpha_n\}$. Then the second-order substitution $\sigma \colon [X \backslash \lambda \bar{\alpha}.A]$ is a solution of $S^\sim$ if $S^\sim \sigma$ is a quasi-tautology; in this case $S^\sim \sigma$ is called an* extended Herbrand sequent. *The instantiation complexity of $S^\sim \sigma$ is defined as $|S^\sim|_i$.*

Theorem 5 in Section 4.2 shows that, from a solution of a schematic extended Herbrand sequent $S^\sim$ of S, we can define a proof ψ of S with a $\forall$-cut and $|\psi|_q = |S^\sim|_i$. The question remains whether every schematic extended Herbrand sequent is solvable. We show below that this is indeed the case.

Let $S^\sim$ as in Definition 7. We define

$$F[l] = \bigwedge \bigcup_{i=1}^{p} \mathcal{F}_i' \text{ and } F[r] = \bigvee \bigcup_{i=p+1}^{q} \mathcal{F}_i'.$$

Definition 9. *Let $S^\sim$ be a schematic extended Herbrand sequent of S as in Definition 7. We define the* canonical formula *$C(S^\sim)$ of $S^\sim$ as $F[l] \wedge \neg F[r]$. The substitution $[X \backslash \lambda \bar{\alpha}.C(S^\sim)]$ is called the* canonical substitution *of $(S, S^\sim)$.*

Theorem 3. *Let S be a Σ_1-sequent, and $S^\sim$ be a schematic extended Herbrand sequent of S. Then the canonical substitution is a solution of $S^\sim$.*

Proof. Let $S^\sim$ be a schematic extended Herbrand sequent as in Definition 7 and $C(S^\sim)$ be the canonical formula of $S^\sim$. We have to prove that

$$S_1 \colon \quad C(S^\sim)(\bar{\alpha}) \supset \bigwedge_{i=1}^{k} C(S^\sim)(\bar{w}_i), \; F[l] \to F[r]$$

is a quasi-tautology. But, by definition of $C(S^\sim)$, S_1 is equivalent to

$$S_2 \colon \quad (F[l] \wedge \neg F[r]) \supset \bigwedge_{i=1}^{k} (F[l] \wedge \neg F[r])(\bar{w}_i), (F[l] \wedge \neg F[r]) \to .$$

Clearly S_2 is a quasi-tautology if the sequent S_3, defined as

$$S_3 \colon \quad \bigwedge_{i=1}^{k} (F[l] \wedge \neg F[r])(\bar{w}_i) \to$$

is a quasi-tautology. But, by $D = U \circ W$ being a decomposition of H, S_3 is logically equivalent to the Herbrand sequent S^* defined over H, which (by definition) is a quasi-tautology.

Example 4. Let

$$S \colon \quad P(f^4 a, a), \forall x. fx = s^2 x, \forall xy (P(sx, y) \supset P(x, sy)) \to P(a, f^4 a)$$

like in Example 1 and D be the decomposition $U \circ W$ of H constructed in Example 3. We have

$$U = (U_2, U_3),$$
$$U_2 = \{\alpha_1, f\alpha_1, \alpha_2, f\alpha_2\},$$
$$U_3 = \{(s^3 \alpha_1, \alpha_2), (s^2 \alpha_1, s\alpha_2), (s\alpha_1, s^2 \alpha_2), (\alpha_1, s^3 \alpha_2)\},$$
$$W = \left\{ \begin{pmatrix} f^2 a \\ a \end{pmatrix}, \begin{pmatrix} a \\ f^2 a \end{pmatrix} \right\}.$$

The corresponding schematic extended Herbrand sequent $S^\sim$ is

$X(\alpha_1, \alpha_2) \supset (X(f^2a, a) \wedge X(a, f^2a))$,
$f\alpha_1 = s^2\alpha_1$, $f^2\alpha_1 = s^2f\alpha_1$, $f\alpha_2 = s^2\alpha_2$, $f^2\alpha_2 = s^2f\alpha_2$,
$P(s^4\alpha_1, \alpha_2) \supset P(s^3\alpha_1, s\alpha_2)$, $P(s^3\alpha_1, s\alpha_2) \supset P(s^2\alpha_1, s^2\alpha_2)$,
$P(s^2\alpha_1, s^2\alpha_2) \supset P(s\alpha_1, s^3\alpha_2), P(s\alpha_1, s^3\alpha_2) \supset P(\alpha_1, s^4\alpha_2), P(f^4a, a) \to P(a, f^4a)$.

Its canonical formula $C(S^\sim)$ which we write as $A(\alpha_1, \alpha_2)$ is

$$\bigwedge_{j=1}^{2}(f\alpha_i = s^2\alpha_i \wedge f^2\alpha_i = s^2f\alpha_i)\wedge$$
$$\bigwedge_{i=0}^{3}(P(s^{4-i}\alpha_1, s^i\alpha_2) \supset P(s^{4-i-1}\alpha_1, s^{i+1}\alpha_2)) \wedge P(f^4a, a) \wedge \neg P(a, f^4a).$$

The canonical solution is $[X\backslash\lambda\alpha_1\alpha_2.A(\alpha_1, \alpha_2)]$ and the corresponding extended Herbrand sequent S' is like $S^\sim$ with $X(\alpha_1, \alpha_2) \supset (X(f^2a, a) \wedge X(a, f^2a))$ replaced by $A(\alpha_1, \alpha_2) \supset (A(f^2a, a) \wedge A(a, f^2a))$. Note that $|S'|_i = 10$, while $|S^*|_i = 12$. So we obtained a compression of quantifier complexity.

4.1   Improving the Solution

In the last section, we have shown that, given a decomposition D of the termset of a cut-free proof of a Σ_1-sequent S, there exists a canonical solution to the schematic extended Herbrand sequent induced by S, D, which gives rise to a proof with a $\forall$-cut. Furthermore, as we will show in Theorem 5, all solutions are equivalent from the point of view of the $|\cdot|_q$ measure. On the other hand, the quality of solutions can be distinguished by other properties, for example by the length of the proof with cut they induce. Since the best known general upper bound on the length of proofs in propositional logic (which corresponds to our setting once a decomposition is fixed) depends on the size of the theorem to be proven, our approach is to search for solutions that have *smaller size than the canonical solution.*

We will consider E-validity of quantifier-free formulas F containing free variables; by "F is E-valid" we mean to say "the universal closure of F is E-valid". Throughout this section, we consider a fixed Σ_1-sequent S using the notation of Definition 1, a fixed decomposition $D = (U_1, \ldots, U_q) \circ W$, with $W = \{\bar{w}_i \mid 1 \leq i \leq k\}$, of a Herbrand structure H of S, along with the schematic extended Herbrand sequent $S^\sim$ induced by S, H, D, using the notation of Definition 7. We will abbreviate $\mathcal{F}'_1 \cup \cdots \cup \mathcal{F}'_p$ by Γ and $\mathcal{F}'_{p+1} \cup \cdots \cup \mathcal{F}'_q$ by Δ, and write "A is a solution" for "$[X\backslash\lambda\bar{x}.A]$ is a solution for $S^\sim$" (note that we will consider the names $\bar{x}$ fixed). In this section, we will focus our attention on solutions in conjunctive normal form (CNF), which always exist since the solution property is semantic (if A is a solution and $A \Leftrightarrow B$ is E-valid, then B is a solution). A clause C is said to be $\bar{x}$-free if it contains no symbol from $\bar{x}$.

The algorithm we will present will involve generating E-consequences of formulas. Although in principle an abstract analysis of our algorithm based on a notion of *E-consequence generator* can be performed, we have chosen, for lack of space, to present only the concrete E-consequence generator used in our implementation.

Our E-consequence generator is based on *forgetful reasoning*. Let C_1, C_2 be two clauses, then denote the set of propositional resolvents of C_1, C_2 by $\text{res}(C_1, C_2)$ and the set of clauses that can be obtained from C_1, C_2 by ground paramodulation by $\text{para}(C_1, C_2)$. Letting F be a formula with CNF $\{C_i\}_{i \in I}$ we define

$$\mathcal{F}(F) = \{C \wedge \bigwedge_{i \in I \setminus \{j,k\}} C_i \mid C \in \text{res}(C_j, C_k) \cup \text{para}(C_j, C_k)\}.$$

Using $\mathcal{F}$, we can now present Algorithm 2: the solution-finding algorithm $\text{SF}_{\mathcal{F}}$. It prunes a solution A of $\bar{x}$-free clauses, then recurses upon those consequences of the pruned A generated by $\mathcal{F}$ which pass a certain E-validity check, finally returning a set of formulas (which will all be solutions).

Algorithm 2. $\text{SF}_{\mathcal{F}}$

---

   function $\text{SF}_{\mathcal{F}}(A$: solution in CNF$)$
      $A \leftarrow A$ without $\bar{x}$-free clauses
      $S \leftarrow \{A\}$
      for $B \in \mathcal{F}(A)$ **do**
         if $B[\bar{x} \backslash \bar{w}_1], \ldots, B[\bar{x} \backslash \bar{w}_k], \Gamma \rightarrow \Delta$ is E-valid **then**   ▷ B is a solution
            $S \leftarrow S \cup \text{SF}_{\mathcal{F}}(B)$
         end if
      end for
      return S
   end function

Theorem 4 (Soundness & Termination). *Let A be any solution in CNF. Then $\text{SF}_{\mathcal{F}}$ terminates on A and, for all $B \in \text{SF}_{\mathcal{F}}(A)$, B is a solution.*

Proof. Termination is trivial since CNFs get smaller under $\mathcal{F}$. If $C(S^{\sim}) \supset B$ is E-valid then B is a solution iff $B[\bar{x} \backslash \bar{w}_1], \ldots, B[\bar{x} \backslash \bar{w}_k], \Gamma \rightarrow \Delta$ is E-valid. Hence the E-validity check in $\text{SF}_{\mathcal{F}}$ suffices. Let A be a solution in CNF and A' be A without $\bar{x}$-free clauses. The fact that A' is a solution follows from the fact that $A \supset C[\bar{x} \backslash \bar{w}_k]$ is E-valid for all $\bar{x}$-free clauses C of A, together with the assumption that A is a solution.

The algorithm $\text{SF}_{\mathcal{F}}$ can be used to lift a compression in quantifier-complexity to a compression w.r.t. proof length. Indeed, in the setting of first-order logic without equality, $\text{SF}_{\mathcal{F}}$ has been applied in [7] to obtain an exponential speed-up result w.r.t. proof length.

Example 5. Consider the canonical formula $C(S^{\sim})$ of Example 4. Then $\text{SF}_{\mathcal{F}}$ generates the CNF

$$F(\alpha_1, \alpha_2): \ f^2 \alpha_1 = s^4 \alpha_1 \wedge f^2 \alpha_2 = s^4 \alpha_2 \wedge (\neg P(s^4 \alpha_1, \alpha_2) \vee P(\alpha_1, s^4 \alpha_2))$$

for the CNF of $C(S^{\sim})$ by applying paramodulation twice to equational atoms and resolution thrice to the clauses corresponding to the implications between the P-atoms. It can be checked that $\lambda \alpha_1 \alpha_2. F(\alpha_1, \alpha_2)$ is a solution for $S^{\sim}$ which is smaller than the canonical solution.

4.2   Proof with Cut

Theorem 5. *Let $S^\sim$ be an extended Herbrand sequent of a Σ_1-sequent S. Then S has a proof φ with a $\forall$-cut s.t. $|\varphi|_q = |S^\sim|_i$.*

Proof. As in [7] (page nr. 12, Theorem 7). Note that the quantifier-blocks in the cut and the equality rules do not change the measured number of weak quantifier-block introductions analyzed in the paper above. The main steps in the proof are the following ones: let S' be an extended Herbrand sequent obtained by the solution $[X\backslash \lambda \bar{\alpha}.A]$. Then a proof with cut formula $\forall \bar{x}.A[\bar{\alpha}\backslash \bar{x}]$ can be constructed where the quantifier substitution blocks for the cut formula on the right-hand-side are $[\bar{x}\backslash \bar{w}]$ for $\bar{w} \in W$ while the cut formula on the left-hand-side gets the substitution $[\bar{x}\backslash \bar{\alpha}]$. The substitutions $[\bar{x}_i\backslash \bar{t}]$ for $\bar{t} \in U_i$ are inserted to introduce the quantifiers of the formula F_i in the end-sequent.

Example 6. Let $\Gamma = P(f^4a, a), \forall x.fx = s^2x, \forall xy(P(sx, y) \supset P(x, sy))$ be the left-hand-side of S. Then, to the canonical solution corresponds an **LK**-proof ψ of the form

$$
\cfrac{
  \cfrac{(\psi_1)}{\cfrac{\Gamma \to P(a, f^4a), A(\alpha_1, \alpha_2)}{\Gamma \to P(a, f^4a), \forall xy.A(x, y)}\ \forall_r^*}
  \qquad
  \cfrac{(\psi_2)}{\cfrac{\Gamma, A(f^2a, a), A(a, f^2a) \to P(a, f^4a)}{\Gamma, \forall xy.A(x, y) \to P(a, f^4a)}\ \forall_l^*}
}{\Gamma \to P(a, f^4a)}\ \text{cut} + \text{c}^*
$$

where ψ_1 and ψ_2 are cut-free and ψ_2 contains only structural and propositional inferences (in ψ_2 only $P(f^4a, a)$ is needed from Γ). The quantifier inferences in ψ_1 use exactly the 8 substitutions encoded in U_1 and U_2, ψ_2 uses the 2 substitutions represented by W. So we have $|\psi|_q = 10$.

5   Implementation and Experiments

Summing up the previous sections, the structure of our cut-introduction algorithm is the following:

Algorithm 3. Cut-Introduction

Require: π: cut-free proof
$\quad T \leftarrow \text{extractTermSet}(\pi)$
$\quad D \leftarrow \text{getMinimalDecomposition}(T)$
$\quad C(\bar{x}) \leftarrow \text{getCanonicalSolution}(D)$
$\quad F(\bar{x}) \leftarrow \text{improveSolution}(C(\bar{x}))$
$\quad \textbf{return } \text{constructProof}(F(\bar{x}))$

Depending on whether the input proof π contains equality reasoning or not we either work modulo quasi-tautologies as described in this paper or modulo tautologies (as described in [8,7]) in improveSolution and constructProof.

In getMinimalDecomposition we can either compute decompositions with a single variable as in [8,7] or with an unbounded number of variables as described in Section 3. We denote these two variants with CI^1 and CI^* respectively.

These algorithms have been implemented in the gapt-system[1] which is a framework for transforming and analyzing formal proofs. It is implemented in Scala and contains data structures such as formulas, sequents, resolution and sequent calculus proofs and algorithms like unification, skolemization, cut-elimination as well as backends for several external solvers and provers. For deciding whether a quantifier-free formula is a tautology we use MiniSat[2]. We use veriT[3] for deciding whether a quantifier-free formula is a quasi-tautology and prover9[4] for the actual proof construction based on the import described in [9].

We have conducted experiments on the prover9-part of the TSTP-library (Thousands of Solutions of Theorem Provers, see [17]). The choice of prover9 was motivated by the simple and clean proof output format Ivy which makes proof import (comparatively) easy. This library contains 6341 resolution proofs. Of those, 5254 can be parsed and transformed into a sequent calculus proof using the transformation described in [9]. Of those, 2849 have non-trivial termsets (we call a term set trivial if every quantified formula in the end-sequent is instantiated at most once).

The input data we have used for our experiments is this collection of proofs with non-trivial term sets. In this collection 66% use equality reasoning and hence must be treated with the method introduced in this paper. The average term set size is 37,1 but 46% have a term set of size ≤ 10. The experiments have been conducted with version 1.6 of gapt on an Intel i5 QuadCore with 3,33GHz with an allocation of 2GB heap space and a timeout of 60 seconds for the cut-introduction algorithm.

On 19% of the input proofs our algorithm terminates with finding a compression, i.e. a non-trivial decomposition (of size at most that of the original termset) and a proof with cut that realizes this decomposition. On 49% it terminates determining that the proof is uncompressible, more precisely: that there is no proof with a single $\forall$-cut which (by cut-elimination) reduces to the given input term set and is of smaller quantifier complexity, see [7]. Figure 1 depicts the return status (in percent) depending on the size of the term set. When reading this figure one should keep in mind the relatively high number of small proofs (see above). One can observe that proofs with term sets up to a size of around 50 can be treated well by our current implementation, beyond that the percentage of timeouts is very large. Small proofs – unsurprisingly – tend to be uncompressible.

In Figure 2 we restrict our attention to runs terminating with a compression. As one can see from the diagram on the left, a significant reduction of quantifier-complexity can be achieved by our method. The diagram on the right

---

[1] Generic Architecture for Proof Transformations, http://www.logic.at/gapt/
[2] http://minisat.se/
[3] http://www.verit-solver.org/
[4] http://www.cs.unm.edu/~mccune/prover9/

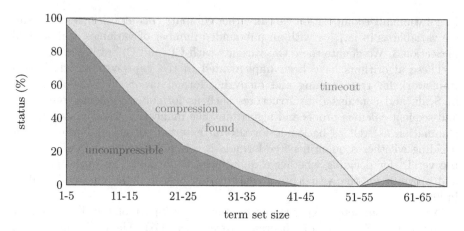

Fig. 1. CI*: return status by term set size

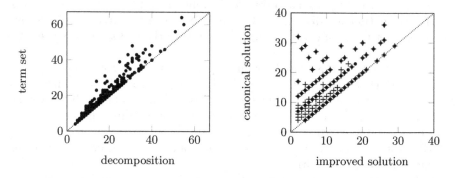

Fig. 2. Size Comparison

demonstrates that forgetful reasoning is highly useful for improving the canonical solution. The points plotted as ● are the result after using forgetful resolution only, the points plotted as + are the result after forgetful resolution and paramodulation.

Our experiments also show that the generalization to the introduction of a block of quantifiers introduced in this paper has a strong effect: of the 548 proofs on which CI* finds a compression, 22% are found to be uncompressible by CI^1.

6   Conclusion

We have introduced a cut-introduction method that works modulo equality and is capable of generating cut-formulas containing a block of quantifiers. We have implemented our new method and have conducted a large-scale empirical evaluation which demonstrates its feasibility on realistic examples. Lessons learned from these experiments include that blocks of quantifiers allow for significantly more proofs to be compressed and that forgetful reasoning methods, while rough in theory, are highly useful for our application in practice.

As future work we plan to extend our method to work modulo (suitably specified) equational theories. We also plan to evaluate our method on proofs produced by Tableaux-provers and SMT-solvers. Another important, and non-trivial, extension will be to cope with cuts that contain quantifier-alternations.

Acknowledgements. The authors would like to thank Pascal Fontaine for help with the veriT-solver and Geoff Sutcliffe for providing the prover9-TSTP test set.

References

1. Bundy, A.: The Automation of Proof by Mathematical Induction. In: Voronkov, A., Robinson, J.A. (eds.) Handbook of Automated Reasoning, vol. 1, pp. 845–911. Elsevier (2001)
2. Bundy, A., Basin, D., Hutter, D., Ireland, A.: Rippling: Meta-Level Guidance for Mathematical Reasoning. Cambridge Tracts in Theoretical Computer Science. Cambridge University Press (2005)
3. Colton, S.: Automated Theory Formation in Pure Mathematics. Ph.D. thesis, University of Edinburgh (2001)
4. Colton, S.: Automated Theory Formation in Pure Mathematics. Springer (2002)
5. Finger, M., Gabbay, D.: Equal Rights for the Cut: Computable Non-analytic Cuts in Cut-based Proofs. Logic Journal of the IGPL 15(5-6), 553–575 (2007)
6. Gentzen, G.: Untersuchungen über das logische Schließen. Mathematische Zeitschrift 39, 176–210, 405–431 (1934-1935)
7. Hetzl, S., Leitsch, A., Reis, G., Weller, D.: Algorithmic Introduction of Quantified Cuts (2013), http://arxiv.org/abs/1401.4330 (submitted)
8. Hetzl, S., Leitsch, A., Weller, D.: Towards Algorithmic Cut-Introduction. In: Bjørner, N., Voronkov, A. (eds.) LPAR-18 2012. LNCS, vol. 7180, pp. 228–242. Springer, Heidelberg (2012)
9. Hetzl, S., Libal, T., Riener, M., Rukhaia, M.: Understanding Resolution Proofs through Herbrand's Theorem. In: Galmiche, D., Larchey-Wendling, D. (eds.) TABLEAUX 2013. LNCS, vol. 8123, pp. 157–171. Springer, Heidelberg (2013)
10. Ireland, A., Bundy, A.: Productive Use of Failure in Inductive Proof. Journal of Automated Reasoning 16(1-2), 79–111 (1996)
11. Johansson, M., Dixon, L., Bundy, A.: Conjecture synthesis for inductive theories. Journal of Automated Reasoning 47(3), 251–289 (2011)
12. Orevkov, V.: Lower bounds for increasing complexity of derivations after cut elimination. Zapiski Nauchnykh Seminarov Leningradskogo Otdeleniya Matematicheskogo Instituta 88, 137–161 (1979)
13. Shoenfield, J.R.: Mathematical Logic, 2nd edn. Addison Wesley (1973)
14. Sorge, V., Colton, S., McCasland, R., Meier, A.: Classification results in quasigroup and loop theory via a combination of automated reasoning tools. Commentationes Mathematicae Universitatis Carolinae 49(2), 319–339 (2008)
15. Sorge, V., Meier, A., McCasland, R., Colton, S.: Automatic Construction and Verification of Isotopy Invariants. Journal of Automated Reasoning 40(2-3), 221–243 (2008)

16. Statman, R.: Lower bounds on Herbrand's theorem. Proceedings of the American Mathematical Society 75, 104–107 (1979)
17. Sutcliffe, G.: The TPTP World - Infrastructure for Automated Reasoning. In: Clarke, E.M., Voronkov, A. (eds.) LPAR-16 2010. LNCS (LNAI), vol. 6355, pp. 1–12. Springer, Heidelberg (2010)
18. Tapolczai, J.: Cut-Introduction with Multiple Universal Quantifiers. Technical report, http://www.logic.at/staff/hetzl/deltavector.pdf
19. Vyskočil, J., Stanovský, D., Urban, J.: Automated Proof Compression by Invention of New Definitions. In: Clarke, E.M., Voronkov, A. (eds.) LPAR-16 2010. LNCS (LNAI), vol. 6355, pp. 447–462. Springer, Heidelberg (2010)
20. Woltzenlogel Paleo, B.: Atomic Cut Introduction by Resolution: Proof Structuring and Compression. In: Clarke, E.M., Voronkov, A. (eds.) LPAR-16 2010. LNCS (LNAI), vol. 6355, pp. 463–480. Springer, Heidelberg (2010)

Quati: An Automated Tool for Proving Permutation Lemmas

Vivek Nigam[1], Giselle Reis[2], and Leonardo Lima[1]

[1] Universidade Federal da Paraíba, Brazil
[2] Technische Universität Wien, Austria

Abstract. The proof of many foundational results in structural proof theory, such as the admissibility of the cut rule and the completeness of the focusing discipline, rely on permutation lemmas. It is often a tedious and error prone task to prove such lemmas as they involve many cases. This paper describes the tool Quati which is an automated tool capable of proving a wide range of inference rule permutations for a great number of proof systems. Given a proof system specification in the form of a theory in linear logic with subexponentials, Quati outputs in LaTeX the permutation transformations for which it was able to prove correctness and also the possible derivations for which it was not able to do so. As illustrated in this paper, Quati's output is very similar to proof derivation figures one would normally find in a proof theory book.

1 Introduction

Permutation lemmas play an important role in proof theory. Many foundational results about proof systems rely on the fact that some rules permute over others. For instance, permutation lemmas are used in Gentzen-style cut-elimination proofs [4], the completeness proof of focusing disciplines [1,7], and the proof of Herbrand's theorem [5].

Proving permutation lemmas, however, is often a tedious and error-prone task as there are normally many cases to consider. As an example, consider the case of permuting $\vee_l$ over $\rightarrow_l$ in the intuitionistic calculus LJ. In order to show whether these two rules permute, one needs to check *every possible case* in which $\rightarrow_l$ occurs above $\vee_l$ in a derivation. When using a multiplicative calculus, there are four possibilities for such derivation, two allow a permutation of the rules while the other two do not. Here's one of each:

$$
\cfrac{\varphi_1 \qquad \cfrac{\cfrac{\varphi_2}{\Gamma' \vdash A} \quad \cfrac{\varphi_3}{\Gamma'', Q, B \vdash F}}{\Gamma', \Gamma'', A \rightarrow B, Q \vdash F} \rightarrow_l}{\Gamma, \Gamma', \Gamma'', A \rightarrow B, P \vee Q \vdash F} \vee_l \qquad \rightsquigarrow \qquad \cfrac{\cfrac{\varphi_2}{\Gamma' \vdash A} \quad \cfrac{\varphi_1 \qquad \cfrac{\varphi_3}{\Gamma, P \vdash F \quad \Gamma'', B, Q \vdash F}}{\Gamma, \Gamma'', P \vee Q, B \vdash F} \vee_l}{\Gamma, \Gamma', \Gamma'', P \vee Q, A \rightarrow B \vdash F} \rightarrow_l
$$

$$
\cfrac{\varphi_1 \qquad \cfrac{\cfrac{\varphi_2}{\Gamma', Q \vdash A} \quad \cfrac{\varphi_3}{\Gamma'', B \vdash F}}{\Gamma', \Gamma'', A \rightarrow B, Q \vdash F} \rightarrow_l}{\Gamma, \Gamma', \Gamma'', A \rightarrow B, P \vee Q \vdash F} \vee_l \qquad \rightsquigarrow \quad ?
$$

The combinatorial nature of proving permutation lemmas can be observed in this example. While there are "only" four cases to consider for this pair of rules, for proving

S. Demri, D. Kapur, and C. Weidenbach (Eds.): IJCAR 2014, LNAI 8562, pp. 255–261, 2014.

the completeness of the focusing discipline, one needs to study which permutations are allowed and therefore all pairs of rules need to be considered [7]. Moreover, the fact that the cases are rarely documented makes it hard for others to check the correctness of the transformations. For instance, the cut-elimination result for bi-intuitionistic logic given by Rauszer [14] was later found to be incorrect [2] exactly because one of the permutation lemmas was not true. Therefore, an automated tool to check for these lemmas would be of great help. This paper introduces such a tool called Quati.[1]

While here we will restrict ourselves to simply illustrate Quati's functionalities and implementation design, we observe that its underlying theory is described in the papers [8,13,9]. We briefly review this body of work.

In [13], we show how to reduce the problem of proving permutation lemmas to solving an answer-set program [3]. That is, given a proof system $\mathcal{P}$ satisfying some properties, we reduce the problem of checking whether a rule r_1 in $\mathcal{P}$ always permutes over r_2 in $\mathcal{P}$ to solving an answer-set program. Each solution of this program corresponds to one possible permutation case. This result sets the foundations for Quati.

However, the exact language in which proof systems are specified was not dealt in [13]. It was subject of the paper [8] which shows that a great number of proof systems for different logics (*e.g.*, linear, intuitionistic, classical, modal logics) can be specified as theories in linear logic with subexponentials (SELL) [11]. These specifications are shown to have a strong adequacy, namely, *on the level of derivations* [12], meaning that there is a one to one correspondence of derivations in the specified logic (object logic) to derivations in linear logic with subexponentials. Moreover, [8] also shows how to check whether proof systems specified in SELL admit cut-elimination. This lead to the tool TATU[2]. Therefore, SELL is a suitable framework for specifying proof systems.

Finally, in the workshop paper [9], we show how to integrate the material in [13] and [8]. Given a proof system specified in SELL, we reduce the problem of checking whether a rule permutes over another to an answer-set program. In the same paper, we also discuss how to extract proof derivation figures similar to those shown in a standard proof theory book [15] from the solutions of the generated answer-set programs.

Quati is the result of this series of papers. This paper is organized as follows: Section 2 describes Quati's syntax and its features, while Section 3 describes its implementation. In Section 4 we end by pointing out future work.

2 Quati at Work

Throughout this section, we will use the specification for the intuitionistic logic's multi-conclusion calculus MLJ [6] as our running example. First we specify Quati's syntax and then its features.

2.1 Syntax

Quati's underlying logic, linear logic with subexponentials (SELL) [11], is a powerful framework for the specification of proof systems. Subexponentials, written $!^\ell$, $?^\ell$, arise

---

[1] Quati is a mammal from the raccoon family native to South America. Its name comes from the Tupi-guarani, a language spoken by native indians in Brazil, and means "long nose".

[2] https://www.logic.at/staff/giselle/tatu/

$Side ::= \text{lft} \mid \text{rght} \quad CtxType ::= \text{many} \mid \text{single} \quad SubType ::= \text{unb} \mid \text{lin}$

$SubSig ::= SubDecl \; SubSpec \; SubRel$

$SubDecl ::= \text{subexp}\langle String\rangle\langle SubType\rangle.$

$SubSpec ::= \text{subexpctx}\langle String\rangle\langle CtxType\rangle\langle Side\rangle.$

$SubRel ::= \text{subexprel}\langle String\rangle\text{<}\langle String\rangle.$

$Bipoles ::= (\text{not}\langle Atoms\rangle)\text{*}\langle BodyPos\rangle.$

$BodyPos ::= \text{one} \mid BodyNeg \mid [\langle String\rangle]\text{bang}BodyNeg \mid$
$\qquad BodyPos\text{*}BodyPos \mid BodyPos\text{+}BodyPos$

$BodyNeg ::= \text{top} \mid \text{bot} \mid \langle MarkAtoms\rangle \mid \langle BodyNeg\rangle|\langle BodyNeg\rangle \mid$
$\qquad \langle BodyNeg\rangle\text{\&}\langle BodyNeg\rangle$

$Atoms ::= \langle Side\rangle\langle Form\rangle \qquad MarkAtoms ::= [\langle String\rangle]?\langle Atoms\rangle$

Fig. 1. Here $Form$ is a term of type form

$\text{*}:\otimes \quad \text{+}:\oplus \quad \text{\&}:\text{\&} \quad |:\bindnasrepma \quad [i]\text{bang}:!^i \quad \text{one}:1 \quad \text{top}:\top \quad \text{bot}:\bot \quad [i]?:?^i$

Fig. 2. Syntax for the linear logic connectives

from the observation that the linear logic exponentials are not canonical (see [8] for an extensive discussion). It is known that these operators greatly increase the expressiveness of the system when compared to linear logic. For instance, subexponentials can be used to represent contexts of proof systems [8], to mark the epistemic state of agents [10], or to specify locations in sequential computations [11]. The main feature of subexponentials is that they are organized in a pre-order, $\preceq$, which specifies the provability relation among them. In [8], we have shown that a great number of proof systems for linear, classic, intuitionistic and modal logics can be specified in SELL with a strong level of adequacy. Another important reason for using SELL as specification language is that one can also use other available tools, such as the tool TATU which is capable of checking whether a proof system specified in SELL admits cut-elimination.

A Quati program is a SELL theory with some more annotations. Its syntax is given in Figure 1 and explained in detail by using our running example MLJ. A Quati program consists of two files: (1) a *type signature* file, with suffix .sig and (2) a specification file with suffix .pl consisting of two parts: (a) a *subexponential signature* and (b) the rules' specifications or *bipoles*.

Type signature This file contains type and kind declarations of the object logic's elements. The kind form is built-in and represent the type of formulas of the object logic. In general, only the connectives' types need to be declared in this file:

```
%%%%%%%%%%%%%%%%%% Signature %%%%%%%%%%%%%%%%%%%%%
type imp form -> form -> form.
```

Subexponential signature The following subexponential signature is used for specifying the proof system MLJ:

```
%%%%%%%%%%%%%% Subexponential Signature %%%%%%%%%%%%%%%%%
subexp l unb.                    subexp r unb.
```

```
subexpctx l many lft.                subexpctx r many rght.
subexprel l > r
```

Intuitively, one subexponential corresponds to one context of the object logic se-quent.[3] MLJ has only two contexts, one to the left and another to the right side of the sequent, thus we use two subexponentials l and r. Moreover, as both contexts (to the left and right) behave classically in MLJ, we specify l and r to be unbounded, denoted by unb. In contrast, the specification of LJ would specify the subexponential r to be linear, as the right side of LJ's sequents behaves linearly.

The commands subexpctx l many lft. and subexpctx r many rght. are not formally needed for specifying proof systems, but as discussed in [9], they are needed in order to improve the visualization of the proof rules. In particular, the former specifies that the context corresponding to the subexponential l contains only formulas of the left side of the sequent, denoted by lft, and may contain many formulas, de-noted by many. In contrast, as the context to the right side of LJ sequents has only one formula, the subexponential r for that system would be annotated with single.

The pre-order among the subexponentials is specified on the last line using the key-word subexprel.

Bipoles The second part of the .pl file is composed by *bipoles*. The concrete syntax for SELL connectives is depicted in Figure 2. The class of *bipole* formulas often appear in proof theory literature due to its good focusing behaviour [1]. The following bipoles specify, respectively, the left and right implication introduction rules [8]. The capital letters are assumed to be existentially quantified.

```
%%%%%%%%%%%%%%%%%%%%%%%%%% Bipoles %%%%%%%%%%%%%%%%%%%%%%%%%%%%%%
% Implication
(not (lft (imp A B))) * (([r]? (rght A)) * ([l]? (lft B))).
(not (rght (imp A B))) * [l]bang ((([l]? (lft A)) | ([r]? (rght B))).
```

The head of these bipoles, formulas (not (lft (imp A B))) and (not (rght (imp A B))), specify that an implication formula to, respectively, the left and right-hand-side is introduced. The body specifies the premises of these rules. For in-stance, the first bipole specifies that its corresponding inference rules has two premises because of the branching caused by the tensor * appearing in the body of its rule, while the second has only one premise as no branching is required. The interesting bit is the $!^l$ ([l]bang) in the second bipole specifying that the context of the subexponential r should be weakened as l > r. In fact, by using advanced proof theoretic machinery, namely focusing [1], we can make this intuition precise in the sense. We refer to [8] for more details on encodings.

2.2 Features

Quati has two main features: (1) It can construct the corresponding inference rule(s) associated to a SELL formula; and (2) it can prove permutation lemmas. We illustrate these features with the specification of MLJ implication introduction rules shown above.

---

[3] There are some specifications where a subexponential is used to capture the structural proper-ties of the proof system and therefore does not necessarily correspond to a context in the object logic. See [8] for more on this.

Rule Construction Proving the adequacy theorems for a given SELL specification is also error-prone. As detailed in [8], to prove (strong) adequacy we need to show that all the possible focused derivations that introduce a formula in the specification correspond to an inference rule of the proof system being specified. Quati automates the proof of such adequacy theorems by constructing from a bipole the corresponding inference rule. To do so, Quati uses the machinery described in [13,9] reducing this problem to the problem of solving answer-set programs.

For the MLJ specification given above, one can use the command #rule in the command line and select a SELL bipole in the loaded specification. Then Quati generates a LaTeX document containing all possible inference rules that correspond to that bipole. If we select the bipole used to specify MLJ's implication right rule, Quati outputs the LaTeX code for the following figure:

$$\frac{i\ \Gamma_l^0, a \vdash_{\dot{r}} b}{i\ \Gamma_l^0 \vdash_{\dot{r}} \Delta_r^0, imp(a)(b)}\ imp_R$$

Notice that this rule looks very similar to MLJ's implication right introduction rule shown in any proof theory textbook. The context Δ_r^0 is erased in the premise. The i and $\dot{r}$ are used to delimit the contexts for the subexponentials 1 and r, respectively. Quati uses the subexponential specification to infer that the context for 1 (resp. for r) should only be on the left-hand-side (resp. right-hand-side) of the sequent.

Under the hood, Quati is constructing the focused derivation [1] that introduces such a SELL bipole as described in [8]. This can be observed by using the command #bipole. For the same SELL bipole used above, Quati returns the LaTeX code for the following figure, corresponding to its focused derivation:

$$\frac{\dfrac{\dfrac{\dfrac{\dfrac{\Gamma_{gamma}^5;\Gamma_r^7;\Gamma_l^5;\Gamma_{infty}^1;\Uparrow}{\Gamma_{gamma}^5;\Gamma_r^5;\Gamma_l^5;\Gamma_{infty}^1;\Uparrow?^r rght(b)}}{\Gamma_{gamma}^5;\Gamma_r^5;\Gamma_l^3;\Gamma_{infty}^1;\Uparrow?^l lft(a) ::?^r rght(b)}}{\dfrac{\Gamma_{gamma}^5;\Gamma_r^5;\Gamma_l^3;\Gamma_{infty}^1;\Uparrow?^l lft(a)\mathbin{\invamp}?^r rght(b)}{\Gamma_{gamma}^4;\Gamma_r^4;\Gamma_l^3;\Gamma_{infty}^1;\Downarrow!^l?^l lft(a)\mathbin{\invamp}?^r rght(b)}}}{\Gamma_{gamma}^5;\Gamma_r^4;\Gamma_l^3;\Gamma_{infty}^1;\Downarrow \neg rght(imp(a)(b))\otimes!^l?^l lft(a)\mathbin{\invamp}?^r rght(b)}}{\Gamma_{gamma}^3;\Gamma_r^4;\Gamma_l^3;\Gamma_{infty}^1;\Uparrow}$$

Rule Permutation. As described in the Introduction, Quati can be used to prove permutation lemmas. The command #permute checks whether the permutation of two selected rules is always allowed or not. Quati outputs, again in LaTeX, the cases for which it was able to find the permutation and the cases for which it was not able to find a permutation. For example, when Quati checks whether MLJ's implication left introduction rule permutes over MLJ's implication right introduction, it correctly finds two possible permutation cases and it cannot find one of the cases for which is indeed not possible. We show one of the cases (reformatted to fit the page margins):

$$\frac{\dfrac{i\ \Gamma_l^0, imp(a)(b), c \vdash_r d}{i\ \Gamma_l^0, imp(a)(b) \vdash_r \Delta_r^0, imp(c)(d), a}\ imp_R \quad i\ \Gamma_l^0, imp(a)(b), b \vdash_r \Delta_r^0, imp(c)(d)}{i\ \Gamma_l^0, imp(a)(b) \vdash_r \Delta_r^0, imp(c)(d)}\ imp_L$$

$$\leadsto \qquad \frac{\dfrac{i\ \Gamma_l^7, imp(a)(b), c \vdash_r a, d \quad i\ \Gamma_l^7, imp(a)(b), c, b \vdash_r d}{i\ \Gamma_l^7, imp(a)(b), c \vdash_r d}\ imp_L}{i\ \Gamma_l^7, imp(a)(b) \vdash_r \Delta_r^0, imp(c)(d)}\ imp_R$$

Once again, this proof figure is very similar to the proof figure that one would find in a standard proof theory textbook. Notice that it uses the fact that the contexts are unbounded, i.e. formulas can be contracted or weakened, to infer the permutation above (see [13] for more discussion on how this works).

3   Implementation Details

Quati is implemented in OCaml[4] and makes use of DLV[5] externally to compute minimal models for the answer-set programs generated. It is part of a bigger project, called sellf[6] which also includes the machinery for TATU mentioned above. The following diagram provides an overview of the main modules in sellf used by Quati for checking permutations.

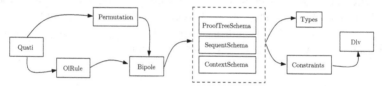

The basic data structure, defined in the module Types, is linear logic formulas with subexponentials. The bipoles in Quati are represented by *proof tree schemas*, defined in the module ProofTreeSchema, which uses the modules SequentSchema and ContextSchema. As the name suggests, these are schematic representations of proof trees, sequents and contexts that use generic contexts [13] to represent possibly non-empty sets of formulas. The constraints that will later compose the answer-set program are implemented in the module Constraints. The application of linear logic rules with constraints is implemented in the ProofTreeSchema module. The computation of possible bipoles of a formula is in the module Bipole. The Permutation module makes use of the bipole generation to construct the derivations of two rules. Given the constraints of a derivation, module Dlv contains the code for executing DLV externally, parsing the result and returning the minimal models. The translation of a proof tree schema and constraints into an object logic derivation is done in the OlRule module. It contains data structures to represent proof trees, sequents and contexts of an object logic and the rewriting algorithm described in [13] (module Derivation).

---

[4] http://ocaml.org/
[5] http://www.dlvsystem.com/dlv/
[6] https://code.google.com/p/sellf/

Quati was tested using some proof systems including LK, LJ, MLJ, LL, S4, G1m and LAX. On most cases, each permutation lemma can be checked in less than one second. The implementation can be downloaded at

http://www.logic.at/staff/giselle/quati.

4   Conclusions and Future Work

This paper introduced Quati, an automated tool for proving permutation lemmas. Besides briefly commenting on its implementation, we illustrated its syntax, usage and features. Besides MLJ, in the download one can find the specification of all proof systems tested, as well as system requirements and installation instructions.

There are several directions we are currently investigating for continuing this work. One is to come up with more graphical ways of writing proof systems and how to translate such representations into SELL specifications. Another possibility is the derivation of completeness of focusing strategies in an automated fashion, since such theorems rely heavily on permutation lemmas. Finally, we are investigating ways to construct machine-readable proof objects for permutation lemmas.

References

1. Andreoli, J.-M.: Logic programming with focusing proofs in linear logic. J. of Logic and Computation 2(3), 297–347 (1992)
2. Crolard, T.: Subtractive logic. Theor. Comput. Sci. 254(1-2), 151–185 (2001)
3. Gelfond, M., Lifschitz, V.: Logic programs with classical negation. In: ICLP (1990)
4. Gentzen, G.: Investigations into logical deductions. The Collected Papers of Gerhard Gentzen (1969)
5. Herbrand, J.: Recherches sur la Théorie de la Démonstration. PhD thesis (1930)
6. Maehara, S.: Eine darstellung der intuitionistischen logik in der klassischen. Nagoya Mathematical Journal, 45–64 (1954)
7. Miller, D., Saurin, A.: From proofs to focused proofs: a modular proof of focalization in linear logic. In: Duparc, J., Henzinger, T.A. (eds.) CSL 2007. LNCS, vol. 4646, pp. 405–419. Springer, Heidelberg (2007)
8. Nigam, V., Pimentel, E., Reis, G.: An extended framework for specifying and reasoning about proof systems. Accepted to Journal of Logic and Computation, http://www.nigam.info/docs/modal-sellf.pdf
9. Nigam, V., Reis, G., Lima, L.: Quati: From linear logic specifications to inference rules (extended abstract). In: Brazilian Logic Conference, EBL (2014), http://www.nigam.info/docs/ebl14.pdf
10. Nigam, V.: On the complexity of linear authorization logics. In: LICS (2012)
11. Nigam, V., Miller, D.: Algorithmic specifications in linear logic with subexponentials. In: PPDP (2009)
12. Nigam, V., Miller, D.: A framework for proof systems. J. Autom. Reasoning 45(2), 157–188 (2010)
13. Nigam, V., Reis, G., Lima, L.: Checking proof transformations with ASP. In: ICLP (Technical Communications) (2013)
14. Rauszer, C.: A formalization of the propositional calculus h-b logic. Studia Logica (1974)
15. Troelstra, A.S., Schwichtenberg, H.: Basic Proof Theory (1996)

A History-Based Theorem Prover
for Intuitionistic Propositional Logic Using
Global Caching: IntHistGC System Description

Rajeev Goré, Jimmy Thomson, and Jesse Wu

Research School of Computer Science, The Australian National University,
Canberra, Australia

Abstract. We describe an implementation of a new theorem prover
for Intuitionistic Propositional Logic based on a sequent calculus with
histories due to Corsi and Tassi. The main novelty of the prover lies in its
use of dependency directed backtracking for global caching. We analyse
the performance of the prover, and various optimisations, in comparison
to current state of the art theorem provers and show that it produces
competitive results on many classes of formulae.

1 Preliminaries

IntHistGC is a theorem prover for Intuitionistic Propositional Logic (Int), based
on a sound and cut-free complete sequent calculus which uses histories to guarantee termination. The key element behind the prover's efficiency is the use
of global caching to reduce search space. This system description provides an
overview of IntHistGC's proof strategy, implementation, optimisations and an
analysis of performance in comparison to the current best provers for Int.

Sequent Calculus. Figure 1 presents the standard multiple conclusioned sequent system for Int using sets of formulae. These rules form the basic elements
of the calculus used in our implementation. Note that we assume $\neg\varphi := (\varphi \to \bot)$.

The $\to$L rule may result in infinite looping since the principal formula $\varphi \to \psi$
must be copied into the left premise for completeness. Various approaches are
known to ensure termination: explicit loop-checking; Vorob'ev's method [1]; and
the use of complex histories to track loops [2, 3].

We take a different approach, by replacing the basic $\to$L and $\to$R rules with
those shown in Figure 2. The $\to$L rule is replaced by a "blocking" version which
prevents any further (backward) applications until "unblocked" by a $\to$R-first
rule. The $\to$R-rest rule can only (and must) be applied (upwards) if we have
previously applied $\to$R-first on the same implication on the same branch.

Rules in this calculus either delete a connective or block an implication. Implications are only unblocked after the use of a $\to$R-first rule, which is prevented
from being applied to the same formula twice. Hence there cannot be an infinite
sequence of blocking and unblocking, so backward proof search in this calculus
always terminates. Although this calculus was developed independently, recently
we have learnt that it is functionally identical to the sequent calculus IG [4].

Lemma 1. $\top$, $\bot$, $\wedge$, $\vee$, *id*, $\to$L-*blocking and* $\to$R-*rest are invertible rules.*

S. Demri, D. Kapur, and C. Weidenbach (Eds.): IJCAR 2014, LNAI 8562, pp. 262–268, 2014.
© Springer International Publishing Switzerland 2014

$$\frac{\Gamma \vdash \Delta}{\Gamma, \top \vdash \Delta} \ \top L \qquad \frac{}{\Gamma, \bot \vdash \Delta} \ \bot L \qquad \frac{\Gamma \vdash \Delta}{\Gamma \vdash \bot, \Delta} \ \bot R \qquad \frac{}{\Gamma \vdash \top, \Delta} \ \top R$$

$$\frac{}{\Gamma, \varphi \vdash \varphi, \Delta} \ id \qquad \frac{\Gamma, \varphi, \psi \vdash \Delta}{\Gamma, \varphi \wedge \psi \vdash \Delta} \ \wedge L \qquad \frac{\Gamma \vdash \varphi, \Delta \qquad \Gamma \vdash \psi, \Delta}{\Gamma \vdash \varphi \wedge \psi, \Delta} \ \wedge R$$

$$\frac{\Gamma, \varphi \vdash \Delta \qquad \Gamma, \psi \vdash \Delta}{\Gamma, \varphi \vee \psi \vdash \Delta} \ \vee L \qquad \frac{\Gamma \vdash \varphi, \psi, \Delta}{\Gamma \vdash \varphi \vee \psi, \Delta} \ \vee R$$

$$\frac{\Gamma, \varphi \to \psi \vdash \varphi, \Delta \qquad \Gamma, \psi \vdash \Delta}{\Gamma, \varphi \to \psi \vdash \Delta} \ \to L \qquad \frac{\Gamma, \varphi \vdash \psi}{\Gamma \vdash \varphi \to \psi, \Delta} \ \to R$$

Fig. 1. Basic sequent calculus rules

$$\frac{\Gamma, \varphi \to_b \psi \vdash \varphi, \Delta \qquad \Gamma, \psi \vdash \Delta}{\Gamma, \varphi \to \psi \vdash \Delta} \ \to\text{L-blocking} \qquad \frac{\Gamma \vdash \psi, \Delta}{\Gamma \vdash \varphi \to \psi, \Delta} \ \to\text{R-rest}$$

$$\frac{\boldsymbol{p}, \varphi_1 \to \psi_1, \cdots, \varphi_n \to \psi_n, \varphi \vdash \psi}{\boldsymbol{p}, \varphi_1 \to_b \psi_1, \cdots, \varphi_n \to_b \psi_n \vdash \varphi \to \psi, \boldsymbol{q}, \varphi_{n+1} \to \psi_{n+1}, \cdots, \varphi_{n+m} \to \psi_{n+m}} \ \to\text{R-first}$$

Fig. 2. New rules where $\boldsymbol{p}$ and $\boldsymbol{q}$ represent disjoint sets of atomic propositions

Backward Proof Search Strategy. Our proof strategy is as below. Note that backtracking is required at →R-first jumps, as the rule is not semantically invertible:

while some rule is applicable to a leaf sequent **do**
  stop: apply any applicable termination rule (*id*, ⊥L, ⊤R) to that leaf
  saturate: else apply any applicable static rule (⊤L, ⊥R, ∧L, ∨R, →L-blocking, →R-rest) to that leaf
  step: else apply the transitional rule (→R-first) to that leaf

Theorem 1. *Our rules are sound and our strategy is complete with respect to* Int *[4], and produces* $O(n^2)$ *depth proofs.*

Related Work. We now describe the main techniques behind three state of the art theorem provers for Int. All use some form of pre-processing: see Section 2.

BDDIntKt checks the validity of a formula φ_0 by constructing a finite closure $cl(\varphi_0)$ and checking if any of the subsets in the closure can lead to a model which falsifies φ_0 [5]. While the closure generally contains exponentially many subsets, by using Binary Decision Diagrams (BDDs) many of these do not necessarily have to be explicitly created.

fCube is a Prolog prover based on a signed tableaux calculus for Int [6]. The prover makes heavy use of a variety of simplification rules, whereby formulae with known sign (as well as atoms of a certain polarity) may be replaced under certain conditions with ⊤ or ⊥. These rules produce an equivalent but simpler sequent, and can significantly reduce unnecessary branching and backtracking.

Imogen uses formula polarity to restrict applicable inference rules [7]. The prover is based on a focused inverse method, which reduces the search space in a sequent calculus based on the polarity of connectives and atomic propositions. Proof search is applied in a forward manner, unlike our strategy above.

2   Implementation and Optimisations

The sequent rules described in Figures 1 and 2, and the general search strategy of Section 1, were implemented in Ocaml. Search is conducted in a depth-first manner, fully exploring one branch before considering others. While breadth-first search is also possible, a depth-first strategy allows for a natural implementation of global caching [8] as described next. IntHistGC includes a (slower) configuration which produces graphs for both valid and invalid derivations.

Pre-processing. A simple processing function which replaces inputs with equivalent formulae can significantly reduce the complexity of a proof. By imposing a standard ordering on all formulae, one can completely remove any complications due to variable ordering at commutative operators. For example, the formula $(\varphi \wedge \psi) \rightarrow (\psi \wedge \varphi)$ can be rewritten as $(\varphi \wedge \psi) \rightarrow (\varphi \wedge \psi)$ which simplifies to $\top$. IntHistGC also makes use of standard logical identities, such as $\varphi \wedge \top = \top \wedge \varphi = \varphi \wedge \varphi = \varphi$, and removes nested implications using the identity $\varphi_0 \rightarrow (\varphi_1 \rightarrow \varphi_2) = (\varphi_0 \wedge \varphi_1) \rightarrow \varphi_2$ before applying backwards proof search.

Dependency directed backtracking and Caching. The application of sequent rules often leads to proofs which only require a subset of formulae within a sequent. If two valid sequents differ only by formulae which are not necessary for a proof of validity, then a proof for one is sufficient to prove the other. *Thrashing* occurs when branches with the same important formulae are redundantly explored. Backjumping is a technique which avoids this phenomenon, by backtracking to the last point which contains all formulae relevant for a proof [9]. This technique is applied at all branching points, to reduce search space as much as possible.

We extend backjumping by also implementing dependency directed backtracking [9] which stores sets of formulae used to close branches. This is done by placing minimal sequents, like those in backjumping, into a cache. Thereafter, any sequent containing formulae which are a superset of a cached sequent can be proved by applying the same rules. This removes the need to continue search within that branch, and so the prover can return to the latest backtracking point.

In addition to a cache for provable sequents, we also implement a cache for unprovable sequents. In contrast to the minimal sequents constructed for the provable cache, here we include as many formulae as possible; as long as the overall sequent is still not provable. If any sequent is a subset of a previously unproven sequent, then it also cannot be proven - at best, all the same rules will be applied, which we know from previous experience cannot lead to a proof.

Both caches are global and thus search space can potentially be reduced across all branches, by substituting previous proofs. However, caching introduces its own problems. Caching sequents after all rule applications will result in an exponential growth in cache size. To somewhat mitigate unnecessary growth in cache size, cache additions are only considered at $\rightarrow$R-first jumps and branching points where backjumping has been applied. As each branch terminates without loops, we do not need the full complications of global caching for modal logics [8].

Data Structures. Formulae are represented using (positive) integers. After pre-processing, each formula (including subformulae) is given a unique identifier, and a number of tables are used to map integers to their respective formula, their formula type, as well as the integers corresponding to any subformulae.

We use a set-trie data structure for caching, and implement an adaptation of the algorithms presented in [10]. As formulae are integers, sequents can be represented by an ordered list of integers. A set-trie takes advantage of such an ordering, allowing efficient storage of sequents and set containment searches.

Heuristics. For efficiency, non-branching rules are prioritised over branching rules. Formulae are also expanded in a lazy fashion: only the top-most connective of a formula is ever considered. Thus the prover can avoid some memory issues when expanding bi-implications, interpreted as $\varphi \leftrightarrow \psi := (\varphi \rightarrow \psi) \wedge (\psi \rightarrow \varphi)$.

Some branches are inherently easier to prove (or disprove) than others. For example, by changing the order of branch choice in a naïve depth-first search, formulae of the form $(\neg(\neg a_1) \leftrightarrow a_2) \leftrightarrow (a_2 \leftrightarrow a_1))$ can take either a few milliseconds or more than ten minutes to solve - a clear indication of the importance of branch choice. Currently our implementation takes an approach based on formula order as defined in the pre-processing stage.

Rule choice is similarly important, and is considered in implementation. For example, when more than one $\rightarrow$L-blocking rule is applicable, our prover attempts to pick one which corresponds to an application of modus ponens. Such a choice essentially allows the rule *id* to be immediately applied to one branch.

3   Experimental Results

Performance is evaluated using the Intuitionistic Logic Theorem Proving (ILTP) Library [11], which allows a maximum of 600 seconds for each individual problem instance. We present results in Figure 3 for our prover (IH) with varying optimisations, along with results for BDDIntKt -autoreorder -nary -assumimp (BDD), fCube (no options) and Imogen -h optimize. We also include experiments on an unofficial extended set of the ILTP problem classes, kindly provided by the authors of the library, Jens Otten and Thomas Raths (http://www.iltp.de/download/SYJ2xx-50/SYJ2xx-50.tar.gz) in Figure 4. All tests were conducted on a machine with an Intel i7-3770 CPU @3.40GHz with 8GB memory. The configurations for IntHistGC are as follows:

Configuration	Optimisations involved
naïve	The baseline prover without any optimisations
b	Backjumping
c	Global caching with additions made at $\rightarrow$R-first jumps
c2	Global caching with additions also at backjumping points

Figure 3 clearly shows that backjumping and caching both improve significantly upon our naïve implementation, allowing the prover to complete the majority of the ILTP benchmark. In fact, global caching by itself, without backjumping,

Class	naïve	IHb	IHc	IHbc	IHbc2	BDD	fCube	Imogen	Out of
LCL	2	2	2	2	2	2	2	2	2
SYN	20	20	20	20	20	20	20	20	20
SYJ10*	12	12	12	12	12	12	12	12	12
SYJ201	2	20	20	20	20	20	20	20	20
SYJ202	3	5	3	5	8	13	9	8	20
SYJ203	10	20	20	20	20	20	20	20	20
SYJ204	20	20	20	20	20	20	20	20	20
SYJ205	10	20	20	20	20	20	20	20	20
SYJ206	5	20	20	20	20	20	20	20	20
SYJ207	2	3	20	20	20	20	20	20	20
SYJ208	10	10	20	20	20	11	20	20	20
SYJ209	9	10	20	20	20	20	20	20	20
SYJ210	20	20	20	20	20	20	20	20	20
SYJ211	3	4	20	20	20	20	20	20	20
SYJ212	20	20	20	20	20	20	20	20	20
Total	148	206	257	259	262	258	263	262	274

Fig. 3. Performance of various versions of IntHistGC on the ILTP benchmark

manages to solve all problems except that of the valid pigeonhole class SYJ202. Note that IntHistGC's 'c' configuration has no positive effect on this class, as there is only one top-level →R-first jump.

While caching clearly has benefits, profiling reveals that the majority of the time is generally spent on cache look-ups. Therefore we implemented a 'c3' option which performs cache look-ups every second node, rather than at all branching points. While such a heuristic may result in cache misses, Figures 4 and 6 shows that this reduction in look-up times often outweighs the extra branching involved since such branching is often mitigated by backjumping from a later cache hit.

We remark that the ILTP library is limited to 12 main classes of formulae, half of which are only slight modifications of the other classes. In light of this, we also benchmarked on other families of formulae: Portia, which are encodings of a "real world puzzle"; Nishimura [12], previously used by the authors of fCube; and randomly generated formulae, noting that these tend to be non-valid. Figures 5 and 6 presents timing performance on these classes, in addition to hard instances of the ILTP problems. Particularly noteworthy are the provable Portia examples, which show that both IntHistGC and fCube seem to be susceptible to input ordering. Nevertheless, while using very different techniques to improve speed, fCube and IntHistGC currently seem to be the best provers for Int (Figures 4, 5 and 6). The two trade wins in pigeonhole and Nishimura, although fCube is superior on (typically non-valid) random formulae (Figure 5).

Ideally, a prover which utilises the best of both caching and simplification would perform very well. Unfortunately the two techniques are somewhat conflicting, as simplification introduces new formulae which often result in cache misses. Indeed, we implemented the replacement and basic permanence rules used by fCube [13] but these resulted in worse performance than caching alone.

Class	IHc	IHbc	IHbcc3	IHbc2	BDD	fCube	Imogen	Out of
SYJ201	50	50	50	50	50	50	36	50
SYJ202	3	5	6	8	13	9	8	38
SYJ203	50	50	50	50	50	50	50	50
SYJ204	50	50	50	50	50	50	50	50
SYJ205	50	50	50	50	50	50	50	50
SYJ206	50	50	50	50	50	50	50	50
SYJ207	50	50	50	50	49	50	26	50
SYJ208	38	38	38	38	11	38	37	38
SYJ209	50	50	50	50	50	50	50	50
SYJ210	50	50	50	50	50	50	50	50
SYJ211	50	50	50	50	45	50	50	50
SYJ212	50	50	50	50	50	50	46	50
Total	541	543	544	546	518	547	503	576

Fig. 4. Performance on extended ILTP formulae

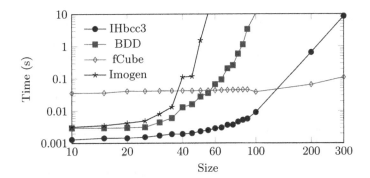

Fig. 5. Average times for random formula of increasing size

Problem	IHbc	IHbc2	IHbcc3	IHbc2c3	BDD	fCube	Imogen
SYJ201+1.050	5.69	10.13	4.58	6.87	31.794	159.92	–
SYJ202+1.008	–	190.16	–	208.21	0.33	27.86	70.76
SYJ207+1.050	301.23	454.69	161.73	242.08	188.36	17.90	–
SYJ208+1.038	181.58	185.99	113.32	112.70	–	1.76	–
SYJ211+1.050	2.18	1.37	10.15	6.17	–	0.02	0.27
SYJ212+1.050	0.26	0.26	0.22	0.22	50.91	0.22	–
p-portia.100	429.97	–	238.00	–	–	334.02	–
n-portia.100	5.32	295.84	4.52	14.64	–	385.12	–
n-portiaV2.100	443.25	–	239.08	–	–	97.62	–
nishimura.035	0.51	0.51	0.51	0.51	7.05	–	–

Fig. 6. Timing comparisons (all times in seconds, – indicates a timeout)
'p' means provable, 'n' is non-provable and V2 is a version with reordered formulae

4 Conclusion and Further Work

We have introduced a new prover, which demonstrates the practicality of a history based sequent calculus using global caching. We find that various optimisation techniques make a significant difference in terms of performance. In particular, global caching was extremely effective on the ILTP benchmark. To our knowledge, there are no other ATPs for Int which similarly utilise caching.

Further research into the interaction between simplification and caching techniques merits investigation. The development of a technique which allows caching and simplification to fully complement each other should have a significant impact on prover efficiency. Other than this, it may also be interesting to examine the performance of global caching using a different underlying calculus.

References

[1] Dyckhoff, R.: Contraction-free sequent calculi for intuitionistic logic. J. Symb. Log. 57, 795–807 (1992)

[2] Howe, J.M.: Two loop detection mechanisms: A comparison. In: Galmiche, D. (ed.) TABLEAUX 1997. LNCS, vol. 1227, pp. 188–200. Springer, Heidelberg (1997)

[3] Heuerding, A., Seyfried, M., Zimmermann, H.: Efficient loop-check for backward proof search in some non-classical propositional logics. In: Miglioli, P., Moscato, U., Ornaghi, M., Mundici, D. (eds.) TABLEAUX 1996. LNCS, vol. 1071, pp. 210–225. Springer, Heidelberg (1996)

[4] Corsi, G., Tassi, G.: Intuitionistic logic freed of all metarules. J. Symb. Log. 72, 1204–1218 (2007)

[5] Goré, R., Thomson, J.: BDD-based automated reasoning for propositional bi-intuitionistic tense logics. In: Gramlich, B., Miller, D., Sattler, U. (eds.) IJCAR 2012. LNCS (LNAI), vol. 7364, pp. 301–315. Springer, Heidelberg (2012)

[6] Ferrari, M., Fiorentini, C., Fiorino, G.: fCube: An efficient prover for intuitionistic propositional logic. In: Fermüller, C.G., Voronkov, A. (eds.) LPAR-17. LNCS, vol. 6397, pp. 294–301. Springer, Heidelberg (2010)

[7] McLaughlin, S., Pfenning, F.: Imogen: Focusing the polarized inverse method for intuitionistic propositional logic. In: Cervesato, I., Veith, H., Voronkov, A. (eds.) LPAR 2008. LNCS (LNAI), vol. 5330, pp. 174–181. Springer, Heidelberg (2008)

[8] Goré, R., Nguyen, L.A.: Exptime tableaux for ALC using sound global caching. J. Autom. Reasoning 50, 355–381 (2013)

[9] Hustadt, U., Schmidt, R.A.: Simplification and backjumping in modal tableau. In: de Swart, H. (ed.) TABLEAUX 1998. LNCS (LNAI), vol. 1397, pp. 187–201. Springer, Heidelberg (1998)

[10] Savnik, I.: Index data structure for fast subset and superset queries. In: Cuzzocrea, A., Kittl, C., Simos, D.E., Weippl, E., Xu, L. (eds.) CD-ARES 2013. LNCS, vol. 8127, pp. 134–148. Springer, Heidelberg (2013)

[11] Raths, T., Otten, J., Kreitz, C.: The ILTP problem library for intuitionistic logic. J. Autom. Reasoning 38, 261–271 (2007)

[12] Fiorentini, C.: All intermediate logics with extra axioms in one variable, except eight, are not strongly ω-complete. J. Symb. Log. 65, 1576–1604 (2000)

[13] Ferrari, M., Fiorentini, C., Fiorino, G.: Simplification rules for intuitionistic propositional tableaux. ACM Trans. Comput. Log. 13, 14 (2012)

MleanCoP: **A Connection Prover for First-Order Modal Logic**

Jens Otten

Institut für Informatik, University of Potsdam
August-Bebel-Str. 89, 14482 Potsdam-Babelsberg, Germany
jeotten@cs.uni-potsdam.de

Abstract. MleanCoP is a fully automated theorem prover for first-order modal logic. The proof search is based on a prefixed connection calculus and an additional prefix unification, which captures the Kripke semantics of different modal logics. MleanCoP is implemented in Prolog and the source code of the core proof search procedure consists only of a few lines. It supports the standard modal logics D, T, S4, and S5 with constant, cumulative, and varying domain conditions. The most recent version also supports heterogeneous multimodal logics and outputs a compact prefixed connection proof. An experimental evaluation shows the strong performance of MleanCoP.

1   Introduction

Modal logics extend the language of classical logic with the unary modal operators $\Box$ and $\Diamond$. They are used to represent the modalities "it is necessarily true that" and "it is possibly true that", respectively. The *Kripke semantics* of the standard *unimodal* logics are defined by a set of worlds and a single binary accessibility relation between these worlds. *Multimodal* logics consider a finite set of distinct modal operators $\Box_1, \ldots, \Box_n$ and $\Diamond_1, \ldots, \Diamond_n$, and the Kripke semantics is specified by a set of n accessibility relations. *First-order* modal logics extend propositional modal logics by *domains*, i.e. sets of objects that are associated with each world, and the standard universal and existential quantifiers [5,8]. Modal logics have applications in, e.g., planning, natural language processing, and program verification. Multimodal logics are in particular suitable for representing knowledge and beliefs. Popular multimodal logics include temporal and epistemic logic, which are used for program verification and representing dynamic knowledge of different agents [7]. Even though many of these applications would benefit from a higher degree of automation, the development of *efficient* fully automated theorem provers for first-order modal logic is still in its infancy.

This paper presents one of the first theorem provers for first-order (multi)modal logic. It is based on a modal connection calculus (Section 2). Whereas the underlying *connection calculus* provides a basis for an efficient proof search [4,11], *prefixes* are used to directly encode sequences of accessible worlds of the Kripke semantics. The calculus for the different modal logics differ only in the prefix unification, which respects the accessibility relation of the modal logic under consideration. The modal connection calculus is implemented in a very compact Prolog program (Section 3), which shows a strong performance on the problems in the QMLTP library (Section 4).

S. Demri, D. Kapur, and C. Weidenbach (Eds.): IJCAR 2014, LNAI 8562, pp. 269–276, 2014.
© Springer International Publishing Switzerland 2014

2   The Modal Connection Calculus

Syntax and Semantics. *First-order modal formulae F* are composed of atomic formulae, the standard (classical) connectives ¬, ∧, ∨, ⇒, the modal operators □, ◇, and the standard quantifiers ∀ and ∃. For *multimodal* logic, sets of modal operators $\{\Box_i, \Diamond_i \mid i \in I\!N\}$ are considered. The *Kripke semantics* of the standard modal logics are defined by a set of worlds W and a binary *accessibility relation* $R_i \subseteq W \times W$ between these worlds [5,8]. In each single world $w \in W$ the classical semantics applies to the standard connectives and quantifiers, e.g.

$\forall x F / \exists x F$ is true in world w iff F is true in world w for *all/some* object(s) x,
whereas the modal operators are interpreted with respect to accessible worlds, i.e.,

$\Box_i F / \Diamond_i F$ is true in world w iff F is true in *all/some* world(s) w', with $(w, w') \in R_i$.
The properties of the accessibility relation R_i determine the particular modal logic. In this paper the modal logics D, T, S4, and S5 are considered. Their accessibility relation is serial (D)[1], reflexive (T), reflexive and transitive (S4), or an equivalence relation (S5). The standard semantics is considered with rigid term designation, i.e. every term denotes the same object in every world, and terms are local, i.e. any ground term denotes an existing object in every world.

Using Prefixes. A *prefix* is used to name a sequence of accessible worlds and is assigned to each literal L and each subformula of a given formula F. E.g., the prefixed formula $F : w_1 w_2$ denotes the fact that F is true in world w_2 that is accessible from a world w_1. Similarly to free variables and Skolem terms used for quantified variables, free "world variables" and "Skolem worlds" are used within prefixes [13]. This can be explained by the fact that the semantics of the quantifiers resembles the semantics of the modal operators (see the definitions given above). In the negation normal form $\Box_i$ adds a Skolem world *(prefix constant)* to the prefix, whereas $\Diamond_i$ adds a world variable *(prefix variable)*. Depending on the modal logic (D, T, S4, or S5) and its accessibility relation, variables can be substituted by exactly one prefix variable or constant (D), by at most one prefix variable or constant (T), or by any sequence of prefix variables and constants (S4). For the modal logic S5 only the last element of every prefix is considered.

In Fitting's modal tableau calculi [5,8], prefixes of literals that close a branch need to denote the same world, i.e., they need to be identical. Similarly to term unification for (first-order) terms, this is achieved by a *prefix unification* during the proof search. This unification problem is a special case of string unification that takes the *prefix property*, i.e. the form of the prefixes, and the accessibility relation of the modal logic into account. For D and S5 the prefix unification is straightforward and there is only one most general unifier. For T and S4 prefix unification procedures that calculate minimal (finite) *sets* of most general unifiers were developed as well [10,13].

For (heterogeneous) *multimodal* logics each prefix constant and variable is marked with the index i of the corresponding modal operator $\Box_i$ or $\Diamond_i$. Prefix constants and variables can only be assigned to variables with the same index, and the modal logic assigned to each index i has to be taken into account. Modal operators with different indices are independent from each other, i.e. interaction axioms must be added explicitly.

---

[1] A relation $R \subseteq W \times W$ is *serial* iff for all $w_1 \in W$ there is some $w_2 \in W$ with $(w_1, w_2) \in R$.

The Modal Connection Calculus. The *connection calculus* [4] is already successfully used for automated theorem proving in first-order classical and first-order intuitionistic logic [10,11]. In order to adapt the calculus to modal logic, prefixes are added to all literals. The axiom and the rules of the *modal connection calculus* are given in Figure 1. $M = \{C_1, \ldots, C_m\}$ is a *prefixed matrix*, i.e., a set of clauses where each clause $C_i = \{L_1 : p_1, \ldots, L_n : p_n\}$ is a set of prefixed literals, i.e., p_i is the prefix of the literal L_i. The *subgoal clause* C and the *active path Path* are sets of (prefixed) literals or ε; C_1 and C_2 are clauses. A *connection* $\{L_1 : p_1, L_2 : p_2\}$ is σ-complementary for a *term substitution* σ_Q and a *prefix substitution* σ_M iff $\sigma_Q(L_1) = \sigma_Q(\overline{L_2})$ and $\sigma_M(p_1) = \sigma_M(p_2)$, where $\overline{L_2}$ is the complement of L_2. These substitutions are *rigid*, i.e. they are applied to the whole derivation, and calculated by algorithms for term and prefix unification.

A *modal connection proof* for the prefixed matrix M is a derivation for $\varepsilon, M, \varepsilon$, with admissible substitutions σ_Q and σ_M. Substitutions are *admissible* if they respect the accessibility relation and the domain condition. The accessibility relation depends on the logic and is captured in the specific prefix unification for each modal logic. The *domain condition* ensures that if a Skolem term t is assigned to a variable x, then t and x need to exist in the same world. This property holds if the prefix of (the quantifier of) t is an initial string of the prefix of (the quantifier of) x for *cumulative* domains, or if these prefixes are equal for *varying* domains; there is no restriction for *constant* domains.

For example, the prefixed matrix of the modal formula $\Box\forall x Px \Rightarrow \Box\forall y\Box Py$ is $M_1 = \{\{\neg Px : W_1\}, \{Pc : w_2 w_3\}\}$ in which c is a Skolem term and w_2 and w_3 are prefix constants. The following derivation for M_1 is a modal connection proof for the modal logics S4 and S5 with constant and cumulative domains (the arc marks the only connection).

$$\cfrac{\cfrac{}{\{\},M_1,\{Pc\colon w_2w_3\}}\ axiom \quad \cfrac{}{\{\},M_1,\{\}}\ axiom}{\cfrac{\{Pc\colon w_2w_3\},\{\{\neg Px\colon W_1\},\{Pc\colon w_2w_3\}\},\{\}}{\varepsilon,\{\{\neg Px\colon W_1\},\{Pc\colon w_2w_3\}\},\varepsilon}\ start}\ extension$$

$\sigma_Q(x) = c$
$\sigma_M(W_1) = w_2 w_3 \ (= w_3 \text{ for S5})$
(the prefix of x is $w_2 w_3$ and the prefix of c is w_2)

The modal connection calculus is based on a *clausal matrix characterization* of logical validity [13], which is a slightly adapted version of the original (non-clausal) matrix characterization [15]. In order to simplify the implementation a *Skolemization* is used not only for eigenvariables but also for prefix constants. A similar approach is already used for intuitionistic logic [10]. Thus, the irreflexivity test of the reduction ordering [15] is realized by the occurs check of the term and prefix unification procedures.

axiom	$\cfrac{}{\{\},M,Path}$	*start* $\cfrac{C_2,M,\{\}}{\varepsilon,M,\varepsilon}$ and C_2 is copy of $C_1 \in M$
reduction	$\cfrac{C,M,Path\cup\{L_2\colon p_2\}}{C\cup\{L_1\colon p_1\},M,Path\cup\{L_2\colon p_2\}}$	$\{L_1\colon p_1,L_2\colon p_2\}$ is σ-complementary
extension	$\cfrac{C_2\backslash\{L_2\colon p_2\},M,Path\cup\{L_1\colon p_1\}\quad C,M,Path}{C\cup\{L_1\colon p_1\},M,Path}$	C_2 is a copy of $C_1\in M$, $L_2\colon p_2\in C_2$, $\{L_1\colon p_1,L_2\colon p_2\}$ is σ-complementary

Fig. 1. The connection calculus for first-order modal logic

3 The Implementation

MleanCoP implements the modal connection calculus presented in Section 2. Version 1.3 of MleanCoP features the following enhancements compared to version 1.2 [2,13]: support for heterogeneous multimodal logics, output of a compact modal connection proof, support for the modal TPTP syntax, integration of the strategy scheduling into the shell script, and an additional check of the domain condition in the core prover. Furthermore, version 1.3 of MleanCoP does not only support ECLiPSe Prolog, but also SWI and SICStus Prolog. The total size of the shell script and the four files containing the Prolog source code is less than 29 KB. MleanCoP is available under the GNU General Public License and can be downloaded at `http://www.leancop.de/mleancop/` .

Invoking and Preprocessing. The MleanCoP prover is invoked by the command

    ./mleancop.sh <problem file> [<time limit>]

which starts the proof search for the modal formula in the file `<problem file>`. The optional `<time limit>` is used to control the *fixed strategy scheduling*. If the problem file contains a formula in the modal TPTP syntax [14], it is translated into the MleanCoP syntax. Afterwards, the formula is translated into a prefixed (clausal) matrix, i.e., prefixes are added to all literals in the matrix; *no* other simplifications are carried out in this step. The prefixed matrix is stored in Prolog's database and represented by the predicate `lit/3`. An optional *definitional clausal form* translation reduces the number of possible connections and might prune the search space significantly.

The Modal Connection Calculus. The implementation of the core proof search procedure extends the automated theorem prover leanCoP for first-order classical logic [10,11] by adding prefixes to literals and a prefix unification algorithm for each considered modal logic. Furthermore, each clause is annotated with a list that contains term variables together with their prefixes in order to check the domain condition. The Prolog source code of the MleanCoP 1.3 core prover is shown in Figure 2. The underlined code was added to leanCoP 2.1; no other modifications were done. The open subgoal C and the active path *Path* in the modal connection calculus of Figure 1 are represented by the Prolog lists `Cla` and `Path`, respectively. Atoms are represented by Prolog atoms, term (and prefix) variables by Prolog variables and negation by "`-`". The substitutions σ_Q and σ_M are stored implicitly by Prolog.

The predicate `prove(PathLim,Set,Proof)` (lines a–g) implements the start rule. `PathLim` is the maximum size of the active path used for *iterative deepening*, `Set` is a list of options used to control the proof search, and `Proof` contains the returned connection proof. First, MleanCoP performs a classical proof search, afterwards, the domain condition is checked (`domain_cond/1`) and the collected prefixes are unified (`prefix_unify/1`) (line g). These are the only external predicates called during the actual proof search. The implementations of the prefix unifications for the modal logics D, T, S4, and S5 need between 2 to 17 lines of Prolog code; the domain condition is implemented by another 15 lines of code. For multimodal logic, prefix constants and variables are marked with the index of the corresponding modal operator. Prefixes are divided into sections and unified according to the modal logic assigned to their indices.

```
(a)  prove(PathLim,Set,Proof) :-
(b)    ( \+member(scut,Set) ->
(c)      prove([(-(#)):(-[])],[],PathLim,[],PreSet,FreeV1,Set,[Proof]) ;
(d)      lit((#):_,FV:C,_) ->
(e)      prove(C,[(-(#)):(-[])],PathLim,[],PreSet,FreeV,Set,Proof1),
(f)      Proof=[C|Proof1], append(FreeV,FV,FreeV1) ),
(g)      domain_cond(FreeV1), prefix_unify(PreSet).
(h)  prove(PathLim,Set,Proof) :-
(i)    member(comp(Limit),Set), PathLim=Limit -> prove(1,[],Proof) ;
(j)    (member(comp(_),Set);retract(pathlim)) ->
(k)    PathLim1 is PathLim+1, prove(PathLim1,Set,Proof).

(1)  prove([],_,_,_,[],[],_,[]).
(2)  prove([Lit:Pre|Cla],Path,PathLim,Lem,[PreSet,FreeV],Set,Proof) :-
(3)    Proof=[[[NegLit:PreN|Cla1]|Proof1]|Proof2],
(4)    \+ (member(LitC,[Lit:Pre|Cla]), member(LitP,Path), LitC==LitP),
(5)    (-NegLit=Lit;-Lit=NegLit) ->
(6)    ( member(LitL,Lem), Lit:Pre==LitL, Cla1=[], Proof1=[],
(7)      PreSet3=[], FreeV3=[]
(8)    ;
(9)      member(NegL:PreN,Path), unify_with_occurs_check(NegL,NegLit),
(10)     Cla1=[], Proof1=[],
(11)     \+ \+ prefix_unify([Pre=PreN]), PreSet3=[Pre=PreN], FreeV3=[]
(12)   ;
(13)     lit(NegLit:PreN,FV:Cla1,Grnd1),
(14)     ( Grnd1=g -> true ; length(Path,K), K<PathLim -> true ;
(15)       \+ pathlim -> assert(pathlim), fail ),
(16)     \+ \+ ( domain_cond(FV), prefix_unify([Pre=PreN]) ),
(17)     prove(Cla1,[Lit:Pre|Path],PathLim,Lem,PreSet1,FreeV1,Set,Proof1),
(18)     PreSet3=[Pre=PreN|PreSet1], append(FreeV1,FV,FreeV3)
(19)   ),
(20)   ( member(cut,Set) -> ! ; true ),
(21)   prove(Cla,Path,PathLim,[Lit:Pre|Lem],PreSet2,FreeV2,Set,Proof2),
(22)   append(PreSet3,PreSet2,PreSet), append(FreeV2,FreeV3,FreeV).
```

Fig. 2. Source code of the MleanCoP core prover

The predicate `prove(Cla,Path,PathLim,Lem,[PreSet,FreeV],Set,Proof)` implements the axiom (line 1), the reduction rule (lines 9–11, 21–22) and the extension rule (lines 13, 16–18, 21–22) of the modal connection calculus in Figure 1. A *weak* prefix unification (and domain check) is carried out for the current connection (line 11 and 16); double negation prevents any variable bindings. If the proof search for the current path limit fails and this limit was actually reached (lines 14–15), then `PathLim` is increased and the proof search restarts with an increased path limit (lines *h–k*). MleanCoP uses a few additional effective techniques already used in the classical prover leanCoP: *regularity* (line 4), *lemmata* (lines 6–7), and *restricted backtracking* [11] (line 20). For the example formula from Section 2 the MleanCoP core prover is invoked by

```
prove((# all X: p(X) => # all Y: # p(Y)),Proof).
```

which is (internally) translated into the prefixed matrix

```
[[]:[p(4^[]^[3^[]]):[3^[],5^[]]],[[X,[W]]:[-(p(X)):-([W])]]]
```

and returns the modal prefixed connection proof (for S4 with cumulative domains)

```
Proof = [[p(4^[]^[3^[]]):[3^[], 5^[]]],
           [[-p(4^[]^[3^[]]): -[[3^[], [5^[]]]]]]]
```

where X is a term variable, $4^\wedge[]^\wedge[3^\wedge[]]$ is a Skolem term; W is a prefix variable for the world W_1, $3^\wedge[]$ and $5^\wedge[]$ are prefix constants for the worlds w_2 and w_3, respectively.

4   Experimental Evaluation

The modal connection prover MleanCoP described in Section 3 was tested on all 580 unimodal and all 20 multimodal problems of version 1.1 of the QMLTP library [14]. All tests were conducted on a 3.4 GHz Xeon system with 4 GB of RAM running Linux 2.6.24 and ECLiPSe Prolog 5.10. The CPU time limit for all proof attempts was set to 100 seconds.

Table 1. Results on the unimodal problems (varying/cumul./constant) of the QMLTP library

Logic	MleanSeP (proved)	MleanTAP (proved)	Satallax (proved)	MleanCoP (proved)	MleanCoP (< 1 sec)	MleanCoP (refuted)
D	– /130/129	100/120/135	113/133/159	186/207/224	160/178/193	273/247/222
T	– /163/165	138/162/175	169/192/212	223/250/270	211/236/253	159/132/114
S4	– /190/189	169/205/220	206/237/258	288/349/364	259/304/320	127/96/83
S5	– / – / –	219/272/272	245/294/301	359/436/436	321/388/388	94/41/41

Table 1 shows the results for *unimodal* logic for the theorem provers MleanSeP 1.2, MleanTAP 1.3, Satallax 2.2, and MleanCoP 1.3. The columns contain the number of proved problems (proved), and for MleanCoP also the number of problems proved within 1 second (< 1 sec) and the number of refuted problems (refuted). For each logic the results are given for the varying/cumulative/constant domain conditions.

MleanSeP implements the standard modal sequent calculus for several unimodal logics with cumulative domains.[2] It performs an analytic proof search and uses free variables with a dynamic Skolemization. For the constant domain variants the *Barcan formulae* are added. MleanTAP is a compact implementation of a prefixed tableau calculus for several unimodal logics.[3] Similarly to MleanCoP it uses prefixes and an additional prefix unification procedure. Hence, MleanTAP can easily be extended to multimodal logic by integrating the multimodal prefix unification of MleanCoP 1.3. Satallax [6] is a theorem prover for higher-order logic (HOL) and is used in combination with an embedding of first-order modal logic into simple type theory [2,3]. These are currently the only available theorem provers for first-order modal logic. Instead of Satallax, other theorem provers for HOL can be used as well, but Satallax shows the strongest performance when using the embedding into HOL [2].

MleanCoP 1.3 proves significantly more problems than any of the other theorem provers for first-order modal logic. This is true, even if the time limit for MleanCoP is reduced to one second. Satallax comes second, proving more problems than MleanSeP and MleanTAP; it also refutes a high number of problems and can deal with many more modal logics, such as the modal logic K [2].

MleanCoP 1.3 solves 17 of the 20 *multimodal* problems included in the QMLTP library; all of these problems are solved within a fraction of a second.

---

[2] MleanSeP can be obtained at http://www.leancop.de/mleansep/

[3] MleanTAP can be obtained at http://www.leancop.de/mleantap/

5   Conclusion

Despite the fact that modal logics are considered as some of the most important non-classical logics and numerous calculi were developed, the availability of actual *implementations* of fully automated theorem provers for *first-order* modal logic is very limited. Extending existing theorem provers for *propositional* modal logic, e.g. mod-leanTAP [1] or MSPASS [9], to *first-order* modal logic is not straightforward [2].

The modal connection calculus extends the classical clausal connection calculus by prefixes and additional prefix unifications, which directly encode the accessibility relations of the different modal logics. MleanCoP is based on the classical connection prover leanCoP and extended by prefix unifications for the unimodal logics D, T, S4, S5 and for the (normal) multimodal logics. The returned modal connection proof contains all necessary information to translate it back into a more readable form.

Future work includes the extension of the classical *non-clausal* connection calculus [12] to first-order modal logic, optimizing the prefix unifications, and extending the prefix unification to other standard modal logics.

References

1. Beckert, B., Goré, R.: Free Variable Tableaux for Propositional Modal Logics. In: Galmiche, D. (ed.) TABLEAUX 1997. LNCS (LNAI), vol. 1227, pp. 91–106. Springer, Heidelberg (1997)
2. Benzmüller, C., Otten, J., Raths, T.: Implementing and Evaluating Provers for First-order Modal Logics. In: De Raedt, L., et al. (eds.) 20th European Conference on Artificial Intelligence, ECAI 2012, pp. 163–168. IOS Press, Amsterdam (2012)
3. Benzmüller, C., Raths, T.: HOL Based First-Order Modal Logic Provers. In: McMillan, K., Middeldorp, A., Voronkov, A. (eds.) LPAR-19 2013. LNCS, vol. 8312, pp. 127–136. Springer, Heidelberg (2013)
4. Bibel, W.: Automated Theorem Proving. Vieweg, Wiesbaden (1987)
5. Blackburn, P., van Bentham, J., Wolter, F.: Handbook of Modal Logic. Elsevier, Amsterdam (2006)
6. Brown, C.: Reducing Higher-Order Theorem Proving to a Sequence of SAT Problems. In: Bjørner, N., Sofronie-Stokkermans, V. (eds.) CADE 2011. LNCS (LNAI), vol. 6803, pp. 147–161. Springer, Heidelberg (2011)
7. Carnielli, W., Pizzi, C.: Modalities and Multimodalities. Springer, Heidelberg (2008)
8. Fitting, M., Mendelsohn, R.L.: First-Order Modal Logic. Kluwer, Dordrecht (1998)
9. Hustadt, U., Schmidt, R.: MSPASS: Modal Reasoning by Translation and First-Order Resolution. In: Dyckhoff, R. (ed.) TABLEAUX 2000. LNCS (LNAI), vol. 1847, pp. 67–81. Springer, Heidelberg (2000)
10. Otten, J.: leanCoP 2.0 and ileanCoP 1.2: High Performance Lean Theorem Proving in Classical and Intuitionistic Logic. In: Armando, A., Baumgartner, P., Dowek, G., et al. (eds.) IJCAR 2008. LNCS (LNAI), vol. 5195, pp. 283–291. Springer, Heidelberg (2008)
11. Otten, J.: Restricting backtracking in connection calculi. AI Communications 23, 159–182 (2010)
12. Otten, J.: A Non-clausal Connection Calculus. In: Brünnler, K., Metcalfe, G. (eds.) TABLEAUX 2011. LNCS (LNAI), vol. 6793, pp. 226–241. Springer, Heidelberg (2011)

13. Otten, J.: Implementing Connection Calculi for First-order Modal Logics. In: Korovin, K., et al. (eds.) IWIL 2012. EPiC, vol. 22, pp. 18–32. EasyChair (2012)
14. Raths, T., Otten, J.: The QMLTP Problem Library for First-order Modal Logics. In: Gramlich, B., Miller, D., Sattler, U. (eds.) IJCAR 2012. LNCS (LNAI), vol. 7364, pp. 454–461. Springer, Heidelberg (2012)
15. Wallen, L.A.: Automated Deduction in Nonclassical Logics. MIT Press, Cambridge (1990)

Optimal Tableaux-Based Decision Procedure for Testing Satisfiability in the Alternating-Time Temporal Logic ATL$^+$

Serenella Cerrito[1], Amélie David[1], and Valentin Goranko[2,3]

[1] Laboratoire IBISC - Université Evry Val-d'Essonne, France
{serena.cerrito,adavid}@ibisc.univ-evry.fr
[2] Department of Applied Mathematics and Computer Science
Technical University of Denmark
[3] University of Johannesburg, South Africa

Abstract. We develop a sound, complete and practically implementable tableaux-based decision method for constructive satisfiability testing and model synthesis in the fragment ATL$^+$ of the full Alternating time temporal logic ATL*. The method extends in an essential way a previously developed tableaux-based decision method for ATL and works in 2EXP-TIME, which is the optimal worst case complexity of the satisfiability problem for ATL$^+$. We also discuss how suitable parameterizations and syntactic restrictions on the class of input ATL$^+$ formulae can reduce the complexity of the satisfiability problem.

1 Introduction

The Alternating-time temporal logic ATL* was introduced and studied in [1] as a multi-agent extension of the branching time temporal logic CTL*, where the path quantifiers are generalized to "strategic quantifiers", indexed with coalitions of agents A and ranging over all computations enabled by a given collective strategy of A. ATL* was proposed as logical framework for specification and verification of properties of open systems modeled as concurrent game models, in which all agents effect state transitions collectively, by taking simultaneous actions at each state. The language of ATL* allows expressing statements of the type *"Coalition A has a collective strategy to guarantee the satisfaction of the objective Φ on every play enabled by that strategy"*. The syntactic fragment ATL of ATL* allows only state formulae, where all occurrences of temporal operators must be immediately preceded by strategic quantifiers. The fragment ATL$^+$ of ATL* extends ATL by allowing any Boolean combinations of ATL objectives in the scope of a strategic quantifier. It is considerably more expressive than ATL, which is reflected in the high – 2EXPTIME – worst case complexity lower bound of the satisfiability problem for ATL$^+$ (inherited from the lower bound for CTL$^+$, see [8]) as opposed to the EXPTIME-completeness of the satisfiability problem for ATL [4,11]. The matching 2EXPTIME upper bound is provided by the automata-based method for deciding satisfiability in the full ATL*, developed in [10].

S. Demri, D. Kapur, and C. Weidenbach (Eds.): IJCAR 2014, LNAI 8562, pp. 277–291, 2014.
© Springer International Publishing Switzerland 2014

The contribution of this paper is the development of a sound, complete and terminating tableaux-based decision method for constructive satisfiability testing of ATL$^+$ formulae, which we also claim to be intuitive, conceptually simple and transparent, as well as practically implementable and even manually usable, despite the inherently high worst-case complexity of the problem. The tableaux method presented here is based on the general methodology going back to [9] and [12]. It was further developed for ATL in [7] to which the reader is referred for more details, and a recent implementation is reported in [3]. The tableaux method for ATL$^+$ is an essential extension of the one for ATL, as it has to deal with much more complex (and computationally expensive) path objectives that can be assigned to the agents. It is also rather different from the above mentioned automata-based method in [10].

The paper is structured as follows. In Section 2 we offer brief technical preliminaries on concurrent game models, syntax and semantics of ATL* and ATL$^+$. Section 3 develops the technical machinery needed for the presentation of the tableaux method itself in Section 4. Section 5 contains the main results related to termination, soundness, completeness and complexity of the procedure. In Section 6 we offer a brief comparison with the automata-based method in [10].

For lack of space, we only provide here very brief sketches of the proofs of the soundness, completeness and some other technical claims. A full version of this paper, including detailed proofs, is available as a technical report [2].

2   Preliminaries

We assume that the reader has basic familiarity with the branching time logic CTL*, see e.g. [5]. Also, basic knowledge on ATL* [1] and the tableaux-based decision procedure for ATL in [7], on which this paper builds, would be beneficial.

2.1   Concurrent Game Models, Strategies and Co-strategies

A **concurrent game model** [1] (CGM) is a tuple
$\mathcal{M} = (\mathbb{A}, \mathsf{St}, \{\mathsf{Act}_\mathsf{a}\}_{\mathsf{a} \in \mathbb{A}}, \{\mathsf{act}_\mathsf{a}\}_{\mathsf{a} \in \mathbb{A}}, \mathsf{out}, \mathsf{Prop}, \mathsf{L})$ comprising:

- a finite, non-empty set of *players (agents)* $\mathbb{A} = \{1, \ldots, k\}$
- a non-empty set of *states* St,
- a set of actions $\mathsf{Act}_\mathsf{a} \neq \emptyset$ for each $\mathsf{a} \in \mathbb{A}$.
  For any $A \subseteq \mathbb{A}$ we denote $\mathsf{Act}_A := \prod_{\mathsf{a} \in A} \mathsf{Act}_\mathsf{a}$ and use σ_A to denote a tuple from Act_A. In particular, $\mathsf{Act}_\mathbb{A}$ is the set of all possible *action profiles* in $\mathcal{M}$.
- for each $\mathsf{a} \in \mathbb{A}$, a map $\mathsf{act}_\mathsf{a} : \mathsf{St} \to \mathcal{P}(\mathsf{Act}_\mathsf{a}) \setminus \{\emptyset\}$ defining for each state s the actions available to a at s,
- a partial *transition function* $\mathsf{out} : \mathsf{St} \times \mathsf{Act}_\mathbb{A} - \to \mathsf{St}$ that assigns deterministically a *successor (outcome) state* $\mathsf{out}(s, \sigma_\mathbb{A})$ to every state s and action profile $\sigma_\mathbb{A} = \langle \sigma_1, \ldots, \sigma_k \rangle$, such that $\sigma_\mathsf{a} \in \mathsf{act}_\mathsf{a}(s)$ for every $\mathsf{a} \in \mathbb{A}$,
- a set of *atomic propositions* Prop, and a *labelling function* $\mathsf{L} : \mathsf{St} \to \mathcal{P}(\mathsf{Prop})$.

Concurrent game models represent multi-agent transition systems that function as follows: at any moment the system is in a given state, where each agent selects an action from those available to him at that state. All agents execute their actions synchronously and the combination of these actions together with the current state determine a transition to a unique successor state in the model. A *play* in a CGM is an infinite sequence of subsequent successor states, i.e., an infinite sequence $s_0 s_1 ... \in \mathsf{St}^\omega$ of states such that for each $i \geq 0$ there exists an action profile $\sigma_\mathbb{A} = \langle \sigma_1, \ldots, \sigma_k \rangle$ such that $\mathsf{out}(s_i, \sigma_\mathbb{A}) = s_{i+1}$. A *history* is a finite prefix of a play. We denote by $\mathsf{Plays}_\mathcal{M}$ and $\mathsf{Hist}_\mathcal{M}$ respectively the set of plays and set of histories in $\mathcal{M}$. For a state $s \in \mathsf{St}$ we define $\mathsf{Plays}_\mathcal{M}(s)$ and $\mathsf{Hist}_\mathcal{M}(s)$ as the set of plays and set of histories with initial state s. Given a sequence of states λ, we denote by λ_0 its initial state, by λ_i its $(i+1)$th state, by $\lambda_{\leq i}$ the prefix $\lambda_0 ... \lambda_i$ of λ and by $\lambda_{\geq i}$ the suffix $\lambda_i \lambda_{i+1} ...$ of λ. When $\lambda = \lambda_0 ... \lambda_\ell$ is finite, we say that it has length ℓ and write $|\lambda| = \ell$. Further, we put $\mathsf{last}(\lambda) = \lambda_\ell$.

A *(perfect recall) strategy* for an agent a in $\mathcal{M}$ is a mapping $F_\mathsf{a} : \mathsf{Hist}_\mathcal{M} \to \mathsf{Act}_\mathsf{a}$ such that for all $h \in \mathsf{Hist}_\mathcal{M}$ we have $F_\mathsf{a}(h) \in \mathsf{act}_\mathsf{a}(\mathsf{last}(h))$. Intuitively, it assigns an admissible action for agent a after any history h of the game. We denote by $\mathsf{Strat}_\mathcal{M}(\mathsf{a})$ the set of strategies of agent a. A *(collective) strategy* of a set *(coalition)* of agents $A \subseteq \mathbb{A}$ is a tuple $(F_\mathsf{a})_{\mathsf{a} \in A}$ of strategies, one for each agent in A. When $A = \mathbb{A}$ this is called a *strategy profile*. We denote by $\mathsf{Strat}_\mathcal{M}(A)$ the set of collective strategies of coalition A. A play $\lambda \in \mathsf{Plays}_\mathcal{M}$ is *consistent with a collective strategy* $F_A \in \mathsf{Strat}_\mathcal{M}(A)$ if for every $i \geq 0$ there exists an action profile $\sigma_\mathbb{A} = \langle \sigma_1, \ldots, \sigma_k \rangle$ such that $\mathsf{out}(\lambda_i, \sigma_\mathbb{A}) = \lambda_{i+1}$ and $\sigma_\mathsf{a} = F_\mathsf{a}(\lambda_{\leq i})$ for all $\mathsf{a} \in A$. The set of plays with initial state s that are consistent with F_A is denoted $\mathsf{Plays}_\mathcal{M}(s, F_A)$. For any coalition $A \subseteq \mathbb{A}$ and a given CGM $\mathcal{M}$ and state $s \in \mathsf{St}$, an *A-co-move* at s in $\mathcal{M}$ is a mapping $\mathsf{Act}_A^c : \mathsf{Act}_A \to \mathsf{Act}_{\mathbb{A} \setminus A}$ that assigns to every collective action of A at the state s a collective action at s for the complementary coalition $\mathbb{A} \setminus A$. Likewise, an *A-co-strategy* in $\mathcal{M}$ is a mapping $F_A^c : \mathsf{Strat}_\mathcal{M}(A) \times \mathsf{St} \to \mathsf{Act}_{\mathbb{A} \setminus A}$ that assigns to every collective strategy of A and a state $s \in \mathsf{St}$ a collective action at s for $\mathbb{A} \setminus A$.

2.2 The Logic ATL* and Fragments

The logic ATL* is a multi-agent extension of CTL* with *strategic quantifiers* $\langle\!\langle A \rangle\!\rangle$ indexed with coalitions A of agents. There are two types of formulae in ATL*: *state formulae*, that are evaluated at states, and *path formulae*, that are evaluated on plays. To simplify the presentation we will work with formulae in negation normal form over a fixed set Prop of atomic propositions and primitive temporal operators *Always* $\square$ and *Until* $\mathcal{U}$. The syntax of the full language ATL* and its fragments ATL$^+$ and ATL can then be defined as follows, where $l \in \mathsf{Prop} \cup \{ \neg p \mid p \in \mathsf{Prop} \}$ is a literal, $\mathbb{A}$ is a fixed set of agents and $A \subseteq \mathbb{A}$:

$$\text{State formulae: } \varphi := l \mid (\varphi \vee \varphi) \mid (\varphi \wedge \varphi) \mid \langle\!\langle A \rangle\!\rangle \Phi \mid [\![A]\!] \Phi \tag{1}$$

$$\text{ATL*-path formulae: } \Phi := \varphi \mid \bigcirc \Phi \mid \square \Phi \mid (\Phi \mathcal{U} \Phi) \mid (\Phi \vee \Phi) \mid (\Phi \wedge \Phi) \tag{2}$$

$$\text{ATL}^+\text{-path formulae: } \Phi := \varphi \mid \bigcirc \varphi \mid \square \varphi \mid (\varphi \mathcal{U} \varphi) \mid (\Phi \vee \Phi) \mid (\Phi \wedge \Phi) \tag{3}$$

$$\text{ATL-path formulae: } \Phi := \quad \bigcirc \varphi \mid \square \varphi \mid (\varphi \mathcal{U} \varphi) \tag{4}$$

Note that the state formulae have the same definition but define different sets in all 3 cases. To keep the notation lighter, we will list the members of the set A in $\langle\langle A \rangle\rangle$ without using $\{\}$. When the length of a formula is measured, A will be assumed given by a bit vector. Parentheses will be omitted whenever safe, but they will be important when conjunctions and disjunctions are composed.

Hereafter, we use φ, ψ, η to denote arbitrary state formulae and Φ, Ψ to denote path formulae. By an ATL^+ formula we will mean by default a *state* formula of ATL^+; likewise for ATL. We define $\top := p \vee \neg p$, $\bot := \neg\top$ and the temporal operators *Sometime* $\Diamond$ by $\Diamond\varphi := \top\mathcal{U}\varphi$ and *Release* $\mathcal{R}$ by $\varphi\mathcal{R}\psi := \Box\varphi \vee \varphi\mathcal{U}(\varphi \wedge \psi)$. Note, that $\langle\langle A \rangle\rangle\varphi\mathcal{R}\psi$ and $[\![A]\!]\varphi\mathcal{R}\psi$ are ATL^+ state formulae.

CTL^* can be regarded as the fragment of ATL^* where $\langle\langle\emptyset\rangle\rangle$ represents the path quantifier $\forall$ and $\langle\langle\mathbb{A}\rangle\rangle$ represents $\exists$. The semantics of ATL^* (inherited by ATL^+) is defined in a given CGM $\mathcal{M}$, state $s \in \mathcal{M}$ and a path λ in $\mathcal{M}$ just like the semantics of CTL^*, with the added clauses for the strategic quantifiers:

- $\mathcal{M}, s \models \langle\langle A \rangle\rangle\Phi$ iff there exists an A-strategy F_A such that, for all computations λ consistent with F_A, $\mathcal{M}, \lambda \models \Phi$.
- $\mathcal{M}, s \models [\![A]\!]\Phi$ iff there exists an A-co-strategy F_A^c such that, for all computations λ consistent with F_A^c, $\mathcal{M}, \lambda \models \Phi$.

Valid, satisfiable and equivalent formulae in ATL^* are defined as usual. Here are some important equivalences in LTL [5] and in ATL^* [1,6], used further:

- $\Box\Psi \equiv \Psi \wedge \bigcirc\Box\Psi$; $\quad \Phi\mathcal{U}\Psi \equiv \Psi \vee (\Phi \wedge \bigcirc(\Phi\mathcal{U}\Psi))$;
- $\langle\langle C \rangle\rangle\Box\Psi \equiv \Psi \wedge \langle\langle C \rangle\rangle \bigcirc \langle\langle C \rangle\rangle\Box\Psi$; $\quad \langle\langle C \rangle\rangle\Phi\mathcal{U}\Psi \equiv \Psi \vee (\Phi \wedge \langle\langle C \rangle\rangle \bigcirc \langle\langle C \rangle\rangle\Phi\mathcal{U}\Psi)$;
- $[\![C]\!]\Box\Psi \equiv \Psi \wedge [\![C]\!] \bigcirc [\![C]\!]\Box\Psi$; $\quad [\![C]\!]\Phi\mathcal{U}\Psi \equiv \Psi \vee (\Phi \wedge [\![C]\!] \bigcirc [\![C]\!]\Phi\mathcal{U}\Psi)$;
- $[\![A]\!] \bigcirc \varphi \equiv \neg\langle\langle A \rangle\rangle \bigcirc \neg\varphi \equiv \langle\langle\emptyset\rangle\rangle \bigcirc \varphi$; $\quad \langle\langle A \rangle\rangle\langle\langle B \rangle\rangle\Phi \equiv \langle\langle B \rangle\rangle\Phi$;
- For every state formula φ: $\langle\langle A \rangle\rangle(\varphi \wedge \Psi) \equiv \varphi \wedge \langle\langle A \rangle\rangle\Psi$, $\langle\langle A \rangle\rangle(\varphi \vee \Psi) \equiv \varphi \vee \langle\langle A \rangle\rangle\Psi$.

Remark 1. It is known [1] that, when restricted to ATL formulae, the semantics above (based on perfect-recall strategies) is equivalent to the semantics based on positional (or memoryless) strategies, where the prescribed actions only depend on the current state, not on the whole history. This is no longer the case for ATL^+. For example, the formula $\langle\langle 1 \rangle\rangle\Diamond(p \wedge \langle\langle 1 \rangle\rangle\Diamond q) \to \langle\langle 1 \rangle\rangle(\Diamond p \wedge \Diamond q)$ is valid in the semantics with perfect-recall strategies (which can be freely composed) but not in the semantics with positional strategies (which cannot be freely composed).

Hereafter, we assume that the semantics is based on perfect-recall strategies.

Here we deal with the *(constructive) satisfiability decision problem* for ATL^+:

Given a state formula φ in ATL^+, does there exist a CGM $\mathcal{M}$ and a state s in $\mathcal{M}$ such that $\mathcal{M}, s \models \varphi$? If so, construct such a satisfying pair $(\mathcal{M}, s)$.

Remark 2. There are two variants of this satisfiability problem: *tight*, where it is assumed that all agents in the model are mentioned in the formula, and *loose*, where additional agents, not mentioned in the formula, are allowed in the model. These variants are really different, but the latter one is immediately reducible to the former, by adding just one extra agent a to the language. Furthermore, this extra agent can be easily added superfluously to the formula, e.g., by adding a conjunct $\langle\langle a \rangle\rangle \bigcirc \top$, so we hereafter only consider the tight satisfiability version. For further details and discussion on this issue, see e.g., [11,7].

3   Decomposition and Closure of ATL$^+$ Formulae

We partition the set of ATL$^+$ formulae into *primitive* and *non-primitive* formulae. The primitive formulae are $\top, \bot$, the literals and all ATL$^+$ *successor formulae*, of the form $\langle\!\langle A \rangle\!\rangle \bigcirc \psi$ or $[\![A]\!] \bigcirc \psi$, each with *successor component* ψ. The non-primitive formulae are classified as α-, β- and γ-formulae. An α-formula in our syntax is a conjunction $\varphi \wedge \psi$ with (conjunctive) α-*components* φ and ψ; a β-formula is a disjunction $\varphi \vee \psi$ with (disjunctive) β-*components* φ and ψ. The rest of the non-primitive formulae are classified as γ-formulae. That is, a γ-formula is one of the form $[\![A]\!]\Phi$ or $\langle\!\langle A \rangle\!\rangle\Phi$, where Φ is an ATL$^+$ path formula whose main operator is not $\bigcirc$. We note that, unlike [7], here we do not treat $\langle\!\langle A \rangle\!\rangle\square\varphi$ as an α-formula nor $\langle\!\langle A \rangle\!\rangle\varphi\mathcal{U}\psi$ as a β-formula; both are γ-formulae.

The α- and β-formulae will be decomposed in the tableau as usual, while the case of γ-formulae $\langle\!\langle A \rangle\!\rangle\Phi$ and $[\![A]\!]\Phi$ is special and needs extra work, because their tableau decomposition will depend on the structure of Φ.

3.1 γ-Decomposition and γ-Components of γ-Formulae

We denote the set of ATL$^+$ state formulae by ATL^+_s and the set of ATL$^+$ path formulae by ATL^+_p. We will define a γ-*decomposition function* dec : ATL$^+_p \to \mathcal{P}(\text{ATL}^+_s \times \text{ATL}^+_p)$ with the following intuitive meaning: for any $\Phi \in \text{ATL}^+_p$ and pair $\langle \psi, \Psi \rangle \in \text{dec}(\Phi)$, ψ is a state formula true at the current state and Ψ is a path formula expressing what must be true at the next state of a possible play starting at the current state. Thus, the set $\text{dec}(\Phi)$ is interpreted as a disjunction describing all possible 'types of paths' starting from the current state and satisfying Φ. The definition of dec is recursive on ATL$^+$ path formulae, as follows.

$\star$ $\text{dec}(\varphi) = \{\langle \varphi, \top \rangle\}$, $\text{dec}(\bigcirc\varphi) = \{\langle \top, \varphi \rangle\}$ for any ATL$^+$ state formula φ. The other base cases derive from the well-known LTL equivalences listed in 2.2:

$\star$ $\text{dec}(\square\varphi) = \{\langle \varphi, \square\varphi \rangle\}$ and $\text{dec}(\varphi\mathcal{U}\psi) = \{\langle \varphi, \varphi\mathcal{U}\psi \rangle, \langle \psi, \top \rangle\}$.

$\star$ $\text{dec}(\Phi_1 \wedge \Phi_2) = \text{dec}(\Phi_1) \otimes \text{dec}(\Phi_2)$, where
$\text{dec}(\Phi_1)\otimes\text{dec}(\Phi_2) := \{\langle \psi_i \wedge \psi_j, \Psi_i \wedge \Psi_j \rangle \mid \langle \psi_i, \Psi_i \rangle \in \text{dec}(\Phi_1), \langle \psi_j, \Psi_j \rangle \in \text{dec}(\Phi_2)\}$.

$\star$ $\text{dec}(\Phi_1 \vee \Phi_2) = \text{dec}(\Phi_1) \cup \text{dec}(\Phi_2) \cup (\text{dec}(\Phi_1) \oplus \text{dec}(\Phi_2))$,
where $\text{dec}(\Phi_1) \oplus \text{dec}(\Phi_2) :=$
$\{\langle \psi_i \wedge \psi_j, \Psi_i \vee \Psi_j \rangle \mid \langle \psi_i, \Psi_i \rangle \in \text{dec}(\Phi_1), \langle \psi_j, \Psi_j \rangle \in \text{dec}(\Phi_2), \Psi_i \neq \top, \Psi_j \neq \top\}$.

The conjunctive case is clear: every path satisfying $\Phi_1 \wedge \Phi_2$ combines a type of path satisfying Φ_1 with a type of path satisfying Φ_2. To understand the disjunctive case, first note that the use of $\text{dec}(\Phi_1) \oplus \text{dec}(\Phi_2)$ in the above union reflects the case of those plays where it is not decided yet which disjunct of $\Phi_1 \vee \Phi_2$ will hold, so we have to keep both disjuncts true at the present state and delay the choice. This is why the state formulae ψ_i and ψ_j are connected by $\wedge$ but the path formulae Ψ_i and Ψ_j are connected by $\vee$. Moreover, the $\oplus$ operation avoids the construction of a pair $\langle \psi_i \wedge \psi_j, \Psi_i \vee \Psi_j \rangle$ where either Ψ_i or Ψ_j is $\top$, because in that case we would be in a situation already included in $\text{dec}(\Phi_1)$ or in $\text{dec}(\Phi_2)$. The three cases for paths satisfying the disjunction $\Phi_1 \vee \Phi_2$ can be illustrated by the picture in Figure 1.

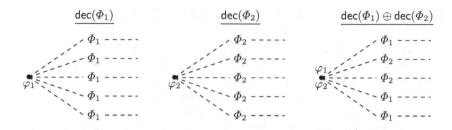

Fig. 1. The 3 cases for disjunctive path objectives in a γ-formula

Now, let $\zeta = \langle\!\langle A \rangle\!\rangle \Phi$ or $\zeta = [\![A]\!]\Phi$ be a γ-formula to be decomposed. Each pair $\langle \psi, \Psi \rangle \in \text{dec}(\Phi)$ is then converted to a γ-*component* $\gamma(\psi, \Psi)$ as follows:

$$\gamma(\psi, \Psi) = \psi \quad \text{if } \Psi = \top \tag{5}$$

$$\gamma(\psi, \Psi) = \psi \wedge \langle\!\langle A \rangle\!\rangle \bigcirc \langle\!\langle A \rangle\!\rangle \Psi \quad \text{if } \zeta \text{ is of the form } \langle\!\langle A \rangle\!\rangle \Phi, \tag{6}$$

$$\gamma(\psi, \Psi) = \psi \wedge [\![A]\!]\bigcirc[\![A]\!]\Psi \quad \text{if } \zeta \text{ is of the form } [\![A]\!]\Phi \tag{7}$$

The following key lemma claims that every γ-formula is equivalent to the disjunction of its γ-components. For the (long and non-trivial) proof see [2].

Lemma 1. *For any* ATL^+ γ-*formula* $\Theta = \langle\!\langle A \rangle\!\rangle \Phi$ *or* $\Theta = [\![A]\!]\Phi$:

1. $\Phi \equiv \bigvee\{\psi \wedge \bigcirc\Psi \mid \langle \psi, \Psi \rangle \in \text{dec}(\Phi)\}$.
2. $\langle\!\langle A \rangle\!\rangle \Phi \equiv \bigvee\{\langle\!\langle A \rangle\!\rangle(\psi \wedge \bigcirc\Psi) \mid \langle \psi, \Psi \rangle \in \text{dec}(\Phi)\}$, *and respectively,*
   $[\![A]\!]\Phi \equiv \bigvee\{[\![A]\!](\psi \wedge \bigcirc\Psi) \mid \langle \psi, \Psi \rangle \in \text{dec}(\Phi)\}$.
3. $\Theta \equiv \bigvee\{\gamma(\psi, \Psi) \mid \langle \psi, \Psi \rangle \in \text{dec}(\Phi)\}$.

Example 1. We will use 2 syntactically similar, yet different, running examples: $\theta = \langle\!\langle 1 \rangle\!\rangle(p\mathcal{U}q \vee \Box q) \wedge \langle\!\langle 2 \rangle\!\rangle(\Diamond p \wedge \Box\neg q)$ and $\vartheta = \langle\!\langle 1 \rangle\!\rangle(p\mathcal{U}q \vee \Box q) \wedge [\![2]\!](\Diamond p \wedge \Box\neg q)$.

First, we consider θ. It is an α-formula with conjunctive components $\theta_1 = \langle\!\langle 1 \rangle\!\rangle(p\mathcal{U}q \vee \Box q)$ and $\theta_2 = \langle\!\langle 2 \rangle\!\rangle(\Diamond p \wedge \Box\neg q)$. Further, θ_1 is a γ-formula of the form $\langle\!\langle A \rangle\!\rangle \Phi$ where the main connective of Φ is $\vee$. So $\text{dec}(\theta_1) = \text{dec}(p\mathcal{U}q) \cup \text{dec}(\Box q) \cup (\text{dec}(p\mathcal{U}q) \oplus \text{dec}(\Box q))$, where $\text{dec}(p\mathcal{U}q) = \{\langle p, p\mathcal{U}q \rangle, \langle q, \top \rangle\}$ and $\text{dec}(\Box q) = \{\langle q, \Box q \rangle\}$. Thus, $\text{dec}(\theta_1) = \{\langle p, p\mathcal{U}q \rangle, \langle q, \top \rangle, \langle q, \Box q \rangle, \langle p \wedge q, p\mathcal{U}q \vee \Box q \rangle\}$, hence $\theta_1 \equiv (p \wedge \langle\!\langle 1 \rangle\!\rangle\bigcirc\langle\!\langle 1 \rangle\!\rangle p\mathcal{U}q) \vee (q) \vee (q \wedge \langle\!\langle 1 \rangle\!\rangle\bigcirc\langle\!\langle 1 \rangle\!\rangle\Box q) \vee (p \wedge q \wedge \langle\!\langle 1 \rangle\!\rangle\bigcirc\langle\!\langle 1 \rangle\!\rangle(p\mathcal{U}q \vee \Box q))$. Likewise, θ_2 is a γ-formula of the form $\langle\!\langle A \rangle\!\rangle \Phi$ where the main connective of Φ is $\wedge$. So $\text{dec}(\theta_2) = \text{dec}(\Diamond p) \otimes \text{dec}(\Box\neg q)$, with $\text{dec}(\Diamond p) = \{\langle \top, \Diamond p \rangle, \langle p, \top \rangle\}$ and $\text{dec}(\Box\neg q) = \{\langle \neg q, \Box\neg q \rangle\}$. Thus, $\text{dec}(\theta_2) = \{\langle \top \wedge \neg q, \Diamond p \wedge \Box\neg q \rangle, \langle p \wedge \neg q, \top \wedge \Box\neg q \rangle\} = \{\langle \neg q, \Diamond p \wedge \Box\neg q \rangle, \langle p \wedge \neg q, \Box\neg q \rangle\}$ and $\theta_2 \equiv (\neg q \wedge \langle\!\langle 2 \rangle\!\rangle\bigcirc\langle\!\langle 2 \rangle\!\rangle(\Diamond p \wedge \Box\neg q)) \vee (p \wedge \neg q \wedge \langle\!\langle 2 \rangle\!\rangle\bigcirc\langle\!\langle 2 \rangle\!\rangle\Box\neg q)$.

For ϑ, the γ-decomposition is similar, we only replace $\langle\!\langle 2 \rangle\!\rangle$ by $[\![2]\!]$. Thus, we obtain $\vartheta_1 \equiv (p \wedge \langle\!\langle 1 \rangle\!\rangle\bigcirc\langle\!\langle 1 \rangle\!\rangle p\mathcal{U}q) \vee (q) \vee (q \wedge \langle\!\langle 1 \rangle\!\rangle\bigcirc\langle\!\langle 1 \rangle\!\rangle\Box q) \vee (p \wedge q \wedge \langle\!\langle 1 \rangle\!\rangle\bigcirc\langle\!\langle 1 \rangle\!\rangle(p\mathcal{U}q \vee \Box q))$ and $\vartheta_2 \equiv (\neg q \wedge [\![2]\!]\bigcirc[\![2]\!](\Diamond p \wedge \Box\neg q)) \vee (p \wedge \neg q \wedge [\![2]\!]\bigcirc[\![2]\!]\Box\neg q)$.

The *closure* $cl(\psi)$ of an ATL^+ state formula ψ is the least set of ATL^+ formulae such that $\psi, \top, \bot \in cl(\psi)$ and $cl(\psi)$ is closed under taking of successor-, α-, β- and γ-components. For any set of state formulae Γ we define $cl(\Gamma) := \bigcup\{cl(\psi) \mid \psi \in \Gamma\}$. We denote by $|\psi|$ the length of ψ and by $\|\Gamma\|$ the cardinality of Γ.

Example 2. The closures of the formulae θ and ϑ from Example 1 are:

$\mathsf{cl}(\theta) = \{\theta, \theta_1, \theta_2, p \wedge q \wedge \langle\langle 1\rangle\rangle \bigcirc \langle\langle 1\rangle\rangle (p\,\mathcal{U}q \vee \Box q), p \wedge q, p, q, \langle\langle 1\rangle\rangle \bigcirc \langle\langle 1\rangle\rangle (p\,\mathcal{U}q \vee$
$\Box q), q \wedge \langle\langle 1\rangle\rangle \bigcirc \langle\langle 1\rangle\rangle \Box q, \langle\langle 1\rangle\rangle \bigcirc \langle\langle 1\rangle\rangle \Box q, \langle\langle 1\rangle\rangle \Box q, p \wedge \langle\langle 1\rangle\rangle \bigcirc \langle\langle 1\rangle\rangle p\,\mathcal{U}q, \langle\langle 1\rangle\rangle \bigcirc \langle\langle 1\rangle\rangle p\,\mathcal{U}q,$
$\langle\langle 1\rangle\rangle p\,\mathcal{U}q, p \wedge \neg q \wedge \langle\langle 2\rangle\rangle \bigcirc \langle\langle 2\rangle\rangle \Box \neg q, p \wedge \neg q, \neg q, \langle\langle 2\rangle\rangle \bigcirc \langle\langle 2\rangle\rangle \Box \neg q, \langle\langle 2\rangle\rangle \Box \neg q, \neg q \wedge \langle\langle 2\rangle\rangle \bigcirc$
$\langle\langle 2\rangle\rangle \Box \neg q, \neg q \wedge \langle\langle 2\rangle\rangle \bigcirc \langle\langle 2\rangle\rangle (\Diamond p \wedge \Box \neg q), \langle\langle 2\rangle\rangle \bigcirc \langle\langle 2\rangle\rangle (\Diamond p \wedge \Box \neg q), \top\}.$

$\mathsf{cl}(\vartheta) = \{\vartheta, \vartheta_1, \vartheta_2, p \wedge q \wedge \langle\langle 1\rangle\rangle \bigcirc \langle\langle 1\rangle\rangle (p\,\mathcal{U}q \vee \Box q), p \wedge q, p, q, \langle\langle 1\rangle\rangle \bigcirc \langle\langle 1\rangle\rangle (p\,\mathcal{U}q \vee$
$\Box q), q \wedge \langle\langle 1\rangle\rangle \bigcirc \langle\langle 1\rangle\rangle \Box q, \langle\langle 1\rangle\rangle \bigcirc \langle\langle 1\rangle\rangle \Box q, \langle\langle 1\rangle\rangle \Box q, p \wedge \langle\langle 1\rangle\rangle \bigcirc \langle\langle 1\rangle\rangle p\,\mathcal{U}q, \langle\langle 1\rangle\rangle \bigcirc \langle\langle 1\rangle\rangle p\,\mathcal{U}q,$
$\langle\langle 1\rangle\rangle p\,\mathcal{U}q, p \wedge \neg q \wedge [2] \bigcirc [2] \Box \neg q, p \wedge \neg q, \neg q, [2] \bigcirc [2] \Box \neg q, [2] \Box \neg q, \neg q \wedge [2] \bigcirc$
$[2] \Box \neg q, \neg q \wedge [2] \bigcirc [2] (\Diamond p \wedge \Box \neg q), [2] \bigcirc [2] (\Diamond p \wedge \Box \neg q), \top\}.$

Lemma 2. *For any* ATL$^+$ *state formula* φ, $\|cl(\varphi)\| < 2^{|\varphi|^2}$.

Proof. Every formula in $cl(\varphi)$ has length less than $2|\varphi|$ and is built from symbols in φ, so there can be at most $|\varphi|^{2|\varphi|} = 2^{2|\varphi|\log_2|\varphi|} < 2^{|\varphi|^2}$ such formulae. $\square$

The estimate above is rather crude, but $\|cl(\varphi)\|$ *can* reach size exponential in $|\varphi|$. Indeed, consider the formulae $\phi_k = \langle\langle 1\rangle\rangle (p_1\,\mathcal{U}q_1 \wedge (p_2\,\mathcal{U}q_2 \wedge (\ldots \wedge p_k\,\mathcal{U}q_k)\ldots)$ for $k = 1, 2, \ldots$ and distinct $p_1, q_1, \ldots, p_k, q_k, \ldots \in \mathsf{Prop}$. Then $|\phi_k| = O(k)$, while the number of different γ-components of ϕ_k is 2^k, hence $\|cl(\phi_k)\| > 2^k$.

3.2   Full Expansions of Sets of ATL$^+$ Formulae

As part of the tableaux construction we will need a procedure that, for any given finite set of ATL$^+$ state formulae Γ, produces all "full expansions" (called in [7] "downward saturated extensions") defined below.

Definition 1. *Let* Γ, Δ *be sets of* ATL$^+$ *state formulae and* $\Gamma \subseteq \Delta \subseteq cl(\Gamma)$.

1. Δ *is* patently inconsistent *if it contains* $\bot$ *or a pair of formulae* φ *and* $\neg\varphi$.
2. Δ *is a* full expansion *of* Γ *if it is not patently inconsistent and satisfies the following closure conditions:*
   - *if* $\varphi \wedge \psi \in \Delta$ *then* $\varphi \in \Delta$ *and* $\psi \in \Delta$;
   - *if* $\varphi \vee \psi \in \Delta$ *then* $\varphi \in \Delta$ *or* $\psi \in \Delta$;
   - *if* $\varphi \in \Delta$ *is a* γ-*formula, then at least one* γ-*component of* φ *is in* Δ *and exactly one of these* γ-*components in* Δ, *denoted* $\gamma(\varphi, \Delta)$, *is designated as the* γ-*component in* Δ *linked to the* γ-*formula* φ, *as explained below.*

The family of all full expansions of Γ will be denoted by $FE(\Gamma)$. It can be constructed by a simple iterative procedure that starts with $\{\Gamma\}$ and repeatedly, until saturation, takes a set X from the currently constructed family, selects a formula $\varphi \in X$ and: if φ is a conjunction, then adds both conjunctive components of φ to X; if φ is a disjunction, then creates two extensions of X by adding respectively each disjunctive component of φ; and if φ is a γ-formula, then creates an extension of X with each γ-component ψ of φ and designates ψ as the γ-component of φ linked to φ in every full expansion of Γ eventually produced by further extending $X \cup \{\psi\}$. In case when such an extension becomes patently inconsistent it is discarded from the family. Clearly, this procedure terminates on every finite input set of formulae Γ and produces a family of at most $2^{\|cl(\Gamma)\|}$ sets. Furthermore, due to Lemma 1, we have the following:

Proposition 1. *For any finite set of* ATL$^+$ *state formulae* Γ:

$$\bigwedge \Gamma \equiv \bigvee \left\{ \bigwedge \Delta \mid \Delta \in FE(\Gamma) \right\}.$$

4   Tableau-Based Decision Procedure for ATL$^+$

The tableaux procedure consists of three major phases: *pretableau construction*, *prestate elimination*, and *state elimination*. It constructs a directed graph $\mathcal{T}^\eta$ (called a *tableau*) with nodes labelled by finite sets of formulae and directed edges between nodes relating them to successor nodes.

The pretableau construction phase produces the so-called *pretableau* $\mathcal{P}^\eta$ for the input formula η, with two kinds of nodes: *states* and *prestates*. States are fully expanded sets, meant to represent states of a CGM, while prestates can be any finite sets of formulae from $cl(\eta)$ and only play a temporary role in the construction of $\mathcal{P}^\eta$. States and prestates are labelled uniquely, so they can be identified with their labels. The prestate elimination phase creates a smaller graph $\mathcal{T}_0^\eta$ out of $\mathcal{P}^\eta$, called the *initial tableau for* η, by eliminating all the prestates from $\mathcal{P}^\eta$ and accordingly redirecting its edges. Finally, the state elimination phase removes, step-by-steps, all the states (if any) that cannot be satisfied in a CGM, because they lack necessary successors or because they contain unrealized eventualities. Eventually, the elimination procedure produces a (possibly empty) subgraph $\mathcal{T}^\eta$ of $\mathcal{T}_0^\eta$, called the *final tableau for* η. If some state Δ of $\mathcal{T}^\eta$ contains η, the procedure declares η satisfiable and a partly defined CGM (called *Hintikka game frame*) satisfying η can be extracted from it; otherwise it declares η unsatisfiable.

4.1   Pretableau Construction Phase

The pretableau construction phase for an input formula η starts with an initial prestate (with label) $\{\eta\}$ and consists of alternating application of two construction rules, until saturation: **(SR)**, expanding prestates into states, and **(Next)**, creating successor prestates from states. This phase closely resembles the corresponding one for ATL tableaux in [7], with the only essential difference being the γ-decomposition of γ-formulae used here by the rule **(SR)**, which causes, as we will see, a possibly exponential blow-up of the size of the tableaux, and eventually of the entire worst case time complexity, as compared to the ATL tableaux. Another (minor) difference with respect to [7] is in the formulation of both rules, because here we work with formulae in negation normal form.

Rule (SR). Given a prestate Γ, do the following:

1. For each full expansion Δ of Γ add to the pretableau a state with label Δ.
2. For each of the added states Δ, if Δ does not contain any formulae of the form $\langle\!\langle A \rangle\!\rangle \bigcirc \varphi$ or $[\![A]\!] \bigcirc \varphi$, add the formula $\langle\!\langle \mathbb{A} \rangle\!\rangle \bigcirc \top$ to it;
3. For each state Δ obtained at steps 1 and 2, link Γ to Δ via a $\Longrightarrow$ edge;
4. If, however, the pretableau already contains a state Δ' with label Δ, do not create another copy of it but only link Γ to Δ' via a $\Longrightarrow$ edge.

Example 3. For the formula θ from Example 1 the initial prestate is $\Gamma_0 = \{\langle\!\langle 1\rangle\!\rangle(p\,\mathcal{U}q \vee \Box q) \wedge \langle\!\langle 2\rangle\!\rangle(\Diamond p \wedge \Box\neg q)\}$. It has 2 full expansions:
$\Delta_1 = \{\theta, \theta_1, \theta_2, p, \neg q, \langle\!\langle 1\rangle\!\rangle\bigcirc\langle\!\langle 1\rangle\!\rangle p\,\mathcal{U}q, \langle\!\langle 2\rangle\!\rangle\bigcirc\langle\!\langle 2\rangle\!\rangle(\Diamond p \wedge \Box\neg q)\}$, and
$\Delta_2 = \{\theta, \theta_1, \theta_2, p, p \wedge \neg q, \neg q, \langle\!\langle 1\rangle\!\rangle\bigcirc\langle\!\langle 1\rangle\!\rangle p\,\mathcal{U}q, \langle\!\langle 2\rangle\!\rangle\bigcirc\langle\!\langle 2\rangle\!\rangle\Box\neg q\}$.

Likewise, for the formula ϑ: $\Gamma_0 = \{\langle\!\langle 1\rangle\!\rangle(p\,\mathcal{U}q \vee \Box q) \wedge [\![2]\!](\Diamond p \wedge \Box\neg q)\}$ is the initial prestate and it has 2 full expansions:
$\Delta_1 = \{\vartheta, \vartheta_1, \vartheta_2, p, \neg q, \langle\!\langle 1\rangle\!\rangle\bigcirc\langle\!\langle 1\rangle\!\rangle p\,\mathcal{U}q, [\![2]\!]\bigcirc[\![2]\!](\Diamond p \wedge \Box\neg q)\}$, and
$\Delta_2 = \{\vartheta, \vartheta_1, \vartheta_2, p, p \wedge \neg q, \neg q, \langle\!\langle 1\rangle\!\rangle\bigcirc\langle\!\langle 1\rangle\!\rangle p\,\mathcal{U}q, [\![2]\!]\bigcirc[\![2]\!]\Box\neg q\}$.

In the following, by *enforceable successor formula* we mean a formula of the form $\langle\!\langle A\rangle\!\rangle\bigcirc\psi$ and by *unavoidable successor formula* one of the form $[\![A]\!]\bigcirc\psi$.

Rule (Next). Given a state Δ do the following, where σ is a shorthand for σ_A:

1. List all primitive successor formulae of Δ in such a way that all enforceable successor formulae precede all unavoidable ones; let the result be the list

$$\mathbb{L} = \langle\!\langle A_0\rangle\!\rangle\bigcirc\varphi_0, \ldots, \langle\!\langle A_{m-1}\rangle\!\rangle\bigcirc\varphi_{m-1}, [\![A_0']\!]\bigcirc\psi_0, \ldots, [\![A_{l-1}']\!]\bigcirc\psi_{l-1}$$

Let $r_\Delta = m + l$; denote by $D(\Delta)$ the set $\{0, \ldots, r_\Delta - 1\}^{|A|}$. Then, for every $\sigma \in D(\Delta)$, denote $N(\sigma) := \{i \mid \sigma_i \geq m\}$, where σ_i is the ith component of the tuple σ, and let $\mathbf{co}(\sigma) := [\Sigma_{i \in N(\sigma)}(\sigma_i - m)] \mod l$.

2. For each $\sigma \in D(\Delta)$ create a prestate:

$$\Gamma_\sigma = \{\varphi_p \mid \langle\!\langle A_p\rangle\!\rangle\bigcirc\varphi_p \in \Delta \text{ and } \sigma_a = p \text{ for all } a \in A_p\}$$
$$\cup \{\psi_q \mid [\![A_q']\!]\bigcirc\psi_q \in \Delta, \mathbf{co}(\sigma) = q, \text{ and } \mathbb{A} - A_q' \subseteq N(\sigma)\}$$

If Γ_σ is empty, add $\top$ to it. Then connect Δ to Γ_σ with $\overset{\sigma}{\longrightarrow}$.
If, however, $\Gamma_\sigma = \Gamma$ for some prestate Γ that has already been added to the pretableau, only connect Δ to Γ with $\overset{\sigma}{\longrightarrow}$.

For intuition on the rule **(Next)** see [7] and [2]. The rules **(SR)** and **(Next)** are applied alternatively until saturation, which is bound to occur because every label is a subset of $cl(\eta)$. Then the construction phase is over. The graph built in that phase is called *pretableau* for the input formula η and denoted by $\mathcal{P}^\eta$.

Example 4. Continuation of Example 3 for θ: For Δ_1, the list of successor formulae is $\mathbb{L} = \langle\!\langle 1\rangle\!\rangle\bigcirc\langle\!\langle 1\rangle\!\rangle p\,\mathcal{U}q, \langle\!\langle 2\rangle\!\rangle\bigcirc\langle\!\langle 2\rangle\!\rangle(\Diamond p \wedge \Box\neg q)$, so $m = 2, l = 0$ and $r_{\Delta_1} = 2$. As there are no unavoidable successor formulae, we do not need to compute $N(\sigma)$ and $\mathbf{co}(\sigma)$. Then, $\Gamma_{(0,0)} = \{\langle\!\langle 1\rangle\!\rangle p\,\mathcal{U}q\} = \Gamma_1$, $\Gamma_{(0,1)} = \{\langle\!\langle 1\rangle\!\rangle p\,\mathcal{U}q, \langle\!\langle 2\rangle\!\rangle(\Diamond p \wedge \Box\neg q)\} = \Gamma_2$, $\Gamma_{(1,0)} = \{\top\} = \Gamma_3$ and $\Gamma_{(1,1)} = \{\langle\!\langle 2\rangle\!\rangle(\Diamond p \wedge \Box\neg q)\} = \Gamma_4$.

For Δ_2, the list of successor formulae is $\mathbb{L} = \langle\!\langle 1\rangle\!\rangle\bigcirc\langle\!\langle 1\rangle\!\rangle p\,\mathcal{U}q, \langle\!\langle 2\rangle\!\rangle\bigcirc\langle\!\langle 2\rangle\!\rangle\Box\neg q$, so $m = 2, l = 0$ and $r_{\Delta_2} = 2$. Here again, we do not compute $N(\sigma)$ and $\mathbf{co}(\sigma)$. Then $\Gamma_{(0,0)} = \{\langle\!\langle 1\rangle\!\rangle p\,\mathcal{U}q\} = \Gamma_1$, $\Gamma_{(0,1)} = \{\langle\!\langle 1\rangle\!\rangle p\,\mathcal{U}q, \langle\!\langle 2\rangle\!\rangle\Box\neg q\} = \Gamma_5$, $\Gamma_{(1,0)} = \{\top\} = \Gamma_3$ and $\Gamma_{(1,1)} = \{\langle\!\langle 2\rangle\!\rangle\Box\neg q\} = \Gamma_6$.

Applying rule **(SR)** to the so-obtained prestates, we have:
states$(\Gamma_1) = \{\Delta_3 : \{\langle\!\langle 1\rangle\!\rangle p\,\mathcal{U}q, p, \langle\!\langle 1\rangle\!\rangle\bigcirc\langle\!\langle 1\rangle\!\rangle p\,\mathcal{U}q\}, \Delta_4 : \{\langle\!\langle 1\rangle\!\rangle p\,\mathcal{U}q, q, \langle\!\langle 1, 2\rangle\!\rangle\bigcirc\top\}\}$,
states$(\Gamma_2) = \Delta_5 : \{\langle\!\langle 1\rangle\!\rangle p\,\mathcal{U}q, \langle\!\langle 2\rangle\!\rangle(\Diamond p \wedge \Box\neg q), p, \neg q, \langle\!\langle 1\rangle\!\rangle\bigcirc\langle\!\langle 1\rangle\!\rangle p\,\mathcal{U}q, \langle\!\langle 2\rangle\!\rangle\bigcirc$

$\langle\!\langle 2\rangle\!\rangle(\Diamond p \wedge \Box\neg q)\}, \Delta_6$: $\{\langle\!\langle 1\rangle\!\rangle pUq, \langle\!\langle 2\rangle\!\rangle(\Diamond p \wedge \Box\neg q), p, p \wedge \neg q, \neg q, \langle\!\langle 1\rangle\!\rangle \bigcirc$
$\langle\!\langle 1\rangle\!\rangle pUq, \langle\!\langle 2\rangle\!\rangle\bigcirc\langle\!\langle 2\rangle\!\rangle\Box\neg q\}\}$;
states$(\Gamma_3) = \{\Delta_7 : \{\top, \langle\!\langle 1, 2\rangle\!\rangle\bigcirc\top\}\}$;
states$(\Gamma_4) = \{\Delta_8 : \{\langle\!\langle 2\rangle\!\rangle(\Diamond p \wedge \Box\neg q), \neg q, \langle\!\langle 2\rangle\!\rangle\bigcirc\langle\!\langle 2\rangle\!\rangle(\Diamond p \wedge \Box\neg q)\}, \Delta_9 : \{\langle\!\langle 2\rangle\!\rangle(\Diamond p \wedge \Box\neg q), p \wedge \neg q, \neg q, \langle\!\langle 2\rangle\!\rangle\bigcirc\langle\!\langle 2\rangle\!\rangle\Box\neg q\}\}$;
states$(\Gamma_5) = \{\Delta_{10} : \{\langle\!\langle 1\rangle\!\rangle pUq, \langle\!\langle 2\rangle\!\rangle\Box\neg q, p, \neg q, \langle\!\langle 1\rangle\!\rangle\bigcirc\langle\!\langle 1\rangle\!\rangle pUq, \langle\!\langle 2\rangle\!\rangle\bigcirc\langle\!\langle 2\rangle\!\rangle\Box\neg q\}\}$;
states$(\Gamma_6) = \{\Delta_{11} : \{\langle\!\langle 2\rangle\!\rangle\Box\neg q, \neg q, \langle\!\langle 2\rangle\!\rangle\bigcirc\langle\!\langle 2\rangle\!\rangle\Box\neg q\}\}$.

The pretableau for θ is given in Figure 2.

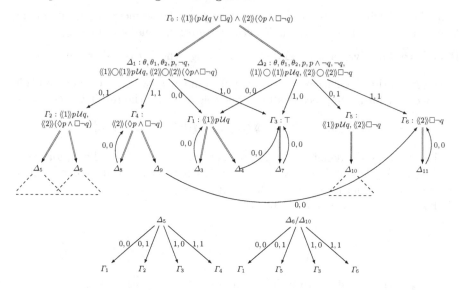

Fig. 2. The pretableau for θ

Example 5. Continuation of Example 3 for ϑ: For Δ_1, the list of successor formulae is $\mathbb{L} = \langle\!\langle 1\rangle\!\rangle\bigcirc\langle\!\langle 1\rangle\!\rangle pUq, [2]\bigcirc[2](\Diamond p \wedge \Box\neg q)$, so $m = 1$, $l = 1$ and $r_{\Delta_1} = 2$. Therefore, $N(0,0) = \emptyset$, $N(0,1) = \{2\}$, $N(1,0) = \{1\}$, $N(1,1) = \{1,2\}$ and also $\mathbf{co}(0,0) = \mathbf{co}(0,1) = 0 = \mathbf{co}(0,1) = \mathbf{co}(0,1) = 0$. Then, $\Gamma_{(0,0)} = \Gamma_{(0,1)} = \{\langle\!\langle 1\rangle\!\rangle pUq\} = \Gamma_1$, and $\Gamma_{(1,0)} = \Gamma_{(1,1)} = \{[2](\Diamond p \wedge \Box\neg q)\} = \Gamma_2$.

For Δ_2, the list of successor formulae is $\mathbb{L} = \langle\!\langle 1\rangle\!\rangle\bigcirc\langle\!\langle 1\rangle\!\rangle pUq, [2]\bigcirc[2]\Box\neg q$, so $m = 1$, $l = 1$ and $r_{\Delta_2} = 2$. Here also $N(0,0) = \emptyset$, $N(0,1) = \{2\}$, $N(1,0) = \{1\}$, $N(1,1) = \{1,2\}$, and $\mathbf{co}(0,0) = \mathbf{co}(0,1) = \mathbf{co}(0,1) = 0$. Then, $\Gamma_{(0,0)} = \Gamma_{(0,1)} = \{\langle\!\langle 1\rangle\!\rangle pUq\} = \Gamma_1$, and $\Gamma_{(1,0)} = \Gamma_{(1,1)} = \{[2]\Box\neg q\} = \Gamma_3$.

In the same way, we obtain:
states$(\Gamma_1) = \{\Delta_3 : \{\langle\!\langle 1\rangle\!\rangle pUq, p, \langle\!\langle 1\rangle\!\rangle\bigcirc\langle\!\langle 1\rangle\!\rangle pUq\}, \Delta_4 : \{\langle\!\langle 1\rangle\!\rangle pUq, q, \langle\!\langle 1, 2\rangle\!\rangle\bigcirc\top\}\}$;
states$(\Gamma_2) = \{\Delta_5 : \{[2](\Diamond p \wedge \Box\neg q), \neg q, [2]\bigcirc[2](\Diamond p \wedge \Box\neg q)\}, \Delta_6 : \{[2](\Diamond p \wedge \Box\neg q), p \wedge\neg q, p, \neg q, [2]\bigcirc[2]\Box\neg q\}\}$; **states**$(\Gamma_3) = \{\Delta_7 : \{[2]\Box\neg q, \neg q, [2]\bigcirc[2]\Box\neg q\}\}$.

4.2 The Prestate and State Elimination Phases. Eventualities

First, we remove from $\mathcal{P}^\eta$ all the prestates and the $\Longrightarrow$ edges, as follows. For every prestate Γ in $\mathcal{P}^\eta$ put $\Delta \xrightarrow{\sigma} \Delta'$ for all states Δ in $\mathcal{P}^\eta$ with $\Delta \xrightarrow{\sigma} \Gamma$ and all

$\Delta' \in \mathbf{states}(\Gamma)$; then, remove Γ from $\mathcal{P}^\eta$. The graph obtained after eliminating all prestates is called the *initial tableau*, denoted by $\mathcal{T}_0^\eta$. The initial tableau for the formula θ in our running example is given on Figure 3.

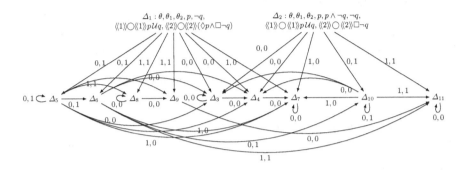

Fig. 3. The initial tableau for θ

The elimination phase starts with $\mathcal{T}_0^\eta$ and goes through stages. At stage $n+1$ we remove exactly one state from the tableau $\mathcal{T}_n^\eta$ obtained at the previous stage, by applying one of the elimination rules described below, thus obtaining the tableau $\mathcal{T}_{n+1}^\eta$. The set of states of $\mathcal{T}_m^\eta$ is noted S_m^η.

The first elimination rule **(ER1)**, defined below, is used to eliminate all states with missing successors for some move vectors determined by the rule **(Next)**. If, due to a previous state elimination, any state has an outgoing move vector for which the corresponding successor state is missing, we delete the state. The reason is clear: if Δ is to be satisfiable, then for each $\sigma \in D(\Delta)$ there should exist a satisfiable Δ' that Δ reaches via σ. Formally, the rule is stated as follows, where $D(\Delta)$ is defined in the rule **(Next)**:

Rule (ER1): If, for some $\sigma \in D(\Delta)$, all states Δ' with $\Delta \xrightarrow{\sigma} \Delta'$ have been eliminated at earlier stages, then obtain $\mathcal{T}_{n+1}^\eta$ by eliminating Δ from $\mathcal{T}_n^\eta$.

The aim of the next elimination rule is to make sure that there are no *unrealized eventualities*. In ATL there are only two kinds of eventualities : $\langle\!\langle A \rangle\!\rangle \varphi \mathcal{U} \psi$ and $[\![A]\!]\Diamond\varphi$. The situation is more complex in ATL$^+$. For instance, should the formula $\langle\!\langle A \rangle\!\rangle (\Box\varphi \vee \psi_1 \mathcal{U} \psi_2)$ be considered an eventuality? Our solution for ATL$^+$ is to consider all γ-formulae as *potential eventualities*. In order to properly define the notion of *realization of a potential eventuality* we first define a function *Real* associating to a γ-formula ψ and a set of ATL$^+$-formulae (representing a state of the current tableau) a Boolean value indicating whether the potential eventuality represented by ψ has been 'realized' at that state:

- $Real(\Phi \wedge \Psi, \Theta) = Real(\Phi, \Theta) \wedge Real(\Psi, \Theta)$
- $Real(\Phi \vee \Psi, \Theta) = Real(\Phi, \Theta) \vee Real(\Psi, \Theta)$
- $Real(\varphi, \Theta) = true$ if $\varphi \in \Theta$, $false$ otherwise
- $Real(\bigcirc\varphi, \Theta) = false$

- $Real(\Box\varphi, \Theta) = true$ if $\varphi \in \Theta$, $false$ otherwise
- $Real(\varphi\mathcal{U}\psi, \Theta) = true$ if $\psi \in \Theta$, $false$ otherwise

Definition 2 (Descendant potential eventualities). *Let $\xi \in \Delta$ be a potential eventuality of the form $\langle\!\langle A\rangle\!\rangle\Phi$ or $[\![A]\!]\Phi$. Suppose the γ-component $\gamma(\xi, \Delta)$ in Δ linked to ξ is, respectively, of the form $\psi \wedge \langle\!\langle A\rangle\!\rangle\bigcirc\langle\!\langle A\rangle\!\rangle\Psi$ or $\psi \wedge [\![A]\!]\bigcirc[\![A]\!]\Psi$. Then the successor potential eventuality of ξ w.r.t. $\gamma(\xi, \Delta)$ is the γ-formula $\langle\!\langle A\rangle\!\rangle\Psi$ (resp. $[\![A]\!]\Psi$) and it will be denoted by ξ^1_Δ. The notion of descendant potential eventuality of ξ of degree d, for $d > 1$, is defined inductively as follows:*
- *any successor eventuality of ξ (w.r.t. some γ-component of ξ) is a descendant eventuality of ξ of degree 1;*
- *any successor eventuality of a descendant eventuality ξ^n of ξ of degree n is a descendant eventuality of ξ of degree $n + 1$.*

We will also consider ξ to be a descendant eventuality of itself of degree 0.

Example 6. (Continuation of Example 5) In Δ_1 we have $\xi = \langle\!\langle 1\rangle\!\rangle(p\mathcal{U}q \vee \Box q)$ with $Real(p\mathcal{U}q \vee \Box q, \Delta_1) = Real(p\mathcal{U}q, \Delta_1) \vee Real(\Box q, \Delta_1) = false \vee false = false$, since $q \notin \Delta_1$, and $\xi' = \langle\!\langle 2\rangle\!\rangle(\Diamond p \wedge \Box\neg q)$ with $Real(\Diamond p \wedge \Box\neg q, \Delta_1) = Real(\Diamond p, \Delta_1) \wedge Real(\Box\neg q, \Delta_1) = true \wedge true = true$ since $p, \neg q \in \Delta_1$.

The successor eventuality of $\xi = \langle\!\langle 1\rangle\!\rangle(p\mathcal{U}q \vee \Box q)$ w.r.t $\gamma(\xi, \Delta_5)$ is $\xi^1_{\Delta_5} = \langle\!\langle 1\rangle\!\rangle p\mathcal{U}q$ in $\Delta_3, \Delta_4, \Delta_5, \Delta_6$. For each $n > 1$, the descendant eventuality of degree n of ξ w.r.t $\gamma(\xi, \Delta_5)$ is $\xi^n_{\Delta_5} = \xi^1_{\Delta_5}$ in $\Delta_3, \Delta_4, \Delta_5, \Delta_6, \Delta_{10}$. The successor eventuality of $\xi' = \langle\!\langle 2\rangle\!\rangle(\Diamond p \wedge \Box\neg q)$ w.r.t $\gamma(\xi', \Delta_5)$ is $\xi'^1_{\Delta_5} = \langle\!\langle 2\rangle\!\rangle(\Diamond p \wedge \Box\neg q)$ in $\Delta_5, \Delta_6, \Delta_8$. For each $n > 1$, the descendant eventualities of degree n of ξ' w.r.t $\gamma(\xi', \Delta_5)$ are $\xi'^n_{\Delta_5} = \xi'^1_{\Delta_5}$ in $\Delta_5, \Delta_6, \Delta_8$ and Δ_9; and $\xi'^n_{\Delta_5} = \langle\!\langle 2\rangle\!\rangle\Box\neg q$ in Δ_{10} and Δ_{11}.

Now, let $\mathbb{L} = \langle\!\langle A_0\rangle\!\rangle\bigcirc\varphi_0, \ldots, \langle\!\langle A_{m-1}\rangle\!\rangle\bigcirc\varphi_{m-1}, [\![A'_0]\!]\bigcirc\psi_0, \ldots, [\![A'_{l-1}]\!]\bigcirc\psi_{l-1}$ be the list of all primitive successor formulae of $\Delta \in S^\eta_0$, induced as part of application of **(Next)**. We will use the following notation:
$D(\Delta, \langle\!\langle A_p\rangle\!\rangle\bigcirc\varphi) := \{\sigma \in D(\Delta) \mid \sigma_a = p$ for every $a \in A_p\}$
$D(\Delta, [\![A'_q]\!]\bigcirc\psi) := \{\sigma \in D(\Delta) \mid \mathbf{co}(\sigma) = q$ and $\mathbb{A} - A'_q \subseteq N(\sigma)\}$

Next, we will define recursively what it means for an eventuality ξ to be realized at a state Δ of a tableau $\mathcal{T}^\eta_n$, followed by our second elimination rule.

Definition 3 (Realization of potential eventualities). *Let $\Delta \in S^\eta_n$ and $\xi \in \Delta$ be a potential eventuality of the form $\langle\!\langle A\rangle\!\rangle\Phi$ or $[\![A]\!]\Phi$. Then:*

1. *If $Real(\Phi, \Delta) = true$ then ξ is realized at Δ in $\mathcal{T}^\eta_n$.*
2. *Else, let ξ^1_Δ be the successor potential eventuality of ξ w.r.t. $\gamma(\xi, \Delta)$. If for every $\sigma \in D(\Delta, \langle\!\langle A\rangle\!\rangle\bigcirc\xi^1_\Delta)$ (resp. $\sigma \in D(\Delta, [\![A]\!]\bigcirc\xi^1_\Delta)$), there exists $\Delta' \in \mathcal{T}^\eta_n$ with $\Delta \xrightarrow{\sigma} \Delta'$ and ξ^1_Δ is realized at Δ' in $\mathcal{T}^\eta_n$, then ξ is realized at Δ in $\mathcal{T}^\eta_n$.*

Rule (ER2) : If $\Delta \in S^\eta_n$ contains a potential eventuality that is not realized at $\Delta \in \mathcal{T}^\eta_n$, then obtain $\mathcal{T}^\eta_{n+1}$ by removing Δ from S^η_n.

Example 7. (Cont. of Example 6) The potential eventuality $\xi'' = \langle\!\langle 1\rangle\!\rangle(p\mathcal{U}q)$ is not realized in Δ_5, so by Rule **(ER2)** we remove the state Δ_5 from $\mathcal{T}^\theta_0$ and obtain the tableau $\mathcal{T}^\theta_1$. The same applies to Δ_6 for ξ'', so we also remove Δ_6

from $\mathcal{T}_1^\theta$ and obtain $\mathcal{T}_2^\theta$ with Rule **(ER2)**. In $\mathcal{T}_2^\theta$ there is no more move vector $(0,1)$ for the state Δ_1, so by Rule **(ER1)** we remove Δ_1 from $\mathcal{T}_2^\theta$ and obtain $\mathcal{T}_3^\theta$. In the same way, Δ_{10} is removed by Rule **(ER2)** and Δ_2 by Rule **(ER1)**.

As for the case of ϑ, it is easy to see that no states get eliminated, so the final tableau is the same as the initial one.

The elimination phase is completed when no more applications of elimination rules are possible. Then we obtain the *final tableau* for η, denoted by $\mathcal{T}^\eta$. It is declared *open* if η belongs to some state in it, otherwise *closed*. The procedure for deciding satisfiability of η returns "No" if $\mathcal{T}^\eta$ is closed, "Yes" otherwise.

Example 8. (Continuation of Example 7) At the end of the elimination phase, Δ_1 and Δ_2 are no longer in $\mathcal{T}^\theta$. Thus $\mathcal{T}^\theta$ is closed and we deduce that the formula $\theta = \langle\!\langle 1 \rangle\!\rangle (p\,\mathcal{U}q \vee \Box q) \wedge \langle\!\langle 2 \rangle\!\rangle (\Diamond p \wedge \Box \neg q)$ is declared unsatisfiable. The final tableaux for θ is given on Figure 4.

Respectively, the final tableau for ϑ is open, hence ϑ is declared satisfiable. Indeed, a CGM can be extracted from the final tableau.

Fig. 4. The final tableau for θ

5   Termination, Soundness, Completeness and Complexity

The termination of the tableaux procedure is straightforward, as there are only finitely many states and prestates that can be added in the construction phase.

Theorem 1. *The tableaux method for* ATL$^+$ *is sound.*

Soundness of the tableaux method means that if the input formula is satisfiable, then the procedure will indeed produce an open tableau. The argument in a nutshell is that if the input formula η is satisfiable, then, due to Proposition 1, there is a satisfiable state Δ in **states**$(\{\eta\})$. The key claim, proved by induction on the number of steps in the elimination phase, is that the elimination rules only remove states with unsatisfiable labels, so the Δ 'survives' the elimination phase and remains in the final tableau. A detailed proof can be found in [2].

Theorem 2. *The tableaux method for* ATL$^+$ *is complete.*

Completeness of the procedure means that an open tableau implies existence of a CGM model. This is proved by first introducing the notion of *Hintikka game structure*, which is essentially a partially defined CGM, and showing that every open tableau provides a Hintikka game structure containing the input formula η in the label of a state in it, which is equivalent of η being satisfiable. Again, a detailed proof can be found in [2].

Theorem 3. *The tableaux procedure for* ATL$^+$ *runs in* 2EXPTIME.

Proof. The argument generally follows the calculations computing the complexity of the tableaux method for ATL in Section 4.7 of [7], with one essential difference: $\|cl(\eta)\|$ for any ATL formula η is linear in its length $|\eta|$, whereas $\|cl(\eta)\|$ for an ATL$^+$ formula η can be exponentially large in $|\eta|$, as shown after Lemma 2. This exponential blowup, combined with the worst-case exponential in $\|cl(\eta)\|$ number of states in the tableaux, accounts for the 2EXPTIME worst-case complexity of the tableaux method for ATL$^+$, which is the expected optimal lower bound. It is also an upper bound for the tableaux method, because no further exponential blowups occur in the prestate- and state-elimination phases.   □

There are various ways to restrict or parameterize the set of ATL$^+$ formulae in order to avoid the exponential blowup of their closure sets. As suggested by the example after Lemma 2, the main cause for that blowup of the number of γ-components of a γ-formulae $\varphi = \langle\!\langle A \rangle\!\rangle \Phi$ or $\varphi = [\![A]\!]\Phi$ in ATL$^+$ is the nesting of conjunctions and disjunctions in the path formula Φ which are not in the scope of temporal operators. Let us call that number the *superficial Boolean depth* of Φ and denote it by $\delta_0(\Phi)$. Then, let the *nested Boolean depth* of any ATL$^+$ formula Ψ, denoted $\delta(\Psi)$, be the maximal superficial Boolean depth $\delta_0(\Phi)$ of a path subformula Φ of Ψ. For instance, $\delta(\langle\!\langle 1 \rangle\!\rangle \bigcirc \langle\!\langle 1 \rangle\!\rangle ((p \vee q)\,\mathcal{U} \neg q)) = 0$, $\delta(\langle\!\langle 1 \rangle\!\rangle (\Box p \vee ((q \wedge p)\,\mathcal{U} \neg q)) = 1$, $\delta(\langle\!\langle 1 \rangle\!\rangle (\Diamond q \wedge (\Box p \wedge (q\,\mathcal{U} \neg q))) = 2$. Now, if this number for a formula η is bounded above, the size of the closure $\|\eta\|$ becomes polynomially bounded in $|\eta|$ because the nesting of $\wedge$ and $\vee$ when they are separated by a temporal operator does not have multiplicative effect on the number of γ-components. Consequently, the complexity of the tableaux method is reduced to single exponential time, caused only by the maximal possible number of states in the tableaux, just like in ATL. Thus, we have the following.

Proposition 2. *The tableaux procedure for* ATL$^+$ *applied to a class of* ATL$^+$ *formulae of bounded nested Boolean depth runs in* EXPTIME.

6   Concluding Remarks

Here we have developed sound, complete and terminating tableaux-based decision method for constructive satisfiability testing of ATL$^+$ formulae and have argued for its practical usability and implementability. The method is amenable to further extension to the full ATL*, but this is left to future work.

Some comparison with the automata-based method for satisfiability testing in ATL*, presented in [10] are in order. The two methods appear to be quite different and, though eventually working in the same worst case complexity, the double exponential blowups seem to occur in different ways, namely, in the automata-based method, one exponential blowup occurs in converting the formula into an automaton, while the other is in the time complexity of checking non-emptiness of the resulting automaton. It would be instructive to compare the practical implications and efficiency of both methods and we leave such systematic comparison to the future, when (hopefully) both methods are implemented. For now, we only mention that the formula θ from our running example, the tableau for which is worked out explicitly and in detail in this paper, is translated with the method from [10] into an automaton with 2^{12} alphabet symbols and over 100 states. Of course, this comparison cannot serve as an argument for general practical superiority in efficiency of the tableaux-based method. Still, the technical details of both methods, illustrated in that example, indicate that, while the worst case exponential blowups are bound to occur in both methods, they seem to be more controllable and avoidable in the tableaux-based method, at the expense of its lesser automaticity and higher degree of user control. Thus, we would argue that both methods have generally incomparable pros and cons, and consequently are of independent interest, both theoretically and practically.

Acknowledgments. We thank the anonymous reviewers for their helpful remarks and suggestions and for some corrections. We also gratefully acknowledge the financial support for this research from Université Évry Val-d'Essonne and the laboratory IBISC by funding visits of V. Goranko to Université Évry Val-d'Essonne and of A. David to the Technical University of Denmark.

References

1. Alur, R., Henzinger, T.A., Kuperman, O.: Alternating-time temporal logic. Journal of the ACM 49(5), 672–713 (2002)
2. Cerrito, S., David, A., Goranko, V.: Optimal tableaux-based decision procedure for testing satisfiability in the alternating-time temporal logic ATL$^+$. Tech. rep., Laboratoire IBISC - Université Evry Val-d'Essonne (2014), https://www.ibisc.univ-evry.fr/~adavid/fichiers/ IJCAR_Extended_Draft_Tableau_ATLplus.pdf
3. David, A.: TATL: Implementation of ATL tableau-based decision procedure. In: Galmiche, D., Larchey-Wendling, D. (eds.) TABLEAUX 2013. LNCS, vol. 8123, pp. 97–103. Springer, Heidelberg (2013)
4. van Drimmelen, G.: Satisfiability in alternating-time temporal logic. In: Proc. of LICS 2003, pp. 208–217 (2003)
5. Emerson, E.A.: Temporal and modal logics. In: van Leeuwen, J. (ed.) Handbook of Theoretical Computer Science, vol. B, pp. 995–1072. MIT Press (1990)
6. Goranko, V., van Drimmelen, G.: Complete axiomatization and decidablity of Alternating-time temporal logic. Theor. Comp. Sci. 353, 93–117 (2006)
7. Goranko, V., Shkatov, D.: Tableau-based decision procedures for logics of strategic ability in multiagent systems. ACM Trans. Comput. Log. 11(1) (2009)
8. Johannsen, J., Lange, M.: CTL+ is Complete for Double Exponential Time. In: Baeten, J.C.M., Lenstra, J.K., Parrow, J., Woeginger, G.J. (eds.) ICALP 2003. LNCS, vol. 2719, pp. 767–775. Springer, Heidelberg (2003)
9. Pratt, V.R.: A near optimal method for reasoning about action. Journal of Computer and System Sciences 20, 231–254 (1980)
10. Schewe, S.: ATL* satisfiability is 2EXPTIME-complete. In: Aceto, L., Damgård, I., Goldberg, L.A., Halldórsson, M.M., Ingólfsdóttir, A., Walukiewicz, I. (eds.) ICALP 2008, Part II. LNCS, vol. 5126, pp. 373–385. Springer, Heidelberg (2008)
11. Walther, D., Lutz, C., Wolter, F., Wooldridge, M.: ATL satisfiability is indeed ExpTime-complete. Journal of Logic and Computation 16(6), 765–787 (2006)
12. Wolper, P.: The tableau method for temporal logic: an overview. Logique et Analyse 28(110-111), 119–136 (1985)

dTL²: Differential Temporal Dynamic Logic with Nested Temporalities for Hybrid Systems

Jean-Baptiste Jeannin and André Platzer

Computer Science Department, Carnegie Mellon University, Pittsburgh, PA, USA

Abstract. The differential temporal dynamic logic dTL² is a logic to specify temporal properties of hybrid systems. It combines differential dynamic logic with temporal logic to reason about the intermediate states reached by a hybrid system. The logic dTL² supports some linear time temporal properties of LTL. It extends differential temporal dynamic logic dTL with nested temporalities. We provide a semantics and a proof system for the logic dTL², and show its usefulness for nontrivial temporal properties of hybrid systems. We take particular care to handle the case of alternating universal dynamic and existential temporal modalities and its dual, solving an open problem formulated in previous work.

1 Introduction

A major task of computer science is to program objects of our physical world: cars, trains, airplanes, robots, etc. — often grouped under the denomination of cyber-physical systems (CPS). A CPS is governed by its programmable controllers, but also by the laws of physics. To fully verify it, one thus needs to model the controllers and their software as well as the relevant laws of physics in the same system. Such a system then becomes hybrid: the controllers are *discrete* while the laws of physics are *continuous*.

In recent years, a number of systems have been explored to reason about such *hybrid systems*. In particular, this paper is based on *differential dynamic logic* [14], [16, chapter 4], a logic based on dynamic logic [15,17], [16, chapter 2] and including programs enabling discrete assignments and discrete control structures, but also execution of differential equations. Differential dynamic logic comes with a semantics as well as a proof system, which is sound and relatively complete.

Based on dynamic logic, differential dynamic logic only reasons about the end state of a system. However, to ensure that a system always stays within some structural limits, or always accomplishes a certain task, one needs to reason about its intermediate states as well. CPSs that are safe when their systems terminate but have been unsafe in the middle of the program run are still not safe to use. The idea is to use both dynamic logic — to quantify over possible executions — and temporal logic — to quantify over the states in the trace of each execution. This is not a new idea, but previous work [1,14] focuses only on the non-alternating cases: "some property is always verified during all executions" and "something happens during some execution."

In this paper, we are developing a *differential temporal dynamic logic* dTL² inspired from LTL, and we are focusing on correctly handling the more complex alternating cases: "something happens during all executions" and "there is an execution where

S. Demri, D. Kapur, and C. Weidenbach (Eds.): IJCAR 2014, LNAI 8562, pp. 292–306, 2014.

some property is always verified," as well as nested temporal modalities. In particular, a property checking that a task is always accomplished can now be checked. This logic is an important stepping stone towards full dTL*, the differential analog of CTL*.

As a simple example, let us look at a satellite with position x trying to leave the solar system, avoiding planets. To simplify, let us consider only two planets with radiuses r_1 and r_2, at (evolving) positions p_1 and p_2. The satellite can be controlled either by a pilot who can set its steering ω to left or right then let x evolve according to differential equation flight(ω), or by an autopilot following a PID controller with target direction set to d. During each evolution, the positions of the planets continue to evolve, following differential equation planets(p_1, p_2). The program of the satellite and its safety property ϕ — expressing that there exists a steering avoiding all planets — can be expressed as:

$$\text{satellite} ::= (((\omega := \text{left} \cup \omega := \text{right}); x' = \text{flight}(\omega), (p_1', p_2') = \text{planets}(p_1, p_2))$$
$$\cup\, (d := *; x' = \text{PID}(d), (p_1', p_2') = \text{planets}(p_1, p_2)))^*;$$
$$\text{control} := \text{lost}; d := *; x' = \text{PID}(d), (p_1', p_2') = \text{planets}(p_1, p_2)$$
$$\phi ::= \langle\text{satellite}\rangle\Box(\text{dist}(x, p_1) > r_1 \wedge \text{dist}(x, p_2) > r_2 \wedge \text{control} \neq \text{lost})$$

This example shows several features of hybrid programs and the logic dTL^2. Under the pilot's command, the variable ω can be *assigned* to either left or right, following a *nondeterministic choice* $\cup$. Then x, p_1 and p_2 follow a *differential equation* modeling the continuous evolution of the system, including movement of the planets. Under the autopilot's command, d is *nondeterministically assigned* ($d := *$). There is a nondeterministic choice between the two commands, followed by a star $*$ representing *repetition*. In case of mechanical or communication failure, control could be lost, which we represent by a variable assignment, and the system continues to evolve. The formula ϕ says that there exists a possible evolution ($\langle\text{satellite}\rangle$) such that throughout this evolution ($\Box$), the satellite does not hit any planet; namely, the evolution avoiding planets where control is never lost. The formula ϕ is expressible in dTL^2, and shows how dTL^2 handles alternating and nested program ($\langle\text{satellite}\rangle$) and temporal modalities ($\Box$ and $\Diamond$). The focus of this paper is to create a semantics and a proof calculus for dTL^2.

There are three main contributions to this paper. First, we show how to correctly handle the alternating cases of a universal dynamic modality followed by an existential temporal modality, and its dual an existential dynamic modality followed by a universal temporal modality. This solves an open problem identified in 2001 [1] and identified as a problem for hybrid systems in 2007 [14], [16, chapter 4]. Second, we offer a treatment where programs are not duplicated by proof rules, solving another open problem formulated in [14], [16, chapter 4]. This is significant for proving hybrid systems in practice, because previous approaches led to a duplication of proof effort, once for intermediate and once for final states. Third and finally, we extend the logic to nested temporal quantifiers, show that all formulas of interest are equivalent to formulas containing at most two quantifiers — thus the name dTL^2 — by identifying the resemblance to modal system S4.2, and develop a logic and proof calculus for the new temporal formulas.

The paper is organized as follows. After presenting the syntax and semantics of Differential Temporal Dynamic Logic dTL^2 in Section 2, we show how to normalize trace formulas and how to axiomatize dTL^2 in Section 3. We study alternative proof systems in Section 4 and related work in Section 5, before concluding in Section 6.

2   Differential Temporal Dynamic Logic dTL^2

This section defines the syntax and semantics of hybrid programs and trace formulas formally. The development mostly follows and extends previous work on differential temporal dynamic logic [14], [16, chapter 4]; we explicitly point out differences and extensions from the previous work.

2.1   Hybrid Programs

We use *hybrid programs* (HP) [15,17], [16, chapter 2] α, β to model hybrid systems. Syntactically, hybrid programs can be *atomic* hybrid programs or *compound* hybrid programs. Atomic hybrid programs can be discrete jump *assignments* ($x := \theta$), *tests* ($?\chi$) and *differential equations* evolving within an evolution domain constraint χ — meaning that the system can evolve following a solution of the differential equation as long as χ remains true ($x' = \theta \,\&\, \chi$). Terms θ are polynomials with rational coefficients, and conditions χ are first-order formulas of real arithmetic.[1] Compound hybrid programs are *nondeterministic choice* ($\alpha \cup \beta$), *sequential composition* ($\alpha; \beta$) and *nondeterministic finite repetition* (α^*):

$$\alpha, \beta ::= x := \theta \mid ?\chi \mid x' = \theta \,\&\, \chi \mid \alpha \cup \beta \mid \alpha; \beta \mid \alpha^*$$

The *trace semantics* of hybrid programs assigns to each program α a set of *traces* $\tau(\alpha)$. The set of *states* Sta is the set of (total) functions from variables to the reals $\mathbb{R}$. In addition, we consider a separate state Λ (not in Sta) denoting a failure of the system. For $v \in$ Sta or $v = \Lambda$, we denote by $\hat{v}$ the function $\sigma : \{0\} \to \{v\}, 0 \mapsto v$, defined only on the singleton interval $[0, 0]$. A *trace* is a (nonempty) finite sequence $\sigma = (\sigma_0, \sigma_1, ..., \sigma_n)$ of functions σ_i. For $0 \le i < n$, the piece σ_i is a function $\sigma_i : [0, r_i] \to$ Sta, where $r_i \ge 0$ is the duration of this step. For $i = n$, the piece σ_n is either a function:

- $\sigma_n : [0, r_i] \to$ Sta; we then say that σ is a *terminating* trace; or
- $\sigma_n : [0, +\infty) \to$ Sta; we then say that σ is an *infinite* trace; or
- $\sigma_n : \{0\} \to \{\Lambda\}, 0 \mapsto \Lambda$, for $n \ge 1$;[2] we then say that σ is an *error* trace.

We often collectively refer to infinite and error traces as *nonterminating*; thus when we refer to terminating traces, we only refer to those traces that terminate but not with an error state Λ. We write Tra for the set of all traces. A *position* of σ is a pair (i, ζ) with $0 \le i \le n$ and ζ in the domain of definition of σ_i; the state of σ at (i, ζ) is $\sigma_i(\zeta)$. For any trace σ, we denote by first σ the state $\sigma_0(0)$; we informally say that "σ starts with v" to say that $v =$ first σ. If $\sigma = (\sigma_0, ..., \sigma_n)$ terminates (and only in that case), we also denote by last σ the state $\sigma_n(r_n)$; when σ does not terminate, last σ is undefined. We denote by $\text{val}(v, \theta)$ the value of term θ in state v, and by $v[x \mapsto r]$ the valuation assigning variable x to $r \in \mathbb{R}$ and matching with v on all other variables. We also write $v \vDash \chi$ if state v satisfies condition χ, and $v \nvDash \chi$ otherwise.

---

[1] Using first-order formulas or real arithmetic results in a poor-test version of the logic. Our results generalize to a rich-test version, where a condition χ is instead defined as any formula ϕ of dTL^2 (see Section 2.2).

[2] We impose $n \ge 1$ so that $(\hat{\Lambda})$ is not considered a trace

Given two traces $\sigma = (\sigma_0, \ldots, \sigma_n)$ and $\rho = (\rho_0, \ldots, \rho_m)$, we say that ρ is a *prefix* of σ if it describes the trace σ truncated at some position. Formally, ρ is a prefix of σ if and only if $\rho = \sigma$ — a condition ensuring that nonterminating traces are also suffixes of themselves — or there exists a position (i, ζ) of σ such that:

- traces $(\sigma_0, \ldots, \sigma_{i-1})$ and $(\rho_0, \ldots, \rho_{m-1})$ are identical.[3] In particular this imposes that $i = m$; and
- the domain of definition of ρ_m is exactly $[0, \zeta]$ and is included in the domain of definition of σ_m, and for all $d \in [0, \zeta]$, $\sigma_m(d) = \rho_m(d)$.

Symmetrically, we say that ρ is a *suffix* of σ if it starts at some position of σ then follows σ. Formally, ρ is a suffix of σ if and only if there exists a position (i, ζ) of σ such that:

- if σ_i has domain of definition $[0, r_i]$, then the domain of definition of ρ_0 is exactly $[0, r_i - \zeta]$ and for all $d \in [\zeta, r_i]$, $\sigma_i(d) = \rho_0(d - \zeta)$; and in the case where σ_i has domain of definition $[0, +\infty)$, the domain of definition of ρ_0 is also $[0, +\infty)$ and for all $d \in [\zeta, +\infty)$, $\sigma_i(d) = \rho_0(d - \zeta)$; and
- $(\sigma_{i+1}, \ldots, \sigma_n)$ and $(\rho_1, \ldots, \rho_m)$ are identical, which imposes that $n - i = m$.

Definition 1 (Trace Semantics of Hybrid Programs). *The* trace semantics $\tau(\alpha) \subseteq 2^{\mathsf{Tra}}$ *of a hybrid program α is then defined inductively as follows:*

- $\tau(x := \theta) = \{(\hat{v}, \hat{w}) \mid w = v[x \mapsto \mathsf{val}(v, \theta)]\};$
- $\tau(x' = \theta \,\&\, \chi) = \{(\sigma) \;:\; \sigma \text{ is a state flow of order 1 [15] defined on } [0, r] \text{ or } [0, +\infty) \text{ solution of } x' = \theta, \text{ and for all } t \text{ in its domain of definition}, \sigma(t) \vDash \chi\}$ $\cup \{(\hat{v}, \hat{\Lambda}) \;:\; v \nvDash \chi\};$[4]
- $\tau(?\chi) = \{(\hat{v}) \;:\; v \vDash \chi\} \cup \{(\hat{v}, \hat{\Lambda}) \;:\; v \nvDash \chi\};$
- $\tau(\alpha \cup \beta) = \tau(\alpha) \cup \tau(\beta);$
- $\tau(\alpha; \beta) = \{\sigma \circ \rho \;:\; \sigma \in \tau(\alpha), \rho \in \tau(\beta) \text{ when } \sigma \circ \rho \text{ is defined}\};$
  where the composition $\sigma \circ \rho$ of $\sigma = (\sigma_0, \ldots, \sigma_n)$ and $\rho = (\rho_0, \ldots, \rho_m)$ is
  - $\sigma \circ \rho = (\sigma_0, \ldots, \sigma_n, \rho_0, \ldots, \rho_m)$ *if σ terminates and* last $\sigma = $ first ρ *(since σ terminates,* last σ *is well-defined);*
  - σ *if σ does not terminate;*
  - *undefined otherwise;*
- $\tau(\alpha^*) = \bigcup_{n \in \mathbb{N}} \tau(\alpha^n)$, *where α^0 is defined as ?true, α^1 is defined as α and α^{n+1} is defined as $\alpha^n; \alpha$ for $n \geq 1$.*

An important property of this trace semantics is that for all programs α and states v, there exists a trace σ of α starting with v (even if it might be an error trace). This property will be key to proving the soundness of assignment rules.

Aside from the correction on $\tau(x' = \theta \,\&\, \chi)$, this definition is slightly different from [14], [16, chapter 4] in two ways: these previous papers also consider infinite sequences $\sigma = (\sigma_0, \sigma_1, \ldots)$, but infinite sequences are not part of the semantics of any program; and these papers do not consider infinite traces in the semantics. Still, we can prove that the interpretation of trace formulas (Section 2.2) is the same on the subset of trace formulas they consider.

---

[3] If $i = m = 0$, $(\sigma_0, \ldots, \sigma_{i-1})$ and $(\rho_0, \ldots, \rho_{m-1})$ are empty and thus not formally traces, but we still consider the condition fulfilled.

[4] This case is corrected from [14], [16, chapter 4], which wrongly forget the error traces of ordinary differential equations — when χ is initially false.

2.2 State and Trace Formulas

To reason about hybrid programs, we use *state formulas* and *trace formulas*. State formulas express properties about states, while trace formulas express properties about traces; their definitions are mutually inductive. A state formula ϕ, ψ can be a *comparison of terms* ($\theta_1 \geq \theta_2$); a *negation* of a state formula ($\neg\phi$); a *conjunction* ($\phi \wedge \psi$) or a *disjunction* ($\phi \vee \psi$) of state formulas; a *universally quantified* ($\forall x\ \phi$) or *existentially quantified* ($\exists x\ \phi$) state formula — quantification of a variable x is over the set of reals $\mathbb{R}$. Finally, a state formula can also be a *program necessity* ($[\alpha]\pi$) — expressing that all traces of hybrid program α starting at the current state satisfy trace formula π — or its dual, a *program possibility* ($\langle\alpha\rangle\pi$) — expressing that there is a trace of α starting at the current state satisfying trace formula π.

A trace formula π can be a *state formula* (ϕ); a *negation* of a trace formula ($\neg\pi$); a *temporal necessity* of a trace formula ($\Box\pi$) — expressing that every suffix of the current trace satisfies π — or its dual, a *temporal possibility* of a trace formula ($\Diamond\pi$) — expressing that there is a suffix of the current trace satisfying π. The syntax of state and trace formulas is thus given by:

$$\phi, \psi ::= \theta_1 \geq \theta_2 \mid \neg\phi \mid \phi \wedge \psi \mid \phi \vee \psi \mid \forall x\ \phi \mid \exists x\ \phi \mid [\alpha]\pi \mid \langle\alpha\rangle\pi$$
$$\pi ::= \phi \mid \neg\pi \mid \Box\pi \mid \Diamond\pi$$

Additionally, as in classical logic, the implication $\phi \rightarrow \psi$ is defined as $\neg\phi \vee \psi$. When a trace formula also happens to be a state formula ϕ, the formula $\neg\phi$ means the same whether it is seen as a state or trace formula; in the rest of the paper we collude the two. We are now ready to define satisfaction of state and trace formulas.

Definition 2 (Satisfaction of dTL^2 Formulas). *For state formulas, we write $v \vDash \phi$ to say that state $v \in \mathsf{Sta}$ satisfies state formula ϕ. Satisfaction of state formulas with respect to a state v is defined inductively as follows:*

- *$v \vDash \theta_1 \geq \theta_2$ if and only if $\mathsf{val}(v, \theta_1) \geq \mathsf{val}(v, \theta_2)$*
- *$v \vDash \neg\phi$ if and only if $v \vDash \phi$ does not hold.*
- *$v \vDash \phi \wedge \psi$ if and only if $v \vDash \phi$ and $v \vDash \psi$.*
- *$v \vDash \phi \vee \psi$ if and only if $v \vDash \phi$ or $v \vDash \psi$.*
- *$v \vDash \forall x\ \phi$ if and only if $v[x \mapsto d] \vDash \phi$ holds for all $d \in \mathbb{R}$.*
- *$v \vDash \exists x\ \phi$ if and only if $v[x \mapsto d] \vDash \phi$ holds for some $d \in \mathbb{R}$.*
- *for ϕ a state formula, $v \vDash [\alpha]\phi$ if and only if for each trace $\sigma \in \tau(\alpha)$ that starts in first $\sigma = v$, if σ terminates, then last $\sigma \vDash \phi$.*
- *for ϕ a state formula, $v \vDash \langle\alpha\rangle\phi$ if and only if there is a trace $\sigma \in \tau(\alpha)$ starting in first $\sigma = v$ such that σ terminates and last $\sigma \vDash \phi$.*
- *If π is not a state formula, $v \vDash [\alpha]\pi$ if and only $\sigma \vDash \pi$ for each trace $\sigma \in \tau(\alpha)$ that starts in first $\sigma = v$.*
- *If π is not a state formula, $v \vDash \langle\alpha\rangle\pi$ if and only $\sigma \vDash \pi$ for some trace $\sigma \in \tau(\alpha)$ that starts in first $\sigma = v$.*

For trace formulas, we write $\sigma \vDash \pi$ to say that trace $\sigma \in \mathsf{Tra}$ satisfies trace formula π. Satisfaction of trace formulas with respect to a trace σ is defined inductively as follows:

- $\sigma \vDash \phi$ *if and only if* first $\sigma \vDash \phi$.
- $\sigma \vDash \neg \pi$ *if and only if* $\sigma \vDash \pi$ *does not hold.*
- $\sigma \vDash \Box \pi$ *if and only if* $\rho \vDash \pi$ *holds for all suffixes ρ of σ that are different from* $(\hat{\Lambda})$.
- $\sigma \vDash \Diamond \pi$ *if and only if* $\rho \vDash \pi$ *holds for some suffix ρ of σ that is different from* $(\hat{\Lambda})$.

This definition follows the intuition given when presenting the syntax of state and trace formulas, except for one point. Note that in the definitions of $\sigma \vDash \Box \pi$ and $\sigma \vDash \Diamond \pi$, the suffix ρ of σ does not have to be proper, and we can have $\rho = \sigma$. When seen as a trace formula, a state formula ϕ can express a property on a trace σ. We then say that σ satisfies ϕ if and only if the first state of σ satisfies ϕ (condition first $\sigma \vDash \phi$ in the definition of $\sigma \vDash \phi$). However, there is an exception to this definition: when ϕ appears directly after a program necessity (as in $[\alpha]\phi$) or a program possibility (as in $\langle \alpha \rangle \phi$), ϕ only refers to *terminating* traces, and we say that σ satisfies ϕ if and only if the *last* state of σ satisfies ϕ (condition $\sigma \vDash$ last ϕ in the definitions of $\sigma \vDash \langle \alpha \rangle \phi$ and $\sigma \vDash [\alpha]\phi$). This discontinuity in the definition of the satisfaction of ϕ enables following both the usual semantics of dynamic logic and of temporal logic, and was also adopted in previous work [7,14], [16, chapter 4]. It is also useful for proof rules as temporal properties often reduce to what happens after a program.

The syntax of dTL^2 formulas extends the syntax of trace formulas given in [14], [16, chapter 4] by allowing nesting of temporal modalities, and otherwise agrees with it. The satisfaction of dTL^2 formulas given in Def. 2, although presented in a slightly different way, agrees with the definitions given in [14], [16, chapter 4] on trace formulas without nested temporal modalities.

3   Proof Calculus

3.1   Equivalence of Trace Formulas

Trace formulas follow the axioms of modal system S4.2 [9], therefore there are only four proper affirmative modalities $\Box \phi$, $\Diamond \phi$, $\Box \Diamond \phi$ or $\Diamond \Box \phi$. Intuitively, because formulas $\neg \Box \pi$ and $\Diamond \neg \pi$ are equivalent — in the sense that they are satisfied by the same traces — formulas can always be expressed in a way where only state formulas have negations. Similarly, formulas $\Box \pi$ and $\Box \Box \pi$ are equivalent, therefore a trace formula containing exclusively $\Box$ temporalities followed by a state formula ϕ is equivalent to $\Box \phi$. Moreover, a formula containing both $\Box$ and $\Diamond$ temporalities, finishing by a $\Diamond$ temporality followed by a state formula ϕ is equivalent to $\Box \Diamond \phi$. Similar properties are true for their duals. This is formalized by the following lemma, proved in [10].

Lemma 1 (Equivalence of Trace Formulas). *For any trace formula π_1, there exists a trace formula π_2 of the form ϕ, $\Box \phi$, $\Diamond \phi$, $\Box \Diamond \phi$ or $\Diamond \Box \phi$ such that $\sigma \vDash \pi_1$ if and only if $\sigma \vDash \pi_2$. Such a π_2 can be computed from π_1 in linear time in the number of temporal modalities and negations in π_1.*

Remark 1. Lemma 1 tells us that the only interesting trace formulas of our system are those of the form ϕ, $\Box \phi$, $\Diamond \phi$, $\Box \Diamond \phi$ and $\Diamond \Box \phi$. For any trace σ, the intuitive meaning of $\sigma \vDash \pi$ for π of the form ϕ, $\Box \phi$ or $\Diamond \phi$ is clear: we have $\sigma \vDash \phi$ if and only if σ

starts in a state satisfying ϕ; we have $\sigma \vDash \Box\phi$ if and only if all non-error states of the trace σ satisfy ϕ; and we have $\sigma \vDash \Diamond\phi$ if and only if there is a non-error state of trace σ satisfying ϕ. When π is of the form $\Box\Diamond\phi$ and $\Diamond\Box\phi$, we get a better intuition by distinguishing cases:

- if σ is a terminating trace, $\sigma \vDash \Diamond\Box\phi$ if and only if last $\sigma \vDash \phi$, and $\sigma \vDash \Box\Diamond\phi$ if and only if last $\sigma \vDash \phi$ as well;
- if σ is an error trace, σ can be written $(\sigma_0, \ldots, \sigma_{n-1}, \hat{\Lambda})$. Let $\rho = (\sigma_0, \ldots, \sigma_{n-1})$, then ρ is a terminating trace and a prefix of σ. Moreover, both $\sigma \vDash \Diamond\Box\phi$ and $\sigma \vDash \Box\Diamond\phi$ are equivalent to last $\rho \vDash \phi$;
- if σ is an infinite trace, $\sigma \vDash \Diamond\Box\phi$ holds if and only if ϕ holds on all states of σ after some position, and $\sigma \vDash \Box\Diamond\phi$ holds if and only if any state of σ has a later state satisfying ϕ (if we did not have continuous dynamics, this would be the same as ϕ being true infinitely often along σ; but here it is not sufficient).

3.2 Normalization of Trace Formulas

The primary goal of this paper is to establish a proof system for differential temporal dynamic logic dTL^2. As for d$\mathcal{L}$ and dTL, rules typically decompose programs syntactically. Let us look at the state formula $\langle\alpha;\beta\rangle\Box\phi$, and to simplify, let us only consider terminating traces for now. Intuitively, this formula says that there exists a trace in $\tau(\alpha;\beta)$ throughout which ϕ holds. Considering only terminating traces, this is true as long as there exists a trace σ of α throughout which ϕ is true, and a trace ρ of β starting at last σ throughout which ϕ is also true. It is thus tempting to write the following rule:

$$\frac{\langle\alpha\rangle\Box\phi \wedge \langle\alpha\rangle\langle\beta\rangle\Box\phi}{\langle\alpha;\beta\rangle\Box\phi} \text{ (unsound)}$$

This rule is unsound because α is possibly nondeterministic. Its premise says that there is a trace σ of α throughout which ϕ is true, and a trace σ' of α followed by a trace ρ of β throughout which ϕ is true. But σ and σ' do not have to be the same trace; the trick is that ϕ is not necessarily true throughout σ'. To fix this rule, we need to express that traces σ and σ' are the same, thus writing a premise resembling:

$$\langle\alpha\rangle(\Box\phi \wedge \langle\beta\rangle\Box\phi) \tag{1}$$

Unfortunately, this is not directly expressible with dTL^2, without using the program $\alpha;\beta$ again: the missing piece is the expressibility of a conjunction on traces that simultaneously talks about temporal properties like $\Box\phi$ and properties true at the end of the trace. To achieve this expressibility, we extend the logic with *normalized trace formulas* to make conjunction of temporal formulas expressible as needed in (1).

A normalized trace formula ξ can be of different forms: for terminating traces, the formula $\phi\sqcap\Box\psi$ captures the conjunction of ending in a state satisfying ϕ, and satisfying $\Box\psi$; and the formula $\phi \sqcup \Diamond\psi$ captures the disjunction of ending in a state satisfying ϕ, or satisfying $\Diamond\psi$. For nonterminating traces, $\phi \sqcap \Box\psi$ is the same as $\Box\psi$, and $\phi \sqcup \Diamond\psi$ is the same as $\Diamond\psi$, because there is no terminal state in which it makes sense to evaluate ϕ. Additionally, the formula $\phi \blacktriangleleft \Box\Diamond\psi$ captures ending in a state satisfying ϕ if terminating, and satisfying $\square\lozenge\psi$ otherwise; and similarly, the formula $\phi \blacktriangleleft \lozenge\square\psi$ captures ending in a state satisfying ϕ if terminating, and satisfying $\lozenge\square\psi$ otherwise.

Formulas $\phi \blacktriangleleft \lozenge\square\psi$ and $\phi \blacktriangleleft \square\lozenge\psi$ play the same role for formulas $\lozenge\square\psi$ and $\square\lozenge\psi$ as formulas $\phi \sqcap \square\psi$ and $\phi \sqcup \lozenge\psi$ play for formulas $\square\psi$ and $\lozenge\psi$: they allow us to define premises of modular inference rules for sequential composition as in (1). Like standard trace formulas, normalized trace formulas can appear after a program necessity $[\alpha]$ or a program possibility $\langle\alpha\rangle$. We therefore extend state formulas to accept normalized trace formulas, and define normalized trace formulas as:

$$\phi, \psi ::= \dots \mid [\alpha]\xi \mid \langle\alpha\rangle\xi$$
$$\xi ::= \phi \sqcap \square\psi \mid \phi \sqcup \lozenge\psi \mid \phi \blacktriangleleft \square\lozenge\psi \mid \phi \blacktriangleleft \lozenge\square\psi$$

Sometimes we will also use the notation $\phi \blacktriangleleft \pi$, with the understanding that in such cases π can only be of the form $\square\lozenge\psi$ or $\lozenge\square\psi$.

Coming back to our example, a sound rule for $\langle\alpha; \beta\rangle\square\phi$ can be expressed as:

$$\frac{\langle\alpha\rangle(\langle\beta\rangle\square\phi \sqcap \square\phi)}{\langle\alpha; \beta\rangle\square\phi} \ (\langle;\rangle\square)$$

In the form of its dual $[;]\lozenge$, this rule will be discussed later and proved sound in [10]. Observe how $\langle;\rangle\square$ does not even duplicate α and β.

Extending Def. 2, the satisfaction of trace formulas $[\alpha]\xi$ and $\langle\alpha\rangle\xi$ is defined in the same way as trace formulas $[\alpha]\pi$ and $\langle\alpha\rangle\pi$ (if π is not a state formula):

- $v \vDash [\alpha]\xi$ if and only $\sigma \vDash \xi$ for each trace $\sigma \in \tau(\alpha)$ that starts in first $\sigma = v$.
- $v \vDash \langle\alpha\rangle\xi$ if and only $\sigma \vDash \xi$ for some trace $\sigma \in \tau(\alpha)$ that starts in first $\sigma = v$.

Satisfaction of normalized trace formulas carefully distinguishes between terminating and nonterminating traces, and is defined as follows.

Definition 3 (Semantics of Normalized dTL^2 Trace Formulas). *For normalized trace formulas, we write $\sigma \vDash \xi$ to say that trace σ satisfies normalized state formula ξ. Satisfaction of normalized trace formulas with respect to a trace σ is defined inductively:*

$$\sigma \vDash \phi \sqcup \lozenge\psi \ \textit{if and only if} \ \begin{cases} \text{last } \sigma \vDash \phi \text{ or } \sigma \vDash \lozenge\psi & \textit{if } \sigma \textit{ terminates} \\ \sigma \vDash \lozenge\psi & \textit{otherwise} \end{cases}$$

$$\sigma \vDash \phi \sqcap \square\psi \ \textit{if and only if} \ \begin{cases} \text{last } \sigma \vDash \phi \text{ and } \sigma \vdash \square\psi & \textit{if } \sigma \textit{ terminates} \\ \sigma \vDash \square\psi & \textit{otherwise} \end{cases}$$

$$\sigma \vDash \phi \blacktriangleleft \pi \quad \textit{if and only if} \ \begin{cases} \text{last } \sigma \vDash \phi & \textit{if } \sigma \textit{ terminates} \\ \sigma \vDash \pi & \textit{otherwise} \end{cases}$$

Not only can normalized trace formulas help express rules like $\langle;\rangle\square$, they can also, along with state formulas, express all possible trace formulas. In Lemma 1, we have shown how to express any trace formula in the form ϕ, $\square\phi$, $\lozenge\phi$, $\square\lozenge\phi$ or $\lozenge\square\phi$. Building on this result, we now show how to *normalize* every trace formula into a state formula or a normalized trace formula. To this effect, we define a relation $\rightsquigarrow$ between the set of state formulas and trace formulas, and the set of state formulas and normalized trace formulas. This simplifies the axiomatization of dTL^2 by allowing us to only consider cases containing normalized trace formulas.

$$\Box\phi \rightsquigarrow \text{true} \sqcap \Box\phi \; (\rightsquigarrow\sqcap)$$

$$\Box\Diamond\phi \rightsquigarrow \phi \blacktriangleleft \Box\Diamond\phi \; (\rightsquigarrow\blacktriangleleft\Box)$$

$$\phi \rightsquigarrow \phi \; (\rightsquigarrow\phi)$$

$$\Diamond\phi \rightsquigarrow \text{false} \sqcup \Diamond\phi \; (\rightsquigarrow\sqcup)$$

$$\Diamond\Box\phi \rightsquigarrow \phi \blacktriangleleft \Diamond\Box\phi \; (\rightsquigarrow\blacktriangleleft\Diamond)$$

$$\frac{\pi_1 \sim \pi_2 \qquad \pi_2 \rightsquigarrow \xi}{\pi_1 \rightsquigarrow \xi} \; (\sim\rightsquigarrow)$$

Fig. 1. Normalization rules for trace formulas

The normalization is sound, meaning that two related formulas are satisfied by the same trace. Additionally, every trace formula is related to either a state formula or a normalized trace formula, which can be found in linear time.

Lemma 2 (Soundness of Normalization). *If $\pi \rightsquigarrow \xi$ then for all traces σ, $\sigma \vDash \pi$ if and only if $\sigma \vDash \xi$.*

Proof. Soundness of $\rightsquigarrow \phi$ is trivial. Soundness of proof rules $\rightsquigarrow \sqcap$, $\rightsquigarrow \sqcup$, $\rightsquigarrow \blacktriangleleft\Box$ and $\rightsquigarrow \blacktriangleleft\Diamond$ is true by Def. 3, keeping in mind the intuition given in Remark 1. Soundness of proof rule $\sim\rightsquigarrow$ is by induction and using Lemma 1. □

Lemma 3 (Existence of a Normalized Form). *For any trace formula π there exists a state formula ϕ such that $\pi \rightsquigarrow \phi$, or a normalized trace formula ξ such that $\pi \rightsquigarrow \xi$. Such a ϕ or ξ can be computed from π in linear time.*

Proof. This lemma is a direct consequence of Lemma 1, using the identities of Fig. 1. Unless π is itself a state formula ϕ, it is related to a normalized trace formula ξ. □

Lemma 3 concludes our study of normalized forms. Since every trace formula is related (and thus semantically equivalent by Lemma 2) to a state formula or a normalized trace formula, we can limit our axiomatization to the study of state formulas and normalized trace formulas. Formulas of the form $[\alpha]\phi$ or $\langle\alpha\rangle\phi$ involving state formulas have already been axiomatized in d$\mathcal{L}$ [15,17], [16, chapter 2]. The rest of this paper focuses on axiomatizing formulas of the form $[\alpha]\xi$ or $\langle\alpha\rangle\xi$ involving normalized trace formulas. In [10], we come back to trace formulas to study a direct treatment of proof rules for state formulas of the form $[\alpha]\pi$ and $\langle\alpha\rangle\pi$ in order to make the system more efficient.

3.3   Proof Calculus for dTL²

In this section we present a proof calculus for dTL² for verifying temporal properties of hybrid programs specified in the differential temporal dynamic logic dTL². The basic idea of the proof calculus is symbolic decomposition. The calculus progressively transforms formulas to simpler formulas, often by inductively decomposing programs that are in program modalities. In particular, the temporal rules progressively transform temporal formulas to temporal-free formulas, in order to leverage the nontemporal rules of d$\mathcal{L}$. The proof system inherits its nontemporal rules from the d$\mathcal{L}$ proof system [15,17], [16, chapter 2], and adds its own temporal rules. As is the case for d$\mathcal{L}$, the basis of our proof system is real arithmetic, and we integrate it as in d$\mathcal{L}$ [15,17], [16, chapter 2]. We first present how to use the rules, then a brief overview on the inherited nontemporal rules from d$\mathcal{L}$, and finally a detailed account of the new temporal rules of dTL², summarized in Fig. 2.

Usage of the Rules. Rules are to be used in the same way as in the d$\mathcal{L}$ calculus. We do, however, use a new double bar notation by writing some rules in the form

$$\frac{\phi}{\psi}$$

This notation denotes equivalence of the premise and its conclusion. This means that there exists a dual rule, hence the two following rules are true

$$\frac{\phi}{\psi} \qquad\qquad\qquad \frac{\neg\phi}{\neg\psi}$$

For space reasons we do not list dual rules explicitly but give them in [10].

Inherited Nontemporal Rules. On top of the temporal rules presented in Fig. 2, the proof calculus of dTL^2 also inherits the rules of the proof calculus of d$\mathcal{L}$. Since the semantics of dTL^2 conservatively extends the semantics of dTL, which itself conservatively extends the semantics of d$\mathcal{L}$ [14], [16, chapter 4], it is sound to inherit the d$\mathcal{L}$ calculus. While we inherit the nontemporal rules of d$\mathcal{L}$, we do not inherit — but rather reformulate with normalized trace formulas — the temporal rules of dTL [14], [16, chapter 4], thus enabling more efficient proofs by exploiting normalized trace formulas.

Temporal Rules. The temporal rules of the proof calculus of dTL^2 are presented in Fig. 2, in which they are grouped by program construct. Rules $[\,]\rightsquigarrow$ and $\langle\,\rangle\rightsquigarrow$ lift trace formula normalization to program modalities. Rule $[\cup]\xi$ for nondeterministic choice easily extends corresponding rule $[\cup]$ of d$\mathcal{L}$, and assignment rules behave as expected, largely because assignments always terminate.

The sequential composition rules exhibit how nicely the normalized formula interact with sequential composition; remember that sequential composition is one of the main technical difficulties of a calculus handling alternating program and temporal modalities. Normalized trace formulas were designed for these rules, and particular care was taken in considering nonterminating traces. Rule $[;]\sqcap$ expresses that all traces of the composition of two programs α and β satisfies $\phi \sqcap \Box\psi$ if and only if all traces of α satisfy $\Box\psi$, and for terminating traces of α, if all following traces of β satisfy $\phi \sqcap \Box\psi$. In particular, this rule improves on the corresponding rule $[;]\Box$ of dTL by *not* duplicating program modality $[\beta]$, thus eliminating proofs that are exponential in the number of sequential compositions. Rule $[;]\sqcup$ is the main rule for alternating program and temporal modalities in the context of sequential composition. It expresses that all traces of the composition of two programs α and β satisfies $\phi \sqcap \Diamond\psi$ if and only if all traces of α either satisfy $\Diamond\psi$, or are terminating and followed only by traces of β satisfying $\phi\sqcup\Diamond\psi$. Finally, rule $[;]\blacktriangleleft$ similarly handles sequential compositions followed by a $\blacktriangleleft$ operator.

For the test rules, let us remember that a test trace terminates only if the test passes, and is otherwise an error trace. Any trace of test $?\chi$ satisfies $\phi \sqcap \Box\psi$ if and only if its initial state satisfies $\phi \wedge \psi$ when it terminates, or satisfies just ψ when it doesn't terminate; this can be summarized as $(\neg\chi \vee \phi) \wedge \psi$ as in rule $[?]\sqcap$. Rule $[?]\sqcup$ is similar. Any trace of test $?\chi$ satisfies $\phi \blacktriangleleft \Diamond\Box\psi$ if and only if it terminates and its initial state satisfied ϕ, or it doesn't terminate and its initial state satisfied ψ; this can be summarized as $(\chi \wedge \phi) \vee (\neg\chi \wedge \psi)$ as in rule $[?] \blacktriangleleft \Diamond$. Rule $[?] \blacktriangleleft \Box$ is similar.

$$\boxed{\text{Normalization of Trace Formulas}} \qquad \dfrac{\pi \rightsquigarrow \xi \quad [\alpha]\xi}{[\alpha]\pi} \; ([\,]\rightsquigarrow) \qquad \dfrac{\pi \rightsquigarrow \xi \quad \langle\alpha\rangle\xi}{\langle\alpha\rangle\pi} \; (\langle\,\rangle\rightsquigarrow)$$

$\boxed{\text{Sequential Composition}}$

$$\dfrac{[\alpha]([\beta](\phi \sqcap \Box\psi) \sqcap \Box\psi)}{[\alpha;\beta](\phi \sqcap \Box\psi)}([;]\sqcap) \qquad \dfrac{[\alpha]([\beta](\phi \sqcup \Diamond\psi) \sqcup \Diamond\psi)}{[\alpha;\beta](\phi \sqcup \Diamond\psi)}([;]\sqcup) \qquad \dfrac{[\alpha]([\beta](\phi \blacktriangleleft \pi) \blacktriangleleft \pi)}{[\alpha;\beta](\phi \blacktriangleleft \pi)}([;]\blacktriangleleft)$$

$\boxed{\text{Nondeterministic Choice}}$ | $\boxed{\text{Test}}$

$$\dfrac{[\alpha]\xi \wedge [\beta]\xi}{[\alpha \cup \beta]\xi}(\lbrack\cup\rbrack\xi)$$

$$\dfrac{(\neg\chi \vee \phi) \wedge \psi}{[?\chi](\phi \sqcap \Box\psi)}([?]\sqcap) \qquad \dfrac{(\chi \wedge \phi) \vee (\neg\chi \wedge \psi)}{[?\chi](\phi \blacktriangleleft \Diamond\Box\psi)}([?]\blacktriangleleft\Diamond)$$

$$\dfrac{(\chi \wedge \phi) \vee \psi}{[?\chi](\phi \sqcup \Diamond\psi)}([?]\sqcup) \qquad \dfrac{(\chi \wedge \phi) \vee (\neg\chi \wedge \psi)}{[?\chi](\phi \blacktriangleleft \Box\Diamond\psi)}([?]\blacktriangleleft\Box)$$

$\boxed{\text{Assignment}}$

$$\dfrac{\psi \wedge [x := \theta](\phi \wedge \psi)}{[x := \theta](\phi \sqcap \Box\psi)}([:=]\sqcap) \qquad \dfrac{\psi \vee [x := \theta](\phi \vee \psi)}{[x := \theta](\phi \sqcup \Diamond\psi)}([:=]\sqcup) \qquad \dfrac{[x := \theta]\phi}{[x := \theta](\phi \blacktriangleleft \pi)}([:=]\blacktriangleleft)$$

$\boxed{\text{Ordinary Differential Equation}}$

$$\dfrac{\psi \wedge [x' = \theta \,\&\, \chi](\phi \wedge \psi)}{[x' = \theta \,\&\, \chi](\phi \sqcap \Box\psi)}(['\,]\sqcap)$$

$$\dfrac{(\chi \vee \psi) \wedge [x' = \theta \,\&\, (\chi \wedge \neg\psi)]\phi \wedge \langle x' = \theta\rangle(\neg\chi \vee \psi)}{[x' = \theta \,\&\, \chi](\phi \sqcup \Diamond\psi)}(['\,]\sqcup)$$

$$\dfrac{(\chi \vee \psi) \wedge [x' = \theta \,\&\, \chi]\phi \wedge (\langle x' = \theta\rangle(\neg\chi) \vee \langle x' = \theta\rangle[x' = \theta]\psi)}{[x' = \theta \,\&\, \chi](\phi \blacktriangleleft \Diamond\Box\psi)}(['\,]\blacktriangleleft\Diamond)$$

$$\dfrac{(\chi \vee \psi) \wedge [x' = \theta \,\&\, \chi]\phi \wedge (\langle x' = \theta\rangle(\neg\chi) \vee [x' = \theta]\langle x' = \theta\rangle\psi)}{[x' = \theta \,\&\, \chi](\phi \blacktriangleleft \Box\Diamond\psi)}(['\,]\blacktriangleleft\Box)$$

$\boxed{\text{Repetition}}$

$$\dfrac{\phi \wedge [\alpha^*][\alpha](\phi \sqcap \Box\psi)}{[\alpha^*](\phi \sqcap \Box\psi)}([^*]\sqcap) \qquad \dfrac{\psi \vee (\phi \wedge [\alpha;\alpha^*](\phi \sqcup \Diamond\psi))}{[\alpha^*](\phi \sqcup \Diamond\psi)}([^{*n}]\sqcup)$$

$$\dfrac{\forall^{\alpha}(\phi \rightarrow [\alpha](\phi \sqcup \Diamond\psi))}{\phi \rightarrow [\alpha^*](\phi \sqcup \Diamond\psi)}(\text{ind}\sqcup) \qquad \dfrac{\phi \wedge [\alpha^*][\alpha](\phi \blacktriangleleft \pi)}{[\alpha^*](\phi \blacktriangleleft \pi)}([^*]\blacktriangleleft)$$

$$\dfrac{\forall^{\alpha}\forall r > 0 \, (\varphi(r) \rightarrow \langle\alpha\rangle(\varphi(r-1) \sqcap \Box\psi))}{(\exists r \, \varphi(r)) \wedge \psi \rightarrow \langle\alpha^*\rangle((\exists r \leq 0 \, \varphi(r)) \sqcap \Box\psi)}(\text{con}\sqcap)$$

Fig. 2. Rule schemata of the proof calculus for dTL^2

Ordinary differential equations have terminating traces, but also infinite and error traces. Additionally, the execution can exit a differential equation at any moment, even if the evolution constraint domain it still verified; thus formulas like $[x' = \theta \,\&\, \chi]\phi$ and $[x' = \theta \,\&\, \chi]\Box\phi$ are equivalent in a state satisfying χ. Rules for ordinary differential equations transform formulas into temporal-free formulas, on which the d$\mathcal{L}$ proof calculus and in particular differential invariants can be used. In rule $['\,]\sqcap$, the first conjunct ψ is necessary to handle error traces, when χ is initially false. In rule $['\,]\sqcup$, the first conjunct $\chi \vee \psi$ expresses that the differential equation can evolve or has satisfied $\Diamond\psi$ initially. The second conjunct handles traces that never satisfy ψ and thus have to satisfy ϕ, and the third conjunct makes sure there is either no infinite trace ($\langle x' = \theta\rangle\neg\chi$), or that such an infinite trace satisfies $\Diamond\psi$ (condition $\langle x' = \theta\rangle\psi$, equivalent to $\langle x' = \theta\rangle\Diamond\psi$). The first conjunct of rule $[']$ ◄ $\Diamond$ again handles error traces as in rule $[']\sqcup$. The second conjunct ensures all terminating traces finish in a state satisfying ϕ, and its third conjunct handles infinite traces by making sure they don't exist ($\langle x' = \theta\rangle\neg\chi$) or that they satisfy $\Diamond\Box\psi$ (condition $\langle x' = \theta\rangle[x' = \theta]\psi$). Rule $[']$ ◄ $\Box$ is similar.

In some way, repetition rules are easier because as long as a repetition only repeats a terminating trace, it is itself terminating. Rules $[^*]\sqcap$ and $[^*]$◄ are particularly satisfying because their premise no longer contains a temporal property of a loop, but only a non-temporal postcondition of a loop, which is thus provable by ordinary, non-temporal induction. Only the postcondition still has a temporal property but no more loops. That is, these rules reduce temporal properties of loops to nontemporal properties of loops, or more complicated temporal properties on a program without the loop. In rule $[^*]\sqcap$, the first disjunct expresses that $\Diamond\psi$ holds without repeating if ψ holds initially. The first conjunct ϕ of the second disjunct is necessary when α repeats zero times; while the second conjunct executes α any number of times n, then checks that the $(n + 1)$-st execution of α also satisfies $\phi \sqcap \Box\psi$. The treatment of rule $[^*]$◄ is similar. Rule $[^{*n}]$ is less satisfying because it leaves an α^* inside a program modality followed by a normalized trace formula. If ψ is true then the conclusion trivially holds; otherwise the rule relies on the fact that α^* is equivalent to $?true \cup \alpha; \alpha^*$ and just unwinds the loop once. Program $\alpha; \alpha^*$ in the modality could as well be the equivalent $\alpha^*; \alpha$. The same thing is *not* true for rule $[^*] \sqcap$, where $[\alpha^*][\alpha](\phi \sqcap \Box\phi)$ ensures progress of the proof, while writing $[\alpha][\alpha^*](\phi\sqcap\Box\phi)$ would not. Rules ind $\sqcup$ and con $\sqcap$ extend induction (ind) and convergence (con) rules of d$\mathcal{L}$ to normalized trace formulas. As in d$\mathcal{L}$, they are not equivalences; and also as in d$\mathcal{L}$, they use the notation $\forall^\alpha$, which quantifies over all variables possibly assigned by α in assignments or differential equations. Rule ind $\sqcup$ shows that ϕ is inductive with exit clause $\Diamond\psi$, i.e., ϕ holds after all traces of α from any state where ϕ holds, except when exit condition ψ was true at some point during that trace. If ψ was true initially, rule $[^{*n}]$ applies instead. Rule con $\sqcap$ proves that φ is a variant of some trace of α (i.e., its level r decreases) during which ψ always holds true. Then starting from some initial r (assumption of conclusion), an r for which $\varphi(r)$ holds will ultimately be ≤ 0 without having violated when repeating α^* often enough.

3.4 Meta-Results

Soundness. The following result shows that verification with the dTL^2 calculus always produces correct results about the temporal behavior of hybrid systems, i.e., the dTL^2 calculus presented in Fig. 2 is sound. Theorem 1 is proved in [10].

Theorem 1 (Soundness of dTL^2). *The dTL^2 calculus presented in Fig. 2 is sound, i.e., derivable state formulas are valid, i.e., valid in every state.*

Incompleteness of dTL^2. In [15,17], [16, chapter 2] it was shown that the discrete and continuous fragments of d$\mathcal{L}$ are non-axiomatizable. An extension of d$\mathcal{L}$, the logic dTL is also non-axiomatizable [14], [16, chapter 4]. Since dTL^2 is a conservative extension of both d$\mathcal{L}$ and dTL, those results lift to dTL^2. Therefore the discrete and continuous fragments of dTL^2, even if only containing nontemporal formulas are non-axiomatizable. In particular dTL^2 is non-axiomatizable.

Relative Completeness for Star-Free Expressions. We now show how to lift the relative completeness result of d$\mathcal{L}$ [15,17], [16, chapter 2] to dTL^2; this completeness result is relative to first order logic of differential equations (FOD), i.e., first-order real arithmetic augmented with formulas expressing properties of differential equations [15,17], [16, chapter 2].

Theorem 2 (Relative completeness for star-free expressions). *The dTL^2 calculus restricted to *-free programs is complete relative to FOD, i.e., every valid dTL^2 formula with only star-free programs can be derived from FOD tautologies.*

Theorem 2 is proved in [10]. We conjecture that the proof system of dTL^2 is also relatively complete relative to FOD for all expressions, including repetitions.

4   Alternative Proof Systems

Normalizing all temporal formulas before applying the rules of Fig. 2 can sometimes result in longer proofs than necessary. In [10] we study a proof system directly handling (non-normalized) trace formulas. This extended proof system alleviates the need for normalizing all trace formulas, and is thus more efficient.

Another alternative, that we also study in [10], is to suppress all the $[\,]\sqcap$ rules of Fig. 2 (rules $[;]\sqcap$, $[?]\sqcap$, $[:=]\sqcap$, $[']\sqcap$ and $[*]\sqcap$) and replace them by rules directly handling formulas of the form $[\alpha]\Box\phi$, and the following rule:

$$\frac{[\alpha]\phi \wedge [\alpha]\Box\psi}{[\alpha](\phi \sqcap \Box\psi)}([\,]\sqcap)$$

This results in a simpler system, because some of the rules are less complicated. However the system is not as efficient, because it duplicates the symbolic execution of α.

5   Related Work

In this section we study work related specifically to temporal reasoning of hybrid systems. For a more general account of previous work on verification of hybrid systems we refer to [15,17], [16, chapter 2].

This paper is based on work by Platzer introducing a temporal dynamic logic for hybrid systems [14], extending previous work by Beckert and Schlager [1] to hybrid programs. Both papers present a relatively complete calculus; however Beckert and Schlager only consider discrete state spaces, and only study temporal formulas of the form $[\alpha]\Box\phi$ and its dual $\langle\alpha\rangle\Diamond\phi$, leaving out any mixed cases alternating program and temporal modalities $[\alpha]\Diamond\phi$ or $[\alpha]\Box\Diamond\phi$. Platzer proposes to handle mixed cases by non-local program transformation, but does not show how to handle them compositionally.

Process logic [7,12,13,20] initially used temporal logic [6,19] in the context of dynamic logic [8] to reason about temporal behavior of programs. It is well studied, but limited to discrete programs. It also only considers an abstract notion of atomic program, without explicitly considering assignments and tests.

Davoren and Nerode [5] study hybrid systems and their topological aspects in the context of the propositional modal μ-calculus. Davoren, Coulthard, Markey and Moor [4] also give a semantics in general flow systems for a generalization of CTL*. In both [5] and [4], the authors provide Hilbert-style calculi to prove formulas of their systems, but in a propositional — not first-order — system, without specific proof rules to handle ordinary differential equations. Zhou, Ravn and Hansen [21] present a duration calculus extended by mathematical expressions with derivatives of state variables. Their system requires external mathematical reasoning about derivatives and continuity.

Other authors have studied temporal properties of hybrid systems in the context of model checking. Mysore, Piazza and Mishra [11] study model checking of semi-algebraic hybrid systems for TCTL (Timed Computation Tree Logic) properties and prove undecidability. They do bounded model checking for differential equations with polynomial solutions only, while we handle more general polynomial differential equations and unbounded safety verification. Additionally TCTL does not allow nesting of temporal modalities as we do. Cimatti, Roveri and Tonetta [3] present HRELTL, a linear temporal logic with regular expressions for hybrid traces. Their work is inspired by requirements validation for the European Train Control System, and uses bounded model checking and satisfiability modulo theory. More recently, Bresolin [2] develops HyLTL, a temporal logic for model checking hybrid systems, and shows how to solve the model checking problem by translating formulas into equivalent hybrid automata.

6   Conclusion and Future Work

In this paper we have presented a proof calculus for dTL^2, extending dTL by allowing nesting of temporal modalities. We showed proof rules for handling compositionally alternating program and temporal modalities, solving an open problem formulated in 2001 [1] and identified as a problem for hybrid systems in 2007 [14], [16, chapter 4]. We also offered a treatment where programs are not duplicated by proof rules, solving another open problem formulated by [14], [16, chapter 4]. We showed that the system is relatively complete with respect to FOD for *-free hybrid programs. The treatment of infinite traces is crucial to make the logic interesting, as temporal properties on terminating and error traces simplify greatly (Remark 1).

Future work includes proving our conjecture that the system is relatively complete with respect to FOD for all expressions; extending the semantics and the proof system to allow repetition — and not just differential equations — to create infinite traces; and implementing our proof rules in a tool such as KeYmaera [18].

A number of extensions to dTL^2 should be explored, such as inclusion of the temporal Until operator, or nested conjunctions and disjunctions inside temporal formulas. Some of these extensions can be handled by program transformations [14], [16, chapter 4], but a compositional proof system such as the one presented here would be more interesting. The proof system of dTL^2 is an important step towards a more general system dTL^*, extending dTL^2 with formulas of CTL*, and expressing formulas such as $[\alpha]\Box(\Diamond\phi \wedge \psi)$. We would like to develop a semantics and a proof system for dTL^*.

Acknowledgements. We are grateful to Khalil Ghorbal, Dexter Kozen, Sarah Loos, Stefan Mitsch, Ed Morehouse, Jan-David Quesel, Marcus Völp, and the anonymous referees for helpful comments and discussions. This material is based upon work supported by the National Science Foundation under NSF CAREER Award CNS-1054246, NSF EXPEDITION CNS-0926181 and under Grant No. CNS-0931985.

References

1. Beckert, B., Schlager, S.: A sequent calculus for first-order dynamic logic with trace modalities. In: Goré, R.P., Leitsch, A., Nipkow, T. (eds.) IJCAR 2001. LNCS (LNAI), vol. 2083, pp. 626–641. Springer, Heidelberg (2001)
2. Bresolin, D.: HyLTL: a temporal logic for model checking hybrid systems. In: Bortolussi, L., Bujorianu, M.L., Pola, G. (eds.) HAS. EPTCS, vol. 124, pp. 73–84 (2013)
3. Cimatti, A., Roveri, M., Tonetta, S.: Requirements validation for hybrid systems. In: Bouajjani, A., Maler, O. (eds.) CAV 2009. LNCS, vol. 5643, pp. 188–203. Springer, Heidelberg (2009)
4. Davoren, J.M., Coulthard, V., Markey, N., Moor, T.: Non-deterministic temporal logics for general flow systems. In: Alur, R., Pappas, G.J. (eds.) HSCC 2004. LNCS, vol. 2993, pp. 280–295. Springer, Heidelberg (2004)
5. Davoren, J.M., Nerode, A.: Logics for hybrid systems. Proc. IEEE (2000)
6. Emerson, E.A., Halpern, J.Y.: "Sometimes" and "Not Never" revisited: on branching versus linear time temporal logic. J. ACM 33(1), 151–178 (1986)
7. Harel, D., Kozen, D., Parikh, R.: Process logic: Expressiveness, decidability, completeness. J. Comput. Syst. Sci. 25(2), 144–170 (1982)
8. Harel, D., Kozen, D., Tiuryn, J.: Dynamic Logic. MIT Press, Cambridge (2000)
9. Hughes, G., Cresswell, M.: A New Introduction to Modal Logic. Routledge (1996)
10. Jeannin, J.B., Platzer, A.: dTL^2: Differential temporal dynamic logic with nested temporalities for hybrid systems. Tech. Rep. CMU-CS-14-109, School of Computer Science. Carnegie Mellon University, Pittsburgh, PA, 15213 (May 2014), http://reports-archive.adm.cs.cmu.edu/anon/2013/abstracts/14-109.html
11. Mysore, V., Piazza, C., Mishra, B.: Algorithmic algebraic model checking II: Decidability of semi-algebraic model checking and its applications to systems biology. In: Peled, D.A., Tsay, Y.-K. (eds.) ATVA 2005. LNCS, vol. 3707, pp. 217–233. Springer, Heidelberg (2005)
12. Nishimura, H.: Descriptively complete process logic. Acta Inf. 14, 359–369 (1980)
13. Parikh, R.: A decidability result for a second order process logic. In: FOCS, pp. 177–183. IEEE Comp. Soc. (1978)
14. Platzer, A.: A temporal dynamic logic for verifying hybrid system invariants. In: Artemov, S., Nerode, A. (eds.) LFCS 2007. LNCS, vol. 4514, pp. 457–471. Springer, Heidelberg (2007)
15. Platzer, A.: Differential dynamic logic for hybrid systems. J. Autom. Reas. 41(2), 143–189 (2008)
16. Platzer, A.: Logical Analysis of Hybrid Systems: Proving Theorems for Complex Dynamics. Springer, Heidelberg (2010)
17. Platzer, A.: Logics of dynamical systems. In: LICS, pp. 13–24. IEEE (2012)
18. Platzer, A., Quesel, J.-D.: KeYmaera: A hybrid theorem prover for hybrid systems. In: Armando, A., Baumgartner, P., Dowek, G. (eds.) IJCAR 2008. LNCS (LNAI), vol. 5195, pp. 171–178. Springer, Heidelberg (2008)
19. Pnueli, A.: The temporal logic of programs. In: FOCS, pp. 46–57. IEEE Comp. Soc. (1977)
20. Pratt, V.R.: Process logic. In: Aho, A.V., Zilles, S.N., Rosen, B.K. (eds.) POPL, pp. 93–100. ACM (1979)
21. Zhou, C., Ravn, A.P., Hansen, M.R.: An extended duration calculus for hybrid real-time systems. In: Grossman, R.L., Ravn, A.P., Rischel, H., Nerode, A. (eds.) HS 1991 and HS 1992. LNCS, vol. 736, pp. 36–59. Springer, Heidelberg (1993)

Axioms vs Hypersequent Rules with Context Restrictions: Theory and Applications*

Björn Lellmann

TU Vienna, Vienna, Austria

Abstract. We introduce transformations between hypersequent rules with context restrictions and Hilbert axioms extending classical (and intuitionistic) propositional logic and vice versa. The introduced rules are used to prove uniform cut elimination, decidability and complexity results as well as finite axiomatisations for many modal logics given by simple frame properties. Our work subsumes many logic-tailored results and allows for new results. As a case study we apply our methods to the logic of uniform deontic frames.

1 Introduction

The automatic construction of reasoning systems and decision procedures from specifications for various logics is an important emerging area in the field of automated reasoning. Results in this area provide general decision procedures and complexity results applicable to specific logics in the spirit of Logic Engineering [11], and also yield deeper insights into strengths, weaknesses, and fundamental properties of different types of calculi used for reasoning systems. But also from the perspective of producing such systems for specific logics investigating the connections between specifications and different frameworks is important, since this allows choosing the most efficient framework for the logic at hand.

Here we investigate the connection between specifications given as Hilbert axioms and the framework of *hypersequent calculi* for extensions of classical propositional logic. Taking the specifications as Hilbert axioms yields a very flexible and semantics-independent approach and allows to capture non-normal modal logics (unlike e.g. [8]) Also, while often not complexity-optimal, the hypersequent framework is very flexible and captures several logics for which no sequent or tableaux systems seem to exists. Of course correspondence results and general decision procedures demand general results about hypersequent calculi. This necessitates a clarification of which kind of calculi we consider. To this aim we introduce the format of *hypersequent rules with context restrictions* which is general enough to capture many existing calculi, e.g. for modal logics S5 [1] and S4.3 [7] as well as for modal logics without symmetry given by simple frame properties [8]. We obtain sufficient conditions for (syntactic) cut elimination, decidability, and complexity results for such systems. The results apply e.g. to the

---

* Supported by FWF START Y544-N23.

S. Demri, D. Kapur, and C. Weidenbach (Eds.): IJCAR 2014, LNAI 8562, pp. 307–321, 2014.

Table 1. The standard modal rule sets

$$\frac{\mathcal{G} \mid \varphi \Rightarrow \psi}{\mathcal{G} \mid \Box\varphi \Rightarrow \Box\psi} \mathsf{K}_n \qquad \frac{\mathcal{G} \mid \Gamma, \varphi \Rightarrow \Delta}{\mathcal{G} \mid \Gamma, \Box\varphi \Rightarrow \Delta} \mathsf{T}_n \qquad \frac{\mathcal{G} \mid \Box\Gamma, \varphi \Rightarrow \psi}{\mathcal{G} \mid \Box\Gamma, \Box\varphi \Rightarrow \Box\psi} 4_n \quad (|\varphi| = n)$$

$$\mathcal{R}_\mathsf{K} := \{\mathsf{K}_n : n \geq 0\} \qquad \mathcal{R}_\mathsf{KT} := \mathcal{R}_\mathsf{K} \cup \{\mathsf{T}_n : n \geq 1\} \qquad \mathcal{R}_\mathsf{K4} := \mathcal{R}_\mathsf{K} \cup \{4_n : n \geq 0\}$$

calculi for extensions of K or K4 from [8]. We also show a correspondence between rules of our format and axioms of a certain form (Def. 5.16). This yields general decidability and complexity results for modal logics axiomatised this way, and as a byproduct finite axiomatisations for modal logics given by certain simple frame properties. As application we construct a new cut-free hypersequent calculus for the non-normal logic LUDF from [13], entailing a new complexity bound. While for space reasons the results in this article are given for logics with unary connectives based on classical logic, they extend to higher arities and intuitionistic logic as base logic similar to [9]. The extension of these investigations to more general frameworks such as tree-hypersequents will be considered in future work.

2 Preliminaries and Notation

In the following we write $\mathbb{N}$ for $\{0, 1, 2, \dots\}$. We take $\mathcal{V}$ to be a countable set of propositional variables. The set of *boolean connectives* is $\Lambda_\mathsf{B} := \{\wedge, \vee, \rightarrow\}$. For a set $\Lambda \subseteq \Lambda_\mathsf{U} \cup \Lambda_\mathsf{B}$ with Λ_U a set of unary connectives the set $\mathcal{F}(\Lambda)$ of *formulae over* Λ is defined by $\mathcal{F}(\Lambda) \ni \varphi ::= p \mid \bot \mid \heartsuit\varphi \mid \varphi \circ \varphi$ with $p \in \mathcal{V}, \heartsuit \in \Lambda \cap \Lambda_\mathsf{U}$ and $\circ \in \Lambda \cap \Lambda_\mathsf{B}$. The connectives $\leftrightarrow$ and $\neg$ are introduced as abbreviations as usual. Connectives in Λ_U are called *modalities*. The set $\{\Box\} \cup \Lambda_\mathsf{B}$ is denoted $\Lambda_\Box$. For $F \subseteq \mathcal{F}(\Lambda)$ we write $\Lambda(F)$ for $\{\heartsuit\varphi : \heartsuit \in \Lambda \setminus \Lambda_\mathsf{B}$ and $\varphi \in F\} \cup \{\varphi \circ \psi : \circ \in \Lambda \cap \Lambda_\mathsf{B}$ and $\varphi, \psi \in F\}$. The *modal rank* of a formula φ, denoted $\mathsf{mrk}(\varphi)$, is the maximum nesting depth of modalities in φ, and its *complexity* is the number of symbols occurring in it. Sequences $\varphi_1, \dots, \varphi_n$ of formulae are written $\boldsymbol{\varphi}$, and $|\boldsymbol{\varphi}|$ denotes the length of $\boldsymbol{\varphi}$. Similarly $*\varphi_1, \dots, *\varphi_n$ is written $*\boldsymbol{\varphi}$ for $* \in \Lambda$.

A *multiset* Γ over a set F of formulae is a function $F \rightarrow \mathbb{N}$ with finite support, and we write $\varphi \in \Gamma$ for $\Gamma(\varphi) > 0$. The *union* of multisets Γ and Δ is denoted by Γ, Δ and defined by $(\Gamma, \Delta)(\varphi) := \Gamma(\varphi) + \Delta(\varphi)$. We also write $\bigsqcup_{i=1}^n \Gamma_n$ for $\Gamma_1, \dots, \Gamma_n$ and φ for the multiset containing only one occurrence of φ. The set $\mathcal{S}(F)$ of *sequents over the set* F *of formulae* contains all tuples of multisets over F, written as $\Gamma \Rightarrow \Delta$. A *hypersequent* over F is a multiset over $\mathcal{S}(F)$, written as $\Gamma_1 \Rightarrow \Delta_1 \mid \cdots \mid \Gamma_n \Rightarrow \Delta_n$ We write H for the hypersequent version of a standard context-sharing sequent calculus for classical logic [10] with the standard external and internal weakening and contraction rules [1]. The rules of $\mathcal{R}_\mathsf{K}$, $\mathcal{R}_\mathsf{KT}$ and $\mathcal{R}_\mathsf{K4}$ are given in Table 1. The cut rule is denoted Cut.

A Λ-*logic* is a set $\mathcal{L}$ of formulae over Λ closed under modus ponens (if $\varphi \in \mathcal{L}$ and $\varphi \rightarrow \psi \in \mathcal{L}$, then $\psi \in \mathcal{L}$) and uniform substitution (if $\varphi \in \mathcal{L}$, then $\varphi\sigma \in \mathcal{L}$ for every substitution $\sigma : \mathcal{V} \rightarrow \mathcal{F}(\Lambda)$) and containing classical propositional logic. For a set $\mathcal{A}$ of formulae, $\mathcal{L}_\mathcal{A}$ is the smallest Λ-logic containing $\mathcal{A}$. For a Λ-logic

$\mathcal{L}$ and $\varphi \in \mathcal{F}(\Lambda)$ we write $\mathcal{L} \oplus \varphi$ for the smallest Λ-logic $\mathcal{L}'$ with $\mathcal{L} \cup \{\varphi\} \subseteq \mathcal{L}'$. We also write $\models_{\mathcal{L}} \varphi$ for $\varphi \in \mathcal{L}$. For the standard notions of modal logic see [4].

3   Hypersequent Rules with Restrictions

The rule format we consider is an abstraction of the rule format found in many calculi for modal logics. One of the main characteristics is that the format of context formulae which are copied into a premiss can be restricted as in the rule 4_n in Table 1. This is captured by the following notion from [10,9]:

Definition 3.1. *For $F \subseteq \mathcal{F}(\Lambda)$ the set of* context restrictions *over F is $\mathfrak{C}(F) := \{\langle F_1, F_2 \rangle : F_1, F_2 \subseteq F\}$. For a sequent $\Gamma \Rightarrow \Delta$ and a context restriction $\mathcal{C} = \langle F_1, F_2 \rangle$ the* restriction *of $\Gamma \Rightarrow \Delta$ according to $\mathcal{C}$ is the sequent $(\Gamma \Rightarrow \Delta) \upharpoonright_{\mathcal{C}} := \Gamma \upharpoonright_{F_1} \Rightarrow \Delta \upharpoonright_{F_2}$ where for a multiset Σ and $F \subseteq \mathcal{F}(\Lambda)$ the multiset $\Sigma \upharpoonright_F$ contains those formulae from Σ which are substitution instances of formulae in F.*

Example 3.2.   1. The context restriction $\mathcal{C}_{\emptyset} := \langle \emptyset, \emptyset \rangle$ intuitively deletes the whole context, we have $(\Gamma \Rightarrow \Delta) \upharpoonright_{\mathcal{C}_{\emptyset}} = \Rightarrow$ for every sequent $\Gamma \Rightarrow \Delta$.
   2. The context restriction $\mathcal{C}_{\mathrm{id}} := \langle \{p\}, \{p\} \rangle$ intuitively copies the whole context, we have $(\Gamma \Rightarrow \Delta) \upharpoonright_{\mathcal{C}_{\mathrm{id}}} = \Gamma \Rightarrow \Delta$ for every sequent $\Gamma \Rightarrow \Delta$.
   3. The context restriction $\mathcal{C}_{\square} := \langle \{\square p\}, \emptyset \rangle$ copies only the boxed formulae on the left side of the context.

We take the rules to introduce exactly one layer of connectives in the principal formulae, and we assume that every premiss includes a restriction for each component of the principal part.

Definition 3.3. *A* hypersequent rule with context restrictions, *written as*

$$\frac{\{(\Gamma_i \Rightarrow \Delta_i; \mathcal{C}_i) : i \le m\}}{\Sigma_1 \Rightarrow \Pi_1 \mid \cdots \mid \Sigma_n \Rightarrow \Pi_n}$$

is given by a natural number $n > 0$, a sequence $\Sigma_1 \Rightarrow \Pi_1 \mid \cdots \mid \Sigma_n \Rightarrow \Pi_n$ called principal part *with $\Sigma_i \Rightarrow \Pi_i \in \mathcal{S}(\Lambda(\mathcal{V}))$ and a set of* premisses, *where each premiss $(\Gamma_i \Rightarrow \Delta_i; \mathcal{C}_i)$ consists of a sequent of variables and a sequence $\mathcal{C}_i = \langle F_i^1, G_i^1 \rangle, \ldots, \langle F_i^n, G_i^n \rangle$ of context restrictions subject to the* variable condition: *every variable occurs at most once in the principal part and it occurs in the principal part whenever it occurs in the premisses. An* application *of such a rule is given by a substitution $\sigma : \mathcal{V} \to \mathcal{F}(\Lambda)$, a side hypersequent $\mathcal{G}$ and a sequence $\Omega_1 \Rightarrow \Upsilon_1 \mid \cdots \mid \Omega_n \Rightarrow \Upsilon_n$ of context sequents. It is written as*

$$\frac{\{\mathcal{G} \mid \Omega_1 \upharpoonright_{F_i^1}, \ldots, \Omega_n \upharpoonright_{F_i^n}, \Gamma_i \sigma \Rightarrow \Delta_i \sigma, \Upsilon_1 \upharpoonright_{G_i^1}, \ldots, \Upsilon_n \upharpoonright_{G_i^n} : i \le m\}}{\mathcal{G} \mid \Omega_1, \Sigma_1 \sigma \Rightarrow \Pi_1 \sigma, \Upsilon_1 \mid \cdots \mid \Omega_n, \Sigma_n \sigma \Rightarrow \Pi_n \sigma, \Upsilon_n} .$$

The notions of a *derivation* and *derivability* for a set $\mathcal{R}$ of hypersequent rules with restrictions are defined in the usual way, and we write $\vdash_{\mathcal{R}} \mathcal{G}$ if $\mathcal{G}$ is derivable in

$\mathcal{R}$. A rule is *derivable in* $\mathcal{R}$ if for all its applications the conclusion can be derived from the premises in $\mathcal{R}$ and *admissible* if whenever the premises are derivable in $\mathcal{R}$, then so is the conclusion. We stipulate that sets of rules are closed under variable renaming and permutation of the components in the principal part. Rules are written inline using "/" to separate premises and conclusion.

Example 3.4. 1. The standard hypersequent rules for the propositional connectives. E.g. the rule $\wedge_\mathsf{L}$ is the rule $\{(p, q \Rightarrow ; \mathcal{C}_{\mathsf{id}})\}/p \wedge q \Rightarrow .$

2. The standard rules for modal logics from Table 1. E.g. the rule 4_n is the rule $\{(\boldsymbol{p} \Rightarrow q; \mathcal{C}_\square)\}/\square\boldsymbol{p} \Rightarrow \square q$ with $|\boldsymbol{p}| = n$.

3. The modalised splitting rule for S5 from [1] with applications $\mathcal{G} \mid \square\Gamma, \Sigma \Rightarrow \square\Delta, \Pi/\mathcal{G} \mid \square\Gamma \Rightarrow \square\Delta \mid \Sigma \Rightarrow \Pi$ is $\{(\Rightarrow ; \langle\{\square p\}, \{\square p\}\rangle, \mathcal{C}_{\mathsf{id}})\}/ \Rightarrow \mid \Rightarrow .$

4 Cut Elimination and Applications

We obtain sufficient criteria for cut elimination by generalising the cut elimination proof in [5]. The cut-elimination strategy is to permute a cut into the premises of the last applied rule on the left until the cut formula is principal in the last applied rule. Then the cut is permuted into the premises on the right until it is principal here as well, in which case it is reduced to cuts on formulae of smaller complexity. To state the condition used to reduce principal cuts we use the notion of a *cut between rules*, where intuitively a new rule is constructed from two rules by cutting their conclusions on a formula $\heartsuit p$ and eliminating p from the premises by cutting on p in all possible ways. To make this precise, write $\mathcal{C} \cup \mathcal{D}$ for the *union* of two sequences $\mathcal{C}, \mathcal{D} \in \mathfrak{C}^n$ of restrictions, defined component-wise: If the i-th components of $\mathcal{C}$ resp. $\mathcal{D}$ are $\langle F_i, G_i\rangle$ resp. $\langle F_i', G_i'\rangle$, then the i-th component of $\mathcal{C} \cup \mathcal{D}$ is $\langle F_i \cup F_i', G_i \cup G_i'\rangle$. In addition, for permuting the cut into the context on the right we need a condition on the context restrictions which ensures that whenever the cut formula satisfies a context restriction, then so does the whole left premise of the cut.

Definition 4.1. *For sets* $\mathcal{P}_1, \mathcal{P}_2$ *of premises and rules* $R_1 = \mathcal{P}_1/\Sigma_1 \Rightarrow \Pi_1 \mid \cdots \mid \Sigma_{n-1} \Rightarrow \Pi_{n-1} \mid \Sigma_n \Rightarrow \Pi_n, \heartsuit p$ *and* $R_2 = \mathcal{P}_2/\heartsuit p, \Omega_1 \Rightarrow \Theta_1 \mid \Omega_2 \Rightarrow \Theta_2 \mid \cdots \mid \Omega_k \Rightarrow \Theta_k$ *the cut between* R_1 *and* R_2 *on* $\heartsuit p$ *is the rule* $\mathrm{cut}(R_1, R_2, \heartsuit p)$ *given by*

$$\frac{\{(\Gamma, \Gamma' \Rightarrow \Delta, \Delta'; \mathcal{C} \cup \mathcal{D}) : (\Gamma \Rightarrow \Delta, p; \mathcal{C}), (\Gamma' \Rightarrow \Delta'; \mathcal{D}) \in \mathcal{P}\} \qquad \{(\Gamma \Rightarrow \Delta; \mathcal{C}) \in \mathcal{P} : p \notin \Gamma, \Delta\}}{\Sigma_1 \Rightarrow \Pi_1 \mid \cdots \mid \Sigma_{n-1} \Rightarrow \Pi_{n-1} \mid \Sigma_n, \Omega_1 \Rightarrow \Pi_n, \Theta_1 \mid \Omega_2 \Rightarrow \Theta_2 \mid \ldots \Omega_k \Rightarrow \Theta_k}$$

where $\mathcal{P} := \{(\Gamma \Rightarrow \Delta; \mathcal{C}, \mathcal{C}_\emptyset, \overset{(k-1)\text{-}times}{\ldots}, \mathcal{C}_\emptyset) : (\Gamma \Rightarrow \Delta; \mathcal{C}) \in \mathcal{P}_1\} \cup \{(\Gamma \Rightarrow \Delta; \mathcal{D}_\emptyset, \overset{(n-1)\text{-}times}{\ldots}, \mathcal{D}_\emptyset, \mathcal{D}) : (\Gamma \Rightarrow \Delta; \mathcal{C}) \in \mathcal{P}_2\}$. *A set* $\mathcal{R}$ *of rules is* principal-cut closed *if it is closed under the addition of cuts between rules. It is* mixed-cut permuting *if for all* $R_1, R_2 \in \mathcal{R}$: *if* $\Gamma \Rightarrow \Delta, \heartsuit p$ *is a component of the principal part of* R_1 *and* $(\heartsuit p \Rightarrow) \upharpoonright_\mathcal{C} = \heartsuit p \Rightarrow$ *for a restriction* $\mathcal{C}$ *of* R_2, *then* $(\Gamma \Rightarrow \Delta) \upharpoonright_\mathcal{C} = \Gamma \Rightarrow \Delta$ *and* $(\Sigma \Rightarrow \Pi) \upharpoonright_\mathcal{D} \upharpoonright_\mathcal{C} = (\Sigma \Rightarrow \Pi) \upharpoonright_\mathcal{D}$ *for every restriction* $\mathcal{D}$ *for this component and sequent* $\Sigma \Rightarrow \Pi$.

Example 4.2. 1. The cut between $K_n = \{(\boldsymbol{p} \Rightarrow q; \mathcal{C}_\emptyset)\}/\Box \boldsymbol{p} \Rightarrow \Box q$ and $K_{m+1} = \{(q, \boldsymbol{q} \Rightarrow r; \mathcal{C}_\emptyset)\}/\Box q, \Box \boldsymbol{q} \Rightarrow \Box r$ is the rule $\mathsf{cut}(K_n, K_{m+1}, \Box q) = \{(\boldsymbol{p}, \boldsymbol{q} \Rightarrow r); \mathcal{C}_\emptyset\}/\Box \boldsymbol{p}, \Box \boldsymbol{q} \Rightarrow \Box r = K_{n+m}$. Thus the rule set $\mathcal{R}_K$ is principal-cut closed.
2. The cut between the rule $K4_n = \{(\boldsymbol{p} \Rightarrow q; \mathcal{C}_\emptyset)\}/\Box \boldsymbol{p} \Rightarrow \Box q$ and the rule $5 = \{q \Rightarrow ; \mathcal{C}_\emptyset, \mathcal{C}_{\mathsf{id}})\}/\Box q \Rightarrow | \Rightarrow$ on $\Box q$ is the rule $\mathsf{cut}(K4_n, 5, \Box q) = \{(\boldsymbol{p} \Rightarrow \mathcal{C}_\Box, \mathcal{C}_{\emptyset, \emptyset})\}/\Box \boldsymbol{p} \Rightarrow | \Rightarrow$ which we denote 5_n. Its applications have the form $\mathcal{G} \mid \Box \Gamma, \Sigma, \varphi_1, \ldots, \varphi_n \Rightarrow \Pi / \mathcal{G} \mid \Box \Gamma, \Box \varphi_1, \ldots, \Box \varphi_n \Rightarrow | \Sigma \Rightarrow \Pi$. It is straightforward to see that the rule set $\mathcal{R}_{\mathsf{KT4}} \cup \{5_n : n \geq 0\}$ is principal-cut closed.

For sequent rules introducing only one connective, principal-cut closure is known as *coherence* [2], and it corresponds to Belnap's condition $C8$ [3]. The two properties of Def. 4.1 ensure that we can eliminate topmost instances of a restricted version of multicut, where the cut formula occurs only once in the left premiss (and is principal in the last applied rule there), but several times in several components in the right premiss by induction on the maximal complexity of a cut formula occurring in a derivation. Allowing the cut formula to occur more than once on the right is necessary due to the internal and external contraction rules. The fact that several instances of the cut formula in the right premiss of such a restricted multicut can be principal also is the reason why we take the cuts between rules of a principal-cut closed rule set to be *in* the rule set and not just derivable: we need to be able to replace iterated cuts by a rule from the rule set. To avoid also several instances of the cut formula being principal in the left premiss and to deal with external contraction we introduce a further restriction.

Definition 4.3. *A rule set $\mathcal{R}$ is* right-contraction closed *if applications of internal contraction right to the conclusion of a rule are derived by internal contractions followed by one rule from $\mathcal{R}$. It is* single-conclusion right *if the principal part of no rule contains $\Gamma \Rightarrow \Delta, \heartsuit p \mid \Sigma \Rightarrow \Pi, \heartsuit q$ for $\heartsuit \in \Lambda$ and $p, q \in \mathcal{V}$.*

The cut formula is not principal in more than one component in the left premiss if the rule set is single-conclusion right, and right-contraction closure prevents the cut formula occurring twice in a single component of the principal part (proof by induction on the complexity of the contracted formulae):

Lemma 4.4. *Let $\mathcal{R}$ be right-contraction closed and single-conclusion right. Then whenever $\vdash_{\mathcal{R}\mathsf{Cut}} \mathcal{G}$ there is a derivation of $\mathcal{G}$ in which in every application of a rule from $\mathcal{R}$ the right hand sides of the principal part are fully contracted.*

Finally, we impose a further restriction which ensures that cuts with cut formula contextual on the left can be permuted into the premisses on the left.

Definition 4.5. *A rule is* right-substitutive *if all restrictions occurring in it have the form $\langle \{p\}; \{p\} \rangle$ or $\langle F; \emptyset \rangle$ for some $F \subseteq \mathcal{F}(\Lambda)$.*

Theorem 4.6 (Cut elimination). *Let $\mathcal{R}$ be right-substitutive, single-conclusion right, right-contraction closed, principal-cut closed and mixed-cut permuting. Then for every hypersequent $\mathcal{G}$ we have: $\vdash_{\mathcal{R}\mathsf{Cut}} \mathcal{G}$ iff $\vdash_{\mathcal{R}} \mathcal{G}$.*

Proof (Sketch). By double induction on the maximal complexity of a cut formula in a derivation and the number of applications with cut formula of maximal complexity. Topmost cuts of maximal complexity are eliminated using the fact that with right-substitutivity applications of a restricted version of multicut allowing the cut formula to occur several times in several components on the left can be eliminated by permuting them up on the left until exactly one occurrence is principal (by Lem. 4.4 and single-conclusion right), permuting the non-principal cuts into the premisses and using principal-cut closure and mixed-cut closure as above to eliminate the remaining cut with cut formula principal on the left.   □

Corollary 4.7. *The hypersequent calculi* $\mathsf{H}, \mathsf{HR_K}, \mathsf{HR_{K4}}, \mathsf{HR_{KT}}$ *and* $\mathsf{HR_{KT}}\{5_n : n \in \mathbb{N}\}$ *with rules* 5_n *from Ex. 4.2.2 admit cut elimination.*

Proof. Inspection of the rules together with Ex. 4.2 shows that these rule sets satisfy the conditions of Thm. 4.6.   □

Thm. 4.6 together with the next Lemma also provides the basis of the method of *cut elimination by saturation* used in Sec. 6, where cut-free hypersequent calculi are constructed by saturating a rule set under cuts between rules. Of course we still need to check that the remaining conditions of Thm. 4.6 are satisfied.

Lemma 4.8. *Let* R_1, R_2 *be hypersequent rules with context restrictions. Then the rule* $\mathsf{cut}(R_1, R_2, \heartsuit p)$ *is a derivable rule in* $\mathsf{H}R_1 R_2\mathsf{Cut}$.

4.1   Applications: Decision Procedures and Complexity Bounds

For general decision procedures apart from cut elimination we also need to deal with Contraction. The idea is to show admissibility of internal contraction under a modified notion of rule applications, where some principal formulae are copied into the premiss (as in Kleene's G3-systems). Then under a mild assumption only a bounded number of components per hypersequent are relevant in a rule application, hence using the subformula property of the rules the total number of hypersequents occurring in a derivation is bounded and we obtain decidability.

Definition 4.9. *A* modified application *of a hypersequent rule* $R = \{(\Gamma_i \Rightarrow \Delta_i; \mathcal{C}_i) : i \in \mathcal{P}\}/\Sigma_1 \Rightarrow \Pi_1 \mid \cdots \mid \Sigma_n \Rightarrow \Pi_n$ *is given by a side hypersequent* $\mathcal{G}$, *a substitution* $\sigma : \mathcal{V} \to \mathcal{F}$ *and contexts* $\Theta_1 \Rightarrow \Omega_1 \mid \cdots \mid \Theta_n \Rightarrow \Omega_n$ *and written as*

$$\frac{\left\{ \mathcal{G} \mid \mathcal{H} \mid \Gamma_i\sigma, \bigsqcup_{j \leq n}(\Sigma_j\sigma, \Theta_j) \restriction_{\mathcal{C}_i^j} \Rightarrow \Delta_i\sigma, \bigsqcup_{j \leq n}(\Pi_j\sigma, \Omega_j) \restriction_{\mathcal{C}_i^j} : i \in \mathcal{P} \right\}}{\mathcal{G} \mid \Sigma_1\sigma, \Theta_1 \Rightarrow \Pi_1\sigma, \Omega_1 \mid \cdots \mid \Sigma_n\sigma, \Theta_n \Rightarrow \Pi_n\sigma, \Omega_n} R^*$$

with $\mathcal{H} = \Sigma_1\sigma, \Theta_1 \Rightarrow \Pi_1\sigma, \Omega_1 \mid \cdots \mid \Sigma_n\sigma, \Theta_n \Rightarrow \Pi_n\sigma, \Omega_n$.

Thus in addition to the context formulae all principal formulae satisfying the corresponding restriction are copied into the premiss, and all components of the principal part are copied to deal with external contraction. If internal contractions can be permuted with rules this yields admissibility of internal contraction.

Definition 4.10. *A rule set* $\mathcal{R}$ *is* contraction closed *if for every rule* $R \in \mathcal{R}$ *with principal part* $\mathcal{G} \mid \Gamma \Rightarrow \Delta, \heartsuit p, \heartsuit q$ *(resp.* $\mathcal{G} \mid \Gamma, \heartsuit p, \heartsuit q \Rightarrow \Delta$) *there is a rule* $R' \in \mathcal{R}$ *with principal part* $\mathcal{G} \mid \Gamma \Rightarrow \Delta, \heartsuit p$ *(resp.* $\Gamma, \heartsuit p \Rightarrow \Delta$) *whose premises are derivable from those of* R *by renaming* q *to* p *and contractions.*

Lemma 4.11. *For contraction closed* $\mathcal{R}$ *internal contraction is admissible in* $\mathcal{R}^*$.

Proof. By simultaneous double induction on the complexity of φ and the depth of the derivation we show: whenever $\vdash_{\mathcal{R}^*} \mathcal{G} \mid \varphi, \varphi, \Gamma_1 \Rightarrow \Delta_1 \mid \cdots \mid \varphi, \varphi, \Gamma_n \Rightarrow \Delta_n$, then $\vdash_{\mathcal{R}^*} \mathcal{G} \mid \varphi, \Gamma_1 \Rightarrow \Delta_1 \mid \cdots \mid \varphi, \Gamma_n \Rightarrow \Delta_n$ and analogously for φ on the right. Contractions between context and principal formulae are dealt with by modified rule applications and the inner induction hypothesis, those between principal formulae using contraction closure and the outer induction hypothesis.   □

Definition 4.12. *A rule set* $\mathcal{R}$ *is* tractable *if there is an encoding* $\ulcorner . \urcorner$ *of applications of rules from* $\mathcal{R}$ *of size polynomial in the size of the conclusion such that given a hypersequent* $\mathcal{G}$ *and an encoding* $\ulcorner R \urcorner$ *of a rule application it is decidable in time exponential in the size of* $\mathcal{G}$ *whether* $\mathcal{G}$ *is the conclusion of* R *and it is decidable in time exponential in the size of* $\ulcorner R \urcorner$ *whether* $\mathcal{G}$ *is a premiss of* R.

Definition 4.13. *A rule set* $\mathcal{R}$ *is* bounded component *if there is* $n \in \mathbb{N}$ *such that the principal part of every rule in* $\mathcal{R}$ *has at most* n *components.*

Theorem 4.14. *Let* $\mathcal{R}$ *be a contraction closed, bounded conclusion and tractable set of rules. Then derivability in* $\mathcal{R}$ *is decidable in double exponential time.*

Proof. Using Weakening and Contraction derivability in $\mathcal{R}$ is equivalent to derivability in $\mathcal{R}^*$. Moreover, Lem. 4.11 allows us to equivalently work with hypersequents build from *set-set* sequents. Since $\mathcal{R}$ is bounded component, for some k at most k components of a hypersequent contain principal formulae of the last applied rule. Thus w.l.o.g. in a derivation every hypersequent contains at most k copies of the same component. Hence in a derivation of a hypersequent with size n at most $(k+1)^{2^{2n}} = 2^{2^{\mathcal{O}(n)}}$ different hypersequents appear. Thus using the fact that derivability in one step from a set of hypersequents is a monotone operator we compute all derivable hypersequents of this set using tractability of $\mathcal{R}$ and the fact that since the size of an encoding of a rule application is polynomial in the size of its conclusion the number of encodings of rules with a given conclusion is only exponential in the size of the conclusion and check whether the given hypersequent is among these in time doubly exponential in n.   □

5   Axioms and Rules

To translate axioms into rules with context restrictions and vice versa we need to interpret hypersequents as formulae. We do this in an abstract way by viewing an interpretation as a family of formulae, one for each number of components in a hypersequent, compatible with the structural rules. Formally:

Definition 5.1. *An* interpretation *for a Λ-logic $\mathcal{L}$ is a set $\iota = \{\iota_n(p_1,\ldots,p_n) : n \geq 1\}$ of formulae in $\mathcal{F}(\Lambda)$ which respects the structural rules, i.e. for all $n \geq 1$:*

1. *ι respects (external) exchange:* $\models_{\mathcal{L}} \iota_n(\varphi,\psi,\chi,\boldsymbol{\xi})$ *iff* $\models_{\mathcal{L}} \iota_n(\varphi,\chi,\psi,\boldsymbol{\xi})$
2. *ι respects external Weakening: if* $\models_{\mathcal{L}} \iota_n(\varphi)$, *then* $\models_{\mathcal{L}} \iota_{n+1}(\varphi,\psi)$
3. *ι respects external Contraction: if* $\models_{\mathcal{L}} \iota_{n+1}(\varphi,\psi,\psi)$, *then* $\models_{\mathcal{L}} \iota_n(\varphi,\psi)$
4. *ι respects* Cut: *if* $\models_{\mathcal{L}} \iota_n(\varphi,\psi \to \chi)$ *and* $\models_{\mathcal{L}} \iota_m(\chi \to \xi,\boldsymbol{\zeta})$, *then we have* $\models_{\mathcal{L}} \iota_{n+m-1}(\varphi,\psi \to \xi,\boldsymbol{\zeta})$.

The interpretation is regular *if for all $\varphi \in \mathcal{F}$ we have* $\models_{\mathcal{L}} \varphi$ *iff* $\models_{\mathcal{L}} \iota_1(\varphi)$.

An interpretation $\iota = \{\iota_n : n \geq 1\}$ for a logic induces a map $\iota : \mathcal{HS} \to \mathcal{F}$ defined by $\Gamma_1 \Rightarrow \Delta_1 \mid \cdots \mid \Gamma_n \Rightarrow \Delta_n \mapsto \iota_n(\bigwedge \Gamma_1 \to \bigvee \Delta_1, \ldots, \bigwedge \Gamma_n \to \bigvee \Delta_n)$.

Example 5.2. 1. The interpretation $\iota_{\boxplus}$ for normal $\Lambda_{\square}$-logics is given by the formulae $\iota_n^{\boxplus}(\varphi_1,\ldots,\varphi_n) = \bigvee_{i=1}^n (\varphi_i \wedge \square\varphi_i)$. It is an interpretation by normality of $\square$ and obviously regular.

2. The standard interpretation for normal $\Lambda_{\square}$-logics from [1] is $\iota_{\square}$ given by $\iota_n^{\square}(\varphi_1,\ldots,\varphi_n) = \bigvee_{i=1}^n \square\varphi_i$. It is regular for a normal logic iff $\square\varphi/\varphi$ is admissible, in particular if $\square p \to p$ is an axiom. It is not regular for e.g. KB.

Depending on whether we involve the interpretation we obtain different notions of soundness. Regular interpretations link these and imply soundness of H.

Definition 5.3. *Let $\mathcal{R}$ be a set of rules and ι an interpretation for the logic $\mathcal{L}$. Then $\mathcal{R}$ is* hypersequent soundness preserving *(briefly: hssp) for $(\mathcal{L}, \iota)$ if for every application of a rule from $\mathcal{R}$ with premisses $\mathcal{H}_k$ for $k \leq n$ and conclusion $\mathcal{G}$: if $\models_{\mathcal{L}} \iota(\mathcal{H}_k)$ for all $k \leq n$, then $\models_{\mathcal{L}} \iota(\mathcal{G})$. The calculus is* sound *for $\mathcal{L}$, if $\vdash_{\mathsf{HR}} \Rightarrow \varphi$ implies $\models_{\mathcal{L}} \varphi$, and* complete *for $\mathcal{L}$, if $\models_{\mathcal{L}} \varphi$ implies $\vdash_{\mathsf{HR}} \Rightarrow \varphi$.*

Proposition 5.4. 1. *If $\mathcal{R}$ is hssp for $(\mathcal{L}, \iota)$ and ι is a regular interpretation for $\mathcal{L}$, then $\mathcal{R}$ is sound for $\mathcal{L}$.*
2. *If ι is a regular interpretation for $\mathcal{L}$, then H is hssp for $(\mathcal{L}, \iota)$.*

Proof. 1. By induction on the depth of a derivation we have: $\vdash_{\mathcal{R}} \mathcal{H}$ implies $\models_{\mathcal{L}} \iota(\mathcal{H})$. Now regularity of ι gives the statement.
2. Using the fact that $\mathcal{L}$ includes all propositional tautologies, all the modalities have congruence and thus $\models_{\mathcal{L}} \iota_n(\varphi,\psi)$ iff $\models_{\mathcal{L}} \iota_n(\varphi, \top \to \psi)$ and the properties of a regular interpretation. $\square$

The interpretation $\iota_{\square}$ is regular e.g. for normal $\Lambda_{\square}$-logics given by a class of Kripke frames closed under the addition of a predecessor to every world:

Definition 5.5. *A class K of frames is* extensible *if whenever for a frame $\mathfrak{F} = (W,R)$ we have $\mathfrak{F} \in$ K then also $\mathfrak{F}^{\bullet} \in$ K where $\mathfrak{F}^{\bullet} = (W \cup \{x\}, R \cup \{(x,y) : y \in W \cup \{x\}\})$ with $x \notin W$.*

Lemma 5.6. *If $\mathcal{L}$ is a normal $\Lambda_{\square}$-logic defined by an extensible class of frames, then $\iota_{\square}$ is a regular interpretation for $\mathcal{L}$.*

Proof. By normality $\iota_\square$ is an interpretation for $\mathcal{L}$. For regularity suppose that $\neg\varphi$ is satisfiable in $\mathfrak{F} \in \mathsf{K}$ with K the extensible class of frames defining $\mathcal{L}$. Then for some world w of $\mathfrak{F}$ and valuation σ we have $\mathfrak{F}, w, \sigma \not\models \varphi$. Thus for the additional world x in $\mathfrak{F}^\bullet$ we have $\mathfrak{F}^\bullet, x, \sigma \not\models \square\varphi$, and since $\mathfrak{F}^\bullet \in \mathsf{K}$ we have $\not\models_\mathcal{L} \square\varphi$. $\square$

From Rules to Axioms. In the construction of axioms from rules we extend the method from [14,10,9]. The idea is to show projectivity (Lem. 5.10) of a formula corresponding to the premisses of the rule and use a substitution witnessing this property to inject the information of the premisses into a formula corresponding to the conclusion. For the sake of presentation here we only consider the normal modality $\square$ and restrict the context restrictions to $\{\mathcal{C}_\emptyset, \mathcal{C}_{\mathsf{id}}, \mathcal{C}_\square\}$. In general the method also works for monotone or antitone n-ary modalities and arbitrary context restrictions. To show projectivity we need to assume the following for every premiss $(\Gamma \Rightarrow \Delta; \mathcal{C})$:

$$\text{If } \mathcal{C}_{\mathsf{id}} \notin \mathcal{C} \text{ then } \Gamma, \Delta \neq \emptyset \tag{1}$$

For the rest of this section we fix a rule R with this property. In presence of HCut we may assume furthermore w.l.o.g. that the restriction $\mathcal{C}_{\mathsf{id}}$ does not occur in R: If it does occur we simply convert R into a rule of this format by introducing a *dummy modality* $\cdot$ satisfying $\cdot\varphi \leftrightarrow \varphi$ for all formulae and replacing every restriction $\mathcal{C}_{\mathsf{id}}$ by the sequent $\Rightarrow s$ for a fresh variable s in the premisses and by $\Rightarrow \cdot s$ in the corresponding component in the principal part. By Lem. 4.8 the resulting rule is equivalent to the original one modulo $\mathsf{HR}_{\mathsf{dm}}\mathsf{Cut}$ where $\mathcal{R}_{\mathsf{dm}} = \{(p \Rightarrow ; \mathcal{C}_{\mathsf{id}})/\cdot p \Rightarrow, (\Rightarrow p; \mathcal{C}_{\mathsf{id}})/ \Rightarrow \cdot p\}$ states equivalence of p and $\cdot p$. Together with property (1) this means that $\Gamma, \Delta \neq \emptyset$ for every premiss $(\Gamma \Rightarrow \Delta; \mathcal{C})$. Since the number of context formulae might vary, a rule can not be translated into a formula directly. This is avoided by fixing the number of context formulae. For normal modalities and the limited restrictions considered here this gives:

Definition 5.7. *The canonical proto rule for a rule* $R = \{(\Gamma_i \Rightarrow \Delta_i; \mathcal{C}_i^1, \ldots, \mathcal{C}_i^n) : i \leq m\}/\Sigma_1 \Rightarrow \Pi_1 \mid \cdots \mid \Sigma_n \Rightarrow \Pi_n$ *is given by the context sequents* $\Omega_1 \Rightarrow \mid \cdots \mid \Omega_n \Rightarrow$ *with* $\Omega_j = \square p_j$ *if* $\mathcal{C}_i^j = \mathcal{C}_\square$ *for some* i *and empty otherwise, using fresh variables* $\boldsymbol{p}$. *An* application *of the canonical proto rule for* R *given by* $\mathcal{G}$ *and* σ *is the same as the application of* R *given by* $\mathcal{G}, \sigma$ *and the above contexts.*

Example 5.8. 1. The canonical proto rule for 4_n from Tab. 1 is given by the context $\square p \Rightarrow$ and has applications $\mathcal{G} \mid \square\chi, \varphi \Rightarrow \psi / \mathcal{G} \mid \square\chi, \square\varphi \Rightarrow \square\psi$.

2. To treat $R_5 := (\Rightarrow ; \mathcal{C}_\square, \mathcal{C}_{\mathsf{id}})/ \Rightarrow \mid \Rightarrow$ we replace $\mathcal{C}_{\mathsf{id}}$ by the dummy modality, giving $(\Rightarrow s; \mathcal{C}_\square, \mathcal{C}_\emptyset)/ \Rightarrow \mid \Rightarrow \cdot s$. The canonical proto rule for R_5 is given by the contexts $\square p \Rightarrow \mid \Rightarrow$ and has applications $\mathcal{G} \mid \square\varphi \Rightarrow \psi / \mathcal{G} \mid \square\varphi \Rightarrow \mid \Rightarrow \cdot\psi$.

In the non-normal case or for arbitrary context restrictions we would need to consider a set of proto rules with every possible number of context formulae also on the right hand side, compare [10,9]. Using the rules for normal modal logics and HCut it is straightforward to see that the canonical proto rule is enough:

Lemma 5.9. R *and its canonical proto rule are interderivable in* $\mathsf{HR}_\mathsf{K}\mathsf{Cut}$. $\square$

Now suppose we have an interpretation $\iota = \{\iota_n : n \geq 1\}$ and that

$$R = \{(\Gamma_i \Rightarrow \Delta_i; \mathcal{C}_i) : i \leq m\}/\Sigma_1 \Rightarrow \Pi_1 \mid \cdots \mid \Sigma_n \Rightarrow \Pi_n$$

with $\mathcal{C}_i^j = \langle F_i^j, G_i^j \rangle$. The canonical proto rule $\hat{R}$ for R is given by the contexts $\Omega_1 \Rightarrow \mid \cdots \mid \Omega_n \Rightarrow$. The formula corresponding to its premises is

$$\varphi := \bigwedge_{i \leq m} \left(\bigwedge (\Omega_1 \lceil_{F_i^1}, \ldots, \Omega_n \lceil_{F_i^n}, \Gamma_i) \rightarrow \bigvee \Delta_i \right) .$$

Now define a substitution θ by $\theta(x) = \varphi \wedge x$ if $x \in \Gamma_i$ for some $i \leq m$ and $\theta(x) = \varphi \rightarrow x$ if $x \in \Delta_i$ for some $i \leq m$ and $\theta(x) = x$ otherwise. Since by monotonicity w.l.o.g. no variable occurs both in antecedent and succedent of a premiss, θ is well-defined. Straightforward propositional reasoning gives:

Lemma 5.10. *The substitution θ witnesses projectivity of φ, i.e. the following hold:* $\vdash_{\mathsf{HMonCut}} \Rightarrow \varphi\theta$ *and* $\vdash_{\mathsf{HMonCut}} \varphi \Rightarrow p \leftrightarrow p\theta$ *for every* $p \in \mathcal{V}$. $\square$

This gives equivalence of $\hat{R}$ to a *ground hypersequent*, i.e. a set of hypersequents closed under substitution, which we then interpret as an axiom using ι:

Lemma 5.11. $\hat{R}$ *is interderivable over* $\mathsf{HCutMon}$ *with the ground hypersequent* $\mathcal{H}_R := \Rightarrow (\bigwedge(\Omega_1, \Sigma_1) \rightarrow \bigvee \Pi_1)\theta \mid \cdots \mid \Rightarrow (\bigwedge(\Omega_n, \Sigma_n) \rightarrow \bigvee \Pi_n)\theta.$

Proof. By Lem. 5.10 we have $\vdash_{\mathsf{HCutMon}} \Rightarrow \varphi\theta$ and thus $\vdash_{\mathsf{HCutMon}} \Rightarrow \varphi\theta\sigma$ for every substitution σ. Now inverting the propositional rules using Cut and an application of $\hat{R}$ give $\mathcal{H}_R\sigma$. For the other direction, Lem. 5.10 implies $\vdash_{\mathsf{HMonCut}} \varphi \Rightarrow \psi \leftrightarrow \psi\theta$ (by induction on the complexity of ψ). Hence $\vdash_{\mathsf{HMonCut}} \varphi\sigma, \chi_i\theta\sigma \Rightarrow \chi_i\sigma$ with $\chi_i = \bigwedge(\Omega_i, \Sigma_i) \rightarrow \bigvee \Delta_i$. From the premises of an application of $\hat{R}$ we obtain $\Gamma \mid \Rightarrow \varphi\sigma$, and cutting these and the ground hypersequent $\mathcal{H}_R\sigma$ followed by invertibility of H and external Contraction yield the conclusion of this application. $\square$

Theorem 5.12 (Soundness). *If* $\mathsf{HR_KCutR}$ *is hssp for* $(\mathcal{L}, \iota)$, *then* $\iota(\mathcal{H}_R) \in \mathcal{L}$.

Proof. Since $\mathcal{H}_R$ is derivable in $\mathsf{HR_KCutR}$ by Lem. 5.11 and $\mathsf{HR_KCutR}$ is hssp for $(\mathcal{L}, \iota)$, the former is hssp for $(\mathcal{L}, \iota)$ as well. Thus $\iota(\mathcal{H}_R) \in \mathcal{L}$. $\square$

Theorem 5.13 (Completeness). *If for sets $\mathcal{A}$ of axioms and $\mathcal{R}$ of rules HCutR is complete for $\mathcal{L_A}$ and the rule $\Rightarrow \varphi_1 \mid \cdots \mid \Rightarrow \varphi_n/ \Rightarrow \iota_n(\varphi_1, \ldots, \varphi_n)$ is derivable in HCutR, then HCutRR is complete for $\mathcal{L_A} \oplus \iota(\mathcal{H}_R)$.*

Proof. By Lem. 5.11 the ground hypersequent $\mathcal{H}_R$ is derivable in HCutRR, and thus the axiom $\iota(\mathcal{H}_R)$ is derivable in HCutRR as well. Simulating modus ponens by Cut we thus obtain completeness of this calculus for $\mathcal{L_A} \oplus \iota(\mathcal{H}_R)$. $\square$

Example 5.14. The premiss of the canonical proto rule for R_5 from Ex. 5.8.2 is turned into $\varphi = \Box p \rightarrow s$. Then with θ defined by $\theta(p) = p$ and $\theta(s) = \varphi \rightarrow s$ we obtain $\mathcal{H} = \Rightarrow \neg\Box p\theta \mid \Rightarrow \cdot s\theta = \Rightarrow \neg\Box p \mid \Rightarrow \cdot(\varphi \rightarrow s)$. Thus R_5 is equivalent under $\iota_\Box$ to the axiom $\iota_\Box(\mathcal{H}) = \Box\neg\Box p \vee \Box \cdot ((\Box p \rightarrow s) \rightarrow s)$ which modulo propositional reasoning and monotonicity is easily seen to be equivalent (as an axiom) to $\Box\neg\Box p \vee \Box \cdot \Box p$. By idempotency of $\cdot$ this is equivalent to $\Box\neg\Box p \vee \Box\Box p$.

Crucially, Thm. 5.12 also implies that rules stay hssp in extensions of a logic:

Corollary 5.15. *If $\mathcal{L}_1 \subseteq \mathcal{L}_2$, and ι is an interpretation for $\mathcal{L}_1, \mathcal{L}_2$, and $\mathsf{HCut}\mathcal{R}_\mathsf{K}$ is hssp for $(\mathcal{L}_1, \iota)$ and $(\mathcal{L}_2, \iota)$, then if R is hssp for $(\mathcal{L}_1, \iota)$ it is also hssp for $(\mathcal{L}_2, \iota)$.*

Proof. Since R and $\mathcal{H}_R$ are interderivable and $\iota(\mathcal{H}_R) \in \mathcal{L}_1 \subseteq \mathcal{L}_2$. □

From Axioms to Rules. The translation from axioms to rules proceeds similar to that for sequent rules in [10,9], but uses the interpretation to peel away one layer of the formula first. The idea is to treat some subformulae of an axiom as *context formulae* and translate the axiom into a proto rule (i.e. a rule with a fixed number of context formulae). To simplify presentation we assume monotonicity of the modalities.

Definition 5.16. *Let $C_\ell, C_r \subseteq \mathcal{F}(\Lambda)$ and $V \subseteq \mathcal{V}$. The class of* translatable *clauses for (C_ℓ, V, C_r) is defined by the following grammar (starting variable S):*

$$S ::= L \rightarrow R$$
$$L ::= L \wedge L \mid \heartsuit P_r \mid \psi_\ell \mid \top \mid \bot \qquad R ::= R \vee R \mid \heartsuit P_\ell \mid \psi_r \mid \top \mid \bot$$
$$P_r ::= P_r \vee P_r \mid P_r \wedge P_r \mid P_\ell \rightarrow P_r \mid \psi_r \mid p \mid \bot \mid \top$$
$$P_\ell ::= P_\ell \vee P_\ell \mid P_\ell \wedge P_\ell \mid P_r \rightarrow P_\ell \mid \psi_\ell \mid p \mid \bot \mid \top$$

where $\heartsuit \in \Lambda, p \in V$ and $\psi_i \in C_i$ for $i \in \{\ell, r\}$. A formula is hypertranslatable *for an interpretation $\iota = \{\iota_n : n \geq 1\}$ if has the form $\iota_n(\chi_1, \ldots, \chi_n)$ with χ_i a translatable clause for (C_ℓ, V, C_r) where no distinct formulae in $C_\ell \cup V \cup C_r$ share a variable, every formula in $C_\ell \cup C_r$ occurs in the χ_i exactly once not in the scope of a modality and at least once in the scope of a modality.*

A little thought shows that hypersequents $\mathcal{G} \mid \Rightarrow \varphi$ (resp. $\mathcal{G} \mid \varphi \Rightarrow$) with φ generated by taking P_r (resp. P_ℓ) as starting variable in the above grammar can be decomposed using invertibility of the propositional rules into sets of hypersequents $\mathcal{G} \mid \Gamma \Rightarrow \Delta$ with $\Gamma \subseteq C_\ell \cup V$ and $\Delta \subseteq C_r \cup V$. The formulae in C_ℓ (resp. C_r) will play the role of context formulae on the left (resp. right). We now fix a logic $\mathcal{L}$, an interpretation $\iota = \{\iota_n : n \geq 1\}$ and a hypertranslatable formula φ for ι and consider the stages of the translation in detail.

Ground hypersequent stage. We have $\varphi = \iota_n(\varphi_1, \ldots, \varphi_n)$ where $\varphi_i = \bigwedge \psi^i \wedge \bigwedge \chi^i \rightarrow \bigvee \xi^i \vee \bigvee \zeta^i$ with context formulae $\chi^i_j \in C_\ell$, $\zeta^i_j \in C_r$ and formulae ψ^i_j (resp. ξ^i_j) of the form $\heartsuit \delta_j$ with δ_j generated by the above grammar with starting variable P_r (resp. P_ℓ). This is turned into the ground hypersequent $\mathcal{H}_\varphi := \psi^1, \chi^1 \Rightarrow \xi^1, \zeta^1 \mid \cdots \mid \psi^n, \chi^n \Rightarrow \xi^n, \zeta^n$ which by HCut is hssp for $(\mathcal{L}, \iota)$.

Shaping the conclusion. We replace each $\psi^i_j = \heartsuit \delta^i_j$ with $\heartsuit p^i_j$ where $p^i_j \in V$ is fresh and add the premiss $p^i_j \Rightarrow \delta^i_j$. Analogously we replace $\xi^i_j = \heartsuit \gamma^i_j$ with $\heartsuit q^i_j$ and add the premiss $\gamma^i_j \Rightarrow q^i_j$. By monotonicity and Cut this is equivalent to $\mathcal{H}_\varphi$.

Resolving propositional logic. Using invertibility of the propositional rules we replace each of these premisses by a number of sequents $\Gamma \Rightarrow \Delta$ with $\Gamma \subseteq C_\ell \cup V$ and $\Delta \subseteq C_r \cup V$. In presence of HCut this gives an equivalent rule.

Cleaning the premisses. To ensure that every variable occurring in the premisses of the rule also occurs in the conclusion we eliminate the variables from V from the premisses by successively cutting the premisses on all variables in V as in Def.4.1 disregarding context restrictions. Reasoning as in Lem. 4.8 the resulting rule is seen to be equivalent to the original rule (compare also [5]).

Introducing context restrictions. The global condition on the context formulae in Def.5.16 guarantees that every formula in $C_\ell \cup C_r$ occurs exactly once in the conclusion and at least once in the premisses. Moreover, it occurs always on the same side of the sequent. Thus we now have a rule with a fixed number of context formulae. Provided the context formulae are *normal* in the sense that formulae in C_ℓ distribute over $\wedge$ and those in C_r over $\vee$ we may replace them with context restrictions by turning a premiss $\chi_1, \ldots, \chi_m, \Gamma \Rightarrow \Delta, \zeta_1, \ldots, \zeta_k$ with context formulae χ_j and ζ_j occurring in the i_j-th component of the conclusion into the premiss with restriction $(\Gamma \Rightarrow \Delta; \mathcal{C})$ where $\mathcal{C}^i = \langle \{\chi_j : i_j = i\}; \{\zeta_j : i_j = i\} \rangle$ and deleting all context formulae from the conclusion. Call the resulting rule R_φ.

Since all steps in the above construction yield rules interderivable with the original ones using HCut and monotonicity and soundness of these rules is preserved by Cor. 5.15 we immediately obtain soundness and completeness.

Proposition 5.17. *Let ι be a regular interpretation for $\mathcal{L}$ and let $\mathcal{R}$ be hssp and complete for $(\mathcal{L}, \iota)$ with the rule $\Rightarrow p_1 \mid \cdots \mid \Rightarrow p_n / \Rightarrow \iota_n(\boldsymbol{p})$ derivable in $\mathcal{R}$. If φ is hypertranslatable for ι with normal context formulae (C_ℓ, C_r), then $\mathcal{R}R_\varphi$ is hssp and complete for $(\mathcal{L} \oplus \varphi, \iota)$.* □

Example 5.18. Using $\iota_\Box$ the axiom $\Box\neg\Box p \vee \Box \cdot \Box p$ from Ex. 5.14 is converted into the ground hypersequent $\Box p \Rightarrow \mid \Rightarrow \cdot \Box p$. Taking $\Box p$ to be in C_ℓ we introduce a fresh variable q and the corresponding premiss to obtain $\Box p \Rightarrow q / \Box p \Rightarrow \mid \Rightarrow \cdot q$. Using normality of $\Box$ (for $\mathcal{R}_\mathsf{K}$) the formula $\Box p$ is now replaced with the context restriction $\langle \{\Box p\}, \emptyset \rangle = \mathcal{C}_\Box$ resulting in the rule $(\Rightarrow q; \mathcal{C}_\Box, \mathcal{C}_\emptyset) / \Rightarrow \mid \Rightarrow \cdot q$.

The translations show that in general a single axiom corresponds to a proto rule, i.e. a rule with a fixed number of context formulae. Thus in general a rule corresponds to an infinite number of (systematically generated) axioms, see [10,9] for the sequent case. The method also works for non-monotone modalities, where in the second stage we introduce both premisses $p_j^i \Rightarrow \delta_j^i$ and $\delta_j^i \Rightarrow p_j^i$ instead of only one of these. Furthermore, in some cases we still obtain rules with restrictions from axioms with non-normal context formulae, see Sec.6.

6   Case Studies

Logics for simple frame properties. An interesting class of examples are the rules constructed from *simple* frame properties for normal modal logics [8]. A *simple* frame property is a formula $\forall w_1, \ldots, \forall w_n \exists u \varphi_S$ in the frame language, with $\varphi_S = \bigvee_{\langle S_R, S_= \rangle \in S}(\bigwedge_{i \in S_R} w_i R u \wedge \bigwedge_{i \in S_=} w_i = u)$ for some non-empty *description* S consisting of a set of tuples $\langle S_R, S_= \rangle$ with $S_R, S_= \subseteq \{1, \ldots, n\}$ and $S_R \cup S_= \neq \emptyset$. We identify a simple frame property with its description. In [8] hypersequent rules corresponding to simple frame properties based on $\mathsf{K}, \mathsf{K4}$ and KB are given and cut admissibility is shown via the semantics. Here we consider the rules based on K and $\mathsf{K4}$ (those for KB do not fit our rule format). The set of *hypersequent rules induced by S for $\mathcal{R}_\mathsf{K}$* is $\mathcal{R}_S := \{R_{k_1,\ldots,k_n} : k_i \geq 0\}$ with

$$R_{k_1,\ldots,k_n} := \frac{\left\{ (\bigsqcup_{j \in S_R} p_1^j, \ldots, p_{k_j}^j \Rightarrow ; \mathcal{C}^1_{\langle S_R, S_= \rangle}, \ldots, \mathcal{C}^n_{\langle S_R, S_= \rangle}) : \langle S_R, S_= \rangle \in S \right\}}{\Box p_1^1, \ldots, \Box p_{k_1}^1 \Rightarrow | \cdots | \Box p_1^n, \ldots, \Box p_{k_n}^n \Rightarrow}$$

where $\mathcal{C}^j_{\langle S_R, S_= \rangle} = \mathcal{C}_\mathsf{id}$ for $j \in S_=$ and $\mathcal{C}_\emptyset$ otherwise. The set of *hypersequent rules induced by S for $\mathcal{R}_\mathsf{K4}$* is the set $\mathcal{R}_S^4 := \{R_n^4 : n \geq 0\}$ with R_n^4 the rule R_n with $\mathcal{C}^j_{\langle S_R, S_= \rangle} = \mathcal{C}_\mathsf{id}$ for $j \in S_=$ and $\mathcal{C}_\Box$ for $j \in S_R$ and $\mathcal{C}_\emptyset$ otherwise. Inspection of the rule sets constructed in this way shows that together with $\mathsf{H}\mathcal{R}_\mathsf{K}$ (resp. $\mathsf{H}\mathcal{R}_\mathsf{K4}$) they satisfy all conditions given in Thm. 4.6. Thus we obtain a purely syntactic analogue to the semantic cut admissibility proof in [8]:

Corollary 6.1. *If $\mathcal{R}$ is a set of rules induced by simple frame properties for $\mathcal{R}_\mathsf{K}$ (resp. $\mathcal{R}_\mathsf{K4}$), then $\mathsf{H}\mathcal{R}_\mathsf{K}\mathcal{R}$ (resp. $\mathsf{H}\mathcal{R}_\mathsf{K4}\mathcal{R}$) has cut elimination.* $\qquad\square$

Using the translation from rules to axioms we furthermore obtain finite axiomatisations from the so constructed rules, provided we have a regular interpretation and the rules are hssp for this interpretation. While $\iota_\boxplus$ is always regular, the interpretation $\iota_\Box$ gives cleaner axioms. Sometimes regularity of $\iota_\Box$ can be read of the frame properties directly: if $S_R \neq \emptyset \neq S_=$ for all $(S_R, S_=) \in \theta$ for one property θ, then the logic is reflexive, and if $S_= = \emptyset$ for all $(S_R, S_=) \in \theta$ for every θ, then the logic is extensible (Def. 5.5). Under certain conditions we may also adjust the original soundness proof to our setting:

Proposition 6.2 ([8]). *If S is a simple frame property and $\mathcal{L}_S$ resp. $\mathcal{L}_S^4$ are the logics of the class of frames (resp. transitive frames) with this property, then $\mathcal{R}_S^4$ is hssp for $(\mathcal{L}_S^4, \iota_\Box)$ and $(\mathcal{L}_S^4, \iota_\boxplus)$. If $\mathcal{L}_S$ is extensible or if $S_= \neq \emptyset$ for all $(S_R, S_=) \in S$, then $\mathcal{R}_S$ is hssp for $(\mathcal{L}_S, \iota_\Box)$ and $(\mathcal{L}_S, \iota_\boxplus)$.*

To obtain the simplest axioms we observe that given $\mathsf{H}\mathcal{R}_\mathsf{K}\mathsf{Cut}$ (resp. $\mathsf{H}\mathcal{R}_\mathsf{K4}\mathsf{Cut}$) by Lem. 4.8 the set of rules induced by a simple property is equivalent (in both cases!) to a *single* rule $\{(\bigsqcup_{i \in S_R} p_i \Rightarrow ; \mathcal{C}_{\langle S_R, S_= \rangle}) : \langle S_R, S_= \rangle \in S\}/\Box p_1 \Rightarrow | \cdots |$ $\Box p_n$ with $\mathcal{C}^i_{\langle S_R, S_= \rangle} = \mathcal{C}_\mathsf{id}$ for $i \in S_=$ and $\mathcal{C}_\emptyset$ otherwise. Translating this rule gives the corresponding axiom. This restricts the shape of the resulting axioms.

Definition 6.3. *A ι-simple axiom for an interpretation $\iota = \{\iota_n : n \geq 1\}$ is an axiom $\iota_n(\varphi_1, \ldots, \varphi_n)$ where $\mathsf{mrk}(\varphi_i) \leq 1$ and $\Box$ occurs only negatively in the φ_i.*

Proposition 6.4. *Let $\mathcal{L}_S$ (resp. $\mathcal{L}_S^4$) be the logic of the class F of frames (resp. transitive frames) satisfying the simple frame property S. Then $\mathcal{L}_S^4$ is axiomatised over $\mathsf{K4}$ by one $\iota_\boxplus$-simple axiom. The logic $\mathcal{L}_S$ is axiomatised by one $\iota_\Box$-simple axiom if: ($\mathcal{L}_S$ is reflexive or F is extensible or $S_= = \emptyset$ for all $\langle S_R, S_= \rangle \in S$) and ($\mathcal{L}_S$ is transitive or F is extensible or $S_= \neq \emptyset$ for all $\langle S_R, S_= \rangle \in S$).* $\qquad\square$

Table 2. The additional axioms for LUDF and their translations

(UC) $\mathcal{P}A \wedge \mathcal{P}B \to \mathcal{P}(A \vee B)$	(OiP) $\mathcal{O}A \to \mathcal{P}A$	(Unif-O) $\mathcal{O}A \to \Box \mathcal{O}A$
(W-P) $\mathcal{O}A \to (\mathcal{P}B \to \Box(B \to A))$	(OiC) $\mathcal{O}A \to \neg\Box\neg A$	(Unif-P) $\mathcal{P}A \to \Box\mathcal{P}A$

$$\frac{\mathcal{G} \mid p \Rightarrow r \quad \mathcal{G} \mid q \Rightarrow r \quad \mathcal{G} \mid r \Rightarrow p,q}{\mathcal{G} \mid \mathcal{P}p, \mathcal{P}q \Rightarrow \mathcal{P}r} \; \text{UC} \qquad \frac{\mathcal{G} \mid p,q \Rightarrow}{\mathcal{G} \mid \mathcal{O}p, \Box q \Rightarrow} \; \text{OiC} \qquad \frac{\mathcal{G} \mid \mathcal{O}p \Rightarrow q}{\mathcal{G} \mid \mathcal{O}p \Rightarrow \Box q} \; \text{Unif-O}$$

$$\frac{\mathcal{G} \mid p \Rightarrow r \quad \mathcal{G} \mid \Rightarrow q,r}{\mathcal{G} \mid \mathcal{O}p, \mathcal{P}q \Rightarrow \Box r} \; \text{W-P} \qquad \frac{\mathcal{G} \mid p \Rightarrow q \quad \mathcal{G} \mid q \Rightarrow p}{\mathcal{G} \mid \mathcal{O}p \Rightarrow \mathcal{P}q} \; \text{OiP} \qquad \frac{\mathcal{G} \mid \mathcal{P}p \Rightarrow q}{\mathcal{G} \mid \mathcal{P}p \Rightarrow \Box q} \; \text{Unif-P}$$

This extends to finite sets of simple frame properties (if using extensibility to show soundness we need *all* frame classes obtained by successively adding properties to be extensible). While seemingly restrictive, the conditions capture all examples of [8], e.g. directedness, universality, linearity or bounded cardinality.

Example 6.5. The property called *Bounded Acyclic Subgraph* in [8] induces the rule $R_{\mathsf{BAS}} = \{(q_k^k \Rightarrow ; \mathcal{C}^i : k < i \le n\}/\Box q_1^1 \Rightarrow \mid \cdots \mid \Box q_n^n \Rightarrow$ where the i-th component of $\mathcal{C}^i$ is $\mathcal{C}_{\mathsf{id}}$ and all other components are $\mathcal{C}_\emptyset$. Using reflexivity and $\iota_\Box$ the translation of this is $\bigvee_{1 \le k \le n} \Box \left(\ell_k \wedge \Box \left(\bigwedge_{1 \le i < j \le n} (\ell_j \wedge p_i \to r_j) \wedge p_k \right) \to r_k \right)$, which in particular implies the axiom $\mathsf{BAS}_n = \bigvee_{k=1}^n \Box(\Box p_k \to \bigvee_{m=1}^{k-1} p_m)$. Translating the latter back into a rule (taking $C_r = C_\ell = \emptyset$ and introducing the dummy modality writing $\bigvee_{m=1}^{k-1} \cdot p_m$) again gives the rule R_{BAS}. Thus the logic given by the Bounded Acyclic Subgraph property is axiomatised over KT by BAS_n.

The Logic of Uniform Deontic Frames. The logic LUDF of uniform deontic frames [13] is based on the connectives $\Lambda_\Box \cup \{\mathcal{P}, \mathcal{O}\}$ with $\mathcal{P}$ and $\mathcal{O}$ unary nonnormal modalities with intended interpretations "...is permissible" and "...is obligatory" and is axiomatised by the S5-axioms for $\Box$ together with the axioms in Table 2. A hypersequent calculus for the fragment without the axioms (Unif-O) and (Unif-P) based on the calculus for S5 from [12] was given in [6]. We now construct a cut-free calculus for the full logic using the developed methods.

First we convert the axioms into hypersequent rules, building on the calculus for S5 constructed in Ex. 4.2. Since S5 is reflexive, the interpretation $\iota_\Box$ is regular and thus we take it as the underlying interpretation. Under S5 adding an axiom A is equivalent to adding the axiom $\Box A$, hence it suffices to translate the boxed versions of the axioms. Doing this using the methods of Sec. 5 gives the rules in Table 2. Next we saturate the rule set under cuts between rules (Def. 4.1) and (to ensure contraction closure) under contracting principal formulae and the corresponding variables in the premises. Omitting superfluous premises this gives the rules in Table 3, where we turned the set of iterated cuts between instances of (Unif-O) and (Unif-P) and 4_n for $n \in \mathbb{N}$ into the rules $(p_1, \ldots, p_n \Rightarrow ; C)/\Box p_1, \ldots, \Box p_n \Rightarrow$ with context restriction $\mathcal{C} = \langle \{\Box p, \mathcal{O}p, \mathcal{P}p\}, \emptyset \rangle$. By construction these rules are hssp, and clearly the translations of the axioms are derivable rules using $\mathcal{R}_{\mathsf{LUDF}}$. Finally, it is

Table 3. The rules in $\mathcal{R}_{\mathsf{LUDF}}$, where $\mathcal{C} := \langle\{\Box p, \mathcal{O}p, \mathcal{P}p\}, \emptyset\rangle$ and writing $*p$ for $*p_1, \ldots, *p_{|p|}$ with $* \in \{\mathcal{O}, \mathcal{P}, \Box\}$

$$\frac{\{(r \Rightarrow p, q; \mathcal{C}_\emptyset)\} \cup \{(p_i \Rightarrow r; \mathcal{C}_\emptyset), (q_i \Rightarrow r; \mathcal{C}_\emptyset) : p_i \in p, q_i \in q\}}{\mathcal{O}p, \mathcal{P}q \Rightarrow \mathcal{P}r} \ (|p| + |q| \geq 1)$$

$$\frac{(p, r \Rightarrow s; \mathcal{C}), (r \Rightarrow q, s; \mathcal{C})\}}{\mathcal{O}p, \mathcal{P}q, \Box r \Rightarrow \Box s} \qquad \frac{(p, r \Rightarrow ; \mathcal{C}_{\mathsf{id}}), (r \Rightarrow q; \mathcal{C}_{\mathsf{id}})\}}{\mathcal{O}p, \mathcal{P}q, \Box r \Rightarrow} \ |p| \geq 1, |q|, |r| \geq 0$$

$$\frac{(p, r \Rightarrow ; \mathcal{C}, \mathcal{C}_{\mathsf{id}}), (r \Rightarrow q; \mathcal{C}, \mathcal{C}_{\mathsf{id}})\}}{\mathcal{O}p, \mathcal{P}q, \Box r \Rightarrow | \Rightarrow} \ |p| \geq 1, |q|, |r| \geq 0$$

$$\frac{(r \Rightarrow s; \mathcal{C})}{\Box r \Rightarrow \Box s} \ |r| \geq 0 \qquad \frac{(r \Rightarrow ; \mathcal{C}_{\mathsf{id}})}{\Box r \Rightarrow} \ |r| \geq 1 \qquad \frac{(r \Rightarrow ; \mathcal{C}, \mathcal{C}_{\mathsf{id}})}{\Box r \Rightarrow | \Rightarrow} \ |r| \geq 1$$

straightforward to see that $\mathcal{R}_{\mathsf{LUDF}}$ satisfies the conditions for cut elimination and the decision procedure of Sec. 4.1. In particular we obtain an apparently new double exponential complexity bound for LUDF.

References

1. Avron, A.: The method of hypersequents in the proof theory of propositional non-classical logics. In: Logic: From Foundations to Applications. Clarendon (1996)
2. Avron, A., Lev, I.: Canonical propositional Gentzen-type systems. In: Goré, R.P., Leitsch, A., Nipkow, T. (eds.) IJCAR 2001. LNCS (LNAI), vol. 2083, pp. 529–544. Springer, Heidelberg (2001)
3. Belnap, N.D.: Display logic. J. Philos. Logic 11, 375–417 (1982)
4. Blackburn, P., de Rijke, M., Venema, Y.: Modal Logic, Cambridge (2001)
5. Ciabattoni, A., Galatos, N., Terui, K.: From axioms to analytic rules in nonclassical logics. In: LICS 2008, pp. 229–240. IEEE Computer Society (2008)
6. Gratzl, N.: Sequent calculi for multi-modal logic with interaction. In: Grossi, D., Roy, O., Huang, H. (eds.) LORI. LNCS, vol. 8196, pp. 124–134. Springer, Heidelberg (2013)
7. Indrzejczak, A.: Cut-free hypersequent calculus for S4.3. Bull. Sec. Log. 41, 89–104 (2012)
8. Lahav, O.: From frame properties to hypersequent rules in modal logics. In: LICS 2013. IEEE Computer Society (2013)
9. Lellmann, B.: Sequent Calculi with Context Restrictions and Applications to Conditional Logic. Ph.D. thesis, Imperial College London (2013)
10. Lellmann, B., Pattinson, D.: Correspondence between modal Hilbert axioms and sequent rules with an application to S5. In: Galmiche, D., Larchey-Wendling, D. (eds.) TABLEAUX 2013. LNCS, vol. 8123, pp. 219–233. Springer, Heidelberg (2013)
11. Ohlbach, H.J.: Logic engineering: Konstruktion von Logiken. KI 3, 34–38 (1992)
12. Poggiolesi, F.: A cut-free simple sequent calculus for modal logic S5. Rev. Symb. Log. 1(1), 3–15 (2008)
13. Roy, O., Anglberger, A.J., Gratzl, N.: The logic of obligation as weakest permission. In: Ågotnes, T., Broersen, J., Elgesem, D. (eds.) DEON 2012. LNCS (LNAI), vol. 7393, pp. 139–150. Springer, Heidelberg (2012)
14. Schröder, L.: A finite model construction for coalgebraic modal logic. J. Log. Algebr. Program. 73(1-2), 97–110 (2007)

Clausal Resolution for Modal Logics of Confluence

Cláudia Nalon[1,*], João Marcos[2,**], and Clare Dixon[3]

[1] Departament of Computer Science, University of Brasília
C.P. 4466 – CEP:70.910-090 – Brasília – DF – Brazil
nalon@unb.br
[2] LoLITA and Dept. of Informatics and Applied Mathematics, UFRN, Brazil
jmarcos@dimap.ufrn.br
[3] Department of Computer Science, University of Liverpool
Liverpool, L69 3BX – United Kingdom
CLDixon@liverpool.ac.uk

Abstract. We present a clausal resolution-based method for normal multimodal logics of confluence, whose Kripke semantics are based on frames characterised by appropriate instances of the Church-Rosser property. Here we restrict attention to eight families of such logics. We show how the inference rules related to the normal logics of confluence can be systematically obtained from the parametrised axioms that characterise such systems. We discuss soundness, completeness, and termination of the method. In particular, completeness can be modularly proved by showing that the conclusions of each newly added inference rule ensures that the corresponding conditions on frames hold. Some examples are given in order to illustrate the use of the method.

Keywords: normal modal logics, combined logics, resolution method.

1 Introduction

Modal logics are often introduced as extensions of classical logic with two additional unary operators: "$\Box$" and "$\Diamond$", whose meanings vary with the field of application to which they are tailored to apply. In the most common interpretation, formulae "$\Box p$" and "$\Diamond p$" are read as "p is necessary" and "p is possible", respectively. Evaluation of a modal formula depends upon an organised collection of scenarios known as *possible worlds*. Different modal logics assume different *accessibility relations* between such worlds. Worlds and their accessibility relations define a so-called *Kripke frame*. The evaluation of a formula hinges on such structure: given an appropriate accessibility relation and a world w, a formula $\Box p$ is satisfied at w if p is true at all worlds accessible from w; a formula $\Diamond p$ is satisfied at w if p is true at some world accessible from w.

In normal modal logics extending the classical propositional logic, the schema $\Box(\varphi \Rightarrow \psi) \Rightarrow (\Box\varphi \Rightarrow \Box\psi)$ (the distribution axiom **K**), where φ and ψ are well-formed formulae and $\Rightarrow$ stands for classical implication, is valid, and the schematic

---

* C. Nalon was partially supported by CAPES Foundation BEX 8712/11-5.
** J. Marcos was partially supported by CNPq and by the EU-FP7 Marie Curie project PIRSES-GA-2012-318986.

S. Demri, D. Kapur, and C. Weidenbach (Eds.): IJCAR 2014, LNAI 8562, pp. 322–336, 2014.

rule $\varphi/\square\varphi$ (the necessitation rule **Nec**) preserves validity. The weakest of these logics, named $\mathsf{K}_{(1)}$, is semantically characterised by the class of Kripke frames with no restrictions imposed on the accessibility relation. In the multimodal version, named $\mathsf{K}_{(n)}$, Kripke frames are directed multigraphs and modal operators are equipped with indexes over a set of *agents*, given by $\mathcal{A}_n = \{1, 2, \ldots, n\}$, for some positive integer n. Accordingly, in this case classical logic is extended with operators $\boxed{1}, \boxed{2}, \ldots, \boxed{n}$, where a formula as $\boxed{a}p$, with $a \in \mathcal{A}_n$, may be read as "agent a considers p to be necessary". The modal operator $\Diamond$ is the dual of $\boxed{a}$, being introduced as an abbreviation for $\neg\boxed{a}\neg$, where $\neg$ stands for classical negation. The logic $\mathsf{K}_{(n)}$ can be seen as the *fusion* of n copies of $\mathsf{K}_{(1)}$ and its axiomatisation is given by the union of the axioms for classical propositional logic with the axiomatic schemata $\mathbf{K}_a$, namely $\boxed{a}(\varphi \Rightarrow \psi) \Rightarrow (\boxed{a}\varphi \Rightarrow \boxed{a}\psi)$, for each $a \in \mathcal{A}_n$; and the set of inference rules is given by *modus ponens* and the rule schemata $\mathbf{Nec}_a$, namely $\varphi/\boxed{a}\varphi$, for each $a \in \mathcal{A}_n$.

The basic normal multimodal logic $\mathsf{K}_{(n)}$ and its extensions have been widely used to represent and reason about complex systems. Some of the interesting extensions include the normal multimodal logics based on $\mathbf{K}_a$ and (the combination of) axioms as, for instance, $\mathbf{T}_a$ ($\boxed{a}\varphi \Rightarrow \varphi$), $\mathbf{D}_a$ ($\boxed{a}\varphi \Rightarrow \Diamond\varphi$), $\mathbf{4}_a$ ($\boxed{a}\varphi \Rightarrow \boxed{a}\,\boxed{a}\varphi$), $\mathbf{5}_a$ ($\Diamond\varphi \Rightarrow \boxed{a}\Diamond\varphi$), and $\mathbf{B}_a$ ($\Diamond\boxed{a}\varphi \Rightarrow \varphi$). For example, the description logic $\mathcal{ALC}$, which is employed for reasoning about ontologies, is a syntactic variant of $\mathsf{K}_{(1)}$ [22]; the epistemic logic, denoted by $\mathsf{S5}_{(n)}$, which is used in dealing with problems ranging from multi-agency to communication protocols [21,11], can be axiomatised by combining $\mathbf{K}_a$, $\mathbf{T}_a$, and $\mathbf{5}_a$. The addition of those axioms (or their combinations) to $\mathsf{K}_{(n)}$ imposes some restrictions on the class of models where formulae are valid. Thus, a formula valid in a logic containing $\mathbf{T}_a$ is valid only if it is valid in a frame where the accessibility relation for each agent a is *reflexive*. The other axioms, $\mathbf{D}_a$, $\mathbf{4}_a$, $\mathbf{5}_a$, and $\mathbf{B}_a$, demand the accessibility relation for each agent a to be, respectively, *serial*, *transitive*, *Euclidean*, and *symmetric*.

A *logic of confluence* $\mathsf{K}_{(n)}^{p,q,r,s}$ is a modal system axiomatised by $\mathsf{K}_{(n)}$ plus axioms $\mathbf{G}_a^{p,q,r,s}$ of the form

$$\Diamond_a^p \boxed{a}^q \varphi \Rightarrow \boxed{a}^r \Diamond_a^s \varphi$$

where $a \in \mathcal{A}_n$, φ is a well-formed formula, $p, q, r, s \in \mathbb{N}$, where $\boxed{a}^0\varphi \stackrel{\text{def}}{=} \varphi$ and $\boxed{a}^{i+1}\varphi \stackrel{\text{def}}{=} \boxed{a}\,\boxed{a}^i\varphi$, and where $\Diamond^0\varphi \stackrel{\text{def}}{=} \varphi$ and $\Diamond^{i+1}\varphi \stackrel{\text{def}}{=} \Diamond\Diamond^i\varphi$, for $i \in \mathbb{N}$ (the superscript is often omitted if equal to 1). Such axiomatic schemata were notably studied by Lemmon [16]. Using Modal Correspondence Theory, it can be shown that the frame condition on a logic where an instance of $\mathbf{G}_a^{p,q,r,s}$ is valid corresponds to a generalised diamond-like structure representing the *Church-Rosser property* (the philosophical literature sometimes calls such property 'incestual' [9]), as illustrated in Fig. 1 [6]. To be more precise, let $\mathcal{W}$ be a nonempty set of worlds and let $\mathcal{R}_a \subseteq \mathcal{W} \times \mathcal{W}$ be the accessibility relation of agent $a \in \mathcal{A}_n$. By $w\mathcal{R}_a^0 w'$ we mean that $w = w'$, and $w\mathcal{R}_a^{i+1}w'$ means that there is some world w'' such that $w\mathcal{R}_a w''$ and $w''\mathcal{R}_a^i w'$. Thus, $w\mathcal{R}_a^i w'$ holds if there is an i-long $\mathcal{R}_a$-path from w to w'; alternatively, to assert that, we may also write $(w, w') \in \mathcal{R}_a^i$. Given these definitions, the condition on frames that corresponds to the axiom $\mathbf{G}_a^{p,q,r,s}$ is described by $\forall w_0, w_1, w_2 \ (w_0\mathcal{R}_a^p w_1 \wedge w_0\mathcal{R}_a^r w_2 \Rightarrow \exists w_3(w_1\mathcal{R}_a^q w_3 \wedge w_2\mathcal{R}_a^s w_3))$, where $w_0, w_1, w_2, w_3 \in \mathcal{W}$.

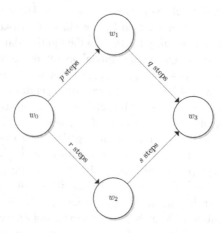

Fig. 1. Church-Rosser property for frames where $\mathbf{G}_a^{p,q,r,s} = \Diamond^p \boxed{a}^q \varphi \Rightarrow \boxed{a}^r \Diamond^s \varphi$ is valid

Many well-known modal axiomatic systems are identified with particular logics of confluence. For instance, $\mathsf{T}_{(n)}$ corresponds to $\mathsf{K}_{(n)}^{0,1,0,0}$, a normal modal logic in which the axiom $\boxed{a}\varphi \Rightarrow \varphi$ is valid, for all $a \in \mathcal{A}_n$ and any formula φ. The axiom $\mathbf{4}_a$ may be written as $\mathbf{G}_a^{0,1,2,0}$, that is, $\boxed{a}^1 \varphi \Rightarrow \boxed{a}^2 \varphi$. The Geach axiom $\mathbf{G1}_a$ is given by $\mathbf{G}_a^{1,1,1,1}$ ($\Diamond \boxed{a} \varphi \Rightarrow \boxed{a} \Diamond \varphi$). Formulae in $\mathsf{K}_{(n)}^{1,1,1,1}$ are satisfiable if, and only if, they are satisfiable in a model with n relations satisfying the so-called 'diamond property', and analogous claims hold for instance concerning formulae of $\mathsf{T}_{(n)}$ and models whose relations are all reflexive, and formulae of $\mathbf{4}_{(n)}$ and models whose relations are all transitive.

Logics of confluence are interesting not only because they encompass a great number of normal modal logics as particular examples, but also in view of their attractive computational behaviour. Indeed, if we think of multimodal frames as abstract rewriting systems, for instance, and think of modal languages as a way of obtaining an internal and local perspective on such frames, then each given notion of confluence ensures that certain different paths of transformation will eventually lead to the same result. Having a decidable proof procedure for a logic underlying such class of frames helps in establishing a direct form of verifying the properties of the structures that they represent.

As a contribution towards a uniform approach to the development of proof methods for logics of confluence, in this work we deal with the logics where $p, q, r, s \in \{0, 1\}$. Table 1 shows the relevant axiomatic schemata, some standard names by which they are known, and the corresponding conditions on frames. The axiom $\mathbf{G}_a^{0,1,1,1}$ seems not to be named in the literature; the corresponding property follows the naming convention given in [5, pg. 127]. Note that $\mathbf{G}_a^{0,0,0,0}$, $\mathbf{G}_a^{0,1,1,0}$, and $\mathbf{G}_a^{1,0,0,1}$ are instances of classical tautologies and are thus not included in Table 1. Also, given the duality between $\boxed{a}$ and $\Diamond$, $\mathbf{G}_a^{p,q,r,s}$ is semantically equivalent to $\mathbf{G}_a^{r,s,p,q}$. Thus, there are in fact eight families of multimodal logics related to the axioms $\mathbf{G}_a^{p,q,r,s}$, where $p, q, r, s \in \{0, 1\}$.

We present a clausal resolution-based method for solving the satisfiability problem in logics axiomatised by $\mathbf{K}_a$ plus $\mathbf{G}_a^{p,q,r,s}$, where $p, q, r, s \in \{0, 1\}$. The resolution

Table 1. Axioms and corresponding conditions on frames

(p,q,r,s)	Name	Axioms	Property	Condition on Frames
$(0,0,1,1)$	$\mathbf{B}_a$	$\varphi \Rightarrow \boxed{a}\Diamond\varphi$	symmetric	$\forall w,w'(w\mathcal{R}_aw' \Rightarrow w'\mathcal{R}_aw)$
$(1,1,0,0)$		$\Diamond\boxed{a}\varphi \Rightarrow \varphi$		
$(0,0,1,0)$	$\mathbf{Ban}_a$	$\varphi \Rightarrow \boxed{a}\varphi$	modally banal	$\forall w,w'(w\mathcal{R}_aw' \Rightarrow w=w')$
$(1,0,0,0)$		$\Diamond\varphi \Rightarrow \varphi$		
$(0,1,0,1)$	$\mathbf{D}_a$	$\boxed{a}\varphi \Rightarrow \Diamond\varphi$	serial	$\forall w\exists w'(w\mathcal{R}_aw')$
$(1,0,1,0)$	$\mathbf{F}_a$	$\Diamond\varphi \Rightarrow \boxed{a}\varphi$	functional	$\forall w,w',w''((w\mathcal{R}_aw' \wedge w\mathcal{R}_aw'') \Rightarrow w'=w'')$
$(0,0,0,1)$	$\mathbf{T}_a$	$\varphi \Rightarrow \Diamond\varphi$	reflexive	$\forall w(w\mathcal{R}_aw)$
$(0,1,0,0)$		$\boxed{a}\varphi \Rightarrow \varphi$		
$(1,0,1,1)$	$\mathbf{5}_a$	$\Diamond\varphi \Rightarrow \boxed{a}\Diamond\varphi$	Euclidean	$\forall w,w',w''((w\mathcal{R}_aw' \wedge w\mathcal{R}_aw'') \Rightarrow w'\mathcal{R}_aw'')$
$(1,1,1,0)$		$\Diamond\boxed{a}\varphi \Rightarrow \boxed{a}\varphi$		
$(1,1,1,1)$	$\mathbf{G1}_a$	$\Diamond\boxed{a}\varphi \Rightarrow \boxed{a}\Diamond\varphi$	convergent	$\forall w,w',w''((w\mathcal{R}_aw' \quad \wedge \quad w\mathcal{R}_aw'') \Rightarrow \exists w'''(w'\mathcal{R}_aw''' \wedge w''\mathcal{R}_aw'''))$
$(0,1,1,1)$	$\mathbf{G}_a^{0,1,1,1}$	$\boxed{a}\varphi \Rightarrow \boxed{a}\Diamond\varphi$	0,1,1,1-convergent	$\forall w,w'(w\mathcal{R}_aw' \Rightarrow \exists w''(w\mathcal{R}_aw'' \wedge w'\mathcal{R}_aw''))$
$(1,1,0,1)$		$\Diamond\boxed{a}\varphi \Rightarrow \Diamond\varphi$		

calculus is based on that of [17], which deals with the logical fragment corresponding to $\mathsf{K}_{(n)}$. The new inference rules to deal with axioms of the form $\mathbf{G}_a^{p,q,r,s}$ add relevant information to the set of clauses: the conclusion of each inference rule ensures that properties related to the corresponding conditions on frames hold, that is, the newly added clauses capture the required properties of a model. We discuss soundness, completeness, and termination. Full proofs can be found in [18].

2   The Normal Modal Logic $\mathsf{K}_{(n)}$

The set $\mathsf{WFF}_{\mathsf{K}_{(n)}}$ of *well-formed formulae* of the logic $\mathsf{K}_{(n)}$ is constructed from a denumerable set of *propositional symbols*, $\mathcal{P} = \{p,q,p',q',p_1,q_1,\ldots\}$, the negation symbol $\neg$, the conjunction symbol $\wedge$, the propositional constant **true**, and a unary connective $\boxed{a}$ for each agent a in the finite set of agents $\mathcal{A}_n = \{1,\ldots,n\}$. When $n=1$, we often omit the index, that is, $\Box\varphi$ stands for $\boxed{1}\varphi$. As usual, $\Diamond$ is introduced as an abbreviation for $\neg\boxed{a}\neg$. A *literal* is either a propositional symbol or its negation; the set of literals is denoted by $\mathcal{L}$. By $\neg l$ we will denote the *complement* of the literal $l \in \mathcal{L}$, that is, $\neg l$ denotes $\neg p$ if l is the propositional symbol p, and $\neg l$ denotes p if l is the literal $\neg p$. A *modal literal* is either $\boxed{a}l$ or $\neg\boxed{a}l$, where $l \in \mathcal{L}$ and $a \in \mathcal{A}_n$.

We present the semantics of $\mathsf{K}_{(n)}$, as usual, in terms of Kripke frames.

Definition 1. *A Kripke frame $\mathcal{S}$ for n agents over $\mathcal{P}$ is given by a tuple $(\mathcal{W}, w_0, \mathcal{R}_1, \mathcal{R}_2, \ldots, \mathcal{R}_n)$, where $\mathcal{W}$ is a set of possible worlds (or states) with a distinguished world w_0, and each $\mathcal{R}_a$ is a binary relation on $\mathcal{W}$. A Kripke model $\mathcal{M} = (\mathcal{S}, \pi)$ equips a Kripke frame $\mathcal{S}$ with a function $\pi : \mathcal{W} \rightarrow (\mathcal{P} \rightarrow \{true, false\})$*

that plays the role of an interpretation that associates to each state $w \in \mathcal{W}$ a truth-assignment to propositional symbols.

The so-called accessibility relation $\mathcal{R}_a$ is a binary relation that captures the notion of relative possibility from the viewpoint of agent a: A pair (w, w') is in $\mathcal{R}_a$ if agent a considers world w' possible, given the information available to her in world w. We write $\langle \mathcal{M}, w \rangle \models \varphi$ (resp. $\langle \mathcal{M}, w \rangle \not\models \varphi$) to say that φ is satisfied (resp. not satisfied) at the world w in the Kripke model $\mathcal{M}$.

Definition 2. *Satisfaction of a formula at a given world w of a model $\mathcal{M}$ is set by:*

- $\langle \mathcal{M}, w \rangle \models$ **true**
- $\langle \mathcal{M}, w \rangle \models p$ *if, and only if,* $\pi(w)(p) = true$, *where* $p \in \mathcal{P}$
- $\langle \mathcal{M}, w \rangle \models \neg\varphi$ *if, and only if,* $\langle \mathcal{M}, w \rangle \not\models \varphi$
- $\langle \mathcal{M}, w \rangle \models (\varphi \wedge \psi)$ *if, and only if,* $\langle \mathcal{M}, w \rangle \models \varphi$ *and* $\langle \mathcal{M}, w \rangle \models \psi$
- $\langle \mathcal{M}, w \rangle \models \boxed{a}\varphi$ *if, and only if* $\langle \mathcal{M}, w' \rangle \models \varphi$, *for all* w' *such that* $w\mathcal{R}_a w'$

The formulae **false**, $(\varphi \vee \psi)$, $(\varphi \Rightarrow \psi)$, and $\diamondsuit\varphi$ are introduced as the usual abbreviations for $\neg$**true**, $\neg(\neg\varphi \wedge \neg\psi)$, $(\neg\varphi \vee \psi)$, and $\neg \boxed{a} \neg\varphi$, respectively. Formulae are interpreted with respect to the distinguished world w_0, that is, satisfiability is defined with respect to pointed-models. A formula φ is said to be *satisfied in the model* $\mathcal{M} = (\mathcal{S}, \pi)$ *of the Kripke frame* $\mathcal{S} = (\mathcal{W}, w_0, \mathcal{R}_1, \ldots, \mathcal{R}_n)$ if $\langle \mathcal{M}, w_0 \rangle \models \varphi$; the formula φ is *satisfiable in a Kripke frame* $\mathcal{S}$ if there is a model $\mathcal{M}$ of $\mathcal{S}$ such that $\langle \mathcal{M}, w_0 \rangle \models \varphi$; and φ is said to be *valid in a class* $\mathcal{C}$ *of Kripke frames* if it is satisfied in any model of any Kripke frame belonging to the class $\mathcal{C}$.

3   Resolution for $\mathsf{K}_{(n)}$

In [17], a sound, complete, and terminating resolution-based method for $\mathsf{K}_{(n)}$, which in this paper we call RES_K, is introduced. As the proof-method for logics of confluence presented here relies on RES_K, in order to keep the present paper self-contained, we reproduce the corresponding inference rules here and refer the reader to [17] for a detailed account of the method. The approach taken in the resolution-based method for $\mathsf{K}_{(n)}$ is clausal: a formula to be tested for (un)satisfiability is first translated into a normal form, explained in Section 3.1, and then the inference rules given in Section 3.2 are applied until either a contradiction is found or no new clauses can be generated.

3.1   A Normal Form for $\mathsf{K}_{(n)}$

Formulae in the language of $\mathsf{K}_{(n)}$ can be transformed into a normal form called Separated Normal Form for Normal Logics (SNF). As the semantics is given with respect to a pointed-model, we add a nullary connective **start** in order to represent the world from which we start reasoning. Formally, given a model $\mathcal{M} = (\mathcal{W}, w_0, \mathcal{R}_1, \ldots, \mathcal{R}_n, \pi)$, we have that $\langle \mathcal{M}, w \rangle \models$ **start** if, and only if, $w = w_0$. A formula in SNF is represented by a conjunction of clauses, which are true at all reachable states, that is, they have the general form $\bigwedge_i \square^* A_i$, where A_i is a clause and $\square^*$, the universal operator, is characterised by (the greatest fixed point of) $\Box^*\varphi \Leftrightarrow \varphi \wedge \bigwedge_{a \in \mathcal{A}_n} \boxed{a}\Box^*\varphi$, for a formula φ. Observe that satisfaction of $\Box^*\varphi$ imposes that φ must hold at the actual world w and at every world reachable from w, where reachability is defined in the usual (graph-theoretic) way. Clauses have one of the following forms:

- Initial clause

$$\mathbf{start} \Rightarrow \bigvee_{b=1}^{r} l_b$$

- Literal clause

$$\mathbf{true} \Rightarrow \bigvee_{b=1}^{r} l_b$$

- Positive a-clause

$$l' \Rightarrow \boxed{a} l$$

- Negative a-clause

$$l' \Rightarrow \neg \boxed{a} l$$

where $l, l', l_b \in \mathcal{L}$. Positive and negative a-clauses are together known as *modal a-clauses*; the index a may be omitted if it is clear from the context.

The translation to SNF uses rewriting of classical operators and the renaming technique [20], where complex subformulae are replaced by new propositional symbols and the truth of these new symbols is linked to the formulae that they replaced in all states. Given a formula φ, the translation procedure is applied to $\Box^*(\mathbf{start} \Rightarrow t_0) \wedge \Box^*(t_0 \Rightarrow \varphi)$, where t_0 is a new propositional symbol. The universal operator, which surrounds all clauses, ensures that the clauses generated by the translation of a formula are true at all reachable worlds. Classical rewriting is used to remove some classical operators from φ (e.g. $\Box^*(t \Rightarrow \psi_1 \wedge \psi_2)$ is rewritten as $\Box^*(t \Rightarrow \psi_1) \wedge \Box^*(t \Rightarrow \psi_2)$). Renaming is used to replace complex subformulae in disjunctions (e.g. if ψ_2 is not a literal, $\Box^*(t \Rightarrow \psi_1 \vee \psi_2)$ is rewritten as $\Box^*(t \Rightarrow \psi_1 \vee t_1) \wedge \Box^*(t_1 \Rightarrow \psi_2)$, where t_1 is a new propositional symbol) or in the scope of modal operators (e.g. if ψ is not a literal, $\Box^*(t \Rightarrow \boxed{a}\psi)$ is rewritten as $\Box^*(t \Rightarrow \boxed{a}t_1) \wedge \Box^*(t_1 \Rightarrow \psi)$, where t_1 is a new propositional symbol). We refer the reader to [17] for details on the transformation rules that define the translation to SNF, their correctness, and examples of their application.

3.2 Inference Rules for $K_{(n)}$

In the following, $l, l', l_i, l_i' \in \mathcal{L}$ ($i \in \mathbb{N}$) and D, D' are disjunctions of literals.

Literal Resolution. This is classical resolution applied to the classical propositional fragment of the combined logic. An initial clause may be resolved with either a literal clause or another initial clause (rules **IRES1** and **IRES2**). Literal clauses may be resolved together (**LRES**).

[IRES1] $\Box^*(\mathbf{true} \Rightarrow D \vee l)$	[IRES2] $\Box^*(\mathbf{start} \Rightarrow D \vee l)$	[LRES] $\Box^*(\mathbf{true} \Rightarrow D \vee l)$
$\Box^*(\mathbf{start} \Rightarrow D' \vee \neg l)$	$\Box^*(\mathbf{start} \Rightarrow D' \vee \neg l)$	$\Box^*(\mathbf{true} \Rightarrow D' \vee \neg l)$
$\Box^*(\mathbf{start} \Rightarrow D \vee D')$	$\Box^*(\mathbf{start} \Rightarrow D \vee D')$	$\Box^*(\mathbf{true} \Rightarrow D \vee D')$

Modal Resolution. These rules are applied between clauses which refer to the same context, that is, they must refer to the same agent. For instance, we may resolve two or more $\boxed{a}$-clauses (rules **MRES** and **NEC2**); or several $\boxed{a}$-clauses and a literal clause (rules **NEC1** and **NEC3**). The modal inference rules are:

[MRES] $\quad \Box^*(l_1 \Rightarrow \boxed{a} l)$ $\qquad\qquad$ **[NEC2]** $\quad \Box^*(l_1' \Rightarrow \boxed{a} l_1)$

$\qquad\qquad \dfrac{\Box^*(l_2 \Rightarrow \neg \boxed{a} l)}{\Box^*(\mathbf{true} \Rightarrow \neg l_1 \vee \neg l_2)}$ $\qquad\qquad\quad \Box^*(l_2' \Rightarrow \boxed{a} \neg l_1)$

$\qquad\qquad\qquad\qquad\qquad\qquad\qquad\qquad\quad \dfrac{\Box^*(l_3' \Rightarrow \neg \boxed{a} l_2)}{\Box^*(\mathbf{true} \Rightarrow \neg l_1' \vee \neg l_2' \vee \neg l_3')}$

[NEC1] $\quad \Box^*(l_1' \Rightarrow \boxed{a} \neg l_1)$ $\qquad\qquad$ **[NEC3]** $\quad \Box^*(l_1' \Rightarrow \boxed{a} \neg l_1)$

$\qquad\qquad\qquad \vdots$ $\qquad\qquad\qquad\qquad\qquad\qquad\qquad\qquad \vdots$

$\qquad\qquad \Box^*(l_m' \Rightarrow \boxed{a} \neg l_m)$ $\qquad\qquad\qquad\quad \Box^*(l_m' \Rightarrow \boxed{a} \neg l_m)$

$\qquad\qquad \dfrac{\Box^*(l' \Rightarrow \neg \boxed{a} l)}{}$ $\qquad\qquad\qquad\qquad \dfrac{\Box^*(l' \Rightarrow \neg \boxed{a} l)}{}$

$\qquad\qquad \dfrac{\Box^*(\mathbf{true} \Rightarrow l_1 \vee \ldots \vee l_m \vee l)}{\Box^*(\mathbf{true} \Rightarrow \neg l_1' \vee \ldots \vee \neg l_m' \vee \neg l')}$ $\qquad\quad \dfrac{\Box^*(\mathbf{true} \Rightarrow l_1 \vee \ldots \vee l_m)}{\Box^*(\mathbf{true} \Rightarrow \neg l_1' \vee \ldots \vee \neg l_m' \vee \neg l')}$

The rule **MRES** is a syntactic variation of classical resolution, as a formula and its negation cannot be true at the same state. The rule **NEC1** corresponds to necessitation (applied to $(\neg l_1 \wedge \ldots \wedge \neg l_m \Rightarrow \neg l)$, which is equivalent to the literal clause in the premises) and several applications of classical resolution. The rule **NEC2** is a special case of **NEC1**, as the parent clauses can be resolved with the tautology $\mathbf{true} \Rightarrow l_1 \vee \neg l_1 \vee l_2$. The rule **NEC3** is similar to **NEC1**, however the negative modal clause is not resolved with the literal clause in the premises. Instead, the negative modal clause requires that resolution takes place between literals on the right-hand side of positive modal clauses and the literal clause. The resolvents in the inference rules **NEC1**–**NEC3** impose that the literals on the left-hand side of the modal clauses in the premises are not all satisfied whenever their conjunction leads to a contradiction in a successor state. Given the syntactic forms of clauses, the three rules are needed for completeness [17]. Note that for **NEC1**, we may have $m = 0$; for **NEC2** the number of premises is fixed; and that for **NEC3**, if $m = 0$, then the literal clause in the premises is $\mathbf{true} \Rightarrow \mathbf{false}$, which cannot be satisfied in any model. Thus, **NEC3** is not applied when $m = 0$.

We define a derivation as a sequence of sets of clauses $\mathcal{T}_0, \mathcal{T}_1, \ldots$, where $\mathcal{T}_i$ results from adding to $\mathcal{T}_{i-1}$ the resolvent obtained by an application of an inference rule of RES_K to clauses in $\mathcal{T}_{i-1}$. A derivation *terminates* if, and only if, either a contradiction, in the form of $\Box^*(\mathbf{start} \Rightarrow \mathbf{false})$ or $\Box^*(\mathbf{true} \Rightarrow \mathbf{false})$, is derived or no new clauses can be derived by further application of the resolution rules of RES_K. We assume standard simplification from classical logic to keep the clauses as simple as possible. For example, $D \vee l \vee l$ on the right-hand side of a clause would be rewritten as $D \vee l$.

Example 1. We wish to check whether the formula $\boxed{1}\boxed{2}(a \wedge b) \Rightarrow \boxed{1}(\boxed{2}a \wedge \boxed{2}b)$ is valid in $\mathsf{K}_{(2)}$. The translation of its negation into the normal form is given by clauses (1)–(9) below. Then the inference rules are applied until **false** is generated. In order to improve readability, the universal operator is suppressed. The full refutation follows.

1. **start** $\Rightarrow t_1$	9. $\quad t_6 \Rightarrow \neg \boxed{2} b$
2. $\quad t_1 \Rightarrow \boxed{1} t_2$	10. **true** $\Rightarrow \neg t_2 \vee \neg t_5$ [**NEC1**, 3, 8, 4]
3. $\quad t_2 \Rightarrow \boxed{2} t_3$	11. **true** $\Rightarrow \neg t_2 \vee \neg t_4 \vee t_6$ [**LRES**, 10, 7]
4. **true** $\Rightarrow \neg t_3 \vee a$	12. **true** $\Rightarrow \neg t_2 \vee \neg t_6$ [**NEC1**, 3, 9, 5]
5. **true** $\Rightarrow \neg t_3 \vee b$	13. **true** $\Rightarrow \neg t_2 \vee \neg t_4$ [**LRES**, 12, 11]
6. $\quad t_1 \Rightarrow \neg \boxed{1} \neg t_4$	14. **true** $\Rightarrow \neg t_1$ [**NEC1**, 2, 6, 13]
7. **true** $\Rightarrow \neg t_4 \vee t_5 \vee t_6$	15. **start** $\Rightarrow$ **false** [**IRES1**, 14, 1]
8. $\quad t_5 \Rightarrow \neg \boxed{2} a$	

Clauses (10) and (12) are obtained by applications of **NEC1** to clauses in the context of agent 2. Clause (14) is obtained by an application of the same rule, but in the context of agent 1. Clauses (11) and (13) result from applications of resolution to the propositional part of the language shared by both agents. Clause (15) shows that a contradiction was found at the initial state. Therefore, the original formula is valid.

4   Clausal Resolution for Logics of Confluence

The inference rules of $\mathsf{RES_K}$, given in Section 3.2, are resolution-based: whenever a set of (sub)formulae is identified as contradictory, the resolvents require that they are not all satisfied together. The extra inference rules for $\mathsf{K}_{(n)}^{p,q,r,s}$, with $p, q, r, s \in \{0, 1\}$, which we are about to present, have a different flavour: whenever we can identify that the set of clauses imply that $\Diamond^p \boxed{a}^q \psi$ holds, we add some new clauses that ensure that $\boxed{a}^r \Diamond^s \psi$ also holds. If this is not the case, that is, if the set of clauses implies that $\neg \boxed{a}^r \Diamond^s \psi$ holds, then a contradiction is found by applying the inference rules for $\mathsf{K}_{(n)}$. Because of the particular normal form we use here, there are, in fact, two general forms for the inference rules for $\mathsf{K}_{(n)}^{p,q,r,s}$, given in Table 2 (where l, l' are literals and C is a conjunction of literals).

Table 2. Inference Rules for $\mathbf{G}_a^{p,q,r,s}$

$[\mathbf{RES}_a^{p,1,r,s}] \quad \dfrac{\Box^*(l \Rightarrow \boxed{a} l')}{\Box^*(\Diamond^p l \Rightarrow \boxed{a}^r \Diamond^s l')}$	$[\mathbf{RES}_a^{p,0,r,s}] \dfrac{\Box^*(C \Rightarrow \Diamond^p l')}{\Box^*(C \Rightarrow \boxed{a}^r \Diamond^s l')}$

Soundness is checked by showing that the transformation of a formula $\varphi \in \mathsf{WFF}_{\mathsf{K}_{(n)}}$ into its normal form is satisfiability-preserving and that the application of the inference rules are also satisfiability-preserving. Satisfiability-preserving results for the transformation into **SNF** are provided in [17]. To extend the soundness results so as to cover the new inference rules, note that the conclusions of the inference rules in Table 2 are derived using the semantics of the universal operator and the distribution axiom, $\mathbf{K}_a$. For $\mathbf{RES}_a^{p,1,r,s}$, we have that the premise $\Box^*(l \Rightarrow \boxed{a} l')$ is semantically equivalent to $\Box^*(\neg \boxed{a} l' \Rightarrow \neg l)$. By the definition of the universal operator, we obtain $\Box^*(\boxed{a}^p(\neg \boxed{a} l' \Rightarrow \neg l))$. Applying the distribution axiom $\mathbf{K}_a$ to this clause results in $\Box^*(\boxed{a}^p \neg \boxed{a} l' \Rightarrow \boxed{a}^p \neg l))$, which is semantically equivalent to $\Box^*(\neg \boxed{a}^p \neg l \Rightarrow$

$\neg\boxed{a}{}^{p}\neg\boxed{a}l')$. As $\Diamond$ is an abbreviation for $\neg\boxed{a}\neg$ and because $\Diamond^{p}\boxed{a}l'$ implies $\boxed{a}^{r}\Diamond^{s}l'$ in $\mathsf{K}_{(n)}^{p,1,r,s}$, by classical reasoning, we have that $\Box^{*}(\neg\boxed{a}{}^{p}\neg l \Rightarrow \neg\boxed{a}{}^{p}\neg\boxed{a}l')$ implies $\Box^{*}(\Diamond^{p}l \Rightarrow \boxed{a}^{r}\Diamond^{s}l')$, the conclusion of $\mathbf{RES}_{a}^{p,1,r,s}$. Soundness of the inference rule $\mathbf{RES}_{a}^{p,0,r,s}$ can be proved in a similar way.

As the conclusions of the above inference rules may contain complex formulae, they might need to be rewritten into the normal form. Thus, we also need to add clauses corresponding to the normal form of $\Diamond^{p}l$ and $\Diamond^{s}l'$, which occur in the conclusions of the inference rules. Let φ be a formula and let $\tau(\varphi)$ be the set of clauses resulting from the translation of φ into the normal form. Let $\mathcal{L}(\tau(\varphi))$ be the set of literals that might occur in the clause set, that is, for all $p \in \mathcal{P}$ such that p occurs in $\tau(\varphi)$, we have that both p and $\neg p$ are in $\mathcal{L}(\tau(\varphi))$. The set of *definition clauses* is given by

$$\Box^{*}(pos_{a,l} \Rightarrow \neg\boxed{a}\neg l)$$

$$\Box^{*}(\neg pos_{a,l} \Rightarrow \boxed{a}\neg l)$$

for all $l \in \mathcal{L}(\tau(\varphi))$, where $pos_{a,l}$ is a new propositional *definition symbol* used for renaming the negative modal literal $\Diamond l$, that is, the definition clauses correspond to the normal form of $pos_{a,l} \Leftrightarrow \neg\boxed{a}\neg l$. Note that we have definition clauses for every propositional symbol and its negation, e.g. for a propositional symbol $p \in \tau(\varphi)$, we have the definition clauses $\Box^{*}(pos_{a,p} \Rightarrow \neg\boxed{a}\neg p)$, $\Box^{*}(\neg pos_{a,p} \Rightarrow \boxed{a}\neg p)$, $\Box^{*}(pos_{a,\neg p} \Rightarrow \neg\boxed{a}p)$, and $\Box^{*}(\neg pos_{a,\neg p} \Rightarrow \boxed{a}p)$, for every $a \in \mathcal{A}_{n}$ occurring in $\tau(\varphi)$. We assume the set of definition clauses to be available whenever those symbols are used. It is also important to note that those new definition symbols and the respective definition clauses can all be introduced at the beginning of the application of the resolution method because we do not need definition clauses applied to definition symbols in the proofs, as given in the completeness proof [18]. As no new propositional symbols are introduced by the inference rules, there is a finite number of clauses that might be expressed (modulo simplification) and, therefore, the clausal resolution method for each modal logic of confluence is terminating.

As discussed above and from the results in [17], we can establish the soundness of the proof method.

Theorem 1. *The resolution-based calculi for logics of confluence are sound.*

Proof (Sketch). The transformation into the normal form is satisfiability preserving [17]. Given a set $\mathcal{T}$ of clauses and a model $\mathcal{M}$ that satisfies $\mathcal{T}$, we can construct a model $\mathcal{M}'$ for the union of $\mathcal{T}$ and the definition clauses, where $\mathcal{M}$ and $\mathcal{M}'$ may differ only in the valuation of the definition symbols. By setting properly the valuations in $\mathcal{M}'$, we have that $\langle\mathcal{M}', w\rangle \models pos_{a,p}$ if and only if $\langle\mathcal{M}, w\rangle \models \Diamond p$, for any $w \in W$. Soundness of the inference rules for $\mathbf{RES}_{\mathsf{K}}$ is also given in [17]. Soundness of $\mathbf{RES}_{a}^{p,1,r,s}$ and $\mathbf{RES}_{a}^{p,0,r,s}$ follow from the axiomatisation of $\mathsf{K}_{(n)}^{p,q,r,s}$.

Table 3 shows the inference rules for each specific instance of $\mathbf{G}_{a}^{p,q,r,s}$, where $p, q, r, s \in \{0, 1\}$, $l, l' \in \mathcal{L}$, and D is a disjunction of literals. As $\mathbf{G}_{a}^{p,q,r,s}$ is semantically equivalent to $\mathbf{G}_{a}^{r,s,p,q}$, the inference rules for both systems are grouped together. Some of the inference rules in Table 3 are obtained directly from Table 2. For instance, the rule for reflexive systems, i.e. where the axiom $\mathbf{G}_{a}^{0,1,0,0}$ is valid, has the

Table 3. Inference Rules for several instances of $\mathbf{G}_a^{p,q,r,s}$

Logic	Inference Rules	Logic	Inference Rules
$\mathbf{T}_a$	$[\text{RES}_a^{0,0,0,1}]\ \dfrac{\Box^*(\textbf{true} \Rightarrow D \vee l)}{\Box^*(\neg D \Rightarrow \neg [a]\neg l)}$ $[\text{RES}_a^{0,1,0,0}]\ \dfrac{\Box^*(l \Rightarrow [a]l')}{\Box^*(\textbf{true} \Rightarrow \neg l \vee l')}$	$\mathbf{G}_a^{0,1,1,1}$	$[\text{RES}_a^{0,1,1,1}]\ \dfrac{\Box^*(l \Rightarrow [a]l')}{\Box^*(l \Rightarrow [a]pos_{a,l'})}$ $[\text{RES}_a^{1,1,0,1}]\ \dfrac{\Box^*(l \Rightarrow [a]l')}{\Box^*(pos_{a,l} \Rightarrow \neg [a]\neg l')}$
$\mathbf{Ban}_a$	$[\text{RES}_a^{0,0,1,0}]\ \dfrac{\Box^*(\textbf{true} \Rightarrow D \vee l)}{\Box^*(\neg D \Rightarrow [a]l)}$ $[\text{RES}_a^{1,0,0,0}]\ \dfrac{\Box^*(l \Rightarrow \neg [a]\neg l')}{\Box^*(\textbf{true} \Rightarrow \neg l \vee l')}$	$\mathbf{F}_a$	$[\text{RES}_a^{1,0,1,0}]\ \dfrac{\Box^*(l \Rightarrow \neg [a]\neg l')}{\Box^*(l \Rightarrow [a]l')}$
$\mathbf{B}_a$	$[\text{RES}_a^{0,0,1,1}]\ \dfrac{\Box^*(\textbf{true} \Rightarrow D \vee l)}{\Box^*(\neg D \Rightarrow [a]pos_{a,l})}$ $[\text{RES}_a^{1,1,0,0}]\ \dfrac{\Box^*(l \Rightarrow [a]l')}{\Box^*(\neg l' \Rightarrow [a]\neg l)}$	$\mathbf{5}_a$	$[\text{RES}_a^{1,0,1,1}]\ \dfrac{\Box^*(l \Rightarrow \neg [a]\neg l')}{\Box^*(l \Rightarrow [a]pos_{a,l'})}$ $[\text{RES}_a^{1,1,1,0}]\ \dfrac{\Box^*(l \Rightarrow [a]l')}{\Box^*(pos_{a,l} \Rightarrow [a]l')}$
$\mathbf{D}_a$	$[\text{RES}_a^{0,1,0,1}]\ \dfrac{\Box^*(l \Rightarrow [a]l')}{\Box^*(l \Rightarrow \neg [a]\neg l')}$	$\mathbf{G1}_a$	$[\text{RES}_a^{1,1,1,1}]\ \dfrac{\Box^*(l \Rightarrow [a]l')}{\Box^*(pos_{a,l} \Rightarrow [a]pos_{a,l'})}$

form $\Box^*(l \Rightarrow [a]l')/\Box^*(\Diamond^0 l \Rightarrow [a]^0 \Diamond^0 l')$ in Table 2; in Table 3, the conclusion is rewritten in its normal form, that is, $\Box^*(\textbf{true} \Rightarrow \neg l \vee l')$. For other systems, the form of the inference rules are slightly different from what would be obtained from a direct application of the general inference rules in Table 2. This is the case, for instance, for the inference rules for symmetric systems, that is, those systems where the axiom $\mathbf{G}_a^{1,1,0,0}$ is valid. From Table 2, in symmetric systems, for a premise of the form $\Box^*(l \Rightarrow [a]l')$, the conclusion is given by $\Box^*(\Diamond l \Rightarrow l')$, which is translated into the normal form as $\Box^*(\textbf{true} \Rightarrow \neg pos_{a,l} \vee l')$. We have chosen, however, to translate the conclusion as $\Box^*(\neg l' \Rightarrow [a]\neg l)$, which is semantically equivalent to the conclusion obtained by the general inference rule, but avoids the use of definition symbols.

The inference rules given in Table 2 provide a systematic way of designing the inference rules for each specific modal logic of confluence. We note, however, that we do not always need both inference rules in order to achieve a complete proof method for a particular logic. In the completeness proofs provided in [18], we show for instance that the inference rules which introduce modalities in their conclusions from literal clauses (that is, the inference rules $\text{RES}_a^{0,0,r,s}$) are not needed for completeness. We also show that we need just one specific inference rule for logics in which $\mathbf{G}_a^{0,1,1,1}$ and $\mathbf{5}_a$ are valid: $\text{RES}_a^{0,1,1,1}$ and $\text{RES}_a^{1,0,1,1}$, respectively.

Given a formula φ in $\mathsf{K}_{(n)}^{p,q,r,s}$, with $p,q,r,s \in \{0,1\}$, the resolution method for $\mathsf{K}_{(n)}$, given in Section 3, and the inference rule $\text{RES}_a^{p,q,r,s}$ are applied to $\tau(\varphi)$ and the set of definition clauses. The extra inference rules for $\mathsf{K}_{(n)}^{p,q,r,s}$ do not need to be applied to clauses if such application generates new nested definition symbols, that is, we do not need definition clauses for definition symbols. For instance, the application of $\mathbf{RES}_a^{1,1,1,1}$ to a clause of the form $\square^*(l \Rightarrow \boxed{a}pos_{a,l'})$ would result in $\square^*(pos_{a,l} \Rightarrow \boxed{a}pos_{a,pos_{a,l'}})$. Although it is not incorrect to apply the inference rules to such a clause, this might cause the method not to terminate. We can show, however, that the application of inference rules to clauses which would result in nested literals is not needed for completeness, as the restrictions imposed by those symbols are already ensured by existing definition symbols and relevant inference rules (see Theorem 3 below). This ensures that no new definition symbols are introduced by the proof method.

Completeness is proved by showing that, for each specific logic of confluence, if a given set of clauses is unsatisfiable, there is a refutation produced by the method presented here. The proof is by induction on the number of nodes of a graph, known as *behaviour graph*, built from a set of clauses. The graph construction is similar to the construction of a canonical model, followed by filtrations based on the set of formulae (or clauses), often used to check completeness for proof methods in modal logics (see [3], for instance, for definitions and examples). Intuitively, nodes in the graph correspond to states and are defined as maximally consistent sets of literals and modal literals occurring in the set of clauses, including those literals introduced by definition clauses. That is, for any literal l occurring in the set of clauses, including definition clauses, and agents $a \in \mathcal{A}_n$, a node contains either l or $\neg l$; and either $\boxed{a}l$ or $\neg\boxed{a}l$. The set of edges correspond to the agents' accessibility relations. Edges or nodes that do not satisfy the set of clauses are deleted from the graph. Such deletions correspond to applications of one or more of the inference rules. We prove that an empty behaviour graph corresponds to an unsatisfiable set of clauses and that, in this case, there is a refutation using the inference rules for $\mathsf{RES_K}$, given in Section 3, and the inference rules for the specific logic of confluence, presented in Table 3.

Theorem 2. *Let $\mathcal{T}$ be an unsatisfiable set of clauses in $\mathbf{G}_a^{p,q,r,s}$, with $p, q, r, s \in \{0, 1\}$. A contradiction can be derived by applying the resolution rules for $\mathsf{RES_K}$, presented in Section 3, and Table 3.*

Proof (Sketch). We construct a behaviour graph and show that the application of rules in Table 3 removes nodes and edges where the corresponding frame condition does not hold. The full proof is provided in [18].

Theorem 3. *The resolution-based calculi for logics of confluence terminate.*

Proof (Sketch). From the completeness proof, the introduction of a literal such as $pos_{a,pos_{a,l}}$ for an agent a and literal l is not needed. We can show that the restrictions imposed by such clauses, together with the resolution rules for each specific logical system, are enough to ensure that the corresponding frame condition already holds. As the proof method does not introduce new literals in the clause set, there is only a finite number of clauses that can be expressed. Therefore, the proof method is terminating.

Example 2. We show that $\varphi \stackrel{\text{def}}{=} p \Rightarrow \boxed{1}\Diamond p$, which is an instance of $\mathbf{B}_1$, is a valid formula in symmetric systems. As symmetry is implied by reflexivity and Euclideanness, instead of using $\mathbf{RES}_1^{1,1,0,0}$, we combine the inference rules for both $\mathbf{T}_1$ and $\mathbf{5}_1$. Clauses (1)–(4) correspond to the translation of the negation of φ into the normal form. Clauses (5)–(8) are the definition clauses used in the proof.

1.	$\mathbf{start} \Rightarrow t_0$		
2.	$\mathbf{true} \Rightarrow \neg t_0 \vee p$		
3.	$t_0 \Rightarrow \neg \boxed{1} \neg t_1$		
4.	$t_1 \Rightarrow \boxed{1} \neg p$		
5.	$\neg pos_{1,t_1} \Rightarrow \boxed{1} \neg t_1$	$[\mathbf{Def.}\, pos_{1,t_1}]$	
6.	$pos_{1,t_1} \Rightarrow \neg \boxed{1} \neg t_1$	$[\mathbf{Def.}\, pos_{1,t_1}]$	
7.	$pos_{1,p} \Rightarrow \neg \boxed{1} \neg p$	$[\mathbf{Def.}\, pos_{1,p}]$	
8.	$\neg pos_{1,p} \Rightarrow \boxed{1} \neg p$	$[\mathbf{Def.}\, pos_{1,p}]$	

9.	$\mathbf{true} \Rightarrow \neg t_0 \vee pos_{1,t_1}$	$[\mathbf{MRES}, 5, 3]$
10.	$\mathbf{true} \Rightarrow \neg t_1 \vee \neg pos_{1,p}$	$[\mathbf{MRES}, 7, 4]$
11.	$pos_{1,p} \Rightarrow \boxed{1} pos_{1,p}$	$[\mathbf{RES}_1^{1,0,1,1}, 7]$
12.	$\mathbf{true} \Rightarrow \neg pos_{1,p} \vee \neg pos_{1,t_1}$	$[\mathbf{NEC1}, 11, 6, 10]$
13.	$\mathbf{true} \Rightarrow \neg p \vee pos_{1,p}$	$[\mathbf{RES}_1^{0,1,0,0}, 8]$
14.	$\mathbf{true} \Rightarrow \neg p \vee \neg pos_{1,t_1}$	$[\mathbf{LRES}, 13, 12]$
15.	$\mathbf{true} \Rightarrow \neg t_0 \vee \neg p$	$[\mathbf{LRES}, 14, 9]$
16.	$\mathbf{true} \Rightarrow \neg t_0$	$[\mathbf{LRES}, 15, 2]$
17.	$\mathbf{start} \Rightarrow \mathbf{false}$	$[\mathbf{IRES1}, 16, 1]$

Clause (11) results from applying the Euclidean inference rule to clause (7). Clause (13) results from applying the reflexive inference rule to (8). The remaining clauses are derived by the resolution calculus for $\mathsf{K}_{(1)}$. As a contradiction is found, given by clause (17), the set of clauses is unsatisfiable and the original formula φ is valid.

5   Closing Remarks

We have presented a sound, complete, and terminating proof method for logics of confluence, that is, normal multimodal systems where axioms of the form

$$\mathbf{G}_a^{p,q,r,s} = \Diamond^p \boxed{a}^q \varphi \Rightarrow \boxed{a}^r \Diamond^s \varphi$$

where $p, q, r, s \in \{0, 1\}$, are valid. The axioms $\mathbf{G}_a^{p,q,r,s}$ provide a general form for axioms widely used in logical formalisms applied to representation and reasoning within Computer Science.

We have proved completeness of the proof method presented in this paper for eight families of logics and their fusions. The inference rules for particular instances of these logics can be systematically obtained and the resulting calculus can be implemented by adding to the existing prover for $\mathsf{K}_{(n)}$ [24] the clauses dependent on the clause-set. Efficiency, of course, depends on several aspects. Firstly, for certain classes of problems, dedicated proof methods might be more efficient. For instance, if the satisfiability problem for a particular logic is in NP (as in the case of $\mathsf{S5}_{(1)}$), then our procedure may be less efficient as the satisfiability problem for $\mathsf{K}_{(1)}$ is already PSPACE-complete [15]. Secondly, efficiency might depend on the inference rules chosen to produce proofs for a specific logic. For instance, for $\mathsf{S5}_{(n)}$, the user can choose the inference rules related to reflexivity and Euclideanness, or choose the inference rules related to seriality, symmetry, and Euclideanness. The number of inference rules used to test the unsatisfiability of a set of clauses for a particular logic might affect the number of clauses generated by the resolution method as well as the size of the proof. As in the case of derived inference rules in other proof methods, using more inference rules might lead to shorter proofs. Thirdly, as in the case of the resolution-based method for propositional logic, efficiency might be affected by strategies used to search for a proof. Future work includes the design of strategies for $\mathsf{RES}_\mathsf{K}(n)$ and for specific logics of confluence. Fourthly, efficiency might also depend on the form of the input problem. For instance, comparisons between tableaux methods and resolution methods [14,13] have shown that there is no overall better approach: for some problems resolution proof methods behave better, for others tableaux based methods behave better. Providing a resolution-based method for the logics axiomatised by $\mathbf{K}_a$ and $\mathbf{G}_a^{p,q,r,s}$ gives the user a choice for automated tools that can be used depending on the type of the input formulae.

There are quite a few dedicated methods for the logics presented in this paper. In general, however, those methods do not provide a systematic way of dealing with logics based on similar axioms or their extensions. Therefore, we restrict attention here to methods related to logics of confluence. Tableaux methods for logics of confluence where the mono-modal axioms $\mathbf{T}$, $\mathbf{D}$, $\mathbf{B}$, $\mathbf{4}$, $\mathbf{5}$, $\mathbf{De}$ (for density, the converse of $\mathbf{4}$), and $\mathbf{G}$ are valid, can be found in [7,8]. For each of those axioms, a tableau inference rule is given. The inference rules can then be combined in order to provide proof methods for modal logics under $\mathbf{S5}_{(1)}$. Whilst the tableaux procedures in [7,8] are designed for mono-modal logics they seem to be extendable to multimodal logics as long as there are no interactions between modalities. Those procedures do not cover all the logics investigated in this paper. In [2], labelled tableaux are given for the mono-modal logics axiomatised by $\mathbf{K}$ and axioms $\mathbf{G}^{p,q,r,s}$ where $q = s = 0$ implies $p = r = 0$. This restriction avoids the introduction of inference rules related to the identity predicate, but also excludes, for instance, functional and modally banal systems, which are treated by the method introduced in the present paper. In [4], hybrid logic tableaux methods for logics of confluence are given: the inference rules create nodes, labelled by nominals. The nominals are used in order to eliminate the Skolem function related to the existential quantifier in the first-order sentence corresponding to the axiom $\mathbf{G}_a^{p,q,r,s}$. This proof method provides tableau rules for all instances of the axiom. Soundness and completeness are discussed, but termination of the method is not dealt with and it is not clear what are the bounds for creating new nodes in the general case. In [12], sound, complete, and terminating display calculi for tense logics and some of its extensions, including those with the axiom $\mathbf{G}_a^{p,q,r,s}$, are presented. It has been shown that these calculi have the property of *separation*, that is, they provide complete proof methods for the component fragments. The paper investigates the relation between the display calculi and deep inference systems (where the sequent rules can be applied at any node of a proof tree). By finding appropriate propagation rules for the fusion of tense logic with either $\mathbf{S4}_{(1)}$, $\mathbf{S5}_{(1)}$, or functional systems, completeness of search strategies are presented. However, propagation rules for the axiom of convergence, $\mathbf{G1}$, or for the combination of path axioms (i.e. axioms of the form $\Diamond^i \varphi \Rightarrow \Diamond^j \varphi$) with seriality are not given. Also related, in [1], prefixed tableaux procedures for confluence logics that validate the multimodal version of the axiom $\Diamond \boxed{b} \varphi \Rightarrow \boxed{c} \Diamond \varphi$, where φ is a formula, are given. Note that the logics in [1] are systems with instances of the axiom $\mathbf{G}_{a,b,c,d}^{1,1,1,1}$, that is, a logic which allows the interaction of the agents $a, b, c, d \in \mathcal{A}_n$, and might lead to undecidable systems.

To the best of our knowledge, there are no resolution-based proof methods for logics of confluence. However, resolution-based methods for modal logics, based on translation into first-order logic, have been proposed for several modal logics. A survey on translation-based approaches for non-transitive modal logics (i.e. modal logics that do not include the axiom $\mathbf{4}$) can be found in [19]. The translation-based approach has the clear advantage of being easily implemented, making use of well-established theorem-provers, and dealing with any logic that can be embedded into first-order, should it be decidable or not. However, first-order provers cannot deal easily with logics that embed some properties which are covered by particular axioms of confluence (e.g. functionality). In order to avoid such problematic fragments within first-order logic, the axiomatic translation principle for modal logic, introduced in [23], besides using the standard translation of a modal formulae into first-order, takes an axiomatisation for a particular modal logic and introduces a set of first-order modal axioms in the form of *schema clauses*. As an example, adapted from [23], in order to prove that $\boxed{a} \neg \boxed{a} p$ is satisfiable in $\mathsf{KT4}_{(n)}$, for each modal subformula (i.e. $\boxed{a} \neg \boxed{a} p$ and $\boxed{a} p$) and for each considered axiom (i.e. **T** and **4**), one schema clause is added, resulting in:

$$\neg Q_{\boxed{a} \neg \boxed{a} p}(x) \vee \neg R(x, y) \vee Q_{\boxed{a} \neg \boxed{a} p}(y) \qquad \neg Q_{\boxed{a} \neg \boxed{a} p}(x) \vee Q_{\neg \boxed{a} p}(y)$$

$$\neg Q_{\boxed{a} p}(x) \vee \neg R(x, y) \vee Q_{\boxed{a} p}(y) \qquad \neg Q_{\boxed{a} p}(x) \vee Q_p(y)$$

where the predicate $Q_\varphi(x)$ can be read as φ holds at world x and R is the predicate symbol to express the accessibility relation for agent a. Note that the clauses on the left are related to transitivity (**4**) and the two clauses on the right are related to reflexivity (**T**). The axiomatic translation approach is similar to the approach taken in the present paper and in [17] as the schema clauses provide a way of talking about properties of the accessibility relation. As in our case, soundness follows easily from the properties of the translation. Termination follows from the fact that only a finite number of schema clauses are needed. However, as in the case of the proof method presented here, general completeness of the method is difficult to be proved and it is given only for particular families of logics. In [10], a translation-based approach for properties which can be expressed by regular grammar logics (including transitivity and Euclideanness) is given. Completeness of the method has been proved for some families of logics.

In the present paper, we have restricted attention to the case where $p, q, r, s \in \{0, 1\}$, but we believe that the proof method can be extended in a uniform way for dealing with the unsatisfiability problem for any values of p, q, r, and s, by adding inference rules of the following form:

$$[\mathbf{RES}_a^{p,q,r,s}] \frac{\Box^*(l \Rightarrow \Diamond^p \boxed{a}^r l')}{\Box^*(l \Rightarrow \boxed{a}^r \Diamond^s l')}$$

which requires search for clauses that correspond to the normal form of the premise and the introduction of as many new definition symbols as the number of modalities occurring in the conclusion. The inference rule $\mathbf{RES}_a^{p,q,r,s}$ is obviously sound, but we have yet to identify the restrictions on the number of new propositional symbols introduced by the method in order to ensure termination. Future work includes this extension, the complexity analysis, the implementation of the proof method, and practical comparisons with other methods.

References

1. Baldoni, M., Giordano, L., Martelli, A.: A tableau calculus for multimodal logics and some (un)decidability results. In: de Swart, H. (ed.) TABLEAUX 1998. LNCS (LNAI), vol. 1397, pp. 44–59. Springer, Heidelberg (1998)

2. Basin, D., Matthews, S., Viganò, L.: Labelled propositional modal logics: Theory and practice. J. Log. Comput 7(6), 685–717 (1997)
3. Blackburn, P., de Rijke, M., Venema, Y.: Modal Logic. Cambridge University Press, Cambridge (2001)
4. Blackburn, P., Dialogue, E.L.E., Cate, B.T.: Beyond pure axioms: Node creating rules in hybrid tableaux. In: Areces, C.E., Blackburn, P., Marx, M., Sattler, U. (eds.) Hybrid Logics, pp. 21–35 (July 25, 2002)
5. Boolos, G.S.: The Logic of Provability. Cambridge University Press (1993)
6. Carnielli, W.A., Pizzi, C.: Modalities and Multimodalities. Logic, Epistemology, and the Unity of Science, vol. 12. Springer (2008)
7. Castilho, M.A., del Cerro, L.F., Gasquet, O., Herzig, A.: Modal tableaux with propagation rules and structural rules. Fundamenta Informaticae 32(3-4), 281–297 (1997)
8. del Cerro, L.F., Gasquet, O.: Tableaux based decision procedures for modal logics of confluence and density. Fundamenta Informaticae 40(4), 317–333 (1999)
9. Chellas, B.: Modal Logic: An Introduction. Cambridge University Press (1980)
10. Demri, S., Nivelle, H.: Deciding regular grammar logics with converse through first-order logic. Journal of Logic, Language and Information 14(3), 289–329 (2005)
11. Fagin, R., Halpern, J., Moses, Y., Vardi, M.: Reasoning About Knowledge. MIT Press (1995)
12. Goré, R., Postniece, L., Tiu, A.: On the correspondence between display postulates and deep inference in nested sequent calculi for tense logics. Logical Methods in Computer Science 7(2) (2011)
13. Goré, R., Thomson, J., Widmann, F.: An experimental comparison of theorem provers for CTL. In: Combi, C., Leucker, M., Wolter, F. (eds.) TIME 2011, Lübeck, Germany, September 12-14, pp. 49–56. IEEE (2011)
14. Hustadt, U., Schmidt, R.A.: Scientific benchmarking with temporal logic decision procedures. In: Fensel, D., Giunchiglia, F., McGuinness, D., Williams, M.-A. (eds.) Proceedings of the KR 2002, pp. 533–544. Morgan Kaufmann (2002)
15. Ladner, R.E.: The computational complexity of provability in systems of modal propositional logic. SIAM J. Comput. 6(3), 467–480 (1977)
16. Lemmon, E.J., Scott, D.: The Lemmon Notes: An Introduction to Modal Logic. Segerberg, K. (ed.). Basil Blackwell (1977)
17. Nalon, C., Dixon, C.: Clausal resolution for normal modal logics. J. Algorithms 62, 117–134 (2007)
18. Nalon, C., Marcos, J., Dixon, C.: Clausal resolution for modal logics of confluence – extended version. Technical Report ULCS-14-001, University of Liverpool, Liverpool, UK (May 2014), http://intranet.csc.liv.ac.uk/research/techreports/?id=ULCS-14-001
19. de Nivelle, H., Schmidt, R.A., Hustadt, U.: Resolution-Based Methods for Modal Logics. Logic Journal of the IGPL 8(3), 265–292 (2000)
20. Plaisted, D.A., Greenbaum, S.A.: A Structure-Preserving Clause Form Translation. Journal of Logic and Computation 2, 293–304 (1986)
21. Rao, A., Georgeff, M.: Modeling Rational Agents within a BDI-Architecture. In: Fikes, R., Sandewall, E. (eds.) Proceedings of KR&R-91, pp. 473–484. Morgan-Kaufmann (April 1991)
22. Schild, K.: A Correspondence Theory for Terminological Logics. In: Proceedings of the 12th IJCAI, pp. 466–471 (1991)
23. Schmidt, R.A., Hustadt, U.: The axiomatic translation principle for modal logic. ACM Transactions on Computational Logic 8(4), 1–55 (2007)
24. Silva, G.B.: Implementação de um provador de teoremas por resolução para lógicas modais normais. Monografia de Conclusão de Curso, Bacharelado em Ciência da Computação, Universidade de Brasília (2013), http://www.cic.unb.br/~nalon/#software

Implementing Tableau Calculi Using BDDs: BDDTab System Description

Rajeev Goré, Kerry Olesen, and Jimmy Thomson

Research School of Computer Science, Australian National University

Abstract. We present a modification of the DPLL-based approach to decide modal satisfiability where we substitute DPLL by BDDs. We demonstrate our method by implementing the standard tableau calculi for automated reasoning in propositional modal logics K and S4, along with extensions to the multiple modalities of $\mathcal{ALC}$. We evaluate our implementation of such a reasoner using several K and S4 benchmark sets, as well as some $\mathcal{ALC}$ ontologies. We show, with comparison to FaCT++, InKreSAT and *SAT, that it can compete with other state of the art methods of reasoning in propositional modal logic. We also discuss how this technique extends to tableau for other propositional logics.

Introduction. Many approaches have been proposed to decide satisfiability and validity in various non-classical logics. In the eager approach the entire problem is encoded into SAT and solved with a SAT-solver, either in one phase [10] or incrementally [7] (e.g. InKreSAT). In the DPLL-based approach [3], DPLL is used to provide propositionally satisfying assignments, which are recursively checked for modal consistency in a master-slave arrangement (e.g. *SAT). Our approach, BDDTab, is a modification on the DPLL-based approach, where we use BDDs instead of DPLL. BDDs have also been used to implement the finite property method directly [4], which is unrelated to the approach we present here.

BDDTab is written in C++, using the BuDDy [2] BDD library. Source and all benchmarks can be found at `http://users.cecs.anu.edu.au/~rpg/BDDTab`

We use a common mapping $[\cdot]$ from propositional modal formulae to BDDs:

$$[p] = \langle v_p, \text{true-node}, \text{false-node} \rangle \quad [\varphi \vee \psi] = [\varphi] \vee [\psi]$$
$$[\neg \varphi] = \neg [\varphi] \qquad\qquad\qquad\quad [\Diamond \varphi] = \neg [\Box \neg \varphi]$$
$$[\varphi \wedge \psi] = [\varphi] \wedge [\psi] \qquad\qquad\quad [\Box \varphi] = \langle v_{\Box \varphi}, \text{true-node}, \text{false-node} \rangle$$

This mapping is both purely propositional and modally shallow: each modal formula $\Box \varphi$ is represented by a BDD variable $v_{\Box \varphi}$, and the mapping "stops" when it reaches a modal formula. We treat modal formulae as if they were atomic propositions by allocating them a BDD variable $v_{\Box \varphi}$. As constructed, the set of all valuations on which $[\varphi]$ is true represents all the open branches of a saturated classical propositional logic (CPL) tableau for φ. However, the correspondence between valuations and tableau leaves is not quite one-to-one, as we demonstrate next.

S. Demri, D. Kapur, and C. Weidenbach (Eds.): IJCAR 2014, LNAI 8562, pp. 337–343, 2014.

$$\frac{a \vee b}{a \quad b} \ (\vee)$$

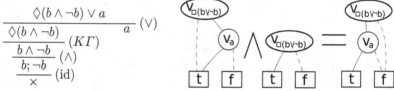

$$(K\Gamma) \ \frac{\Diamond\varphi; \Box X; Z}{\varphi; X; \Gamma} \ \forall \psi.\Box\psi \notin Z$$

On the left is the tableau for $a \vee b$, and in the middle is $[a \vee b]$, where the *low* branch is a dashed line, and the *high* branch is a solid line. Each tableau branch has only one of the atomic propositions, but $[a \vee b]$, with variable order $v_a < v_b$, has the valuation $a = f$, $b = t$, which would be equivalent to the tableau node $\neg a; b$. Thus using BDDs can involve a kind of semantic branching. This is not limited to the local branching demonstrated above; every satisfying valuation in a BDD is mutually disjoint with every other satisfying valuation in that BDD, so the branches in its representation of a saturated tableau are mutually disjoint.

Method. Our method is basically a standard tableau procedure, but we use BDDs to perform the saturation phases, as described above. Thus we compute a CPL saturation phase using BDDs, get a branch from the saturation, then compute modal jumps via $(K\Gamma)$ (shown above) as necessary, where Γ is a finite set of global assumptions (TBox). Branches are explored recursively, depth-first.

An important part of our method is when a modal jump closes, and a new OR-branch needs to be explored. Instead of traversing the BDD for the next satisfying valuation, we refine the BDD to remove the closed leaf, and then ask for a new one. The refinement is of the form $BDD_{final} = BDD_{initial} \wedge \neg(v_0 \wedge v_1 \wedge \ldots)$, where v_i are some of the variables of the branch that closed. This process is analogous to clause learning in SAT procedures.

$$\frac{\dfrac{\dfrac{\dfrac{b \wedge \neg b}{b; \neg b} \ (\wedge)}{\times} \ (id)}{\Diamond(b \wedge \neg b)} \ (K\Gamma)}{\Diamond(b \wedge \neg b) \vee a} \quad a \ (\vee)$$

Above left is the tableau for $\{\Diamond(b \wedge \neg b) \vee a\}$. The first BDD on the right is $[\Diamond(b \wedge \neg b) \vee a]$. In the tableau, exploring the modal jump over $\Diamond(b \wedge \neg b)$ will close, so we would refine $[\Diamond(b \wedge \neg b) \vee a]$ with the second BDD, giving the third BDD, from which we would get a new satisfying valuation, and continue the procedure.

This process has the potential to eliminate many branches at a time. For instance, if a in the previous example were some more complicated formula, we would have also eliminated every other branch that included $\Diamond(b \wedge \neg b)$.

The flip-side of this is that the refinement process may introduce new variables to existing branches. In the above example, every other branch will now explicitly include $\neg\Diamond(b \wedge \neg b)$, even if it didn't mention it previously. If these new variables are "irrelevant" to the branch, they may cause a net increase in the cost of exploring the branch. We refer to this problem as "**irrelevant variables.**"

Optimisations. <u>R</u>esponsible <u>V</u>ariables: Given the choice of refining with $\neg(v_0)$ or $\neg(v_0 \wedge v_1)$, we prefer $\neg(v_0)$ because refining with $\neg(v_0)$ would eliminate at least as many branches as $\neg(v_0 \wedge v_1)$. Thus a significant optimisation of this algorithm is to find small subsets of modal variables over which to refine. We recursively track sets of variables that we deem "responsible" for any *unsatisfiable* result.

This method for minimising the set v_1, v_2 is like modal back-jumping [6], and is the BDD counterpart of the learning technique from standard lazy SMT [1], but we use only an approximation of the dependency graph, as explained next.

The base case is when a modal jump closes immediately because the subsequent saturation returned the false-bdd. In this case we can (relatively cheaply) incrementally reconstruct the saturation phase from the modal variables, and explicitly determine a small subset that still yields the false-bdd.

In general, we will only have the responsible variables from beyond the jump. From this, we can only approximate which modal variables were responsible for those variables, as the details are hidden in the BDDs. Our approximation is to consider all the subformulae of a modal variable. If one of its subformulae is a responsible variable, then we consider it responsible, but we must also consider other modal variables with common subformulae, and so on for those variables.

$$\frac{\Diamond(b \vee \Diamond(a \wedge \neg a)); \Box(\neg b)}{\vdots} \quad \text{(jump 1)}$$

$$\frac{\Diamond(a \wedge \neg a); \neg b}{\vdots} \quad \text{(jump 2)}$$

$$\frac{\vdots}{\times} \quad \text{(id)}$$

Consider the above example in K. Jump 2 closes immediately due to $a \wedge \neg a$. This is our base case, and we would find that $v_{\Diamond(a \wedge \neg a)}$ is the only responsible variable. As there is only one branch, $v_{\Diamond(a \wedge \neg a)}$ would be returned to jump 1. We would then identify $v_{\Diamond(b \vee \Diamond(a \wedge \neg a))}$ as responsible, because its subformula $\Diamond(a \wedge \neg a)$ was returned as a responsible variable from beyond jump 1. We would also identify $v_{\Box(\neg b)}$ as responsible, as it shares the subformula b with $v_{\Diamond(b \vee \Diamond(a \wedge \neg a))}$.

Each BDD corresponds to a saturation phase. If the saturation phase contains an L-satisfiable branch, then we say that the corresponding BDD is L-satisfiable.

Sat Cache: We use a global cache of L-satisfiable BDDs. A new saturation (BDD) is immediately flagged as L-satisfiable if the cache already contains it (as a unique integer). The cache size is limited by a first-in-first-out removal policy.

Unsat Cache: We use a global cache of L-unsatisfiable modal jumps (the refinement BDD $\neg(v_0 \wedge v_1 \wedge \ldots)$ is cached). When creating a saturation, we iterate through this cache, and if the variables in a cached refinement are a subset of the variables of the saturation, we apply that refinement to the saturation. The subset check aids in avoiding introducing irrelevant variables. This cache is also size-limited, with a first-in-first-out removal policy.

Consider again the example shown under **Method**. We made a refinement of $[\neg \Diamond(b \wedge \neg b)]$, which we would cache. So if a future modal jump includes $\Diamond(b \wedge \neg b)$, we would immediately apply this refinement. From this point in time onward, no branches would ever be explored that contain $\Diamond(b \wedge \neg b)$.

BDD Unsat Cache: If we were not concerned about irrelevant variables we could skip the (potentially expensive) subset checks used in the unsatisfiable

Table 1. Results on the LWB benchmarks for K (left) and S4 (right)

subclass	BDDTab	BUC	SUC	NUC	Norm	Reorder	R to L	FaCT++	InKreSAT	*SAT
branch_n	18	18	18	18	18	19	17	10	13	12
branch_p	21	1	21	21	21	21	21	10	16	18
lin_n	21	21	21	21	21	21	21	21	21	14
path_n	21	2	21	21	21	21	21	21	10	21
path_p	21	3	21	21	21	21	21	21	10	21
ph_n	10	10	10	10	10	10	10	13	21	12
ph_p	9	9	9	9	10	9	9	7	9	9
poly_p	21	9	21	21	21	21	21	21	21	21
t4p_p	21	0	21	21	21	21	21	21	21	21

subclass	BDDTab	BUC	SUC	NUC	Norm	Reorder	R to L	FaCT++	InKreSAT
branch_n	19	19	19	19	19	19	18	6	12
grz_n	21	21	21	21	21	21	14	21	21
ipc_n	12	12	12	12	12	12	21	11	10
ipc_p	21	21	21	21	21	21	21	10	9
md_n	15	15	15	15	15	15	15	10	8
md_p	21	21	21	21	21	21	21	4	3
path_n	3	3	3	3	3	3	7	21	4
path_p	4	4	4	4	4	4	8	21	5
ph_n	5	5	5	5	5	5	4	8	14
ph_p	9	9	9	9	9	9	9	6	9
s5_n	21	21	21	21	21	21	21	21	16

cache above. Instead, we could maintain a single BDD of all the refinements and apply that to every modal jump.

Saturation Unsat Cache: Instead of leaves, we could cache saturated tableaux like we do for the satisfiable cache, which would also skip the subset checks.

Normalisation: The mapping from a formula φ to its BDD $[\varphi]$ is syntactic, so two semantically equivalent but syntactically different formulae can exist as two separate BDD variables. To avoid such duplications, we add an extra normalisation step where all modal formulae are constructed as BDDs. Since BDDs are canonical, and equality checking is just integer comparison, duplicate variables can then be easily identified and merged into one variable before the tableau process begins. This is similar to the syntactic normalisations (sorting and associativity) in other approaches [9], but is stronger, as it also catches distributivity.

Reordering: The order of the BDD variables can have a significant effect on the size of the constructed BDDs and thus on how long they take to construct. The BDD package we use, Buddy [2], provides automatic reordering, which periodically attempts to reorder the variables to reduce the number of BDD nodes.

Results. We evaluated BDDTab, against InKreSAT [7], FaCT++ [12] and *SAT [11], all with default settings, as representatives of the state-of-the-art.

We also evaluated several versions of our prover: **BDDTab** uses the optimisations RV, SC and UC, to which the other versions introduce one modification; **BUC** uses a single BDD for the unsatisfiable cache; **SUC** caches unsatisfiable saturations instead of leaves; **NUC** does not do any unsatisfiable caching at all; **Norm** does extra formula normalisation; **Reorder** uses automatic BDD variable reordering; and **R to L** explores saturated BDDs from right to left.

We have followed Kaminski and Tebbi [7] in their choice of benchmarks. Our tests were performed on an Intel 3.4GHz CPU with 8GB of memory.

Table 1 contains results for the Logic Work Bench (LWB) benchmarks for K and S4 [5]. Each subclass consists of a formula shape which takes an argument n

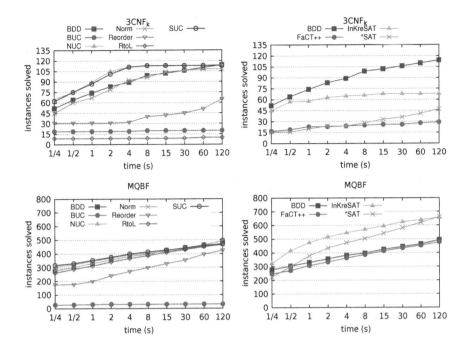

Fig. 1. 3CNF$_K$ and MQBF benchmarks

and generates a formula. As n increases, the formulae become larger. The entries in Table 1 mark the highest problem instance solved by each program, given a 30 second time limit per formula. For each subclass the best result is marked in bold. Classes where all provers solved all problems are omitted.

On the LWB benchmarks, BDDTab has arguably the best results, completing every subclass that two or more other provers complete, and gaining a higher score than all of them for several subclasses, including branch_n and _p for K, and branch_n, ipc_p, md_n and md_p for S4. BDDTab performs quite poorly on only a few subclasses, and at least one other prover also performs badly on these.

For the ph_n subclasses, this poor result is expected since the formulae are primarily propositional, and tableau-based approaches can halt once they find a single open branch. Our approach always computes the entirety of the saturation phase, and so wastes time finding every open branch, when only one is needed.

BDDTab's memory use for the ph subclasses grows exponentially with n, as expected, but is not always a significant factor in its performance on these subclasses. The first unsolved instance of S4 ph_p uses 2.7GB before timing out. On the first unsolved instances of the other ph subclasses it uses at most 50MB.

Figure 1 shows results for a set of randomly generated 3CNF$_K$ [3] formulae of modal depth 2, 4 and 6, as well as for a subset of the TANCS-2000 [8] Unbounded Modal QBF (MQBF) benchmarks for K. On the 3CNF$_K$, BDDTab shows the best performance by a large margin. On the MQBF benchmarks BDDTab performs just as well as FaCT++ and worse than InKreSAT and *SAT.

BUC is catastrophically worse over all the benchmarks than BDDTab, except for LWB S4. It is clear that this is due to irrelevant variables being introduced.

NUC performs similarly to BDDTab over the benchmarks, apart from the $3CNF_K$, where it performs much better; it solves the same instances, but is significantly faster on almost every single one, showing that the overhead of the subset checking can easily outweigh the benefits of using the cache, especially when the cache is large. On the MQBF each version is able to solve instances that the other could not, showing that the caching can be worthwhile.

SUC performs almost identically to NUC, which suggests that even cheap unsatisfiability caching does not provide a net benefit over these benchmarks.

Norm is almost identical to BDDTab. For LWB and MQBF the normalisation does not find any duplicate formulae, so the only difference is the time taken to perform the normalisation. These show that this is not necessarily an expensive procedure. In the $3CNF_K$ the normalisation typically reduces the number of variables by around 30%, but this doesn't translate into any net benefit.

Reorder is much worse than BDDTab on $3CNF_K$ and MQBF. Thus naive automatic reordering can be expensive without providing significant benefit.

R to L is perhaps the most striking result, since it just reverses an arbitrary ordering decision. On the LWB it performs significantly better than BDDTab on a number of classes and worse on some others. On the $3CNF_K$ it performs atrociously. This demonstrates the importance of these ordering decisions.

Description Logics. We tested BDDTab on classifying 15 $\mathcal{ALC}$ ontologies from public online collections. This required a trivial extension to multiple modalities, and we present only general results. Each ontology was translated into a single multi-modal K formula, and then BDDTab was asked to read this formula and perform an exhaustive n^2 check of all possible subsumptions of named classes. FaCT++ was asked to classify the original ontologies through OWL API.

Often, the majority of BDDTab's time is taken up by constructing the BDD representation of the TBox (the global assumptions), after which the actual classification is relatively quick. Reordering is very beneficial here, which is primarily because keeping the size of this BDD small makes it much faster to construct.

BDDTab's worst performance with reordering was two seconds, except for ABA-Adult-Mouse-Brain where, even with reordering, BDDTab failed to construct the global assumptions within 10 minutes. FaCT++ took at most 4 seconds to load and classify each. These results are not conclusive, but suggest it is possible for our method to handle large sets of global assumptions.

Conclusion and Further Work. Determining better ordering schemes and heuristics is needed. These include BDD variable ordering and reordering, modal jump exploration order, saturation exploration order and cache removal policies.

The approach could be extended to include features of other logics such as graded modalities or nominals. Each syntactic element would require its own BDD variable, and saturation phases would always be performed in one hit, but otherwise any high level tableau algorithm could use BDDs in this way.

The refinement process could be modified to only remove paths from the BDD, and avoid introducing new variables to existing paths.

Finally, our notion of "irrelevant variables" can be refined. DPLL-based modal reasoners (and lately SMT solvers) implement a technique called "pure-literal filtering": if a non-Boolean atom (e.g. a box or a diamond formula) occurs only positively [resp. negatively] in the *original* formula (*i.e.* before adding the $\neg(v_0 \wedge v_1 ...)$) and it is assigned negatively [resp. positively] in a branch, then the negated [resp. positive] atom can be ignored when recursively checking the branch. In our example, we can simply consider v_a in the final OBDD ignoring the "$\neg \diamond (...)$" since $\diamond(...)$ occurs only positively in the original formulae.

Our work shows that BDDs can be an effective base data structure for computing tableaux. The results explicitly show this for K, S4 and $\mathcal{ALC}$ with respect to the state-of-the-art as represented by FaCT++, InKreSAT and *SAT.

Acknowledgments. We thank an anonymous reviewer for many suggestions.

References

[1] Barrett, C.W., Sebastiani, R., Seshia, S.A., Tinelli, C.: Satisfiability modulo theories. In: Biere, A., Heule, M., van Maaren, H., Walsh, T. (eds.) Handbook of Satisfiability, Frontiers in Artificial Intelligence and Applications, vol. 185, pp. 825–885. IOS Press (2009)

[2] Buddy (2013), http://sourceforge.net/projects/buddy/

[3] Giunchiglia, F., Sebastiani, R.: Building decision procedures for modal logics from propositional decision procedures - the case study of modal K(m). Information and Computation 162(1/2) (October/November 2000)

[4] Goré, R., Thomson, J.: BDD-based automated reasoning for propositional bi-intuitionistic tense logics. In: Gramlich, B., Miller, D., Sattler, U. (eds.) IJCAR 2012. LNCS, vol. 7364, pp. 301 315. Springer, Heidelberg (2012)

[5] Heuerding, A., Schwendimann, S.: A benchmark method for the propositional modal logics K, KT, S4 (1996)

[6] Horrocks, I., Patel-Schneider, P.F.: Optimizing Description Logic Subsumption. Journal of Logic and Computation 9(3), 267–293 (1999)

[7] Kaminski, M., Tebbi, T.: Inkresat: Modal reasoning via incremental reduction to SAT. In: Bonacina, M.P. (ed.) CADE 2013. LNCS, vol. 7898, pp. 436–442. Springer, Heidelberg (2013)

[8] Massacci, F., Donini, F.M.: Design and results of TANCS-2000 non-classical (modal) systems comparison. In: Dyckhoff, R. (ed.) TABLEAUX 2000. LNCS, vol. 1847, pp. 52–56. Springer, Heidelberg (2000)

[9] Sebastiani, R., Tacchella, A.: SAT techniques for modal and description logics. In: Handbook of Satisfiability, Frontiers in Artificial Intelligence and Applications, vol. 185, pp. 781–824. IOS Press (2009)

[10] Sebastiani, R., Vescovi, M.: Automated Reasoning in Modal and Description Logics via SAT Encoding: the Case Study of K(m)/ALC-Satisfiability. Journal of Artificial Intelligence Research (JAIR) 35, 343–389 (2009)

[11] Tacchella, A.: *sat system description. In: Description Logics (1999)

[12] Tsarkov, D., Horrocks, I.: Fact++ description logic reasoner: System description. In: Furbach, U., Shankar, N. (eds.) IJCAR 2006. LNCS (LNAI), vol. 4130, pp. 292–297. Springer, Heidelberg (2006)

Approximations for Model Construction

Aleksandar Zeljić[1], Christoph M. Wintersteiger[2], and Philipp Rümmer[1]

[1] Uppsala University, Sweden
[2] Microsoft Research

Abstract. We consider the problem of efficiently computing models for satisfiable constraints, in the presence of complex background theories such as floating-point arithmetic. Model construction has various applications, for instance the automatic generation of test inputs. It is well-known that naive encoding of constraints into simpler theories (for instance, bit-vectors or propositional logic) can lead to a drastic increase in size, and be unsatisfactory in terms of memory and runtime needed for model construction. We define a framework for systematic application of approximations in order to speed up model construction. Our method is more general than previous techniques in the sense that approximations that are neither under- nor over-approximations can be used, and shows promising results in practice.

1 Introduction

The construction of satisfying assignments (or, more generally, models) for a set of given constraints is one of the most central problems in automated reasoning. Although the problem has been addressed extensively in research fields including constraint programming, and more lately satisfiability modulo theories (SMT), there are still constraint languages and background theories where effective model construction is challenging. Such theories are, in particular, arithmetic domains such as bit-vectors, nonlinear real arithmetic (or real-closed fields), and floating-point arithmetic (FPA); even when decidable, the high computational complexity of such languages turns model construction into a bottleneck in applications such as bounded model checking, white-box testcase generation, analysis of hybrid systems, and mathematical reasoning in general.

We follow a recent line of research that applies the concept of *abstraction* to model construction (e.g., [3,5,10,19]). In this setting, constraints are usually simplified prior to solving to obtain over- or under-approximations, or some combination thereof (*mixed abstractions*); experiments show that this concept can speed up model construction significantly. However, previous work in this area suffers from the fact that the definition of good over- and under-approximations can be difficult and limiting, for instance in the context of floating-point arithmetic. We argue that the focus on over- and under-approximations is neither necessary nor optimal: as a more flexible alternative, we present a general algorithm that can integrate *any form of approximation* into the model construction process, including approximations that cannot naturally be represented as a

S. Demri, D. Kapur, and C. Weidenbach (Eds.): IJCAR 2014, LNAI 8562, pp. 344–359, 2014.

combination of over- and under-approximation. Our method preserves essential properties like soundness, completeness, and termination.

For the purpose of empirical evaluation, we instantiate our model construction procedure for the domain of floating-point arithmetic, and present an evaluation based on an implementation thereof within the Z3 theorem prover [22]. Experiments with publicly available floating-point benchmarks show a uniform speed-up of about one order of magnitude compared to the naive bit-blasting-based decision procedure that is built into Z3 (on satisfiable benchmarks), and performance that is competitive with other state-of-the-art solvers for floating-point arithmetic.

The contributions of the paper are: 1. a general method for model construction based on approximations, 2. an instantiation of our framework for the theory of floating-point arithmetic, and 3. an experimental evaluation of our approach.

We would like to emphasize that the present paper focuses on the construction of models for satisfiable constraints. Although our framework can in principle show unsatisfiability of constraints, this is neither the goal, nor within the scope of the paper; we believe that further research is necessary to improve reasoning in the unsatisfiable case.

1.1 Motivating Example

We first illustrate our approach by considering a (strongly simplified) PI controller operating on floating-point data:

```
double Kp=1.0; double Ki=0.25; double set_point=20.0;
double integral = 0.0; double error;
for (int i = 0; i < N; ++i) {
  in       = read_input();
  error    = set_point - in;
  integral = integral + err;
  out      = Kp*err + Ki*integral;
  set_output(out);
}
```

All variables in this example range over double precision (64-bit) IEEE-754 floating-point numbers. The PI controller is initialized with the set_point value and the constants Kp and Ki. The controller reads input values via function read_input, and computes output values which control the system using the function set_output. The controller computes the control values (out) so that the input values are as close to set_point as possible. For simplicity, we assume that there is a bounded number N of control iterations.

Suppose we want to prove that if the input values stay within the range $18.0 \leq$ in ≤ 22.0, then the control values will stay within a range that we consider safe, e.g., $-3.0 \leq$ out $\leq +3.0$. This property is true of our controller only for two control iterations, but it can be violated within three iterations.

A bounded model checking approach to this problem produces a series of formulas, one for each N and checks the satisfiability of those formulas (usually in sequence). Today, most (precise) solvers for floating-point formulas implement this satisfiability check by means of *bit-blasting*, i.e., using a bit-precise encoding of FPA semantics as a propositional formula. Due to the complexity of FPA, the resulting formulas grow very quickly, and tend to overwhelm even the fastest SAT/SMT solvers. For example, an unrolling of the PI controller example to 30 steps cannot be solved by Z3 within an hour of runtime:

Bound N		1	2	5	10	20	30	40	50
Clauses	(millions)	0.28	0.66	1.80	3.71	7.53	11.34	15.15	18.97
Variables	"	0.04	0.09	0.25	0.51	1.04	1.57	2.10	2.63
Z3 solving time (s)		4	13	18	213	1068	>1h	...	

The example has the property, however, that full range of FP numbers is not required to find suitable program inputs; essentially a prover just needs to find a sequence of inputs such that the errors add up to a sum that is greater than 3.0. There is no need to consider numbers with large magnitude, or a large number of significant digits/bits. We postulate that this situation is typical for many applications. Since bit-precise treatment of FP numbers is clearly wasteful in this setting, we might consider some of the following alternatives:

- all operations in the program can be evaluated in **real** instead of FP arithmetic. For problems with only linear operations, such as the program at hand, this enables the use of highly efficient LP solvers. However, the encoding ignores the possibility of overflows or rounding errors; bounded model checking will in this way be neither sound nor complete. In addition, little is gained in terms of computational complexity for nonlinear constraints.
- operations can be evaluated in **fixed-point** arithmetic. Again, this encoding does not preserve the overflow- and rounding-semantics of FPA, but it enables solving using more efficient bit-vector encodings and solvers.
- operations can be evaluated in FPA with **reduced precision**: we can use single precision numbers, or even smaller formats.

Strictly speaking, soundness and completeness are lost in all three cases, since the precise nature of overflows and rounding in FPA is ignored. All three methods enable, however, the efficient computation of *approximate models,* which are likely to be "close" to genuine double-precision FPA models. In this paper, we define a general framework for model construction with approximations. In order to establish overall soundness and completeness, the framework contains a *model reconstruction* phase, in which approximate models are translated to precise models. This reconstruction may fail, in which case *refinement* is used to iteratively increase the precision of approximate models.

2   Related Work

Related work to our contribution falls into two categories: general abstraction and approximation frameworks, and specific decision procedures for FPA.

The concept of abstraction is central to software engineering and program verification and it is increasingly employed in general mathematical reasoning and in decision procedures. Usually, and in contrast to our work, only under- and over-approximations are considered, i.e., the formula that is solved either implies or is implied by an approximated formula. Counter-example guided abstraction refinement [7] is a general concept that is applied in many verification tools and decision procedures (e.g., in QBF [18] and MBQI for SMT [13]).

A general framework for abstracting decision procedures is Abstract CDCL, recently introduced by D'Silva et al. [10], which was also instantiated with great success for FPA [11,2]. This approach relies on the definition of suitable abstract domains for constraint propagation and learning. In our experimental evaluation, we compare to the FPA decision procedure in MathSAT, which is an instance of ACDCL. ACDCL could also be integrated with our framework, e.g., to solve approximations. A further framework for abstractions in theorem proving was proposed by Giunchiglia et al. [14]. Again, this work focusses on under- and over-approximations, not on other forms of approximation.

Specific instantiations of abstraction schemes in related areas also include the bit-vector abstractions by Bryant et al. [5] and Brummayer and Biere [4], as well as the (mixed) floating-point abstractions by Brillout et al. [3]. Van Khanh and Ogawa present over- and under-approximations for solving polynomials over reals [19]. Gao et al. [12] present a δ-complete decision procedure for nonlinear reals, considering over-approximations of constraints by means of δ-weakening.

There is a long history of formalising and analysing FPA concerns using proof assistants, among others in Coq by Melquiond [21] and HOL Light by Harrison [15]. Coq has also been integrated with a dedicated floating-point prover called Gappa by Boldo et al. [1], which is based on interval reasoning and forward error propagation to determine bounds on arithmetic expressions in programs [9]. The ASTRÉE static analyzer [8] features abstract interpretation-based analyses for FPA overflow and division-by-zero problems in ANSI-C programs. The SMT solvers MathSAT [6], Z3 [22], and Sonolar [20], all feature (bit-precise) conversions from floating-point to bit-vector constraints.

3   Preliminaries

We establish a formal basis in the context of multi-sorted first-order logic (e.g., [16]). A signature $\Sigma = (S, P, F, \alpha)$ consists of a set of sort symbols S, a set of sorted predicate symbols P, a set of sorted function symbols F, and a sort mapping α. Each predicate $p \in P$ is assigned a k-tuple $\alpha(p)$ of argument sorts (with $k \geq 0$); each function $f \in F$ is assigned a $(k + 1)$-tuple $\alpha(f)$ of sorts. We assume a countably infinite set X of variables, and (by abuse of notation) overload α to assign sorts also to variables. Given a multi-sorted signature Σ and variables X, the notions of well-sorted terms, atoms, literals, clauses, and formulas are defined as usual. The function $fv(\phi)$ denotes the set of free variables in a formula ϕ. In what follows, we assume that formulas are quantifier-free.

A Σ-structure $m = (U, I)$ with underlying universe U and interpretation function I maps each sort $s \in S$ to a non-empty set $I(s) \subseteq U$, each predicate

$p \in P$ of sorts $(s_1, s_2, \ldots, s_k)$ to a relation $I(p) \subseteq I(s_1) \times I(s_2) \times \ldots \times I(s_k)$, and each function $f \in F$ of sort $(s_1, s_2, \ldots, s_k, s_{k+1})$ to a set-theoretic function $I(f) : I(s_1) \times I(s_2) \times \ldots \times I(s_k) \to I(s_{k+1})$. A variable assignment β under a Σ-structure m maps each variable $x \in X$ to an element $\beta(x) \in I(\alpha(x))$. The valuation function $val_{m,\beta}(\cdot)$ is defined for terms and formulas in the usual way. A theory T is a pair (Σ, M) of a multi sorted signature Σ and a class of Σ-structures M. A formula ϕ is T-satisfiable if there is a structure $m \in M$ and a variable assignment β such that ϕ evaluates to *true*; we denote this by $m, \beta \models_T \phi$, and call β a T-solution of ϕ.

4   The Approximation Framework

We describe a decision procedure for problems ϕ over a set of variables X, using a theory T. The goal is to obtain a T-solution of ϕ. The main idea underlying our method is to replace the theory T with an *approximation theory* $\hat{T}$, which enables explicit control over the precision used to evaluate theory operations. In our method, the T-problem ϕ is first lifted to a $\hat{T}$-problem $\hat{\phi}$, then solved in the theory $\hat{T}$, and if a solution is found, it is translated back to a T-solution. The benefit of using the theory $\hat{T}$ is that different levels of approximation may be used during computation. We will use the theory of floating-point arithmetic as a running example for instantiation of the presented framework.

4.1   Approximation Theories

In order to formalize the approach of finding models by means of approximation, we construct the *approximation theory* $\hat{T} = (\hat{\Sigma}, \hat{M})$ from T, by extending function and predicate symbols with a new argument representing the *precision* of the approximation.

Syntax. We introduce a new sort for the precision s_p, and a new predicate symbol $\preceq$ which orders precision values. The signature $\hat{\Sigma} = (\hat{S}, \hat{P}, \hat{F}, \hat{\alpha})$ is obtained from Σ in the following manner: $\hat{S} = S \cup \{s_p\}$; the set of predicate symbols is extended with the new predicate symbol $\preceq$, $\hat{P} = P \cup \{\preceq\}$; the set of function symbols is extended with the new constant ω, representing the maximum precision value, $\hat{F} = F \cup \{\omega\}$; the sort function $\hat{\alpha}$ is defined in the following manner:

$$\hat{\alpha}(g) = \begin{cases} (s_p, s_1, s_2, \ldots, s_n) & \text{if } g \in P \cup F \text{ and } \alpha(g) = (s_1, s_2, \ldots, s_n) \\ (s_p, s_p) & \text{if } g = \preceq \\ (s_p) & \text{if } g = \omega \\ \alpha(g) & \text{otherwise} \end{cases}$$

Semantics. $\hat{\Sigma}$-structures $(\hat{U}, \hat{I})$ enrich the original Σ-structures by providing approximate versions of function and predicate symbols. The resulting operations can be under- or over-approximations, but they can also be approximations that are close to the original operations' semantics by some other metric. The degree of approximation is controlled with the help of the precision argument. We assume that the set $\hat{M}$ of $\hat{\Sigma}$-structures satisfies the following properties:

- for every structure $(\hat{U}, \hat{I}) \in \hat{M}$, the relation $\hat{I}(\preceq)$ is a partial order on $\hat{I}(s_p)$ that satisfies the ascending chain condition (every ascending chain is finite), and that has the unique greatest element $\hat{I}(\omega) \in \hat{I}(s_p)$;
- for every structure $(U, I) \in M$, an approximation structure $(\hat{U}, \hat{I}) \in \hat{M}$ extending (U, I) exists, together with an embedding $h : U \mapsto \hat{U}$ such that, for every sort $s \in S$, function $f \in F$, and predicate $p \in P$:

$$h(I(s)) \subseteq \hat{I}(s)$$

$$(a_1, \ldots, a_n) \in I(p) \iff (\hat{I}(\omega), h(a_1), \ldots, h(a_n)) \in \hat{I}(p) \quad (a_i \in I(\alpha(p)_i))$$

$$h(I(f)(a_1, \ldots, a_n)) = \hat{I}(f)(\hat{I}(\omega), h(a_1), \ldots, h(a_n)) \quad (a_i \in I(\alpha(f)_i))$$

- vice versa, for every approximation structure $(\hat{U}, \hat{I}) \in \hat{M}$ there is a structure $(U, I) \in M$ that can be embedded in $(\hat{U}, \hat{I})$ in the same way.

These properties ensure that every T-model has a corresponding $\hat{T}$-model, i.e. that no models are lost. Interpretations of function and predicate symbols under $\hat{I}$ with maximal precision are isomorphic to their original interpretation under I. The interpretation $\hat{I}$ should interpret the function and predicate symbols in such a way that their interpretations for a given value of the precision argument approximate the interpretations of the corresponding function and predicate symbols under I.

Applied to FPA. The IEEE-754 standard for floating point numbers [17] defines floating point numbers, their representation in bit-vectors, and the corresponding operations. Most crucially, bit-vectors of various sizes are used to represent the significand and the exponent of numbers; e.g., double-precision floating-point numbers are represented by using 11 bits for the exponent and 53 bits for the significand. We denote the set of floating-point numbers that can be represented using s significand bits and e exponent bits by $FP_{s,e}$. Note that FP domains are growing monotonically when increasing e or s, i.e., $FP_{s',e'} \subseteq FP_{s,e}$ provided that $s' \leq s$ and $e' \leq e$.

For fixed values e of exponent bits and s of significant bits, FPA can be modeled as a theory in our sense. We denote this theory by $TF_{s,e}$, and write s_f for the sort of FP numbers, and s_r for the sort of rounding modes. The various FP operations are represented as functions and predicates of the theory; for instance, floating-point addition turns into the function symbol $\oplus$ with signature $\alpha(\oplus) = (s_p, s_r, s_f, s_f)$. The semantics of $TF_{s,e}$ is defined by a unique structure $(U_{s,e}, I_{s,e})$; in particular, $I_{s,e}(s_f) = FP_{s,e}$.

We construct the approximation theory $\hat{TF}_{s,e}$ by introducing the precision sort s_p, predicate symbol $\preceq$, and a constant symbol ω. The function and predicate symbols have their signature changed to include the precision argument. For example, the signature of the floating-point addition symbol $\oplus$ is $\hat{\alpha}(\oplus) = (s_p, s_r, s_f, s_f)$ in the approximation theory.

The semantics of the approximation theory $\hat{TF}_{s,e}$ is again defined through a singleton set $\hat{M} = \{(\hat{U}_{s,e}, \hat{I}_{s,e})\}$ of structures. The universe of the approximation theory extends the original universe with a set of integers which are the domain of the precision sort, i.e., $\hat{U}_{s,e} = U_{s,e} \cup \{0, 1, \ldots, n\}$, $\hat{I}_{s,e}(s_p) = \{0, 1, \ldots, n\}$, and $\hat{I}_{s,e}(\omega) = n$. The embedding h is the identity mapping.

In order to use precision to regulate the semantics of FP operations, we introduce the notation $(s, e) \downarrow p$ to denote the number of bits in reduced precision $p \in \{0, 1, \ldots, n\}$; for instance, the reduced bit-widths can be defined as $(s, e) \downarrow p = (\lceil s \cdot \frac{p}{n} \rceil, \lceil e \cdot \frac{p}{n} \rceil)$. The approximate semantics of functions is derived from the FP semantics for the reduced bit-widths. For example, $\oplus$ in approximation theory $\hat{TF}_{s,e}$ is defined as:

$$\hat{I}_{s,e}(\oplus)(p, r, a, b) = cast_{s,e}(I_{(s,e)\downarrow p}(\oplus)(r, cast_{(s,e)\downarrow p}(a), cast_{(s,e)\downarrow p}(b)))$$

This definition uses a function $cast_{s,e}$ to map any FP number to a number with s significand bits and e exponent bits, i.e., $cast_{s,e}(a) \in FP_{s,e}$ for any $a \in FP_{s',e'}$. The cast performs rounding (if required) using a fixed rounding mode. Note that many occurrences of $cast_{s,e}$ can be eliminated in practice, if they only concern intermediate results. For example, consider $\oplus(c_1, \otimes(c_2, a_1, a_2), a_3)$. The result of $\otimes(c_2, a_1, a_2)$ can be directly cast to precision c_1 without the need of casting up to full precision when calculating the value of the expression.

4.2 Lifting Constraints to Approximate Constraints

In order to solve a constraint ϕ using an approximation theory $\hat{T}$, it is first necessary to lift ϕ to an extended constraint $\hat{\phi}$ that includes explicit variables c_l for the precision of each operation. This is done by means of a simple traversal of ϕ, using a recursive function L that receives a formula (or term) ϕ and a position $l \in \mathbb{N}^*$ as argument. For every position l, the symbol c_l denotes a fresh variable of the precision sort $\alpha(c_l) = s_p$ and we define

$$\begin{aligned}
L(l, \neg\phi) &= \neg L(l.1, \phi) \\
L(l, \phi \circ \psi) &= L(l.1, \phi) \circ L(l.2, \psi) & (\circ \in \{\vee, \wedge\}) \\
L(l, x) &= x & (x \in X) \\
L(l, g(t_1, \ldots, t_n)) &= g(c_l, L(l.1, t_1), \ldots, L(l.n, t_n)) & (g \in F \cup P)
\end{aligned}$$

Then we obtain the lifted formula $\hat{\phi} = L(\epsilon, \phi)$, where ϵ denotes an empty word. Since T-structures can be embedded into $\hat{T}$-structures, it is clear that no models are lost as a result of lifting:

Lemma 1 (Completeness). *If a T-constraint ϕ is T-satisfiable, then the lifted constraint $\hat{\phi} = L(\epsilon, \phi)$ is $\hat{T}$-satisfiable as well.*

An approximate model that chooses full precision for all operations induces a model for the original constraint:

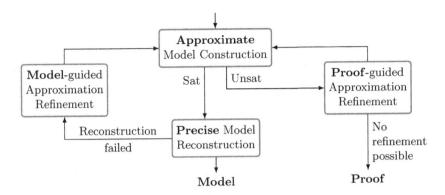

Fig. 1. The model construction process

Lemma 2 (Fully precise operations). *Let $\hat{m} = (\hat{U}, \hat{I})$ be a $\hat{T}$-model, and $\hat{\beta}$ be a variable assignment. If $\hat{m}, \hat{\beta} \models_{\hat{T}} \hat{\phi}$ for an approximate constraint $\hat{\phi} = L(\epsilon, \phi)$, then $m, \beta \models_T \phi$, provided that: 1. there is a T-structure m embedded in $\hat{m}$ via h, and a variable assignment β such that $h(\beta(x)) = \hat{\beta}(x)$ for all variables $x \in fv(\phi)$, and 2. $\hat{\beta}(c_l) = \hat{I}(\omega)$ for all precision variables c_l introduced by L.*

The fully precise case however, is not the only case in which an approximate model is easily translated to a precise model. For instance, approximate operations might still yield a precise result for some arguments. Examples of this are constraints in floating-point arithmetic that have small integer solutions or fixed-point arithmetic solutions.

Theorem 1 (Precise evaluation). *Suppose $\hat{m}, \hat{\beta} \models_{\hat{T}} \hat{\phi}$ for an approximate constraint $\hat{\phi} = L(\epsilon, \phi)$, such that all operations in $\hat{\phi}$ are performed exactly with respect to T. Then $\hat{m}, \hat{\beta} \models_T \phi$.*

5   Model Refinement Scheme

In the following sections, we will use the approximation framework to successively construct more and more precise solutions of given constraints, until eventually either a genuine solution is found, or the constraints are determined to be unsatisfiable. We fix a partially ordered precision domain $(D_p, \preceq_p)$ (where, as before, $\preceq_p$ satisfies the ascending chain condition, and has a greatest element), and consider approximation structures $(\hat{U}, \hat{I})$ such that $\hat{I}(s_p) = D_P$ and $\hat{I}(\preceq) = \preceq_p$.

Given a lifted constraint $\hat{\phi} = L(\epsilon, \phi)$, let $X_p \subseteq X$ be the set of precision variables introduced by the function L. A *precision assignment* $\gamma : X_p \to D_p$ maps the precision variables to precision values. We write $\gamma \preceq_p \gamma'$ if for all variables $c_l \in X_p$ we have $\gamma(c_l) \preceq_p \gamma'(c_l)$. Precision assignments are partially ordered by $\preceq_p$. There is a greatest precision assignment, which maps each precision variable to the greatest element of the precision domain D_p.

The proposed procedure is outlined in Fig. 1. First, an initial precision assignment γ is chosen, depending on the theory T. In *Approximate Model Construction,* the procedure tries to find $(\hat{m}, \hat{\beta})$, a model of the approximated constraint $\hat{\phi}$. If $(\hat{m}, \hat{\beta})$ is found, *Precise Model Reconstruction* tries to translate it to (m, β), a model of the original constraint ϕ. If this succeeds, the procedure stops and returns a model. Otherwise, *Model-guided Approximation Refinement* uses (m, β) and $(\hat{m}, \hat{\beta})$ to increase the precision assignment γ. If Approximate Model Construction cannot find any model $(\hat{m}, \hat{\beta})$, then *Proof-guided Approximation Refinement* decides how to modify the precision assignment γ. If the precision assignment is maximal and cannot be further increased, the procedure has determined unsatisfiability. In the following sections we provide additional details for each of the components of our procedure.

General Properties. Since $\preceq_p$ has the ascending chain property, our procedure is guaranteed to terminate and either produce a genuine precise model, or detect unsatisfiability of the constraints. The potential benefits of this approach are that it often takes less time to solve multiple smaller (approximate) problems than to solve the full problem straight away. The candidate models provide useful hints for the following iterations. The downside is that it might be necessary to solve the whole problem eventually anyway, which is the case, for instance, for unsatisfiable problems. Therefore, our approach is mainly useful when the goal is to obtain a model, e.g., when searching for counter-examples.

5.1   Approximate Model Construction

Once a precision assignment γ has been fixed, existing solvers for the operations in the approximation theory can be used to construct a model $\hat{m}$ and a variable assignment $\hat{\beta}$ s.t. $\hat{m}, \hat{\beta} \models_{\hat{T}} \hat{\phi}$. It is necessary that $\hat{\beta}$ and γ agree on X_p. As an optimisation, the model search can be formulated in various theory-dependent ways which heuristically benefit the Precise Model Reconstruction. For example, search can prefer models with small values of some error criterion, or first attempt to find models that are similar to models found in earlier iterations.

Applied to FPA. Since our FP approximations are again formulated using FP semantics, any solver for FPA can be used for Approximate Model Construction. In our implementation, the lifted constraints $\hat{\phi}$ of $\hat{T}F_{s,e}$ are encoded in bit-vector arithmetic, and then bit-blasted and solved using a SAT solver. The encoding of a particular function or predicate symbol uses the precision argument to determine the floating-point domain of the interpretation. This kind of approximation reduces the size of the encoding of each operation, and results in smaller problems handed over to the SAT solver. An example of theory-specific optimisation of the model search is to look for models where no rounding occurs during the calculation.

Algorithm 1. Model reconstruction

---

1 $(m, h) := \texttt{extract\_Tstructure}(\hat{m})$;

2 $lits := \texttt{extract\_asserted\_literals}(\hat{m}, \hat{\beta}, \hat{\phi})$;

3 **for** $l \in lits$ **do**

4 | $(m, \beta) := \texttt{extend\_model}(l, \beta, h, \hat{\beta}, \hat{m})$;

5 **end**

6 $\texttt{complete}(\beta, \hat{\beta})$;

7 **return** (m, β);

5.2 Reconstructing Precise Models

In the model reconstruction phase, our procedure attempts to produce a model (m, β) for the original formula ϕ from an approximate model $(\hat{m}, \hat{\beta})$ obtained by solving $\hat{\phi}$. Since we consider arbitrary approximations (which might be neither over- nor under-), this translation is non-trivial; for instance, approximate and precise operations might exhibit different rounding behavior. In practice, it might still be possible to 'patch' approximate models that are close to real models, avoiding further refinement iterations.

First, note that by definition it is possible to embed a T-structure m in $\hat{m}$; the structure m and the embedding h are retrieved from $\hat{m}$ via $\texttt{extract\_Tstructure}$ in Alg. 1. The structure m and h will be used to evaluate ϕ using values from $\hat{\beta}$.

The function $\texttt{extract\_asserted\_literals}$ determines a set $lits$ of literals in $\hat{\phi}$ that are true under $(\hat{m}, \hat{\beta})$, such that the conjunction $\bigwedge lits$ implies $\hat{\phi}$. For instance, if $\hat{\phi}$ is in CNF, one literal per clause can be selected that is true under $(\hat{m}, \hat{\beta})$. Any pair (m, β) that satisfies the literals in $lits$ will be a T-model of ϕ.

The procedure then iterates over $lits$, and successively constructs a valuation $\beta : X \to U$ such that (m, β) satisfies all selected literals, and therefore is a model of ϕ ($\texttt{extend\_model}$). During this loop, we assume that β is a *partial* valuation and only defined for some of the variables in X. We use the notation $\beta \uparrow h$ to lift β from m to $\hat{m}$, setting all precision variables to greatest precision, defined by

$$(\beta \uparrow h)(x) = \begin{cases} \hat{I}(\omega) & \text{if } x \in X_p \\ h(\beta(x)) & \text{otherwise .} \end{cases}$$

The precise implementation of $\texttt{extend\_model}$ is theory-specific. In general, the function first attempts to evaluate a literal l as $val_{\hat{m}, \beta \uparrow h}(l)$. If this fails, the valuation β has to be extended, for instance by including values $\hat{\beta}(x)$ for variables x not yet assigned in β.

After all literals have been successfully asserted, β may be incomplete, so we complete it by mapping value assignments from $\hat{\beta}$ and return the model (m, β). Note, that if all the asserted literals already have maximum precision assigned then, by Lemma 2, model reconstruction cannot fail.

Applied to FPA. The function $\texttt{extract\_Tstructure}$ is trivial for our FPA approximations, since m and $\hat{m}$ coincide for the sort s_f of FP numbers. Further, by approximating FPA using smaller domains of FP numbers, all of which are subsets of the original domain, reconstruction of models is easy in some cases and boils down to padding the obtained values with zero bits. The more difficult case concerns literals with rounding in approximate FP semantics, since a significant error emerges when the literal is re-interpreted using higher-precision FP numbers. A useful optimization is special treatment of equalities $x = t$ in which one side is a variable x not assigned in β, and all right-hand side variables are assigned. In this case, the choice $\beta(x) := val_{\hat{m},\beta\uparrow h}(t)$ will satisfy the equation. Use of this heuristic partly mitigates the negative impact of rounding in approximate FP semantics, since the errors originating in the $(\hat{m}, \hat{\beta})$ will not be present in (m, β). The heuristic is not specific to the floating-point theory, and can be carried over to other theories as well.

5.3   Approximation Refinement

The overall goal of the refinement scheme outlined in Fig. 1 is to find a model of the original constraints using a series of approximations defined by precision assignments γ. We usually want γ to be as small as possible in the partial order of precision assignments, since approximations with lower precision can be solved more efficiently. During refinement, the precision assignment is adjusted so that the approximation of the problem in the next iteration is closer to full semantics. Intuitively, this increase in precision should be kept as small as possible, but as large as necessary. Note that two different refinement procedures are required, depending on whether an approximation is satisfiable or not. We refer to these procedures as Model- and Proof-guided Approximation Refinement, respectively.

Model-Guided Approximation Refinement is performed after obtaining a model $(\hat{m}, \hat{\beta})$ of $\hat{\phi}$, together with a reconstructed model (m, β) that does *not* satisfy ϕ. We use the procedure described in Alg. 2 for adjusting γ in this situation. Since the model reconstruction failed, there are literals in $\hat{\phi}$ which are critical for $(\hat{m}, \hat{\beta})$, in the sense that they are satisfied by $(\hat{m}, \hat{\beta})$ and required to satisfy $\hat{\phi}$, but are not satisfied by (m, β). Such literals can be identified through evaluation with both $(\hat{m}, \hat{\beta})$ and (m, β) (choose_critical_literals), and can then be traversed, evaluating each sub-term under both structures. If a term $g(c_l, \bar{t})$ is assigned different values in the two models, it witnesses discrepancies between precise and approximate semantics; in this case, an error is computed using the error function, mapping to some suitably defined error domain (e.g., the real numbers $\mathbb{R}$ for errors represented numerically). The computed errors are then used to select those operations whose precision argument c_l should be assigned a higher value.

Depending on refinement criteria, the rank_terms function can be implemented in different ways. For example, terms can be ordered according to the absolute error which was calculated earlier; if there are too many terms to refine, only a certain number of them will be selected for refinement. An example of a more complex criterion follows:

Algorithm 2. Model-guided Approximation Refinement

---

1 $lits := \texttt{choose\_critical\_literals}(\hat{m}, \hat{\beta}, \beta, \hat{\phi});$
2 **for** $l \in lits$ **do**
3 $\quad$ **for** $g(c_l, \bar{t}) \in \texttt{ordered\_subterms}(l)$ **do**
4 $\quad\quad$ **if** $val_{\hat{m},\hat{\beta}}(g(c_l,\bar{t})) \neq val_{\hat{m},\beta\uparrow h}(g(\omega,\bar{t}))$ **then**
5 $\quad\quad\quad$ $\Delta(c_l) := \texttt{error}(val_{\hat{m},\hat{\beta}}(g(c_l,\bar{t})), val_{\hat{m},\beta\uparrow h}(g(\omega,\bar{t}));$
6 $\quad\quad$ **end**
7 $\quad$ **end**
8 **end**
9 $chosenTerms := \texttt{rank\_terms}(\Delta);$
10 $\gamma := \texttt{refine}(\gamma, chosenTerms);$

Error-based selection aims at refining the terms introducing the greatest imprecision first. The absolute error of an expression is determined by the errors of its sub-terms, and the error introduced by approximation of the operation itself. By calculating the ratio between output and input error, refinement tries to select those operations that cause the biggest *increase* in error. If we assume that theory T is some numerical theory (i.e., it can be mapped to reals in a straightforward manner), then we can define the $\texttt{error}$ function (in Alg. 2) as absolute difference between its arguments. Then $\Delta(c_l)$ represents the *absolute error* of the term $g(c_l, \bar{t})$. This allows us to define the *relative error* $\delta(c_l)$ of the term $g(c_l, \bar{t})$ in the following way:

$$\delta(c_l) = \frac{\Delta(c_l)}{|val_{\hat{m},\beta\uparrow h}(g(\omega,\bar{t}))|}$$

Similar measures can be defined for non-numeric theories.

Since a term can have multiple sub-terms, we calculate the average relative input error; alternatively, minimum or maximum input errors could be used. We obtain a function characterizing increase in error caused by an operation:

$$errInc(c_l) = \frac{\delta(c_l)}{1 + \frac{1}{k}\Sigma_{i=1}^{k}\delta(c_{l.i})} \ ,$$

where $g(c_l, \bar{t})$ represents the term being ranked. The function $\texttt{rank\_terms}$ then selects terms $g(c_l, \bar{t})$ with maximum error increase $errInc(c_l)$.

Applied to FPA. The only difference to the general case is that we define relative error $\delta(c_l)$ to be $+\infty$ if a special value ($\pm\infty$, NaN) from $(\hat{m}, \hat{\beta})$ turns into a normal value under (m, β). Our $\texttt{rank\_terms}$ function ignores terms which have an infinite average relative error of sub-terms. The refinement strategy will prioritize the terms which introduce the largest error, but in the case of special values it will refine the first imprecise terms that are encountered (in bottom up evaluation), because once the special values occur as input error to a term we have no way to estimate its actual error. After ranking the terms using the described criteria $\texttt{rank\_terms}$ returns the top 30% highest ranked terms. The precision of chosen terms is increased by a constant value.

Proof-Guided Approximation Refinement. When no approximate model can be found, some theory solvers may still provide valuable information why the problem could not be satisfied; for instance, proofs of unsatisfiability or unsatisfiable cores. While it may be (computationally) hard to determine which variables absolutely need to be refined in this case (and by how much), in many cases a tight estimate is easy to compute. For instance, it is possible to increase the precision of all variables appearing in the literals of an unsatisfiable core.

6 Experimental Evaluation

To assess the efficacy of our method, we present results of an experimental evaluation obtained through an implementation of the approximation using smaller floating-point numbers. We implemented this approach as a custom tactic [23] within the Z3 theorem prover [22]. All experiments were performed on Intel Xeon 2.5 GHz machines with a time limit of 1200 sec and a memory limit of 2 GB. The symbols T/o and M/o indicate that the time or the memory limit were exceeded.

Implementation Details. For the sake of reproducibility of our experiments, we note that our implementation starts with an initial precision mapping γ that limits the precision of all floating-point operations to $s = 3$ significand and $e = 3$ exponent bits. Upon refinement, operations receive an increase in precision that represents 20% of the width of the full precision. We do not currently implement any sophisticated proof-guided approximation refinement, but simply increase the precision of all operations by a constant when an approximation is determined unsatisfiable.

Evaluation. Our benchmarks are taken from a recent evaluation of the ACDCL-based MathSAT, by Brain et al. [2]. This benchmark set contains 213 benchmarks, both satisfiable and unsatisfiable ones. The benchmarks originate from verification problems of C programs performing numerical computations, where ranges and error bounds of variables and expressions are verified; other benchmarks are randomly generated systems of inequalities over bounded floating-point variables. We compare against Z3 and MathSAT.

The results we obtain are briefly summarized in Table 1, which shows that our approximation solves more satisfiable instances than other solvers, but the least number of unsatisfiable problems. This is expected, as our approximation scheme does not yet in-

Table 1. Evaluation Statistics

	Z3	MathSAT	Approx.
SAT	76	76	**86**
UNSAT	56	**76**	46

corporate any specialized refinement techniques for unsatisfiable formulas. Fig. 2 provides more detailed results, which show that on satisfiable formulas, our approach is about one order of magnitude faster than Z3. In comparison to MathSAT, the picture is less clear (right side of Fig. 2): while our approximation solves a number of satisfiable problems that are hard for MathSAT, it requires more time than MathSAT on other problems.

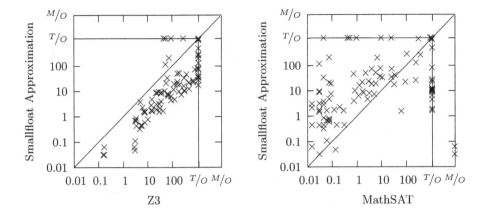

Fig. 2. Comparisons of our method with other tools (on satisfiable instances)

Overall, it can be observed that our approximation method leads to significant improvements in solver performance, especially where satisfiable formulas are concerned. Our method exhibits complementary performance to the ACDCL procedure in MathSAT; one of the aspects to be investigated in future work is a possible combination of the two methods, using an ACDCL solver to solve the constraints obtained through approximation with our procedure.

7 Conclusion

We present a general method for efficient model construction through the use of approximations. By computing a model of a formula interpreted in suitably approximated semantics, followed by reconstruction of a genuine model in the original semantics, scalability of existing decision procedures is improved for complex background theories. Our method uses a refinement procedure to increase the precision of the approximation on demand. Finally, we show that an instantiation of our framework for floating-point arithmetic shows promising results in practice and often outperforms state-of-the-art solvers.

We plan to further extend the procedure presented here, in particular considering other theories, other approximations, and addressing the case of unsatisfiable constraints. Furthermore, it is possible to solve approximations with different precision assignments in parallel, and use the refinement information from multiple models (or proofs) simultaneously. Increases in precision may then be adjusted based on differences in precision between models, or depending on the runtime required to solve each of the approximations.

Acknowledgments. We would like to thank Alberto Griggio for assistance with MathSAT and help with the benchmarks in our experiments, as well as the anonymous referees for insightful comments.

References

1. Boldo, S., Filliâtre, J.-C., Melquiond, G.: Combining Coq and Gappa for certifying floating-point programs. In: Carette, J., Dixon, L., Coen, C.S., Watt, S.M. (eds.) MKM 2009, Held as Part of CICM 2009. LNCS, vol. 5625, pp. 59–74. Springer, Heidelberg (2009)

2. Brain, M., D'Silva, V., Griggio, A., Haller, L., Kroening, D.: Deciding floating-point logic with abstract conflict driven clause learning. In: FMSD (2013)

3. Brillout, A., Kroening, D., Wahl, T.: Mixed abstractions for floating-point arithmetic. In: FMCAD. IEEE (2009)

4. Brummayer, R., Biere, A.: Effective bit-width and under-approximation. In: Moreno-Díaz, R., Pichler, F., Quesada-Arencibia, A. (eds.) EUROCAST 2009. LNCS, vol. 5717, pp. 304–311. Springer, Heidelberg (2009)

5. Bryant, R.E., Kroening, D., Ouaknine, J., Seshia, S.A., Strichman, O., Brady, B.A.: Deciding bit-vector arithmetic with abstraction. In: Grumberg, O., Huth, M. (eds.) TACAS 2007. LNCS, vol. 4424, pp. 358–372. Springer, Heidelberg (2007)

6. Cimatti, A., Griggio, A., Schaafsma, B.J., Sebastiani, R.: The MathSAT5 SMT solver. In: Piterman, N., Smolka, S.A. (eds.) TACAS 2013 (ETAPS 2013). LNCS, vol. 7795, pp. 93–107. Springer, Heidelberg (2013)

7. Clarke, E.M., Grumberg, O., Jha, S., Lu, Y., Veith, H.: Counterexample-guided abstraction refinement. In: Emerson, E.A., Sistla, A.P. (eds.) CAV 2000. LNCS, vol. 1855, Springer, Heidelberg (2000)

8. Cousot, P., Cousot, R., Feret, J., Mauborgne, L., Miné, A., Monniaux, D., Rival, X.: The ASTREÉ analyzer. In: Sagiv, M. (ed.) ESOP 2005. LNCS, vol. 3444, pp. 21–30. Springer, Heidelberg (2005)

9. Daumas, M., Melquiond, G.: Certification of bounds on expressions involving rounded operators. ACM Trans. Math. Softw. 37(1) (2010)

10. D'Silva, V., Haller, L., Kroening, D.: Abstract conflict driven learning. In: POPL. ACM (2013)

11. D'Silva, V., Haller, L., Kroening, D., Tautschnig, M.: Numeric bounds analysis with conflict-driven learning. In: Flanagan, C., König, B. (eds.) TACAS 2012. LNCS, vol. 7214, pp. 48–63. Springer, Heidelberg (2012)

12. Gao, S., Kong, S., Clarke, E.M.: dReal: An SMT solver for nonlinear theories over the reals. In: Bonacina, M.P. (ed.) CADE 2013. LNCS, vol. 7898, pp. 208–214. Springer, Heidelberg (2013)

13. Ge, Y., de Moura, L.: Complete instantiation for quantified formulas in satisfiability modulo theories. In: Bouajjani, A., Maler, O. (eds.) CAV 2009. LNCS, vol. 5643, pp. 306–320. Springer, Heidelberg (2009)

14. Giunchiglia, F., Walsh, T.: A theory of abstraction. Artif. Intell. 57(2-3) (1992)

15. Harrison, J.: Floating point verification in HOL Light: the exponential function. TR 428, University of Cambridge Computer Laboratory (1997), available on the Web as http://www.cl.cam.ac.uk/~jrh13/papers/tang.html

16. Harrison, J.: Handbook of Practical Logic and Automated Reasoning. Cambridge University Press (2009)

17. IEEE Comp. Soc.: IEEE Standard for Floating-Point Arithmetic 754-2008 (2008)

18. Janota, M., Klieber, W., Marques-Silva, J., Clarke, E.: Solving QBF with counterexample guided refinement. In: Cimatti, A., Sebastiani, R. (eds.) SAT 2012. LNCS, vol. 7317, pp. 114–128. Springer, Heidelberg (2012)

19. Khanh, T.V., Ogawa, M.: SMT for polynomial constraints on real numbers. In: TAPAS. Electronic Notes in Theoretical Computer Science, vol. 289 (2012)

20. Lapschies, F., Peleska, J., Gorbachuk, E., Mangels, T.: SONOLAR SMT-solver. In: Satisfiability Modulo Theories Competition; System Description (2012)
21. Melquiond, G.: Floating-point arithmetic in the Coq system. In: Conf. on Real Numbers and Computers. Information & Computation, vol. 216. Elsevier (2012)
22. de Moura, L., Bjørner, N.S.: Z3: An efficient SMT solver. In: Ramakrishnan, C.R., Rehof, J. (eds.) TACAS 2008. LNCS, vol. 4963, pp. 337–340. Springer, Heidelberg (2008)
23. de Moura, L., Passmore, G.O.: The strategy challenge in SMT solving. In: Bonacina, M.P., Stickel, M.E. (eds.) Automated Reasoning and Mathematics. LNCS, vol. 7788, pp. 15–44. Springer, Heidelberg (2013)

A Tool That Incrementally Approximates Finite Satisfiability in Full Interval Temporal Logic*

Rüdiger Ehlers[1,2] and Martin Lange[1]

[1] School of Electrical Engineering and Computer Science, University of Kassel, Germany
[2] University of Bremen and DFKI GmbH, Bremen, Germany

Abstract. Interval Temporal Logic (ITL) is a powerful formalism to reason about sequences of events that can occur simultaneously and in an overlapping fashion. Despite its importance for various application domains, little tool support for automated ITL reasoning is available, possibly also owed to ITL's undecidability.

We consider bounded satisfiability which approximates finite satisfiability and is only NP-complete. We provide an encoding into SAT that is designed to use the power of modern incremental SAT solvers. We present a tool that tests an ITL specification for finite satisfiability.

1 Introduction

Propositional Interval Temporal Logic (ITL) [13,9] is a modal logic that is interpreted over interval structures which enrich the natural numbers with propositional evaluations of all its intervals. Its modalities are obtained from Allen's relations on intervals [1]. Thus, it can make assertions like "every right-neighbouring interval contains an interval which ..." etc.

Despite many claims about the importance of ITL in various areas like hardware verification, A.I. planning etc., there is little tool support for automatically checking the satisfiability of an ITL formula. On one hand this may be caused by ITL's undecidability. This issue has been studied extensively together with questions regarding the expressive power and axiomatisability of ITL and its fragments, naturally obtained by restricting it to a subset of Allen's interval relations [10,12,11]. The complexity of their satisfiability problems varies between NP and undecidability depending on the combination of relations chosen.

There is an implementation of a tableau-based procedure [5] for the Right Neighbourhood fragment only. This is quite a weak fragment featuring a single modality only. The same fragment has also been targeted with an approach based on evolutionary algorithms [4]. This constitutes a sound but incomplete approximation method for the *finite satisfiability problem*, i.e. the question of whether or not a formula is satisfied by an interval structure based on a finite prefix of the natural numbers.

Approximative solutions can help to tackle difficult (like undecidable or just very hard) problems. One such method that is particularly successful in the area of hardware verification is *bounded model checking* [7]. It approximates a PSPACE-hard model

---

* The European Research Council has provided financial support under the European Community's Seventh Framework Programme (FP7/2007-2013) / ERC grant agreement no 259267.

S. Demri, D. Kapur, and C. Weidenbach (Eds.): IJCAR 2014, LNAI 8562, pp. 360–366, 2014.

checking problem through successive calls to a problem in NP, i.e. one that easily reduces to the satisfiability problem for propositional logic (SAT). It has been shown that such approximations can also yield useful approaches to even undecidable problems [2].

Here we report on an implementation of a similar method for the finite satisfiability problem for ITL. We show how to encode the problem of deciding whether or not a given ITL formula φ has a model of length k by a propositional formula of size polynomial in $|\varphi|$ and $|k|$. The approximation then iterates through increasing lengths k, thus being able to report satisfiability but not unsatisfiability. The encoding is *incremental*, i.e. the SAT formula for length $k + 1$ can be obtained using the SAT formula for length k as starting point and does not have to be computed from scratch. This approach bears the following advantages.

- It aims at *efficiency* by using modern SAT solvers and thus benefits from developments in this area.
- It covers the *entire* ITL unlike the few implementations described above.
- It is *extensible*; further logical operators like those of the Duration Calculus [6] or CDT [15] could easily be integrated into the encoding.

2   Interval Temporal Logic

As usual, we write $[i, j]$ when $i \le j$ for the interval of natural numbers between i and j inclusively. For an $n \in \mathbb{N}$ let $I(n) = \{[i, j] \mid 0 \le i \le j < n\}$ be the set of all intervals with upper bound less than n. Let $\mathcal{P} = \{p, q, \ldots\}$ be a set of atomic propositions. A *(finite) interval structure* (over $\mathcal{P}$) is a pair $\mathcal{I} = (n, \vartheta)$ with $n \in \mathbb{N}$ and $\vartheta : I(n) \to 2^{\mathcal{P}}$. We call n the *length* of the interval structure $\mathcal{I}$.

There are twelve relations on intervals over a linear order, known as Allen's relations [1], which describe their relative position on this linear order. Here we consider four of them defined by $[i, j]\ B\ [i', j']$ iff $i = i' \wedge j' < j$ ("started-by"); $[i, j]\ E\ [i', j']$ iff $i < i' \wedge j' = j$ ("finished-by"); as well as their inverses $\bar{B}$ and $\bar{E}$ where $[i, j]\ \bar{r}\ [i', j']$ iff $[i', j']\ r\ [i, j]$.

Formulas of ITL in positive normal form over the set $\mathcal{P}$ of atomic propositions are given by the following grammar.

$$\varphi ::= p \mid \neg p \mid \varphi \wedge \varphi \mid \varphi \vee \varphi \mid \langle r \rangle \varphi \mid [r]\varphi$$

where $p \in \mathcal{P}$ and $r \in \{B, \bar{B}, E, \bar{E}\}$.

We use usual Boolean abbreviations like $\bot := p \wedge \neg p$ and $\top := p \vee \neg p$ for some p. Modal operators for the other eight Allen relations are definable via $\langle A \rangle \varphi := ([E]\bot \wedge \langle \bar{B} \rangle \varphi) \vee \langle E \rangle ([E]\bot \wedge \langle \bar{B} \rangle \varphi)$ ("meets"); $\langle D \rangle \varphi := \langle B \rangle \langle E \rangle \varphi$ ("contains"); $\langle L \rangle \varphi := \langle A \rangle \langle E \rangle \varphi$ ("before"); $\langle O \rangle \varphi := \langle E \rangle \langle \bar{B} \rangle \varphi$ ("overlaps") and similarly for their inverses. The set $Sub(\varphi)$ of subformulas of φ is defined as usual. We measure the size of a formula φ in terms of the number of its different subformulas: $|\varphi| := |Sub(\varphi)|$.

An ITL formula φ is interpreted in an interval $[i,j]$ of a finite interval structure $\mathcal{I} = (n, \vartheta)$ as follows [9].

$\mathcal{I}, [i,j] \models p$ iff $p \in \vartheta([i,j])$

$\mathcal{I}, [i,j] \models \neg p$ iff $p \notin \vartheta([i,j])$

$\mathcal{I}, [i,j] \models \varphi \wedge \psi$ iff $\mathcal{I}, [i,j] \models \varphi$ and $\mathcal{I}, [i,j] \models \psi$

$\mathcal{I}, [i,j] \models \varphi \vee \psi$ iff $\mathcal{I}, [i,j] \models \varphi$ or $\mathcal{I}, [i,j] \models \psi$

$\mathcal{I}, [i,j] \models \langle r \rangle \varphi$ iff   there is $[i',j'] \in I(n)$ s.t. $[i,j] r [i',j']$ and $\mathcal{I}, [i',j'] \models \varphi$

$\mathcal{I}, [i,j] \models [r] \varphi$ iff   for all $[i',j'] \in I(n)$ with $[i,j] r [i',j']$ we have $\mathcal{I}, [i',j'] \models \varphi$

The *finite satisfiability problem* for ITL is defined as follows. Given an ITL formula φ, decide whether or not there is a finite interval structure $\mathcal{I} = (n, \vartheta)$ and an interval $[i,j] \in I(n)$ such that $\mathcal{I}, [i,j] \models \varphi$. A formula that has a model in the above sense is said to be *finitely satisfiable*.

3 Approximating Finite Satisfiability

We develop the notion of bounded ITL satisfiability which approximates finite satisfiability. To keep the presentation short we work with the minimal set of modal operators introduced above. For efficiency purposes it may be useful to treat the other operators as basic and not as abbreviations; this is done for instance in the implementation that is reported on in the next section.

Definition 1. *Let $n \geq 1$. An ITL formula φ is said to be n-bounded satisfiable if there is a finite interval structure $\mathcal{I} = (n, \vartheta)$ such that $\mathcal{I}, [0,0] \models \varphi$. The n-bounded satisfiability problem is to decide, given an ITL formula φ, whether or not it is n-bounded satisfiable.*

First note that φ is finitely satisfiable iff there is a finite interval structure $\mathcal{I}$ such that $\mathcal{I}, [0,0] \models Somewhere(\varphi)$ with $Somewhere(\varphi) := \varphi \vee \langle \bar{B} \rangle (\varphi \vee \langle E \rangle \varphi)$. Thus, when determining finite satisfiability it is possible to restrict the attention to satisfaction in the interval $[0,0]$ at the cost of extending the input formula by 4 additional subformulas. Then we get that φ is finitely satisfiable iff $Somewhere(\varphi)$ is n-bounded satisfiable for some $n \geq 1$.

Note that n-bounded satisfiability is neither monotone nor antitone in n. Consider $\varphi_2 := \langle B \rangle \langle B \rangle \top \wedge [B][B][B]\bot$. The first conjunct is satisfied in an interval of the form $[i,j]$ when $j \geq i+2$, the second one requires $j \leq i+2$. Thus, it is only satisfied by intervals of length 2. Then $\langle A \rangle (\varphi_2 \wedge [\bar{B}]\bot)$ is 3-bounded satisfiable but neither 2- nor 4-bounded satisfiable. This is an important observation for the design of a procedure that successively approximates finite satisfiability by bounded satisfiability for increasing bounds. It means that one has to increase by steps of 1 for the procedure to be complete.

Let $n \geq 1$ be fixed. We reduce the n-bounded ITL satisfiability problem to the propositional satisfiability problem as follows. Given an ITL formula φ we construct a finite set $\mathcal{C}_\varphi^n := \{X_{0,0}^\varphi\} \cup \mathcal{P}_\varphi^n \cup \mathcal{T}_\varphi^n$ of propositional formulas which is satisfiable iff φ is n-bounded satisfiable. The elements of $\mathcal{P}_\varphi^n := \bigcup_{i=0}^{n-1} \bigcup_{j=i}^{n-1} \mathcal{P}_\varphi^n(i,j)$ are called the *main* constraints, and those of $\mathcal{T}_\varphi^n$ are called *temporary* constraints. This distinction is necessary to make the encoding incremental, to be explained in detail below.

$\mathcal{C}_\varphi^n$ is defined over the set of atomic propositions $\{X_{i,j}^\psi\}_{0 \le i \le j \le n, \psi \in Sub(\varphi)}$. Intuitively, a variable $X_{i,j}^\psi$ expresses that ψ is satisfied by $[i,j]$ in the interval structure of length n that is represented by a model of $\mathcal{C}_\varphi^n$.

Each main constraint in $\mathcal{P}_\varphi^n(i,j)$ is associated with a subformula of φ as follows.

$\neg p$	$X_{i,j}^{\neg p} \to \neg X_{i,j}^p$	$\langle E \rangle \psi$	$X_{i,j}^{\langle E \rangle \psi} \to \bigvee_{k=i+1}^j X_{k,j}^\psi$
$\psi_1 \wedge \psi_2$	$X_{i,j}^{\psi_1 \wedge \psi_2} \to X_{i,j}^{\psi_1}$	$\langle \bar{E} \rangle \psi$	$X_{i,j}^{\langle \bar{E} \rangle \psi} \to \bigvee_{k=0}^{i-1} X_{k,j}^\psi$
	$X_{i,j}^{\psi_1 \wedge \psi_2} \to X_{i,j}^{\psi_2}$	$[B]\psi$	$X_{i,j}^{[B]\psi} \to X_{i,k}^\psi, \qquad k = i, \ldots, j-1$
$\psi_1 \vee \psi_2$	$X_{i,j}^{\psi_1 \vee \psi_2} \to X_{i,j}^{\psi_1} \vee X_{i,j}^{\psi_2}$	$[\bar{B}]\psi$	$X_{i,j}^{[\bar{B}]\psi} \to X_{i,k}^\psi, \quad k = j+1, \ldots, n-1$
$\langle B \rangle \psi$	$X_{i,j}^{\langle B \rangle \psi} \to \bigvee_{k=i}^{j-1} X_{i,k}^\psi$	$[E]\psi$	$X_{i,j}^{[E]\psi} \to X_{k,j}^\psi, \qquad k = i+1, \ldots, j$
$\langle \bar{B} \rangle \psi$	$X_{i,j}^{\langle \bar{B} \rangle \psi} \to X_{i,j+1}^\psi \vee X_{i,j+1}^{\langle \bar{B} \rangle \psi}$	$[\bar{E}]\psi$	$X_{i,j}^{[\bar{E}]\psi} \to X_{k,j}^\psi, \qquad k = 0, \ldots, i-1$

Each of them is defined by case distinction on the type of the corresponding subformula. For every subformula of the form in the left column, $\mathcal{P}_\varphi^n(i,j)$ contains *all* the constraints given in the corresponding right column.

The temporary constraints in $\mathcal{T}_\varphi^n$ are defined to be $\{\neg X_{i,n}^{\langle \bar{B} \rangle \psi} \mid 0 \le i < n, \langle \bar{B} \rangle \psi \in Sub(\varphi)\}$. Note that these variables describe the truth value of subformulas on intervals that do not exist with the currently considered length of an interval structure. They are used in the permanent constraints in order to make the encoding for the $\langle \bar{B} \rangle$-operators incremental. Since these intervals do not exist, these variables are forced to be false by the temporary constraints. When increasing the length of considered interval structures, the temporary constraints are deleted and more permanent constraints are being used to describe the truth values of these new intervals of the form $[i,n]$ for $i \le n$.

Theorem 2. *For all ITL formulas φ and all $n \ge 1$ we have that φ is n-bounded satisfiable iff $\mathcal{C}_\varphi^n$ is satisfiable.*

The proof is standard and therefore omitted. The following estimation on the size of $\mathcal{C}_\varphi^n$ is also easy to verify.

Lemma 3. *$\mathcal{C}_\varphi^n$ contains $\mathcal{O}(|\varphi| \cdot n^2)$ many variables and is of size $\mathcal{O}(|\varphi| \cdot n^3)$.*

Also note that $\mathcal{C}_\varphi^n$ consists of propositional clauses and can therefore be given to a SAT solver as it is. Modern SAT solvers support incremental solving meaning that, as long as in between runs of the solver only clauses are added, the next solving process can re-use information gathered in the last one like learnt clauses etc. [8]. Solvers such as picosat [3], which we use in the tool described here can do so even if some variable values are assumed, and satisfiability checking is only performed for assignments that respect these values. Note that all clauses in $\mathcal{T}_\varphi^n$ are single-literal clauses and can thus be used as variable value assumptions during solving.

In order to benefit from incremental SAT solving we need to explain how $\mathcal{C}_\varphi^m$ can be obtained from $\mathcal{C}_\varphi^n$ for $m > n$ adding minimal sets of clauses. Because of the remark on non-monotonicity of bounded ITL satisfiability made at the beginning of this section it makes sense to consider the case of $m = n + 1$ only. The next lemma stating the possibility of using this encoding incrementally is also easily verified. It is straightforward to extend it to the case of $m > n + 1$.

Lemma 4. *For all $n \geq 1$ we have $C_\varphi^{n+1} = (C_\varphi^n \setminus T_\varphi^n) \cup \left(\bigcup_{i=0}^{n} P_\varphi^{n+1}(i, n) \right) \cup T_\varphi^{n+1}$.*

Thus, C_φ^{n+1} can be constructed from C_φ^n by removing the temporary constraints (of a cardinality that is linear in n), and then adding a quadratic number of constraints to it. This is asymptotically better than building C_φ^{n+1} from scratch which would take cubic time in n.

Based on Theorem 2 and Lemma 4 we can devise a simple approximation scheme that tests an ITL formula for finite satisfiability.

procedure ITLFINSATTEST(φ)
$\quad n \leftarrow 0$
$\quad C \leftarrow \{X_{0,0}^\varphi\}$
$\quad$**repeat**
$\quad\quad n \leftarrow n + 1$
$\quad\quad C \leftarrow C \cup \left(\bigcup_{i=0}^{n-1} P_\varphi^n(i, n-1) \right)$
$\quad$**until** C is satisfiable assuming all assignment in T_φ^n to hold
$\quad$extract a model for φ of size n from a satisfying propositional assignment
end procedure

Completeness of this approximation is a direct consequence of the fact that it symbolically tests all interval structures of increasing size for being a model for its input formula. Soundness only holds in the weak sense that this method does not return any false positives. However, it is not able to detect unsatisfiability, i.e. on unsatisfiable inputs it simply does not terminate.

Theorem 5 (Completeness). ITLFINSATTEST(*Somewhere(φ)*) *terminates on an n-bounded satisfiable φ after at most n iterations of its loop and produces a model for Somewhere(φ).*

We remark that ITLFINSATTEST can be made to terminate on fragments of ITL for which the small model property is known. Such fragments are necessarily decidable. In such cases it suffices to run the loop up to the maximal size of a minimal model of the input formula. If none has been found, unsatisfiability can be reported.

4   Experiments

To evaluate the scalability of incremental bounded ITL satisfiability checking using the ideas described in this paper, we implemented the tool ITLFinSat. It is available for download at https://github.com/progirep/ITLFinSat. The tool is written in C++ and uses the SAT solver picosat v.957 [3] as a library for incremental solving. The tool is completely single-threaded.

We consider three benchmark cases: an ITL model of the Fischer Mutex protocol [14], a formalisation of a binary counter, and a classical puzzle as an ITL satisfiability problem.

Table 1. Results of the experiments

benchmark φ	$|\varphi|$	prop. formula size	# prop. variables	model size k	time
Fischer, $n = 2$	414	45,441	31,482	10	0.18s
Fischer, $n = 3$	638	103,987	66,885	12	0.72s
Fischer, $n = 4$	880	201,650	121,800	14	6.2s
Fischer, $n = 5$	1140	352,529	201,501	16	311s
chicken puzzle	215	19,705	15,840	9	0.09s
5-bit counter	195	77,028	40,869	17	0.3s
6-bit counter	236	464,168	171,955	33	3.2s
7-bit counter	277	3,178,956	749,529	65	58.9s
8-bit counter	318	24,087,440	3,312,335	129	31m57s

Fischer Protocol. This protocol orchestrates n agents that want to enter some critical section. Mutual exclusion is achieved through a clever set-wait-and-test phase in which each agent can indicate their intention to enter by setting a common variable to its ID, then wait for a while and enter the critical section only if the variable's value still equals the agent's ID. We formalise the possibility for more than one agent to enter their critical section at the same time as an ITL satisfiability problem, using 4 propositions for each agent to indicate the state that they are in currently, and $n + 1$ propositions for the common variable's values. The ITL formula then expresses that at every time the agents' states and the variable's value are unique, and that the agents can only change states according to the description above, i.e. when the variable's value allows them to do so.

Intervals in a model for this formalisation can be seen as durations for how long the agents need to remain in certain states, and the satisfiability check reveals that mutual exclusion does not hold when the waiting phase is too short for some agents. Note that correctness of Fischer's protocol relies on some phases being longer than others, and general interval length comparisons are not formalisable in plain ITL. Mutual exclusion in this protocol does depend on certain intervals being longer than others, though; this is why the reason for violation in this example would have to be found manually from the output of the satisfiability test. For the Fischer protocol, we model the question if for some value of n if for a setting with n processes, all of them can be in their critical regions at the same time as an ITL formula.

The Chicken Crossing Puzzle. We formulate the classical problem of the farmer trying to get a fox, a chicken and some corn across the river without ever leaving the chicken with the fox or the corn unattended on one side. The existence of a solution can naturally be formalised in ITL using propositions for the locations (i.e. side of the river) of the four protagonists. The ITL formula states that none of them is on both sides at the same time, that the farmer can only take one of them across the river at a time, that they are all on the left side at the beginning and on the right side at the end, etc.

Binary Counter. This benchmark family is used to test the limits of the SAT-based approach. It formalises the evolution of an n-bit counter using propositions for "the i-th bit is set/unset on this interval" by stating that the highest bit is unset and afterwards set, and whenever bit i is set or unset on an interval then this begins with an interval in which bit $i - 1$ is unset and ends in one in which bit $i - 1$ is set. Moreover, we require that phases in which some bit is set resp. unset must not overlap. This formula for n bits is always satisfiable, but its shortest models are of length $2^{n-1} + 1$.

Table 1 presents data collected from satisfiability checks for these benchmarks. All experiments were carried out on a computer with an Intel i5-3230M CPU running at 2.60GHz. A memory limit of 2GB was never exceeded in our experiments. The table shows the size of the underlying ITL formula, the size and number of propositional variables of its SAT encoding when it has been found to be k-bounded satisfiable, the size k of the model that has been found, and the overall time taken for the satisfiability check including the encoding and checks at model sizes less than k.

References

1. Allen, J.F.: Maintaining knowledge about temporal intervals. Communications of the ACM 26(11), 832–843 (1983)
2. Axelsson, R., Heljanko, K., Lange, M.: Analyzing context-free grammars using an incremental SAT solver. In: Aceto, L., Damgård, I., Goldberg, L.A., Halldórsson, M.M., Ingólfsdóttir, A., Walukiewicz, I. (eds.) ICALP 2008, Part II. LNCS, vol. 5126, pp. 410–422. Springer, Heidelberg (2008)
3. Biere, A.: Picosat essentials. JSAT 4(2-4), 75–97 (2008)
4. Bresolin, D., Jiménez, F., Sánchez, G., Sciavicco, G.: Finite satisfiability of propositional interval logic formulas with multi-objective evolutionary algorithms. In: FOGA 2013, pp. 25–36. ACM (2013)
5. Bresolin, D., Della Monica, D., Montanari, A., Sciavicco, G.: A tableau system for right propositional neighborhood logic over finite linear orders: An implementation. In: Galmiche, D., Larchey-Wendling, D. (eds.) TABLEAUX 2013. LNCS, vol. 8123, pp. 74–80. Springer, Heidelberg (2013)
6. Chaochen, Z., Hoare, C.A.R., Ravn, A.P.: A calculus of durations. Information Processing Letters 40(5), 269–276 (1991)
7. Clarke, E.M., Biere, A., Raimi, R., Zhu, Y.: Bounded model checking using satisfiability solving. Formal Methods in System Design 19(1), 7–34 (2001)
8. Eén, N., Sörensson, N.: An extensible SAT-solver. In: Giunchiglia, E., Tacchella, A. (eds.) SAT 2003. LNCS, vol. 2919, pp. 502–518. Springer, Heidelberg (2004)
9. Halpern, J.Y., Shoham, Y.: A propositional modal logic of time intervals. In: LICS 1986, pp. 279–292. IEEE (1986)
10. Hodkinson, I.M., Montanari, A., Sciavicco, G.: Non-finite axiomatizability and undecidability of interval temporal logics with C, D, and T. In: Kaminski, M., Martini, S. (eds.) CSL 2008. LNCS, vol. 5213, pp. 308–322. Springer, Heidelberg (2008)
11. Della Monica, D., Goranko, V., Montanari, A., Sciavicco, G.: Expressiveness of the interval logics of allen's relations on the class of all linear orders: Complete classification. In: IJCAI 2011, pp. 845–850. AAAI (2011)
12. Della Monica, D., Goranko, V., Montanari, A., Sciavicco, G.: Interval temporal logics: a journey. Bulletin of the EATCS 105 (2011)
13. Moszkowski, B.: Reasoning about digital circuits. Ph.D. thesis, Stanford Univ (1983)
14. Peterson, G.L., Fischer, M.J.: Economical solutions to the critical section problem in a distributed system. In: STOC 1977, pp. 91–97. ACM (1977)
15. Venema, Y.: A modal logic for chopping intervals. Journal of Logic and Computation 1(4), 453–476 (1991)

StarExec: A Cross-Community Infrastructure for Logic Solving*

Aaron Stump[1], Geoff Sutcliffe[2], and Cesare Tinelli[1]

[1] Department of Computer Science, The University of Iowa, Iowa City, IA, USA
[2] Department of Computer Science, University of Miami, Miami, FL, USA

Abstract. We introduce StarExec, a public web-based service built to facilitate the experimental evaluation of *logic solvers*, broadly understood as automated tools based on formal reasoning. Examples of such tools include theorem provers, SAT and SMT solvers, constraint solvers, model checkers, and software verifiers. The service, running on a compute cluster with 380 processors and 23 terabytes of disk space, is designed to provide a single piece of storage and computing infrastructure to logic solving communities and their members. It aims at reducing duplication of effort and resources as well as enabling individual researchers or groups with no access to comparable infrastructure. StarExec allows community organizers to store, manage and make available benchmark libraries; competition organizers to run logic solver competitions; and community members to do comparative evaluations of logic solvers on public or private benchmark problems.

1 Introduction

Ongoing breakthroughs in a number of fields depend on continuing advances in the development of high-performance automated reasoning tools such as SAT solvers, SMT solvers, theorem provers, model finders, constraint solvers, rewrite systems, model checkers, and so on, which we generically refer to here as *(logic) solvers*. Typically, application problems are translated into possibly large and complex formal descriptions (logical formulas, rewrite rules, transition systems, ...) for these tools to reason about. Different tradeoffs between linguistic expressiveness and the difficulty of the original problems have led to the adoption of a variety of reasoning approaches and logical formalisms to encode those problems. Distinct research communities have developed their own research infrastructure to spur innovation and ease the adoption of their solver technology. This includes standard input/output formats for solvers, e.g., [3, 14]; libraries of benchmark problems, e.g., [2, 6, 11, 14]; solver execution services, e.g., [12, 13, 15]; and solver competitions, e.g., [1, 4, 5, 7, 8, 9, 10]. By and large, so far these different infrastructures have been developed separately in the various logic communities, at significant and largely duplicated cost in development effort, equipment and support.

StarExec is a solver execution and benchmark library service aimed at facilitating the experimental evaluation of automated reasoning tools. It is designed to provide a single piece of storage and computing infrastructure to all logic solving communities, with the

---

* Work made possible in large part by the support of the National Science Foundation through grants 0957438, 1058748, and 1058925.

S. Demri, D. Kapur, and C. Weidenbach (Eds.): IJCAR 2014, LNAI 8562, pp. 367–373, 2014.

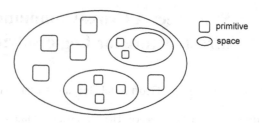

Fig. 1. The StarExec space hierarchy

twofold goal of reducing the effort and resources duplication while also enabling communities and individuals that could not afford developing their own infrastructure. The service allows community organizers to store, manage and make available benchmark libraries, competition organizers to run competitions, and individual community members to run comparative evaluations of logic solvers on benchmark problems. These capabilities are accessible through a web browser interface at the StarExec web site which provides facilities to upload and organize benchmarks and solvers, browse and query the stored benchmark libraries, run user-selected solvers on user-selected benchmarks, and view and analyze execution results and statistics.

This paper gives an overview of StarExec, briefly describing its main design, components, functionality, and usage, and discusses its current status and further development plans. For more details on the service and its usage, we refer the reader to the online documentation on the StarExec web site: http://www.starexec.org.

2   Main Concepts

StarExec is built as a service for logic solving *communities*, groups of logic solver developers and users such as the SAT, SMT, theorem proving, confluence, model checking, and software verification communities. In StarExec, a community is a user group administered by one or more designated *community leaders*, users with special privileges and responsibilities such as accepting new users or uploading and managing benchmarks and solvers. The system is built on top of a few basic concepts: *spaces*, *users*, *solvers*, *benchmarks*, *job pairs*, *jobs*, and *processors*, which are described below.

Spaces. A space is a collection of solvers, benchmarks, jobs, and users—collectively referred to as *primitives*—as well as other spaces—also referred to as *subspaces*. Spaces are similar to folders in a file system and are the means by which StarExec primitives are organized (see Figure 1). The subspace hierarchy has a tree structure: each space other than the root space is contained in exactly one superspace. In contrast, each primitive can be contained in more than one space. The direct subspaces of the root space are *community spaces*. Each logic community has one community space which contains, directly or within subspaces, all benchmarks, solvers, users and jobs in that community. Spaces can be *public* or *private*, *locked* or *unlocked*. A public space is visible to all StarExec users; a private one is visible only to the users contained in it. If a space is locked no one can copy items *out* of it. This is useful to store copy-restricted benchmarks and solvers while still allowing other users to include them in job pairs.

Users. A user is a registered and verified person who belongs to at least one community and is therefore endorsed within the system by a community leader. Every user has visibility over a set of (sub)spaces and has a set of *permissions* for each of these spaces controlling the kind of allowed operations. Users also have *leadership* over some spaces. A leader of a space has the full set of permissions over that space. A community leader is simply the leader of a community space.

Solvers. Solvers are programs that solve logic problems. Specifically, they are Linux programs that take a text file on standard input and may produce text on standard output. Solvers can consist of groups of files and folders—as opposed to a single executable. Each solver is associated with one or more *configurations*, executable scripts that invoke the solver with specific input flags and settings. Configurations and processor scripts (see below) can be written in any of the scripting languages (Bash, Perl, Python, and so on) available by default in Linux distributions. A configuration takes as input a benchmark and feeds it to its solver.

Benchmarks. A benchmark is a single text file. Currently there is no support for multi-file benchmarks; however, StarExec supports benchmark dependencies, the referencing of other benchmark files (e.g., TPTP axioms files [14]) within a benchmark. Each benchmark has a *type* which consists of a (system-wide) unique name and an associated benchmark processor.

Jobs and Job Pairs. A job pair consists of a solver configuration and a benchmark to be run by that configuration. Job pairs are atomic execution units. After execution they are associated to various pieces of information such as results, CPU time, wall clock time, etc. A job is simply a collection of job pairs executed or to be executed.

Processors. Processors are executables that take textual input and produce textual output at certain stages in the job execution pipeline. They are currently of two kinds: benchmark processors and post-processors. The former are run on benchmarks being uploaded to a space. They can be used to perform various checks and validations on a benchmark and to extract meta-data to be stored in a central database. Post-processors are run on solver output produced by executing a job pair. They can be used to extract specific data from this output which too are stored in a database for later access.

3 Functionality and Usage

StarExec was designed as a web service accessible through a web browser interface. Most of the web service functionality is also available through a downloadable client application, called StarExecCommand. That program provides a command line shell and a large number of commands mirroring the operations that can be performed via a web browser. In the following, we focus on the web browser interface.

User Accounts. Using StarExec requires first obtaining a user account. Since every registered user belongs to at least one of the StarExec communities, every new account request requires joining one of them (more communities can be joined later). Such requests are sent to the leaders of the chosen community each of whom has the power to approve them or not. Communities themselves and community leader users can be created only by the StarExec administrators. Registered users must log in to StarExec before being able to use it. External observers can view all public information on the StarExec website by logging in as the special user *guest*.

Spaces and Communities. Using an interface similar to that of file explorers in computer desktop GUIs, users can browse the StarExec space hierarchy or, rather, the subset of it visible to them; view the contents of any spaces they belong to; and inspect details of users, benchmarks, solvers, and jobs in those spaces. As an example, details about benchmarks include name, description, owner, upload date, file size, type, and any user-defined attributes automatically extracted from the benchmark when it was uploaded. Users can also perform a number of stateful operations on a space depending on the set of permissions they have on it. These operations may include creating and deleting subspaces; uploading, copying, linking and deleting benchmarks and solvers; creating, removing jobs; copying, removing users; making other users leaders; changing the space's default permissions; and locking or unlocking the space.

In general, users can copy any visible primitive from an unlocked space to any space where they have permission to add that kind of primitive. While copy operations are done uniformly by a drag-and-drop action, their semantics depends on the copied item. Copying a user U from a space A to a space B amounts to giving U certain permissions on B. Copying a job J from A to B only creates a link to J in B. Copying a benchmark or a solver, on the other hand, can be done either way: as an actual copy or just as a link. All users are by default *leaders* of the spaces they create. The leaders of a space have full permissions on that space, which consist in the ability to add/remove primitives and subspaces as well as make other users space leaders. A space leader can also decide which set of permissions from the above to grant to regular (i.e., non-leader) users in the space. For security and integrity reasons, only StarExec administrators can remove space leaders from a space or demote them to regular users.

Since communities are spaces, users can operate on them as discussed above. However, users can also browse the communities visible to them, request to join or leave a community, and download a community's benchmark processors. Community leaders (i.e., leaders of a community space) can, in addition, set several community-wide defaults such as timeouts or post-processors for job pairs. They can also upload post-processors, define benchmark types, and upload their associated processor.

Solvers. StarExec users can add a solver to a space (in which they have the permission to do so) either by copying/linking the solver from another space or by uploading it. Solvers must be uploaded as tar archives or compressed tar or zip archives. The archives can consist of any collection of folders and files but must contain at least one configuration script—used to establish proper settings and input flags for the solver and then launch the solver. All configuration scripts must reside in a designated top-level directory (bin) in the archive and have a name starting with a designated prefix (starexec_run_) so that they can be easily identified by the StarExec system.

During a job execution, StarExec will run a configuration script in an environment with a number of predefined environment variables. Those variables contain such information as the absolute path to the benchmark input file, the absolute path to a designated output directory for the script, and limits on wall clock time, CPU time, memory, and disk space. Any files written to the output directory during the execution of the script are saved by StarExec for later download by the user. Files written elsewhere are removed after each job run. If a job pair exceeds any of its time or space limits it is terminated.

As with any space item, users can inspect details of any solver visible to them. In particular, they can view the list and the content of each configuration script. If permitted by the owner of the containing space, they can also download the entire solver and its configurations as a compressed archive.

Benchmarks. Users can add benchmarks to a space either by copying/linking them from another space or by uploading an archive file containing benchmarks all of the same type. The user has the option of (i) recreating the archive's directory structure as a space structure in the destination space and place each benchmark in the (sub)space corresponding to the benchmark's source (sub)directory, or (ii) simply placing all the files contained in the archive directly in the destination space. With the first option the new subspaces take the name of the corresponding subdirectory and all get the same set of permissions, specified by the user before uploading the archive. All uploaded files are run through the benchmark processor associated with their type, and are added to their destination space only if accepted by the processor. An upload status page summarizes the results of the upload as it progresses and lists all the discarded files, if any.

Benchmark types are global across all communities. They can be created only by community leaders who must also provide an associated processor. Every benchmark processor is expected to print for its input benchmark a sequence of lines of the form *key=value* where *key* is an attribute name and *value* its value. These attribute-value pairs are stored in StarExec's database and will be shown later when inspecting the benchmarks. Most attribute names are user-defined. A predefined attribute, and the only mandatory one, is `starexec-valid`. The processor is required to print a pair with that key and the value `true` or `false` depending on whether the benchmark was accepted by the processor or not. The system will discard the benchmark if the value of this attribute is `false`. Other predefined but optional attributes allow the processor to specify dependencies on other benchmarks or an expected result for the benchmark.

Compute Cluster. StarExec compute nodes are partitioned into a number of execution queues. Each job is submitted to one of these queues. There is a general queue available to all users. The other queues serve special functions, such as solver competitions, or are reserved for exclusive use by a community. Only community leaders can request the creation and the reservation of a queue. A dedicated page on the web server allows all users to inspect any queue to see pending job pairs for that queue, and inspect any node in a queue to see currently executing job pairs on that node. To assure result reproducibility, each job pair is run in isolation on one compute node processor, i.e., no two job pairs share the same processor at the same time. To assure a basic level of security a job pair's solver is run as a *sandboxed* user with very limited permissions.

Jobs. Users can create and immediately execute a job in a space. At creation time, they set execution parameters specifying timeouts, post-processors, and execution queues. They can choose to have job pairs executed in depth-first or in round-robin fashion. In the first case, StarExec will execute all job pairs in one subspace before moving on to the next; in the second, all job pairs in all subspaces will make progress in the execution concurrently. Users have different ways to generate a job pairs starting from the *root space*, the space in which the job is created. For instance, they can instruct the system to find all subspaces containing solvers and benchmarks, and execute all possible combinations of those benchmarks and available configurations for those solvers. Alternatively, users can manually select which benchmarks and solvers to execute. In that case, they also have different options for how to pair solvers with benchmarks.

After a job has been set up, its job pairs are created automatically and sent to the specified execution queue. A running job can be monitored by looking at its details web page, which gets updated in real time with information on how many pairs have been completed, have been solved, have failed and so on. Specific job pair information includes its status, final runtimes, and user-defined results. After the job completes, its page will also provide various statistics in graphical form, such as scatter and cactus plots. All job data can be also downloaded in real time in CSV format (with one line per job pair) for off-line analysis, process and visualization. The job's web page is assigned a unique, persistent URL that can be used as a reference in publications and the like.

4   Infrastructure and Technologies

The StarExec software infrastructure relies on common web standards and several freely available software applications and libraries. Almost all of the software developed in-house for this service is written in Java, with the rest consisting mostly of shell scripts. We plan to open source the entirety of this software in the near future.

The StarExec hardware infrastructure is located in a dedicated state-of-the-art server hosting facility at the University of Iowa. The service runs on a Red Hat Fedora compute cluster consisting of 3 head nodes and 190 rack-mounted compute nodes. Each node has two 4-core 2.4GHz Intel processors with 256GB of RAM and a 1TB hard drive. Local disk space is used only for caching purposes during job execution. Solvers, benchmarks and all other persistent data and meta-data are stored centrally in a dual NetApp network-attached storage system with a capacity of 23TB.

5   Current Status and Future Development

At the time of this writing, StarExec has been used by two public events: the Confluence Competition (CoCo) 2013, which ran in June, 2013; and the SMT Evaluation (SMT-EVAL) 2013, which ran over several months in 2013. Each event had an execution queue giving it exclusive access to a subset of the compute nodes. Otherwise, they had quite different requirements for StarExec. CoCo 2013 ran during a meeting of the Confluence Workshop, and thus it was very important that the CoCo organizers could initiate the job and monitor its results live. They used StarExecCommand for this. On the other hand, their workload was small, just 509 job pairs. In contrast, SMT-EVAL 2013 had a workload of 1,663,478 job pairs, split over four jobs. This meant that operating at scale was an important criterion for success for SMT-EVAL.

For the FLoC Olympic Games of Summer 2014, many additional events have expressed interest or made plans to run on StarExec. The latter include the CASC, QBF, and SMT-COMP competitions. With its basic functionality now in place, we anticipate that future developments of the StarExec system and service will be driven by the needs of such events or of whole communities. The challenge in adding new features to meet these needs will be to make them general enough to be potentially useful to other communities. For example, in some tracks of the Termination Competition, termination proofs produced by termination checkers for rewriting systems are fed on the fly to a proof checker. To support this we plan to add to StarExec a general facility for pipelining tools, something that we expect will be useful to other competitions as well.

Acknowledgements. We would like to thank several people who have contributed in various capacities to the StarExec project so far. The following people were involved in the development of the software infrastructure at various stages of the project: E. Burns, T. Elvers, T. Jensen, W. Kaiser, B. McCune, M. Nassar, CJ Palmer, V. Sardeshmukh, S. Stark, and R. Zhang. Computer system support and assistance in designing and building the hardware infrastructure was provided by H. Brown, D. Holstad, J. Tisdale, and JJ Ulrich. Several people, from user communities and from the StarExec Advisor Board, provided useful feedback and input. A full list can be found on StarExec website.

References

[1] Barrett, C., Deters, M., de Moura, L., Oliveras, A., Stump, A.: 6 Years of SMT-COMP. Journal of Automated Reasoning 50(3), 243–277 (2012)

[2] Barrett, C., Stump, A., Tinelli, C.: The Satisfiability Modulo Theories Library (SMT-LIB) (2010), http://www.SMT-LIB.org

[3] Barrett, C., Stump, A., Tinelli, C.: The SMT-LIB Standard: Version 2.0. In: Proceedings of the 8th International Workshop on Satisfiability Modulo Theories (2010)

[4] Beyer, D.: Competition on software verification. In: Flanagan, C., König, B. (eds.) TACAS 2012. LNCS, vol. 7214, pp. 504–524. Springer, Heidelberg (2012)

[5] Biere, A., Claessen, K.: Hardware model checking competition. In: Hardware Verification Workshop (2010)

[6] Hoos, H., Stützle, T.: SATLIB: An Online Resource for Research on SAT. In: Proceedings of the 3rd Workshop on the Satisfiability Problem (2001), http://www.satlib.org/

[7] Le Berre, D., Simon, L. (eds.): Special Issue on the SAT 2005 Competitions and Evaluations, vol. 2. JSAT (2006)

[8] Marché, C., Zantema, H.: The termination competition. In: Baader, F. (ed.) RTA 2007. LNCS, vol. 4533, pp. 303–313. Springer, Heidelberg (2007)

[9] Nieuwenhuis, R.: Special Issue: The CADE ATP System Competition. AI Communications 15(2-3) (2002)

[10] Peschiera, C., Pulina, L., Tacchella, A., Bubeck, U., Kullmann, O., Lynce, I.: The seventh QBF solvers evaluation (QBFEVAL'10). In: Strichman, O., Szeider, S. (eds.) SAT 2010. LNCS, vol. 6175, pp. 237–250. Springer, Heidelberg (2010)

[11] Raths, T., Otten, J., Kreitz, C.: The ILTP Problem Library for Intuitionistic Logic - Release v1.1. Journal of Automated Reasoning 38(1-2), 261–271 (2007)

[12] Simon, L., Chatalic, P.: SatEx: A Web-based Framework for SAT Experimentation. In: Proceedings of SAT 2001. ENDM, vol. 9, pp. 129–149 (2001)

[13] Stump, A., Deters, M.: SMT-Exec., http://www.smtexec.org

[14] Sutcliffe, G.: The TPTP Problem Library and Associated Infrastructure. The FOF and CNF Parts, v3.5.0. Journal of Automated Reasoning 43(4), 337–362 (2009)

[15] Sutcliffe, G.: The TPTP World – Infrastructure for Automated Reasoning. In: Clarke, E.M., Voronkov, A. (eds.) LPAR-16 2010. LNCS (LNAI), vol. 6355, pp. 1–12. Springer, Heidelberg (2010)

Skeptik: A Proof Compression System[*]

Joseph Boudou[1], Andreas Fellner[2,3], and Bruno Woltzenlogel Paleo[3]

[1] IRIT, Université de Toulouse, France
joseph.boudou@irit.fr
[2] Free University of Bolzano, Italy
fellner.a@gmail.com
[3] Vienna University of Technology, Austria
bruno@logic.at

Abstract. This paper introduces Skeptik: a system for checking, compressing and improving proofs obtained by SAT- and SMT-solvers.

1 Introduction

There are various reasons why it is desirable for automated reasoning tools to output not only a *yes* or *no* answer to a problem but also *proofs/refutations* or *(counter)models*. Firstly, state-of-the-art tools are complex and heavily optimized. Their code is often long, hard to understand and difficult to automatically verify. Consequently, *yes/no* answers cannot be fully trusted, unless they are accompanied by proofs or (counter)models that serve as independently checkable certificates of their correctness.

Furthermore, for most applications, a *yes/no* answer is inherently insufficient. We often already know in advance whether a problem is expected to be satisfiable or unsatisfiable, and we want more than just a confirmation of this expectation. For satisfiable formulas, the desired information is encoded in the model; while for valid formulas, it is contained in the proof. In case the expectation was wrong, refutations and countermodels can be helpful to explain issues in the encoding of the problem, in order to correct and refine it. Proofs can been used, for example, to obtain unsat cores, interpolants and Herbrand disjunctions [12].

Although current automated deduction tools are very efficient at finding proofs, they do not necessarily find the *best* proofs (e.g. shortest, least spacious, with smallest core... depending on the intended application). The Skeptik tool (http://github.com/Paradoxika/Skeptik/) finds and eliminates redundancies in proofs, in order to improve and compress them according to various metrics. It is licensed under a Creative Commons CC-BY-NC-SA License.

Related Work: CERes (http://www.logic.at/ceres) [8] is another proof transformation system, specialized in cut-elimination for sequent calculus. It was replaced by GAPT (http://code.google.com/p/gapt/) [7], extended with cut-introduction techniques. MINLOG (http://www.mathematik.uni-muenchen.de/~logik/minlog/) extracts functional programs from proofs, employing a refined A-translation.

[*] Funded by Google Summer of Code 2012 and 2013 and FWF project P24300.

S. Demri, D. Kapur, and C. Weidenbach (Eds.): IJCAR 2014, LNAI 8562, pp. 374–380, 2014.
© Springer International Publishing Switzerland 2014

2   Implementation Details

In Skeptik every logical expression is a simply typed lambda expression, implemented by the abstract class E with concrete subclasses Var, App and Abs for, respectively, *variables*, *applications* and *abstractions*. Scala's *case classes* are used to make E behave like an algebraic datatype with (pattern-matchable) constructors Var, App and Abs à la functional programming.

Skeptik is flexible w.r.t. the underlying proof calculus. Every proof node is an instance of the abstract class ProofNode and must contain a judgment of some concrete subclass of Judgment and a (possibly empty) collection of premises (which are other proof nodes). A proof is a directed acyclic graph of proof nodes; it is implemented as the class Proof, which provides higher-order methods for traversing proofs. Thanks to Scala's syntax conventions, these methods can be used as an internal domain specific language that integrates harmoniously with the Scala language itself. A proof calculus is a collection of inference rules, implemented as concrete subclasses of ProofNode. In particular, the main inference rules of the propositional resolution calculus used for representing proofs generated by SAT- and SMT-solvers are Axiom and R (for resolution). They use a Sequent class (a subclass of Judgment) to represent clauses.

Classes for expressions and proof nodes are small and correctness conditions (e.g. typing conditions for expressions and the existence of a pivot for a resolution inference) are checked during object construction. Once constructed, they cannot be changed, because they are *immutable*. Therefore, incorrect expressions and proofs cannot result from the transformations performed by Skeptik. Auxiliary functionality is not implemented in the classes but in their homonymous *companion objects*. Therefore, even though Skeptik has more than 21000 lines of code, its most critical core data structures are less than a few hundred lines long.

3   Supported Proof Formats

Scala's *combinator parsing* library makes it easy to implement parsers for various proof formats. Skeptik uses the extension of a file to determine its proof format. To export proofs in various formats, Skeptik provides many exporter classes that extend Java's java.io.Writer class. Available proof formats are:

TraceCheck Format (".tc"): The TraceCheck format is one of the three formats accepted at the *Certified Unsat* track of the SAT-Competition and is used by SAT-solvers such as PicoSAT [3]. Each line declares a new clause, specifying its name (a fresh positive integer), a space separated list of literals (positive or negative integers, depending on the polarity of the literals), and a list of premises (other clauses, referred by their names) needed to derive the new clause by regular input resolution. Zero is used as a delimiter. Figure 1 shows an example. Other formats accepted at the *Certified Unsat* track are less detailed and hence less convenient to be used by tools that post-process proofs. The omission of premises in the RUP format, for example, throws away information about the

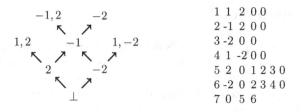

Fig. 1. A proof and its representation in the TraceCheck format

DAG structure of the proofs. Nevertheless, there are tools for converting RUP proofs to the trace-check format [13].

SMT Proof Format (".smt2"):   Although there is a well-established format for SMT problems, there is still no agreement on a format for SMT proofs. Skeptik supports the format used by veriT [5], which is close in style to SMT-Lib's problem format. Other formats could be supported if requested by users. In contrast to the TraceCheck format, in veriT's format, expressions can be arbitrary first-order terms and formulas and not only propositional variables represented as integers. The proofs are purely resolution-based at the bottom (closer to the root) but may contain theory-related and CNF transformation inferences at the top. This clear separation makes the proofs amenable to propositional resolution proof compression techniques. Skeptik currently simply ignores non-resolution inferences, but does keep them in the compressed proof it outputs.

Skeptik's Proof Format (".s"):   Skeptik's own proof format is a propositional resolution proof format meant to be simple and easy to read and write by humans. Each line either declares a new named subproof or deletes a previously declared subproof. Axioms are represented as sequents surrounded by curly braces. The infix resolution operator on subproofs is denoted by the pivot literal surrounded by square brackets or by a single dot (if there is a single pivot candidate and its omission is desired). Subproof names and literals can be arbitrary strings of letters and digits. The last named subproof is considered to be the whole proof.

$$u = (\{\ 1 \vdash 2\ \}\ [2]\ \{\ 2 \vdash\ \})$$
$$q = ((\{\ \vdash 1, 2\ \}\ .\ u)\ .\ (u\ .\ \{\ 2 \vdash 1\ \}))$$

Fig. 2. The proof from Fig. 1 represented in the Skeptik format

4   Proof Compression Algorithms

One of Skeptik's essential design goals is ease of implementation, combination and comparison of proof compression algorithms. This is evidenced by the fact that most algorithms for the compression of propositional resolution proofs described in the literature are available in Skeptik. They are shortly described below:

RecyclePivots (RP) [1,2] compresses a proof by partially regularizing it. A proof is *irregular* [16] if the resolved literal (pivot) of a resolution proof node is resolved again on the path from this node to the root node. RP finds irregular nodes efficiently by traversing the proof from the root to the leaves a single time and memorizing which literals were resolved. When it finds an irregular node, it marks one of its premises for deletion. In a second traversal, from the leaves to the root, irregular nodes are replaced by their non-deleted premises. As full regularization can lead to an exponential blow-up in the proof length [11], it is important to regularize carefully and only partially. RP achieves this by resetting the set of literals for a node to the empty set when it has more than one child (i.e. when it is the premise of more than one node).

RecyclePivotsWithIntersection (RPI) [9] differs from RP in the treatment of a node with more than one child. Instead of resetting its set of literals to the empty set, the intersection of the sets of literals incoming from its children is computed. In this manner, the exponential blow-up is still avoided, but strictly more irregular nodes are detected and regularized.

LowerUnits (LU) [9] partially eliminates a kind of redundancy that is almost orthogonal to irregularity. When a node η appears as premise of many resolutions with the same pivot p, it is desirable to resolve η on p only once instead. This is not always possible, unless η contains a unit clause (a clause with only the literal p). LowerUnits reduces redundancy by lowering all unit nodes. The nodes are removed from their places and reintroduced in the very bottom of the proof, by resolving them (at most once) with the root of the fixed proof.

LowerUnivalents (LUV) [4] generalizes LowerUnits. By keeping track of nodes that have already been lowered and their pivots, it becomes possible to lower a non-unit node if it is *univalent*: all its literals but one (its so-called *valent* literal) can be resolved against the valent literals of the already lowered nodes.

LUVRPI [4] is a non-sequential combination equivalent to the sequential combination of LUV after RPI and currently provides one of the best trade-offs between compression time and compression ratio. Non-sequential combinations with LUV are easy to implement, because LUV has been implemented as a replacement for the `fixProof` function used by some algorithms to reconstruct a proof after deletions.

RPI3LU and RPI3LUV are non-sequential combinations of RPI after LU and LUV, respectively. They consist of three traversals. The first traversal collects subproofs to be lowered. The second traversal computes the sets of safe literals for each node, taking into account the subproofs marked for being lowered. The last traversal actually compresses the proof by removing redundant branches and lowering subproofs. These algorithm are optimizations of the corresponding sequential compositions, achieving the same compression ratio in less time.

Reduce&Reconstruct (RR) [15] applies local transformation rules that either eliminate local redundancies or shuffle the order of resolution steps (similarly to what Gentzen's rank reduction rules do) in order to gradually transform non-local redundancies into local ones. Although the given set of local rules is sufficient to emulate any other compression algorithm, the algorithm may need many traversals to shuffle the order of resolutions steps sufficiently well to eliminate non-local redundancies. This algorithm can achieve very good compression ratio, if executed for long enough. The implementation in Skeptik is very modular, allowing convenient experimentation with various alternative local transformation rules, rule application heuristics and termination criteria.

Split [6] lowers pivot variables in a proof. From a proof with conclusion C, two proofs with conclusions $v \vee C$ and $\neg v \vee C$ are constructed, where the variable v is chosen heuristically. In a first step the positive/negative premises of resolvents with pivot v are removed from the proof. Afterwards the proof is fixed, by traversing it top-down and fixing each proof node. A proof node is fixed by either replacing it by one of its fixed premises or resolving them. The roots of the resulting proofs are resolved, using v as pivot, to obtain a new proof of C. The time-complexity of this algorithm is linear in the proof length, but it has to be repeated many times to obtain significant compression. This can be done iteratively or recursively. Also multiple variables can be chosen in advance. All these variants are implemented in Skeptik.

Tautology Elimination (ET) eliminates proof nodes containing tautological clauses. Although tautological clauses normally do not occur in proofs generated by SAT- and SMT-solvers, they may occur in post-processed proofs.

DAGification (D) finds proof nodes having equal clauses and replaces one of them by the other. *Subsumption* algorithms generalize DAGification by replacing a node containing a clause C_2 by another node containing a clause C_1 if C_1 subsumes C_2. There are three subsumption-based proof compression algorithms implemented in Skeptik, `TopDownSubsumption` (TDS), `BottomUpSubsumption` (BUS) and `RecycleUnits` (RU) [2], all with quadratic worst-case complexity.

Pebbling algorithms compress proofs w.r.t. their *space*, not their length. The space of a proof is the maximum number of proof nodes that have to be kept in memory simultaneously, while reading and checking the proof. Minimizing the space measure is analogous to minimizing the number of pebbles used for playing the *Black Pebbling Game* [10] on the DAG of the proof. This is a hard problem and Skeptik provides many greedy heuristics for reducing space well, though not optimally. Among them, the fastest and most compressive are some heuristic variants of the `BottomUpPebbler` (BUP).

5   Installation and Usage

Skeptik is implemented in Scala and runs on the Java virtual machine (JVM). Therefore, Java (`https://www.java.com/`) must be installed. The easiest way to download Skeptik is via git (`http://git-scm.com/`), by executing [`git clone git@github.com:Paradoxika/Skeptik.git`] in the folder where Skeptik should be downloaded. It is helpful to install SBT (`http://www.scala-sbt.org/`), a build tool that automatically downloads all compilers and libraries on which Skeptik depends. To compile, build and package Skeptik, run [sbt one-jar] in Skeptik's home folder. This generates an executable jar file. SBT and Scala programs may need a lot of memory for compilation and execution. If out-of-memory problems occur, the JVM's maximum available memory can be increased by executing [export JAVA_TOOL_OPTIONS='‘-Xmx1024m -Xss4m -XX:MaxPermSize=256m'’].

The command [java -jar skeptik.jar --help] displays a help message explaining how to use Skeptik. To compress the proof "eq_diamond9.smt2" using the algorithm RPI and write the compressed proof using the 'smt2' proof format, for example, the following command should be executed: [java -jar skeptik.jar -a RPI -f smt2 examples/proofs/VeriT/eq_diamond9.smt2]. Skeptik can be called with an arbitrary number of algorithms and proofs. The following command would compress the proofs "p1.smt2" and "p2.smt2" with two algorithms each (RP and a sequential composition of D, RPI and LU): [java -jar skeptik.jar -a RP -a (D-RPI-LU) p1.smt2 p2.smt2]

Skeptik can also be used as a library of proof data structures and compression algorithms from within any Java or Scala program. In this manner, communication via proof files can be avoided.

6   Conclusions and Future Work

Skeptik's development started around March 2012 and since then it has been used internally to compare various proof compression algorithms and develop new ones. Now its wide collection of algorithms is ready to be released to external users interested in improving the proofs they obtain from SAT- and SMT-solvers. One particularly successful external use of Skeptik was in interpolation-based controller synthesis [14]: proofs with millions of nodes have been compressed by 70% and better interpolants could be extracted from the compressed proof. Compression ratios varying from 10% to 70% (typically around 20%), depending on the benchmark and on the employed algorithm, have been observed. The performance is acceptable; the conveniences of a high-level language such as Scala outweigh its overheads.

In the near future, parsers for other proof formats for propositional resolution proofs could be easily added if requested by users. Moreover, Skeptik was designed to be flexible with respect to the underlying proof system. It is not restricted to propositional resolution. Techniques for compressing natural deduction proofs are currently being investigated and implemented [18]. Data structures for sequent calculus proofs are partially available, and techniques for compressing them [17] could be implemented in Skeptik as well. Many of the propositional resolution proof compression algorithms could be extended to first- or even higher-order resolution, by taking extra care of unification. Features such as the extraction of unsat cores and interpolants could be useful additions to Skeptik as well.

Acknowledgments.   The Vienna Scientific Cluster (http://www.vsc.ac.at) is regularly used to evaluate Skeptik on thousands of proofs.

References

1. Bar-Ilan, O., Fuhrmann, O., Hoory, S., Shacham, O., Strichman, O.: Linear-time reductions of resolution proofs. In: Chockler, H., Hu, A.J. (eds.) HVC 2008. LNCS, vol. 5394, pp. 114–128. Springer, Heidelberg (2009)
2. Bar-Ilan, O., Fuhrmann, O., Hoory, S., Shacham, O., Strichman, O.: Reducing the size of resolution proofs in linear time. STTT 13(3), 263–272 (2011)
3. Biere, A.: Picosat essentials. Journal on Satisfiability, Boolean Modeling and Computation, JSAT (2008)
4. Boudou, J., Woltzenlogel Paleo, B.: Compression of propositional resolution proofs by lowering subproofs. In: Galmiche, D., Larchey-Wendling, D. (eds.) TABLEAUX 2013. LNCS, vol. 8123, pp. 59–73. Springer, Heidelberg (2013)
5. Bouton, T., de Oliveira, D.C.B., Déharbe, D., Fontaine, P.: verit: an open, trustable and efficient smt-solver. In: Schmidt, R.A. (ed.) CADE 2009. LNCS (LNAI), vol. 5663, pp. 151–156. Springer, Heidelberg (2009)
6. Cotton, S.: Two techniques for minimizing resolution proofs. In: Strichman, O., Szeider, S. (eds.) SAT 2010. LNCS, vol. 6175, pp. 306–312. Springer, Heidelberg (2010)
7. Dunchev, C., Leitsch, A., Libal, T., Riener, M., Rukhaia, M., Weller, D., Woltzenlogel Paleo, B.: Prooftool: a gui for the gapt framework. In: UITP, pp. 1–14 (2013)
8. Dunchev, T., Leitsch, A., Libal, T., Weller, D., Woltzenlogel Paleo, B.: System description: The proof transformation system ceres. In: Giesl, J., Hähnle, R. (eds.) IJCAR 2010. LNCS (LNAI), vol. 6173, pp. 427–433. Springer, Heidelberg (2010)
9. Fontaine, P., Merz, S., Woltzenlogel Paleo, B.: Compression of propositional resolution proofs via partial regularization. In: Bjørner, N., Sofronie-Stokkermans, V. (eds.) CADE 2011. LNCS (LNAI), vol. 6803, pp. 237–251. Springer, Heidelberg (2011)
10. Gilbert, J.R., Lengauer, T., Tarjan, R.E.: The pebbling problem is complete in polynomial space. SIAM Journal on Computing 9(3), 513–524 (1980)
11. Goerdt, A.: Comparing the complexity of regular and unrestricted resolution. In: Marburger, H. (ed.) GWAI. Informatik-Fachberichte, vol. 251. Springer (1990)
12. Hetzl, S., Leitsch, A., Weller, D., Woltzenlogel Paleo, B.: Herbrand sequent extraction. In: Autexier, S., Campbell, J., Rubio, J., Sorge, V., Suzuki, M., Wiedijk, F. (eds.) AISC/Calculemus/MKM 2008. LNCS (LNAI), vol. 5144, pp. 462–477. Springer, Heidelberg (2008)
13. Heule, M., Hunt Jr., W.A., Wetzler, N.: Trimming while checking clausal proofs. In: FMCAD, pp. 181–188 (2013)
14. Hofferek, G., Gupta, A., Könighofer, B., Jiang, J.H.R., Bloem, R.: Synthesizing multiple boolean functions using interpolation on a single proof. In: FMCAD, pp. 77–84 (2013)
15. Rollini, S.F., Bruttomesso, R., Sharygina, N.: An efficient and flexible approach to resolution proof reduction. In: Raz, O. (ed.) HVC 2010. LNCS, vol. 6504, pp. 182–196. Springer, Heidelberg (2010)
16. Tseitin, G.S.: On the complexity of derivation in propositional calculus. In: Siekmann, J., Wrightson, G. (eds.) Automation of Reasoning: Classical Papers in Computational Logic 1967-1970, vol. 2. Springer (1983)
17. Woltzenlogel Paleo, B.: Atomic cut introduction by resolution: Proof structuring and compression. In: Clarke, E.M., Voronkov, A. (eds.) LPAR-16 2010. LNCS, vol. 6355, pp. 463–480. Springer, Heidelberg (2010)
18. Woltzenlogel Paleo, B.: Contextual natural deduction. In: Artemov, S., Nerode, A. (eds.) LFCS 2013. LNCS, vol. 7734, pp. 372–386. Springer, Heidelberg (2013)

Terminating Minimal Model Generation Procedures for Propositional Modal Logics*

Fabio Papacchini and Renate A. Schmidt

The University of Manchester, UK

Abstract. Model generation and minimal model generation are useful for tasks such as model checking and for debugging of logical specifications. This paper presents terminating procedures for the generation of models minimal modulo subset-simulation for the modal logic **K** and all combinations of extensions with the axioms **T**, **B**, **D**, **4** and **5**. Our procedures are minimal model sound and complete. Compared with other minimal model generation procedures, they are designed to have smaller search space and return fewer models. In order to make the models more effective for users, our minimal model criterion is aimed to be semantically meaningful, intuitive and contain a minimal amount of information. Depending on the logic, termination is ensured by a variation of equality blocking.

1 Introduction

Automated reasoning methods are often designed to check satisfiability and validity of formulae. In many applications the "yes or no" answer returned by these methods is all the information that is needed, but there are tasks where additional information is required. Model generation methods complement such automated reasoning methods by returning models that explain why a certain answer holds. Examples of tasks where model generation methods are useful are fault analysis, model checking and debugging of logical specifications [18,13]. Even for the most well-behaved, decidable logics, in general, there are uncountably many different models for satisfiable formulae and models can be very large, which makes effective model generation a challenging problem. For these reasons, there have been several studies about the generation of different kinds of *minimal* models for classical and non-classical logics [2,15,14,16,17,13,7].

In this paper we introduce a semantic notion of minimality, similar to the notions used in [17,13]. Our minimality criterion is based on a preorder on models called subset-simulation (i.e., it is a variation of the more common notion of simulation [4]). The criterion is designed so that minimal models are semantically meaningful, more natural than models minimal with respect to other minimality criteria, and contain a minimal amount of information. In this paper we propose

---

* The first author is supported by an EPSRC EU Doctoral Training Award. The research was partially supported by EPSRC research grant EP/H043748/1.

S. Demri, D. Kapur, and C. Weidenbach (Eds.): IJCAR 2014, LNAI 8562, pp. 381–395, 2014.

the first terminating, minimal model sound and complete procedures for the generation of models minimal modulo subset-simulation for all normal modal logics in between **K** and **S5**. If a model generator is minimally sound and complete, not only will it generate one minimal model, but it will generate the complete set of all minimal models. In comparison with other approaches, our procedures benefit from smaller search spaces, and fewer models are returned. In particular, we aim to return the smallest set of all relevant minimal models, so that the user is not swamped with too many similar models.

As modal logics are closely related to description logics, our procedures can be used as methods complementary with respect to ontology debugging methods such as the ones proposed in [19,8]. The usual definition of ontology debugging assumes that an ontology is incoherent (inconsistent). In this case, debugging is the ability to identify the cause of the incoherence and fix it. An ontology can however be considered faulty even when it is coherent, but it does not properly model the domain of interest. This can be because aspects and properties of the domain of interest, that are expected to hold, do not follow from the ontology. In this context, similarly to test-driven software development paradigms, our procedures complement the notion of ontology debugging and provide assistance to model correctly the domain of interest. Minimal model generation procedures can be used to check whether these properties hold at any stage of the life cycle of the ontology, and then corrected based on the computed models.

Another possible use for the generation of models minimal modulo subset-simulation is positive query answering for Horn fragments of modal logics similar to [13]. In [13] the query answering problem is reduced to a model checking problem. Our procedures can be used in the same way, but it is not restricted to Horn fragments of modal logics.

The logics and the main properties of models minimal modulo subset-simulation are presented in Section 2. Section 3 defines minimal model sound and complete procedures for the logics under consideration. As minimal model soundness can be easily shown if the procedures are minimal model complete, we focus on formally proving minimal model completeness (Section 4). Termination results for all the logics under consideration are presented in Section 5, as one of the main contributions of the paper. How our procedures relate to other minimal model generation procedures, what are possible extensions of the procedures, and how it is possible to further improve them is discussed in Section 6. Section 7 summarises the contributions of the paper and mentions directions of future work.

2   Modal Logics and the Minimality Criterion

We work with modal formulae of the propositional modal logic **K** possibly extended with a subset of the well-known axioms **T**, **B**, **D**, **4**, and **5**. Specifically, the logics covered in this paper are all propositional normal modal logics below **S5**, namely, **K**, **KD**, **KDB**, **K4**, **K5**, **KD4**, **KD5**, **K45**, **KD45**, **KB4**, **KT4**, and **KT5**(= **S5**). All these logics are decidable. Table 1 lists the axioms and their semantic characterisations as frame properties.

Table 1. Modalities and their corresponding frame conditions

$\Box$	Axiom	Frame condition	First-order representation
K			
T	$\Box p \to p$	reflexivity	$\forall x R(x,x)$
B	$p \to \Box \Diamond p$	symmetry	$\forall x \forall y(R(x,y) \to R(y,x))$
D	$\Box p \to \Diamond p$	seriality	$\forall x \exists y R(x,y)$
4	$\Box p \to \Box \Box p$	transitivity	$\forall x \forall y \forall z(R(x,y) \wedge R(y,z) \to R(x,z))$
5	$\Diamond p \to \Box \Diamond p$	Euclideanness	$\forall x \forall y \forall z(R(x,y) \wedge R(x,z) \to R(y,z))$

A *modal formula* is a formula of the form $\top$, $\bot$, p_i, $\neg \phi$, $\phi_1 \wedge \phi_2$, $\phi_1 \vee \phi_2$, $\Diamond \phi$, $\Box \phi$, where $\top$ and $\bot$ are two nullary logical operators for, respectively, true and false; $p_i \in \Sigma$ is a propositional symbol belonging to the set Σ of propositional symbols; $\neg$, $\wedge$, $\vee$, $\Diamond$, $\Box$ are, respectively, the logical operators of negation, conjunction, disjunction, diamond and box; and ϕ_1, ϕ_2, ϕ are modal formulae.

We adopt the standard semantics of modal formulae, known as Kripke semantics. A *frame* for a modal logic is a tuple (W, R), where W is a non-empty set of worlds and $R \subseteq W \times W$ is the accessibility relation over W. An *interpretation* $\mathcal{I}$ is a tuple (W, R, V) composed of a frame and an interpretation function V that assigns to each world $u \in W$ a set of propositional symbols, meaning that such propositional symbols hold in u. Given an interpretation $\mathcal{I} = (W, R, V)$ and a world $u \in W$, truth of a modal formula ϕ is inductively defined as follows.

$\mathcal{I}, u \not\models \bot$ \qquad $\mathcal{I}, u \models \top$

$\mathcal{I}, u \models p_i$ \qquad iff $p_i \in V(u)$

$\mathcal{I}, u \models \neg \phi$ \qquad iff $\mathcal{I}, u \not\models \phi$

$\mathcal{I}, u \models \phi_1 \vee [\wedge] \phi_2$ \qquad iff $\mathcal{I}, u \models \phi_1$ or[and] $\mathcal{I}, u \models \phi_2$

$\mathcal{I}, u \models \Box \phi$ \qquad iff for every $v \in W$ if $(u,v) \in R$ then $\mathcal{I}, v \models \phi$

$\mathcal{I}, u \models \Diamond \phi$ \qquad iff there is a $v \in W$ such that $(u,v) \in R$ and $\mathcal{I}, v \models \phi$

Given an interpretation $\mathcal{I}$, a world u and a modal formula ϕ, if $\mathcal{I}, u \models \phi$ holds, then $\mathcal{I}$ is a *model* of ϕ.

A *model graph* $M = (W, R, \mathcal{V})$ is an interpretation except that $\mathcal{V}(u)$ returns the set of formulae true in u. Given a model graph $M = (W, R, \mathcal{V})$ it is possible to obtain the corresponding interpretation $\mathcal{I} = (W, R, V)$, where $V(u) = \mathcal{V}(u) \cap \Sigma$ for all $u \in W$.

Let u and v be two elements of the domain of a model $\mathcal{I}$. If there is a path in the model from u to v, then u is an *ancestor* of v and v is a *descendant* of u.

The *frame closure* of a model $\mathcal{I}$ is the model obtained by computing the closures of the relevant frame properties (e.g., transitive closure).

Let $\mathcal{I} = (W, R, V)$ and $\mathcal{I}' = (W', R', V')$ be two models of a modal formula ϕ. A *bisimulation* is a binary relation $B \subseteq W \times W'$ such that for any two worlds $u \in W$ and $u' \in W'$, if uBu' then the following hold.

– $V(u) = V'(u')$,

- if uRv, then there exists a $v' \in W'$ such that $u'R'v'$ and vBv', and
- if $u'R'v'$, then there exists a $v \in W$ such that uRv and vBv'.

An *auto-bisimulation* is a bisimulation between a model and itself.

Let $\mathcal{I} = (W, R, V)$ and $\mathcal{I}' = (W', R', V')$ be two models of a modal formula ϕ. A *subset-simulation* is a binary relation $S \subseteq W \times W'$ such that for any two worlds $u \in W$ and $u' \in W'$, if uSu' then the following hold.

- $V(u) \subseteq V'(u')$, and
- if uRv, then there exists a $v' \in W'$ such that $u'R'v'$ and vSv'.

If S is such that for all $u \in W$ there is at least one $u' \in W'$ such that uSu', then we call S a *full subset-simulation* from $\mathcal{I}$ to $\mathcal{I}'$. We say a subset-simulation S is a *maximal subset-simulation* if there is no other subset-simulation $S' \neq S$ such that $S \subset S'$. Given two models $\mathcal{I}$ and $\mathcal{I}'$, if there is a full subset-simulation S from $\mathcal{I}$ to $\mathcal{I}'$, we say that $\mathcal{I}'$ *subset-simulates* $\mathcal{I}$, or $\mathcal{I}$ *is subset-simulated* by $\mathcal{I}'$. We write $\mathcal{I} \leq_{\subseteq} \mathcal{I}'$ if $\mathcal{I}$ is subset-simulated by $\mathcal{I}'$.

Subset-simulation is a preorder on models. That is, subset-simulation is a reflexive and transitive relation on models. For this reason it can be used to define the following minimality criterion. A model $\mathcal{I}$ of a modal formula φ is *minimal modulo subset-simulation* iff for any model $\mathcal{I}'$ of φ, if $\mathcal{I}' \leq_{\subseteq} \mathcal{I}$, then $\mathcal{I} \leq_{\subseteq} \mathcal{I}'$.

As bisimulation is more restrictive than subset-simulation, any model $\mathcal{I}_B$ bisimilar to a model $\mathcal{I}$ preserves the original subset-simulation relationship of $\mathcal{I}$. This result is formally expressed in the following lemma, and is used for proving termination of our procedures.

Lemma 1. *Bisimulation preserves subset-simulation. That is, given two models $\mathcal{I}$ and $\mathcal{I}'$, any bisimilar model $\mathcal{I}_B$ of $\mathcal{I}$ is such that if $\mathcal{I} \leq_{\subseteq} \mathcal{I}'$ then $\mathcal{I}_B \leq_{\subseteq} \mathcal{I}'$, and if $\mathcal{I}' \leq_{\subseteq} \mathcal{I}$ then $\mathcal{I}' \leq_{\subseteq} \mathcal{I}_B$.*

3   Procedures for the Generation of Minimal Models

Our procedures for the generation of models minimal modulo subset-simulation are composed of a tableau calculus and a minimality test. Depending on which logic below **S5** is considered, different rules for handling frame properties and different termination techniques are used. The tableau calculus, without the minimality test, is devised to generate minimal models, but it can also generate non-minimal models. We prove minimal model completeness of the calculus in Section 4, and that the use of the minimality test results in minimal model sound and complete procedures.

As the minimality criterion is based on a preorder, it is possible to have symmetry classes of models and minimal models belong to the same symmetry class. Models that belong to the same symmetry class share the same positive information, meaning that all the models entail the same positive formulae. For this reason, we define minimal model completeness as follows. A procedure is *minimal model complete* if it generates at least one witness for each symmetry class of minimal models.

Table 2. Tableau calculus for the generation of minimal models

$$(\Box) \quad \frac{(u,v) : R \quad u : \Box\phi}{v : \phi} \qquad\qquad (\alpha) \quad \frac{u : (\phi_1 \wedge \ldots \wedge \phi_n) \vee \Phi_\alpha^+}{u : \phi_1 \vee \Phi_\alpha^+}$$

$$\vdots$$

$$u : \phi_n \vee \Phi_\alpha^+$$

$$(\beta) \quad \frac{u : \mathcal{A} \vee \Phi^+}{\left. \begin{array}{c} u : \mathcal{A} \\ u : neg(\Phi^+) \end{array} \right| u : \Phi^+} \quad \begin{array}{l} \text{where } \mathcal{A} \text{ is of the form } \Diamond\phi, \Box\phi, \text{ or } p_i, \text{ and} \\ neg(\Phi^+) = \neg p_1 \wedge \ldots \wedge \neg p_n, \text{ where each } p_i \\ \text{is a disjunct of } \Phi^+. \end{array}$$

$$(\Diamond) \quad \frac{u : \Diamond\phi}{\begin{array}{c} (u,v) : R \\ v : \phi \end{array}} \quad \text{where } v \text{ is fresh.}$$

$$(SBR) \quad \frac{u : p_1 \quad \ldots \quad u : p_n \quad u : \neg p_1 \vee \ldots \vee \neg p_n \vee \Phi_\alpha^+}{u : \Phi_\alpha^+}$$

The input to the calculus is a modal formula in negation normal form labelled by an initial world u. Transformation to negation normal form is not essential, but it simplifies the presentation. It also means that there is no need for pre-processing before applying the calculus, and reduces the number of rules in the calculus. Disjunctions and conjunctions are assumed to be flattened (e.g., we write $\phi_1 \vee \phi_2 \vee \phi_3$ instead of $\phi_1 \vee (\phi_2 \vee \phi_3)$). By $\mathcal{A}$ we mean a modal formula of the form p_i, $\Diamond\phi$ or $\Box\phi$. We use Φ^+ to denote a non-empty disjunction, where all disjuncts are of the form $\mathcal{A}$, and use Φ_α^+ to denote a possibly empty disjunction, where all disjuncts are of the form $\mathcal{A}$ or are conjunctions. By $neg(\Phi^+)$ we mean the conjunction $\neg p_1 \wedge \ldots \wedge \neg p_n$, where the p_i are all the positive propositional variables appearing as disjuncts of Φ^+. If Φ^+ does not contain any p_i, then $neg(\Phi^+) = \top$. The exclusive selection of positive propositional variables is crucial for the minimal model completeness of the calculus. An example of this is given in the explanation of the (β) rule.

Table 2 presents the rules of the calculus for the modal logic **K**. Given an input formula $u : \phi$, the rules are exhaustively applied. At most one rule is applied to any formula appearing as the main premise, where the *main premise* of a multi-premise rule is the premise on the right. For fairness, each instance of a rule application is performed exactly once. Given an open branch $\mathcal{B}$ in a tableau derivation, a model $\mathcal{I} = (W, R, V)$ can be extracted from $\mathcal{B}$ as follows. The domain W is the set of all the labels occurring in $\mathcal{B}$, the accessibility relation is composed of all the instances $(u, v) : R$ in $\mathcal{B}$, the interpretation function V is such that $V(u) = \{p_i \mid u : p_i \in \mathcal{B}\}$. A partial model graph M is extracted from a branch $\mathcal{B}$ in a similar way, except that $\mathcal{V}(u) = \{\phi \mid u : \phi \in \mathcal{B}\}$.

The (α) rule is a variation of standard rule for conjunctions. If $\Phi_\alpha^+ = \top$ then it just expands the conjunction, otherwise the application of the (α) rule performs lazy clausification. If such lazy clausification is performed in a clever way, for example, by using a good heuristic for choosing the right conjunction to expand, it can result in the reduction of inferences due to the implicit restriction of Φ_α^+ in the premise of the rule.

The ($\square$) rule and the ($\Diamond$) rule are the common rules for box and diamond formulae. They simply expand formulae in the scope of a modality as required by their semantics.

The (β) rule is the only branching rule of the calculus. Its purpose is to branch over disjunctions without any negated propositional variables, and to close the left branch if it is not minimal. This latter point is achieved by the use of a limited form of complement splitting (more common uses of complement splitting can be found in the literature, e.g. [2]). The reason why complement splitting is applied only on positive propositional variables is that the negation of diamond formulae or box formulae would result in new modal formulae (specifically, box formulae and diamond formulae) that can compromise the minimality of the resulting model. For example, let us assume that the (β) rule is applied to $u : \phi_1 \vee \square\phi_2$. If the complement $\Diamond\neg\phi_2$ of $\square\phi_2$ would have been added to the left branch, the left branch would still be open, and the resulting model would still be a model for the original formula, but the newly introduced diamond formula would generate unnecessary information. The resulting model would not be minimal. A similar example can be given for the case of the negation of diamond formulae.

The (SBR) rule is a selection-based resolution rule. It can be seen as a weaker version of the (SBR) rule in [16], the $PUHR$ rule in [2], or the hyper-tableau rule in [1]. The aim of this rule is twofold. First, it provides the closure rule of the calculus, because atomic closure is sufficient. Second, it allows to remove negative information (i.e., all negative propositional variables) from a disjunction. The rationale for the (SBR) rule is that if a disjunction contains negative information (at least one negated propositional variable) that is not in conflict with any formula on the branch, then any expansion of such a disjunction results in either a minimal model, where the disjunction is true due to the negative information, or in a non-minimal model. Hence, there is no advantage in expanding a disjunction as long as it is not possible to remove all the negative information from it. The (SBR) rule is the reason why other rules, specifically the (β) rule and the (α) rule, can be applied only to disjunctions of the form Φ^+ or Φ_α^+. This decreases the number of required inferences.

Theorem 1. *The tableau calculus in Table 2 is sound and refutationally complete for* **K**.

For reasons of space we omit a formal proof, but the calculus does not differ much from known calculi. All the rules are sound variations of common rules. The rule modifications help in directing the calculus toward the generation of minimal models, for example, the restrictions in Φ^+ or Φ_α^+.

In the next section we show that the calculus is minimal model complete. *Minimal model completeness* means that the calculus generates at least one witness per symmetry class of minimal models. We also want the calculus to be *minimal model sound*, that is, only minimal models are generated.

Table 3. Rules for extending the calculus

$$(\mathbf{T})\ \frac{}{(u,u):R} \qquad\qquad (\mathbf{B})\ \frac{(u,v):R}{(v,u):R}$$

$$(4)\ \frac{(u,v):R\quad (v,w):R}{(u,w):R} \qquad (5)\ \frac{(u,v):R\quad (u,w):R}{(v,w):R}$$

$$(\mathbf{D})\ \frac{}{u:\Diamond\top}$$

To achieve minimal model soundness we define a minimality test to close branches from which non-minimal models can be extracted. The minimality test is called *subset-simulation test*. It consists of two operations. First, let $\mathcal{I}$ be a partial model extracted from an open branch $\mathcal{B}$. If a model $\mathcal{I}'$ such that $\mathcal{I}' \leq_{\subseteq} \mathcal{I}$ has already been found, then close $\mathcal{B}$. Second, let $\mathcal{I}$ be a model newly extracted from an open and fully-expanded branch $\mathcal{B}$. If a model $\mathcal{I}'$ such that $\mathcal{I}' \leq_{\subseteq} \mathcal{I}$ has already been found, then close $\mathcal{B}$, and for any already extracted model $\mathcal{I}'$, if $\mathcal{I} \leq_{\subseteq} \mathcal{I}'$, then close the branch from which $\mathcal{I}'$ was extracted.

Computing subset-simulation relations between finite models is a decidable problem. Our procedure uses the algorithm for computing subset-simulations presented in [17], which is a variation of the algorithm for computing auto-simulation in [6]. The complexity of the algorithm depends directly on the size of the domains and on the number of relations in the involved models.

Our minimal model generation procedure extends to all sublogics of **S5**. Table 3 contains the structural rules enabling the handling of all these logics. Augmenting the extensions with the subset simulation test results in minimal model sound and complete procedures. This is because the subset-simulation test is independent of the logic. What matters is minimal model completeness, and [17] proves that such structural rules preserve minimal model completeness. It is worth noting that some of the extensions, e.g., **K4**, might have minimal models with an infinite domain, and this affects termination. However, minimal model soundness and completeness can be ensured by choosing good branch selection strategies. A suitable branch selection strategy is to always select the branch with the smallest number of labels (i.e., the branch where the extracted partial model has the smallest domain). In Section 5 we show that a better branch selection strategy can be adopted as soon as termination of the procedure is ensured.

4   Minimal Model Completeness

Because our minimal model completeness proof relies on results in [17] for the multi-modal logic $\mathbf{K}_{(m)}$ and its extensions, we recall here some definitions from [17]. A *simulation* relation between models is as a subset-simulation relation, except that the first property is $V(u) = V'(u')$. $\mathcal{I} \leq_{=} \mathcal{I}'$ denotes that $\mathcal{I}$ is simulated by $\mathcal{I}'$. The minimality criterion in [17] is as follows. A model $\mathcal{I}$ of a modal formula φ is *minimal modulo subset-simulation* iff for any model $\mathcal{I}'$ of φ, if $\mathcal{I}' \leq_\subseteq \mathcal{I}$, then $\mathcal{I} \leq_\subseteq \mathcal{I}'$ and for any model $\mathcal{I}''$ of φ belonging to the same symmetry class of $\mathcal{I}$, if $\mathcal{I}'' \leq_= \mathcal{I}$ then $\mathcal{I} \leq_= \mathcal{I}''$. The notion of minimal model completeness used in [17] requires the generation of all minimal models, and not just a witness per symmetry class.

As the logics considered in this paper are a subset of the logics considered in [17], the following lemma is restricted to the logics covered in this paper.

Lemma 2. *Let $\mathcal{I}$ and $\mathcal{I}'$ be models of a modal formula φ such that $\mathcal{I}$ is minimal with respect to the minimality criterion in [17], and $\mathcal{I}'$ is minimal with respect to the minimality criterion used in this paper. Then the following hold.*

- *$\mathcal{I}$ is minimal with respect to the minimality criterion used in this paper, and*
- *there exists a model $\mathcal{I}''$ minimal with respect to the minimality criterion used in [17] such that $\mathcal{I}'' \leq_\subseteq \mathcal{I}'$.*

The lemma explains the relation between the minimality criterion used in [17] and the minimality criterion used in this paper. The first point of Lemma 2 tells us that the minimality criterion used in this paper considers as minimal all the models considered minimal by the minimality criterion in [17], and potentially more than these. The second point tells us, indirectly, that the symmetry classes for the two notions are the same.

As the notion of minimal model completeness in [17] is wider than our notion, and given the relation between the two minimality criteria, the following holds.

Lemma 3. *The procedure in [17] is minimal model complete with respect to our notions of minimal model and minimal model completeness.*

From a procedural perspective, the minimal model generation procedure we propose and the procedure proposed in [17] mainly differ in how diamond formulae are expanded. This is due to the use of different minimality criteria and different notions of minimal model completeness, which force the calculus in [17] to explore all possible expansions of diamond formulae. The ($\Diamond$) rule, simplified to the uni-modal case, used in [17] is the following.

$$(\Diamond) \quad \frac{u : \Diamond \phi}{(u,u_1) : R \Big|...\Big|(u,u_n) : R \Big|(u,v) : R} \qquad \text{where each } u_i \text{ appears on the}$$
$$\frac{}{u_1 : \phi \qquad\qquad u_n : \phi \quad v : \phi} \qquad \text{branch, and } v \text{ is fresh.}$$

Theorem 2. *For any model $\mathcal{I}'$ extracted from an open and fully expanded branch of the procedure in [17], there is a model $\mathcal{I}$ extracted from an open and fully expanded branch $\mathcal{B}$ of our procedure such that $\mathcal{I} \leq_\subseteq \mathcal{I}'$.*

Proof. Let $\mathcal{I} = (W, R, \mathcal{V})$ and $\mathcal{I}' = (W', R', \mathcal{V}')$. We prove the theorem inductively by creating a relation $S \subseteq W \times W'$ during the construction of the branch $\mathcal{B}$, and show that S is a full subset-simulation.

Base Case: Let us assume that the input of the procedure in [17] is $u' : \varphi$, and the input of our procedure is $u : \varphi$. This means that $\mathcal{I}', u' \models \varphi$, and the initial partial model graph $\mathcal{I}$ is $(\{u\}, \emptyset, \mathcal{V}(u) = \{\varphi\})$. As $\mathcal{I}'$ is a complete model graph and $\mathcal{I}', u' \models \varphi$, then $\varphi \in \mathcal{V}'(u')$. This implies that $\mathcal{V}(u) \subseteq \mathcal{V}'(u')$. Let $(u, u') \in S$. Then it is immediate that S is a full subset-simulation from $\mathcal{I}$ to $\mathcal{I}'$.

Induction Step: Let us assume that after n rule applications, for the extracted model $\mathcal{I}$ there is a subset-simulation S from $\mathcal{I}$ to $\mathcal{I}'$. We prove that S, or a variation of it, is still a subset-simulation relation after the application of any rule ρ of our procedure. For reasons of space we give proofs only for three of the rules, but all the other cases are similar. In all the following cases, we assume $\mathcal{I} = (W, R, V)$ is the model extracted before the application of ρ.

ρ is the (α) rule. This means that the expanded formula is a labelled disjunction $u : \varphi$, where at least one disjunct φ_α is a conjunction. Let Φ be the set of labelled formulae representing the conclusion of the (α) rule and $\Psi = \{\psi \mid u : \psi \in \Phi\}$. This means that Φ is on the branch and the new extracted model graph $\mathcal{I}'' = \mathcal{I}$, where $\mathcal{V}'' = \mathcal{V}$ except for $\mathcal{V}''(u) = \mathcal{V}(u) \cup \Psi$. By the inductive hypothesis, there is a $u' \in W'$ such that $(u, u') \in S$, $\mathcal{V}(u) \subseteq \mathcal{V}'(u')$ and $\mathcal{I} \leq_{\subseteq} \mathcal{I}'$. As $\varphi \in \mathcal{V}'(u')$ and $\mathcal{I}'$ is a complete model, then $\Psi \subseteq \mathcal{V}'(u')$. This implies that $\mathcal{V}''(u) \subseteq \mathcal{V}'(u')$. That is, the current S is a full subset-simulation such that $\mathcal{I}'' \leq_{\subseteq} \mathcal{I}'$. In principle there may be more than one conjunction to be selected for the application of the (α) rule. This implies that Φ may be different from the application of the (α) rule applied to generate $\mathcal{I}'$. Even though the two sets of conclusions are syntactically different, they are semantically equivalent. Hence, w.l.o.g. we can assume that the same conjunction is used.

ρ is the $(\Diamond)$ rule. This means that the expanded formula is a labelled diamond formula, let us say $u : \Diamond\varphi$. As a result of the application of the $(\Diamond)$ rule, $\{v : \varphi, (u, v) : R\}$ are on the branch and v is fresh on the branch. The new extracted model $\mathcal{I}''$ is as follows. $W'' = W \cup \{v\}$, $R'' = R \cup \{(u, v)\}$, and $\mathcal{V}'' = \mathcal{V}$ except for $\mathcal{V}''(v) = \{\varphi\}$. By the inductive hypothesis, there is a $u' \in W'$ such that $(u, u') \in S$, $\mathcal{V}(u) \subseteq \mathcal{V}'(u')$ and $\mathcal{I} \leq_{\subseteq} \mathcal{I}'$. As $\Diamond\varphi \in \mathcal{V}'(u')$ and $\mathcal{I}'$ is a complete model, then there is an R-successor $v' \in W'$ of u' such that $\varphi \in \mathcal{V}'(v')$. Let $S' = S \cup \{(v, v')\}$. S' is a full subset-simulation such that $\mathcal{I}'' \leq_{\subseteq} \mathcal{I}'$.

ρ is the (4) rule. This means that there are two labelled relations $(u, v) : R$ and $(v, w) : R$ for which transitivity has not been applied yet. As a result of the application of the (4) rule, $(u, w) : R$ is on the branch and the new extracted model $\mathcal{I}''$ is such that $\mathcal{I}'' = \mathcal{I}$, except for $R'' = R \cup \{(u, w)\}$. By the inductive hypothesis, there are $u', v', w' \in W'$ such that $(u, u'), (v, v'), (w, w') \in S$, $(u', v'), (v', w') \in R$ and $\mathcal{I} \leq_{\subseteq} \mathcal{I}'$. As $\mathcal{I}'$ is a complete model and R is transitive, then $(u', w') \in R$. That is, the current S is a full subset-simulation such that $\mathcal{I}'' \leq_{\subseteq} \mathcal{I}'$. $\qquad\square$

Corollary 1. *For any model $\mathcal{I}'$ minimal with respect to [17], there is a model $\mathcal{I}$ generated by our procedure such that $\mathcal{I} \leq_{\subseteq} \mathcal{I}'$.*

Minimal model completeness of our tableau calculus follows from Lemma 3 and Corollary 1. Minimal model soundness is the result of applying the subset-simulation test to minimal model complete tableaux calculi.

Theorem 3. *The tableau calculus in Table 2 and the extensions with the rules in Table 3 are minimal model complete. That is, the calculi generate at least one witness for each symmetry class of minimal models.*

Theorem 4. *Augmenting the tableau calculus in Table 2 and its extensions with the rules in Table 3 and the subset-simulation test gives us minimal model sound and complete procedures. That is, only minimal models and at least a witness for each symmetry class of minimal models are generated.*

5   Ensuring Termination

The presented calculus is (strongly) terminating for the modal logic **K** and its reflexive and symmetric extensions (i.e., it terminates for **KT**, **KB** and **KTB**). It is known that it is always possible to generate finite models for these logics without using any termination technique. As a reference for this, [16] proves that these logics have finite minimal Herbrand models and presents a tableau calculus that does not require any termination technique.

The same reasoning cannot be used for the other normal modal logics, namely, **KD**, **KDB**, **K4**, **K5**, **KD4**, **KD5**, **K45**, **KD45**, **KB4**, **KT4**, and **KT5**. The main challenge to obtain terminating procedures for these logics is to find blocking techniques preserving minimal model completeness.

For **KD** and **KDB** it is not difficult to achieve termination while preserving minimal model completeness. This is because the seriality condition is what affects termination, forcing all models to have paths where a world in which the only true formula is ⊤ is repeated infinitely many times. It is, therefore, enough to add a reflexive edge as soon as the first such world appears, and the resulting finite model is bisimilar to the original model. Therefore, the following holds.

Theorem 5. *Our procedures to handle the modal logics **KD** and **KDB** are minimal model sound and complete, and terminate if a reflexive loop is added to each occurrence of a fully-expanded world u where $\mathcal{V}(u) = \emptyset$.*

The rule application order is important in a practical implementation. Specifically, the ($\lozenge$) rule needs to be the last rule to be applied.

More interesting are the logics **K4**, **KT4**, and **KD4**. For these logics, the models can be infinite due to the seriality axiom or because of transitivity. The previous termination strategy is not sufficient for these logics (i.e., it is possible to have infinite chains where no fully-expanded world has an empty interpretation). The usual method to obtain a terminating tableau calculus, see, e.g. [9], is to use static subset blocking. Formally, a world u is *subset blocked* if there is a parent v of u such that $\mathcal{V}(u) \subseteq \mathcal{V}(v)$. This kind of blocking, however, is not compatible with our minimality criterion because it might potentially merge a world with less positive information and a world with more positive information. This may lead to non-minimal models being considered minimal, affecting the minimal model soundness of the procedure, or to the non-generation of some minimal model, affecting the minimal model completeness of the procedure. It turns out that

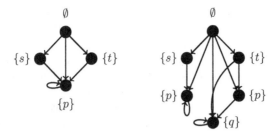

Fig. 1. Example of incompleteness when anywhere equality blocking is used

equality blocking is more suitable. However, not all forms of equality blocking can be used, but *ancestor equality blocking* can. A world u is considered *ancestor equality blocked* if there is an ancestor v of u such that $\mathcal{V}(u) = \mathcal{V}(v)$. We show that anywhere equality blocking (i.e., v can block u even if it is not an ancestor of u) violates minimal model completeness by means of an example. Suppose that $\phi = \Diamond s \wedge \Diamond t \wedge \Box \Diamond (p \vee q)$ is the input formula. While the model on the left in Figure 1 is generated when anywhere equality blocking is used, the model on the right is not and it is a model minimal modulo subset-simulation. However:

Theorem 6. *Our procedures to handle the modal logics* **K4**, **KT4** *and* **KD4** *are minimal model sound and complete, and terminate when ancestor equality blocking is used.*

Proof. Termination follows from known results that ancestor equality blocking is enough to ensure termination for these logics (e.g., [9]). We prove that minimal model completeness is preserved only for the case of **K4**, but the same reasoning can be used to prove it for the other logics. Suppose ϕ is the input modal formula. Let us assume that M is an infinite model graph of ϕ such that $M, u \models \phi$. Hence, there is at least one connected component in the graph with a path composed of an infinite sequence of worlds. Since the tableau calculus only propagates subformulae of ϕ, there are only finitely many formulae per world. This implies that the infinite path must contain a finite number of distinguishable worlds, that can be identified with each other. Due to transitivity each world in the path is connected with all its descendants, meaning we can focus our attention on the infinite repetition of a finite path of distinct worlds. This implies that M subset-simulates another model where a loop over the finite path is created. This reasoning can be iterated for all the infinite components of the graph. Hence, M subset-simulates a finite model. Therefore, as ancestor equality blocking blocks the infinite path at the first appearance of the first repeated world, minimal model completeness is preserved.   □

As in the previous case, the rule application order is important. Specifically, the ($\Diamond$) rule needs to be applied last. In this way it is possible to build the model by exhaustively expanding all formulae world by world. If ancestor equality blocking is checked only before applying the ($\Diamond$) rule, then when equality is detected the set of formulae true in a world cannot change anymore.

For all the remaining cases *anywhere dynamic equality blocking* can be used. This blocking technique blocks a world if there is another world for which the interpretation is the same. The dynamic part is due to the possibility of having false guesses. Specifically, it is possible that two worlds are considered to be the same at some point in the derivation, but they do not have the same interpretation in a subsequent point in the derivation. Making the blocking technique dynamic allows for the possibility of blocking and unblocking pairs of worlds. This technique, or a variation of it, is already used in the literature (e.g., [3,10]).

Theorem 7. *Our procedures to handle the modal logics* **K5**, **KD5**, **K45**, **KD45**, **KB4** *and* **KT5** *are minimal model sound and complete, and terminate when anywhere dynamic equality blocking is used.*

Proof. As for Theorem 6, we only need to prove that minimal model completeness is guaranteed. We prove it only for the case of **K5**, but the same reasoning can be used to prove it for the other logics. Due to frames being Euclidean, the resulting model is a strongly connected graph where the only exception is the root world. Hence, any two non-root worlds u and v are related by a reflexive, symmetric and transitive relation. If $\mathcal{V}(u) = \mathcal{V}(v)$, then an auto-bisimulation of the model would merge them into a single world. This implies that the application of anywhere dynamic equality blocking is equivalent to a bisimulation step. For Lemma 1, if the original model was minimal modulo subset-simulation, then the resulting model is still minimal modulo subset-simulation.   □

From minimal model completeness and the termination results of this section, this theorem follows.

Theorem 8. *All the normal modal logics between* **K** *and* **KT5** *have finitely many symmetry classes of models minimal modulo subset-simulation.*

Having strong termination techniques for all the normal modal logics allows us to vary the branch selection strategy. Specifically, a depth-first left-to-right branch selection strategy can be used. From a practical perspective this is important because it allows for memory efficient implementations.

6   Related Work and Discussion

The most similar approach for the generation of minimal models, for both the methodology used and the minimality criterion involved, is [17]. [17] is the first paper to introduce the notion of models minimal modulo subset-simulation and, hence, first to propose a technique for the generation of such minimal models for the multi-modal logics $\mathbf{K}_{(m)}$ and some of its extensions. Our procedures have however much smaller search spaces than those in [17]. This is because our notion of minimal model completeness requires the generation of one witness per symmetry class, while the notion used in [17] requires the whole symmetry classes to be generated. This is reflected in the rules of the two tableaux calculi. The only branching rule of our tableau calculus is the (β) rule, while in [17] also the ($\Diamond$) rule is a branching rule. The ($\Diamond$) rule in [17] has a high branching factor (i.e., the number of branches is equal to current number of labels in the branch plus one), and it leads to the generation of many similar and, therefore, unnecessary models. In the literature, other notions of minimal model completeness similar to the one we adopt exist. For example, in [5,13] not all minimal models are generated, but only witnesses of a specific kind of equivalences classes are generated.

The used notions of minimality and minimal model completeness allowed us to simplify the subset-simulation test in [17]. The subset-simulation test can be improved even more if for any extracted minimal model the auto-bisimulation is computed. This is because the complexity of checking subset-simulation relations depends on the number of worlds and the number of edges. Using auto-bisimulation can potentially result in minimal models having a smaller domain, making the comparison with other models easier. It is important to note though that the procedure proposed in [17] is designed to cover more expressive modal logics than what we cover in this paper, but, as long as termination is not taken into consideration, our approach can easily be extended to cover exactly the same expressive multi-modal logics while maintaining the results of this paper.

The minimality criterion in [13] has similarities to our minimality criterion and, using our terminology, it can be defined as a minimality criterion based on subset-bisimulation. [13] proposes a method to reduce the problem of answering positive queries for Horn modal formulae to the task of model checking. The creation of a minimal model that preserves all positive entailments simplifies the model checking task. It is interesting to note that any model minimal modulo subset-simulation is also minimal with respect to the minimality criterion proposed in [13]. This means that our approach can be used to address exactly the same problem, even for formulae that are outside of the modal Horn fragment.

Apart from our notion of minimal models, other minimality criteria exist. These can be classified into: syntactic notions of minimality, minimal Herbrand models [2,14,16], and domain minimality [7,11]. The class of minimal Herbrand models has the advantage that it can be ordered by the subset relation. It is thus possible to focus on generating models minimal under this ordering. Generating minimal Herbrand models for classical logics has been studied in [2,14] and for modal logics in [16]. Despite the use of a different minimality criterion, there are similarities between the models considered minimal by our approach and those considered minimal in [16]. As long as termination is not taken into consideration, models minimal modulo subset-simulation are a subset of minimal Herbrand models. As our minimality criterion takes into consideration the semantics of models, some minimal Herbrand model can be considered redundant or not minimal, resulting in a smaller set of minimal models. As soon as termination techniques are necessary, comparing the two notions of minimality becomes more difficult. This is also due to the fact that the approach proposed in [16] cannot be extended easily to cover logics with potentially infinite models, and is restricted to the multi-modal logics $\mathbf{K}_{(m)}$, $\mathbf{KT}_{(m)}$, $\mathbf{KB}_{(m)}$ and $\mathbf{KTB}_{(m)}$.

By contrast, domain minimal models are finite for all logics with the finite model property. Another possibility therefore is to focus on the generation of models with minimised domains [7,11]. Domain minimal models, however, tend to be counter-intuitive because too many worlds are collapsed into a single world. As a result, all the information needed to satisfy the input formula is pushed to the least number of domain elements, making tasks such as verification and debugging harder. Our approach is designed to avoid the creation of domain minimal models while spreading the positive information as much as possible. This results in more meaningful and intuitive models, as is shown in [17].

Description logics are closely related to the modal logics considered in this paper, and all results can be transferred to the corresponding description logics. An important difference is the presence of TBoxes in description logics. This difference can be accommodated by using a calculus for modal logics extended with rules for handling universal modalities. As TBoxes do not need the complete expressiveness of the universal modalities, we can extend our procedures in such a way that only specific patterns of universal modalities are allowed. In this way the procedures can handle description logics such as $\mathcal{ALC}$ with non-empty TBoxes. ABoxes pose no technical challenges. The full expressive power of universal modalities, however, increases the complexity of the procedure, and termination techniques preserving minimal model completeness are needed.

In this paper we used structural rules to accommodate frame conditions. A common alternative to the structural rules are propagation rules (e.g., [12]). The use of propagation rules is possible, but it would require expensive changes to the procedures. It can be proved that if there is a subset-simulation relation between two models obtained by using propagation rules, then the same subset-simulation relation holds also for the frame closures of the models. As the complexity of computing subset-simulation depends on the size of the domain and on the number of edges, the use of propagation rules seems promising. On the other hand, the other direction does not hold. In particular, subset-simulation relations between models where the frame closures are computed are not necessarily transferred to models generated by using a procedure based on propagation rules. As the subset-simulation test is applied many times, the use of propagation rules would require repeated computations of frame closures leading to worse performance.

7   Conclusion

We presented the first terminating, minimal model sound and complete procedures for the generation of models minimal modulo subset-simulation for all the sublogics of **S5**. Compared with other minimal model generation approaches, our procedures greatly benefit from smaller search spaces, fewer models are generated, and the semantically meaningfulness and naturalness of the models make them more effective for debugging purposes. These features of the procedures are really promising from both an implementation and a practical point of view.

We plan to extend our procedures by introducing rules handling more expressive modal logics. Logics we aim to handle are all the extensions from the uni-modal case to the multi-modal case, converse relations, universal modalities and inclusion axioms. These generalisations correspond to expressive logics widely used in real world applications. This is why we are currently working on implementing the procedures. We believe efficient implementations are achievable, and they will have important impact by complementing and improving techniques for debugging and verification.

References

1. Baumgartner, P., Fürbach, U., Niemelä, I.: Hyper tableaux. In: Orłowska, E., Alferes, J.J., Moniz Pereira, L. (eds.) JELIA 1996. LNCS, vol. 1126, pp. 1–17. Springer, Heidelberg (1996)
2. Bry, F., Yahya, A.: Positive unit hyperresolution tableaux and their application to minimal model generation. J. Automat. Reason. 25(1), 35–82 (2000)
3. Cialdea Mayer, M.: A proof procedure for hybrid logic with binders, transitivity and relation hierarchies. In: Bonacina, M.P. (ed.) CADE 2013. LNCS (LNAI), vol. 7898, pp. 76–90. Springer, Heidelberg (2013)
4. Clarke, E.M., Schlingloff, B.: Model checking. In: Robinson, A., Voronkov, A. (eds.) Handbook of Automated Reasoning, pp. 1635–1790. Elsevier (2001)
5. Denecker, M., De Schreye, D.: On the duality of abduction and model generation in a framework for model generation with equality. Theoret. Computer Sci. 122(1&2), 225–262 (1994)
6. Henzinger, M.R., Henzinger, T.A., Kopke, P.W.: Computing simulations on finite and infinite graphs. In: Proc. FCS-36, pp. 453–462. IEEE Comput. Soc. (1995)
7. Hintikka, J.: Model minimization—An alternative to circumscription. J. Automat. Reason. 4(1), 1–13 (1988)
8. Horridge, M., Parsia, B., Sattler, U.: Extracting justifications from bioportal ontologies. In: Cudré-Mauroux, P., Heflin, J., Sirin, E., Tudorache, T., Euzenat, J., Hauswirth, M., Parreira, J.X., Hendler, J., Schreiber, G., Bernstein, A., Blomqvist, E. (eds.) ISWC 2012, Part II. LNCS, vol. 7650, pp. 287–299. Springer, Heidelberg (2012)
9. Horrocks, I., Hustadt, U., Sattler, U., Schmidt, R.A.: Computational modal logic. In: Blackburn, P., van Benthem, J., Wolter, F. (eds.) Handbook of Modal Logic, pp. 181–245. Elsevier (2007)
10. Horrocks, I., Sattler, U.: A description logic with transitive and inverse roles and role hierarchies. J. Logic Comput. 9(3), 385–410 (1999)
11. Lorenz, S.: A tableaux prover for domain minimization. J. Automat. Reason. 13(3), 375–390 (1994)
12. Massacci, F.: Single step tableaux for modal logics. J. Automat. Reason. 24(3), 319–364 (2000)
13. Nguyen, L.A.: Constructing finite least Kripke models for positive logic programs in serial regular grammar logics. Logic J. IGPL 16(2), 175–193 (2008)
14. Niemelä, I.: Implementing circumscription using a tableau method. In: Proc. ECAI 1996, pp. 80–84. Wiley (1996)
15. Niemelä, I.: A tableau calculus for minimal model reasoning. In: Miglioli, P., Moscato, U., Ornaghi, M., Mundici, D. (eds.) TABLEAUX 1996. LNCS, vol. 1071, pp. 278–294. Springer, Heidelberg (1996)
16. Papacchini, F., Schmidt, R.A.: A tableau calculus for minimal modal model generation. Electr. Notes Theoret. Computer Sci. 278(3), 159–172 (2011)
17. Papacchini, F., Schmidt, R.A.: Computing minimal models modulo subset-simulation for propositional modal logics. In: Fontaine, P., Ringeissen, C., Schmidt, R.A. (eds.) FroCoS 2013. LNCS (LNAI), vol. 8152, pp. 279–294. Springer, Heidelberg (2013)
18. Reiter, R.: A theory of diagnosis from first principles. Artificial Intelligence 32(1), 57–95 (1987)
19. Schlobach, S., Huang, Z., Cornet, R., van Harmelen, F.: Debugging incoherent terminologies. J. Automat. Reason. 39(3), 317–349 (2007)

COOL – A Generic Reasoner
for Coalgebraic Hybrid Logics (System Description)[*]

Daniel Gorín[1], Dirk Pattinson[2], Lutz Schröder[1],
Florian Widmann[3], and Thorsten Wißmann[1]

[1] Friedrich-Alexander-Universität Erlangen-Nürnberg, Germany
[2] The Australian National University, Canberra
[3] Imperial College London, UK

Abstract. We describe the Coalgebraic Ontology Logic solver COOL, a generic reasoner that decides the satisfiability of modal (and, more generally, hybrid) formulas with respect to a set of global assumptions – in Description Logic parlance, we support a general TBox and internalize a Boolean ABox. The level of generality is that of coalgebraic logic, a logical framework covering a wide range of modal logics, beyond relational semantics. The core of COOL is an efficient unlabelled tableaux search procedure using global caching. Concrete logics are added by implemening the corresponding (one-step) tableaux rules. The logics covered at the moment include standard relational examples as well as graded modal logic and Pauly's Coalition Logic (the next-step fragment of Alternating-time Temporal Logic), plus every logic that arises as a fusion of the above. We compare the performance of COOL with state-of-the-art reasoners.

1 Introduction

Many modal logics can be interpreted using a Kripke-style relational semantics, but there is a vast array of modal logics that cannot be captured using relational models. Examples include classical and monotone modal logics [5], coalition logic / alternating-time logic [18,1], and probabilistic modal logic [10]. Graded modal logic [11] was originally formulated as a relational logic but is more naturally seen as talking about weighted graphs [7]. Semantically, these logics are captured using *coalgebraic logic* [17], a unifying framework that systematizes semantics, meta-theory and algorithms. Reasoning algorithms harness the syntactical presentation of these logics in terms of *one-step rules*. In tableaux presentation, these rules have the form

$$\frac{\Gamma_0}{\Gamma_1 \mid \cdots \mid \Gamma_n}$$

where $\Gamma_1, \ldots, \Gamma_n$ are sets of literals (read conjunctively) over a set V of propositional variables, and Γ_0 is set of literals over $\Lambda(V) = \{\heartsuit a \mid \heartsuit \in \Lambda, a \in V\}$ where Λ is an abstract set of modal operators. The reading of these rules is standard: to show satisfiability of the premiss, one needs to establish satisfiability of one conclusion, for every applicable rule.

---

[*] Work of the first, third, and fifth author supported by DFG grant GenMod2 (SCHR 1118/5-2).

S. Demri, D. Kapur, and C. Weidenbach (Eds.): IJCAR 2014, LNAI 8562, pp. 396–402, 2014.

Example 1.   1. *Modal Logic K.* A simple one-conclusion example is the standard (unlabeled) tableau rule $\Box a_1, \ldots, \Box a_n, \Diamond b / a_1, \ldots, a_n, b$ for the modal logic K, over modal operators $\Lambda = \{\Box, \Diamond\}$.

2. *Coalition logic:* Pauly's coalition logic [18], or the next-step fragment of alternating-time temporal logic ATL [1], is parametrized by a set $N = \{1, \ldots, n\}$ of *agents*; subsets of N are called *coalitions*. Operators are of the form $[C]$, 'coalition C can force', with duals $\langle C \rangle$ 'coalition C cannot avoid'. In the terminology of [1], the semantics is based on *concurrent game frames*. A complete set of rules is given by

$$\frac{[C_1]a_1, \ldots, [C_n]a_n}{a_1, \ldots, a_n} \qquad \frac{[C_1]a_1, \ldots, [C_n]a_n, \langle D \rangle b, \langle N \rangle c_1, \ldots, \langle N \rangle c_m}{a_1, \ldots, a_n, b, c_1, \ldots, c_m}$$

where $n, m \geq 0$ and the C_i are disjoint and contained in D [18,19,12].

3. *Graded logic:* Fine's graded modal logic [11] counts successor states in relational models; it has found its way into modern description logics [3] in the shape of *qualified number restrictions*. Its operators are $\Diamond_k$, read 'in more than k successors', with duals $\Box_k$ 'in all but at most k successors'. A complete set of rules [19,6] is given by

$$\frac{\Diamond_{k_1} a_1, \ldots, \Diamond_{k_n} a_n, \Box_{l_1} b_1, \ldots, \Box_{l_m} b_m}{\sum_{1 \leq i \leq n} r_i a_i - \sum_{1 \leq j \leq m} s_j (\neg q_j) > 0} \quad \left(\sum_{1 \leq i \leq n} r_i (k_i + 1) - \sum_{1 \leq j \leq m} s_j l_j \geq 1 \right)$$

where $n, m \geq 0$ and $r_i, s_j > 0$, subject to the side condition indicated, and with the sums in the conclusion of the rule referring to arithmetic of characteristic functions, i.e. counting 1 for 'true' and 0 for 'false'. Sufficient tractability of this rule is shown using results from integer linear programming [19].

The one-step rules are combined with propositional rules such as $\Gamma, a \vee b / \Gamma, a \mid \Gamma, b$ and rules that deal with nominal and satisfaction operators. One of the crucial feature of these logics is *compositionality*: the restriction on the rule format allows us to synthesize logics in a modular fashion. This is best understood by thinking of the one-step tableau rules as building blocks for logics that de-construct modal operators of a given type.

Sequencing of Logics. To describe, e.g. simple Segala systems [20] that describe systems that perform non-deterministic actions followed by a (probabilistic) action of the environment, we use a two-sorted syntax

$$L_0 \ni \phi ::= p_0 \mid \neg \phi \mid \phi \wedge \phi \mid \Diamond_a \psi \qquad L_1 \ni \psi ::= p_1 \mid \neg \psi \mid \psi \wedge \psi \mid \langle p \rangle \phi$$

where p_i is a typed propositional variable of the language L_i and $\langle p \rangle$ is an exemplaric operator of probabilistic modal logic 'with probability at least p'. To show satisfiability of $\phi \in L_0$ we deconstruct propositional connectives and apply tableaux rules for Hennessy-Milner logic. This leaves us with formulae in L_1 that are deconstructed in the same way, but using the rules of propositional modal logic, recursively yielding a formula in L_0 to which the same procedure is applied.

Fusion of Logics. To ensure the same typing discipline we describe the fusion of two logics over modal operators Λ_1 and Λ_2 in the same typed framework using three sorts and two new operators $[\pi_1]$ and $[\pi_2]$:

$$L_0 \ni \phi ::= p_0 \mid \neg \phi \mid \phi \wedge \phi \mid [\pi_1] \psi \mid [\pi_2] \sigma$$
$$L_1 \ni \psi ::= p_1 \mid \neg \psi \mid \psi \wedge \psi \mid \heartsuit_1 \phi \qquad L_2 \ni \sigma ::= p_2 \mid \neg \sigma \mid \sigma \wedge \sigma \mid \heartsuit_2 \phi$$

where $\heartsuit_i \in \Lambda_i$ is an operator of type i, and p_i is a propositional variable of type i. It is straightforward to embed the standard (language of the) fusion into the language L_0. We have two modal operators together with the tableau rules

$$\frac{\neg[\pi_i]a_1, \dots, \neg[\pi_i]a_i, [\pi_i]b_1, \dots, [\pi_i]b_k}{\neg a_1, \dots, \neg a_n, b_1, \dots, b_k}$$

for $i = 1, 2$. This allows us to reason about fusion using the same, typed, reasoning algorithm as described above for sequencing.

Choice. Choice allows us to axiomatise, e.g., the alternating systems of Hansson and Jonsson [16] where a successor state either originates from a labelled transition, or from a probabilistic action of the environment. Like fusion, we describe choice by means of a multi-sorted language that introduces one new modal operator, $+$, described by a one-step tableau rule. For alternating systems, we have

$$L_0 \ni \phi ::= p_0 \mid \neg\phi \mid \phi \wedge \phi \mid \psi + \sigma$$
$$L_1 \ni \psi ::= p_1 \mid \neg\psi \mid \psi \wedge \psi \mid \heartsuit_1\phi \qquad L_2 \ni \sigma ::= p_2 \mid \neg\sigma \mid \sigma \wedge \sigma \mid \heartsuit_2\phi$$

where we read the binary operator $\psi + \sigma$ as 'ψ for labelled successors and σ for probabilistic ones'. Reasoning over logics defined using choice is governed by the rules

$$\frac{\neg(a_1 + c_1), \dots, \neg(a_n + c_n), (b_1 + d_1), \dots, (b_k + d_k)}{\neg a_1, \dots, \neg a_n, b_1, \dots, b_n \mid \neg c_1, \dots, \neg c_n, d_1, \dots, d_k}$$

that induce type-correct formulae of sort L_1 (left) and L_2 (right).

A range of generic reasoning procedures of optimal complexity has been developed for coalgebraic logics with various additional features, including global assumptions, nominals, and fixpoints that all support modular combinations as described. The most basic of these, the generic PSPACE algorithm for satisfiability in next-step logics [19], has been implemented (already supporting modularity) in the CoLoSS tool [4]. Here, we present the *Coalgebraic Ontology Logic* Reasoner (CooL), available at https://www8.cs.fau.de/research/cool, which supports modular combinations of logics, global assumptions, and nominals, and uses global caching.

2   The CooL Solver: Supported Features

CooL implements a global caching algorithm for coalgebraic hybrid logic with global assumptions [13]. In description logic parlance, we support terminological reasoning (TBoxes) as well as nominals and satisfaction operators (thus internalizing Boolean ABoxes in concepts). These features are orthogonal to the underlying base logic which is constructed in a modular way from a number of basic building blocks, and the effort of adding a new logic is typically quite limited. Global caching combines theoretical optimality (i.e. an exponential time upper bound) with amenability to heuristic optimization [15]. In more detail, CooL supports the following.

– *Global assumptions*, or, in description logic parlance, a *general TBox*: one can restrict the class of models to ones in which all states/worlds/individuals satisfy a given finite set of formulas. In knowledge representation, such global assumptions serve to express background knowledge about the terminological domain.

– *Nominals:* we incorporate two key features of hybrid logic [2], *nominals* and *satisfaction operators*. Here, a nominal is a name i for an individual state in the model; as a formula, i is satisfied precisely in the unique state named by i. The satisfaction operator $@_i$ lets the evaluation point of a formula jump to i. Reasoning with these features encompasses DL-style ABox reasoning: recall that an *ABox* (*assertional box*) contains statements of the forms $\phi(i)$ or $R(i, j)$, respectively read 'individual i satisfies formula ϕ' and 'individuals i and j are in relation R'. In hybrid logic, these statements can be expressed as $@_i\phi$ and $@_i\Diamond_R j$, respectively.

For reasoning with these features, we use the global caching algorithm introduced in [13]. Global caching for relational modal logics (phrased in DL terminology) goes back to [14]; the principle has been generalized to coalgebraic logic in [12]. The basic idea of global caching is to regard a tableau as a directed (possibly cyclic) graph rather than a tree, thus enabling sharing of nodes. This allows one to visit each label (i.e. finite set of subformulas) at most once, ensuring at most exponential (hence in most cases asymptotically optimal) run time. The algorithm partitions the set of currently created tableau nodes into *unexpanded* (X), *undecided* (U), *satisfiable* (E), and *unsatisfiable* (A) nodes. It consists in applying the following two types of steps in near-arbitrary sequence, until either the root node is marked A or E or no further steps are applicable:

– *Expand:* Apply all matching rules to an unexpanded node (then moved from X to U), creating either new successor nodes (initially marked X) or links to existing nodes.

– *Propagate:* Mark expanded nodes as unsatisfiable if there is a matching tableau rule with only unsatisfiable conclusions, and as satisfiable if all matching rules have some satisfiable conclusion. Here, the recursion is understood as a *least* fixed point for unsatisfiability, and as a *greatest* fixed point for satisfiability.

After the final propagation step, all nodes marked U are reported as satisfiable. Note that the algorithm may leave nodes marked X, thus allowing for quick answers in many cases; the apparent non-determinism works in favour of the implementer, as *any* terminating sequence of steps will yield a correct result, thus leaving room for heuristics.

The novel global caching algorithm for coalgebraic hybrid logic [13] deals with the global demands arising from satisfaction operators ($@_i\phi$ holds everywhere or nowhere) by means of a dedicated second type of nodes called $@$-*constraints*. An $@$-constraint records $@$-formulas to be satisfied for a given standard node to be satisfiable. It is linked to standard nodes in a new type of step called $@$-*expansion* (essentially, having $@_i\phi$ in an $@$-constraint requires a standard node satisfying i and ϕ). In $@$-*propagation* steps, the $@$-constraints are updated throughout the model using greatest fixed points, essentially following a winning strategy of the player advocating satisfiability.

3 The COOL Solver: Implementation Details

The implementation of COOL focuses on modal rules, and uses the minisat sat-solver [9] for reasoning in classical propositional logic, more precisely for expanding the propositional part of nodes in the tableau graph. The SAT solver is used as a black box, and no optimisations that concern propositional reasoning are implemented on top of those performed by minisat. Rules for graded and probabilistic modal logics are generated using the GNU Linear Programming Toolkit [1]. We refer to [21] for the details of generating rules for propositional and graded logics on the basis of linear inequalities.

Compositionality of Logics. The underlying modal logic (semantically: the branching type of systems) is described using an algebraic term with one free variable S (for state) where each base logic is represented by a unary function symbol and two binary ones for choice and fusion. These terms are best read as equalities, and e.g. the alternating systems of Hansson and Jonsson mentioned in the introduction can be specified by

$$S = \mathsf{Ch}(\mathsf{HM}(S), \mathsf{P}(S))$$

where Ch represents choice, HM represents Hennessy-Milner logic (multi-modal K) and P is probabilistic modal logic. Semantically, this expression defines the system type: given a state, we either see labelled successors, or a probability distribution over states, i.e. we observe an external choice between both. Similarly, simple Segala systems are modelled by $S = \mathsf{HM}(\mathsf{P}(S))$ and the fusion of probabilistic modal logic and Hennessy-Milner logic would be $S = \mathsf{Fus}(\mathsf{HM}(S)), \mathsf{P}(S))$ where Fus is a binary function representing fusion.

Generic Reasoning and Tableau Rules. Our reasoner is *generic* in that the reasoning algorithm is conceptually independent of the underlying modal logic. This is achieved by isolating the modal rules into OCAML functions that – given a premiss – compute the set of all (instances of) applicable tableau rules The underlying reasoning algorithm then invokes the respective rules in sequence, following the construction of any given particular logic. The treatment of nominals, satisfaction operators and global assumptions is identical for all logics, and is hard-coded into the reasoning algorithm.

Optimisations. As mentioned above, we do not implement any optimisations on the propositional level, but we use global caching for dealing both with nominals and modal tableaux. The only conceptual implementation supported by COOL (and implemented for K) is backjumping [3]: each logical feature provides a hook by which a subset of literals that cause a clash can be passed back to the reasoner.

4   Experimental Evaluation

The COOL reasoner is still on its early stages of development and, having finished a robust implementation of the generic core, our main focus at the moment is on adding support for more logics. Still, our measurements suggest that it already offers a fair performance: even for the logic K (that is, the baseline logic $\mathcal{ALC}$ of the DL community), the response times of COOL are often within those of a long-established DL reasoner, implementing many advanced optimization strategies, such as FACT++.[22]

The experiments we report here are based on random formula generation in clausal form. This methodology allows one to compare different reasoners without risking an *optimization bias* (testing w.r.t. a set of problems for which one of the reasoners was specifically tuned) and give reasonable expectations regarding the capacity of a reasoner to handle large problems. We acknowledge that on real data-sets the difference in

---
[1] http://www.gnu.org/s/glpk/

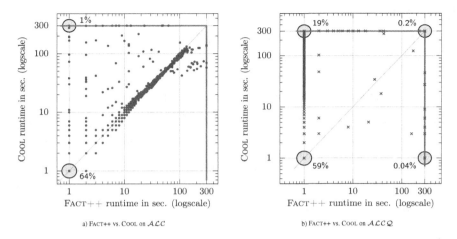

a) FACT++ vs. COOL on $\mathcal{ALC}$ b) FACT++ vs. COOL on $\mathcal{ALCQ}$

Fig. 1. Comparative evaluation of COOL vs. FACT++ on random formulas in $\mathcal{ALC}$ and $\mathcal{ALCQ}$, with 5 atoms and 0.25 chance of occurring; up to 6 dis-/conjuncts per dis-/conjunction; TBox formulas with modal depth of 2. Percentages shown refer to percentage of samples represented by indicated points. Times correspond to the USER TIME field as reported by the GNU TIME command. Test conducted on a heterogeneous cluster of computers with similar load.

performance between reasoners can become more noticeable and that random testing may oversample trivial formulas, but we defer alternative measurements (benchmarking with dedicated formula series) until more substantial sets of benchmark formulas are available also for non-relational logics.

We report on three comparisons: i) COOL vs. FACT++ on $\mathcal{ALC}$ (with a TBox), ii) COOL vs. FACT++ on $\mathcal{ALCQ}$ (with and without a TBox), and iii) COOL vs. TATL (a tableau reasoner for full ATL [8]) on coalition logic (with and without a TBox, encoded using ATL temporal operators for TATL). We kept fixed a number of parameters such as number of atoms, average number of conjuncts/disjuncts etc., and gradually increased the modal depth. In coalition logic, COOL answered consistently and substantially faster and with fewer timeouts than TATL, especially in the presence of a TBox (a scatter plot of the comparison reveals no additional information). Scatter plots for COOL vs. FACT++ are shown in Fig. 1. On $\mathcal{ALC}$, COOL shows a behaviour comparable to that of FACT++. Contrastingly, FACT++ is still substantially faster on $\mathcal{ALCQ}$, possibly due to the fact that COOL does not yet implement backjumping for $\mathcal{ALCQ}$.

5   Conclusions

Based on generic results from coalgebraic logic, the COOL reasoner supports a simple implementation and automatic combination of a wide spectrum of logics; very few other reasoners support any logic outside the standard relational setup. A preliminary empirical evaluation suggests that, while there is still plenty of room for optimizations, the implementation of the core global-caching algorithm is robust and efficient.

Acknowledgments. The authors wish to thank Erwin R. Catesbeiana for helpful comments on consistency checking.

References

1. Alur, R., Henzinger, T.A., Kupferman, O.: Alternating-time temporal logic. J. ACM 49, 672–713 (2002)
2. Areces, C., ten Cate, B.: Hybrid logics. In: Blackburn, P., van Benthem, J., Wolter, F. (eds.) Handbook of Modal Logic, pp. 821–868. Elsevier (2007)
3. Baader, F., Calvanese, D., McGuinness, D.L., Nardi, D., Patel-Schneider, P.F. (eds.): The Description Logic Handbook. Cambridge University Press (2003)
4. Calin, G., Myers, R., Pattinson, D., Schröder, L.: Coloss: The coalgebraic logic satisfiability solver. In: Methods for Modalities, M4M-5. ENTCS, vol. 231, pp. 41–54. Elsevier (2009)
5. Chellas, B.: Modal Logic. Cambridge University Press (1980)
6. Cîrstea, C., Kupke, C., Pattinson, D.: EXPTIME tableaux for the coalgebraic μ-calculus. In: Grädel, E., Kahle, R. (eds.) CSL 2009. LNCS, vol. 5771, pp. 179–193. Springer, Heidelberg (2009)
7. D'Agostino, G., Visser, A.: Finality regained: A coalgebraic study of Scott-sets and multisets. Arch. Math. Logic 41, 267–298 (2002)
8. David, A.: TATL: Implementation of ATL tableau-based decision procedure. In: Galmiche, D., Larchey-Wendling, D. (eds.) TABLEAUX 2013. LNCS, vol. 8123, pp. 97–103. Springer, Heidelberg (2013)
9. Eén, N., Sörensson, N.: An extensible SAT-solver. In: Giunchiglia, E., Tacchella, A. (eds.) SAT 2003. LNCS, vol. 2919, pp. 502–518. Springer, Heidelberg (2004)
10. Fagin, R., Halpern, J.Y.: Reasoning about knowledge and probability. J. ACM 41, 340–367 (1994)
11. Fine, K.: In so many possible worlds. Notre Dame J. Formal Logic 13, 516–520 (1972)
12. Goré, R., Kupke, C., Pattinson, D.: Optimal tableau algorithms for coalgebraic logics. In: Esparza, J., Majumdar, R. (eds.) TACAS 2010. LNCS, vol. 6015, pp. 114–128. Springer, Heidelberg (2010)
13. Goré, R., Kupke, C., Pattinson, D., Schröder, L.: Global caching for coalgebraic description logics. In: Giesl, J., Hähnle, R. (eds.) IJCAR 2010. LNCS (LNAI), vol. 6173, pp. 46–60. Springer, Heidelberg (2010)
14. Goré, R., Nguyen, L.: EXPTIME tableaux for $\mathcal{ALC}$ using sound global caching. In: Description Logics, DL 2007. CEUR Workshop Proceedings, vol. 250 (2007)
15. Goré, R.P., Postniece, L.: An experimental evaluation of global caching for $\mathcal{ALC}$ (system description). In: Armando, A., Baumgartner, P., Dowek, G. (eds.) IJCAR 2008. LNCS (LNAI), vol. 5195, pp. 299–305. Springer, Heidelberg (2008)
16. Hansson, H., Jonsson, B.: A calculus for communicating systems with time and probabilities. In: Real-Time Systems, RTSS 1990, pp. 278–287. IEEE Computer Society (1990)
17. Pattinson, D.: Coalgebraic modal logic: Soundness, completeness and decidability of local consequence. Theoret. Comput. Sci. 309, 177–193 (2003)
18. Pauly, M.: A modal logic for coalitional power in games. J. Log. Comput. 12, 149–166 (2002)
19. Schröder, L., Pattinson, D.: PSPACE bounds for rank-1 modal logics. ACM Trans. Comput. Log. 13, 1–13 (2009)
20. Segala, R.: Modelling and Verification of Randomized Distributed Real-Time Systems. PhD thesis, Massachusetts Institute of Technology (1995)
21. Snell, W., Pattinson, D., Widmann, F.: Solving graded/probabilistic modal logic via linear inequalities (system description). In: Bjørner, N., Voronkov, A. (eds.) LPAR-18 2012. LNCS, vol. 7180, pp. 383–390. Springer, Heidelberg (2012)
22. Tsarkov, D., Horrocks, I.: FaCT++ description logic reasoner: System description. In: Furbach, U., Shankar, N. (eds.) IJCAR 2006. LNCS (LNAI), vol. 4130, pp. 292–297. Springer, Heidelberg (2006)

The Complexity of Theorem Proving in Circumscription and Minimal Entailment

Olaf Beyersdorff* and Leroy Chew**

School of Computing, University of Leeds, UK

Abstract. We provide the first comprehensive proof-complexity analysis of different proof systems for propositional circumscription. In particular, we investigate two sequent-style calculi: *MLK* defined by Olivetti [28] and *CIRC* introduced by Bonatti and Olivetti [8], and the tableaux calculus *NTAB* suggested by Niemelä [26]. In our analysis we obtain exponential lower bounds for the proof size in *NTAB* and *CIRC* and show a polynomial simulation of *CIRC* by *MLK*. This yields a chain $NTAB <_p CIRC <_p MLK$ of proof systems for circumscription of strictly increasing strength with respect to lengths of proofs.

1 Introduction

Circumscription is one of the main formalisms for non-monotonic reasoning. It uses reasoning with minimal models, the key idea being that minimal models have as few exceptions as possible. Therefore circumscription embodies common sense reasoning. Indeed, circumscription is known to be equivalent to reasoning under the extended closed world assumption, one of the main formalisms for reasoning with incomplete information. Apart from its foundational relation to human reasoning, circumscription has wide-spread applications, e.g. in AI, description logics [7] and SAT solving [21]. Circumscription is used both in first-order as well as in propositional logic, and we concentrate in this paper on the propositional case.

The semantics and complexity of circumscription have been the subject of intense research (see e.g. the recent articles [7, 14, 29]). In particular, deciding circumscriptive inference is harder than for propositional logic as it is complete for Π_2^p, the second level of the polynomial hierarchy [11, 16]. Likewise, from the proof-theoretic side there are a number of formal systems for circumscription ranging from sequent calculi [8, 28] to tableau methods [25, 26, 28].

The contribution of the present paper is a comprehensive analysis of these formal systems from the perspective of proof complexity. The main objective in proof complexity is a precise understanding of lengths of proofs. The two main tools for this are *lower bound methods* for the size of proofs for specific proof systems as well as *simulations* between proof systems. While lower bounds provide exact information on proof size, simulations compare the relative strength

---

* Supported by a grant from the John Templeton Foundation.
** Supported by a Doctoral Training Grant from EPSRC.

S. Demri, D. Kapur, and C. Weidenbach (Eds.): IJCAR 2014, LNAI 8562, pp. 403–417, 2014.

of proof systems and determine whether proofs can be efficiently translated between different formalisms. In this paper our results will employ both of these paradigms. While the bulk of research in proof complexity has concentrated on propositional proofs the last decade has seen ever increasing interest in proof complexity of non-classical logics (cf. [4] for a survey). In particular, very impressive results have been obtained for modal and intuitionistic logics [20, 22].

Prior to this paper, very little was known about the proof complexity of propositional circumscription. Our analysis concentrates on three of the main formalisms for circumscription: the tableau system $NTAB$ introduced by Niemelä [26], the analytic sequent calculus $CIRC$ by Bonatti and Olivetti [8], and the sequent calculus MLK by Olivetti [28]. Our main results are exponential lower bounds for the proof size in the tableau system $NTAB$ and the sequent calculus $CIRC$ (Theorems 6 and 19) as well as an efficient simulation of $CIRC$ by MLK (Theorem 13). Together with the simulation of $NTAB$ by $CIRC$ shown by Bonatti and Olivetti [8] this gives a hierarchy of proof systems $NTAB <_p CIRC <_p MLK$. Moreover, this hierarchy is strict as our results provide separations between the proof systems (Theorems 8 and 19). While the systems $NTAB$ and MLK only work for minimal entailment — the most important special case of circumscription — we also extend the results on MLK to the calculus $DMLK$ from [28] for general circumscription (Theorem 16).

In related research, Egly and Tompits [15] investigated the proof-theoretic strength of circumscription in a first-order version of Bonatti and Olivetti's sequent calculus. They showed that for some formulas, first-order $CIRC$ has much shorter proofs than classical first-order LK. Also in [1,5] the authors investigated the proof complexity of propositional default logic and autoepistemic logic, two other main approaches to non-monotonic reasoning. Although there are several translations between the different non-monotonic logics, we stress that none of these previous results imply lower bounds or simulations for circumscription.

This paper is organised as follows. In Sect. 2 we review background information and notation about circumscription and proof complexity. In particular, we discuss the antisequent calculus AC. Section 3 contains our first main result: the exponential lower bound for $CIRC$. In Sect. 4 we prove the simulation of $CIRC$ by MLK for minimal entailment; and this is extended to full circumscription and the calculus $DMLK$ in Sect. 5. Section 6 then contains the comparison to Niemelä's tableau calculus $NTAB$, obtaining a separation between this tableau and $CIRC$. We conclude in Sect. 7 with a discussion and some open problems. Due to space restrictions some proofs are omitted or briefly sketched.

2   Preliminaries

Our propositional language contains the logical symbols $\bot, \top, \neg, \rightarrow, \vee, \wedge$. The notation $A[x/y]$ indicates that in the formula A every occurrence of formula x is replaced by formula y. For a set of formulae Σ, $\mathrm{VAR}(\Sigma)$ is the set of all atoms that occur in Σ. For a set P of atoms we set $\neg P = \{\neg p \mid p \in P\}$. Disjoint union of two sets A and B is denoted by $A \sqcup B$.

Circumscription is a non-monotonic logic introduced by McCarthy [24]. It looks at finding the 'minimal' situations that can occur, given our assumptions (cf. McCarthy's famous example of the "missionaries and cannibals" problem [24]). For circumscription, the propositional atoms are partitioned into three sets: P is the set of all atoms that are *minimised*, R is the set of *fixed* atoms, and Z denotes all remaining atoms, which may vary from the minimisation but are not themselves minimised. We usually only display P and R in the notation.

A *model* is a subset of the propositional atoms Σ_{Prop}. We define a pre-order $\leq_{P;R}$ on models I, J as follows: $I \leq_{P;R} J \Leftrightarrow I \cap P \subseteq J \cap P$ and $I \cap R = J \cap R$. The relation $\leq_{P;R}$ is transitive and minimality can be defined for models. Let $I \models \Gamma$. We say that I is a $(P;R)$-*minimal model* of Γ (and denote it by $I \models_{P;R} \Gamma$) if and only if for any model J, if $J \models \Gamma$ then $(J \leq_{P;R} I) \Rightarrow (I \leq_{P;R} J)$.

If ϕ is a formula, then $\Gamma \models_{P;R} \phi$ means that ϕ holds in all $(P;R)$-minimal models of Γ. This is the notion of semantic entailment in circumscription. A few special cases can be noted. When $P = \emptyset$ then $\models_{P;R}$ coincides with $\models$, the classical entailment. When P is the set of all variables appearing in the formulae of either the antecedent or the succedent then entailment is known as *minimal entailment*, and we denote it with the symbol $\models_M$.

Proof Complexity. A *proof system* (Cook, Reckhow [12]) for a language L over alphabet Γ is a polynomial-time computable partial function $f : \Gamma^\star \to \Gamma^\star$ with $rng(f) = L$. An f-*proof* of string y is a string x such that $f(x) = y$.

From this we can start defining proof size. For f a proof system for language L and string $x \in L$ we define $s_f(x) = \min(|w| : f(w) = x)$. Thus the partial function s_f tells us the minimum proof size of a theorem. We can overload the notation by setting $s_f(n) = \max(s_f(x) : |x| \leq n)$ where $n \in \mathbb{N}$. For a function $t : \mathbb{N} \to \mathbb{N}$, a proof system f is called t-*bounded* if $\forall n \in \mathbb{N}, s_f(n) \leq t(n)$.

Proof systems are compared by simulations. We say that a proof system f *simulates* g ($g \leq f$) if there exists a polynomial p such that for every g-proof π_g there is an f-proof π_f with $f(\pi_f) = g(\pi_g)$ and $|\pi_f| \leq p(|\pi_g|)$. If π_f can even be constructed from π_g in polynomial time, then we say that f p-*simulates* g ($g \leq_p f$). Two proof systems f and g are *(p-)equivalent* ($g \equiv_{(p)} f$) if they mutually (p-)simulate each other.

Gentzen's system LK is one of the historically first and best studied proof systems [18]. It operates with sequents. Formally, a *sequent* is a pair (Γ, Δ) with Γ and Δ finite sets of formulae. A sequent is usually written in the form $\Gamma \vdash \Delta$. In classical logic $\Gamma \vdash \Delta$ is true if every model for $\bigwedge \Gamma$ is also a model of $\bigvee \Delta$, where the disjunction of the empty set is taken as $\perp$ and the conjunction as $\top$. When considering LK in proof complexity we treat sequents as strings in binary, built from binary strings representing atoms and connectives. The system can be used both for propositional and first-order logic; the propositional rules are displayed in Fig. 1. Notice that the rules here do not contain structural rules for contraction or exchange. These come for free as we chose to operate with sets of formulae rather than sequences. Note the soundness of rule ($\bullet \vdash$), which gives us monotonicity of classical propositional logic.

$$\overline{A \vdash A}\ (\vdash) \qquad \overline{\bot \vdash}\ (\bot \vdash) \qquad \overline{\vdash \top}\ (\vdash \top)$$

$$\frac{\Gamma \vdash \Sigma}{\Delta, \Gamma \vdash \Sigma}\ (\bullet \vdash) \qquad \frac{\Gamma \vdash \Sigma}{\Gamma \vdash \Sigma, \Delta}\ (\vdash \bullet) \qquad \frac{\Gamma \vdash \Sigma, A}{\neg A, \Gamma \vdash \Sigma}\ (\neg \vdash)$$

$$\frac{A, \Gamma \vdash \Sigma}{\Gamma \vdash \Sigma, \neg A}\ (\vdash \neg) \qquad \frac{A, \Gamma \vdash \Sigma}{B \wedge A, \Gamma \vdash \Sigma}\ (\bullet \wedge \vdash) \qquad \frac{A, \Gamma \vdash \Sigma}{A \wedge B, \Gamma \vdash \Sigma}\ (\wedge \bullet \vdash)$$

$$\frac{\Gamma \vdash \Sigma, A \qquad \Gamma \vdash \Sigma, B}{\Gamma \vdash \Sigma, A \wedge B}\ (\vdash \wedge) \qquad \frac{A, \Gamma \vdash \Sigma \qquad B, \Gamma \vdash \Sigma}{A \vee B, \Gamma \vdash \Sigma}\ (\vee \vdash)$$

$$\frac{\Gamma \vdash \Sigma, A}{\Gamma \vdash \Sigma, B \vee A}\ (\vdash \bullet \vee) \qquad \frac{\Gamma \vdash \Sigma, A}{\Gamma \vdash \Sigma, A \vee B}\ (\vdash \vee \bullet) \qquad \frac{A, \Gamma \vdash \Sigma, B}{\Gamma \vdash \Sigma, A \to B}\ (\vdash \to)$$

$$\frac{\Gamma \vdash \Sigma, A \qquad B, \Delta \vdash \Lambda}{A \to B, \Gamma, \Delta \vdash \Sigma, \Lambda}\ (\to \vdash) \qquad \frac{\Gamma \vdash \Sigma, A \qquad A, \Gamma \vdash \Sigma}{\Gamma \vdash \Sigma}\ (cut)$$

Fig. 1. Rules of the sequent calculus LK [18]

A useful ingredient for working towards a calculus for non-monotonic logics is the notion of *underivability*. We use $\Gamma \nvdash \phi$ to denote that "there is a model M that satisfies all formulae in Γ but for which $\neg\phi$ holds". An *antisequent* is a pair of sets Γ, Δ of formulae, denoted $\Gamma \nvdash \Delta$. Semantically, an antisequent $\Gamma \nvdash \Sigma$ is true if there is some model $M \models \Gamma$ so that for all ϕ in Σ we have $M \models \neg\phi$. This is equivalent to saying that we cannot derive $\Gamma \vdash \Sigma$.

Bonatti [6] devised an *antisequent calculus* AC (cf. also [30]; rules of AC are given in Fig. 2. Correctness and completeness of AC was proven by Bonatti.

Theorem 1. *(Bonatti [6])* An antisequent is true if and only if it is derivable in the antisequent calculus AC.

While the truth of an antisequent tells us of the existence of a model that satisfies the left hand side but contradicts the right hand side, this does not point immediately to the model itself. The model, however, can be constructed from an AC-proof.

Proposition 2. *Given an AC-proof of an antisequent $\Gamma \nvdash \Delta$ we can construct in polynomial-time a model M that satisfies Γ and falsifies Δ.*

We mention that Proposition 2 implies that AC is presumably not *automatizable*, *i.e.*, it is not possible to construct AC-proofs in polynomial time (even though AC-proofs are always of quadratic size [5]). In fact, using Proposition 2 it can be shown that automatizability of AC is equivalent to a complexity assumption Q, studied in [17] and shown to be equivalent to the p-optimality of the standard proof system for SAT in [3].

$$\frac{}{\Gamma \not\vdash \Sigma} \ (\not\vdash) \qquad \text{where } \Gamma \text{ and } \Sigma \text{ are disjoint sets of propositional variables}$$

$$\frac{\Gamma \not\vdash \Sigma, \alpha}{\Gamma, \neg\alpha \not\vdash \Sigma} \ (\neg \not\vdash) \qquad\qquad \frac{\Gamma, \alpha \not\vdash \Sigma}{\Gamma \not\vdash \Sigma, \neg\alpha} \ (\not\vdash \neg)$$

$$\frac{\Gamma, \alpha, \beta \not\vdash \Sigma}{\Gamma, \alpha \wedge \beta \not\vdash \Sigma} \ (\wedge \not\vdash) \qquad \frac{\Gamma \not\vdash \Sigma, \alpha}{\Gamma \not\vdash \Sigma, \alpha \wedge \beta} \ (\not\vdash \bullet\wedge) \qquad \frac{\Gamma \not\vdash \Sigma, \beta}{\Gamma \not\vdash \Sigma, \alpha \wedge \beta} \ (\not\vdash \wedge\bullet)$$

$$\frac{\Gamma \not\vdash \Sigma, \alpha, \beta}{\Gamma \not\vdash \Sigma, \alpha \vee \beta} \ (\not\vdash \vee) \qquad \frac{\Gamma, \alpha \not\vdash \Sigma}{\Gamma, \alpha \vee \beta \not\vdash \Sigma} \ (\bullet\vee \not\vdash) \qquad \frac{\Gamma, \beta \not\vdash \Sigma}{\Gamma, \alpha \vee \beta \not\vdash \Sigma} \ (\vee\bullet \not\vdash)$$

$$\frac{\Gamma, \alpha \not\vdash \Sigma, \beta}{\Gamma \not\vdash \Sigma, \alpha \rightarrow \beta} \ (\not\vdash \rightarrow) \qquad \frac{\Gamma \not\vdash \Sigma, \alpha}{\Gamma, \alpha \rightarrow \beta \not\vdash \Sigma} \ (\bullet \rightarrow \not\vdash) \qquad \frac{\Gamma, \beta \not\vdash \Sigma}{\Gamma, \alpha \rightarrow \beta \not\vdash \Sigma} \ (\rightarrow \bullet \not\vdash)$$

Fig. 2. Inference rules of the antisequent calculus AC by Bonatti [6]

3   A Lower Bound for the Sequent Calculus *CIRC*

Bonatti and Olivetti [8] devised sequent calculi for several non-monotonic logics, among them was circumscription in a sequent calculus referred to as *CIRC*. A new item Σ known as a *constraint* has been added to the sequent. Σ is a set of atoms disjoint from R, so the *circumscriptive sequents* are of form $\Sigma; \Gamma \vdash_{P;R} \Delta$ (which may be regarded as a 5-tuple). As defined by Bonatti and Olivetti [8], the sequent $\Sigma; \Gamma \vdash_{P;R} \Delta$ is true when: "In every $(P \cup \Sigma; R)$-minimal model of Γ that satisfies Σ there is a formula $\phi \in \Delta$ that holds."

When Σ is empty we omit it from the notation, and these are the circumscriptive sequents we are primarily interested in. The rules of the calculus *CIRC* comprise the rules given in Fig. 3 together with all rules from *LK* and *AC*. Bonatti and Olivetti proved the correctness and completeness of *CIRC*:

Theorem 3. *(Bonatti, Olivetti [8])* A sequent $\Sigma; \Gamma \vdash_{P;R} \Delta$ *is true if and only if it is derivable in CIRC.*

To start a proof-theoretic investigation of *CIRC* we need the following notion:

Definition 4. *Let π be a CIRC-proof of a circumscriptive sequent $\Gamma \vdash_{P;R} \Delta$ and let s be a sequent occurring in π (we will also call this a line of π). We call s involved in π if either s is $\Gamma \vdash_{P;R} \Delta$ or is used as premise for some rule whose conclusion is an involved sequent. We call s intermediate if s is involved in π and occurs in π as a conclusion of any of rules (C1)–(C4).*

Thus the intermediate sequents form the "essential *CIRC*-part" of the proof on which we will focus our analysis. The whole proof can be much larger due to *LK* and *AC*-derivations. The next lemma shows that intermediate sequences are always of a special form.

$$\frac{\Gamma, \neg P \nvdash q}{q, \Sigma; \Gamma \vdash_{P;\emptyset} \Delta} \ (C1) \qquad\qquad \frac{\Sigma, \Gamma \vdash \Delta}{\Sigma; \Gamma \vdash_{P;R} \Delta} \ (C2)$$

$$\frac{q, \Sigma; \Gamma \vdash_{P;R} \Delta \qquad \Sigma; \Gamma, \neg q \vdash_{P;R} \Delta}{\Sigma; \Gamma \vdash_{P,q;R} \Delta} \ (C3)$$

$$\frac{\Sigma; \Gamma, q \vdash_{P;R} \Delta \qquad \Sigma; \Gamma, \neg q \vdash_{P;R} \Delta}{\Sigma; \Gamma \vdash_{P;R,q} \Delta} \ (C4)$$

In all rules q is atomic and does not occur in P or R.

Fig. 3. Inference rules of the circumscription calculus $CIRC$ of Bonatti & Olivetti [8]

Lemma 5. *Let π be a proof of the minimal entailment formula $\Gamma \vdash_{\mathrm{VAR}(\Gamma \cup \Delta);\emptyset} \Delta$. Then every intermediate line in π (in the sense of Definition 4) is of the form $P^+; \Gamma, \neg P^- \vdash_{P^0;\emptyset} \Delta$, where $\mathrm{VAR}(\Gamma \cup \Delta) = P^0 \sqcup P^+ \sqcup P^-$.*

Our first result shows an exponential lower bound to the proof size of $CIRC$. We do this by forcing the $CIRC$-proof to enumerate all minimal models, however in general a $CIRC$-proof may not be required to do so. For an easy example, consider $\bigwedge_{1 \leq i \leq n} p_i \vee q_i \vdash_M \bigwedge_{1 \leq i \leq n} p_i \vee q_i$, which has exponentially many minimal models, but can be derived in two lines from ($\vdash$) and (C2).

Theorem 6. *$CIRC$ needs exponential-size proofs, i.e., $s_{CIRC}(n) \in 2^{\Omega(n/\log n)}$.*

Proof. The idea is to construct a class of formulae which are of size $O(n \log n)$, but whose proof size grows exponentially. We use propositional variables $P_n = \{p_i, q_i : 1 \leq i \leq n\}$ and define antecedent $\Gamma_n := \{p_i \vee q_i : 1 \leq i \leq n\}$ and succedent $\Delta_n := \bigwedge_{1 \leq i \leq n} (p_i \wedge \neg q_i) \vee (q_i \wedge \neg p_i)$. We consider the class of sequents $\Gamma_n \vdash_{P_n;\emptyset} \Delta_n$.

Intuitively the sequents express $\bigwedge_{1 \leq i \leq n} p_i \vee q_i \models_M \bigwedge_{1 \leq i \leq n} p_i \oplus q_i$, which is not classically true. But they are true circumscriptive sequents, because every minimal model of Γ_n will include p_i or q_i but cannot include both as these models are not minimal. Notice that the size of the sequents is bounded by $O(n \log n)$ because to represent each of the n variables we need $O(\log n)$ bits.

Let now π be a $CIRC$-proof of $\emptyset; \Gamma_n \vdash_{P_n;\emptyset} \Delta_n$. We now argue inductively.

Induction Hypothesis (on k for $k \leq n$): Let $P^+; \Gamma_n, \neg P^- \vdash_{P^0;\emptyset} \Delta_n$ be an intermediate sequent of π (we know it is of this form by Lemma 5) with $k = n - |P^- \sqcup P^+|$. Then the sub-proof of $P^+; \Gamma_n, \neg P^- \vdash_{P^0;\emptyset} \Delta_n$ in π contains at least 2^k lines of the form $B; \Gamma_n, \neg A \vdash_{C;\emptyset} \Delta_n$, where A, B, C are sets of atoms, with $P^+ \subseteq B$, $P^- \subseteq A$, and with B, A disjoint in any line.

Base Case (when $k = 0$): A single line is needed to state the end result $P^+; \Gamma_n, \neg P^- \vdash_{P^0;\emptyset} \Delta_n$, and it suffices to take $B = P^+$, $A = P^-$.

Inductive Step: Assume the induction hypothesis holds for $k - 1$. Our aim is to show that if $1 \leq k \leq n$, then $P^+; \Gamma_n, \neg P^- \vdash_{P^0;\emptyset} \Delta_n$ can only be inferred in $CIRC$ by using (C3) in the form of

$$\frac{s, P^+; \Gamma_n, \neg P^- \vdash_{P^0\setminus\{s\};\emptyset} \Delta_n \qquad P^+; \Gamma_n, \neg P^-, \neg s \vdash_{P^0\setminus\{s\};\emptyset} \Delta_n}{P^+; \Gamma_n, \neg P^- \vdash_{P^0;\emptyset} \Delta_n}$$

for some s in P^0. Lemma 5 tells us that $P^+ \sqcup P^- \sqcup P^0 = P_n$. As $k < n$ there is some i, $1 \le i \le n$, such that $p_i, q_i \notin P^+ \sqcup P^-$ and so $p_i, q_i \in P^0$.

Suppose that $P^+; \Gamma_n, \neg P^- \vdash_{P^0;\emptyset} \Delta_n$ is inferred via (C1). Then, for some $p \in P^+$, the sequent $\Gamma_n, \neg P^-, \neg P^0 \nvdash p$ must be obtainable in the antisequent calculus. But as $p_i, q_i \in P^0$ and $p_i \vee q_i \in \Gamma_n$ the set $\Gamma_n, \neg P^-, \neg P^0$ is inconsistent and has no models. Hence $\Gamma_n, \neg P^-, \neg P^0 \vDash p$ and $\Gamma_n, \neg P^-, \neg P^0 \nvdash p$ is not derivable in AC.

Suppose instead that it is inferred via (C2). Then $P^+, \Gamma_n, \neg P^- \vDash \Delta_n$ must be true. However, as $p_i, q_i \notin P^+ \sqcup P^-$ the model which takes p_i, q_i as both true is consistent with the antecedent but not the succedent; so (C2) cannot be used.

Rule (C4) cannot be used either as the resulting sequent always has an element in R. Hence, (C3) is used to infer $P^+; \Gamma_n, \neg P^- \vdash_{P^0;\emptyset} \Delta_n$.

The inductive case needs proofs of both $s, P^+; \Gamma_n, \neg P^- \vdash_{P^0\setminus\{s\};\emptyset} \Delta_n$ and $P^+; \Gamma_n, \neg P^-, \neg s \vdash_{P^0\setminus\{s\};\emptyset} \Delta_n$ to construct the full proof. By the induction hypothesis each takes at least 2^{n-k-1} many lines of our desired form. Atom s is either in B or in A but not both. Therefore the lines are all distinct and there are $2 \cdot 2^{n-k-1}$ many lines, hence at least 2^{n-k} lines for the inductive step.

Finally, when $k = n$ we get that the full proof π of $\emptyset; \Gamma_n \vdash_{P_n;\emptyset} \Delta_n$ contains at least 2^n applications of (C3). □

In fact the proof even shows an exponential lower bound to the number of lines, *i.e.*, the proof length, which is a stronger statement.

4   Separating the Sequent Calculi *CIRC* and *MLK*

We now focus our attention on minimal entailment. In particular we will discuss Olivetti's sequent calculus *MLK* from [28] and compare its proof complexity with *CIRC*. *MLK* operates with sequents $\Gamma \vdash_M \Delta$. Semantically, $\Gamma \vdash_M \Delta$ is true if $\bigvee \Delta$ holds in all (VAR$(\Gamma \cup \Delta); \emptyset$)-minimal models of Γ.

To introduce derivability we use the property of a *positive* atom in a formula from [28], defined inductively as follows. Atom p is positive in formula p. Atom p is positive in formula ϕ if and only if it is negative in $\neg\phi$. If atom p is positive in formula ϕ or χ, it is positive in $\phi \wedge \chi$ and $\phi \vee \chi$. If atom p is negative in formula ϕ or positive in χ then it is positive in $\phi \rightarrow \chi$.

The *MLK* calculus comprises all rules detailed in Fig. 4 together with all rules from *LK*. Olivetti showed soundness and completeness of *MLK*.

Theorem 7. *(Olivetti [28]) A sequent $\Gamma \vdash_M \Delta$ is true if and only if it is derivable in MLK.*

We first show that for minimal entailment, *CIRC* is not better than *MLK*.

Theorem 8. *CIRC does not p-simulate MLK for minimal entailment.*

$$\frac{}{\Gamma \vdash_M \neg p}\ (\vdash_M) \qquad \frac{\Gamma \vdash \Delta}{\Gamma \vdash_M \Delta}\ (\vdash\vdash_M)$$

for p atomic and not positive in any formula in Γ

$$\frac{\Gamma \vdash_M \Sigma, A \qquad A, \Gamma \vdash_M \Lambda}{\Gamma \vdash_M \Sigma, \Lambda}\ (\text{M-cut}) \qquad \frac{\Gamma \vdash_M \Sigma \qquad \Gamma \vdash_M \Delta}{\Gamma, \Sigma \vdash_M \Delta}\ (\bullet \vdash_M)$$

$$\frac{\Gamma \vdash_M \Sigma, A \qquad \Gamma \vdash_M \Sigma, B}{\Gamma \vdash_M \Sigma, A \wedge B}\ (\vdash_M \wedge) \qquad \frac{A, \Gamma \vdash_M \Sigma \qquad B, \Gamma \vdash_M \Sigma}{A \vee B, \Gamma \vdash_M \Sigma}\ (\vee \vdash_M)$$

$$\frac{\Gamma \vdash_M \Sigma, A}{\Gamma \vdash_M \Sigma, B \vee A}\ (\vdash_M \bullet\vee) \qquad \frac{\Gamma \vdash_M \Sigma, A}{\Gamma \vdash_M \Sigma, A \vee B}\ (\vdash_M \vee\bullet)$$

$$\frac{A, \Gamma \vdash_M \Sigma}{\Gamma \vdash_M \Sigma, \neg A}\ (\vdash_M \neg) \qquad \frac{A, \Gamma \vdash_M \Sigma, B}{\Gamma \vdash_M \Sigma, A \to B}\ (\vdash_M \to)$$

Fig. 4. Rules of the sequent calculus *MLK* for minimal entailment (Olivetti [28])

Proof. We use the hard examples from Theorem 6 and show that they can be proved in *MLK* in polynomial size. Using the same notation as in the proof of Theorem 6 we define Γ^i as $\Gamma_n \backslash \{p_i \vee q_i\}$. Consider the following *MLK* derivation.

$$\dfrac{\dfrac{\dfrac{\dfrac{\dfrac{\dfrac{}{p_i \vdash p_i}\,(\vdash)}{\Gamma^i, p_i \vdash p_i}\,(\bullet\vdash)}{\Gamma^i, p_i \vdash_M p_i}\,(\vdash\vdash_M) \quad \dfrac{}{\Gamma^i, p_i \vdash_M \neg q_i}\,(\vdash_M)}{\Gamma^i, p_i \vdash_M p_i \wedge \neg q_i}\,(\vdash_M \wedge)}{\Gamma^i, p_i \vdash_M (p_i \wedge \neg q_i) \vee (q_i \wedge \neg p_i)}\,(\vdash_M \vee\bullet) \quad \dfrac{\dfrac{\dfrac{\dfrac{\dfrac{}{q_i \vdash q_i}\,(\vdash)}{\Gamma^i, q_i \vdash q_i}\,(\bullet\vdash)}{\Gamma^i, q_i \vdash_M q_i}\,(\vdash\vdash_M) \quad \dfrac{}{\Gamma^i, q_i \vdash_M \neg p_i}\,(\vdash_M)}{\Gamma^i, q_i \vdash_M q_i \wedge \neg p_i}\,(\vdash_M \wedge)}{\Gamma^i, q_i \vdash_M (p_i \wedge \neg q_i) \vee (q_i \wedge \neg p_i)}\,(\vdash_M \bullet\vee)}{\Gamma_n \vdash_M (p_i \wedge \neg q_i) \vee (q_i \wedge \neg p_i)}\,(\vee \vdash_M)$$

This proof tree shows that $\Gamma_n \vdash_M (p_i \wedge \neg q_i) \vee (q_i \wedge \neg p_i)$ can be proved in linear length. By repeated use (at most a linear number of times) of rule $(\vdash_M \wedge)$ we build the big conjunction and obtain $\Gamma_n \vdash_M \Delta_n$ in polynomial size. □

The next lemma provides a translation of intermediate *CIRC*-sequents to *MLK*-sequents, which is easy to verify model-theoretically.

Lemma 9. *Let* $\text{VAR}(\Gamma, \Delta) = P^0 \sqcup P^+ \sqcup P^-$. *Then* $P^+; \Gamma, \neg P^- \vdash_{P^0;\emptyset} \Delta$ *is true if and only if* $\Gamma, \neg P^- \vdash_M \Delta, \neg P^+$ *is true.*

Given a minimal entailment sequent $\Lambda \vdash_{\text{VAR}(\Lambda,\Delta);\emptyset} \Delta$ and its proof $(t_i)_{0 \leq i \leq n}$ in *CIRC* we define a map τ that acts on intermediate sequents of the form $\Sigma; \Gamma \vdash_{P;\emptyset} \Delta$ and maps them to the *MLK*-sequent $\Gamma \vdash_M \Delta, \neg\Sigma$. This map is well defined as Lemma 5 guarantees that all intermediate sequents are exactly of the form that allow the translation in Lemma 9.

To compare *MLK* with *CIRC* we need a few facts on *LK*.

Lemma 10. *1. For sets of formulae Γ, Δ and disjoint sets of atoms Σ^+, Σ^- with $VAR(\Gamma \cup \Delta) = \Sigma^+ \sqcup \Sigma^-$ we can efficiently construct quadratic-size LK-proofs of $\Sigma^+, \neg\Sigma^-, \Gamma \vdash \Delta$ when the sequent is true.*
2. For formulae ϕ, χ we have $s_{LK}(\phi \vdash \phi[\chi/\bot]) \in O(|\chi| + |\phi|)$.

Lemma 11. *Let Σ, Γ, Δ be sets of formulae. From a sequent $\Sigma, \bigwedge \Gamma \vdash_M \Delta$ of size n we can derive $\Sigma, \Gamma \vdash_M \Delta$ in an $O(n^3)$ size MLK proof.*

Proof (Sketch). Informally, the idea is that writing a conjunction or a list of formulae is semantically the same thing, but must be treated as different objects in a proof. The lemma demonstrates the ability of *MLK* to prove one direction of the equivalence in polynomial size. The strategy used is to inductively prove $\Sigma, \bigwedge \Gamma, \Gamma' \vdash_M \Delta$ for $\Gamma' \subseteq \Gamma$. We use proof by induction on the number of elements r of Γ'. We then use M-cut to remove the conjunction from the antecedent. □

Remark 12. As can be seen, the M-cut rule is very powerful and allows to manipulate the minimal entailment sequents, by using classical sequents. In fact, even when omitting all rules $(\vdash_M \wedge), (\vee \vdash_M), (\vdash_M \bullet\vee), (\vdash_M \vee\bullet), (\vdash_M \neg), (\vdash_M \rightarrow)$ from *MLK* we still obtain a calculus that is complete for minimal entailment and p-simulates the original *MLK*. An example illustrating this for $(\vdash_M \neg)$ is given below.

$$\frac{\dfrac{\dfrac{\dfrac{\overline{A \vdash A}\;(\vdash)}{\vdash A, \neg A}\;(\vdash \neg)}{\Gamma \vdash A, \neg A}\;\text{(repeated use of }\bullet\vdash)}{\Gamma \vdash_M A, \neg A}\;(\vdash\vdash_M) \qquad A, \Gamma \vdash_M \Sigma}{\Gamma \vdash_M \Sigma, \neg A}\;\text{(M-cut)}$$

The next theorem is the main result in this section. Together with Theorem 8 it shows that *MLK* is strictly stronger than *CIRC* for minimal entailment.

Theorem 13. *MLK p-simulates CIRC for minimal entailment.*

Proof (Sketch). Let π be a proof in *CIRC* of the minimal entailment sequent $\Lambda \vdash_{VAR(\Lambda, \Delta); \emptyset} \Delta$. We will show that there exist constants a and b (independent of π and the sequent) such that there is a proof $\pi^\star$ of $\Lambda \vdash_M \Delta$ in *MLK* with $|\pi^\star| \leq a|\pi|^3 + b$. The induction argument is based on translating each line of the *CIRC*-proof using τ defined after Lemma 9 and deriving it in *MLK*.

Induction Hypothesis (on the number r of applications of (C3) and (C4)): Let $\Lambda \vdash_{VAR(\Lambda, \Delta); \emptyset} \Delta$ be a minimal entailment sequent with *CIRC* proof π. Let $\Sigma; \Gamma \vdash_{P; \emptyset} \Delta$ be an intermediate sequent of π (as in Definition 4), which is preceded by r applications of rules (C3) and (C4) in π, and the sub-proof up to that line is of size k. Then $\tau(\Sigma; \Gamma \vdash_{P; \emptyset} \Delta)$ can be derived in an $(ak^3 + b)$-size *MLK* proof.

Base Case ($r = 0$): For the base cases we only have to consider conclusions of rules (C1) and (C2).

C1: What makes (C1) the most difficult case is that it uses the antisequent calculus, which is not incorporated in MLK. When using (C1) in $CIRC$ proof π we would start with premise $\Gamma, \neg P \nvdash q$ and end with conclusion $q, \Sigma; \Gamma \vdash_{P;\emptyset} \Delta$, so we have to find an MLK proof starting with the axioms of the MLK calculus that is cubic in size and reaches conclusion $\tau(q, \Sigma; \Gamma \vdash_{P;\emptyset} \Delta) = \Gamma \vdash_M \Delta, \neg q, \neg \Sigma$.

Suppose that the intermediate sequent $q, \Sigma; \Gamma \vdash_{P;\emptyset} \Delta$ is inferred via (C1) in the $CIRC$ proof π. Then $\Gamma, \neg P \nvdash q$ holds; so there is some model N in which $\Gamma, \neg P$ and $\neg q$ hold. Moreover, since we have the AC-proof we can efficiently construct this N by Proposition 2, which is needed to get a p-simulation.

Consider the sets of atoms $\Sigma^+ = \text{VAR}(\Gamma) \cap N$ and $\Sigma^- = \text{VAR}(\Gamma) \setminus (N \cup \{q\} \cup P)$. We claim that $\Sigma^+ \subseteq \Sigma \subseteq \Sigma^+ \sqcup \Sigma^-$ (but must omit the proof here). Therefore we can find $\Sigma^\star \subseteq \Sigma^-$ such that $\Sigma = \Sigma^+ \sqcup \Sigma^\star$.

For set of atoms $A = \{a_1, \ldots, a_l\}$ let us define $\hat{\Gamma}(A) = \bigwedge \Gamma[a_1/\bot, \ldots, a_l/\bot]$. This notation allows us to replace the variables with their assigned value, and treat the antecedent as a single formula. Let $m = |\Lambda \vdash_{\text{VAR}(\Lambda,\Delta);\emptyset} \Delta|$. We will let U and Q be arbitrary sets of atoms such that $U \sqcup Q = \Sigma^- \cup P$. Then $\Sigma^+ \vdash_M \hat{\Gamma}(U)$ is true. This is because all atoms in Q and U are minimised to not true, and the remaining positive atoms of N are all true, hence the minimal model is N and so Γ is satisfied. We incorporate these sequents in a proof by induction where we replace $\bot$ with atoms in $\hat{\Gamma}$ one by one (we omit details of this induction). For $Q = \Sigma^- \cup P$ we obtain from this induction an MLK-proof of $\Sigma^+ \vdash_M \bigwedge \Gamma$ of size $O(m^2)$. We proceed extending the proof with

$$\frac{\Sigma^+ \vdash_M \bigwedge \Gamma \qquad \dfrac{\Sigma^+ \vdash_M \neg q}{} \, (\vdash_M)}{\Sigma^+, \bigwedge \Gamma \vdash_M \neg q} \, (\bullet \vdash_M)$$

Using Lemma 11 we can add a cubic size proof to get $\Sigma^+, \Gamma \vdash_M \neg q$. Now we wish to weaken the right hand side. To do this we start with the axiom $\neg q \vdash \neg q$. Then use the weakening rules of LK to get $\Sigma^+, \Gamma, \neg q \vdash \neg q, \neg \Sigma^\star, \Delta$. We then continue with

$$\frac{\Sigma^+, \Gamma \vdash_M \neg q \qquad \dfrac{\Sigma^+, \Gamma, \neg q \vdash \neg q, \neg \Sigma^\star, \Delta}{\Sigma^+, \Gamma, \neg q \vdash_M \neg q, \neg \Sigma^\star, \Delta} \, (\vdash \vdash_M)}{\Sigma^+, \Gamma \vdash_M \neg q, \neg \Sigma^\star, \Delta} \, (\text{M-cut})$$

Repeated use of rule $(\vdash_M \neg)$ on sequents derives $\Gamma \vdash_M \Delta, \neg q, \neg \Sigma$, which is equivalent to the conclusion in (C1) under translation τ.

C2: We start with the classical sequent $\Sigma, \Gamma \vdash \Delta$ and then continue with

$$\frac{\dfrac{\Sigma, \Gamma \vdash \Delta}{\Sigma, \Gamma \vdash_M \Delta} \, (\vdash \vdash_M)}{\Gamma \vdash_M \Delta, \neg \Sigma} \text{ repeated use of } (\vdash_M \neg)$$

to obtain $\Gamma \vdash_M \Delta, \neg \Sigma = \tau(\Sigma; \Gamma \vdash_{P;\emptyset} \Delta)$.

Inductive Step: In our overall induction we still need to consider the cases of applications of rules (C3) and (C4).

C3: For (C3), because of Lemma 5 our premises translated under τ must be $\Lambda, \neg P^- \vdash_M \Delta, \neg P^+, \neg p$ and $\Lambda, \neg P^-, \neg p \vdash_M \Delta, \neg P^+$, yielding

$$\frac{\Lambda, \neg P^- \vdash_M \Delta, \neg P^+, \neg p \qquad \Lambda, \neg P^-, \neg p \vdash_M \Delta, \neg P^+}{\Lambda, \neg P^- \vdash_M \Delta, \neg P^+} \text{ (M-cut)}$$

C4: Since we have no fixed elements there are no applications of (C4).

Finally, the inductive claim for the entire proof gives us a cubic size proof of the sequent $\tau(\Lambda \vdash_{\mathrm{VAR}(\Lambda, \Delta); \emptyset} \Delta)$, and this is $\Lambda \vdash_M \Delta$ as required. Since our proof is constructive we even obtain a p-simulation. □

5 Extending the Simulation to Full Circumscription

While MLK only works for minimal entailment Olivetti [28] also augmented this calculus to obtain a sequent calculus for full circumscription. The rules of this calculus $DMLK$ are shown in Figure 5. To distinguish between the different sequent calculi we use the notation $\Gamma \rhd_{P;R} \Delta$ for derivability in $DMLK$.

$$\frac{}{\Gamma \rhd_{P;R} \neg p} \text{ (P-int)} \qquad \frac{\Gamma, N(U) \rhd_{P;R} \Delta}{\Gamma, N(z), U \to z \rhd_{P;R} \Delta} \text{ (Z-int)} \qquad \frac{\Gamma \vdash \Delta}{\Gamma \rhd_{P;R} \Delta} \text{ (} \vdash \rhd \text{)}$$

for $p \in P$ and not positive in any formula in Γ

for $z \in Z$ and $z \notin \Gamma, \Delta, U$ and formula U occurring negatively in $N(U)$

$$\frac{\Gamma \rhd_{P;R} \Sigma, A \qquad A, \Gamma \rhd_{P;R} \Lambda}{\Gamma \rhd_{P;R} \Sigma, \Lambda} \text{ (} \rhd \text{-cut)} \qquad \frac{\Gamma \rhd_{P;R} \Sigma \qquad \Gamma \rhd_{P;R} \Delta}{\Gamma, \Sigma \rhd_{P;R} \Delta} \text{ (} \bullet \rhd \text{)}$$

$$\frac{\Gamma \rhd_{P;R} \Sigma, A \qquad \Gamma \rhd_{P;R} \Sigma, B}{\Gamma \rhd_{P;R} \Sigma, A \wedge B} \text{ (} \rhd \wedge \text{)} \qquad \frac{A, \Gamma \rhd_{P;R} \Sigma \qquad B, \Gamma \rhd_{P;R} \Sigma}{A \vee B, \Gamma \rhd_{P;Z} \Sigma} \text{ (} \vee \rhd \text{)}$$

$$\frac{\Gamma \rhd_{P;R} \Sigma, A}{\Gamma \rhd_{P;R} \Sigma, B \vee A} \text{ (} \rhd \bullet \vee \text{)} \qquad \frac{\Gamma \rhd_{P;R} \Sigma, A}{\Gamma \rhd_{P;R} \Sigma, A \vee B} \text{ (} \rhd \vee \bullet \text{)}$$

$$\frac{A, \Gamma \rhd_{P;R} \Sigma}{\Gamma \rhd_{P;R} \Sigma, \neg A} \text{ (} \rhd \neg \text{)} \qquad \frac{A, \Gamma \rhd_{P;R} \Sigma, B}{\Gamma \rhd_{P;R} \Sigma, A \to B} \text{ (} \rhd \to \text{)}$$

Fig. 5. Rules of the sequent calculus $DMLK$ for circumscription (Olivetti [28])

Theorem 14. *(Olivetti [28]) DMLK is sound and complete for circumscription.*

If we want to prove a p-simulation of $CIRC$ by $DMLK$ it is necessary to make use of the (Z-int) rule. This seems problematic as the (Z-int) rule is syntactically quite restrictive and specialised for Olivetti's proof of Theorem 14. We therefore alternatively suggest to incorporate the antisequent calculus, adding rules of AC and the following new rule

$$\frac{\Gamma, R^+, \neg R^-, \neg P^-, \neg P^0 \nvdash p}{\Gamma, R^+, \neg R^-, \neg P^- \rhd_{P;R} \neg P^+} \ (\nvdash \rhd)$$

for $p \in P^+$, $P^- \sqcup P^0 \sqcup P^+ = P$, and $R^+ \sqcup R^- = R$. This still yields a sequent calculus $DMLK + (\nvdash \rhd)$ which is sound and complete for circumscription.

Similarly to Lemmas 5 and 9, the next lemma provides a translation of circumscriptive sequents to $\rhd$-sequents.

Lemma 15. *Let $\Gamma \vdash_{P;R} \Delta$ be a circumscriptive sequent with a CIRC-proof π.*

1. *Every intermediate sequent of π is of form $P^+; \Gamma, \neg P^-, R^+, \neg R^- \vdash_{P^0;R^0} \Delta$, where P is partitioned into sets P^+, P^-, P^0; R is partitioned analogously.*
2. *Let σ be the function that takes intermediate sequents of π of the form $P^+; \Gamma, \neg P^-, R^+, \neg R^- \vdash_{P^0;R^0} \Delta$ to sequents $\Gamma, \neg P^-, R^+, \neg R^- \rhd_{P;R} \Delta, \neg P^+$. Let A be an intermediate sequent of π, then $\sigma(A)$ is a true sequent.*

We can now state the simulation.

Theorem 16. *$DMLK + (\nvdash \rhd)$ p-simulates CIRC.*

6   Comparison to Niemelä's Tableau Calculus

We now discuss the relations of these sequent calculi to a tableau calculus for minimal entailment. This tableau works for clausal theories and was introduced by Niemelä [26]. In this paper we will refer to this tableau calculus as *NTAB*.

For clausal theory Γ and formula ϕ, a Niemelä-tableau is defined as follows. We start the construction of the tableau T with a single branch $(C_i)_{0 \leq i \leq k}$ containing all the clauses of $\Gamma \cup \Delta$, where Δ is $\neg \phi$ expressed in CNF (conjunctive normal form). There are two rules for extending a branch, where the premises must occur earlier in the branch. Figure 6 gives these two rules where those clauses above the line indicate the premises needed to use the rule, and the clauses below indicate the extensions.

Niemelä's tableau *NTAB* uses the following conditions to close branches.

$$\frac{\{a_1, a_2, \ldots, a_m, \neg b_1, \neg b_2, \ldots, \neg b_n\}, \{b_1\}, \ldots, \{b_n\},}{\{a_j\}} \ (N1)$$
$$\{\neg a_1\}, \ldots, \{\neg a_{j-1}\}, \{\neg a_{j+1}\}, \ldots, \{\neg a_m\}$$

$$\frac{\{a_1, a_2, \ldots, a_m, \neg b_1, \neg b_2, \ldots, \neg b_n\}, \{b_1\}, \ldots, \{b_n\}}{\{a_j\} \mid \{\neg a_j\}} \ (N2)$$

Fig. 6. Rules of Niemelä's tableau *NTAB* [26]. The notation $\{a_j\} \mid \{\neg a_j\}$ indicates that the branch splits.

1. A branch B is *(classically) closed* when for some atoms $b_1, \ldots, b_n$ the clauses $\{\neg b_1, \ldots, \neg b_n\}, \{b_1\}, \ldots, \{b_n\}$ occur in the same branch.
2. Let $N_\Gamma(B) = \{\neg c \mid c$ is an atom, $\{c\}$ does not occur in B, and $\exists C \in \Gamma$ s.t. $c \in C\}$. A branch B is *ungrounded* when B contains a unit clause $\{a\}$, for which $N_\Gamma(B) \cup \Gamma \not\models a$.
3. A branch is *MM-closed* if it is either closed or ungrounded.

The correctness and completeness of $NTAB$ was shown by Niemelä:

Theorem 17. *(Niemelä [26]) For clausal Γ and arbitrary ϕ there is an $NTAB$ proof for Γ, ϕ with all its branches MM-closed if and only if $\Gamma \models_M \phi$.*

In the same work [8], where Bonatti and Olivetti introduce $CIRC$, they also compare it to $NTAB$, showing that tableaux in $NTAB$ can be efficiently translated into $CIRC$-proofs.

Theorem 18. *(Bonatti, Olivetti [8]) $CIRC$ p-simulates $NTAB$.*

We will now show that the converse simulation does not hold, *i.e.*, we will prove a separation between $NTAB$ and $CIRC$. This separation uses the well-known *pigeonhole principle* PHP_n^{n+1}. This an elementary, but famous principle for which a wealth of lower bounds is known in proof complexity (cf. [2, 19]). PHP_n^{n+1} uses variables $x_{i,j}$ with $i \in [n+1]$ and $j \in [n]$, indicating that pigeon i goes into hole j. PHP_n^{n+1} consists of the clauses $\bigvee_{j \in [n]} x_{i,j}$ for all pigeons $i \in [n+1]$ and $\neg x_{i_1,j} \vee \neg x_{i_2,j}$ for all choices of distinct pigeons $i_1, i_2 \in [n+1]$ and holes $j \in [n]$. We use these formulas to obtain an exponential separation between $NTAB$ and $CIRC$.

Theorem 19. *$NTAB$ does not simulate $CIRC$ for minimal entailment.*

Proof. We first show that $s_{NTAB}(PHP_n^{n+1} \vdash \bot) \geq 2^{\Omega(n)}$. The crucial observation is that any tableau in $NTAB$ for the pigeonhole principle, is in fact a refutation using the DPLL algorithm [13]. This can be seen as follows. The formula $\neg \bot$ in conjunctive normal form is just the empty set. So each tableau has as starting nodes just the clauses of PHP_n^{n+1}. In any MM-closed tableau for this sequent, every branch must be closed. This holds as PHP_n^{n+1} is inconsistent; so the antisequent $N_\Gamma(B), \Gamma \not\vdash a$ is untrue and the ungrounded condition never holds for any branch.

The only clauses that can be derived by (N1) and (N2) are unit clauses. The unit clauses being derived by rule (N2) can be interpreted as the branching labels in the DPLL algorithm. Using (N1) is a restricted form of unit propagation; this step can be done at any point in the DPLL algorithm, and normally it is done automatically between each branching step. Using (N2) is equivalent to branching on a variable. When a branch is (classically) closed this means that the empty clause can be inferred by unit propagation in a constant number of steps. Therefore each proof of $PHP_n^{n+1} \vdash \bot$ in $NTAB$ can be efficiently turned into a DPLL execution.

It is well known that runs of the DPLL algorithm can be efficiently translated into resolution refutations. Therefore the exponential lower bound for PHP_n^{n+1}

of Haken [19] applies and each $NTAB$-proof of $PHP_n^{n+1} \vdash \perp$ must be of exponential size. On the other hand, Buss [10] showed that the pigeonhole formulas admit polynomial-size Frege proofs; and Frege systems are known to be p-equivalent to LK (cf. [23]). As LK is part of $CIRC$ we obtain polynomial-size $CIRC$-proofs of $PHP_n^{n+1} \vdash_M \perp$. $\qquad\qquad\qquad\qquad\qquad\qquad\qquad\qquad\qquad\qquad$ $\square$

7 Conclusion

Combining results from this paper together with earlier results from [8] we obtain the p-simulations $NTAB \leq_p CIRC \leq_p MLK$ of proof systems for propositional circumscription. Moreover, all these systems are exponentially separated. While this tells us that MLK is the best proof system with respect to size of proofs, this might be different when it comes to proof search. In fact, $NTAB$ and $CIRC$ are both analytic[1], which enables efficient proof search strategies (cf. [8]), whereas for MLK the restricted cut rule is very powerful, making the system highly non-analytic. This is in line with the experience from classical proof complexity and SAT solving where strong proof systems are known to be *not automatizable* under suitable assumptions (cf. [9]); and modern SAT solvers all build on rather weak proof systems [27].

In terms of proof complexity, the main question left open by this paper is to show lower bounds for MLK. Clearly, as circumscription is complete for the second level Π_2^p of the polynomial hierarchy [11,16], there exist at least super-polynomial lower bounds for MLK assuming $\mathsf{NP} \neq \Pi_2^p$. However, it might be very hard to show such bounds unconditionally. We note that for default logic and autoepistemic logic it is even known that showing lower bounds for the sequent calculi of these logics from [8] is as hard as showing lower bounds for classical LK [1,5], which is the main open problem in propositional proof complexity. We leave open whether a similar connection as in [1,5] can also be shown between LK and MLK.

References

1. Beyersdorff, O.: The complexity of theorem proving in autoepistemic logic. In: Järvisalo, M., Van Gelder, A. (eds.) SAT 2013. LNCS, vol. 7962, pp. 365–376. Springer, Heidelberg (2013)
2. Beyersdorff, O., Galesi, N., Lauria, M.: A lower bound for the pigeonhole principle in tree-like resolution by asymmetric prover-delayer games. Inf. Process. Lett. 110(23), 1074–1077 (2010)
3. Beyersdorff, O., Köbler, J., Messner, J.: Nondeterministic functions and the existence of optimal proof systems. Theor. Comput. Sci. 410(38-40), 3839–3855 (2009)
4. Beyersdorff, O., Kutz, O.: Proof complexity of non-classical logics. In: Bezhanishvili, N., Goranko, V. (eds.) ESSLLI 2010/2011, Lectures. LNCS, vol. 7388, pp. 1–54. Springer, Heidelberg (2012)
5. Beyersdorff, O., Meier, A., Müller, S., Thomas, M., Vollmer, H.: Proof complexity of propositional default logic. Archive for Mathematical Logic 50(7), 727–742 (2011)

---

[1] The $CIRC$-rules in Fig. 3 are analytic, but cut is available in the LK-part. If we replace LK by cut-free LK, we obtain a fully analytic sequent calculus for circumscription.

6. Bonatti, P.A.: A Gentzen system for non-theorems. Technical Report CD/TR 93/52, Christian Doppler Labor für Expertensysteme (1993)
7. Bonatti, P.A., Lutz, C., Wolter, F.: The complexity of circumscription in DLs. J. Artif. Intell. Res (JAIR) 35, 717–773 (2009)
8. Bonatti, P.A., Olivetti, N.: Sequent calculi for propositional nonmonotonic logics. ACM Transactions on Computational Logic 3(2), 226–278 (2002)
9. Bonet, M.L., Pitassi, T., Raz, R.: On interpolation and automatization for Frege systems. SIAM Journal on Computing 29(6), 1939–1967 (2000)
10. Buss, S.R.: Polynomial size proofs of the propositional pigeonhole principle. The Journal of Symbolic Logic 52, 916–927 (1987)
11. Cadoli, M., Lenzerini, M.: The complexity of propositional closed world reasoning and circumscription. J. Comput. Syst. Sci. 48(2), 255–310 (1994)
12. Cook, S.A., Reckhow, R.A.: The relative efficiency of propositional proof systems. The Journal of Symbolic Logic 44(1), 36–50 (1979)
13. Davis, M., Logemann, G., Loveland, D.W.: A machine program for theorem-proving. Commun. ACM 5(7), 394–397 (1962)
14. Durand, A., Hermann, M., Nordh, G.: Trichotomies in the complexity of minimal inference. Theory Comput. Syst. 50(3), 446–491 (2012)
15. Egly, U., Tompits, H.: Proof-complexity results for nonmonotonic reasoning. ACM Transactions on Computational Logic 2(3), 340–387 (2001)
16. Eiter, T., Gottlob, G.: Propositional circumscription and extended closed world reasoning are Π_2^p-complete. Theor. Comput. Sci. 114(2), 231–245 (1993)
17. Fenner, S.A., Fortnow, L., Naik, A.V., Rogers, J.D.: Inverting onto functions. Information and Computation 186(1), 90–103 (2003)
18. Gentzen, G.: Untersuchungen über das logische Schließen. Mathematische Zeitschrift 39, 68–131 (1935)
19. Haken, A.: The intractability of resolution. Theor. Comput. Sci. 39, 297–308 (1985)
20. Hrubeš, P.: On lengths of proofs in non-classical logics. Annals of Pure and Applied Logic 157(2-3), 194–205 (2009)
21. Janota, M., Marques-Silva, J.: cmMUS: A tool for circumscription-based MUS membership testing. In: Delgrande, J.P., Faber, W. (eds.) LPNMR 2011. LNCS (LNAI), vol. 6645, pp. 266–271. Springer, Heidelberg (2011)
22. Jeřábek, E.: Substitution Frege and extended Frege proof systems in non-classical logics. Annals of Pure and Applied Logic 159(1-2), 1–48 (2009)
23. Krajíček, J.: Bounded Arithmetic, Propositional Logic, and Complexity Theory. Cambridge University Press (1995)
24. McCarthy, J.: Circumscription – a form of non-monotonic reasoning. Artificial Intelligence 13, 27–39 (1980)
25. Niemelä, I.: Implementing circumscription using a tableau method. In: ECAI, pp. 80–84 (1996)
26. Niemelä, I.: A tableau calculus for minimal model reasoning. In: Miglioli, P., Moscato, U., Ornaghi, M., Mundici, D. (eds.) TABLEAUX 1996. LNCS, vol. 1071, pp. 278–294. Springer, Heidelberg (1996)
27. Nieuwenhuis, R.: SAT and SMT are still resolution: Questions and challenges. In: Gramlich, B., Miller, D., Sattler, U. (eds.) IJCAR 2012. LNCS (LNAI), vol. 7364, pp. 10–13. Springer, Heidelberg (2012)
28. Olivetti, N.: Tableaux and sequent calculus for minimal entailment. J. Autom. Reasoning 9(1), 99–139 (1992)
29. Thomas, M.: The complexity of circumscriptive inference in Post's lattice. Theory of Computing Systems 50(3), 401–419 (2012)
30. Tiomkin, M.L.: Proving unprovability. In: Proc. 3rd Annual Symposium on Logic in Computer Science, pp. 22–26 (1988)

Visibly Linear Temporal Logic[*]

Laura Bozzelli[1] and César Sánchez[2,3]

[1] Technical University of Madrid (UPM), Madrid, Spain
[2] IMDEA Software Institute, Madrid, Spain
[3] Institute for Information Security, CSIC, Spain

Abstract. We introduce a robust and tractable temporal logic, we call *Visibly Linear Temporal Logic* (VLTL), which captures the full class of Visibly Pushdown Languages. The novel logic avoids fix points and provides instead natural temporal operators with simple and intuitive semantics. We prove that the complexities of the satisfiability and visibly pushdown model checking problems are the same as for other well known logics, like CaRet and the nested word temporal logic NWTL, which in contrast are strictly more limited in expressive power than VLTL. Moreover, formulas of CaRet and NWTL can be easily and inductively translated in linear-time into VLTL.

1 Introduction

Visibly Pushdown Languages (VPL), introduced by Alur et al. [5,6], are a subclass of context-free languages that is similar in tractability and robustness to the less expressive class of regular languages. A VPL consists of *nested words*, that is words over an alphabet (pushdown alphabet) which is partitioned into three disjoint sets of calls, returns, and internal symbols. This partition induces a nested hierarchical structure in a given word obtained by associating to each call the corresponding matching return (if any) in a well-nested manner. VPL are accepted by Nondeterministic Visibly Pushdown Automata (NVPA) [5,6], a subclass of pushdown automata where the input symbol controls the kind of operations permissible on the stack. Alternative characterizations of VPL have been given in terms of operational and declarative formalisms. Here, we recall alternating automata-based characterizations [7,11], like the class of parity alternating visibly pushdown automata and the more tractable class of parity two-way *alternating finite-state jump automata* (AJA) [11], which extend standard alternating finite-state automata (AFA) with non local moves for navigating the nested structure of words in VPL.

VPL have applications in the formal verification of recursive programs with finite data modeled by pushdown systems [9,6,4]. VPL turn out to be useful also in the streaming processing of semi-structured data, such as XML documents, where each open-tag is matched with a closing-tag in a well-nested manner (see e.g. [18,2]).

---

[*] This work was funded in part by Spanish MINECO Project "TIN2012-39391-C04-01 STRONGSOFT" and by Spanish MINECO Project "TIN2012-38137-C02 VIVAC".

S. Demri, D. Kapur, and C. Weidenbach (Eds.): IJCAR 2014, LNAI 8562, pp. 418–433, 2014.
© Springer International Publishing Switzerland 2014

The theory of VPL is connected to the theory of regular *tree*-languages since nested words can be encoded by labeled binary trees satisfying some regular constraints, and there are translations from VPL into regular tree languages over tree-encodings of nested words, and vice versa. However, as shown in [18,2], NVPA are often more natural (and sometimes exponentially more succinct) than tree automata, and preferable in the streaming processing of XML documents.

Linear Temporal Logics for VPL-Properties. Well-known and tractable linear temporal logics for VPL are the logic CaRet [4] and its extension NWTL$^+$ [3], which in turn are context-free extensions of standard linear temporal logic LTL. Like LTL, which does *not* allow to specify all the linear-time ω-regular properties, the logics CaRet and NWTL$^+$ can only express a strict subclass of VPL. Known logical frameworks which capture the full class of VPL are an extension of standard MSO over nested words with a binary matching-predicate (MSO$_\mu$) [5] and a fixpoint calculus [11], where for the latter, satisfiability and visibly pushdown model checking are EXPTIME-complete [11]. One drawback is that MSO$_\mu$ is not elementarily decidable. Additionally, fixpoint logics are considered in some sense low-level logics, making them "unfriendly" as specification languages. In the setting of regular languages, some tractable formalisms allow to avoid fixpoint binders and still obtain full expressivity, like ETL [23] and fragments of the industrial-strength logic PSL [1], like the regular linear temporal logic RLTL [16,20], which fuses regular expressions and LTL modalities. Merging regular expressions and temporal operators in the linear-time setting has been motivated by the need of human readable specification languages, as witnessed by the widespread adoption of ForSpec, PSL, SVA in industry (see e.g. [8]). Our work follows this direction for visibly-pushdown languages. We recently introduced [12] an algebraic characterization of VPL over finite nested words in terms of *visibly rational expressions* (VRE). VRE extend regular expressions with two novel operators which capture in a natural way the nested relation between calls and matching returns in nested words. These two operators, when applied to languages $\mathcal{L}$ of *well-matched* words (i.e., nested words without pending calls and pending returns), correspond to classical tree substitution and Kleene closure applied to the tree language encoding of $\mathcal{L}$ (in accordance with the encoding of well-matched words by ordered unranked finite trees [2]). However, as observed in [2] when comparing *well-matched* words with ordered unranked trees, "word operations such as prefixes, suffixes, and concatenation [...] do not have analogous tree operations." This is explicitly witnessed by VRE having both word-like concatenation and tree-like substitution (and their Kleene closures), so allowing to describe both the linear structure and the hierarchical structure of nested words.

Our Contribution. We investigate a new linear temporal logic for VPL specifications, which merges in a convenient way VRE and LTL modalities. The task of combining language operators (such as concatenation and Kleene closure) and logical modalities is in general not easy, since allowing unrestricted complementation (corresponding to logical negation) in regular expressions already leads to a non-elementary decidable declarative formalism [22]. Thus, we propose a generalization of RLTL with past that we call *Visibly Linear Temporal Logic* (VLTL), which is obtained by replacing regular expressions for VRE expressions as building blocks for the temporal modalities. Our natural choice leads to a unifying and convenient logical framework for specifying VPL-properties because:

- VLTL is closed under Boolean combinations including negation and captures the full class of VPL. Moreover, VLTL avoids fix points and only offers temporal operators with simple and intuitive semantics.
- VLTL is elementarily decidable. In particular, satisfiability and visibly pushdown model checking have the same complexity as for the strictly less expressive logics CaRet and NWTL$^+$: i.e. are EXPTIME-complete.

Another advantage of VLTL is that CaRet and NWTL$^+$ can be inductively translated in linear-time into VLTL. In particular, the temporal modalities of CaRet and NWTL$^+$ can be viewed as derived operators of VLTL, and, in principle, one can introduce additional "user-friendly" temporal modalities as VLTL derived operators. Thus, VLTL can be also used as a common unifying setting for obtaining efficient decision procedures for other "simple-to-use" logics for VPL.

In order to tackle the decision problems for VLTL, we propose an elegant and unifying framework which extends in a non-trivial and sophisticated way the efficient alternating automata-theoretic approach recently proposed for future RLTL [21]. The technique for future RLTL makes use of a translation of the logic into parity AFA, which is crucially based on the well-known linear-time translation of regular expressions into *nondeterministic* finite-state automata. A direct generalization of this construction based on the use of parity alternating visibly pushdown automata would lead to *doubly* exponential time decision procedures. Instead, our approach exploits as an intermediate step a compositional polynomial-time translation of VLTL formulas into a subclass of parity two-way alternating AJA with index 2, that we call *stratified AJA with main states* (SAJA). Moreover, we identify a subclass of VRE such that the corresponding fragment of VLTL has the same expressiveness as full VLTL and admits a *linear-time* translation into SAJA. Hence, we obtain a translation for this fragment of VLTL into equivalent Büchi NVPA of size $2^{O(|\varphi| \log m)}$, where m is the size of the largest VRE used in φ. Full proofs are omitted due to space limitations.

Related Work. Combining modal logic and regular expressions is also the main feature of the branching temporal logic PDL. In [17], an extension of PDL for recursive programs, has been investigated, where low-level operational aspects are allowed in the form of path expressions given by NVPA. This logic is incomparable with VLTL and the related satisfiability and visibly pushdown model-checking problems are 2-EXPTIME-complete. In [10], a linear temporal framework for VPL has been introduced which allows PDL-like path regular expressions extended with the binary matching-predicate μ of MSO$_\mu$. The setting is parameterized by a finite set of MSO-definable temporal modalities, which leads to an infinite family of linear temporal logics having the same complexity as VLTL and subsuming the logics CaRet and NWTL$^+$. However, it seems clear (even if this issue is not discussed in [10]) that each of these logics does not capture the full class of VPL. Moreover, the complexity analysis in [10], based on the use of two-way alternating tree automata, is not fine-grained and it just allows to obtain a generic polynomial in the exponent of the complexity upper bound.

2   Preliminaries

We recall Visibly Pushdown Automata [5] and Visibly Rational Expressions [12].

In the rest of the paper, we fix a *pushdown alphabet* $\Sigma = \Sigma_{call} \cup \Sigma_{ret} \cup \Sigma_{int}$, that is a finite alphabet Σ which is partitioned into a set Σ_{call} of *calls*, a set Σ_{ret} of *returns*, and a set Σ_{int} of *internal actions*.

Visibly Pushdown Automata [5]. *Nondeterministic Visibly Pushdown Automata* (NVPA) are standard Pushdown Automata operating on finite words over a *pushdown alphabet* Σ satisfying the following "visibly" restriction: (*i*) on reading a call, one symbol is pushed onto the stack, (*ii*) on reading a return, one symbol is popped from the stack (if the stack is empty, the stack content remains unchanged), and (*iii*) on reading an internal action, no stack operation is performed. The languages of finite words accepted by NVPA are called *visibly pushdown languages* (VPL). We also consider *Büchi ω-NVPA* [5], which are standard Büchi Pushdown Automata on infinite words over Σ satisfying the above "visibly" restriction. The ω-languages accepted by Büchi NVPA are called ω-*visibly pushdown languages* (ω-VPL). For details on the syntax and semantics of NVPA and Büchi ω-NVPA, see [5].

Matched Calls and Returns. For a word w on Σ, $|w|$ is the length of w (we set $|w| = \omega$ if w is infinite). For all $1 \leq i \leq j \leq |w|$, $w(i)$ is the ith symbol of w, and $w[i,j]$ is the word $w(i)w(i+1)\ldots w(j)$. The empty word is denoted by ε. The set $WM(\Sigma)$ of *well-matched words* is the subset of Σ^* inductively defined as follows: (i) $\varepsilon \in WM(\Sigma)$ (ii) $\square \cdot w \in WM(\Sigma)$ if $\square \in \Sigma_{int}$ and $w \in WM(\Sigma)$, and (iii) $c \cdot w \cdot r \cdot w' \in WM(\Sigma)$ if $c \in \Sigma_{call}$, $r \in \Sigma_{ret}$, and $w, w' \in WM(\Sigma)$. Let i be a call position of a word w. If there is $j > i$ such that j is a return position of w and $w(i+1)\ldots w(j-1)$ is a well-matched word (note that j is uniquely determined if it exists), we say that j is the *matching return* of i along w. The set $MWM(\Sigma)$ of *minimally well-matched words* is the set of well-matched words of the form $c \cdot w \cdot r$ such that c is a call, r is a return, and w is well-matched. For a language $\mathcal{L} \subseteq \Sigma^*$, we define $MWM(\mathcal{L}) \overset{\text{def}}{=} \mathcal{L} \cap MWM(\Sigma)$, that is the set of words in $\mathcal{L}$ which are minimally well-matched.

Visibly Rational Expressions (VRE) [12]. We recall the classes of pure VRE and pure ω-VRE [12], here called simply VRE and ω-VRE. VRE extend regular expressions (RE) with two non-regular operators: the binary M-substitution operator and the unary S-closure operator.[1] Given $\mathcal{L} \subseteq \Sigma^*$ and a language $\mathcal{L}'$ of

---

[1] The origin of the name M-substitution is *minimally well-matched substitution*, while S-closure stands for *Strict Mimimally Well-Matched Closure*, see [12].

finite or infinite words on Σ, we use $\mathcal{L} \cdot \mathcal{L}'$ for the concatenation of $\mathcal{L}$ and $\mathcal{L}'$, $\mathcal{L}^*$ for the Kleene closure of $\mathcal{L}$, and $\mathcal{L}^\omega$ for the ω-Kleene closure of $\mathcal{L}$.

Definition 1 (M-substitution [12]). *Let $w \in \Sigma^*$, $\square \in \Sigma_{int}$, and $\mathcal{L} \subseteq \Sigma^*$. The M-substitution of $\square$ by $\mathcal{L}$ in w, denoted by $w \curvearrowright_\square \mathcal{L}$, is the language of finite words over Σ obtained by replacing occurrences of $\square$ in w by minimally well-matched words in $\mathcal{L}$. Formally, $w \curvearrowright_\square \mathcal{L}$ is inductively defined as follows:*

- $\varepsilon \curvearrowright_\square \mathcal{L} \overset{def}{=} \{\varepsilon\}$;
- $(\square \cdot w') \curvearrowright_\square \mathcal{L} \overset{def}{=} \left(MWM(\mathcal{L}) \cdot (w' \curvearrowright_\square \mathcal{L}) \right) \cup \left((\{\square\} \cap \mathcal{L}) \cdot (w' \curvearrowright_\square \mathcal{L}) \right)$
- $(\sigma \cdot w') \curvearrowright_\square \mathcal{L} \overset{def}{=} \{\sigma\} \cdot (w' \curvearrowright_\square \mathcal{L})$ *for each $\sigma \in \Sigma \setminus \{\square\}$.*

For two languages $\mathcal{L}, \mathcal{L}' \subseteq \Sigma^*$ and $\square \in \Sigma_{int}$, the M-substitution of $\square$ by $\mathcal{L}'$ in $\mathcal{L}$, written $\mathcal{L} \curvearrowright_\square \mathcal{L}'$, is defined as $\mathcal{L} \curvearrowright_\square \mathcal{L}' \overset{def}{=} \bigcup_{w \in \mathcal{L}} w \curvearrowright_\square \mathcal{L}'$. Note that $\curvearrowright_\square$ is associative, and $\{\square\} \curvearrowright_\square \mathcal{L} = MWM(\mathcal{L})$ if $\{\square\} \cap \mathcal{L} = \emptyset$.

Definition 2 (S-closure [12]). *Given $\mathcal{L} \subseteq \Sigma^*$ and $\square \in \Sigma_{int}$, the S-closure of $\mathcal{L}$ through $\square$, denoted by $\mathcal{L}^{\circlearrowright\square}$, is defined as follows:*

$$\mathcal{L}^{\circlearrowright\square} \overset{def}{=} \bigcup_{n \geq 0} MWM(\mathcal{L}) \underbrace{\curvearrowright_\square (\mathcal{L} \cup \{\square\}) \curvearrowright_\square \ldots \curvearrowright_\square (\mathcal{L} \cup \{\square\})}_{n \text{ occurrences of } \curvearrowright_\square}.$$

Example 1. Let $\Sigma_{call} = \{c_1, c_2\}$, $\Sigma_{ret} = \{r_1, r_2\}$, and $\Sigma_{int} = \{\square\}$. Let us consider the languages $\mathcal{L} = \{c_1 \square r_1, c_2 \square r_2\}$ and $\mathcal{L}' = \{c_1 r_1, c_2 r_2\}$. Then, $\mathcal{L}^{\circlearrowright\square} \curvearrowright_\square \mathcal{L}' = \{c_{i_1} c_{i_2} \ldots c_{i_n} r_{i_n} \ldots r_{i_2} r_{i_1} \mid n \geq 2, \ i_1, \ldots, i_n \in \{1,2\}\}$.

Definition 3. *The syntax of VRE α and ω-VRE β over Σ is defined as follows:*

$$\alpha := \varepsilon \mid int \mid call \mid ret \mid \sigma \mid cr \mid c\square r \mid \alpha \cup \alpha \mid \alpha \cdot \alpha \mid \alpha^* \mid \alpha \curvearrowright_\square \alpha \mid \alpha^{\circlearrowright\square}$$
$$\beta := \alpha^\omega \mid \beta \cup \beta \mid \alpha \cdot \beta$$

where $\sigma \in \Sigma$, $c \in \Sigma_{call}$, $r \in \Sigma_{ret}$, and $\square \in \Sigma_{int}$. The basic expressions int, call, ret are used to denote in a succinct way the languages Σ_{int}, Σ_{call}, and Σ_{ret}, while the redundant basic expressions cr and $c\square r$ in the syntax of VRE are used for defining subclasses of VRE. A VRE α (resp., ω-VRE β) denotes a language of finite words (resp., infinite words) over Σ, written $\mathcal{L}(\alpha)$ (resp., $\mathcal{L}(\beta)$), which is inductively defined in the obvious way.

Note that ω-VRE are defined in terms of VRE in the same way as ω-regular expressions are defined in terms of regular expressions. A VRE is *well-matched* if it does *not* contain basic subexpressions in $\Sigma_{call} \cup \Sigma_{ret} \cup \{call, ret\}$. A VRE α is *well-formed* if each subexpression of α of the form $(\alpha_1 \curvearrowright_\square \alpha_2)$ or $\alpha_1^{\circlearrowright\square}$ is well-matched, and an ω-VRE β is *well-formed* if each VRE occurring in β is well-formed. As usual, the size $|\alpha|$ of a VRE α is the length of the string describing α.

Theorem 1 (from [12]). *(Well-formed) VRE and (well-formed) ω-VRE capture the classes of VPL and ω-VPL, respectively.*

Proof. The results for VRE and ω-VRE were established in [12]. Moreover, a straightforward adaptation of the translations from NVPA to VRE and from Büchi ω-NVPA to ω-VRE in [12] show that *well-formed* VRE and *well-formed* ω-VRE are sufficient to capture the classes of VPL and ω-VPL, respectively.   □

3   Visibly Linear Temporal Logic (VLTL)

In this section, we introduce the Visibly Linear Temporal Logic (VLTL), an extension of Regular Linear Temporal Logic (RLTL) with past (see [16,20]) obtained by replacing regular expressions in the temporal modalities of RLTL with VRE.

The syntax of VLTL formulas φ over the pushdown alphabet Σ is as follows:

$$\varphi := \mathtt{true} \mid \varphi \vee \varphi \mid \neg\varphi \mid \alpha;\varphi \mid \varphi;\alpha \mid \varphi|\alpha\rangle\!\rangle\varphi \mid \varphi|\alpha\rangle\varphi \mid \varphi\langle\!\langle\alpha|\varphi \mid \varphi\langle\alpha|\varphi$$

where α is a VRE over Σ, the symbol ; is the sequencing operator, $|\rangle\!\rangle$ and $\langle\!\langle|$ are the *(future) power operator* and the *past power operator*, and $|\rangle$ and $\langle|$ are the *(future) weak power operator* and the *past weak power operator*. The power formulas $\varphi_1|\alpha\rangle\!\rangle\varphi_2$, $\varphi_1\langle\!\langle\alpha|\varphi_2$, $\varphi_1|\alpha\rangle\varphi_2$, and $\varphi_1\langle\alpha|\varphi_2$ are built from three elements: φ_2 (the *attempt*), φ_1 (the *obligation*), and α (the *delay*). Informally, for $\varphi_1|\alpha\rangle\!\rangle\varphi_2$ (resp., $\varphi_1\langle\!\langle\alpha|\varphi_2$) to hold, either the attempt holds, or the obligation is met and the whole formula evaluates successful after (resp., before) the delay; additionally, the attempt must be eventually met. The weak formulas $\varphi_1|\alpha\rangle\varphi_2$ and $\varphi_1\langle\alpha|\varphi_2$ do not require the attempt to be eventually met. For a VLTL formula φ, φ is *well-formed* if every VRE occurring in φ is well-formed. Let $\|\varphi\|$ be the integer 1 if either $\varphi = \mathtt{true}$ or φ has a Boolean connective at its root; otherwise, $\|\varphi\|$ is the size of the VRE associated with the root operator of φ. The size $|\varphi|$ of φ is defined as $\sum_{\psi\in SF(\varphi)}\|\psi\|$, where $SF(\varphi)$ is the set of subformulas of φ.

VLTL formulas φ are interpreted over *infinite pointed words* (w, i) over Σ, where $w \in \Sigma^\omega$ and $i \geq 1$ is a position along w. The satisfaction relation $(w, i) \models \varphi$ is defined by induction as follows (we omit the rules for Boolean connectives):

$$(w, i) \models \alpha;\varphi \quad\Leftrightarrow\quad \text{for some } j > i,\ (w, j) \models \varphi \text{ and } w[i, j] \in \mathcal{L}(\alpha)$$

$$(w, i) \models \varphi;\alpha \quad\Leftrightarrow\quad \text{for some } j < i,\ (w, j) \models \varphi \text{ and } w[j, i] \in \mathcal{L}(\alpha)$$

$$(w, i) \models \varphi_1|\alpha\rangle\!\rangle\varphi_2 \Leftrightarrow \text{for some sequence } i = j_1 < \ldots < j_n,\ (w, j_n) \models \varphi_2$$
$$\text{and for all } 1 \leq k < n,\ w[j_k, j_{k+1}] \in \mathcal{L}(\alpha) \text{ and } (w, j_k) \models \varphi_1$$

$$(w, i) \models \varphi_1\langle\!\langle\alpha|\varphi_2 \Leftrightarrow \text{for some sequence } j_1 < \ldots < j_n = i,\ (w, j_1) \models \varphi_2$$
$$\text{and for all } 1 < k \leq n,\ w[j_{k-1}, j_k] \in \mathcal{L}(\alpha) \text{ and } (w, j_k) \models \varphi_1$$

$$(w, i) \models \varphi_1|\alpha\rangle\varphi_2 \Leftrightarrow (w, i) \models \varphi_1|\alpha\rangle\!\rangle\varphi_2,$$
$$\text{or for some infinite sequence } i = j_1 < j_2 < \ldots,$$
$$w[j_k, j_{k+1}] \in \mathcal{L}(\alpha) \text{ and } (w, j_k) \models \varphi_1 \text{ for all } k \geq 1$$

$$(w, i) \models \varphi_1\langle\alpha|\varphi_2 \Leftrightarrow (w, i) \models \varphi_1\langle\!\langle\alpha|\varphi_2,$$
$$\text{or for some sequence } 1 = j_1 < \ldots < j_n = i,\ (w, j_n) \models \varphi_1$$
$$\text{and } w[j_k, j_{k+1}] \in \mathcal{L}(\alpha) \text{ and } (w, j_k) \models \varphi_1 \text{ for all } 1 \leq k < n$$

The ω-pointed language $\mathcal{L}_p(\varphi)$ of φ is the set of infinite pointed words (w, i) over Σ satisfying φ (i.e. $(w, i) \models \varphi$). The ω-language $\mathcal{L}(\varphi)$ of φ is the set of infinite words w over Σ such that $(w, 1) \in \mathcal{L}_p(\varphi)$. Two formulas φ_1 and φ_2 are *globally equivalent* if $\mathcal{L}_p(\varphi_1) = \mathcal{L}_p(\varphi_2)$. The *satisfiability* problem for VLTL is checking for a VLTL formula φ, whether $\mathcal{L}(\varphi) \neq \emptyset$. The *visibly pushdown model checking* problem for VLTL is checking for a VLTL formula φ over Σ and a *pushdown system* $\mathcal{P}$ (defined as a Büchi NVPA $\mathcal{P}$ over the same pushdown alphabet Σ and with all states accepting), whether $\mathcal{L}(\mathcal{P}) \subseteq \mathcal{L}(\varphi)$.

Note that the VLTL operators generalize both the operators of standard LTL with past (in particular, the next, previous, until, and since modalities) and the operators of ω-visibly rational expressions. For example, the until formula $\varphi_1\mathcal{U}\varphi_2$ requires that either φ_2 holds (attempt) or otherwise φ_1 holds (obligation) and the formula is reevaluated after a delay of a single step. Similarly, the ω-visibly rational expression α^ω has no possible escape, a trivially fulfilled obligation, with a delay indicated by α.

In the rest of this section, we use some VRE of constant size (where $\square \in \Sigma_{int}$):
- $\alpha_{ONE} := int \cup ret \cup call$, $\alpha_{MWM} := \square \curvearrowright_\square (call \cdot (\alpha_{ONE})^* \cdot ret)$,
  $\alpha_{WM} := (int^* \cdot (\alpha_{MWM})^*)^*$

Note that $\mathcal{L}(\alpha_{ONE}) = \Sigma$, $\mathcal{L}(\alpha_{MWM}) = MWM(\Sigma)$, and $\mathcal{L}(\alpha_{WM}) = WM(\Sigma)$. Moreover, we use some shortcuts in VLTL. The formula $(\sigma \cdot \alpha_{ONE})$; true is satisfied by words that begin with letter $\sigma \in \Sigma$. We abbreviate this formula by σ. Additionally, we use $\mathbf{G}\varphi$ to stand for $\varphi \,|\alpha_{ONE} \cdot \alpha_{ONE}\rangle \,\neg$true (the LTL *always* operator), and $\ominus\varphi$ to stand for $\varphi; (\alpha_{ONE} \cdot \alpha_{ONE})$ (the LTL *previous* operator).

Expressiveness of VLTL. First, we observe that (well-formed) ω-VRE can be translated in linear-time into language-equivalent (well-formed) VLTL formulas by the mapping f from ω-VRE to VLTL inductively defined as follows.
- $f(\alpha^\omega) := $ true$|\alpha \cdot \alpha_{ONE}\rangle \neg$true
- $f(\beta \cup \beta') := f(\beta) \vee f(\beta')$ and $f(\alpha \cdot \beta) := (\alpha \cdot \alpha_{ONE}); f(\beta)$.

Thus, by Theorem 1, (well-formed) VLTL formulas can express every ω-VPL (note that past temporal modalities are not required to capture ω-VPL). The converse direction holds as well (see Section 5). Hence, we obtain the following.

Theorem 2. (*Well-formed*) *VLTL formulas capture the class of* ω-VPL.

Comparison with Known Context-Free Extensions of LTL. We compare now VLTL with some known context-free extensions of LTL: CaRet [4], NWTL [3], and NWTL$^+$ [3]. NWTL and NWTL$^+$ are expressively complete for the first-order fragment FO$_\mu$ of MSO$_\mu$ [3], while it is an open question whether the same holds for CaRet [3], the latter being subsumed by NWTL$^+$. In the analysis of recursive programs, CaRet and NWTL$^+$ allow to express in a natural way LTL properties over non-regular patterns such as (*) the stack content at a given position, and (**) the local computations of procedures which skip over nested procedure invocations. Theorem 3 below shows that these logics can be easily translated in linear time into VLTL. Additionally, VLTL can specify more expressive regular properties over the patterns (*) and (**) such as the following requirement for a given $N \geq 1$, "whenever the procedure A is invoked, the depth of the stack content is a multiple of N", which can be expressed by the following VLTL formula (where the call c_A denotes the invocation of procedure A),

$$\mathbf{G}(c_A \longrightarrow (\neg \ominus \mathbf{true}); \alpha_N) \qquad \alpha_N := [\underbrace{(\alpha_{WM} \cdot call \cdots \cdot \alpha_{WM} \cdot call \cdot \alpha_{WM})}_{N \text{ times}}]^*$$

Theorem 3. *For a CaRet, NWTL or NWTL$^+$ formula φ, one can build in linear-time a VLTL formula with* constant-size VRE *which is globally equivalent to φ.*

Proof. We sketch only the translation of CaRet into VLTL. CaRet extends LTL with non-regular versions of the temporal modalities: the *abstract* next and until modalities and their past counterparts, and the *caller* modalities. Here, we focus on the abstract modalities $\bigcirc^a$ and $\mathcal{U}^a$ which correspond to the standard next and until modalities interpreted on *abstract paths*. Formally, for an infinite pointed word (w, i) on Σ, *the abstract path of w from i* is a maximal (possibly infinite) sequence of positions $i = j_1 < j_2 < \ldots < j_n < \ldots$ such that for all pairs of adjacent positions j_k and j_{k+1}: *either j_k is a call with matching return j_{k+1}, or j_k is not a call, j_{k+1} is not a return, and $j_{k+1} = j_k + 1$.* To translate $\bigcirc^a$ and $\mathcal{U}^a$ into VLTL, we use the following constant-size VRE: $\alpha_a := \alpha_{MWM} \cup ((int \cup ret) \cdot (int \cup call))$. Then, the VLTL formula $\alpha_a; \varphi_1$ is globally equivalent to $\bigcirc^a \varphi_1$, and $\varphi_1 \, |\alpha_a\rangle\!\rangle \, \varphi_2$ is globally equivalent to $\varphi_1 \, \mathcal{U}^a \, \varphi_2$. $\square$

4   Subclasses of Alternating Jump Automata

Alternating Jump Automata (AJA) over finite and infinite words [11] are an alternative automata-theoretic characterization of VPL and ω-VPL. In this section, in order to capture compositionally and efficiently VLTL formulas, we introduce a subclass of two-way parity AJA with index 2, called *two-way stratified AJA with main states* (SAJA). Then, we show how to translate (well-formed) VRE into a subclass of AJA over finite words; this result is used in Section 5 to handle the temporal operators in the translation of VLTL formulas into SAJA. Note that a naive approach based on the use of unrestricted two-way parity AJA would lead to decision procedures for VLTL that are computationally more expensive. More concretely, following [13], two-way parity AJA with n states and index k can be translated into equivalent Büchi NVPA with $2^{O((nk)^2)}$ states and stack symbols. We show that SAJA with n states can be more efficiently translated into equivalent Büchi NVPA with $2^{O(n \log m)}$ states and stack symbols, where m is the size of the largest non-trivial coBüchi stratum. Another technical issue is the efficient handling of logical negation. Like for RLTL, VLTL does not have a positive normal form. Hence, a construction for the negation operator must be given explicitly. Like for standard parity AFA, complementation of parity two-way AJA is easy: one only has to dualize the transition function and to complement the acceptance condition. However, the classical complementation for the parity condition increases in one unit the color assigned to every state, so that the total number of colors could grow linearly in the size of the formula (by alternating the constructions for complementation with those related to other modalities that reintroduce the lowest color). Instead, we show that SAJA (which only use three colors) are closed under complementation.

AJA operate on words over a pushdown alphabet and extend standard alternating finite-state automata by also allowing non-local moves: when the current input position is a matched call, a copy of the automaton can move (jump) in a single step to the matched-return position. We also allow ε-moves and local and non-local backward moves. We first give the notion of *Alternating Jump Transition Tables* (AJT), which represent AJA without acceptance conditions. Let $DIR = \{\varepsilon, \rightarrow, \leftarrow, \curvearrowright, \curvearrowleft\}$. Intuitively, the symbols $\rightarrow$ and $\leftarrow$ are used to denote forward and backward *local* moves and $\curvearrowright$ and $\curvearrowleft$ are for *non-local* moves which lead from a matched call to the matching return, and vice-versa. For a set X, $\mathcal{B}^+(X)$ denotes the set of positive Boolean formulas over X built from elements in X using $\vee$ and $\wedge$ (we also allow the formulas $\mathtt{true}$ and $\mathtt{false}$). For a formula $\theta \in \mathcal{B}^+(X)$, a *model* Y of θ is a subset Y of X which satisfies θ. The model Y of θ is minimal if no strict subset of Y satisfies θ. The *dual formula* $\tilde{\theta}$ of θ is obtained from θ by switching $\vee$ and $\wedge$, and switching $\mathtt{true}$ and $\mathtt{false}$.

Two-Way AJT. A two-way AJT $\mathcal{T}$ over Σ is a tuple $\mathcal{T} = \langle Q, q_0, \delta \rangle$, where Q is a finite set of states, $q_0 \in Q$ is the initial state, and $\delta : Q \times \Sigma \rightarrow \mathcal{B}^+(DIR \times Q \times Q)$ is a transition function. Now, we give the notion of run. We restrict ourselves to *memoryless* runs, in which the behavior of the automaton depends only on the current input position and current state. Since later we will deal only with parity acceptance conditions, memoryless runs are sufficient (see e.g. [24]). Formally, given a finite or infinite pointed word (w, i) on Σ and a state $p \in Q$, a (i, p)-*run of* $\mathcal{T}$ *over* w is a directed graph $\langle V, E, v_0 \rangle$ with set of vertices $V \subseteq \{0, \ldots, |w|+1\} \times Q$ and initial vertex $v_0 = (i, p)$. Intuitively, a vertex (j, q) describes a copy of the automaton which is in state q and reads the jth input position. Additionally, we require that the set of edges E is consistent with the transition function δ. Formally, for every vertex $(j, q) \in V$ such that $1 \leq j \leq |w|$, there is a *minimal model* $X = \{(dir_1, q_1, q_1'), \ldots, (dir_n, q_n, q_n')\}$ of $\delta(q, w(j))$ such that the set of successors of (j, q) is $\{v_1, \ldots, v_n\}$ and for all $1 \leq k \leq n$, the following holds:

- $dir_k = \varepsilon$: $v_k = (j, q_k)$.
- $dir_k = \rightarrow$: $v_k = (j+1, q_k)$ if $j+1 \leq |w|$, and $v_k = (j+1, q_k')$ otherwise.
- $dir_k = \leftarrow$: $v_k = (j-1, q_k)$ if $j-1 > 0$, and $v_k = (j-1, q_k')$ otherwise.
- $dir_k = \curvearrowright$: $v_k = (j_r, q_k)$ if j is a call with matching return j_r; otherwise $v_k = (j+1, q_k')$.
- $dir_k = \curvearrowleft$: $v_k = (j_c, q_k)$ if j is a return with matching call j_c; otherwise $v_k = (j-1, q_k')$.

An infinite path π of a run is *eventually strictly-forward* whenever π has a suffix of the form $(i_1, q_1), (i_2, q_2), \ldots$ such that: (*i*) $i_j \leq i_{j+1}$ for all $j \geq 1$ and (*ii*) for infinitely many j, $i_j < i_{j+1}$.

A two-way AJT $\mathcal{T} = \langle Q, q_0, \delta \rangle$ is an AJT *with main states* if:
- the set of states is partitioned into a set M of *main states* and into a set S of *secondary states* such that $q_0 \in$ M.
- there are no moves from secondary states to main states. Hence, every path starting from a secondary state visits only secondary states.

Two-Way Stratified AJA with Main States (SAJA). We introduce now the class of SAJA as a two-way and non-regular extension of one-way hesitant AFA over infinite words introduced in [14]. Intuitively, the ability to combine both forward and backward moves is syntactically restricted in such a way to ensure that every infinite path in a run is eventually strictly-forward. Moreover, for efficiency issues, we distinguish between main states and secondary states. Intuitively, in the translation of VLTL formulas into SAJA, main states are associated with the regular part of the formula, while secondary states (whose number can be quartic in the number of main states) are associated with the non-regular part (the M-substitution and S-closure operators in the VRE of the formula). Formally, a SAJA $\mathcal{A}$ is a tuple $\mathcal{A} = \langle Q, q_0, \delta, \mathsf{F} \rangle$ with $Q = \mathsf{M} \cup \mathsf{S}$, where $\langle Q, q_0, \delta \rangle$ is a two-way AJT with main states and F is a *strata family* of the form $\mathsf{F} = \{ \langle \rho_1, Q_1, F_1 \rangle, \ldots, \langle \rho_k, Q_k, F_k \rangle \}$, where $Q_1, \ldots, Q_k$ is a partition of the set of states Q, and for all $1 \leq i \leq k$, $\rho_i \in \{-, \mathsf{t}, \mathsf{B}, \mathsf{C}\}$ and $F_i \subseteq Q_i$, such that $F_i = \emptyset$ whenever $\rho_i = \mathsf{t}$. A stratum $\langle \rho_i, Q_i, F_i \rangle$ is called a *negative* stratum if $\rho_i = -$, a *transient* stratum if $\rho_i = \mathsf{t}$, a Büchi stratum (with Büchi acceptance condition F_i) if $\rho_i = \mathsf{B}$, and a coBüchi stratum (with coBüchi acceptance condition F_i) if $\rho_i = \mathsf{C}$. Additionally, there is a partial order $\leq$ on the sets $Q_1, \ldots, Q_k$ such that:

R1. Moves from states in Q_i lead to states in components Q_j such that $Q_j \leq Q_i$; additionally, if Q_i belongs to a transient stratum, there are no moves from Q_i leading to Q_i.
R2. For all $q \in Q_i$ and atoms (dir, q, q') or (dir, q', q) occurring in δ, the following holds: (i) $dir \in \{\leftarrow, \curvearrowleft, \varepsilon\}$ if the stratum of Q_i is negative, and $dir \in \{\rightarrow, \curvearrowright, \varepsilon\}$ otherwise, and (ii) if $dir = \varepsilon$, then there are no ε-moves from q.
R3. For every Büchi or coBüchi stratum $\langle \rho_i, Q_i, F_i \rangle$, $F_i \cap \mathsf{S} = \emptyset$.

R1 is the *stratum order requirement* and it ensures that every infinite path π of a run gets trapped in the component Q_i of some non-transient stratum. R2 is the *eventually syntactical requirement* and it ensures that Q_i belongs to a Büchi or coBüchi stratum and that π is eventually strictly-forward. Moreover, note that R2 also ensures that for all runs and vertices of the form $(0, q)$ reachable from the initial vertex, q belongs to a negative stratum.

Now we define when a run is accepting. Let π be an infinite path of a run, $\langle \rho_i, Q_i, F_i \rangle$ be the Büchi or coBüchi stratum in which π gets trapped, and $Inf(\pi)$ be the states from Q that occur infinitely many times in π. The path π is *accepting* whenever $Inf(\pi) \cap F_i \neq \emptyset$ if $\rho_i = \mathsf{B}$ and $Inf(\pi) \cap F_i = \emptyset$ otherwise (i.e. π satisfies the corresponding Büchi or coBüchi requirement). Note that R3 in the definition of SAJA ensures that whenever π starts at a vertex associated with a secondary state (hence, π visits only secondary states), then π is accepting if the stratum $\langle \rho_i, Q_i, F_i \rangle$ is a coBüchi stratum, and it is rejecting otherwise. A run is *accepting* if: (i) all its infinite paths are accepting and (ii) for each vertex $(0, q)$ reachable from the initial vertex such that q is in the stratum $\mathcal{S} = \langle \rho_i, Q_i, F_i \rangle$ (recall that $\mathcal{S}$ is ensured to be a negative stratum), it holds that $q \in F_i$. Note that this last condition is necessary to allow complementation of SAJA by dualization. The ω-*pointed language* $\mathcal{L}_p(\mathcal{A})$ of $\mathcal{A}$ is the set of infinite pointed words (w, i) over

Σ such that there is an accepting (i, q_0)-run of $\mathcal{A}$ on w. The ω-language $\mathcal{L}(\mathcal{A})$ of $\mathcal{A}$ is the set of infinite words w over Σ such that $(w, 1) \in \mathcal{L}_p(\mathcal{A})$.

The *dual automaton* $\widetilde{\mathcal{A}}$ of the SAJA $\mathcal{A}$ is defined as $\widetilde{\mathcal{A}} = \langle \mathsf{M} \cup \mathsf{S}, q_0, \widetilde{\delta}, \widetilde{\mathsf{F}} \rangle$, where $\widetilde{\delta}(q, \sigma)$ is the dual formula of $\delta(q, \sigma)$, and $\widetilde{\mathsf{F}}$ is obtained from F by converting a Büchi stratum $\langle \mathsf{B}, Q_i, F_i \rangle$ into the coBüchi stratum $\langle \mathsf{C}, Q_i, F_i \rangle$, a coBüchi stratum $\langle \mathsf{C}, Q_i, F_i \rangle$ into the Büchi stratum $\langle \mathsf{B}, Q_i, F_i \rangle$, and a negative stratum $\langle -, Q_i, F_i \rangle$ into the negative stratum $\langle -, Q_i, Q_i \setminus F_i \rangle$. Following standard arguments (see e.g. [24]) we obtain the following lemma, which is crucial for handling, compositionally and efficiently, negation in VLTL formulas.

Lemma 1. *The dual automaton $\widetilde{\mathcal{A}}$ of a SAJA $\mathcal{A}$ is a SAJA whose ω-pointed language $\mathcal{L}_p(\widetilde{\mathcal{A}})$ is the complement of $\mathcal{L}_p(\mathcal{A})$.*

From SAJA to Büchi NVPA. The *size* of a SAJA stratum $\langle \rho_i, Q_i, F_i \rangle$ is the number of *main states* in Q_i (we do not take into account the number of secondary states in Q_i). A coBüchi stratum $\langle \rho_i, Q_i, F_i \rangle$ is *trivial* whenever $F_i = \emptyset$.

Theorem 4. *For a SAJA $\mathcal{A} = \langle \mathsf{M} \cup \mathsf{S}, q_0, \delta, \mathsf{F} \rangle$, one can build in singly exponential time a Büchi NVPA $\mathcal{P}$ accepting $\mathcal{L}(\mathcal{A})$ with $2^{O(|\mathsf{S}| + |\mathsf{M}| \cdot \log(k))}$ states and stack symbols, where k is the size of the largest non-trivial coBüchi stratum of $\mathcal{A}$.*

Sketched Proof. Our approach is a refinement of a non-trivial variation of the method used in [13] to convert parity two-way AJA into equivalent Büchi NVPA. First, we give a characterization of the fulfillment of the acceptance condition for a non-trivial coBüchi stratum along a run in terms of the existence of an *odd ranking function*; the latter generalizes the notion of odd ranking function for standard coBüchi alternating finite-state automata [15] which intuitively, allows to convert a coBüchi acceptance condition into a Büchi-like acceptance condition. Then, by exploiting the above result and a non-trivial generalization of the Miyano-Hayashi construction [19], we give a characterization of the words in $\mathcal{L}(\mathcal{A})$ in terms of infinite sequences of finite sets (called *regions*) satisfying determined requirements which can be easily checked by a Büchi NVPA, where the control states and stack symbols range over the set of regions. The number of regions is at most $2^{O(|\mathsf{S}| + |\mathsf{M}| \cdot \log(k))}$, where k is the size of the largest non-trivial coBüchi stratum of the given SAJA $\mathcal{A}$. $\square$

Translation of VRE in Subclasses of AJA on Finite Words. In the translation of VLTL formulas into SAJA, we use two subclasses of AJA over finite words (for which we give different acceptance notions) in order to handle the VRE associated with the future and past temporal operators. Note that the proposed approach substantially differs from the alternating automata-theoretic approach for RLTL, the latter being crucially based on the use of *nondeterministic* automata for handling the regular expressions of the temporal modalities.[2]

Definition 4. *A forward (resp., backward) AJA with main states is an AJT with main states $\mathcal{A} = \langle \mathsf{M} \cup \mathsf{S}, q_0, \delta, Acc \rangle$ augmented with a set Acc of accepting*

---

[2] AJA are strictly more expressive than their nondeterministic counterpart [11].

states and such that no moves (dir, q, q') *with* $dir \in \{\varepsilon, \leftarrow, \frown\}$ *(resp.,* $dir \in \{\varepsilon, \rightarrow, \frown\}$) *are allowed, and* $\delta(q, \sigma) = \texttt{false}$ *for all accepting main states* q *and* $\sigma \in \Sigma$. *If* $\mathcal{A}$ *is forward (resp., backward), then a run of* $\mathcal{A}$ *over a finite word* w *is* accepting *if for all vertices of the form* $(|w| + 1, q)$ *(resp.,* $(0, q)$), $q \in Acc$. *Moreover, the language* $\mathcal{L}(\mathcal{A})$ *of* $\mathcal{A}$ *is the set of* finite *words* w *on* Σ *such that there is an accepting* $(1, q_0)$-*run (resp., accepting* $(|w|, q_0)$-*run) on* w.

In order to correctly handle the VRE expressions in the translation of VLTL formulas into SAJA, we need to impose additional restrictions on the above two classes of AJA (which intuitively allow to simulate the behavior of nondeterministic automata), ensuring at the same time that these restrictions still allow to (efficiently) capture VRE. These restrictions in their *semantic form* are the following ones, where a *pseudo* run is defined as a run but for all accepting main states q and $\sigma \in \Sigma$, we replace the value $\texttt{false}$ of $\delta(q, \sigma)$ with $\texttt{true}$.

J1. In each (pseudo) run starting from a main state, there is exactly one maximal path (the *main path*) from the initial vertex which visits only main states. Moreover, each vertex of the run which is not visited by the main path is associated with a secondary state.

J2. In a pseudo run over an infinite word, the main path cannot end at a vertex (j, q) such that $j > 0$ and q is not accepting.

J3. Let the given AJA $\mathcal{A}$ be forward (resp., backward). Then, for all infinite words w on Σ and $1 \leq i \leq j$, $w[i, j] \in \mathcal{L}(\mathcal{A})$ *iff* there is a pseudo (i, q_0)-run (resp., pseudo (j, q_0)-run) of $\mathcal{A}$ over the infinite word w whose main path visits position $j + 1$ (resp., $i - 1$) in an accepting main state, the latter being obtained by a local move.

Intuitively, the main path simulates the unique path of a run in a nondeterministic automaton. The notion of pseudo run is used just to ensure that runs of AJA with main states over infinite words whose main path visits an accepting main state exist. Moreover, the semantic requirements J2 and J3 crucially allow to deal with the sequencing and power operators in the translation of VLTL formulas into SAJA. Interestingly, we can show that the semantic requirements J1–J3 can be *syntactically* captured. These syntactical constraints also ensure that in a (pseudo) run, the secondary vertices are associated with positions inside minimally well-matched subwords of the input word. The forward (resp., backward) AJA with main states satisfying these syntactical requirements (ensuring J1–J3) are called *forward* (resp., *backward*) AJA *with main paths* (MAJA). It is worth noting that MAJA with no secondary states correspond to standard finite-state nondeterministic automata. For the class of MAJA, we show the following result.

Theorem 5 (From VRE to MAJA). *Given a VRE* α, *one can build in polynomial time a forward (resp., backward) MAJA* $\mathcal{A}$ *with* $O(|\alpha|)$ *main states and* $O(|\alpha|^4)$ *secondary states such that* $\mathcal{L}(\mathcal{A}) = \mathcal{L}(\alpha) \setminus \{\varepsilon\}$. *Moreover, if* α *is well-formed, then* $\mathcal{A}$ *can be* compositionally *constructed in* linear time.

Sketched Proof. The result for the general case of unrestricted VRE is an adaptation of two known results: VRE can be translated in quadratic time into equivalent NVPA [12], and NVPA can be translated in quadratic time into equivalent AJA over finite words [11]. The proof of the surprising result that well-formed VRE can be compositionally translated in *linear time* into forward and backward MAJA is instead non-trivial. This proof exploits an additional syntactical subclass of MAJA that captures more efficiently the restricted class of *well-matched* VRE (the additional syntactical constraints are used to implement in an efficient way M-substitution and S-closure in well-matched VRE). Note that thanks to the fulfillment of the semantic requirements J1–J3, the concatenation and the Kleene closure operators can be handled in a way analogous to the standard translation of regular expressions in nondeterministic automata.   □

5   Decision Procedures for the Logic VLTL

In this section, we study the satisfiability and visibly pushdown model checking problems for VLTL. Based on Lemma 1 and Theorem 5, we derive a polynomial-time compositional translation of VLTL formulas into SAJA, which provides an automata-theoretic approach to these decision problems. The translation is described by induction on the structure of the given VLTL formula φ. The base case $\varphi = \texttt{true}$ is immediate. For the induction step, given two VLTL formulas φ_1 and φ_2, assume that $\mathcal{A}_1 = \langle M_1 \cup S_1, q_1^0, \delta_1, F_1 \rangle$ and $\mathcal{A}_2 = \langle M_2 \cup S_2, q_2^0, \delta_2, F_2 \rangle$ are the SAJA associated with the VLTL formulas φ_1 and φ_2, accepting the ω-pointed languages $\mathcal{L}_p(\varphi_1)$ and $\mathcal{L}_p(\varphi_2)$, respectively. We illustrate now how to build the SAJA $\mathcal{A} = \langle M \cup S, q^0, \delta, F \rangle$ accepting $\mathcal{L}_p(\varphi)$ for formulas φ built using a single VLTL operator applied to φ_1 and φ_2. For $\varphi = \neg\varphi_1$, $\mathcal{A}$ is the dual automaton of $\mathcal{A}_1$, and the correctness directly follows from Lemma 1. For the other operators, here, we focus on the future power operator and the future weak power operator. Thus, let $\varphi = \varphi_1|\alpha\rangle\!\rangle\varphi_2$ or $\varphi = \varphi_1|\alpha\rangle\varphi_2$. Moreover, let $\mathcal{A}_\alpha = \langle M_\alpha \cup S_\alpha, q_\alpha, \delta_\alpha, Acc_\alpha \rangle$ be the *forward* MAJA of Theorem 5 for the VRE α and such that $(M_\alpha \cup S_\alpha) \cap (M_1 \cup S_1) = \emptyset$ and $(M_\alpha \cup S_\alpha) \cap (M_2 \cup S_2) = \emptyset$. Then, the initial state q^0 of $\mathcal{A}$ is a fresh state and:

$$M = M_1 \cup M_2 \cup M_\alpha \cup \{q^0\} \text{ and } S = S_1 \cup S_2 \cup S_\alpha$$

$$\delta(q,\sigma) = \begin{cases} \delta_2(q_2^0,\sigma) \vee (\delta_1(q_1^0,\sigma) \wedge \delta_\alpha(q_\alpha,\sigma)) & \text{if } q = q^0 \\ \delta_1(q,\sigma) & \text{if } q \in M_1 \cup S_1 \\ \delta_2(q,\sigma) & \text{if } q \in M_2 \cup S_2 \\ \delta_\alpha(q,\sigma) & \text{if } q \in S_\alpha \text{ or } q \not\to_\sigma Acc_\alpha \\ \delta_\alpha(q,\sigma) \vee (\varepsilon, q^0, q^0) & \text{if } q \to_\sigma Acc_\alpha \end{cases}$$

$$F = \begin{cases} F_1 \cup F_2 \cup \{\langle B, M_\alpha \cup S_\alpha \cup \{q^0\}, \emptyset\rangle\} & \text{if } \varphi = \psi_1|\alpha\rangle\!\rangle\psi_2 \\ F_1 \cup F_2 \cup \{\langle B, M_\alpha \cup S_\alpha \cup \{q^0\}, \{q^0\}\rangle\} & \text{if } \varphi = \psi_1|\alpha\rangle\psi_2 \end{cases}$$

where the notation $q \to_\sigma Acc_\alpha$ (resp., $q \not\to_\sigma Acc_\alpha$) means that $q \in M_\alpha$ and there is a (resp., there is no) local move in $\mathcal{A}_\alpha$ from q on reading σ which leads to an accepting main state. Note that the construction adds a new Büchi stratum above all strata from previous stages, so paths that move to the automaton of a subformula do not visit the newly added stratum. Moreover, the number of main states (resp., secondary states) of the new stratum is at most $|M_\alpha| + 1$ (resp., $|S_\alpha|$). Also, the SAJA for formulas φ_1 and φ_2 share the strata belonging to the SAJA of common subformulas of φ_1 and φ_2.[3] Thus, since a MAJA satisfies the semantic requirements J1–J3 at the end of Section 4, by Theorem 5, we obtain the following theorem and its immediate corollary (combined with Theorem 4).

Theorem 6. *For a VLTL formula φ, one can build in polynomial time a SAJA $\mathcal{A}$ such that: $\mathcal{L}_p(\mathcal{A}) = \mathcal{L}_p(\varphi)$, $\mathcal{A}$ has $O(|\varphi|)$ main states and $O(|\varphi|^4)$ secondary states in the general case, and just $O(|\varphi|)$ states if φ is well-formed or has constant-size VRE. Also, the size of the largest non-trivial coBüchi stratum of $\mathcal{A}$ is linear in the size of the largest VRE associated with a weak future power operator in φ which is in the scope of an odd number of negations.*

Corollary 1. *For a well-formed VLTL formula φ, one can build a Büchi NVPA $\mathcal{P}$ accepting $\mathcal{L}(\varphi)$ with $2^{O(|\varphi|\cdot\log(k))}$ states and stack symbols, k being the size of the largest VRE associated with a weak future power operator in φ which is in the scope of an odd number of negations.*

Checking whether $\mathcal{L}(\mathcal{P}) \subseteq \mathcal{L}(\varphi)$ for a pushdown system $\mathcal{P}$ and a VLTL formula φ, reduces to check emptiness of $\mathcal{L}(\mathcal{P}) \cap \mathcal{L}(\neg\varphi)$. Thus, since checking emptiness for the intersection of ω-VPL by Büchi NVPA is in PTIME [5], and satisfiability and visibly pushdown model checking for CaRet are EXPTIME-complete [4], by Theorems 3, 4, and 6, we obtain the following.

Corollary 2. *Satisfiability and visibly pushdown model checking for VLTL are EXPTIME-complete.*

6   Concluding Remarks

Our automata-theoretic approach, based on the use of SAJA as an intermediate step, can be conveniently used also for less expressive logical frameworks. In particular, by Theorems 3, 4, and 6, CaRet and NWTL$^+$ formulas φ can be translated into equivalent Büchi NVPA of size $2^{O(|\varphi|)}$, which matches the upper bounds for the known direct translations [4,3]. Analogously, our approach can also be used to convert formulas φ of RLTL with past into equivalent Büchi nondeterministic finite-state automata of size $2^{O(|\varphi|\cdot\log(k))}$, where k is the size of the largest regular expression associated with a weak future power operator in φ (which follows from Theorem 4 and the fact that the SAJA obtained from φ has only local moves and no secondary states). The recent upper bounds for RLTL [21] tackled only future operators leaving RLTL with past as an open problem.

Future work includes to adapt our automata-based constructions to alphabets based on atomic propositions, and to explore whether alternative formalisms like ETL [23] – adapted to VPL– can be efficiently integrated in the VLTL framework. Other interesting problems are to explore the relative expressive power of fragments of VLTL and to capture *minimal* expressively complete VLTL fragments.

---

[3] In fact, for a given subformula, we need to distinguish between the occurrences which are in the scope of an even number of negations from those which are in the scope of an odd number of negations.

References

1. IEEE Standard for Property Specification Language (PSL). IEEE Standard 1850–2010 (April 2010)
2. Alur, R.: Marrying words and trees. In: Proc. 26th PODS, pp. 233–242. ACM (2007)
3. Alur, R., Arenas, M., Barceló, P., Etessami, K., Immerman, N., Libkin, L.: First-order and temporal logics for nested words. In: Proc. 22nd LICS, pp. 151–160. IEEE Computer Society (2007)
4. Alur, R., Etessami, K., Madhusudan, P.: A temporal logic of nested calls and returns. In: Jensen, K., Podelski, A. (eds.) TACAS 2004. LNCS, vol. 2988, pp. 467–481. Springer, Heidelberg (2004)
5. Alur, R., Madhusudan, P.: Visibly pushdown languages. In: Proc. 36th STOC, pp. 202–211. ACM (2004)
6. Alur, R., Madhusudan, P.: Adding nesting structure to words. J. ACM 56(3) (2009)
7. Arenas, M., Barceló, P., Libkin, L.: Regular languages of nested words: Fixed points, automata, and synchronization. In: Arge, L., Cachin, C., Jurdziński, T., Tarlecki, A. (eds.) ICALP 2007. LNCS, vol. 4596, pp. 888–900. Springer, Heidelberg (2007)
8. Armoni, R., Fix, L., Flaisher, A., Gerth, R., Ginsburg, B., Kanza, T., Landver, A., Mador-Haim, S., Singerman, E., Tiemeyer, A., Vardi, M.Y., Zbar, Y.: The ForSpec temporal logic: A new temporal property-specification language. In: Katoen, J.-P., Stevens, P. (eds.) TACAS 2002. LNCS, vol. 2280, pp. 296–311. Springer, Heidelberg (2002)
9. Ball, T., Rajamani, S.K.: Bebop: a symbolic model checker for boolean programs. In: Havelund, K., Penix, J., Visser, W. (eds.) SPIN 2000. LNCS, vol. 1885, pp. 113–130. Springer, Heidelberg (2000)
10. Bollig, B., Cyriac, A., Gastin, P., Zeitoun, M.: Temporal logics for concurrent recursive programs: Satisfiability and model checking. In: Murlak, F., Sankowski, P. (eds.) MFCS 2011. LNCS, vol. 6907, pp. 132–144. Springer, Heidelberg (2011)
11. Bozzelli, L.: Alternating automata and a temporal fixpoint calculus for visibly push-down languages. In: Caires, L., Vasconcelos, V.T. (eds.) CONCUR 2007. LNCS, vol. 4703, pp. 476–491. Springer, Heidelberg (2007)
12. Bozzelli, L., Sánchez, C.: Visibly rational expressions. In: Proc. FSTTCS. LIPIcs, vol. 18, pp. 211–223 (2012)
13. Dax, C., Klaedtke, F.: Alternation elimination for automata over nested words. In: Hofmann, M. (ed.) FOSSACS 2011. LNCS, vol. 6604, pp. 168–183. Springer, Heidelberg (2011)
14. Kupferman, O., Vardi, M., Wolper, P.: An Automata-Theoretic Approach to Branching-Time Model Checking. J. ACM 47(2), 312–360 (2000)
15. Kupferman, O., Vardi, M.Y.: Weak alternating automata are not that weak. ACM Transactions on Computational Logic 2(3), 408–429 (2001)
16. Leucker, M., Sánchez, C.: Regular linear temporal logic. In: Jones, C.B., Liu, Z., Woodcock, J. (eds.) ICTAC 2007. LNCS, vol. 4711, pp. 291–305. Springer, Heidelberg (2007)
17. Löding, C., Serre, O.: Propositional dynamic logic with recursive programs. In: Aceto, L., Ingólfsdóttir, A. (eds.) FOSSACS 2006. LNCS, vol. 3921, pp. 292–306. Springer, Heidelberg (2006)
18. Madhusudan, P., Viswanathan, M.: Query automata for nested words. In: Královič, R., Niwiński, D. (eds.) MFCS 2009. LNCS, vol. 5734, pp. 561–573. Springer, Heidelberg (2009)

19. Miyano, S., Hayashi, T.: Alternating finite automata on ω-words. Theoretical Computer Science 32, 321–330 (1984)
20. Sánchez, C., Leucker, M.: Regular linear temporal logic with past. In: Barthe, G., Hermenegildo, M. (eds.) VMCAI 2010. LNCS, vol. 5944, pp. 295–311. Springer, Heidelberg (2010)
21. Sánchez, C., Samborski-Forlese, J.: Efficient regular linear temporal logic using dualization and stratification. In: Proc. 19th TIME, pp. 13–20 (2012)
22. Stockmeyer, L.J., Meyer, A.R.: Word problems requiring exponential time: Preliminary report. In: Proc. 5th STOC, pp. 1–9. ACM (1973)
23. Vardi, M.Y., Wolper, P.: Reasoning about infinite computations. Information and Computation 115, 1–37 (1994)
24. Zielonka, W.: Infinite games on finitely coloured graphs with applications to automata on infinite trees. Theoretical Computer Science 200(1-2), 135–183 (1998)

Count and Forget: Uniform Interpolation of $\mathcal{SHQ}$-Ontologies

Patrick Koopmann* and Renate A. Schmidt

The University of Manchester, UK

Abstract. We propose a method for forgetting concept symbols and non-transitive roles symbols of $\mathcal{SHQ}$-ontologies, or for computing uniform interpolants in $\mathcal{SHQ}$. Uniform interpolants restrict the symbols occuring in an ontology to a specified set, while preserving all logical entailments that can be expressed using this set in the description logic under consideration. Uniform interpolation has applications in ontology reuse, information hiding and ontology analysis, but so far no method for computing uniform interpolants for expressive description logics with number restrictions has been developed. Our results are not only interesting because they allow to compute uniform interpolants of ontologies using a more expressive language. Using number restrictions also allows to preserve more information in uniform interpolants of ontologies in less complex logics, such as $\mathcal{ALC}$ or $\mathcal{EL}$. The presented method computes uniform interpolants on the basis of a new resolution calculus for $\mathcal{SHQ}$. The output of our method is expressed using $\mathcal{SHQ}\mu$, which is $\mathcal{SHQ}$ extended with fixpoint operators, to always enable a finite representation of the uniform interpolant. If the uniform interpolant uses fixpoint operators, it can be represented in $\mathcal{SHQ}$ without fixpoints operators using additional concept symbols or by approximation.

1 Introduction

Ontologies are at the center of the semantic web and knowledge-based systems in an increasing number of domains. They model terminological domain knowledge and are usually represented using a description logic to allow reasoning to be performed automatically. Uniform interpolation and forgetting deal with the problem of reducing the vocabulary used in an ontology in such a way that entailments expressed in this reduced vocabulary are preserved. Eliminating concepts or relations from an ontology is referred to as *forgetting* them, and the result is a *uniform interpolant* for the reduced vocabulary.

Uniform interpolation has applications in a range of areas. **(i) Ontology Reuse and Distributed Ontologies.** Big ontologies such as the National Cancer Institute Thesaurus often cover a huge amount of terms, whereas for applications often only a subset is needed. A uniform interpolant can provide a basis in applications where too many symbols in the ontology that users are unfamiliar

---

* Patrick Koopmann is supported by an EPSRC doctoral training award.

S. Demri, D. Kapur, and C. Weidenbach (Eds.): IJCAR 2014, LNAI 8562, pp. 434–448, 2014.

with could be harmful [20]. **(ii) Information Hiding.** In a lot of applications, an ontology may be used by a number of people with different privileges. For such an environment it is crucial to have safe techniques to hide confidential information from users that are not privileged to access them [4]. Uniform interpolation provides a way to remove confidential concepts and relations from an ontology without affecting the entailments over the remaining terminology. **(iii) Understanding concept relations.** Relations between concepts in big ontologies are often indirect and hard to understand with growing complexity of the ontology. Uniform interpolation can be used to compute an ontology that only uses a small number of symbols of interest, to get a direct representation of the relations between them [7]. **(iv) Ontology Maintenance.** A related task is understanding how changes to an ontology, for example the addition of new concept definitions, affect the meaning of other concepts. Uniform interpolants can be used to determine whether the meaning of certain concepts changed, and to get a direct representation of these changes [12]. Further applications of uniform interpolation can be found in [7,14].

So far, the only expressive description logics for which methods for computing uniform interpolants exists are $\mathcal{ALC}$ and $\mathcal{ALCH}$ [10,8,12,19]. In this paper we extend the methods of [10,8] to the description logic $\mathcal{SHQ}$, which extends $\mathcal{ALCH}$ with transitive roles and number restrictions. This way, we broaden the application of uniform interpolants to ontologies that use a more expressive description logic. But the expressivity of the underlying description logic also determines what information is included in the uniform interpolant. Consider for example the following simple $\mathcal{ALC}$-ontology $\mathcal{T}_{bike}$.

$$\text{Bicycle} \sqsubseteq \exists\text{hasWheel.FrontWheel} \sqcap \exists\text{hasWheel.RearWheel}$$
$$\text{FrontWheel} \sqsubseteq \text{Wheel} \sqcap \neg\text{RearWheel}$$
$$\text{RearWheel} \sqsubseteq \text{Wheel} \sqcap \neg\text{FrontWheel}$$

This TBox states that every bicycle has a front wheel and a rear wheel, and that those are disjoint types of wheels. Assume we are not interested in the distinction between front wheels and rear wheels. If we want to preserve all logical entailments in $\mathcal{ALC}$ over the remaining symbols Bicycle, hasWheel and Wheel, this can be done by the single TBox axiom Bicycle $\sqsubseteq \exists\text{hasWheel.Wheel}$, which states that every bicycle has a wheel, and which is the $\mathcal{ALC}$-uniform interpolant of $\mathcal{T}_{bike}$ for {Bicycle, hasWheel, Wheel}. We do lose however the indirectly expressed information that a bicycle has at least two wheels, since we cannot express this in $\mathcal{ALC}$ without using at least one of the concepts FrontWheel and RearWheel. Using number restrictions however, we can express this. The $\mathcal{SHQ}$-uniform interpolant of $\mathcal{T}_{bike}$ consists therefore of the axiom Bicycle $\sqsubseteq \geq 2\text{hasWheel.Wheel}$, which states that every bicycle has at least two wheels.

The results in [10,8,12] suggest that resolution-based approaches allow for an efficient computation of uniform interpolants in a lot of cases, since they make it possible to derive consequences for a specified symbol in a goal-oriented manner. Motivated by this, we follow a similar approach as in [10,8]. In Section 4, we present a new resolution calculus for $\mathcal{SHQ}$. Based on this calculus, we present respectively two methods for forgetting concept symbols and non-transitive role symbols in Sections 5 and 6. Since a finite representation of uniform interpolants is not always possible in pure $\mathcal{SHQ}$, the result may involve the use of fixpoint operators. This way uniform interpolants can always be represented finitely. For this reason, the output of our method is at worst represented in $\mathcal{SHQ}\mu$, which is $\mathcal{SHQ}$ extended with fixpoint operators. If fixpoint operators are not desired, it is possible to obtain a finite presentation in $\mathcal{SHQ}$ using additional symbols, or to approximate the uniform interpolant.

All proofs, some examples and an empirical evaluation of our method are provided in the long version of this paper [11].

2   Definition of $\mathcal{SHQ}\mu$ and Uniform Interpolation

To begin with, we define the description logic $\mathcal{SHQ}\mu$, which is $\mathcal{SHQ}$ extended with fixpoint operators.

Let N_r be a set of *role symbols*. An *RBox* $\mathcal{R}$ is a set of *role axioms* of the form $r \sqsubseteq s$ (*role inclusion*), $r \equiv s$ (*role equivalence*) and $\mathsf{trans}(r)$ (*transitivity axiom*), where $r, s \in N_r$. $r \equiv s$ is defined as abbreviation for the two role inclusions $r \sqsubseteq s$ and $s \sqsubseteq r$. Given an RBox $\mathcal{R}$, we denote by $\sqsubseteq_\mathcal{R}$ the reflexive transitive closure of the role inclusions in $\mathcal{R}$. A role r is *transitive* in $\mathcal{R}$ if $\mathsf{trans}(r) \in \mathcal{R}$. r is *simple* in $\mathcal{R}$ if there is no role s with $s \sqsubseteq_\mathcal{R} r$ and $\mathsf{trans}(s) \in \mathcal{R}$.

Let N_c and N_v be two sets of respectively *concept symbols* and *concept variables*. $\mathcal{SHQ}\mu$-concepts have the following form:

$$\bot \mid A \mid X \mid \neg C \mid C \sqcup D \mid {\geq} nr.C \mid \nu X.C[X],$$

where $A \in N_c$, $X \in N_v$, $r \in N_r$, C and D are arbitrary concepts, n is a non-zero natural number, and $C[X]$ is a concept expression in which X occurs under an even number of negations. We define further concept expressions as abbreviations: $\top = \neg\bot$, $C \sqcap D = \neg(\neg C \sqcup \neg D)$, ${\leq} mr.C = \neg({\geq} nr.C)$ with $m = n - 1$, $\exists r.C = {\geq} 1r.C$, $\forall r.C = {\leq} 0r.\neg C$ and $\mu X.C[X] = \neg\nu X.\neg C[X/\neg X]$, where $C[E_1/E_2]$ denotes the concept obtained by replacing every E_1 in C by E_2. Concepts of the form ${\geq} nr.C$ and ${\leq} nr.C$ are called *number restrictions*, and concepts of the form $\nu X.C[X]$ and $\mu X.C[X]$ are called *fixpoint expressions*. $\nu X.C[X]$ and $\mu X.C[X]$ denote respectively the *greatest* and the *least fixpoint* of $C[X]$, and ν and μ are respectively the *greatest* and *least fixpoint operator*. A concept variable X is *bound* if it occurs in the scope $C[X]$ of a fixpoint expression $\nu X.C[X]$ or $\mu X.C[X]$. Otherwise it is *free*. A concept is *closed* if it does not contain any free variables, otherwise it is *open*.

A *TBox* $\mathcal{T}$ is a set of *concept axioms* of the forms $C \sqsubseteq D$ (*concept inclusion*) and $C \equiv D$ (*concept equivalence*), where C and D are closed concepts. $C \equiv D$ is short-hand for the two concept axioms $C \sqsubseteq D$ and $D \sqsubseteq C$. An *ontology* $\mathcal{O} = \langle \mathcal{T}, \mathcal{R} \rangle$ consists of a TBox $\mathcal{T}$ and an RBox $\mathcal{R}$ with the additional restriction that non-simple roles in $\mathcal{R}$ occur only in number restrictions of the form ${\leq} 0r.C$ or ${\geq} 1r.C$ in $\mathcal{T}$. This restriction in necessary to ensure decidability of common $\mathcal{SHQ}$

reasoning tasks [6], and our method for uniform interpolation assumes that it is satisfied.

Next, we define the semantics of $\mathcal{SHQ}\mu$. An *interpretation* $\mathcal{I}$ is a pair $\langle \Delta^{\mathcal{I}}, \cdot^{\mathcal{I}} \rangle$ of the *domain* $\Delta^{\mathcal{I}}$ is a nonempty set and the *interpretation function* $\cdot^{\mathcal{I}}$ assigns to each concept symbol $A \in N_c$ a subset of $\Delta^{\mathcal{I}}$ and to each role symbol $r \in N_r$ a subset of $\Delta^{\mathcal{I}} \times \Delta^{\mathcal{I}}$. The interpretation function is extended to $\mathcal{SHQ}\mu$-concepts as follows.

$$\bot^{\mathcal{I}} = \emptyset \quad (\neg C)^{\mathcal{I}} = \Delta^{\mathcal{I}} \setminus C^{\mathcal{I}} \quad (C \sqcup D)^{\mathcal{I}} = C^{\mathcal{I}} \cup D^{\mathcal{I}}$$

$$(\geq nr.C)^{\mathcal{I}} = \{x \in \Delta^{\mathcal{I}} \mid \#\{(x,y) \in r^{\mathcal{I}} \mid y \in C^{\mathcal{I}}\} \geq n\}$$

The semantics of fixpoint expressions is defined following [2]. Whereas concept symbols are assigned fixed subsets of the domain, concept variables range over arbitrary subsets, which is why only closed concepts have a fixed interpretation. Open concepts are interpreted using *valuations* ρ that map concept variables to subsets of $\Delta^{\mathcal{I}}$. Given a valuation ρ and a set $W \subseteq \Delta^{\mathcal{I}}$, $\rho[X \mapsto W]$ denotes a valuation identical to ρ except that $\rho[X \mapsto W](X) = W$. Given an interpretation $\mathcal{I}$ and a valuation ρ, the function $\cdot^{\mathcal{I}}_{\rho}$ is $\cdot^{\mathcal{I}}$ extended with the cases $X^{\mathcal{I}}_{\rho} = \rho(X)$ and

$$(\nu X.C)^{\mathcal{I}}_{\rho} = \bigcup \{W \subseteq \Delta^{\mathcal{I}} \mid W \subseteq C^{\mathcal{I}, \rho[X \mapsto W]}\}.$$

If C is closed, we define $C^{\mathcal{I}} = C^{\mathcal{I}}_{\rho}$ for any valuation ρ. Since C does not contain any free variables in this case, this defines $C^{\mathcal{I}}$ uniquely.

A concept inclusion $C \sqsubseteq D$ is *true* in an interpretation $\mathcal{I}$ iff $C^{\mathcal{I}} \subseteq D^{\mathcal{I}}$, a role inclusion $r \sqsubseteq s$ is true in $\mathcal{I}$ iff $r^{\mathcal{I}} \subseteq s^{\mathcal{I}}$ and a transitivity axiom $\mathsf{trans}(r)$ is true in $\mathcal{I}$ if for any domain elements $x, y, z \in \Delta^{\mathcal{I}}$ we have $(x,z) \in r^{\mathcal{I}}$ if $(x,y), (y,z) \in r^{\mathcal{I}}$. $\mathcal{I}$ is a *model* of an ontology $\mathcal{O}$ if all axioms in $\mathcal{O}$ are true in $\mathcal{I}$. An ontology $\mathcal{O}$ is *satisfiable* if there exists a model for $\mathcal{O}$, otherwise it is *unsatisfiable*. Two TBoxes $\mathcal{T}_1$ and $\mathcal{T}_2$ are *equi-satisfiable* if every model of $\mathcal{T}_1$ can be extended to a model of $\mathcal{T}_2$, and vice versa. $\mathcal{T} \models C \sqsubseteq D$ holds iff in every model $\mathcal{I}$ of $\mathcal{T}$ we have $C^{\mathcal{I}} \subseteq D^{\mathcal{I}}$. If an axiom α is true in all models of $\mathcal{O}$, we write $\mathcal{O} \models \alpha$. Observe that $\mathcal{O} \models r \sqsubseteq s$ iff $r \sqsubseteq_{\mathcal{R}} s$. Interestingly, allowing number restrictions and fixpoint operators does not affect the complexity of deciding satisfiability of ontologies: for $\mathcal{SHQ}$ as well as for $\mathcal{SHQ}\mu$ it is ExpTime-complete [17,2].[1]

Let $sig(E)$ denote the concept and role symbols occurring in E, where E can denote a concept, an axiom, a TBox, an RBox or an ontology.

Definition 1 (Uniform Interpolation). *Given an ontology $\mathcal{O}$ and a set of concept and role symbols $\mathcal{S}$, an ontology $\mathcal{O}^{\mathcal{S}}$ is a uniform interpolant of $\mathcal{O}$ for $\mathcal{S}$ iff the following conditions are satisfied:*

1. *$sig(\mathcal{O}^{\mathcal{S}}) \subseteq \mathcal{S}$, and*
2. *$\mathcal{O}^{\mathcal{S}} \models \alpha$ iff $\mathcal{O} \models \alpha$ for every $\mathcal{SHQ}$-axiom α with $sig(\alpha) \subseteq \mathcal{S}$.*

---

[1] [2] proves only ExpTime-completeness for $\mathcal{ALCQ}\mu$, but the result can be easily extended to incorporate role hierarchies and transitive roles using the technique proposed in [16].

3 The Normal Form

Our method for computing uniform interpolants in $\mathcal{SHQ}$ is based on a new resolution calculus $Res_{\mathcal{SHQ}}$ which provides a decision procedure for satisfiability of $\mathcal{SHQ}$-ontologies, and which allows for goal-oriented elimination of concept symbols. Before this calculus can be applied to an ontology, its TBox has to be normalised into a set of clauses using structural transformation or flattening. Let $N_d \subseteq N_c$ be a set of *definer (concept) symbols* that is disjoint with the signature of the given TBox.

Definition 2 (Normal form). *An $\mathcal{SHQ}$-literal is a concept description of the form A, $\neg A$, $\geq nr.\mathcal{D}$ or $\leq mr.\neg\mathcal{D}$, where $A \in N_c$, $r \in N_r$, $n \geq 1$, $m \geq 0$ are natural numbers and $\mathcal{D} = D_1 \sqcup \ldots \sqcup D_n$ is a disjunction of definer symbols. A literal of the form $\neg D, D \in N_d$, is called* negative definer literal. *An $\mathcal{SHQ}$-clause is an unordered set of $\mathcal{SHQ}$-literals $l_1, \ldots, l_n$, represented as $l_1 \sqcup \ldots \sqcup l_2$. The empty clause and the empty disjunction are represented as $\bot$. $\mathcal{SHQ}$-clauses are abbreviations for TBox axioms of the respective forms $\top \sqsubseteq l_1 \sqcup \ldots \sqcup l_2$ and $\top \sqsubseteq \bot$. A TBox is in $\mathcal{SHQ}$-clausal form if every axiom in it is an $\mathcal{SHQ}$-clause.*

Number restrictions of the form $\leq nr.C$ contain a hidden negation of the concept under the restriction (they are equivalent to $\neg \geq (n+1)r.C$). Hence C occurs negatively in $\leq nr.C$. The normal form ensures that every concept under a role restriction occurs positively. This is why $\leq$-literals have the form $\leq nr.\neg\mathcal{D}$.

A TBox is converted into $\mathcal{SHQ}$-clausal form as follows. First we replace existential and universal role restrictions $\exists r.C$ and $\forall r.C$ by corresponding number restrictions $\geq 1r.C$ and $\leq 0r.\neg C$. Then every axiom is converted into negation normal form (every axiom is of the form $\top \sqsubseteq C$, and in C negation only occurs in front of concept symbols or directly under $\leq$-restrictions, and every $\leq$-restriction is of the form $\leq nr.\neg C$). Next, we replace each concept C that occurs under a role restriction of the form $\geq nr.C$ or $\leq nr.\neg C$ by a new concept definer symbol D and add the new axiom $\neg D \sqcup C$ to the TBox. This flattens the TBox, which means every role restriction is of the form $\geq nr.D$ or $\leq nr.\neg D$, where D is a definer symbol. The flattened TBox can be converted into $\mathcal{SHQ}$-clausal form using standard CNF transformations.

Observe that the definition of the $\mathcal{SHQ}$-clausal form allows for disjunctions of arbitrary length under role restrictions $\geq nr.\mathcal{D}$ and $\leq nr.\neg\mathcal{D}$. These disjunctions are not introduced by the initial transformation of the TBox, but may be produced by the rules of the calculus.

Example 1 ($\mathcal{SHQ}$-normal form). Consider the TBox $\mathcal{T} = \{A_1 \sqsubseteq \geq 5r.(A \sqcup B), A_2 \sqsubseteq \leq 3r.A\}$. Observe that the normal form requires a negation under each $\leq$-restriction. The $\mathcal{SHQ}$-clausal form of $\mathcal{T}$ consists of the following clauses.

1. $\neg A_1 \sqcup \geq 5r.D_1$ 2. $\neg D_1 \sqcup A \sqcup B$
3. $\neg A_2 \sqcup \leq 3r.\neg D_2$ 4. $\neg D_2 \sqcup \neg A$

Since the normal form has to be preserved, some rule applications of the calculus require the introduction of new definer symbols that represent the conjunction of existing definer symbols. In particular, during the derivation a new definer symbol D_{12} is introduced for the conjunction $D_1 \sqcap D_2$ by adding two clauses $\neg D_{12} \sqcup D_1$ and $\neg D_{12} \sqcup D_2$, which are equivalent to the concept inclusion $D_{12} \sqsubseteq D_1 \sqcap D_2$. As exemplified here, throughout the paper we indicate which conjunction an introduced definer symbol represents using its index. To avoid the infinite introduction of new definer symbols, we check whether a definer symbol representing this conjunction already exists. This way the number of introduced definer symbols is limited to 2^k, where k is the number of definer symbols introduced by the initial transformation of the TBox.

4   The Underlying Calculus

We now introduce a sound and refutationally complete calculus $Res_{\mathcal{SHQ}}$ that decides satisfiability of TBoxes in $\mathcal{SHQ}$-normal form. This calculus serves as the basis for the method of computing uniform interpolants.

The calculus consists of the rules shown in Figure 1. Most of the rules are motivated by the tautology $(C_1 \sqcup L_1) \sqcap (C_1 \sqcup L_2) \sqsubseteq (C_1 \sqcup C_2 \sqcup (L_1 \sqcap L_2))$. Therefore, the conclusion often contains literals entailed by $L_1 \sqcap L_2$, where L_1 and L_2 occur in the premises. In the case of the resolution rule, which is known from propositional resolution calculi, we have that $(A \sqcap \neg A)$ entails $\bot$.

For the transitivity rule, observe that $\leq 0r.\neg D$ is equivalent to $\forall r.D$, and due to the restrictions on $\mathcal{SHQ}$ ontologies, roles with transitive sub-roles do not occur in number restrictions of the form $\leq nr.\neg D$, where $n > 0$. If a domain element a satisfies $\forall r_1.D$, and we have a transitive role $r_2 \sqsubseteq r_1$, the transitive closure of r_2-successors of a are all r_1-successors of a, and they all have to satisfy D. We put this information into clausal form by adding a new cyclic definer symbol D' that is subsumed by D, and by stating that every r_2-successor of a and every r_2-successor of an D'-instance has to satisfy D' (this is similar to what is done in [16] to incorporate transitivity axioms into formulae).

For the $\geq$-combination rule, observe that our normal form does not allow for conjunctions under number restrictions. We can however express the conjunction $\mathcal{D}_1 \sqcap \mathcal{D}_2$ using a disjunction $\mathcal{D}_{12}$ of possibly new definer symbol symbols that represent the conjunctions of each pair of definer symbols from $\mathcal{D}_1$ and $\mathcal{D}_2$. The $\geq$-combination rule becomes more intuitive by interpreting the last two literals of each conclusion as an implication. For example, $\geq (n_1+n_2)r.(\mathcal{D}_1 \sqcup \mathcal{D}_2) \sqcup \geq 1r.\mathcal{D}_{12}$ is equivalent to $\leq (n_1+n_2-1)r.(\mathcal{D}_1 \sqcup \mathcal{D}_2) \rightarrow \geq 1r.\mathcal{D}_{12}$. Figure 2 illustrates the idea. Every column represents an r-successor. If an upper cell is light, it satisfies $\mathcal{D}_1$, if a lower cell is light, it satisfies $\mathcal{D}_2$. The two columns in the middle represent r-sucessors satisfying both $\mathcal{D}_1$ and $\mathcal{D}_2$, that is, satisfying $\mathcal{D}_{12}$. All except the rightmost column represent r-successors satisfying the union $\mathcal{D}_1 \sqcup \mathcal{D}_2$. Depending on how many r-successors satisfy $\mathcal{D}_1 \sqcup \mathcal{D}_2$, the set of r-successors in $\mathcal{D}_{12}$ gets smaller or bigger according to the conclusions of the $\geq$-rule.

For the $\geq \leq$-combination rule, observe that in Figure 2, if there are more elements in $\mathcal{D}_1$ than in $\neg \mathcal{D}_2$, $\mathcal{D}_1$ and $\mathcal{D}_2$ have to overlap. If $\mathcal{D}_1$ contains at least n_1 elements and the complement of $\mathcal{D}_2$ contains at most n_2 elements, the intersection $\mathcal{D}_{12}$ must contain at least $n_1 - n_2$ elements.

Resolution:

$$\frac{C_1 \sqcup A \qquad C_2 \sqcup \neg A}{C_1 \sqcup C_2}$$

Transitivity:

$$\frac{C \sqcup \leq 0 r_1.\neg D \qquad \text{trans}(r_2) \in \mathcal{R} \qquad r_2 \sqsubseteq_{\mathcal{R}} r_1}{C \sqcup \leq 0 r_2.\neg D' \qquad \neg D' \sqcup D \qquad \neg D' \sqcup \leq 0 r_2.\neg D'}$$

where D' is a new definer symbol.

$\geq$-Combination:

$$\frac{C_1 \sqcup \geq n_1 r_1.\mathcal{D}_1 \qquad C_2 \sqcup \geq n_2 r_2.\mathcal{D}_2 \qquad r_1 \sqsubseteq_{\mathcal{R}} r \qquad r_2 \sqsubseteq_{\mathcal{R}} r}{C_1 \sqcup C_2 \sqcup \geq (n_1 + n_2) r.(\mathcal{D}_1 \sqcup \mathcal{D}_2) \sqcup \geq 1 r.\mathcal{D}_{12}}$$

$$\vdots$$

$$C_1 \sqcup C_2 \sqcup \geq (n_1 + 1) r.(\mathcal{D}_1 \sqcup \mathcal{D}_2) \sqcup \geq n_2 r.\mathcal{D}_{12}$$

where $\mathcal{D}_{12} = \bigsqcup_{D_i \in \mathcal{D}_1, D_j \in \mathcal{D}_2} D_{ij}$ represents the conjunction of $\mathcal{D}_1$ and $\mathcal{D}_2$.

$\geq\leq$-Combination:

$$\frac{C_1 \sqcup \geq n_1 r_1.(D_1 \sqcup \ldots \sqcup D_m) \qquad C_2 \sqcup \leq n_2 r_2.\neg D_a \qquad r_1 \sqsubseteq_{\mathcal{R}} r_2}{C_1 \sqcup C_2 \sqcup \geq (n_1 - n_2) r_2.(D_{1a} \sqcup \ldots \sqcup D_{ma})} \qquad n_1 > n_2$$

$\geq$-Resolution:

$$\frac{C \sqcup \geq n r.(\mathcal{D} \sqcup D) \qquad \neg D}{C \sqcup \geq n r.\mathcal{D}}$$

$\geq$-Elimination:

$$\frac{C \sqcup \geq n r.\bot}{C}$$

Fig. 1. Inference rules of $Res_{\mathcal{SHQ}}$

The $\geq$-resolution rule is a variant of the classical resolution rule, and the $\geq$-elimination rule eliminates unsatisfiable literals. The six rules form a sound and refutationally complete calculus for ontologies in $\mathcal{SHQ}$-clausal form, as the following theorem shows.

Theorem 1. *$Res_{\mathcal{SHQ}}$ is sound and refutationally complete. Given any set $\boldsymbol{N}$ of $\mathcal{SHQ}$-clauses and any RBox $\mathcal{R}$, the saturation of $\boldsymbol{N}$ using the rules of $Res_{\mathcal{SHQ}}$ contains the empty clause iff the ontology $\mathcal{O} = \langle \boldsymbol{N}, \mathcal{R} \rangle$ is unsatisfiable.*

Observe that the $\geq$-combination rule can be applied arbitrarily often, resulting in clauses with larger and larger numbers occurring in the number restrictions. For this reason, $Res_{\mathcal{SHQ}}$ on its own is not a decision procedure, since we can derive infinitely many clauses. In order to achieve termination, we need to add redundancy elimination. This is also essential to make the uniform interpolation method practical. Our notion of redundancy is close to the one introduced in [8],

$\mathcal{D}_1$								$\neg\mathcal{D}_1$
$\neg\mathcal{D}_2$						$\mathcal{D}_2$		

Fig. 2. Diagram illustrating the $\geq$- and the $\geq\leq$-combination rules

but is extended to incorporate number restriction literals and disjunctions under role restrictions.

Definition 3 (Subsumption and Reduction). *A definer symbol D_1 is subsumed by a definer symbol D_2 ($D_1 \sqsubseteq_d D_2$), if either $D_1 = D_2$ or there is a clause $\neg D_1 \sqcup D_2$ in the current clause set. A disjunction $\mathcal{D}_1$ of definer symbols is subsumed by a disjunction $\mathcal{D}_2$ of definer symbols ($\mathcal{D}_1 \sqsubseteq_d \mathcal{D}_2$) if every definer symbol in $\mathcal{D}_1$ is subsumed by a definer symbol in $\mathcal{D}_2$. A literal l_1 is subsumed by a literal l_2 ($l_1 \sqsubseteq_l l_2$) if one of the following is satisfied: (i) $l_1 = l_2$, (ii) $l_1 = \geq n_1 r_1.\mathcal{D}_1$ and $l_2 = \geq n_2 r_2.\mathcal{D}_2$, where $n_1 \geq n_2$, $r_1 \sqsubseteq_{\mathcal{R}} r_2$ and $\mathcal{D}_1 \sqsubseteq_d \mathcal{D}_2$, or (iii) $l_1 = \leq n_1 r_1.\neg\mathcal{D}_1$ and $l_2 = \leq n_2 r_2.\neg\mathcal{D}_2$, where $n_1 \leq n_2$, $r_2 \sqsubseteq_{\mathcal{R}} r_1$ and $\mathcal{D}_1 \sqsubseteq_d \mathcal{D}_2$. A clause C_1 is subsumed by a clause C_2 ($C_1 \sqsubseteq_c C_2$) if every literal in C_1 is subsumed by a literal in C_2. A clause C is redundant with respect to a clause set $\mathbf{N}$, if $\mathbf{N}$ contains a clause C' with $C' \sqsubseteq_c C$.*

The reduction of a disjunction $\mathcal{D}$, denoted by $red(\mathcal{D})$, is obtained from $\mathcal{D}$ by removing every definer symbol from $\mathcal{D}$ that is subsumed by another definer symbol in $\mathcal{D}$. The reduction of a clause C, denoted by $red(C)$, is obtained from C by removing every literal that is subsumed by another literal in C and reducing every disjunction that occurs under a number restriction in the remaining literals.

Observe that the roles and the numbers for $\leq$-restrictions are compared in the other direction as for $\geq$-restrictions. This is due to the hidden negation present in $\leq$-restrictions.

Example 2 (Subsumption and reduction). Assume D_{12} represents $D_1 \sqcap D_2$, which means we have the clauses $\neg D_{12} \sqcup D_1$ and $\neg D_{12} \sqcup D_2$, and we have $r \sqsubseteq s \in \mathcal{R}$. Then $\geq 3r.D_{12}$ is subsumed by $A \sqcup \geq 2s.D_1$ and $\leq 2s.\neg D_{12}$ is subsumed by $B \sqcup \leq 3r.\neg D_2$ ($A \sqcup \geq 2s.D_1$ and $B \sqcup \leq 3r.\neg D_2$ are redundant). The reduction of $\geq 1r.(D_{12} \sqcup D_1)$ is $\geq 1r.D_1$ and the reduction of $\geq 1r.(D_1 \sqcup D_2) \sqcup \geq 2r.D_1$ is $\geq 1r(D_1 \sqcup D_2)$.

In addition to subsumption deletion and reduction, we also remove tautological clauses which contain pairs of contradictory literals. This leads to the set of simplification rules shown in Figure 3. Our use of the terminology for subsumption follows the traditional use in description logics. This means that it is C_2 that is deleted when C_1 is subsumed by C_2 ($C_1 \sqsubseteq_c C_2$), and not vice versa. We denote the calculus $Res_{\mathcal{SHQ}}$ extended with these rules by $Res^s_{\mathcal{SHQ}}$.

Theorem 2. *$Res^s_{\mathcal{SHQ}}$ is sound and refutationally complete, and provides a decision procedure for $\mathcal{SHQ}$-ontology satisfiability.*

$$\textbf{Tautology deletion:} \quad \frac{N \cup \{C \sqcup A \sqcup \neg A\}}{N}$$

$$\textbf{Subsumption deletion:} \quad \frac{N \cup \{C_1,\, C_2\}}{N \cup \{C_1\}} \qquad \text{provided } C_1 \sqsubseteq_c C_2$$

$$\textbf{Reduction:} \quad \frac{N \cup \{C\}}{N \cup \{red(C)\}}$$

Fig. 3. Simplification rules of $Res^s_{\mathcal{SHQ}}$

$\leq$-Combination:

$$\frac{C_1 \sqcup \leq n_1 r_1 . \neg D_1 \qquad C_2 \sqcup \leq n_2 r_2 . \neg D_2 \qquad r \sqsubseteq r_1 \qquad r \sqsubseteq r_2}{C_1 \sqcup C_2 \sqcup \leq (n_1 + n_2) r . \neg D_{12}}$$

$\leq\geq$-Combination:

$$\frac{C_1 \sqcup \leq n_1 r_1 . \neg D_1 \qquad C_2 \sqcup \geq n_2 r_2 . D_2 \qquad r_2 \sqsubseteq_{\mathcal{R}} r_1 \qquad n_1 \geq n_2}{C_1 \sqcup C_2 \sqcup \leq (n_1 - n_2) r_1 . \neg(D_1 \sqcup D_2) \sqcup \geq 1 r_1 . D_{12}}$$

$$\vdots$$

$$C_1 \sqcup C_2 \sqcup \leq (n_1 - 1) r_1 . \neg(D_1 \sqcup D_2) \sqcup \geq n_2 r_1 . D_{12}$$

Fig. 4. Additional inference rules of $Forget_{\mathcal{SHQ}}$

5 Forgetting Concept Symbols

We reduce the problem of computing uniform interpolants to the problem of forgetting single symbols. We denote the result of forgetting a single symbol x from an ontology $\mathcal{O}$ by $\mathcal{O}^{-x}$, where x can be a role or a concept symbol. $\mathcal{O}^{-x}$ is the uniform interpolant of $\mathcal{O}$ over $sig(\mathcal{O}) \setminus \{x\}$.

The general idea for forgetting a concept symbol A is to saturate the clausal representation of $\mathcal{O}$ in such a way that every clause that cannot be represented in an $\mathcal{SHQ}$-ontology in the signature $sig(\mathcal{O}) \setminus \{A\}$ becomes superfluous. In addition to the rules of $Res^s_{\mathcal{SHQ}}$, we need two more inference rules for the forgetting procedure, which are shown in Figure 4. It turns out that we do not have to consider number restrictions with disjunctions in our rules, which is the situation with all number restrictions after the transformation into $\mathcal{SHQ}$-clausal form.

Assume a domain element has maximally n_1 r-successors satisfying $\neg D_1$ and maximally n_2 r-successors satisfying $\neg D_2$. If we sum them up without further knowledge, we have that there are at most $n_1 + n_2$ r-successors satisfying either $\neg D_1$ or $\neg D_2$. Since formulae under $\leq$-restrictions are negated, and because $(\neg D_1 \sqcup \neg D_2) \equiv \neg(D_1 \sqcap D_2)$, we can verify that the $\leq$-combination rule is sound.

$\mathcal{D}_1$							$\neg\mathcal{D}_1$
$\neg\mathcal{D}_2$							$\mathcal{D}_2$

Fig. 5. Diagram illustrating the $\leq\geq$-combination rule

For the $\leq\geq$-combination rule, we again interpret the last two literals of each conclusion as an implication. For example for the first conclusion, the implication is $\leq 0r.(D_{12}) \rightarrow \leq(n_1 - n_2)r.\neg(D_1 \sqcup D_2)$. That this implication follows from $\leq n_1 r.D_1$ and $\geq n_2 r.D_2$ if $n_1 \geq n_2$ is illustrated in the diagram in Figure 5.

Let $\mathbf{N}$ be the $\mathcal{SHQ}$-normal form of the TBox of $\mathcal{O}$. In order to forget A, or to compute $\mathcal{O}^{-A}$, we saturate $\mathbf{N}$, where we apply resolution only on the symbol A we want to forget, or on definer symbols. The combination rules are only applied if they lead to the introduction of new clauses that make further resolution steps on A possible. For example, if we have the clauses $\leq 5r.\neg D_1$, $\geq 3r.D_2$, $\neg D_1 \sqcup A$ and $\neg D_2 \sqcup \neg A$, we apply the $\leq\geq$-combination rule, since this leads to the introduction of a new definer symbol D_{12}, and, after resolving on the definer symbols, the clauses $\neg D_{12} \sqcup A$ and $\neg D_{12} \sqcup \neg A$. These two clauses can be resolved on A. If we do not have $\neg D_1 \sqcup A$ or $\neg D_2 \sqcup \neg A$, we do not have to apply the $\leq\geq$-combination rule in order to compute the uniform interpolant.

After this saturation is computed, we can remove all clauses that contain the symbol A we want to forget, or that are of the form $\neg D_1 \sqcup D_2$ or $\neg D_1 \sqcup \neg D_2 \sqcup C$. Clauses of the form $\neg D_1 \sqcup D_2$ become superfluous since we computed all resolvents on D_2. Clauses of the form $\neg D_1 \sqcup \neg D_2 \sqcup C$ can also be discarded, as is proved in the long version of this paper.

$\mathbf{N}^{-A}$ is the clausal representation of the result of forgetting A from $\mathbf{N}$, as the following lemma shows.

Lemma 1. *Given a set of clauses $\mathbf{N}$ and an RBox $\mathcal{R}$, $\mathbf{N}^{-A}$ does not contain A and we have $\langle \mathbf{N}^{-A}, \mathcal{R} \rangle \models \alpha$ iff $\langle \mathbf{N}, \mathcal{R} \rangle \models \alpha$ for all $\mathcal{SHQ}$-axioms α that do not contain A.*

It remains to eliminate all introduced definer symbols, so that the ontology is completely represented in the desired signature. Since every clause in $\mathbf{N}^{-A}$ contains at most one negative definer literal $\neg D$, we can compute for each definer symbol D a unique concept inclusion $D \sqsubseteq C_1 \sqcap \ldots \sqcap C_n$, where $\neg D \sqcup C_1, \ldots, \neg D \sqcup C_n$ are the clauses in which $\neg D$ occurs outside of a role restriction. We call this concept inclusion the *definition of D*. $D \sqsubseteq C_1 \sqcap \ldots \sqcap C_n$ is equivalent to the set of clauses $\neg D \sqcup C_1, \ldots, \neg D \sqcup C_n$, and we obtain therefore an equivalent TBox by replacing these clauses by the corresponding definitions. We denote the result of this transformation by $\mathcal{T}_D^{-A}$.

In order to compute a TBox representation of $\mathcal{T}_D^{-A}$ without definer symbols, we apply the definer elimination rules shown in Figure 6, where $\mathcal{T}^{[D \mapsto C]}$ denotes the TBox obtained by replacing D with C. If a definer symbol occurs only on the left-hand side of its definition, we can replace all positive occurrences of it using

Non-cyclic definer elimination:

$$\frac{\mathcal{T} \cup \{D \sqsubseteq C\}}{\mathcal{T}^{[D \mapsto C]}} \qquad \text{provided } D \notin sig(C)$$

Definer purification:

$$\frac{\mathcal{T}}{\mathcal{T}^{[D \mapsto \top]}} \qquad \text{provided } D \text{ occurs in } \mathcal{T} \text{ only under number restrictions}$$

Cyclic definer elimination:

$$\frac{\mathcal{T} \cup \{D \sqsubseteq C[D]\}}{\mathcal{T}^{[D \mapsto \nu X.C[X]]}} \qquad \text{provided } D \in sig(C[D])$$

Fig. 6. Rules for eliminating definer symbols

the non-cyclic definer elimination rule. If there is no definition of D, D occurs only positively and we can replace all its occurrences by $\top$. If a definer symbol occurs on both sides of its definition, applying the non-cyclic definer elimination rule would lead to an infinite derivation. Instead we apply the cyclic-definer elimination rule, which introduces a greatest fixpoint operator.

Since ontologies can have cyclic definitions, it is in general not always possible to find a finite uniform interpolant that does not use fixpoint operators. If we want to compute a representation of the uniform interpolant that is completely in $\mathcal{SHQ}$ and does not use fixpoint operators, we can either keep the cyclic definer symbols, or approximate the uniform interpolant. Keeping the cyclic definer symbols has the advantage that we preserve all entailments of the uniform interpolant and obtain an ontology that can be processed by any common reasoner supporting $\mathcal{SHQ}$. The remaining definer symbols can be seen as "helper concepts" that make a finite representation possible.

If we want an ontology without fixpoints that is completely in the desired signature, we can in general only approximate the uniform interpolant, since it might be infinite. This approximation can be performed by replacing the cyclic definer symbols a finite number of times following their definitions, and then replacing them by $\top$ (see also [10] for this).

Theorem 3. *Given any ontology $\mathcal{O} = \langle \mathcal{T}, \mathcal{R} \rangle$ and concept symbol A, $\mathcal{O}^{-A} = \langle \mathcal{T}^{-A}, \mathcal{R} \rangle$ is a uniform interpolant of $\mathcal{O}$ for $sig(\mathcal{O}) \setminus \{A\}$ in $\mathcal{SHQ}\mu$. If $\mathcal{O}^{-A}$ does not use any fixpoint operators, it is the uniform interpolant of $\mathcal{O}$ in $\mathcal{SHQ}$ for $sig(\mathcal{O}) \setminus \{A\}$.*

We conducted a small evaluation of forgetting concept symbols from real-life ontologies, details of which can be found in the long version of this paper [11]. Our results suggest that at least for smaller ontologies of up to 700 axioms, forgetting half of the concept symbols in the signature can be performed in a few minutes in the majority of cases.

Role hierarchy:	$\leq$-Monotonicity:	$\geq$-Monotonicity:
$\dfrac{s \sqsubseteq r \quad r \sqsubseteq t}{s \sqsubseteq t}$	$\dfrac{C \sqcup \leq nr.\neg D \qquad s \sqsubseteq r}{C \sqcup \leq ns.\neg D}$	$\dfrac{C \sqcup \geq nr.D \qquad r \sqsubseteq s}{C \sqcup \geq ns.D}$

Fig. 7. Inference rules for forgetting the role symbol r

6 Forgetting Role Symbols

We can adapt the method from [8] to obtain a procedure for forgetting role symbols from $\mathcal{SHQ}$ ontologies, provided the role to be forgotten is not transitive. For forgetting role symbols, we have to process the RBox as well, and act differently depending on what we can make of the role hierarchy.

Forgetting transitive roles is not possible if we want to express the uniform interpolant in $\mathcal{SHQ\mu}$, as the following theorem shows.

Theorem 4. *There are ontologies $\mathcal{O}$ and role symbols r, where r is transitive in $\mathcal{O}$, without a finite uniform interpolant of $\mathcal{O}$ for $sig(\mathcal{O}) \setminus \{r\}$ in $\mathcal{SHQ\mu}$.*

Proof. Consider an ontology $\mathcal{O}$ with an RBox $\mathcal{R} = \{s \sqsubseteq r, r \sqsubseteq t, \text{trans}(r)\}$. We have the following infinite number of entailments of $\mathcal{O}$, where C is any concept: $\mathcal{O} \models \forall t.C \sqsubseteq \forall s.C, \forall t.C \sqsubseteq \forall s.\forall s.C, \dots$. Since neither t nor s are transitive, there can be no finite ontology in $\mathcal{SHQ\mu}$ defined over $sig(\mathcal{O}) \setminus \{r\}$ and entails all these consequences.

In order to forget a non-transitive role symbol r, we have to apply the combination rules on number restrictions with r exhaustively. As for forgetting concept symbols, the other rules have to be applied only if they lead to the introduction of new definer symbols and clauses that make further derivations on r possible. On the resulting clause set, we apply the monotonicity rules shown in Figure 7. We denote the result by $\mathbf{N}^{*r}$.

If r has a super-role s (that is, $r \sqsubseteq s$), we can afterwards filter out all clauses that contain r and obtain a clausal representation of the uniform interpolant. The $\geq$-monotonicity rule ensures that all information regarding $\geq$-restrictions in the desired signature is preserved. If r has no super-role, we have to check which clauses of the form $C \sqcup \geq nr.D$ can be filtered out. If $\mathbf{N}^{*r} \models \neg D$, we replace $C \sqcup \geq nr.D$ with C, otherwise, we can remove $C \sqcup \geq nr.D$ from the clause set, since all derivations on r have already been performed. To decide whether $\mathbf{N}^{*r} \models \neg D$, we can either use an external reasoner or the calculus $Res^s_{\mathcal{SHQ}}$ presented in Section 4. We also remove all clauses that are of the form $\neg D \sqcup D$ or $\neg D_1 \sqcup \neg D_2 \sqcup C$. The resulting set $\mathbf{N}^{-r}$ of clauses is transformed into an $\mathcal{SHQ\mu}$- or $\mathcal{SHQ}$-ontology using the definer elimination techniques described in the previous section. The resulting TBox $\mathcal{T}^{-r}$ is the TBox of the uniform interpolant. The RBox $\mathcal{R}^{-r}$ is computed by applying the role hierarchy rule from Figure 7 on the RBox and filtering out role inclusions containing r. We have the following result.

Theorem 5. *Given any ontology* $\mathcal{O} = \langle \mathcal{T}, \mathcal{R} \rangle$ *and any role symbol* r *such that* $\text{trans}(r) \notin \mathcal{R}$, $\mathcal{O}^{-r} = \langle \mathcal{T}^{-r}, \mathcal{R}^{-r} \rangle$ *is a uniform interpolant of* $\mathcal{O}$ *for* $\text{sig}(\mathcal{O}) \setminus \{r\}$ *in* $\mathcal{SHQ}\mu$. *If* $\mathcal{O}^{-r}$ *does not contain any fixpoint operators, it is a uniform interpolant of* $\mathcal{O}$ *for* $\text{sig}(\mathcal{O}) \setminus \{r\}$ *in* $\mathcal{SHQ}$.

7   Discussion and Related Work

There has been research in developing methods that deal with uniform interpolation in several description logics, starting from simple ones, such as DL-Lite [20] and $\mathcal{EL}$ [7,15,13], to more expressive ones such as $\mathcal{ALC}$ and $\mathcal{ALCH}$ [19,12,10,9,8]. Forgetting in more expressive description logics was first investigated by [18] and [14]. In [14], it was shown that deciding the existence of uniform interpolants that can be represented finitely in $\mathcal{ALC}$ without fixpoints is 2-EXPTIME and that these uniform interpolants can in the worst case have a size triple exponential with respect to size of the original ontology. It can be shown that, using fixpoint operators, this bound can be reduced to a double-exponential complexity. [18,19] were first to consider the computation of uniform interpolants in $\mathcal{ALC}$ and presented a tableau-based approach. A more goal-oriented method was presented in [12], following a resolution approach based on a different calculus than our method. [12] also included first experimental results, showing the practicality for a lot of applications.

The first resolution-based method incorporating fixpoints, using ideas from the area of second-order quantifier elimination [3], was presented in [10]. This method was implemented and evaluated on a large set of real-life ontologies [9], showing that the worst case complexity of uniform interpolants is hardly reached in reality, and that uniform interpolants can often be computed in a few seconds. In [8], this method was extended by redundancy elimination techniques and the ability to forget role symbols, and an evaluation showed even better results with respect to the size and use of fixpoint operators in the result.

Forgetting for description logics with number restrictions was first considered in [20], where a method for the description logic DL-Lite$^{\mathcal{N}}_{bool}$ is presented. DL-Lite$^{\mathcal{N}}_{bool}$ extends DL-Lite with unqualified number restrictions and Boolean operators [1]. The logic allows for inverse roles, but does not allow concepts under number restrictions. More specifically, it cannot express universal restrictions or qualified existential role restrictions. This makes it possible to implement forgetting in DL-Lite$^{\mathcal{N}}_{bool}$ using propositional resolution.

Apart from number restrictions, there are more extensions to $\mathcal{SHQ}$ that have not been investigated yet, such as inverse roles or nominals. Whereas it is possible that a method for computing uniform interpolants in an expressive description logic with inverse roles will be discovered in the future, for nominals this will likely not be the case. This follows from result for module extraction from [5]. Given two ontologies $\mathcal{O}$ and $\mathcal{M}$ and a signature $\mathcal{S}$, $\mathcal{M}$ is an $\mathcal{S}$-*module* of $\mathcal{O}$, if $\mathcal{M}$ is a subset of $\mathcal{O}$ and has the same logical entailments over $\mathcal{S}$ as $\mathcal{O}$. Different from uniform interpolants, modules can contain symbols that are not in $\mathcal{S}$. Uniform interpolation can be used to test whether an ontology $\mathcal{M}$ is an $\mathcal{S}$-module of another ontology $\mathcal{O}$. More specifically, $\mathcal{M}$ is an $\mathcal{S}$-module of $\mathcal{O}$ iff $\mathcal{M} \subseteq \mathcal{O}$ and $\mathcal{M} \models \mathcal{O}^{\mathcal{S}}$, where $\mathcal{O}^{\mathcal{S}}$ is the uniform interpolant of $\mathcal{O}$ over $\mathcal{S}$. But in [5] it was shown that determining whether $\mathcal{M}$ is an $\mathcal{S}$-module of $\mathcal{O}$ is undecidable already for the description logic $\mathcal{ALCO}$, which extends $\mathcal{ALC}$ with nominals. From this follows that there can be no general method for computing uniform interpolants of $\mathcal{ALCO}$-ontologies that are represented in a decidable logic.

8 Conclusion and Future Work

We have presented a method for uniform interpolation of $\mathcal{SHQ}$-ontologies. The method allows to compute uniform interpolants for ontologies in more expressive description logics than previous approaches, and also to preserve indirect cardinality information from ontologies that do not use number restrictions. The method makes use of a new sound and refutationally complete resolution-calculus for the description logic $\mathcal{SHQ}$. If a finite representation cannot be computed in pure $\mathcal{SHQ}$, fixpoint operators are used in the result, which can be simulated using helper concepts. The result of forgetting transitive roles cannot be represented in $\mathcal{SHQ}\mu$, and is not computed by our method. A solution might be to use a description logic that allows for transitive closures of roles.

Results of a preliminary evaluation of our method indicate that at least for smaller ontologies, forgetting of concept symbols can be performed in short amounts of time [11]. Even for input ontologies that do not contain number restrictions, we sometimes obtained interesting results where further entailments using $\geq$-number restrictions where derived, that would not have been part of an $\mathcal{ALCH}$ uniform interpolant. It is likely that for smaller signatures (100–2000 symbols, depending on the ontology), uniform interpolants of large ontologies can in most cases still be computed in short times, if only concept symbols have to be forgotten, and there are optimisations we have not investigated yet. Forgetting role symbols is likely a much more expensive task, since our combination rules derive a lot of consequences. An evaluation of this is future work.

We have not investigated the complexity of our method and uniform interpolation in $\mathcal{SHQ}$ in general. It is therefore open whether our method is optimal. Additionally, we are currently investigating uniform interpolation for description logics with inverse roles, such as $\mathcal{SHI}$ and $\mathcal{SHIQ}$. Besides extending the expressivity of the supported description logic, there are further ways in which the framework of uniform interpolation can be extended. So far, most work on uniform interpolation for more expressive description logics focused on the TBox and the RBox. An open problem is uniform interpolation of ontologies in $\mathcal{SHQ}$ or more expressive logics that also have an ABox, which would be useful in many applications, including privacy and ontology analysis.

References

1. Artale, A., Calvanese, D., Kontchakov, R., Zakharyaschev, M.: DL-Lite in the Light of First-Order Logic. In: AAAI 2007, pp. 361–366. AAAI Press (2007)

2. Calvanese, D., Giacomo, G.D., Lenzerini, M.: Reasoning in Expressive Description Logics with Fixpoints based on Automata on Infinite Trees. In: Proc. IJCAI 1999, pp. 84–89. Morgan Kaufmann (1999)
3. Gabbay, D.M., Schmidt, R.A., Szałas, A.: Second Order Quantifier Elimination: Foundations, Computational Aspects and Applications. College Publ. (2008)
4. Grau, B.C.: Privacy in ontology-based information systems: A pending matter. Semantic Web 1(1-2), 137–141 (2010)
5. Grau, B.C., Horrocks, I., Kazakov, Y., Sattler, U.: Modular reuse of ontologies: theory and practice. J. Artif. Intell. Res. 31(1), 273–318 (2008)
6. Horrocks, I., Sattler, U., Tobies, S.: Practical Reasoning for Very Expressive Description Logics. Logic J. IGPL 8(3), 239–264 (2000)
7. Konev, B., Walther, D., Wolter, F.: Forgetting and Uniform Interpolation in Large-Scale Description Logic Terminologies. In: Proc. IJCAI 2009, pp. 830–835. AAAI Press (2009)
8. Koopmann, P., Schmidt, R.A.: Forgetting Concept and Role Symbols in $\mathcal{ALCH}$-Ontologies. In: McMillan, K., Middeldorp, A., Voronkov, A. (eds.) LPAR-19 2013. LNCS, vol. 8312, pp. 552–567. Springer, Heidelberg (2013)
9. Koopmann, P., Schmidt, R.A.: Implementation and Evaluation of Forgetting in $\mathcal{ALC}$-Ontologies. In: Proc. WoMO 2013. CEUR-WS.org (2013)
10. Koopmann, P., Schmidt, R.A.: Uniform Interpolation of $\mathcal{ALC}$-Ontologies Using Fixpoints. In: Fontaine, P., Ringeissen, C., Schmidt, R.A. (eds.) FroCoS 2013. LNCS, vol. 8152, pp. 87–102. Springer, Heidelberg (2013)
11. Koopmann, P., Schmidt, R.A.: Count and Forget: Uniform Interpolation of $\mathcal{SHQ}$-Ontologies—Long Version. Tech. Rep., The University of Manchester (2014), http://www.cs.man.ac.uk/~koopmanp/IJCAR_KoopmannSchmidt2014_long.pdf
12. Ludwig, M., Konev, B.: Towards Practical Uniform Interpolation and Forgetting for $\mathcal{ALC}$ TBoxes. In: Proc. DL 2013, pp. 377–389. CEUR-WS.org (2013)
13. Lutz, C., Seylan, I., Wolter, F.: An Automata-Theoretic Approach to Uniform Interpolation and Approximation in the Description Logic $\mathcal{EL}$. In: Proc. KR 2012, pp. 286–296. AAAI Press (2012)
14. Lutz, C., Wolter, F.: Foundations for Uniform Interpolation and Forgetting in Expressive Description Logics. In: Proc. IJCAI 2011, pp. 989–995. AAAI Press (2011)
15. Nikitina, N.: Forgetting in General $\mathcal{EL}$ Terminologies. In: Proc. DL 2011, pp. 345–355. CEUR-WS.org (2011)
16. Schmidt, R.A., Hustadt, U.: A principle for incorporating axioms into the first-order translation of modal formulae. In: Baader, F. (ed.) CADE 2003. LNCS (LNAI), vol. 2741, pp. 412–426. Springer, Heidelberg (2003)
17. Tobies, S.: Complexity results and practical algorithms for logics in knowledge representation. Ph.D. thesis, RWTH-Aachen, Germany (2001)
18. Wang, K., Wang, Z., Topor, R., Pan, J.Z., Antoniou, G.: Concept and Role Forgetting in $\mathcal{ALC}$ Ontologies. In: Bernstein, A., Karger, D.R., Heath, T., Feigenbaum, L., Maynard, D., Motta, E., Thirunarayan, K. (eds.) ISWC 2009. LNCS, vol. 5823, pp. 666–681. Springer, Heidelberg (2009)
19. Wang, K., Wang, Z., Topor, R., Pan, J.Z., Antoniou, G.: Eliminating concepts and roles from ontologies in expressive descriptive logics. Comput. Intell. (to appear)
20. Wang, Z., Wang, K., Topor, R.W., Pan, J.Z.: Forgetting for knowledge bases in DL-Lite. Ann. Math. Artif. Intell. 58(1-2), 117–151 (2010)

Coupling Tableau Algorithms for Expressive Description Logics with Completion-Based Saturation Procedures

Andreas Steigmiller[1,*], Birte Glimm[1], and Thorsten Liebig[2]

[1] University of Ulm, Ulm, Germany
{Andreas.Steigmiller,Birte.Glimm,Thorsten.Liebig}@uni-ulm.de
[2] derivo GmbH, Ulm, Germany
liebig@derivo.de

Abstract. Nowadays, saturation-based reasoners for the OWL EL profile are able to handle large ontologies such as SNOMED very efficiently. However, saturation-based reasoning procedures become incomplete if the ontology is extended with axioms that use features of more expressive Description Logics, e.g., disjunctions. Tableau-based procedures, on the other hand, are not limited to a specific OWL profile, but even highly optimised reasoners might not be efficient enough to handle large ontologies such as SNOMED. In this paper, we present an approach for tightly coupling tableau- and saturation-based procedures that we implement in the OWL DL reasoner Konclude. Our detailed evaluation shows that this combination significantly improves the reasoning performance on a wide range of ontologies.

1 Introduction

The current version of the Web Ontology Language (OWL 2) [19] is based on the very expressive Description Logic (DL) $\mathcal{SROIQ}$ [6]. To handle (standard) reasoning tasks, sound and complete tableau algorithms are typically used, which are easily extensible and adaptable. Moreover, the use of a wide range of optimisation techniques allows for handling many expressive, real-world ontologies. Since standard reasoning tasks for $\mathcal{SROIQ}$ have N2EXPTIME-complete worst-case complexity [9], it is, however, not surprising that larger ontologies easily become unpractical for existing systems.

In contrast, the OWL 2 profiles define language fragments of $\mathcal{SROIQ}$ for which reasoning tasks can be realised efficiently, e.g., within polynomial worst-case complexity. For example, the OWL 2 EL profile is based on the DL $\mathcal{EL}^{++}$ which can be handled very efficiently by variants of saturation-based reasoning procedures [2,10]. These saturation algorithms have also been pushed to more expressive DLs (e.g., Horn-$\mathcal{SHIQ}$ [10] or $\mathcal{ALCH}$ [12]) for which they are often able to outperform the more general tableau algorithms. In particular, they allow a fast one-pass handling of several reasoning tasks such as classification (i.e., the task of arranging the named concepts of an ontology in a subsumption hierarchy), whereas tableau-based procedures perform classification by a pairwise comparison of the named concepts. Handling cardinality restrictions with saturation procedures is, however, still an open question.

---

* The author acknowledges the support of the doctoral scholarship under the Postgraduate Scholarships Act of the Land of Baden-Wuerttemberg (LGFG).

S. Demri, D. Kapur, and C. Weidenbach (Eds.): IJCAR 2014, LNAI 8562, pp. 449–463, 2014.

Recently, new approaches have been proposed to also improve the reasoning performance for ontologies of more expressive DLs by combining saturation procedures and fully-fledged tableau reasoners in a black box manner [1,14]. These approaches try to delegate as much work as possible to the specialised and more efficient reasoner, which allows for reducing the workload of the fully-fledged tableau algorithm, and often results in a better pay-as-you-go behaviour than using a tableau reasoner alone.

In this paper, we present a much tighter coupling between saturation- and tableau-based algorithms, whereby further performance improvements are achieved. After introducing some preliminaries (Section 2), we present a saturation procedure that is adapted to the data structures of a tableau algorithm (Section 3). This allows for easily passing information between the saturation and the tableau algorithm within the same reasoning system. Moreover, the saturation partially handles features of more expressive DLs in order to efficiently derive as many consequences as possible (Section 3.1). We then show how parts of the ontology can be identified for which the saturation procedure is possibly incomplete and where it is necessary to fall-back to the tableau procedure (Section 3.2). Subsequently, we present several optimisations that are based on passing information from the saturation to the tableau algorithm (Section 4) and back (Section 5). Finally, we present the results of a detailed evaluation (Section 6) before we conclude (Section 7). Further details, proofs, an extended evaluation, and comparisons with other reasoners are available in a technical report [16].

2 Preliminaries

For brevity, we do not introduce DLs (see, e.g., [3]) and we only present our approach for the DL $\mathcal{ALCHOIQ}$. However, the approach can easily be extended to $\mathcal{SROIQ}$ (see [16]), e.g., by encoding role chains and adding appropriate rules for the remaining features such as $\exists r.\mathsf{Self}$ concepts.

2.1 Tableau Algorithm

For ease of presentation, we assume in the remainder of the paper that all concepts are in negation normal form (NNF) and we use $\dot{\neg}$ to denote the negation of a concept in NNF. Moreover, we assume that all ABox axioms are internalised into the TBox of a knowledge base.[1]

A tableau algorithm decides the consistency of a knowledge base $\mathcal{K}$ by trying to construct an abstraction of a model for $\mathcal{K}$, a so-called "completion graph". A completion graph G is a tuple $(V, E, \mathcal{L}, \dot{\neq})$, where each node $v \in V$ (edge $\langle v, w \rangle \in E$) represents one or more (pairs of) individuals. Each node v (edge $\langle v, w \rangle$) is labelled with a set of concepts (roles), $\mathcal{L}(v)$ ($\mathcal{L}(\langle v, w \rangle)$), which the individuals represented by v ($\langle v, w \rangle$) are instances of. The relation $\dot{\neq}$ records inequalities between nodes.

The algorithm works by initialising the graph with one node for each nominal in the input knowledge base. Complex concepts are then decomposed using a set of expansion rules, where each rule application can add new concepts to node labels and/or new

---

[1] In the presence of nominals, this can easily be realised, e.g., by expressing a concept assertion $C(a)$ (role assertion $r(a, b)$) as $\{a\} \sqsubseteq C$ ($\{a\} \sqsubseteq \exists r.\{b\}$).

nodes and edges to the completion graph, thereby explicating the structure of a model. The rules are applied until either the graph is fully expanded (no more rules are applicable), in which case the graph can be used to construct a model that is a *witness* to the consistency of $\mathcal{K}$, or an obvious contradiction (called a *clash*) is discovered (e.g., both C and $\dot{\neg} C$ in a node label), proving that the completion graph does not correspond to a model. The input knowledge base $\mathcal{K}$ is *consistent* if the rules (some of which are non-deterministic) can be applied such that they build a fully expanded, clash-free completion graph. A cycle detection technique called *blocking* ensures the termination of the algorithm.

Typically, lazy unfolding rules are used in the tableau algorithm to process axioms of the form $A \sqsubseteq C$, where the concept C is added to the label of a node if it contains the atomic concept A. Axioms that are not directly supported by this lazy unfolding approach must be internalised, which can be realised by expressing a general concept inclusion (GCI) axiom $C \sqsubseteq D$ by $\top \sqsubseteq \dot{\neg} C \sqcup D$. Given that $\top$ is satisfied at each node, the disjunction is then also added to all node labels.

2.2   (Binary) Absorption

Absorption is used as a preprocessing step in order to reduce the non-determinism in the tableau algorithm. Basically, axioms are rewritten into (possibly several) simpler concept inclusion axioms such that lazy unfolding rules in the tableau algorithm can be used and, therefore, internalisation of axioms is often not required. Algorithms based on binary absorption [8] allow for and create axioms of the form $(A_1 \sqcap A_2) \sqsubseteq C$, whereby also more complex axioms can be absorbed. To efficiently support a binary absorption axiom $(A_1 \sqcap A_2) \sqsubseteq C$ in the tableau algorithm, a separate unfolding rule is used, which adds C only to node labels if A_1 and A_2 are already present. More sophisticated absorption algorithms, such as partial absorption [15], further improve the handling of knowledge bases for more expressive DLs since the non-determinism that is caused by disjunctions on the right-hand side of axioms is further reduced. Roughly speaking, the non-absorbable disjuncts are partially used as conditions on the left-hand side of additional inclusion axioms such that the processing of the disjunctions can further be delayed. Many state-of-the-art reasoning systems are at least using some kind of binary absorption, which makes the processing of simple ontologies (e.g., EL ontologies) also with the tableau algorithm deterministic. In the following, we assume that knowledge bases are, at least, preprocessed with a variant of binary absorption and we also use the syntax of binary absorption axioms to illustrate the algorithms and examples.

3   Saturation Compatible with Tableau Algorithms

In this section, we describe a saturation method that is an adaptation of the completion-based procedure [2] such that it generates data structures that are compatible for further usage within a fully-fledged tableau algorithm for more expressive DLs. Similarly to completion graphs, this saturation generates nodes that are labelled with sets of concepts and, therefore, it directly allows for transferring results from the saturation to the tableau algorithm. For example, the saturated labels can be used to initialise the labels of new nodes in the completion graph or to block the processing of nodes. In some cases, it is directly possible to extract completion graphs from the data structures of the saturation, which makes an explicit model construction with the tableau algorithm unnecessary.

Note, the adapted saturation method is not designed to cover a certain OWL 2 profile or a specific DL language. In contrast, we saturate those parts of a knowledge base that can easily be supported with an efficient algorithm (see Section 3.1). Unsupported concept constructors are (partially) ignored by the saturation, but we dynamically detect which parts have not been completely handled afterwards (see Section 3.2). Hence, the results of the saturation are possibly incomplete, but since we know how and where they are incomplete, we can use the results from the saturation appropriately.

3.1 Saturation Based on Tableau Rules

The adapted saturation method generates so-called saturation graphs, which approximate completion graphs in a compressed form (e.g., it allows for "reusing" nodes).

Definition 1 (Saturation Graph). *Let* $\mathsf{Rols}(\mathcal{K})$ *($\mathsf{fclos}(\mathcal{K})$) denote the roles (concepts) that occur in* $\mathcal{K}$ *(in completion graphs for* $\mathcal{K}$*). A saturation graph for* $\mathcal{K}$ *is a directed graph* $S = (V, E, \mathcal{L})$ *with the nodes* $V \subseteq \{v_C \mid C \in \mathsf{fclos}(\mathcal{K})\}$. *Each node* $v_C \in V$ *is labelled with a set* $\mathcal{L}(v_C) \subseteq \mathsf{fclos}(\mathcal{K})$ *such that* $\mathcal{L}(v_C) \supseteq \{\top, C\}$. *We call* v_C *the* representative node *for the concept* C. *Each edge* $\langle v, v' \rangle \in E$ *is labelled with a set* $\mathcal{L}(\langle v, v' \rangle) \subseteq \mathsf{Rols}(\mathcal{K})$. *We say that a node* $v \in V$ *is clashed if* $\bot \in \mathcal{L}(v)$.

A major difference to a completion graph is the missing $\not\doteq$ relation, which can be omitted since the saturation is not designed to completely handle cardinality restrictions and, therefore, we also do not need to keep track of inequalities between nodes in the saturation graph. Furthermore, each node in the saturation graph is the representative node for a specific concept, which allows for reusing nodes. For example, instead of creating new successors for existential restrictions, we reuse the representative node for the existentially restricted concept as a successor.

In principle, the nodes, edges, and labels are used as in completion graphs (cf. [6]) and, therefore, we also use the (r-)neighbour, (r-)successor, (r-)predecessor, ancestor and descendant relations analogously. Please note, however, that a node in the saturation graph can have several predecessors due to the reuse of nodes.

We initialise the saturation graph with the representative nodes for all concepts that have to be saturated. For example, if the satisfiability of the concept C has to be tested, then we are interested in the saturation of the concept C and, therefore, we add the node v_C with the label $\mathcal{L}(v_C) = \{\top, C\}$ to the saturation graph. Note that we only build one saturation graph, i.e., if we are later also interested in the saturation of a concept D that is not already saturated, then we simply extend the existing saturation graph by v_D. For knowledge bases that contain nominals, we also add a node $v_{\{a\}}$ with $\mathcal{L}(v_{\{a\}}) = \{\top, \{a\}\}$ for each nominal $\{a\}$ occurring in the knowledge base.

For the initialised saturation graph, we apply the saturation rules depicted in Table 1. Note that if a saturation rule refers to the representative node for a concept C and the node v_C does not yet exist, then we assume that the saturation graph is automatically extended by this node. Although the saturation rules are very similar to the

Table 1. Saturation rules for the (partial) handling of $\mathcal{ALCHOIQ}$ knowledge bases

$\sqsubseteq_1$-rule: if $H \in \mathcal{L}(v), H \sqsubseteq C \in \mathcal{K}$ with $H = A, H = \{a\}$, or $H = \top$, and $C \notin \mathcal{L}(v)$,   then $\mathcal{L}(v) \longrightarrow \mathcal{L}(v) \cup \{C\}$
$\sqsubseteq_2$-rule: if $\{A, B\} \subseteq \mathcal{L}(v), (A \sqcap B) \sqsubseteq C \in \mathcal{K}$, and $C \notin \mathcal{L}(v)$,   then $\mathcal{L}(v) \longrightarrow \mathcal{L}(v) \cup \{C\}$
$\sqcap$-rule: if $C_1 \sqcap C_2 \in \mathcal{L}(v)$ and $\{C_1, C_2\} \nsubseteq \mathcal{L}(v)$,   then $\mathcal{L}(v) \longrightarrow \mathcal{L}(v) \cup \{C_1, C_2\}$
$\exists$-rule: if $\exists r.C \in \mathcal{L}(v)$ and $r \notin \mathcal{L}(\langle v, v_C \rangle)$,   then $\mathcal{L}(\langle v, v_C \rangle) \longrightarrow \mathcal{L}(\langle v, v_C \rangle) \cup \{r\}$
$\forall$-rule: if $\forall r.C \in \mathcal{L}(v)$, there is an $\mathsf{inv}(r)$-predecessor v' of v, and $C \notin \mathcal{L}(v')$,   then $\mathcal{L}(v') \longrightarrow \mathcal{L}(v') \cup \{C\}$
$\sqcup$-rule: if $C_1 \sqcup C_2 \in \mathcal{L}(v)$, there is some $D \in \mathcal{L}(v_{C_1}) \cap \mathcal{L}(v_{C_2})$, and $D \notin \mathcal{L}(v)$,   then $\mathcal{L}(v) \longrightarrow \mathcal{L}(v) \cup \{D\}$
$\geqslant$-rule: if $\geqslant n r.C \in \mathcal{L}(v)$ with $n \geq 1$ and $r \notin \mathcal{L}(\langle v, v_C \rangle)$,   then $\mathcal{L}(\langle v, v_C \rangle) \longrightarrow \mathcal{L}(\langle v, v_C \rangle) \cup \{r\}$
o-rule: if $\{a\} \in \mathcal{L}(v)$, there is some $D \notin \mathcal{L}(v)$, and $\qquad D \in \mathcal{L}(v_{\{a\}})$ or there is a descendant v' of v with $\{\{a\}, D\} \subseteq \mathcal{L}(v')$,   then $\mathcal{L}(v) \longrightarrow \mathcal{L}(v) \cup \{D\}$
$\bot$-rule: if $\bot \notin \mathcal{L}(v)$, and $\quad$ 1. $\{C, \neg C\} \subseteq \mathcal{L}(v)$, or $\quad$ 2. $\{\geqslant n r.C, \leqslant m s.D\} \subseteq \mathcal{L}(v)$ with $n > m, r \sqsubseteq^* s$ and $D \in \mathcal{L}(v_C)$, or $\quad$ 3. $\geqslant n r.C \in \mathcal{L}(v)$ with $n > 1$, and $\{a\} \in \mathcal{L}(v_C)$, or $\quad$ 4. there exist a successor node v' of v with $\bot \in \mathcal{L}(v')$, or $\quad$ 5. there exist a node $v_{\{a\}}$ with $\bot \in \mathcal{L}(v_{\{a\}})$,   then $\mathcal{L}(v) \longrightarrow \mathcal{L}(v) \cup \{\bot\}$

corresponding expansion rules in the tableau algorithm, there are some differences. For example, the number of nodes is limited by the number of (sub-)concepts occurring in the knowledge base due to the reuse of nodes for satisfying existentially restricted concepts. Consequently, the saturation is terminating since the rules are only applied when they can add new concepts or roles to node or edge labels. Moreover, a cycle detection technique such as blocking is not required, which makes the rule application very fast. Note also that the $\forall$-rule propagates concepts only to the predecessors of a node, which is necessary in order to allow the reuse of nodes for existentially restricted concepts. Language features of more expressive DLs are only partially supported. For instance, the $\sqcup$-rule adds only those concepts that are implied by both disjuncts. In order to (partially) handle a nominal $\{a\}$ in the label of a node v, we use an o-rule that adds those concepts that are derived for $v_{\{a\}}$ or for descendant nodes that also have $\{a\}$ in their label (instead of merging such nodes as in tableau procedures). This enables a very efficient implementation and is sufficient for many ontologies.

The rules also include a $\bot$-rule, which adds the concept $\bot$ to the label of those nodes for which a clash can be discovered. Furthermore, it propagates $\bot$ to the ancestor nodes. In case $\bot$ occurs in the label of a representative node for a nominal, the knowledge base is inconsistent and $\bot$ is propagated to every node label in the saturation graph; otherwise $\bot$ in the label of a node v_C indicates the unsatisfiability of C. Although it is, in principle, possible to detect also several other kinds of clashes for the incompletely handled parts

$$\mathcal{L}(v_A) = \left\{ \top, A, \exists s^-.B, B \sqcup \{a\}, C \right\} \quad \mathcal{L}(v_B) = \left\{ \top, B, C, \exists s.\{a\}, \leqslant 1\, s.C \right\}$$

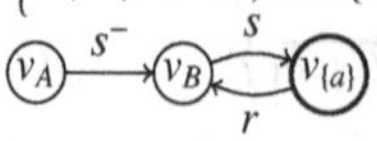

$$\mathcal{L}(v_{\{a\}}) = \left\{ \top, \{a\}, C, \geqslant 2\, r.B \right\}$$

Fig. 1. Generated saturation graph for testing the satisfiability of A_1 for Example 1

in the saturation (e.g., for a concept C that has to be propagated to a successor node v, where v has already the negation of C in its label), the presented conditions of the $\bot$-rule are already sufficient to show the completeness. Hence, we omit further clash conditions for ease of presentation. Note, the use of a $\bot$-rule is typical for saturation procedures since we are interested in associating clashes with specific nodes instead of entire completion graphs. As a consequence, the saturation allows for handling several independent concepts within the same saturation graph, while unsatisfiable nodes can nevertheless be distinguished from nodes that are (possibly) still satisfiable.

Example 1. Let us assume that the TBox $\mathcal{T}$ contains the following axioms:

$$A \sqsubseteq \exists s^-.B \qquad A \sqsubseteq B \sqcup \{a\} \qquad B \sqsubseteq C \qquad B \sqsubseteq \exists s.\{a\}$$
$$B \sqsubseteq \leqslant 1\, s.C \qquad \{a\} \sqsubseteq C \qquad \{a\} \sqsubseteq \geqslant 2\, r.B$$

In order to test the satisfiability of the concept A, we initialise the saturation graph with the representative node for A and the nominal $\{a\}$. Applying the rules of Table 1 yields the saturation graph depicted in Figure 1. Note that the procedure creates new nodes on demand, e.g., for the processing of disjunctions, existential restrictions, and at-least cardinality restrictions. Although the concept C is added to node labels, a node for C is not created since C is not used in a way that requires this. Also note that the $\sqcup$-rule application adds C to the label of v_A, because C is in the label of the representative nodes for both disjuncts of the disjunction $B \sqcup \{a\}$ (i.e., $C \in \mathcal{L}(v_B) \cap \mathcal{L}(v_{\{a\}})$).

With a suitable absorption technique, the saturation is usually able to derive and add the majority of those concepts that would also be added by the tableau algorithm for an equivalent node. This is especially the case for ontologies that primarily use features of the DL $\mathcal{EL}^{++}$. Since $\mathcal{EL}^{++}$ covers many important and often used constructors (e.g., $\sqcap$, $\exists$), the saturation does already the majority of the work for many ontologies (as confirmed by our evaluation in Section 6).

3.2 Saturation Status Detection

If used alone, the presented saturation procedure easily becomes incomplete for more expressive DLs, similarly to other saturation-based procedures. Our aim is, however, to gain as much information as possible from the saturation, i.e., we would like to detect more precisely for which nodes the saturation was incomplete. In principle, this can easily be approximated by testing for which nodes the actual tableau expansion rules are applicable. However, since we partially saturate some more expressive concept constructors, this approach is often too conservative. For example, consider a saturation graph without nominals and at-most cardinality restrictions, but with an at-least cardinality restriction $\geqslant n\, r.C$ with $n > 1$ in some node label. When constructing a model from the saturation graph, we could create the required n successors by "copying" the node v_C. Nevertheless, the tableau expansion rule for this at-least cardinality restriction is still applicable since we only have one successor. It would, however, be sufficient to check whether the number of successors is possibly limited by at-most cardinality restrictions or nominals. Similar relaxations are also possible for other concept constructors, which is exploited by the approach described in this section.

In order to identify nodes for which the saturation procedure might be incomplete, we first identify nodes that depend (directly or indirectly) on nominals and nodes that have tight at-most restrictions.

Definition 2. *Let $S = (V, E, \mathcal{L})$ be a saturation graph and $v \in V$ a node. We say that v is* directly nominal dependent *if $\{a\} \in \mathcal{L}(v)$; v is* nominal dependent *if v is directly nominal dependent or v has a successor node v' such that v' is nominal dependent.*

For a role s and a concept D, the number of merging candidates *for v w.r.t. s and D is defined as $\sum_{\geqslant n\,r.C \in G} n$ with*

$$G = \{\geqslant n\,r.C \in \mathcal{L}(v) \mid r \sqsubseteq^* s \text{ and } D \in \mathcal{L}(v_C)\} \cup$$
$$\{\geqslant 1\,r.C \mid \exists r.C \in \mathcal{L}(v), r \sqsubseteq^* s \text{ and } D \in \mathcal{L}(v_C)\}.$$

The node v has tight at-most restrictions *if there is an at-most cardinality restriction $\leqslant m\,s.D \in \mathcal{L}(v)$ and the number of merging candidates for v w.r.t. s and D is exactly m.*

For nodes with tight at-most restrictions, it is not necessary to merge some of its merging candidates, but every additional candidate might require merging and, therefore, these nodes cannot be used arbitrarily.

We can now identify critical nodes that are possibly incompletely handled by the saturation as follows:

Definition 3. *Let $S = (V, E, \mathcal{L})$ be a saturation graph and $v \in V$ a node. We say that v is* directly critical, *if*

C1 $\forall r.C \in \mathcal{L}(v)$ *and there is an r-successor v' of v such that $C \notin \mathcal{L}(v')$;*

C2 $C \sqcup D \in \mathcal{L}(v)$ *and $C, D \notin \mathcal{L}(v)$;*

C3 $\leqslant m\,s.D \in \mathcal{L}(v)$ *and there is an s-successor v' of v such that $\mathcal{L}(v') \cap \{D, \dot{\neg}D\} = \emptyset$;*

C4 $\leqslant m\,s.D \in \mathcal{L}(v)$ *and the number of merging candidates for v w.r.t. s and D is greater than m;*

C5 v *has an $\mathsf{inv}(s)$-successor v' with $\leqslant m\,s.D \in \mathcal{L}(v')$ and $\mathcal{L}(v) \cap \{D, \dot{\neg}D\} = \emptyset$;*

C6 v *has an $\mathsf{inv}(s)$-successor v' with $\leqslant m\,s.D \in \mathcal{L}(v')$, $D \in \mathcal{L}(v)$, and the number of merging candidates for v' w.r.t. s and D is m;*

C7 $\{a\} \in \mathcal{L}(v)$ *and there is some $v' \in V$ with $\{a\} \in \mathcal{L}(v')$ and $\mathcal{L}(v) \not\sqsubseteq \mathcal{L}(v')$;*

C8 v *is nominal dependent and for some nominal $\{a\}$ the node $v_{\{a\}}$ is critical; or*

C9 v *has an $\mathsf{inv}(s)$-successor v' with $\leqslant m\,s.D \in \mathcal{L}(v')$ and $\{a\} \in \mathcal{L}(v')$.*

We say that v is critical *if v is directly critical or v has a critical successor v'.*

Conditions **C1**, **C2**, and **C3** identify nodes as critical for which the $\forall$-, the $\sqcup$-, or the ch-rule of the tableau algorithm is applicable. Note that in Condition **C1** it is only necessary to check whether the concept can be propagated to successor nodes since the propagation to predecessors is ensured by the saturation procedure. Condition **C4** identifies nodes as critical for which at-most restrictions might not be satisfied. Conditions **C5** and **C6** work analogously to **C3** and **C4**, but check this from the perspective of a predecessor node. Note that C6 only has to check whether the number of merging candidates is equal to m since nodes with an at-most cardinality restriction $\leqslant m\ s.D$ and more merging candidates than m are already critical due to C4. Condition C7 checks whether merging different nodes in the saturation graph that have the same nominal in their label could lead to problems, while C8 marks nominal dependent nodes as critical if representative nodes for nominals are critical since it cannot be excluded that more consequences are propagated to these nodes over the nominals. Finally, Condition C9 identifies nodes as critical for which an interaction between at-most restrictions, nominals and inverse roles could occur and thus the NN-rule of the tableau algorithm could be applicable.

A concept C is obviously unsatisfiable if its representative node is clashed (i.e., $\perp \in \mathcal{L}(v_C)$), whereas the satisfiability of C can only be guaranteed (for the general case) if v_C is not critical, v_C does not depend on a nominal, and the knowledge base is consistent. Consistency is explicitly required, because a concept is satisfiable only if the knowledge base is consistent, which, however, cannot always be determined by the saturation procedure since it might not be able to completely handle all representative nodes for nominals. In particular, if the saturation graph contains a critical representative node for a nominal, then only the nominal dependent nodes are also marked as critical. Thus, for the remaining nodes, we have to require that the knowledge base is consistent in order to be able to guarantee the satisfiability of their associated concepts. In addition, if a node v_C is nominal dependent, then the consequences that are propagated to v_C obviously depend on the labels of the corresponding representative nodes for these nominals. Therefore, we cannot generally guarantee the satisfiability of C without knowing the status of the representative nodes for those nominals on which v_C depends.

Please also note that a critical representative node for a nominal also makes all nominal dependent nodes critical, which can obviously be very problematic in practice. In Section 5, we show how we can use information from a completion graph, e.g., from the initial consistency check, to improve the status of the saturation graph.

Example 1 (continued). For the saturation graph depicted in Figure 1, v_A, v_B, and $v_{\{a\}}$ are nominal dependent, v_B has a tight at-most restriction, and only v_A is critical: First, C6 applies to v_A since v_B is an s^--successor of v_A due to $\exists s^-.B \in \mathcal{L}(v_A)$, $\leqslant 1\ s.C \in \mathcal{L}(v_B)$ and the number of merging candidates for v_B w.r.t. s and C is 1. Second, C2 applies to v_A since none of the disjuncts of $B \sqcup \{a\} \in \mathcal{L}(v_A)$ occurs in $\mathcal{L}(v_A)$.

4   Assisting Tableau Algorithms

In this section, we show how we can use the saturation graph to improve the tableau algorithm such that existing optimisations can still be used. For example, to further support the important dependency directed backtracking [3,18], which allows for evaluating only relevant non-deterministic alternatives, we have to correctly manage the dependencies for all results that we transfer from the saturation into a completion graph.

4.1   Transfer of Saturation Results to Completion Graphs

Since the saturation uses compatible data structures, we can directly transfer the saturation results into the completion graph. For example, if we create a new successor node v

due to an existential restriction $\exists r.C$, then we can directly initialise v with the concepts from $\mathcal{L}(v_C)$ and record that the added concepts deterministically depend on C. The most notable advantage of the transferred consequences is that they often allow for blocking much earlier. Basically, concepts that would be propagated back from successor nodes are already present in the node label and, thus, a block can often be established even without creating and processing the required successors.

Furthermore, the successors of a node v in the completion graph can be blocked if there is a node v' in the saturation graph such that v and v' are labelled with the same concepts and v' is neither clashed, critical nor nominal dependent. If v' is nominal dependent and we would block the successors of v, then we might miss the handling of new consequences if the dependent nominal nodes are modified in this completion graph. If v' does not have a tight at-most restriction, then we can directly block v since merging with a predecessor can be excluded. Of course, if new concepts are propagated to v, then the block becomes invalid and the processing of the successors has to be reactivated unless another node can be used for the blocking.

4.2 Subsumer Extraction

Higher level reasoning tasks such as classification often exploit information that can be extracted from the constructed completion graphs [5]. Obviously, we can also use the saturation graph to improve classification. For example, if a node v_A is neither clashed nor critical, then A is satisfiable and $\mathcal{L}(v_A)$ contains all of its subsumers. In particular, if no nodes are critical (which is the case for many EL ontologies), only a transitive reduction is necessary for classification and, thus, we automatically get a one-pass classification for simple ontologies. Otherwise, the subsumers identified by the saturation can be used to initialise the tableau-based classification algorithm, which is more accurate than the often used told subsumers extracted from the ontology axioms.

4.3 Model Merging

Many ontologies contain axioms of the form $C \equiv D$, which can be seen as an abbreviation for $C \sqsubseteq D$ and $D \sqsubseteq C$. Treating axioms of the form $A \equiv D$ with A an atomic concept as $A \sqsubseteq D$ and $D \sqsubseteq A$ can, however, downgrade the performance of tableau algorithms since absorption might not apply to $D \sqsubseteq A$, i.e., the axiom is internalised into $\top \sqsubseteq \dot{\neg} D \sqcup A$. To avoid this, many implemented tableau algorithms explicitly support $A \equiv D$ axioms by an additional unfolding rule, where the concept A in the label of a node is unfolded to D and $\neg A$ to $\dot{\neg} D$ (exploiting that $D \sqsubseteq A$ is equivalent to $\neg A \sqsubseteq \dot{\neg} D$) [7].[2] Unfortunately, using such an unfolding rule also comes at a price since the tableau algorithm is no longer forced to add either A or $\dot{\neg} D$ to each node in the completion graph, i.e., we might not know for some nodes whether they represent instances of A or $\neg A$. This means that we cannot exclude A as possible subsumer for other (atomic) concepts if the nodes in the completion graph (or in the saturation graph) do not contain A, which is an important optimisation for classification procedures (cf. Section 4.2).

---

[2] Note that this only works as long as there are no other axioms of the form $A \sqsubseteq D'$ or $A \equiv D'$ with $D' \neq D$ in the knowledge base.

To compensate this, we can create a "candidate concept" A^+ for A, for example by partially absorbing D, which is then automatically added to a node in the completion graph if the node is possibly an instance of A. Hence, if A^+ is not added to a node label, then we know that A is not a possible subsumer of the concepts in the label of this node. Although the candidate concepts already allow a significant pruning of subsumption tests, there are still ontologies where we have to add these candidate concepts to many node labels, especially if only a limited absorption of D is possible. Hence, A can still be a possible subsumer for many concepts.

The saturation graph can, however, again be used to further improve the identification of (more or less obvious) non-subsumptions. Basically, if a candidate concept A^+ for $A \equiv D$ is in the label of a node v in the completion graph, then we test whether we can merge v with the saturated node $v_{\neg D}$. Since D is often a conjunction, we can also try to merge v with the representative node for a disjunct of $\neg D$. If the "models" can be "merged" as defined below, then v is obviously not an instance of A.

Definition 4 (Model Merging). *Let $S = (V, E, \mathcal{L})$ be a fully saturated saturation graph and $G = (V', E', \mathcal{L}', \neq)$ be a fully expanded and clash-free completion graph for a knowledge base $\mathcal{K}$. A node $v \in V$ is mergeable with a node $v' \in V'$ if*

- *v is not critical, not nominal dependent, and not clashed;*
- *$\{C, \neg C\} \cap (\mathcal{L}(v) \cup \mathcal{L}'(v')) = \emptyset$ for some concept C;*
- *if $\{A_1, A_2\} \subseteq (\mathcal{L}(v) \cup \mathcal{L}'(v'))$ and $(A_1 \sqcap A_2) \sqsubseteq C \in \mathcal{K}$, then $C \in (\mathcal{L}(v) \cup \mathcal{L}'(v'))$;*
- *if $\forall r.C \in \mathcal{L}(v)$ ($\leqslant m\, r.C \in \mathcal{L}(v)$), then $C \in \mathcal{L}'(w')$ ($\neg C \in \mathcal{L}'(w')$) for every r-neighbour w' of v';*
- *if $\forall r.C \in \mathcal{L}'(v')$ ($\leqslant m\, r.C \in \mathcal{L}'(v')$), then $C \in \mathcal{L}(w)$ ($\neg C \in \mathcal{L}'(w')$) for every r-successor w of v.*

The conditions that guarantee that the models are mergeable can be checked very efficiently. Note that it is possible to relax some of the conditions. For instance, it is not necessary to enforce that v is not nominal dependent as long as we can ensure that there is no interaction with the generated completion graph. This can, for example, be guaranteed if the completion graph does not use nominals.

5   Saturation Improvements

Obviously, the tableau algorithm can benefit more from the saturation, if few nodes are critical. In the following, we present different approaches for improving the saturation by reducing the number of critical nodes.

5.1   Extending Saturation to More Language Features

One way to improve the saturation is to extend the rules to cover more language features, e.g., as in the consequence-based reasoning procedure for Horn-$\mathcal{SHIF}$ [10]. Although we cannot directly modify existing r-successors to support universal restrictions of the form $\forall r.C$, we can easily create a new r-successor whose label additionally contains C. By further removing the previous r-successor, the node is no longer critical due to **C1**. Analogously, to (partially) support at-most restrictions of the form $\leqslant 1\,r.\top$, several r-successors can be merged into a new node. To completely cover the DL $\mathcal{EL}^{++}$, it would be necessary to integrate a more sophisticated handling of nominals. Currently, we are, however, not aware of real-world ontologies where this would result in significant improvements. Note that it is possible to limit the number of additionally created nodes for these extensions and to consider parts that are not handled as critical, thus the overhead of the saturation can be managed.

5.2 Improving Saturation with Results from Completion Graphs

As already mentioned, even if there is only one critical representative node for a nominal, all nominal dependent nodes have to be considered critical. Analogously, nodes with incompletely handled concepts (e.g., disjunctions) are considered critical and also all nodes that indirectly refer to other critical nodes, even if all concepts in their labels can be handled completely. Extending the saturation rules only also has its limits since we are not aware of saturation-based procedures that cover very expressive DLs such as $\mathcal{SROIQ}$. Hence, we can still get many critical nodes in knowledge bases that use unsupported features.

An approach to overcome this issue is to "patch" the saturation graph with results from fully expanded and clash-free completion graphs, e.g., from consistency or satisfiability checks. Roughly speaking, we replace the labels of critical nodes in the saturation graph with corresponding labels from a completion graph, where we know that they are completely handled. Applying the saturation rules again, then hopefully results in a saturation graph with less critical nodes. However, since the completion graph contains deterministically and non-deterministically derived consequences, we also have to distinguish them for the saturation. An interesting way to achieve this is to simultaneously manage two saturation graphs: one where only the deterministically derived concepts are added and a second one, where also the non-deterministically derived concepts and consequences are considered. If the non-deterministic consequences have only a locally limited influence, i.e., the non-deterministically added concepts propagate new consequences only to a limited number of ancestor nodes, then, by comparing both saturation graphs, we can possibly identify ancestor nodes that are not further influenced by non-deterministic consequences and, thus, do not have to be considered critical.

6 Implementation and Evaluation

We extended Konclude[3] [17] with the presented saturation procedure and optimisations. Konclude is a tableau-based reasoner for $\mathcal{SROIQ}$ [6] with extensions for the handling of nominal schemas [15]. It integrates many state-of-the-art optimisations such as lazy unfolding, dependency directed backtracking, caching, etc. Moreover, Konclude uses partial absorption in order to significantly reduce the non-determinism in ontologies, which makes Konclude very suitable for the integration of saturation procedures.

The saturation algorithm integrated in Konclude almost covers the DL Horn-$\mathcal{SRIF}$ by using the extensions described in Section 5.1 for universal restrictions and functional

---

[3] Available at http://www.konclude.com/

Table 2. Statistics of ontology metrics for the evaluated ontology repositories (Ø stands for average and M for median)

Repository	# Ontologies	Axioms Ø	Axioms M	Classes Ø	Classes M	Properties Ø	Properties M	Individuals Ø	Individuals M
Gardiner	276	6,143	95	1,892	16	36	7	90	3
NCBO BioPortal	403	25,561	1,068	7,617	339	47	13	1,782	0
NCIt	185	178,818	167,667	69,720	68,862	116	123	0	0
OBO Foundry	422	44,424	1,990	8,033	839	28	6	24,868	66
Oxford	383	74,248	4,249	8,789	544	52	13	18,798	12
TONES	200	7,697	337	2,907	100	28	5	66	0
Google Crawl	413	6,282	194	1,122	38	69	15	830	1
OntoCrawler	544	1,876	119	125	18	56	12	638	0
OntoJCrawl	1,680	5,848	218	1,641	43	29	8	810	0
Swoogle Crawl	1,635	2,529	109	420	21	26	8	888	0
ALL	6,141	18,583	252	4,635	50	39	9	3,674	0

at-most restrictions (only merging with predecessors is not implemented). The number of nodes that are additionally processed for the handling of these saturation extensions is mainly limited by the number of concepts occurring in the knowledge base. However, the saturation in Konclude only supports a very limited handling of individuals since the individuals also have to be handled by the tableau algorithm (at least in the worst-case) and several representations of the individuals easily multiply the memory consumption. To compensate this, Konclude primarily handles individuals with the tableau algorithm and uses patches from completion graphs (as presented in Section 5.2) to improve those parts in the saturation graph that depend on nominals.

In the following, we present a detailed evaluation that shows the improvement of Konclude due to the integrated saturation procedure. The evaluation uses a large test corpus of ontologies which have been obtained by collecting all downloadable and parseable ontologies from the Gardiner ontology suite [4], the NCBO BioPortal,[4] the National Cancer Institute thesaurus (NCIt) archive,[5] the Open Biological Ontologies (OBO) Foundry [13], the Oxford ontology library,[6] the TONES repository,[7] and those subsets of the OWLCorpus [11] that were gathered by the crawlers Google, OntoCrawler, OntoJCrawl, and Swoogle.[8] All ontologies were parsed and converted to self-contained OWL/XML files with the OWL API. For the 1,380 ontologies with imports we created a version with resolved imports and another one without the imports (for testing the reasoning performance on the main ontology content without imports, which are frequently shared by many ontologies). Since Konclude does not yet support datatypes, we removed all data

---

[4] http://bioportal.bioontology.org/

[5] http://ncit.nci.nih.gov/

[6] http://www.cs.ox.ac.uk/isg/ontologies/; We ignored repositories that are redundantly contained in the Oxford ontology library (e.g., the Gardiner ontology suite).

[7] http://owl.cs.manchester.ac.uk/repository/

[8] In order to avoid too many redundant ontologies, we only used those subsets of the OWLCorpus which were gathered with the crawlers OntoCrawler, OntoJCrawl, Swoogle, and Google.

properties and we replaced all data property restrictions with *owl:Thing* in all ontologies. Table 2 shows an overview of our obtained test corpus with overall 6,141 ontologies including statistics of ontology metrics for the source repositories. Please note that 34.9 % of all ontologies are not even in the OWL 2 DL profile, which is, however, mainly due to undeclared entities.

The evaluation was carried out on a Dell PowerEdge R420 server running with two Intel Xeon E5-2440 hexa core processors at 2.4 GHz with Hyper-Threading and 48 GB RAM under a 64bit Ubuntu 12.04.2 LTS. Our evaluation focuses on classification, which is a central reasoning task that is supported by many reasoners and, thus, it is ideal for the comparison of results. In principle, we only measured the wall clock time for classification, i.e., the times spent for parsing and loading ontologies as well as for writing classification output to files are not included. Each test was executed with a time limit of 5 minutes, but without any limitation of memory allocation. Although Konclude supports parallelisation, we only used one worker thread, which allows for a comparison independent of the number of CPU cores and facilitates the presentation of the improvements through saturation.

Table 3 shows a comparison of the accumulated classification times for the evaluated repositories (in seconds) between the following versions of Konclude:

- NONE, where none of the saturation optimisations are activated,
- NONE+RT, where only the transfer of results from the saturation into the completion graph (as presented in Section 4.1) is activated;
- NONE+SE, where only the extraction of subsumers from the saturation (as presented in Section 4.2) is activated;
- NONE+MM, where only the model merging with the saturation graph (as presented in Section 4.3) is activated;
- ALL−SI, where the saturation improvements (as presented in Section 5) are deactivated (i.e., all the saturation optimisations presented in Setion 4 are activated),
- ALL, where all saturation optimisations and saturation improvements are activated.

In addition, the column on the right side shows the performance gains (in percent) from NONE to the version ALL. Please note that the saturation improvements are optimisations to further improve the saturation procedure and, therefore, a separate evaluation of these techniques does not make sense.

It can be observed that the most significant improvements are achieved with the model merging optimisation (cf. NONE+MM), which is due to the large amount of NCI-Thesaurus ontologies in the NCIt archive, where this optimisation significantly reduces the classification effort. In contrast, if only the transfer of the saturation results (NONE+RT) or the extraction of subsumers from the saturation (NONE+SE) is activated, then only minor improvements with respect to the version NONE are possible. However, the combined activation of these optimisations (cf. ALL−SI) again leads to a significant performance gain for the repositories, which indicates that there is a synergy effect from the combination of these optimisation. Since all of these optimisations are based on the saturation procedure, which also requires a significant amount of processing time for large ontologies (approximately $1,953$ s for all repositories), this synergy effect is not very surprising. By further activating the saturation improvements (cf. ALL), we obtain another performance gain. Considering all repositories, the

Table 3. Accumulated classification times (in seconds) with separately activated saturation optimisations for the evaluated ontology repositories

Repository	NONE	NONE+RT	NONE+SE	NONE+MM	ALL–SI	ALL	↓ [%]
Gardiner	531	611	469	535	558	559	−5.2
NCBO BioPortal	2,071	1,947	971	2,156	988	793	61.7
NCIt	28,639	28,538	28,276	3,223	2,496	2,457	91.4
OBO Foundry	879	821	979	1,078	741	649	26.2
Oxford	6,623	5,006	6,012	6,510	3,429	2,743	58.6
TONES	1,756	1,456	1,413	494	321	337	80.8
Google	465	428	448	467	363	138	70.3
OntoCrawler	26	25	24	25	23	22	14.7
OntoJCrawl	1,417	923	715	1,427	517	548	61.4
Swoogle	2,501	2,502	2,493	1,402	1,248	1,343	46.3
ALL	44,910	42,256	41,800	17,317	10,684	9,589	78.6

combined activation of all saturation optimisations and improvements reduces the accumulated reasoning time by 78.6 %. It is also worth pointing out that the version NONE timed out for 128 ontologies, whereas the version ALL only reached the time limit for 10 ontologies. Thus, an evaluation with an increased time limit would show even better performance gains. For example, the Oxford ontology library contains the SCT-SEP ontology,[9] which can be classified by the version ALL in 181.2 s, whereas the version NONE requires 1709.4 s. SCT-SEP is a SNOMED extension that intensively uses disjunctions and disjointness and, thus, is clearly outside the OWL EL fragment. Nevertheless, large parts of the ontology have an EL structure and, therefore, our optimisations are able to improve the reasoning performance by almost one order of magnitude.

Table 3 also reveals that some saturation optimisations are not really relevant for some repositories. For instance, the activated result transfer yields worse reasoning times for the ontologies in the Gardiner ontology suite. Moreover, the model merging optimisation causes significant performance losses for some repositories (e.g., for OBO Foundry), which indicates that further optimisation is possible; e.g., one could learn statistics about the success of model merging with certain nodes in the saturation graph and automatically skip a merging test if there is a high likelihood that it will fail.

7 Conclusions

In this paper, we have presented a technique for tightly coupling saturation- and tableau-based procedures. Unlike standard consequence-based procedures, the approach is applicable on arbitrary OWL 2 DL ontologies. Furthermore, it has a very good pay-as-you-go behaviour, i.e., if only few axioms use features that are problematic for saturation-based procedures (e.g., disjunction), then the tableau procedure can still benefit significantly from the saturation. This seems to be confirmed by our evaluation over several thousand ontologies, where the integration of the presented saturation optimisations into the reasoning system Konclude significantly improves the classification performance.

---

[9] Originally from https://code.google.com/p/condor-reasoner/

References

1. Armas Romero, A., Cuenca Grau, B., Horrocks, I.: MORe: Modular combination of OWL reasoners for ontology classification. In: Cudré-Mauroux, P., Heflin, J., Sirin, E., Tudorache, T., Euzenat, J., Hauswirth, M., Parreira, J.X., Hendler, J., Schreiber, G., Bernstein, A., Blomqvist, E. (eds.) ISWC 2012, Part I. LNCS, vol. 7649, pp. 1–16. Springer, Heidelberg (2012)
2. Baader, F., Brandt, S., Lutz, C.: Pushing the $\mathcal{EL}$ envelope. In: Proc. 19th Int. Joint Conf. on Artificial Intelligence (IJCAI 2005), pp. 364–369. Professional Book Center (2005)
3. Baader, F., Calvanese, D., McGuinness, D., Nardi, D., Patel-Schneider, P. (eds.): The Description Logic Handbook: Theory, Implementation, and Applications, 2nd edn. Cambridge University Press (2007)
4. Gardiner, T., Horrocks, I., Tsarkov, D.: Automated benchmarking of description logic reasoners. In: Proc. 19th Int. Workshop on Description Logics (DL 2006), vol. 198. CEUR (2006)
5. Glimm, B., Horrocks, I., Motik, B., Shearer, R., Stoilos, G.: A novel approach to ontology classification. J. of Web Semantics 14, 84–101 (2012)
6. Horrocks, I., Kutz, O., Sattler, U.: The even more irresistible $\mathcal{SROIQ}$. In: Proc. 10th Int. Conf. on Principles of Knowledge Representation and Reasoning (KR 2006), pp. 57–67. AAAI Press (2006)
7. Horrocks, I., Tobies, S.: Reasoning with axioms: Theory and practice. In: Proc. 7th Int. Conf. on Principles of Knowledge Representation and Reasoning (KR 2000), pp. 285–296. Morgan Kaufmann (2000)
8. Hudek, A.K., Weddell, G.E.: Binary absorption in tableaux-based reasoning for description logics. In: Proc. 19th Int. Workshop on Description Logics (DL 2006), vol. 189. CEUR (2006)
9. Kazakov, Y.: $\mathcal{RIQ}$ and $\mathcal{SROIQ}$ are harder than $\mathcal{SHOIQ}$. In: Proc. 11th Int. Conf. on Principles of Knowledge Representation and Reasoning (KR 2008), pp. 274–284. AAAI Press (2008)
10. Kazakov, Y.: Consequence-driven reasoning for Horn-$\mathcal{SHIQ}$ ontologies. In: Proc. 21st Int. Conf. on Artificial Intelligence (IJCAI 2009), pp. 2040–2045. IJCAI (2009)
11. Matentzoglu, N., Bail, S., Parsia, B.: A corpus of OWL DL ontologies. In: Proc. 26th Int. Workshop on Description Logics (DL 2013), vol. 1014. CEUR (2013)
12. Simančík, F., Kazakov, Y., Horrocks, I.: Consequence-based reasoning beyond Horn ontologies. In: Proc. 22nd Int. Joint Conf. on Artificial Intelligence (IJCAI 2011), pp. 1093–1098. IJCAI/AAAI (2011)
13. Smith, B., Ashburner, M., Rosse, C., Bard, J., Bug, W., Ceusters, W., Goldberg, L.J., Eilbeck, K., Ireland, A., Mungall, C.J., The, O.B.I., Consortium, L.N., Rocca-Serra, P., Ruttenberg, A., Sansone, S.A., Scheuermann, R.H., Shah, N., Whetzeland, P.L., Lewis, S.: The OBO Foundry: coordinated evolution of ontologies to support biomedical data integration. Nature Biotechnology 25, 1251–1255 (2007)
14. Song, W., Spencer, B., Du, W.: WSReasoner: A prototype hybrid reasoner for $\mathcal{ALCHOI}$ ontology classification using a weakening and strengthening approach. In: Proc. 1st Int. Workshop on OWL Reasoner Evaluation (ORE 2012), vol. 858. CEUR (2012)
15. Steigmiller, A., Glimm, B., Liebig, T.: Nominal schema absorption. In: Proc. 23rd Int. Joint Conf. on Artificial Intelligence (IJCAI 2013), pp. 1104–1110. AAAI Press (2013)
16. Steigmiller, A., Glimm, B., Liebig, T.: Coupling tableau algorithms for the DL $\mathcal{SROIQ}$ with completion-based saturation procedures. Tech. Rep. UIB-2014-02, University of Ulm, Ulm, Germany (2014), http://www.uni-ulm.de/fileadmin/website_uni_ulm/iui/Ulmer_Informatik_Berichte/2014/UIB-2014-02.pdf
17. Steigmiller, A., Liebig, T., Glimm, B.: Konclude: system description. J. of Web Semantics (accepted, 2014)
18. Tsarkov, D., Horrocks, I., Patel-Schneider, P.F.: Optimizing terminological reasoning for expressive description logics. J. of Automated Reasoning 39, 277–316 (2007)
19. W3C OWL Working Group: OWL 2 Web Ontology Language: Document Overview. W3C Recommendation (October 27, 2009)

$\mathcal{EL}$-ifying Ontologies*

David Carral[1], Cristina Feier[2], Bernardo Cuenca Grau[2],
Pascal Hitzler[1], and Ian Horrocks[2]

[1] Department of Computer Science, Wright State University, Dayton US
[2] Department of Computer Science, University of Oxford, Oxford UK

Abstract. The OWL 2 profiles are fragments of the ontology language
OWL 2 for which standard reasoning tasks are feasible in polynomial
time. Many OWL ontologies, however, contain a typically small number
of out-of-profile axioms, which may have little or no influence on reasoning outcomes. We investigate techniques for rewriting axioms into the
EL and RL profiles of OWL 2. We have tested our techniques on both
classification and data reasoning tasks with encouraging results.

1 Introduction

Description Logics (DLs) are a family of knowledge representation formalisms
underpinning the W3C standard ontology languages OWL and OWL 2. State-
of-the-art DL reasoners such as Pellet [18], JFact, FaCT^{++} [21], RacerPro [9],
and HermiT [15] are highly-optimised for classification (i.e., the problem of computing all subsumption relationships between atomic concepts in an ontology)
and have been exploited successfully in many applications. In a recent large-
scale evaluation campaign, these reasoners exhibited excellent performance on a
corpus of more than 1,000 ontologies, as they were able to classify 75%-85% of
the corpus in less than 10 seconds when running on stock hardware [8,3].

However, notwithstanding extensive research into optimisation techniques, DL
reasoning remains a challenge in practice. Indeed, the aforementioned evaluation
also revealed that many ontologies are still hard for reasoners to classify. Furthermore, due to the high worst-case complexity of reasoning, systems are inherently
not robust, and even minor changes to ontologies can have a significant effect on
performance. Finally, the limitations of DL reasoners become even more apparent when reasoning with ontologies and large datasets.

These issues have motivated a growing interest in lightweight DLs: weaker
logics that enjoy more favourable computational properties. OWL 2 specifies
several profiles (language fragments) based on lightweight DLs [14]: OWL 2 EL
(or just EL) is based on the $\mathcal{EL}$ family of DLs; OWL 2 RL (or just RL) is based on
Datalog; and OWL 2 QL (or just QL) is based on DL-Lite. Standard reasoning
tasks, including classification and fact entailment (checking whether an ontology
and a dataset entail a given ground atom), are feasible in polynomial time for
all profiles, and many highly scalable reasoners have been developed [22,11,2,4].

* Work supported by the Royal Society, the EPSRC project Score! and the National
Science Foundation under the award TROn: Tractable Reasoning with Ontologies.

S. Demri, D. Kapur, and C. Weidenbach (Eds.): IJCAR 2014, LNAI 8562, pp. 464–479, 2014.
© Springer International Publishing Switzerland 2014

Unfortunately, many ontologies fall outside the OWL 2 profiles, and we are forced to resort to a fully-fledged reasoner if a completeness guarantee is required. Even in such cases, the majority of axioms typically still fall within one of the profiles, and the out-of-profile axioms may have little or no influence on the results of classification or query answering. Effectively detecting cases where the additional expressivity is used in a "harmless" way is, however, challenging, since even a single axiom can have a dramatic effect on reasoning outcomes.

In this paper we investigate techniques for rewriting out-of-profile axioms so as to improve reasoner performance. All rewritings are polynomial and preserve classification and fact entailment reasoning outcomes. In Section 3, we consider rewritings that are applicable to $\mathcal{SHOIQ}$—a DL that covers OWL DL and most of OWL 2 [10]—and that can transform non-EL axioms into EL by elimination of inverse roles and universal restrictions. If all non-EL axioms can be rewritten, then we can provide completeness guarantees using only an EL reasoner. Otherwise, the rewritings can still improve the performance of fully-fledged reasoners (e.g., by enabling the use of optimisation techniques that are applicable only in the absence of certain constructs) and/or the effectiveness of modular reasoners that combine profile-specific with OWL 2 reasoners, such as MORe [1].

In Section 4, we focus on Horn ontologies and consider rewritings into OWL 2 RL. The RL profile is tightly connected to Datalog, and hence existential restrictions $\exists R.C$ occurring positively in axioms are disallowed, unless C is a singleton nominal $\{o\}$. We show that when R fulfills certain conditions, such concepts $\exists R.C$ can be rewritten into existential restrictions over nominals as accepted in OWL 2 RL; we call such roles R *reuse-safe*. In the limit case where all roles are reuse-safe, the ontology can be polynomially rewritten into RL; if, additionally, the ontology contains no cardinality constraints, it can also be rewritten into EL. Furthermore, if only some roles are reuse-safe, they can be treated by (hyper-)tableau reasoners in an optimised way, potentially reducing the size of the constructed pre-models and improving reasoning times.

We have implemented our rewriting techniques and evaluated their effect on reasoning times over a large repository of ontologies. Our experiments reveal that our EL-ification techniques can lead to substantial improvements in classification times for both standard and modular reasoners. Furthermore, we show that many ontologies contain only reuse-safe roles and hence can be rewritten into RL; thus, highly scalable RL triple stores can be exploited for large-scale data reasoning.

This paper is accompanied by an online technical report.[1]

2   Preliminaries

A *signature* consists of disjoint countably infinite sets of *individuals* N_I, *atomic concepts* N_C and *atomic roles* N_R. A *role* is an element of $N_R \cup \{R^- | R \in N_R\}$. The function $\mathsf{Inv}(\cdot)$ is defined over the set of roles as follows, where $R \in N_R$: $\mathsf{Inv}(R) = R^-$ and $\mathsf{Inv}(R^-) = R$. An *RBox* $\mathcal{R}$ is a finite set of *RIAs* $R \sqsubseteq R'$ and *transitivity axioms* $\mathsf{Tra}(R)$, with R and R' roles. We denote with $\sqsubseteq_\mathcal{R}$ the minimal

---

[1] http://www.cs.ox.ac.uk/isg/TR/safeshoiq.pdf

relation over roles in $\mathcal{R}$ s.t. $R \sqsubseteq_{\mathcal{R}} S$ and $\mathsf{Inv}(R) \sqsubseteq_{\mathcal{R}} \mathsf{Inv}(S)$ hold if $R \sqsubseteq S \in \mathcal{R}$. We define $\sqsubseteq_{\mathcal{R}}^{*}$ as the reflexive-transitive closure of $\sqsubseteq_{\mathcal{R}}$. A role R is *transitive* in $\mathcal{R}$ if there is a role S such that $S \sqsubseteq_{\mathcal{R}}^{*} R$, $R \sqsubseteq_{\mathcal{R}}^{*} S$ and either $\mathsf{Tra}(S) \in \mathcal{R}$ or $\mathsf{Tra}(\mathsf{Inv}(S)) \in \mathcal{R}$. A role R is *simple* in $\mathcal{R}$ if no transitive role S exists s.t. $S \sqsubseteq_{\mathcal{R}}^{*} R$. The set of $\mathcal{SHOIQ}$ *concepts* is the smallest set containing A (atomic concept), $\top$ (top), $\bot$ (bottom), $\{o\}$ (nominal), $\neg C$ (negation), $C \sqcap D$ (conjunction), $C \sqcup D$ (disjunction), $\exists R.C$ (existential restriction), $\forall R.C$ (universal restriction), $\leqslant nS.C$ (at-most restriction), and $\geqslant nR.C$ (at-least restriction), for $A \in N_\mathsf{C}$, C and D $\mathcal{SHOIQ}$ concepts, $o \in N_\mathsf{I}$, R a role and S a simple role, and n a nonnegative integer. A *literal concept* is either atomic or the negation of an atomic concept. A *TBox* $\mathcal{T}$ is a finite set of GCIs $C \sqsubseteq D$ with C, D concepts. An *ABox* $\mathcal{A}$ is a finite set of assertions $C(a)$ (concept assertion), $R(a, b)$ (role assertion), $a \approx b$ (equality assertion), and $a \not\approx b$ (inequality assertion), with C a concept, R a role and a, b individuals. A *fact* is either a concept assertion $A(a)$ with A atomic, a role assertion, an equality assertion, or an inequality assertion. A knowledge base is a triple $\mathcal{K} = (\mathcal{R}, \mathcal{T}, \mathcal{A})$. The semantics is standard [10].

We assume familiarity with standard conventions for naming DLs, and we just provide here a definition of the OWL 2 profiles. A $\mathcal{SHOIQ}$ KB is:

- EL if *(i)* it does not contain inverse roles, negation, disjunction, at-most restrictions and at-least restrictions; and *(ii)* every universal restriction appears in a GCI of the form $\top \sqsubseteq \forall R.C$.
- RL if each GCI $C \sqsubseteq D$ satisfies *(i)* C does not contain negation as well as universal, at-least, and at-most restrictions; *(ii)* D does not contain negation (other than $\bot$), union, existential restrictions (other than of the form $\exists R.\{o\}$), at-least restrictions, and at-most restrictions with $n > 1$.
- QL if it does not contain transitivity and for each GCI $C \sqsubseteq D$ *(i)* C is either atomic or $\exists R.\top$; *(ii)* D is of the form $\prod_{i=1}^{n} B_i$ with each B_i either a literal concept, or $\bot$, or of the form $\exists R.A$ with R a role and A either atomic or $\top$.

Classification of $\mathcal{K}$ is the task of computing all subsumptions $\mathcal{K} \models A \sqsubseteq B$ with $A \in N_\mathsf{C} \cup \{\top\}$, and $B \in N_\mathsf{C} \cup \{\bot\}$. Fact entailment is to check whether $\mathcal{K} \models \alpha$, for α a fact. Both problems are reducible to knowledge base unsatisfiability.

3   EL-ification of $\mathcal{SHOIQ}$ Ontologies

In this section, we propose techniques for transforming non-EL axioms into EL. Whenever possible, inverse roles are replaced with fresh symbols and the knowledge base is extended with axioms simulating their possible effects. At the same time, we attempt to transform positive occurrences of universal restrictions into negative occurrences of existential restrictions while inverting the relevant role. Note that our techniques do not rewrite disjunctions and cardinality restrictions; thus, ontologies containing such constructs will not be fully rewritten into EL.

3.1   Preprocessing

Before attempting to rewrite a $\mathcal{SHOIQ}$ knowledge base $\mathcal{K}$ into EL, we first bring $\mathcal{K}$ into a suitable normal form. Normalisation facilitates further rewriting

$$\Theta(\mathcal{T}) = \bigcup_{\alpha \in \mathcal{T}} \Theta(\alpha)$$

$$\Theta(\mathbb{C} \sqsubseteq \mathbb{D} \sqcup \forall R.B) = \Theta(\mathbb{C} \sqsubseteq \mathbb{D} \sqcup \alpha_B) \cup \{\alpha_B \sqsubseteq \forall R.B\}$$

$$\Theta(\mathbb{C} \sqsubseteq \mathbb{D} \sqcup \forall R.\neg B) = \Theta(\mathbb{C} \sqcap \alpha_B \sqsubseteq \mathbb{D}) \cup \{\exists R.B \sqsubseteq \alpha_B\}$$

$$\Theta(\mathbb{C} \sqsubseteq \mathbb{D} \sqcup \bowtie nR.B) = \Theta(\mathbb{C} \sqsubseteq \mathbb{D} \sqcup \alpha_B) \cup \{\alpha_B \sqsubseteq \bowtie nR.B\}$$

$$\Theta(\mathbb{C} \sqsubseteq \mathbb{D} \sqcup \geqslant nR.\neg B) = \Theta(\mathbb{C} \sqsubseteq \mathbb{D} \sqcup \geqslant nR.\alpha_B) \cup \{\alpha_B \sqcap B \sqsubseteq \top\}$$

$$\Theta(\mathbb{C} \sqsubseteq \mathbb{D} \sqcup \leqslant nR.\neg B) = \Theta(\mathbb{C} \sqsubseteq \mathbb{D} \sqcup \leqslant nR.\alpha_B) \cup \{\top \sqsubseteq \alpha_B \sqcup B\}$$

$$\Theta(\mathbb{C} \sqsubseteq \mathbb{D} \sqcup \neg B) = \Theta(\mathbb{C} \sqcap B \sqsubseteq \mathbb{D})$$

$$\Theta(\alpha) = \alpha \text{ for any other axiom } \alpha.$$

Fig. 1. $\mathbb{C}$ is a conjunction of atomic concepts or $\top$, $\mathbb{D}$ is a disjunction of concepts C, $\forall R.C, \bowtie nR.C$ ($\bowtie \in \{\leqslant, \geqslant\}$) or $\bot$, with C literal, B atomic, and α_B is fresh

steps, and it allows us to identify axioms with a direct correspondence in EL. For example, $A \sqcup B \sqsubseteq \neg\forall R.\neg C$ is equivalent to the EL axioms $A \sqsubseteq \exists R.C$ and $B \sqsubseteq \exists R.C$. Furthermore, although $A \sqsubseteq \exists R.\neg B$ is not equivalent to an EL axiom, it can be trivially transformed into the EL axioms $A \sqsubseteq \exists R.X$ and $X \sqcap B \sqsubseteq \bot$ by introducing a fresh symbol X. We therefore introduce a normal form that makes explicit those axioms that are neither logically equivalent to EL axioms, nor can be transformed into EL by means of the trivial introduction of fresh symbols.

Definition 1. *A GCI is normalised if it is of either of the following forms, where each $A_{(i)}$ is atomic or $\top$, B is atomic, each $C_{(j)}$ is atomic, $\bot$, or a nominal, R is a role, $n \geqslant 2$, and $m \geqslant 1$:*

(N1) $\displaystyle\prod_{i=1}^{n} A_i \sqsubseteq \bigsqcup_{j=1}^{m} C_j;$ **(N2)** $A \sqsubseteq \exists R.A_i;$ **(N3)** $\exists R.A \sqsubseteq A_i$

(N4) $A \sqsubseteq \geqslant n\,R.A_i;$ **(N5)** $A \sqsubseteq \forall R.B;$ **(N6)** $A \sqsubseteq \leqslant m\,R.A_i$

A knowledge base $\mathcal{K} = (\mathcal{R}, \mathcal{T}, \mathcal{A})$ is normalised if $\mathcal{A}$ has only facts and each GCI in $\mathcal{T}$ is normalised. Finally, $\mathcal{K}$ is Horn if $m = 1$ in each axiom N1 or N6.

Note that axioms of type **N2** and **N3**, as well as Horn axioms of type **N1**, are EL. To normalise a knowledge base $\mathcal{K}$, we proceed in two steps. First, we translate $\mathcal{K}$ into the following disjunctive normal form [15].

Definition 2. *A GCI is in disjunctive normal form (DNF) if it is of the form $\top \sqsubseteq \bigsqcup_{i=1}^{n} C_i$, where each C_i is of the form B, $\{o\}$, $\exists R.B$, $\forall R.B$, $\geqslant n\,R.B$, or $\leqslant n\,R.B$, for B a literal concept, R a role, and n a nonnegative integer. A knowledge base $\mathcal{K} = (\mathcal{R}, \mathcal{T}, \mathcal{A})$ is in DNF if all roles in $\mathcal{A}$ are atomic, all concept assertions in $\mathcal{A}$ contain only a literal concept, and each GCI in $\mathcal{T}$ is in DNF.*

DNF normalisation can be seen as a variant of the structural transformation, in which all complex concepts are "flattened" and negations are made explicit (see [15] for details). Once $\mathcal{K}$ is in DNF, we can further normalise by replacing concepts $\neg B$ in restrictions $\forall R.\neg B$, $\exists R.\neg B$, $\geqslant n\,R.\neg B$ and $\leqslant n\,R.\neg B$ with fresh symbols, bringing the remaining negated concepts to the left in GCIs, and introducing fresh symbols for all restrictions occurring in disjunctions.

Definition 3. *Let $\mathcal{K}$ be a KB. Then, $\Upsilon(\mathcal{K})$ is computed from $\mathcal{K}$ as follows:* (i) *apply the transformation in [15] to obtain $\mathcal{K}' = (\mathcal{R}', \mathcal{T}', \mathcal{A}')$ in DNF;* (ii) *replace each assertion $\alpha = \neg A(a)$ in $\mathcal{A}'$ with a fact $X_\alpha(a)$, with X_α fresh, and extend $\mathcal{T}'$ with $X_\alpha \sqcap A \sqsubseteq \bot$; and* (iii) *apply to $\mathcal{T}'$ the transformation Θ in Figure 1.*

The following proposition establishes the properties of normalisation.

Proposition 1. *Let $\mathcal{K}$ be a KB, then $\Upsilon(\mathcal{K})$ is normalised and can be computed in polynomial time in the size of $\mathcal{K}$. Furthermore, if $\mathcal{K}$ is EL, then so is $\Upsilon(\mathcal{K})$. Finally, $\mathcal{K}$ is satisfiable iff $\Upsilon(\mathcal{K})$ is satisfiable.*

3.2   Rewritable Inverse Roles

Satisfiability of $\mathcal{SHOIQ}$ KBs is NExpTime-complete, whereas for $\mathcal{SHOQ}$ it is ExpTime-complete; thus, in general, inverse roles cannot be faithfully eliminated from $\mathcal{SHOIQ}$ KBs by means of a polynomial transformation. The following example illustrates that an obstacle to rewritability is the interaction between inverses and at-most restrictions.

Example 1. Consider $\mathcal{K} = (\mathcal{R}, \mathcal{T}, \mathcal{A})$, with $\mathcal{R} = \emptyset$, $\mathcal{A} = \{A(a)\}$, and $\mathcal{T}$ as follows:

$$\mathcal{T} = \{A \sqsubseteq \exists R^-.B; \quad B \sqsubseteq \exists R.C; \quad B \sqsubseteq \,\leqslant 1\,R.\top\}$$

Note that $\mathcal{K} \models C(a)$. In every model $(\Delta_\mathcal{I}, \cdot^\mathcal{I})$, object $a^\mathcal{I}$ must be R^--connected to some $x \in B^\mathcal{I}$ (due to the first axiom in $\mathcal{T}$); also, x must be R-connected to some $y \in C^\mathcal{I}$ (due to the second axiom). Then, for the last axiom to be satisfied, $a^\mathcal{I}$ and y must be identical; thus, $a^\mathcal{I} \in C^\mathcal{I}$. Figure 2 a) depicts such a model. Consider now $\mathcal{K}'$ obtained from $\mathcal{K}$ by replacing R^- with a fresh atomic role N_{R^-}. Then, $\mathcal{K}' \not\models C(a)$, and Figure 2 b) depicts a model of $\mathcal{K}'$ not satisfying $C(a)$. Extending $\mathcal{K}'$ with EL axioms to simulate the interaction between inverses and cardinality restrictions (and thus recover the missing entailment) seems infeasible. $\Diamond$

   We next propose sufficient conditions for inverse roles to be rewritable in the presence of cardinality constraints. Our conditions ensure existence of a one-to-one correspondence between the *canonical* forest-shaped models of the original and rewritten KBs, and hence disallow cases such as Example 2.[2]

Definition 4. *Let $\mathcal{K} = (\mathcal{R}, \mathcal{T}, \mathcal{A})$ be a normalised $\mathcal{SHOIQ}$ knowledge base. A (possibly inverse) role R is generating in $\mathcal{K}$ if there exists a role R' occurring in $\mathcal{T}$ in an axiom of type $N2$ or $N4$ such that $R' \sqsubseteq^*_\mathcal{R} R$.*

---

[2] Roughly speaking, a forest-shaped model of a (normalised) knowledge base is canonical if every fact that holds in the model is "justified" by an axiom or assertion in the knowledge base. In particular, the result of unravelling a pre-model constructed by a (hyper-)tableau algorithm is a canonical forest-shaped model.

Fig. 2. A situation where rewriting away inverse roles leads to missing entailments

An inverse role S^- is *rewritable* if for each $X \in \{S, S^-\}$ occurring in an axiom of type **N6** we have that $\mathsf{Inv}(X)$ is not generating in $\mathcal{K}$.

Intuitively, roles R' in axioms **N2** or **N4** are those "inducing" the edges between individuals and their successors in a canonical model; then, a role R is generating if it is a super-role of one such R'. Our condition ensures that "backwards" edges in a canonical model of $\mathcal{K}$ (i.e., those induced by an inverse role) cannot invalidate an at-most cardinality restriction. In the limit case where all inverse roles in a $\mathcal{SHOIQ}$ KB are rewritable, we can faithfully eliminate inverses and rewrite the KB into $\mathcal{SHOQ}$ by means of a polynomial transformation.

Theorem 1. *Let $\mathcal{C}$ be the class of all normalised $\mathcal{SHOIQ}$ ontologies containing only rewritable inverse roles. Then, there exists a polynomial transformation mapping each $\mathcal{K} \in \mathcal{C}$ to an equisatisfiable $\mathcal{SHOQ}$ knowledge base.*[3]

Theorem 1 identifies a class of $\mathcal{SHOIQ}$ ontologies for which standard reasoning is feasible in ExpTime (in contrast to NExpTime). This result can also be exploited for optimisation: tableaux reasoners employ pairwise blocking techniques over $\mathcal{SHOIQ}$ ontologies, while they rely on more aggressive single blocking techniques for $\mathcal{SHOQ}$ inputs, which can reduce the size of pre-models.

3.3 The EL-ification Transformation

Before presenting our transformation formally, we provide two motivating examples. First, we show how a rewritable inverse role can be eliminated in the presence of cardinality constraints.

Example 2. Let $\mathcal{K} = (\mathcal{R}, \mathcal{T}, \mathcal{A})$ be the following knowledge base:

$\mathcal{R} = \{R \sqsubseteq T^-; \ \ S \sqsubseteq T^-\}$
$\mathcal{T} = \{A \sqsubseteq \exists R.B; \ \ A \sqsubseteq \exists S.C; \ \ A \sqsubseteq {\leqslant} 1\,T^-.\top; \ \ B \sqcap C \sqsubseteq D; \ \ \exists R.D \sqsubseteq B\}$
$\mathcal{A} = \{A(a); \ \ T(b, a)\}$

Figure 3(a) depicts a canonical model for $\mathcal{K}$. The facts entailed by $\mathcal{K}$ are precisely those that hold in the canonical model. By Definition 4, T^- is rewritable

---

[3] Theorem 1 is given here for presentation purposes: it follows as a corollary of Theorem 3, which we state only after presenting our transformations.

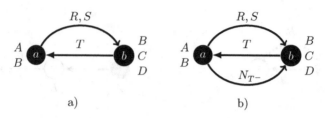

Fig. 3. Rewriting away inverse roles in a KB with cardinality constraints

since T is not generating; however, it does not suffice to replace T^- with a fresh N_{T^-} since the resulting KB will no longer entail the facts $R(a,b)$, $S(a,b)$, $B(b)$, $C(b)$, and $D(b)$. Instead, we can extend $\mathcal{A}$ with $T^-(a,b)$, and only then replace T^- with N_{T^-}. The canonical model of the resulting KB is given in Figure 3(b).◇

Next, we show how axioms of type **N5**, which involve a universal restriction, can be replaced with EL axioms of type **N3** if the relevant roles are not generating.

Example 3. Consider $\mathcal{K} = (\mathcal{R}, \mathcal{T}, \mathcal{A})$ where $\mathcal{R} = \{R \sqsubseteq S^-\}$, $\mathcal{A} = \{A(a); S(a,b)\}$, and $\mathcal{T}$ is defined as follows:

$$\mathcal{T} = \{A \sqsubseteq \forall S.B; \quad B \sqsubseteq \exists R.C; \quad \exists S.B \sqsubseteq D; \quad C \sqcap D \sqsubseteq \bot\}$$

Clearly, $\mathcal{K}$ is unsatisfiable. Furthermore, it does not contain axioms **N6**, and hence S^- is rewritable. In a first step, we extend $\mathcal{K}$ with logically redundant axioms, which make explicit information that may be lost when replacing inverses with fresh symbols. Thus, we extend $\mathcal{T}$ with $\exists S^-.A \sqsubseteq B$, and $B \sqsubseteq \forall S^-.D$; furthermore, we extend $\mathcal{R}$ with $R^- \sqsubseteq S$; and finally, $\mathcal{A}$ with the assertion $S^-(b,a)$.

An important observation is that S is not generating. As a result, we can dispense with axiom $A \sqsubseteq \forall S.B$. Then we replace S^- with a fresh symbol N_{S^-} and R^- with N_{R^-}. The resulting $\mathcal{K}' = (\mathcal{R}', \mathcal{T}', \mathcal{A}')$ is as follows:

$$\mathcal{R}' = \{R \sqsubseteq N_{S^-}; \quad N_{R^-} \sqsubseteq S\}$$
$$\mathcal{T}' = \{\exists N_{S^-}.A \sqsubseteq B; \quad B \sqsubseteq \exists R.C; \quad \exists S.B \sqsubseteq D; \quad B \sqsubseteq \forall N_{S^-}.D; \quad C \sqcap D \sqsubseteq \bot\}$$
$$\mathcal{A}' = \{A(a); \quad S(a,b); \quad N_{S^-}(b,a)\}$$

$\mathcal{K}'$ is unsatisfiable; furthermore it is in EL except for axiom $B \sqsubseteq \forall N_{S^-}.D$. This axiom cannot be dispensed with since S^- is generating, and hence it is needed to propagate information along N_{S^-}-edges in a canonical model.   ◇

We next present our transformation. For simplicity, we first restrict ourselves to $\mathcal{ALCHOIQ}$ KBs; later on, we discuss issues associated with transitivity axioms and show how our techniques extend to $\mathcal{SHOIQ}$.

Definition 5. *Let $\mathcal{K} = (\mathcal{R}, \mathcal{T}, \mathcal{A})$ be a normalised $\mathcal{ALCHOIQ}$ knowledge base. The knowledge base $\Xi(\mathcal{K}) = (\mathcal{R}', \mathcal{T}', \mathcal{A}')$ is obtained as follows:*

1. Extension: *the knowledge base* $\mathcal{K}_e = (\mathcal{R}_e, \mathcal{T}_e, \mathcal{A}_e)$ *is defined as follows:*
   - $\mathcal{R}_e$ *extends* $\mathcal{R}$ *with an axiom* $\mathsf{Inv}(R) \sqsubseteq \mathsf{Inv}(S)$ *for each* $R \sqsubseteq S$ *in* $\mathcal{R}$;
   - $\mathcal{T}_e$ *extends* $\mathcal{T}$ *with the following axioms:*
     - *an axiom* $\exists\mathsf{Inv}(R).A \sqsubseteq B$ *for each axiom* $A \sqsubseteq \forall R.B$ *in* $\mathcal{T}$ *where either* $\mathsf{Inv}(R)$ *is generating, or* R *is not generating; and*
     - *an axiom* $A \sqsubseteq \forall\mathsf{Inv}(R).B$ *for each axiom* $\exists R.A \sqsubseteq B$ *in* $\mathcal{T}$ *where* $\mathsf{Inv}(R)$ *is generating;*
   - $\mathcal{A}_e$ *extends* $\mathcal{A}$ *with an assertion* $R^-(b,a)$ *for each* $R(a,b) \in \mathcal{A}$.
2. $\mathcal{EL}$-ification: $\Xi(\mathcal{K}) = (\mathcal{R}', \mathcal{T}', \mathcal{A}')$ *is obtained from* $\mathcal{K}_e$ *by first removing all axioms* $A \sqsubseteq \forall R.B$ *in* $\mathcal{T}_e$ *where* R *is not generating in* $\mathcal{T}$ *and then replacing each occurrence of an inverse role that is rewritable in* $\mathcal{K}_e$ *with a fresh role.*

The extension step only adds redundant information, and hence $\mathcal{K}$ and $\mathcal{K}_e$ are equivalent. Making such information explicit is crucial for the subsequent EL-ification step, where ineffectual axioms involving universal restrictions are removed, and rewritable inverse roles are replaced with fresh atomic roles. The following theorem extablishes the properties of the transformation.

Theorem 2. *Let* $\mathcal{K}' = \Xi(\mathcal{K})$. *The following conditions hold:*

1. $\mathcal{K}'$ *is satisfiable iff* $\mathcal{K}$ *is satisfiable;*
2. $\mathcal{K}'$ *is of size polynomial in the size of* $\mathcal{K}$;
3. *If* $\mathcal{K}$ *satisfies all of the following properties, then* $\mathcal{K}'$ *is EL:*
   - $\mathcal{K}$ *is Horn and does not contain axioms of type* **N4** *or* **N6**;
   - *each axiom* **N5** *satisfies either* $A = \top$, *or* R *is not generating.*
   - *each axiom* **N3** *satisfies either* $A = \top$, *or* $\mathsf{Inv}(R)$ *is not generating.*

Note that the third condition in the theorem establishes sufficient conditions on $\mathcal{K}$ for the transformed knowledge base $\mathcal{K}'$ to be in EL. A simple case is when $\mathcal{K}$ is in the QL profile of OWL 2, in which case the transformed KB is guaranteed to be in EL. An interesting consequence of this result is that highly optimised EL reasoners, such as ELK, can be exploited for classifying QL ontologies.

Corollary 1. *If* $\mathcal{K}$ *is a normalised QL knowledge base, then* $\Xi(\mathcal{K})$ *is in EL.*

In many cases our transformation may only succeed in partially rewriting a knowledge base into EL (c.f. Example 3). Even in these cases, our techniques can have substantial practical benefits (see Evaluation section). As discussed in Section 3.2, in the absence of inverse roles (hyper-)tableau reasoners may exploit more aggressive blocking techniques. Furthermore, modular reasoning systems such as MORe, which are designed to behave better for ontologies with a large EL subset, are substantially enhanced by our transformations.

3.4   Dealing with Transitivity Axioms

As shown by the following example, the transformation in Definition 5 is not applicable to knowledge bases containing transitivity axioms in the RBox.

Example 4. Consider $\mathcal{K} = (\mathcal{R}, \mathcal{T}, \mathcal{A})$ with $\mathcal{R} = \{R \sqsubseteq R^-; \mathsf{Tra}(R)\}$, $\mathcal{A} = \{A(a)\}$, and $\mathcal{T} = \{A \sqsubseteq \exists R.B; A \sqsubseteq C; \exists R^-.C \sqsubseteq D\}$. Let $\mathcal{K}' = \Xi(\mathcal{K})$, where we assume that the transitivity axiom $\mathsf{Tra}(R)$ stays unmodified in $\mathcal{K}'$. More precisely, $\mathcal{A}' = \mathcal{A}$, and $\mathcal{R}' = \{R \sqsubseteq N_{R^-}; N_{R^-} \sqsubseteq R; \mathsf{Tra}(R)\}$, and $\mathcal{T}' = \{A \sqsubseteq \exists R.B; A \sqsubseteq C; \exists N_{R^-}.C \sqsubseteq D; C \sqsubseteq \forall R.D\}$. It can be checked that $\mathcal{K} \models D(a)$, but $\mathcal{K}' \not\models D(a)$; thus, a relevant entailment is lost. An attempt to recover this entailment by making N_{R^-} transitive does not solve the problem.   ◇

To address this issue, we eliminate transitivity before applying our transformation in Definition 5. Standard techniques for eliminating transitivity axioms in DLs (e.g., [15]) have the effect of introducing non-Horn axioms. As a result, a Horn knowledge base may not remain Horn after eliminating transitivity. Therefore, we propose a modification of the standard technique that preserves Horn axioms and which is compatible with our transformation in Definition 5.

Definition 6. *Let $\mathcal{K} = (\mathcal{R}, \mathcal{T}, \mathcal{A})$ be a normalised $\mathcal{SHOIQ}$ knowledge base. For each axiom of the form $A \sqsubseteq \forall R.B$ in $\mathcal{T}$ and each transitive sub-role S of R in $\mathcal{R}$, let $X^S_{R,B}$ be an atomic concept uniquely associated to R, B, S. Furthermore, for each axiom $\exists R.A \sqsubseteq B$ in $\mathcal{T}$ and each transitive sub-role S of R in $\mathcal{R}$, let $Y^S_{R,B}$ be a fresh atomic concept uniquely associated to R, B, S.*

The knowledge base $\Omega(\mathcal{K}) = (\mathcal{R}', \mathcal{T}', \mathcal{A}')$ is defined as follows: (i) $\mathcal{R}'$ is obtained from $\mathcal{R}$ by removing all transitivity axioms; (ii) $\mathcal{T}'$ is obtained from $\mathcal{T}$ by adding axioms $A \sqsubseteq \forall S.X^S_{R,B}$, $X^S_{R,B} \sqsubseteq \forall S.X^S_{R,B}$, and $X^S_{R,B} \sqsubseteq \forall S.B$ for each concept $X^S_{R,B}$, and axioms $\exists S.A \sqsubseteq Y^S_{R,B}$, $\exists S.Y^S_{R,B} \sqsubseteq Y^S_{R,B}$, and $\exists S.Y^S_{R,B} \sqsubseteq B$ for each concept $Y^S_{R,B}$; finally, (iii) $\mathcal{A}' = \mathcal{A}$.

Lemma 1 establishes the properties of transitivity elimination, and Theorem 3 shows that our techniques extend to a $\mathcal{SHOIQ}$ knowledge base $\mathcal{K}$ by first applying Ω to $\mathcal{K}$ and then Ξ to the resulting KB.

Lemma 1. *Let $\mathcal{K}$ be a normalised $\mathcal{SHOIQ}$ KB. The following holds:*

1. *$\Omega(\mathcal{K})$ is satisfiable iff $\mathcal{K}$ is .*
2. *$\Omega(\mathcal{K})$ is a normalised $\mathcal{ALCHOIQ}$; furthermore, $\Omega(\mathcal{K})$ is Horn iff $\mathcal{K}$ is Horn.*
3. *$\Omega(\mathcal{K})$ can be computed in time polynomial in the size of $\mathcal{K}$.*
4. *if $\mathcal{K}$ is EL, then so is $\Omega(\mathcal{K})$.*
5. *If an inverse role R^- is rewritable in $\mathcal{K}$, then it is also rewritable in $\Omega(\mathcal{K})$.*

Theorem 3. *Let $\mathcal{K} = (\mathcal{R}, \mathcal{T}, \mathcal{A})$ be a normalised $\mathcal{SHOIQ}$ knowledge base, and let $\mathcal{K}' = \Xi(\Omega(\mathcal{K}))$. Then, $\mathcal{K}'$ satisfies all properties $1 - 3$ in Theorem 2.*

4   Reuse-Safe Roles

We next focus on Horn ontologies, and show how to further optimise reasoning by identifying roles that are "reuse-safe", and which can thus be treated by (hyper-)tableau reasoners in a more optimised way. Each application of an axiom **N2** or **N4** triggers the generation of fresh individuals in a (hyper-)tableau. If these

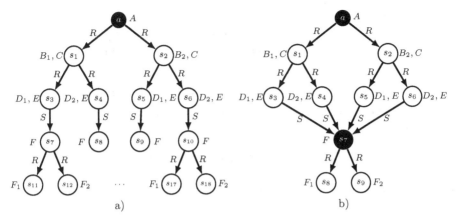

a)   b)

Fig. 4. Decreasing model size by reusing individuals

axioms involve a reuse-safe role, however, we show that reasoners can associate with each such axiom a single fresh nominal, which can be deterministically "reused" whenever the axiom is applied during construction of a pre-model. This may reduce the size of pre-models, and improve reasoning times. Our technique extends the results in [16], which show that for EL ontologies all roles admit reuse, and pre-model size can be bounded polynomially.

Example 5. Consider the following knowledge base $\mathcal{K} = (\mathcal{R}, \mathcal{T}, \mathcal{A})$ where $\mathcal{R} = \emptyset$, $\mathcal{A} = \{A(a)\}$, and $\mathcal{T}$ consists of the following axioms:

$$A \sqsubseteq \exists R.B_1 \quad C \sqsubseteq \exists R.D_1 \quad A \sqsubseteq \forall R.C \quad E \sqsubseteq \exists S.F \quad F \sqsubseteq \exists R.F_2 \quad B_1 \sqcap B_2 \sqsubseteq \bot$$
$$A \sqsubseteq \exists R.B_2 \quad C \sqsubseteq \exists R.D_2 \quad C \sqsubseteq \forall R.E \quad F \sqsubseteq \exists R.F_1 \quad F_1 \sqcap F_2 \sqsubseteq \bot \quad D_1 \sqcap D_2 \sqsubseteq \bot$$

Since R is generating and $\mathcal{K}$ has no inverses, we have $\Xi(\mathcal{K}) = \mathcal{K}$. Figure 4 a) depicts a canonical model of $\mathcal{K}$. Role S is reuse-safe since it is not "affected" by non-EL axioms involving universal restrictions. We can exploit this fact to "fold" the model by identifying all nodes with an S-predecessor to a single fresh nominal, as in Figure 4 b). In this way, we can reduce model size. $\Diamond$

Definition 7. *Let $\mathcal{K} = (\mathcal{R}, \mathcal{T}, \mathcal{A})$ be a normalised Horn KB. A role R in $\mathcal{K}$ is reuse-safe if either R is not generating or the following conditions hold:*

- *Each axiom $A \sqsubseteq \,\leqslant 1\,S.B$ in $\mathcal{K}$ satisfies $R \not\sqsubseteq_{\mathcal{R}}^* S$ and $R \not\sqsubseteq_{\mathcal{R}}^* \mathsf{Inv}(S)$;*
- *Each axiom $A \sqsubseteq \forall S.B$ in $\mathcal{K}$ with $A \neq \top$ satisfies $R \not\sqsubseteq_{\mathcal{R}}^* S$; and*
- *Each axiom $\exists S.A \sqsubseteq B$ in $\mathcal{K}$ with $A \neq \top$ satisfies $R \not\sqsubseteq_{\mathcal{R}}^* \mathsf{Inv}(S)$.*

If a generating role R is reuse-safe, we can ensure that R-edges in a canonical model of $\mathcal{K}$ are irrelevant to the satisfaction of non-EL axioms in $\mathcal{K}$. To ensure that (hyper-)tableau algorithms will exploit reuse-safety, and construct succinct "folded" canonical models such as the one in Example 5, we provide the following transformation, which makes the relevant nominals explicit.

Definition 8. Let $\mathcal{K} = (\mathcal{R}, \mathcal{T}, \mathcal{A})$ be a normalised Horn knowledge base.

For each each positive occurrence of a concept $\exists R.B$ (resp. $\geqslant nR.B$) in $\mathcal{K}$ with R reuse-safe, let $c_{R,B}$ (resp. $c_{i,R,B}$ for $1 \leqslant i \leqslant n$) be fresh individual(s). Then, $\Psi(\mathcal{K})$ is KB obtained from $\mathcal{K}$ by:

- replacing each axiom in $\mathcal{T}$ of the form $A \sqsubseteq \exists R.B$, where R is safe, by $A \sqsubseteq \exists R.\{c_{R,B}\}$ and adding the fact $B(c_{R,B})$ to $\mathcal{A}$, and by
- replacing each axiom of the form $A \sqsubseteq \geqslant nR.B$, where R is safe, by all $\alpha \in \{A \sqsubseteq \exists R.\{c_{i,R,B}\}, \{c_{j,R,B}\} \sqcap \{c_{k,R,B}\} \sqsubseteq \bot \mid 1 \leqslant i \leqslant n \text{ and } 1 \leqslant j < k \leqslant n\}$ and adding the facts $B(c_{i,R,B})$, for $1 \leqslant i \leqslant n$, to $\mathcal{A}$.

The following theorem establishes the correctness of our transformation.

Theorem 4. $\mathcal{K}$ is satisfiable iff $\Psi(\mathcal{K})$ is satisfiable.

In practice, system developers can achieve the same goal as our transformation by making their implementations sensitive to reuse-safe roles: to satisfy an axiom involving existential or an at-least restrictions over such role, a system should reuse a suitable distinguished individual instead of generating a fresh one.

We next analyse the case where all roles in a Horn KB $\mathcal{K} = (\mathcal{R}, \mathcal{T}, \mathcal{A})$ are reuse-safe. In this case, we can show that $\Psi(\mathcal{K})$ is in RL. Furthermore, we can identify a new efficiently-recognisable class of DL knowledge bases that contains both EL and RL, and for which both classification and fact entailment are feasible in polynomial time.

Theorem 5. Let $\mathcal{C}$ be the class of Horn knowledge bases $\mathcal{K}$ such that all roles in $\mathcal{K}$ are reuse-safe. Then, the following conditions hold:

1. Checking whether a $\mathcal{SHOIQ}$ KB $\mathcal{K}$ is in $\mathcal{C}$ is feasible in polynomial time;
2. Every EL and RL knowledge base is contained in $\mathcal{C}$;
3. $\Psi(\mathcal{K})$ is an RL knowledge base for each $\mathcal{K} \in \mathcal{C}$; and
4. Classification and fact entailment in $\mathcal{C}$ are feasible in polynomial time.

Finally, it is worth emphasising that, although the transformations Ψ in Definition 8 and Ξ in Section 3 are very different and serve rather orthogonal purposes, they are connected in the limit case where all roles are reuse-safe and the ontology does not contain cardinality restrictions.

Proposition 2. Let $\mathcal{K}$ be a normalised Horn KB that does not contain axioms **N4** or **N6**. Then, $\Xi(\Omega(\mathcal{K}))$ is EL iff all roles in $\mathcal{K}$ are reuse-safe.

5   Evaluation

We have implemented the transformations described in Sections 3 and 4, and we have performed a range of classification and data reasoning experiments over both realistic ontologies and standard benchmarks.

Table 1. Classification times for representative ontologies (in seconds)

Ontology ID	00018	00352	00448	00461	00463	00470	00660	Fly
Original (HermiT)	76.787	18.679	68.545	2.260	t-out	286.89	102.80	840.014
Normalised (HermiT)	30.730	7.235	41.529	11.768	t-out	318.60	123.71	807.167
EL-ified (HermiT)	9.006	7.953	21.395	1.801	651.884	54.40	17.62	17.361
Original (MORe)	42.292	15.095	5.949	2.515	t-out	258.53	99.93	844.639
Normalised (MORe)	10.521	3.195	5.061	11.442	t-out	293.55	85.42	819.640
EL-ified (MORe)	3.0792	2.650	5.019	1.310	694.046	3.48	17.58	17.409

5.1 Classification Experiments

For our input data, we used the OWL 2 ontologies in the Oxford Ontology Repository,[4] which contains 793 realistic ontologies, as well as a "hard" version of the FlyAnatomy ontology, which is not yet in the repository. Several of the test ontologies contain a small number of axioms exploiting constructs (such as complex RIAs) not available in $\mathcal{SHOIQ}$; in these cases we tested filtered versions of the ontologies where such axioms have been removed.

We tested classification times for the latest versions of HermiT (v.1.3.8) and MORe (v.0.1.5) using their standard settings. All experiments were performed on a laptop with 16 GB RAM and Intel Core 2.9 GHz processor running Java v.1.7.0_21, with a timeout set to 3,000s.

EL-ification Experiments. Out of the 793 ontologies in the corpus, we selected those 70 that contain inverse roles, and which HermiT takes at least 1s to classify. For each test ontology $\mathcal{K}$ we have computed a normalised version $\varUpsilon(\mathcal{K})$ and an EL-ified version $\mathcal{K}'$ (see Section 3), and have compared classification times for HermiT and MORe on each version.

We found that 50 out of the 70 test ontologies contained only rewritable inverse roles, which could be successfully eliminated using our transformations, and 4 of these ontologies could be fully rewritten into EL. Of these 50 ontologies, 6 could not be classified by HermiT even after EL-ification; however, HermiT succeeded on 2 EL-ified ontologies that could not be classified in their original form. For the remaining 42 ontologies, normalisation alone leads to a slight deterioration in average performance due to the introduction of new class names (which HermiT must classify); however, EL-ification improves HermiT's performance by an average factor of approximately 3. We believe that this improvement is due to HermiT being able to use single blocking instead of pairwise blocking.

Like HermiT, MORe failed on 8 of the original ontologies, but succeeded on two of these after EL-ification. With the remaining 42, as for HermiT, normalisation alone leads to a slight deterioration in performance, but EL-ification improves performance by an average factor of approximately 6. The larger improvement can be explained by the fact that many axioms are rewritten into EL, and hence MORe can delegate a greater part of the computational work to ELK. Table 1 presents results for some representative cases.

---

[4] http://www.cs.ox.ac.uk/isg/ontologies/

Finally, as already mentioned, our test corpus contains 20 ontologies with non-rewritable inverse roles. As expected, in these cases we obtained no consistent improvement since the presence of inverses forces HermiT to use pairwise blocking; furthermore, in some cases the transformation negatively impacts performance, as it adds a substantial number of axioms to simulate the effect of inverse roles. Hence, it seems that our techniques are clearly beneficial only when all inverse roles are rewritable.

Reuse Safety. From the 793 ontologies in the corpus, we identified 174 Horn ontologies that do not fall within any of the OWL 2 profiles. We have applied our transformation in Definition 8 to these ontologies and found that 53 do not contain unsafe roles and hence are rewritten into RL. Furthermore, we found that in the remaining ontologies 89% of the roles were reuse-safe, on average. We have tested classification times with HermiT over the transformed ontologies, but found that the transformation had a negative impact on performance. This is explained by the fact that our transformation introduces nominals. In the presence of nominals, HermiT disables *anywhere blocking*—a powerful technique that makes nodes blockable by any other node in the tableau (and not just by its ancestors). As mentioned in Section 4, it would be more effective to implement safe reuse as a modification of HermiT's calculus; this, however, implies non-trivial modifications to the core of the reasoner, which is left for future work.

5.2   Data Reasoning Experiments

We have used the standard LUBM benchmark, which comes with an ontology about academic departments and a dataset generator parameterised by the number of universities for which data is generated (LUBM(n) denotes the dataset for n universities). The LUBM ontology is not in RL, as it contains axioms of type **N2**; however, all roles in LUBM are reuse-safe and hence we rewrote it into RL using the transformation in Definition 8. For each dataset, we recorded the times needed to compute the instances of all atomic concepts in the ontology. We compared HermiT over the original ontology and the RL reasoner RD-Fox[5] over the transformed ontology. HermiT took 3.7s for LUBM(1), and timed out for LUBM(5). In contrast RDFox only required 0.2s for LUBM(1), 1.5s for LUBM(10), and 7.4s for LUBM(20). These results suggest the clear benefits of transforming an ontology to RL and exploiting highly scalable reasoners such as RDFox.

6   Related Work

The observation that many ontologies consist of a large EL "backbone" and a relatively small number of non-EL axioms is exploited by the modular reasoner MORe [1] to delegate the bulk of the classification work to EL reasoner ELK [11]. Modular reasoning techniques, however, are sensitive to syntax and all

---

[5] http://www.cs.ox.ac.uk/isg/tools/RDFox/

non-EL axioms (as well as those "depending" on them) must be processed by a fully-fledged OWL reasoner. Ren et al. propose a technique for approximating an OWL ontology into EL [17]; this approximation, however, is incomplete for classification and hence valid subsumptions might be lost.

Several techniques for inverse role elimination in DL ontologies have been developed. Ding et al. [7] propose a polynomial reduction from $\mathcal{ALCI}$ into $\mathcal{ALC}$, which is then extended in [6] to $\mathcal{SHOI}$. Similarly, Song et al. [19] propose a polynomial reduction from $\mathcal{ALCHI}$ to $\mathcal{ALCH}$ KBs to optimise classification. In all of these approaches inverse roles are replaced with fresh symbols and new axioms are introduced to compensate for the loss of implicit inferences. These approaches, however, are not applicable to KBs with cardinality restrictions; furthermore, inverse role elimination heavily relies on the introduction of universal restrictions, and hence they are not well-suited for EL-ification. Calvanese et al. [5] propose a transformation from $\mathcal{ALCFI}$ knowledge bases to $\mathcal{ALC}$ which is sound and complete for classification; this technique exhaustively introduces universal restrictions to simulate at-most cardinality restrictions and inverse roles, and hence it is also not well-suited for EL-ification; furthermore, this technique is not applicable to knowledge bases with transitive roles or nominals. Finally, Lutz et al. study rewritability of first-order formulas into EL as a decision problem [13]; the rewritings studied in [13], however, require preservation of logical equivalence, whereas ours preserve satisfiability.

The techniques described in Section 4 extend the so-called combined approach to query answering in EL [12,20]. They are also related to are strongly related to individual reuse optimisations [16], where to satisfy existential restrictions a (hyper-)tableau reasoner tries to reuse an individual from the model constructed thus far. Individual reuse, however, may introduce non-determinism in exchange for a smaller model: if the reuse fails (i.e., a contradiction is derived), the reasoner must backtrack and introduce a fresh individual. In contrast, in the case of reuse-safe roles reuse can be done *deterministically* and hence model size is reduced without the need of backtracking.

Finally, Zhou et. al use a very similar transformation as ours to strengthen ontologies and overestimate query answers [23]. It follows from Theorem 5 that the technique in [23] leads to exact answers to atomic queries for Horn ontologies where all roles are reuse-safe.

7   Conclusions and Future Work

In this paper, we have proposed novel techniques for rewriting ontologies into the OWL 2 profiles. Our techniques are easily implementable as preprocessing steps in DL reasoners, and can lead to substantial improvements in reasoning times. Furthermore, we have established sufficient conditions for ontologies to be polynomially rewritable into the EL and RL profiles. Thus, for the class of ontologies satisfying our conditions reasoning becomes feasible in polynomial time. There are many avenues to explore for future work. For example, we will investigate extensions of our EL-ification techniques that are capable of rewriting away disjunctive axioms. Furthermore, we are planning to implement safe reuse in HermiT and evaluate the impact of this optimisation on classification.

References

1. Armas Romero, A., Cuenca Grau, B., Horrocks, I.: MORe: modular combination of OWL reasoners for ontology classification. In: Cudré-Mauroux, P., et al. (eds.) ISWC 2012, Part I. LNCS, vol. 7649, pp. 1–16. Springer, Heidelberg (2012)
2. Baader, F., Lutz, C., Suntisrivaraporn, B.: CEL — A polynomial-time reasoner for life science ontologies. In: Furbach, U., Shankar, N. (eds.) IJCAR 2006. LNCS (LNAI), vol. 4130, pp. 287–291. Springer, Heidelberg (2006)
3. Bail, S., Glimm, B., Gonçalves, R.S., Jiménez-Ruiz, E., Kazakov, Y., Matentzoglu, N., Parsia, B. (eds.): ORE. CEUR, vol. 1015 (2013)
4. Bishop, B., Kiryakov, A., Ognyanoff, D., Peikov, I., Tashev, Z., Velkov, R.: OWLim: A family of scalable semantic repositories. Semantic Web J. 2(1), 33–42 (2011)
5. Calvanese, D., De Giacomo, G., Rosati, R.: A note on encoding inverse roles and functional restrictions in $\mathcal{ALC}$ knowledge bases. In: Proceedings of the 1998 Description Logic Workshop (DL 1998), pp. 69–71. CEUR (1998)
6. Ding, Y.: Tableau-based Reasoning for Description Logics with Inverse Roles and Number Restrictions. Ph.D. thesis, Concordia University, Canada (2008)
7. Ding, Y., Haarslev, V., Wu, J.: A new mapping from $\mathcal{ALCI}$ to $\mathcal{ALC}$. In: Calvanese, D., Franconi, E., Haarslev, V., Lembo, D., Motik, B., Turhan, A., Tessaris, S. (eds.) DL 2007. CEUR Workshop Proceedings, vol. 250 (2007)
8. Gonçalves, R.S., Matentzoglu, N., Parsia, B., Sattler, U.: The empirical robustness of Description Logic classification. In: DL, pp. 197–208 (2013)
9. Haarslev, V., Hidde, K., Möller, R., Wessel, M.: The racerpro knowledge representation and reasoning system. Semantic Web J. 3(3), 267–277 (2012)
10. Horrocks, I., Sattler, U.: A tableau decision procedure for $\mathcal{SHOIQ}$. J. of Automated Reasoning 39(3), 249–276 (2007)
11. Kazakov, Y., Krötzsch, M., Simančík, F.: Concurrent classification of EL ontologies. In: Aroyo, L., Welty, C., Alani, H., Taylor, J., Bernstein, A., Kagal, L., Noy, N., Blomqvist, E. (eds.) ISWC 2011, Part I. LNCS, vol. 7031, pp. 305–320. Springer, Heidelberg (2011)
12. Kontchakov, R., Lutz, C., Toman, D., Wolter, F., Zakharyaschev, M.: The Combined Approach to Ontology-Based Data Access. In: IJCAI, pp. 2656–2661 (2011)
13. Lutz, C., Piro, R., Wolter, F.: Description logic tboxes: Model-theoretic characterizations and rewritability. In: IJCAI, pp. 983–988 (2011)
14. Motik, B., Cuenca Grau, B., Horrocks, I., Wu, Z., Fokoue, A., Lutz, C. (eds.): OWL 2 Web Ontology Language: Profiles. W3C Recommendation (October 27, 2009), http://www.w3.org/TR/owl2-profiles/
15. Motik, B., Shearer, R., Horrocks, I.: Hypertableau reasoning for description logics. J. Artificial Intelligence Research (JAIR) 36(1), 165–228 (2009)
16. Motik, B., Horrocks, I.: Individual reuse in description logic reasoning. In: Armando, A., Baumgartner, P., Dowek, G. (eds.) IJCAR 2008. LNCS (LNAI), vol. 5195, pp. 242–258. Springer, Heidelberg (2008)
17. Ren, Y., Pan, J.Z., Zhao, Y.: Soundness preserving approximation for TBox reasoning. In: AAAI (2010)
18. Sirin, E., Parsia, B., Cuenca Grau, B., Kalyanpur, A., Katz, Y.: Pellet: A practical OWL-DL reasoner. J. Web Semantics (JWS) 5(2), 51–53 (2007)

19. Song, W., Spencer, B., Du, W.: A transformation approach for classifying $\mathcal{ALCHI}(\mathcal{D})$ ontologies with a consequence-based $\mathcal{ALCH}$ reasoner. In: ORE. CEUR, vol. 1015, pp. 39–45 (2013)

20. Stefanoni, G., Motik, B., Horrocks, I.: Introducing Nominals to the Combined Query Answering Approaches for EL. In: AAAI (2013)

21. Tsarkov, D., Horrocks, I.: FaCT++ description logic reasoner: System description. In: Furbach, U., Shankar, N. (eds.) IJCAR 2006. LNCS (LNAI), vol. 4130, pp. 292–297. Springer, Heidelberg (2006)

22. Wu, Z., Eadon, G., Das, S., Chong, E.I., Kolovski, V., Annamalai, M., Srinivasan, J.: Implementing an inference engine for RDFS/OWL constructs and user-defined rules in Oracle. In: ICDE, pp. 1239–1248 (2008)

23. Zhou, Y., Cuenca Grau, B., Horrocks, I., Wu, Z., Banerjee, J.: Making the most of your triple store: query answering in OWL 2 using an RL reasoner. In: WWW (2013)

The Bayesian Description Logic $\mathcal{BEL}$

İsmail İlkan Ceylan[1,*] and Rafael Peñaloza[1,2,**]

[1] Theoretical Computer Science, TU Dresden, Germany
[2] Center for Advancing Electronics Dresden
{ceylan,penaloza}@tcs.inf.tu-dresden.de

Abstract. We introduce the probabilistic Description Logic $\mathcal{BEL}$. In $\mathcal{BEL}$, axioms are required to hold only in an associated context. The probabilistic component of the logic is given by a Bayesian network that describes the joint probability distribution of the contexts. We study the main reasoning problems in this logic; in particular, we (i) prove that deciding positive and almost-sure entailments is not harder for $\mathcal{BEL}$ than for the BN, and (ii) show how to compute the probability, and the most likely context for a consequence.

1 Introduction

Description Logics (DLs) [2] are a family of knowledge representation formalisms originally designed for representing the terminological knowledge of a domain in a precise and well-understood manner. They have been successfully employed for creating large knowledge bases, representing real application domains. For instance, they are the logical formalism underlying prominent bio-medical ontologies such as SNOMED CT, GALEN, or the Gene Ontology.

Description logic ontologies are usually composed of axioms that restrict the class of possible interpretations. As these are hard restrictions, DL ontologies can only encode absolute, immutable knowledge. For some application domains, however, knowledge depends on the situation (or context) in which it is considered. For example, the notion of a *luxury hotel* in a small rural center will be different from the one in a large cosmopolitan city. When building an ontology for hotels, it makes sense to contextualize the axioms according to location, and possibly other factors like season, type of weather, etc. Since these contexts refer to notions that are external to the domain of interest, it is not always desirable, or even possible, to encode them directly into the classical DL axioms.

We follow a different approach for handling contextual knowledge. We label every axiom with the context in which it is valid. For example, we could have statements like ⟨LuxuryHotel ⊑ ∃hasFeature.MeetingRoom : city⟩ stating that in the context of a city, every luxury hotel has a meeting room. This axiom imposes no restriction in case the context is not a city: it might still hold, or not, depending on other factors.

---

* Supported by DFG within the Research Training Group "RoSI" (GRK 1907).
** Partially supported by DFG within the Cluster of Excellence 'cfAED'.

S. Demri, D. Kapur, and C. Weidenbach (Eds.): IJCAR 2014, LNAI 8562, pp. 480–494, 2014.
© Springer International Publishing Switzerland 2014

Labeling the axioms in an ontology allows us to give a more detailed description of the knowledge domain. Reasoning in these cases can be used to infer knowledge that is guaranteed to hold in any given context. While the knowledge *within* this context is precise, there might be a level of uncertainty regarding the current context. To model this uncertainty, we attach a probability to each of the possible contexts. Since we cannot assume that the contexts are (probabilistically) independent, we need to describe the joint probability distribution over the space of all contexts. Thus, we consider knowledge bases that are composed of an ontology labeled with contextual information, together with a joint probability distribution over the space of contexts.

To represent the probabilistic component of the knowledge base, we use Bayesian networks (BNs) [14], a well-known probabilistic graphical model that allows for a compact representation of the probability distribution, with the help of conditional independence assumptions. For the logical component, we focus on $\mathcal{EL}$ [1], a light-weight DL that allows for polynomial-time reasoning. These formalisms together yield the Bayesian DL $\mathcal{BEL}$.

We study classical and probabilistic reasoning problems in $\mathcal{BEL}$. Not surprisingly, reasoning in this logic is intractable in general, as is reasoning in BNs already. However, we show that hardness arises exclusively from the probabilistic component: the parameterized complexity of reasoning is polynomial, if the size of the BN is considered as a parameter.

The choice of $\mathcal{EL}$ as underlying logical formalism is meant as a simple prototypical case. It allows us to understand the subtleties of combining BNs with DLs, as a first step towards more expressive formalisms. For a preliminary discussion on more expressive Bayesian DLs, and additional details and examples for $\mathcal{BEL}$, see [9].

2   The Description Logic $\mathcal{BEL}$

The DL $\mathcal{BEL}$ is a probabilistic extension of the light-weight DL $\mathcal{EL}$, where probabilities are encoded using a Bayesian network [14]. Formally, a *Bayesian network* (BN) is a pair $\mathcal{B} = (G, \Phi)$, where $G = (V, E)$ is a finite directed acyclic graph (DAG) whose nodes represent Boolean random variables,[1] and Φ contains, for every node $x \in V$, a conditional probability distribution $P_{\mathcal{B}}(x \mid \pi(x))$ of x given its parents $\pi(x)$. If V is the set of nodes in G, we say that $\mathcal{B}$ is a BN *over* V.

The idea behind BNs is that $G = (V, E)$ encodes a series of conditional independence assumptions between the random variables. More precisely, every variable $x \in V$ is conditionally independent of its non-descendants given its parents. Thus, every BN $\mathcal{B}$ defines a unique joint probability distribution (JPD) over V given by

$$P_{\mathcal{B}}(V) = \prod_{x \in V} P_{\mathcal{B}}(x \mid \pi(x)).$$

---

[1] In their general form, BNs allow for arbitrary discrete random variables. We restrict w.l.o.g. to Boolean variables for ease of presentation.

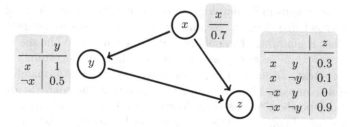

Fig. 1. The BN $\mathcal{B}_0$ over $V_0 = \{x, y, z\}$

A very simple BN is shown in Figure 1. From this network we can derive e.g.
$P(x, \neg y, z) = P(z \mid x, \neg y) \cdot P(\neg y \mid x) \cdot P(x) = 0.1 \cdot 0 \cdot 0.7 = 0$.

As with classical DLs, the main building blocks in $\mathcal{BEL}$ are *concepts*, which are syntactically built as $\mathcal{EL}$ concepts. Given two disjoint sets N_C and N_R of *concept names* and *role names*, respectively, $\mathcal{BEL}$ concepts are defined through the syntactic rule

$$C ::= A \mid \top \mid C \sqcap C \mid \exists r.C$$

where $A \in N_C$ and $r \in N_R$. In DLs, the domain knowledge is typically encoded as a finite set of general concept inclusions (GCIs), called a TBox. $\mathcal{BEL}$ generalizes classical TBoxes by annotating the GCIs with a context, defined by a set of literals belonging to a BN.

Definition 1 (KB). *Let V be a finite set of Boolean variables. A V-literal is an expression of the form x or $\neg x$, where $x \in V$; a V-context is a consistent set of V-literals.*

A V-restricted general concept inclusion (V-GCI) is an expression of the form $\langle C \sqsubseteq D : \kappa \rangle$ where C and D are $\mathcal{BEL}$ concepts and κ is a V-context. A V-TBox is a finite set of V-GCIs.

A $\mathcal{BEL}$ knowledge base (KB) over V is a pair $\mathcal{K} = (\mathcal{B}, \mathcal{T})$ where $\mathcal{B}$ is a BN over V and $\mathcal{T}$ is a V-TBox.

Intuitively, a V-GCI is an axiom that is only guaranteed to hold when its context is enforced. The semantics of this logic is defined with the help of interpretations that map concept and role names to unary and binary predicates, respectively; additionally, these interpretations evaluate the random variables from the BN.

Definition 2 (interpretation). *Given a finite set of Boolean variables V, a V-interpretation is a tuple $\mathcal{I} = (\Delta^{\mathcal{I}}, \cdot^{\mathcal{I}}, \mathcal{V}^{\mathcal{I}})$ where $\Delta^{\mathcal{I}}$ is a non-empty set called the domain, $\mathcal{V}^{\mathcal{I}} : V \to \{0, 1\}$ is a valuation of the variables in V, and $\cdot^{\mathcal{I}}$ is an interpretation function that maps every concept name A to a set $A^{\mathcal{I}} \subseteq \Delta^{\mathcal{I}}$ and every role name r to a binary relation $r^{\mathcal{I}} \subseteq \Delta^{\mathcal{I}} \times \Delta^{\mathcal{I}}$.*

When there is no danger of ambiguity, we will usually ignore the parameter V and speak simply of e.g. a *TBox*, a *KB*, or an *interpretation*.

The interpretation function $\cdot^{\mathcal{I}}$ is extended to arbitrary $\mathcal{BEL}$ concepts by the following rules.

- $\top^{\mathcal{I}} := \Delta^{\mathcal{I}}$
- $(C \sqcap D)^{\mathcal{I}} := C^{\mathcal{I}} \cap D^{\mathcal{I}}$
- $(\exists r.C)^{\mathcal{I}} := \{d \in \Delta^{\mathcal{I}} \mid \exists e \in \Delta^{\mathcal{I}} : (d, e) \in r^{\mathcal{I}} \text{ and } e \in C^{\mathcal{I}}\}$

The valuation $\mathcal{V}^{\mathcal{I}}$ is extended to contexts by defining, for every $x \in V$, $\mathcal{V}^{\mathcal{I}}(\neg x) = 1 - \mathcal{V}^{\mathcal{I}}(x)$, and for every context κ,

$$\mathcal{V}^{\mathcal{I}}(\kappa) = \min_{\ell \in \kappa} \mathcal{V}^{\mathcal{I}}(\ell),$$

where $\min_{\ell \in \emptyset} \mathcal{V}^{\mathcal{I}}(\ell) := 1$. Intuitively, a context κ can be thought as a conjunction of literals, which is evaluated to 1 iff each literal is so and 0 otherwise. We say that the V-interpretation $\mathcal{I}$ is a *model* of the GCI $\langle C \sqsubseteq D : \kappa \rangle$, denoted as $\mathcal{I} \models \langle C \sqsubseteq D : \kappa \rangle$, iff (i) $\mathcal{V}^{\mathcal{I}}(\kappa) = 0$, or (ii) $C^{\mathcal{I}} \subseteq D^{\mathcal{I}}$. It is a *model* of the TBox $\mathcal{T}$ iff it is a model of all the GCIs in $\mathcal{T}$. The idea is that the restriction $C \sqsubseteq D$ is only required to hold whenever the context κ is satisfied. Thus, any interpretation that violates the context trivially satisfies the whole axiom.

Example 3. Let $V_0 = \{x, y, z\}$, and consider the V_0-TBox

$$\mathcal{T}_0 := \{\langle A \sqsubseteq C : \{x, y\}\rangle, \ \langle A \sqsubseteq B : \{\neg x\}\rangle, \ \langle B \sqsubseteq C : \{\neg x\}\rangle\}.$$

The interpretation $\mathcal{I}_0 = (\{d\}, \cdot^{\mathcal{I}_0}, V_0)$ where $\mathcal{V}_0(\{x, \neg y, z\}) = 1$, $A^{\mathcal{I}_0} = \{d\}$, and $B^{\mathcal{I}_0} = C^{\mathcal{I}_0} = \emptyset$ is a model of $\mathcal{T}_0$, but is not a model of the GCI $\langle A \sqsubseteq B : \{x\}\rangle$.

The classical DL $\mathcal{EL}$ can be seen as a special case of $\mathcal{BEL}$ in which all GCIs are associated with an empty context; that is, are of the form $\langle C \sqsubseteq D : \emptyset \rangle$. Notice that every valuation satisfies the empty context $\emptyset$. Thus, a V-interpretation $\mathcal{I}$ satisfies the GCI $\langle C \sqsubseteq D : \emptyset \rangle$ iff $C^{\mathcal{I}} \subseteq D^{\mathcal{I}}$. We say that $\mathcal{T}$ *entails* $\langle C \sqsubseteq D : \emptyset \rangle$, denoted by $\mathcal{T} \models C \sqsubseteq D$, if every model of $\mathcal{T}$ is also a model of $\langle C \sqsubseteq D : \emptyset \rangle$. For a valuation $\mathcal{W}$ of the variables in V, we can define a TBox containing all axioms that must be satisfied in any V-interpretation $\mathcal{I} = (\Delta^{\mathcal{I}}, \cdot^{\mathcal{I}}, \mathcal{V}^{\mathcal{I}})$ with $\mathcal{V}^{\mathcal{I}} = \mathcal{W}$.

Definition 4 (restriction). *Let $\mathcal{K} = (\mathcal{B}, \mathcal{T})$ be a KB. The* restriction *of $\mathcal{T}$ to a valuation $\mathcal{W}$ of the variables in V is the TBox*

$$\mathcal{T}_{\mathcal{W}} := \{\langle C \sqsubseteq D : \emptyset \rangle \mid \langle C \sqsubseteq D : \kappa \rangle \in \mathcal{T}, \mathcal{W}(\kappa) = 1\}.$$

So far, our semantics have focused on the evaluation of the Boolean variables and the interpretation of concepts, ignoring the probabilistic information provided by the BN. To handle these probabilities, we introduce multiple-world semantics next. Intuitively, a V-interpretation describes a possible world; by assigning a probabilistic distribution over these interpretations, we describe the required probabilities, which should be consistent with the BN.

Definition 5 (probabilistic model). *A probabilistic interpretation is a pair $\mathcal{P} = (\mathfrak{I}, P_{\mathfrak{I}})$, where $\mathfrak{I}$ is a set of V-interpretations and $P_{\mathfrak{I}}$ is a probability distribution over $\mathfrak{I}$ such that $P_{\mathfrak{I}}(\mathcal{I}) > 0$ only for finitely many interpretations $\mathcal{I} \in \mathfrak{I}$.*

This probabilistic interpretation is a model of the TBox $\mathcal{T}$ if every $\mathcal{I} \in \mathfrak{I}$ is a model of $\mathcal{T}$. $\mathcal{P}$ is consistent with the BN $\mathcal{B}$ if for every possible valuation $\mathcal{W}$ of the variables in V it holds that

$$\sum_{\mathcal{I} \in \mathfrak{I}, \mathcal{V}^{\mathcal{I}} = \mathcal{W}} P_{\mathfrak{I}}(\mathcal{I}) = P_{\mathcal{B}}(\mathcal{W}).$$

The probabilistic interpretation $\mathcal{P}$ is a model of the KB $(\mathcal{B}, \mathcal{T})$ iff it is a (probabilistic) model of $\mathcal{T}$ and consistent with $\mathcal{B}$.

One simple consequence of this semantics is that probabilistic models preserve the probability distribution of $\mathcal{B}$ for subsets of literals; i.e., contexts. The proof follows from the fact that a context corresponds to a partial valuation. Hence, the probability of a context κ is the sum of the probabilities of all valuations that extend κ.

Theorem 6. *Let $\mathcal{K} = (\mathcal{B}, \mathcal{T})$ be a KB, and κ a context. For every model $\mathcal{P}$ of $\mathcal{K}$ it holds that*

$$\sum_{\mathcal{I} \in \mathfrak{I}, \mathcal{V}^{\mathcal{I}}(\kappa) = 1} P_{\mathfrak{I}}(\mathcal{I}) = P_{\mathcal{B}}(\mathcal{V}^{\mathcal{I}}(\kappa)).$$

For the following sections it will be useful for proving our results to consider a special kind of interpretations, which we call *pithy* . These interpretations contain at most one V-interpretation for each valuation of the variables in V. Each of these V-interpretations provides the essential information associated to the corresponding valuation.

Definition 7 (pithy). *The probabilistic interpretation $\mathcal{P} = (\mathfrak{I}, P_{\mathfrak{I}})$ is called pithy if for every valuation $\mathcal{W}$ of the variables in V there exists at most one V-interpretation $\mathcal{I} = (\Delta^{\mathcal{I}}, \cdot^{\mathcal{I}}, \mathcal{V}^{\mathcal{I}}) \in \mathfrak{I}$ such that $\mathcal{V}^{\mathcal{I}} = \mathcal{W}$.*

We now study classical and probabilistic reasoning problems in $\mathcal{BEL}$, and analyse their complexity.

3   Reasoning in $\mathcal{BEL}$

In the previous section we have described how probabilistic knowledge can be represented using a $\mathcal{BEL}$ KB. We now focus our attention to reasoning with this knowledge. The most basic decision problem in any DL is whether an ontology is consistent. It turns out that, as for classical $\mathcal{EL}$, this problem is trivial in $\mathcal{BEL}$.

Theorem 8. *Every $\mathcal{BEL}$ KB is consistent.*

Proof (Sketch). Let $\mathcal{K} = (\mathcal{B}, \mathcal{T})$ be a $\mathcal{BEL}$ KB. Let $\Delta^{\mathcal{I}} = \{a\}$ and $\cdot^{\mathcal{I}}$ be such that $A^{\mathcal{I}} = \{a\}$ and $r^{\mathcal{I}} = \{(a, a)\}$ for all $A \in \mathsf{N_C}$ and $r \in \mathsf{N_R}$. For every valuation $\mathcal{W}$, define the V-interpretation $\mathcal{I}_{\mathcal{W}} = (\Delta^{\mathcal{I}}, \cdot^{\mathcal{I}}, \mathcal{W})$. Then, the probabilistic interpretation $\mathcal{P} = (\mathfrak{I}, P_{\mathfrak{I}})$ where $\mathfrak{I} = \{\mathcal{I}_{\mathcal{W}} \mid \mathcal{W}$ is a valuation$\}$ and $P_{\mathfrak{I}}(\mathcal{I}_{\mathcal{W}}) = P_{\mathcal{B}}(\mathcal{W})$ is a model of $\mathcal{K}$.

A more interesting reasoning problem is subsumption: decide whether a concept is interpreted as a subclass of another one. We generalize this problem to consider also the contexts and probabilities provided by the BN.

Definition 9 (subsumption). *Let C, D be two $\mathcal{BEL}$ concepts, κ a context, and $\mathcal{K}$ a KB. C is contextually subsumed by D in κ w.r.t. $\mathcal{K}$, denoted as $\langle C \sqsubseteq_\kappa D : \kappa \rangle$, if every probabilistic model of $\mathcal{K}$ is also a model of the TBox $\{\langle C \sqsubseteq D : \kappa\rangle\}$. For a probabilistic interpretation $\mathcal{P} = (\mathfrak{I}, P_\mathfrak{I})$, we define the probability of a consequence $P(\langle C \sqsubseteq_\mathcal{P} D : \kappa\rangle) := \sum_{\mathcal{I} \in \mathfrak{I}, \mathcal{I} \models \langle C \sqsubseteq D : \kappa\rangle} P_\mathfrak{I}(\mathcal{I})$. The probability of $\langle C \sqsubseteq D : \kappa \rangle$ w.r.t. $\mathcal{K}$ is defined as*

$$P(\langle C \sqsubseteq_\mathcal{K} D : \kappa\rangle) := \inf_{\mathcal{P} \models \mathcal{K}} P(\langle C \sqsubseteq_\mathcal{P} D : \kappa\rangle).$$

We say that C is positively subsumed *by D in κ if $P(\langle C \sqsubseteq_\mathcal{K} D : \kappa\rangle) > 0$, and C is p-subsumed by D in κ, for $p \in (0, 1]$ if $P(\langle C \sqsubseteq_\mathcal{K} D : \kappa\rangle) \geq p$. We sometimes refer to 1-subsumption as* almost-sure *subsumption.*

Clearly, if C is subsumed by D in κ w.r.t. a KB $\mathcal{K}$, then $P(\langle C \sqsubseteq_\mathcal{K} D : \kappa\rangle) = 1$. The converse, however, may not hold since the subsumption relation might be violated in V-interpretations of probability zero.

Example 10. Consider the KB $\mathcal{K}_0 = (\mathcal{B}_0, \mathcal{T}_0)$, where $\mathcal{B}_0$ is the BN depicted in Figure 1 and $\mathcal{T}_0$ the TBox from Example 3. It follows that $P(\langle A \sqsubseteq_{\mathcal{K}_0} C : \emptyset\rangle) = 1$ and $P(\langle C \sqsubseteq_{\mathcal{K}_0} B : \{x, y\}\rangle) = 0$. Moreover, for any two concepts E, F, it holds that $P(\langle E \sqsubseteq_{\mathcal{K}_0} F : \{x, \neg y\}\rangle) = 1$ since $\langle E \sqsubseteq_{\mathcal{K}_0} F : \{x, \neg y\}\rangle$ can only be violated in V-interpretations that have probability 0. However, in general the consequence $\langle E \sqsubseteq_{\mathcal{K}_0} F : \{x, \neg y\}\rangle$ does not hold.

3.1   Probabilistic Subsumption

We consider first the problem of computing the probability of a subsumption, or deciding positive, p-subsumption, and almost-sure subsumption. As an intermediate step, we show that it is possible w.l.o.g. to restrict reasoning to pithy models.

Lemma 11. *Let $\mathcal{K}$ be a KB. If $\mathcal{P}$ is a probabilistic model of $\mathcal{K}$, then a pithy model $\mathcal{Q}$ of $\mathcal{K}$ can be computed such that for every two concepts C, D and context κ it holds that $P(\langle C \sqsubseteq_\mathcal{Q} D : \kappa\rangle) \leq P(\langle C \sqsubseteq_\mathcal{P} D : \kappa\rangle)$.*

Proof (Sketch). Let $\mathcal{W}$ be a valuation and $\mathcal{I}, \mathcal{I}' \in \mathfrak{I}$ two V-interpretations such that $\mathcal{V}^\mathcal{I} = \mathcal{V}^{\mathcal{I}'} = \mathcal{W}$. Construct a new interpretation $\mathcal{J}$ as the disjoint union of $\mathcal{I}$ and $\mathcal{I}'$. The probabilistic interpretation $(\mathfrak{H}, P_\mathfrak{H})$ with $\mathfrak{H} = (\mathfrak{I} \cup \{\mathcal{J}\}) \setminus \{\mathcal{I}, \mathcal{I}'\}$ and

$$P_\mathfrak{H}(\mathcal{H}) := \begin{cases} P_\mathfrak{I}(\mathcal{H}) & \mathcal{H} \neq \mathcal{J} \\ P_\mathfrak{I}(\mathcal{I}) + P_\mathfrak{I}(\mathcal{I}') & \mathcal{H} = \mathcal{J} \end{cases}$$

is a model of $\mathcal{K}$. Moreover, $\mathcal{J} \models \langle C \sqsubseteq D : \kappa\rangle$ iff both $\mathcal{I} \models \langle C \sqsubseteq D : \kappa\rangle$ and $\mathcal{I}' \models \langle C \sqsubseteq D : \kappa\rangle$. $\qquad\square$

As we show next, the probability of a consequence can be computed by reasoning over the restrictions $\mathcal{T}_W$ of $\mathcal{T}$.

Theorem 12. *Let $\mathcal{K} = (\mathcal{B}, \mathcal{T})$ be a KB, C, D two concepts and κ a context.*

$$P(\langle C \sqsubseteq_{\mathcal{K}} D : \kappa \rangle) = 1 - P_{\mathcal{B}}(\kappa) + \sum_{\substack{\mathcal{T}_W \models C \sqsubseteq D \\ W(\kappa)=1}} P_{\mathcal{B}}(W).$$

Proof. For every valuation W construct the V-interpretation $\mathcal{I}_W$ as follows. If $\mathcal{T}_W \models C \sqsubseteq D$, then $\mathcal{I}_W$ is any model $(\Delta^{\mathcal{I}_W}, \cdot^{\mathcal{I}_W}, W)$ of $\mathcal{T}_W$; otherwise, $\mathcal{I}_W$ is any model $(\Delta^{\mathcal{I}_W}, \cdot^{\mathcal{I}_W}, W)$ of $\mathcal{T}_W$ that does not satisfy $\langle C \sqsubseteq D : \kappa \rangle$, which must exist by definition. The probabilistic interpretation $\mathcal{P}_{\mathcal{K}} = (\mathfrak{J}, P_{\mathfrak{J}})$ such that $\mathfrak{J} = \{\mathcal{I}_W \mid W \text{ a valuation of } V\}$ and $P_{\mathfrak{J}}(\mathcal{I}_W) = P_{\mathcal{B}}(W)$ for all W is a model of $\mathcal{K}$ and

$$P(\langle C \sqsubseteq_{\mathcal{P}_{\mathcal{K}}} D : \kappa \rangle) = \sum_{\mathcal{I}_W \models \langle C \sqsubseteq D : \kappa \rangle} P_{\mathfrak{J}}(\mathcal{I}_W)$$

$$= \sum_{W(\kappa)=0} P_{\mathfrak{J}}(\mathcal{I}_W) + \sum_{\substack{W(\kappa)=1, \\ \mathcal{I}_W \models \langle C \sqsubseteq D : \kappa \rangle}} P_{\mathfrak{J}}(\mathcal{I}_W)$$

$$= 1 - P_{\mathcal{B}}(\kappa) + \sum_{\substack{\mathcal{T}_W \models C \sqsubseteq D \\ W(\kappa)=1}} P_{\mathcal{B}}(W).$$

Thus, $P(\langle C \sqsubseteq_{\mathcal{K}} D : \kappa \rangle) \leq 1 - P_{\mathcal{B}}(\kappa) + \sum_{\mathcal{T}_W \models C \sqsubseteq D, W(\kappa)=1} P_{\mathcal{B}}(W)$. Suppose now that the inequality is strict, then there exists a probabilistic model $\mathcal{P} = (\mathfrak{J}, P_{\mathfrak{J}})$ of $\mathcal{K}$ such that $P(\langle C \sqsubseteq_{\mathcal{P}} D : \kappa \rangle) < P(\langle C \sqsubseteq_{\mathcal{P}_{\mathcal{K}}} D : \kappa \rangle)$. By Lemma 11, we can assume w.l.o.g. that $\mathcal{P}$ is pithy, and hence for every valuation W with $P_{\mathcal{B}}(W) > 0$ there exists exactly one $\mathcal{J}_W \in \mathfrak{J}$ with $V^{\mathcal{J}_W} = W$. We thus have

$$\sum_{\mathcal{J}_W \models \langle C \sqsubseteq D : \kappa \rangle, W(\kappa)=1} P_{\mathfrak{J}}(\mathcal{J}_W) < \sum_{\mathcal{I}_W \models \langle C \sqsubseteq D : \kappa \rangle, W(\kappa)=1} P_{\mathfrak{J}}(\mathcal{I}_W).$$

Since $P_{\mathfrak{J}}(\mathcal{I}_W) = P_{\mathfrak{J}}(\mathcal{J}_W)$ for all W, then there must exist a valuation V such that $\mathcal{I}_V \models \langle C \sqsubseteq D : \kappa \rangle$ but $\mathcal{J}_V \not\models \langle C \sqsubseteq D : \kappa \rangle$. Since $\mathcal{J}_V$ is a model of $\mathcal{T}_V$ it follows that $\mathcal{T}_V \not\models C \sqsubseteq D$. By construction, then we have that $\mathcal{I}_V \not\models \langle C \sqsubseteq D : \kappa \rangle$, which is a contradiction. □

Based on this theorem, we can compute the probability of a subsumption as described in Algorithm 1. The algorithm simply verifies for all possible valuations W, whether $\mathcal{T}_W$ entails the desired axiom. Clearly, the **for** loop is executed $2^{|V|}$ times; that is, once for each possible valuation of the variables in V. Each of these executions needs to compute the probability $P_{\mathcal{B}}(W)$ and, possibly, decide whether $\mathcal{T}_W \models C \sqsubseteq D$. The former can be done in polynomial time on the size of $\mathcal{B}$, using the standard chain rule [14], while deciding entailment from an $\mathcal{EL}$ TBox is polynomial on $\mathcal{T}$ [8]. Overall, Algorithm 1 runs in time exponential on

---

Algorithm 1. Probability of Subsumption

Input: KB $\mathcal{K} = (\mathcal{B}, \mathcal{T})$, GCI $\langle C \sqsubseteq D : \kappa \rangle$
Output: $P(\langle C \sqsubseteq_{\mathcal{K}} D : \kappa \rangle)$
1: $P \leftarrow 0, Q \leftarrow 0$
2: **for all** valuations $\mathcal{W}$ **do**
3:     **if** $\mathcal{W}(\kappa) = 0$ **then**
4:         $Q \leftarrow Q + P_{\mathcal{B}}(\mathcal{W})$
5:     **else if** $\mathcal{T}_{\mathcal{W}} \models C \sqsubseteq D$ **then**
6:         $P \leftarrow P + P_{\mathcal{B}}(\mathcal{W})$
7: **return** $1 - Q + P$

$\mathcal{B}$ but polynomial on $\mathcal{T}$. Moreover, the algorithm requires only polynomial space since the different valuations can be enumerated using only $|V|$ bits. Thus, we obtain the following result.

Theorem 13. *The problem of deciding p-subsumption is in* PSPACE *and fixed-parameter tractable where $|V|$ is the parameter.*[2]

As a lower bound, unsurprisingly, p-subsumption is at least as hard as deciding probabilities from the BN. Since this latter problem is hard for the class PP [19], we get the following result.

Theorem 14. *Deciding p-subsumption is* PP-*hard.*

If we are interested only in deciding positive or almost-sure subsumption, then we can further improve these upper bounds to NP and coNP, respectively.

Theorem 15. *Deciding positive subsumption is* NP-*complete. Deciding almost-sure subsumption is* coNP-*complete.*

Proof. To decide positive subsumption, we can simply guess a valuation $\mathcal{W}$ and check in polynomial time that (i) $P_{\mathcal{B}}(\mathcal{W}) > 0$ and (ii) either $\mathcal{W}(\kappa) = 0$ or $\mathcal{T}_{\mathcal{W}} \models C \sqsubseteq D$. The correctness of this algorithm is given by Theorem 12. Thus the problem is in NP.

To show hardness, we recall that deciding, given a BN $\mathcal{B}$ and a variable $x \in V$, whether $P_{\mathcal{B}}(x) > 0$ is NP-hard [11]. Consider the KB $\mathcal{K} = (\mathcal{B}, \emptyset)$ and A, B two arbitrary concept names. It follows from Theorem 12 that $P_{\mathcal{B}}(x) > 0$ iff $P(\langle A \sqsubseteq_{\mathcal{K}} B : \{\neg x\}\rangle) > 0$. Thus positive subsumption is NP-hard. The coNP-completeness of almost-sure subsumption can be shown analogously. $\square$

Notice once again that the non-determinism needed to solve these problems is limited to the number of random variables in $\mathcal{B}$. More precisely, exactly $|V|$ bits need to be non-deterministically guessed, and the rest of the computation runs in polynomial time. In practical terms this means that subsumption is tractable

---

[2] Recall that a problem is fixed-parameter tractable if it can be solved in polynomial time, assuming that the parameter is fixed [15].

as long as the DAG remains small. On the other hand, Algorithm 1 shows that the probabilistic and the logical components of the KB can be decoupled while reasoning. This is an encouraging result as it means that one can apply the optimized methods developed for BN inference and for DL reasoning directly in $\mathcal{BEL}$ without major modifications.

3.2   Contextual Subsumption

We now turn our attention to deciding whether a contextual subsumption relation follows from all models of the KB in a classical sense; that is, whether $\langle C \sqsubseteq_\mathcal{K} D : \kappa \rangle$ holds. Contrary to classical $\mathcal{EL}$, subsumption in $\mathcal{BEL}$ is already intractable, even if we consider only the empty context.

Theorem 16. *Let $\mathcal{K}$ be a KB and C, D two concepts. Deciding $\langle C \sqsubseteq_\mathcal{K} D : \emptyset \rangle$ is coNP-hard.*

Proof. We present a reduction from validity of DNF formulas, which is known to be coNP-hard [10]. Let $\phi = \sigma_1 \vee \ldots \vee \sigma_n$ be a DNF formula where each σ_i is a conjunctive clause and let V be the set of all variables appearing in ϕ. For each variable $x \in V$, we introduce the concept names B_x and $B_{\neg x}$ and define the TBox $\mathcal{T}_x := \{\langle A \sqsubseteq B_x : \{x\} \rangle, \langle A \sqsubseteq B_{\neg x} : \{\neg x\} \rangle\}$. For every conjunctive clause $\sigma = \ell_1 \wedge \ldots \wedge \ell_m$ define the TBox $\mathcal{T}_\sigma := \{\langle B_{\ell_1} \sqcap \ldots \sqcap B_{\ell_m} \sqsubseteq C : \emptyset \rangle\}$. Let now $\mathcal{K} = (\mathcal{B}, \mathcal{T})$ where $\mathcal{B}$ is an arbitrary BN over V and $\mathcal{T} = \bigcup_{x \in V} \mathcal{T}_x \cup \bigcup_{1 \leq i \leq n} \mathcal{T}_{\sigma_i}$. It is easy to see that ϕ is valid iff $\langle A \sqsubseteq_\mathcal{K} C : \emptyset \rangle$.   □

The main reason for this hardness is that the interaction of contexts might produce consequences that are not obvious at first sight. For instance, a consequence might follow in context κ not because the axioms from κ entail the consequence, but rather because any valuation satisfying κ will yield it. That is the main idea in the proof of Theorem 16; the axioms that follow directly from the empty context never entail the subsumption $A \sqsubseteq C$, but if ϕ is valid, then this subsumption follows from all valuations. We obtain the following result.

Lemma 17. *Let $\mathcal{K} = (\mathcal{B}, \mathcal{T})$ be a KB. Then $\langle C \sqsubseteq_\mathcal{K} D : \kappa \rangle$ iff for every valuation $\mathcal{W}$ with $\mathcal{W}(\kappa) = 1$, it holds that $\mathcal{T}_\mathcal{W} \models C \sqsubseteq D$.*

It thus suffices to identify all valuations that define TBoxes entailing the consequence. To do this, we will take advantage of techniques developed in the area of axiom-pinpointing [6], access control [3], and context-based reasoning [4]. It is worth noticing that subsumption relations depend only on the TBox and not on the BN. For that reason, for the rest of this section we focus only on the terminological part of the KB.

We can think of every context κ as the conjunctive clause $\chi_\kappa := \bigwedge_{\ell \in \kappa} \ell$. In this view, the V-TBox $\mathcal{T}$ is a labeled TBox over the (distributive) lattice $\mathbb{B}$ of all Boolean formulas over the variables V, modulo equivalence. Each formula ϕ in this lattice defines a sub-TBox $\mathcal{T}_\phi$ which contains all axioms $\langle C \sqsubseteq D : \kappa \rangle \in \mathcal{T}$ such that $\chi_\kappa \models \phi$.

Using the terminology from [4], we are interested in finding a boundary for a consequence. Given a TBox $\mathcal{T}$ labeled over the lattice $\mathbb{B}$ and concepts C, D, a *boundary* for $C \sqsubseteq D$ w.r.t. $\mathcal{T}$ is an element $\phi \in \mathbb{B}$ such that for every join-prime element $\psi \in \mathbb{B}$ it holds that $\psi \models \phi$ iff $\mathcal{T}_\psi \models C \sqsubseteq D$ (see [4] for further details). Notice that the join-prime elements of $\mathbb{B}$ are exactly the valuations of variables in V. Using Lemma 17 we obtain the following result.

Theorem 18. *Let ϕ be a boundary for $C \sqsubseteq D$ w.r.t. $\mathcal{T}$ in $\mathbb{B}$. Then, for any context κ we have that $\langle C \sqsubseteq_\kappa D : \kappa \rangle$ iff $\chi_\kappa \models \phi$.*

While several methods have been developed for computing the boundary of a consequence, they are based on a *black-box* approach that makes several calls to an external reasoner. We present a *glass-box* approach that computes a compact representation of the boundary directly. This method, based on the standard completion algorithm for $\mathcal{EL}$ [8], can in fact compute the boundaries for all subsumption relations between concept names that follow from the KB.

For our completion algorithm we assume that the TBox is in normal form; i.e., all GCIs are of the form $\langle A_1 \sqcap A_2 \sqsubseteq B : \kappa \rangle$, $\langle A \sqsubseteq \exists r.B : \kappa \rangle$, or $\langle \exists r.A \sqsubseteq B : \kappa \rangle$, where $A, A_1, A_2, B \in \mathsf{N_C} \cup \{\top\}$. It is easy to see that every V-TBox can be transformed into an equivalent one in normal form in linear time.

Given a TBox in normal form, the completion algorithm uses rules to label a set of assertions until no new information can be added. Assertions are tuples of the form (A, B) or (A, r, B) where $A, B \in \mathsf{N_C} \cup \{\top\}$ and $r \in \mathsf{N_R}$ are names appearing in the TBox. The function lab maps every assertion to a Boolean formula ϕ over the variables in V. Intuitively, $\mathsf{lab}(A, B) = \phi$ expresses that $\mathcal{T}_\mathcal{W} \models A \sqsubseteq B$ in all valuations $\mathcal{W}$ that satisfy ϕ; and $\mathsf{lab}(A, r, B) = \phi$ expresses that $\mathcal{T}_\mathcal{W} \models A \sqsubseteq \exists r.B$ in all valuations $\mathcal{W}$ that satisfy ϕ. The algorithm is initialized with the labeling of assertions

$$\mathsf{lab}(\alpha) := \begin{cases} \mathsf{t} & \alpha \text{ is of the form } (A, \top) \text{ or } (A, A) \text{ for } A \in \mathsf{N_C} \cup \{\top\} \\ \mathsf{f} & \text{otherwise,} \end{cases}$$

where t is a tautology and f a contradiction in $\mathbb{B}$. This function is modified by applying the rules from Table 1 where for brevity, we denote $\mathsf{lab}(\alpha) = \phi$ by α^ϕ. Every rule application changes the label of one assertion for a more general formula. The number of assertions is polynomial on $\mathcal{T}$ and the depth of the lattice $\mathbb{B}$ is exponential on $|V|$. Thus, in the worst case, the number of rule applications is bounded exponentially on $|V|$, but polynomially on $\mathcal{T}$.

Clearly, all the rules are sound; that is, at every step of the algorithm it holds that $\mathcal{T}_\mathcal{W} \models A \sqsubseteq B$ for all concept names A, B and all valuations $\mathcal{W}$ that satisfy $\mathsf{lab}(A, B)$, and analogously for (A, r, B). It can be shown using techniques from axiom-pinpointing (see e.g. [7,4]) that after termination the converse also holds; i.e., for every valuation $\mathcal{W}$, if $\mathcal{T}_\mathcal{W} \models A \sqsubseteq B$, then $\mathcal{W} \models \mathsf{lab}(A, B)$. Thus, we obtain the following result.

Theorem 19. *Let lab be the labelling function obtained through the completion algorithm. For every two concept names A, B appearing in $\mathcal{T}$, $\mathsf{lab}(A, B)$ is a boundary for $A \sqsubseteq B$ w.r.t. $\mathcal{T}$.*

Table 1. Completion rules for subsumption in $\mathcal{BEL}$

If $\left\{\begin{array}{l}\langle A_1 \sqcap A_2 \sqsubseteq B : \kappa\rangle \in \mathcal{T}, \\ (X, A_1)^{\phi_1}, (X, A_2)^{\phi_2}, (X, B)^\psi \\ \chi_\kappa \wedge \phi_1 \wedge \phi_2 \not\models \psi\end{array}\right\}$	**then** $\mathsf{lab}(X, B) := (\chi_\kappa \wedge \phi_1 \wedge \phi_2) \vee \psi$
If $\left\{\begin{array}{l}\langle A \sqsubseteq \exists r.B : \kappa\rangle \in \mathcal{T} \\ (X, A)^\phi, (X, r, B)^\psi \\ \chi_\kappa \wedge \phi \not\models \psi\end{array}\right\}$	**then** $\mathsf{lab}(X, r, B) := (\chi_\kappa \wedge \phi) \vee \psi$
If $\left\{\begin{array}{l}\langle \exists r.A \sqsubseteq B : \kappa\rangle \in \mathcal{T} \\ (X, r, Y)^{\phi_1}, (Y, A)^{\phi_2}, (X, B)^\psi \\ \chi_\kappa \wedge \phi_1 \wedge \phi_2 \not\models \psi\end{array}\right\}$	**then** $\mathsf{lab}(X, B) := (\chi_\kappa \wedge \phi_1 \wedge \phi_2) \vee \psi$

Once we know a boundary ϕ for $A \sqsubseteq B$ w.r.t. $\mathcal{T}$, we can decide whether $\langle A \sqsubseteq_\kappa B : \kappa\rangle$: we need only to verify whether $\chi_\kappa \models \phi$. This decision is in NP on $|V|$. Although the algorithm is described exclusively for concept names A, B, it can be used to compute a boundary for $C \sqsubseteq D$, for arbitrary $\mathcal{BEL}$ concepts C, D, simply by adding the axioms $\langle A_0 \sqsubseteq C : \emptyset\rangle$ and $\langle D \sqsubseteq B_0 : \emptyset\rangle$, where A_0, B_0 are new concept names, to the TBox, and then computing a boundary for $A_0 \sqsubseteq B_0$ w.r.t. the extended TBox. This yields the following result.

Corollary 20. *Subsumption in $\mathcal{BEL}$ can be decided in exponential time, and is fixed-parameter tractable where $|V|$ is the parameter.*

Clearly, the boundary for $C \sqsubseteq D$ provides more information than necessary for deciding whether the subsumption holds in a *given* context κ. It encodes *all* contexts that entail the desired subsumption. We can use this knowledge to deduce the most likely context.

3.3   Most Likely Context

The problem of finding the most likely context for a consequence can be seen as the dual of computing the probability of this consequence. Intuitively, we are interested in finding the most likely explanation for an event; assuming that a consequence holds, we are interested in finding an explanation for it, in the form of a context, that has the maximal probability of occurring.

Definition 21 (most likely context). *Given a KB $\mathcal{K} = (\mathcal{B}, \mathcal{T})$ and concepts C, D, the context κ is called a* most likely context *for $C \sqsubseteq D$ w.r.t. $\mathcal{K}$ if (i) $\langle C \sqsubseteq_\mathcal{K} D : \kappa\rangle$, and (ii) for every context κ', if $\langle C \sqsubseteq_\mathcal{K} D : \kappa'\rangle$ holds, then $P_\mathcal{B}(\kappa') \leq P_\mathcal{B}(\kappa)$.*

Notice that we are not interested in maximizing $P(\langle C \sqsubseteq_\mathcal{K} D : \kappa\rangle)$ but rather $P_\mathcal{B}(\kappa)$. Indeed, these two problems can be seen as dual, since $P(\langle C \sqsubseteq_\mathcal{K} D : \kappa\rangle)$ depends inversely, but not exclusively, on $P_\mathcal{B}(\kappa)$ (see Theorem 12).

Algorithm 2. Compute all most likely contexts

Input: KB $\mathcal{K} = (\mathcal{B}, \mathcal{T})$, concepts C, D
Output: The set Λ of most likely contexts for $C \sqsubseteq D$ w.r.t. $\mathcal{K}$ and probability $p \in [0, 1]$
1: $\Lambda \leftarrow \emptyset$, $p \leftarrow 0$
2: $\phi \leftarrow \mathsf{boundary}(C \sqsubseteq D, \mathcal{T})$ ▷ compute a boundary for $C \sqsubseteq D$ w.r.t. $\mathcal{T}$
3: **for all** contexts κ **do**
4:    **if** $\chi_\kappa \models \phi$ **then**
5:       **if** $P_\mathcal{B}(\kappa) > p$ **then**
6:          $\Lambda \leftarrow \{\kappa\}$
7:          $p \leftarrow P_\mathcal{B}(\kappa)$
8:       **else if** $P_\mathcal{B}(\kappa) = p$ **then**
9:          $\Lambda \leftarrow \Lambda \cup \{\kappa\}$
10: **return** Λ, p

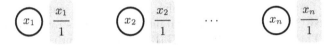

Fig. 2. The BN $\mathcal{B}_n$ over $\{x_1, \ldots, x_n\}$

Algorithm 2 computes the set of all most likely contexts for $C \sqsubseteq D$ w.r.t. $\mathcal{K}$, together with their probability. It maintains a value p of the highest known probability for a context, and a set Λ with all the contexts that have probability p. The algorithm first computes a boundary for the consequence, which is used to test, for every context κ whether $\langle C \sqsubseteq_\mathcal{K} D : \kappa \rangle$ holds. In that case, it compares $P_\mathcal{B}(\kappa)$ with p. If the former is larger, then the highest probability is updated to this value, and the set Λ is restarted to contain only κ. If they are the same, then κ is added to the set of most likely contexts.

Computing a boundary requires exponential time on $\mathcal{T}$. Likewise, the number of contexts is exponential on $\mathcal{B}$, and for each of them we have to test propositional entailment, which is also exponential on $\mathcal{B}$. Overall, we have the following.

Theorem 22. *Algorithm 2 computes all most likely contexts for $C \sqsubseteq D$ w.r.t. $\mathcal{K}$ in exponential time.*

In general, it is not possible to lower this exponential upper bound, since a simple consequence may have exponentially many most likely contexts. For example, given a natural number $n \geq 1$, let $\mathcal{B}_n = (G_n, \Phi_n)$ be the BN where $G = (\{x_1, \ldots, x_n\}, \emptyset)$, i.e., G contains n nodes and no edges connecting them, and for each $i, 1 \leq i \leq n$ Φ_n contains the distribution with $P_\mathcal{B}(x_i) = 1$ (see Figure 2). For every context $\kappa \subseteq \{x_1, \ldots, x_n\}$, we have that $P_\mathcal{B}(\kappa) = 1$ which means that there are 2^n most likely contexts for $A \sqsubseteq A$ w.r.t. the KB $(\mathcal{B}_n, \emptyset)$.

Algorithm 2 can be adapted to compute *one* most likely context in a more efficient way. The main idea is to order the calls in the **for** loop by decreasing probability. Once one context κ with $\chi_\kappa \models \phi$ has been found, it is guaranteed to be a most likely context and the algorithm may stop. This approach would still require exponential time in the worst case. However, recall that simply *verifying* whether κ is a context for $C \sqsubseteq D$ is already coNP-hard (Theorem 16), and hence deciding whether it is a most likely context is arguably hard for the second level of the polynomial hierarchy. On the other hand, this exponential bound depends exclusively on $|V|$. Hence, as before, we have that deciding whether a context is a most likely context for a consequence is fixed-parameter tractable over $|V|$.

4 Related Work

The amount of work on handling uncertain knowledge with description logics is too vast to cover in detail here. Many probabilistic description logics have been defined, which differ not only in their syntax but also in their use of the probabilities and their application. These logics were recently surveyed in [17]. We discuss here only those logics most closely related to ours.

One of the first attempts for combining BNs and DLs was P-CLASSIC [16], which extended CLASSIC through probability distributions over the interpretation domain. The more recent PR-OWL [12] uses multi-entity BNs to describe the probability distributions of some domain elements. In both cases, the probabilistic component is interpreted providing individuals with a probability distribution; this differs greatly from our multiple-world semantics, in which we consider a probability distribution over a set of classical DL interpretations.

Perhaps the closest to our approach are the Bayesian extension of DL-Lite [13] and DISPONTE [18]. The latter allows for so-called epistemic probabilities that express the uncertainty associated to a given axiom. Their semantics are based, as ours, on a probabilistic distribution over a set of interpretations. The main difference with our approach is that in [18], the authors assume that all probabilities are independent, while we provide a joint probability distribution through the BN. Another minor difference is that in DISPONTE it is impossible to obtain classical consequences, as we do.

Abstracting from the different logical constructors used, the logic in [13] looks almost identical to ours. There is, however, a subtle but important difference. In our approach, an interpretation $\mathcal{I}$ satisfies an axiom $\langle C \sqsubseteq D : \kappa \rangle$ if $\mathcal{V}^{\mathcal{I}}(\kappa) = 1$ implies $C^{\mathcal{I}} \subseteq D^{\mathcal{I}}$. In [13], the authors employ a closed-world assumption over the contexts, where this implication is substituted for an equivalence; i.e., $\mathcal{V}^{\mathcal{I}}(\kappa) = 0$ also implies $C^{\mathcal{I}} \not\subseteq D^{\mathcal{I}}$. The use of such semantics can easily produce inconsistent KBs, which is impossible in $\mathcal{BEL}$.

5 Conclusions

We have introduced the probabilistic DL $\mathcal{BEL}$, which extends the classical $\mathcal{EL}$ to express uncertainty. Our basic assumption is that we have *certain* knowledge, which depends on an *uncertain* situation, or context. In practical terms, this means that every axiom is associated to a context with the intended meaning that, if the context holds, then the axiom must be true. Uncertainty is represented through a BN that encodes the probability distribution of the contexts.

The advantage of using Bayesian networks relies in their capacity of describing conditional independence assumptions in a compact manner.

We have studied the complexity of reasoning in this probabilistic logic. Contrary to classical $\mathcal{EL}$, reasoning in $\mathcal{BEL}$ is in general intractable. More precisely, we have shown that positive subsumption is NP-complete, and almost-sure subsumption is coNP-complete. For the other reasoning problems we have not found tight complexity bounds, but we proved that p-subsumption is NP-hard and in PSPACE, while contextual subsumption and deciding most likely contexts are between coNP and EXPTIME.

In contrast to these negative complexity results, we have shown that the complexity can be decoupled between the probabilistic and the logical components of the KB. Indeed, all these problems are fixed-parameter tractable over the parameter $|V|$. This means that, if we have a fixed number of contexts, then all these problems can be solved in polynomial time. It is not unreasonable, moreover, to assume that the number of contexts is quite small in comparison to the size of the TBox. Finally notice that reasoning with the BN itself is already intractable. What we have shown is that intractability is a consequence of the contextual and probabilistic components of the KB, and not of the logical one.

There are several directions for future work. First, we would like to tighten our complexity results. Notice that the main bottleneck in our algorithms for deciding contextual subsumption and computing the most likely contexts is the computation of the boundary, which requires exponential time. It has been argued that a compact representation of the pinpointing formula, which is a special case of the boundary, can be computed in polynomial time for $\mathcal{EL}$ using an automata-based approach [5]. If making logical inferences over this compact encoding is not harder than for the formula itself, then we would automatically obtain a Σ_2^P algorithm for deciding contextual subsumption. Likewise, a context could be verified to be a most likely context for a consequence in PSPACE.

A different direction will be to extend our semantics to more expressive logics. In particular, we will include assertion and role axioms into our knowledge bases. Since many of our algorithms depend only on the existence of a reasoner for the logic, such extension should not be a problem. Our complexity results, on the other hand, would be affected by these changes. From the probabilistic side, we can also consider other probabilistic graphical models to encode the JPD of the contexts. Finally, we would like to consider problems that tighten the relationship between the probabilistic and the logical components. One of such problems would be to update the BN according to evidence attached to the TBox.

References

1. Baader, F.: Terminological cycles in a description logic with existential restrictions. In: Proc. 18th International Joint Conference on Artificial Intelligence (IJ-CAI 2003), pp. 325–330. Morgan Kaufmann (2003)

2. Baader, F., Calvanese, D., McGuinness, D.L., Nardi, D., Patel-Schneider, P.F. (eds.): The Description Logic Handbook: Theory, Implementation, and Applications, 2nd edn. Cambridge University Press (2007)
3. Baader, F., Knechtel, M., Peñaloza, R.: A generic approach for large-scale ontological reasoning in the presence of access restrictions to the ontology's axioms. In: Bernstein, A., Karger, D.R., Heath, T., Feigenbaum, L., Maynard, D., Motta, E., Thirunarayan, K. (eds.) ISWC 2009. LNCS, vol. 5823, pp. 49–64. Springer, Heidelberg (2009)
4. Baader, F., Knechtel, M., Peñaloza, R.: Context-dependent views to axioms and consequences of semantic web ontologies. J. of Web Semantics, 12–13, 22–40 (2012)
5. Baader, F., Peñaloza, R.: Automata-based axiom pinpointing. J. of Automated Reasoning 45(2), 91–129 (2010)
6. Baader, F., Peñaloza, R.: Axiom pinpointing in general tableaux. J. of Logic and Computation 20(1), 5–34 (2010)
7. Baader, F., Peñaloza, R., Suntisrivaraporn, B.: Pinpointing in the description logic $\mathcal{EL}^+$. In: Hertzberg, J., Beetz, M., Englert, R. (eds.) KI 2007. LNCS (LNAI), vol. 4667, pp. 52–67. Springer, Heidelberg (2007)
8. Brandt, S.: Polynomial time reasoning in a description logic with existential restrictions, GCI axioms, and—what else? In: Proc. 16th European Conference on Artificial Intelligence (ECAI 2004), pp. 298–302. IOS Press (2004)
9. Ceylan, İ.İ.: Context-Sensitive Bayesian Description Logics. Master's thesis, Dresden University of Technology, Germany (2013)
10. Cook, S.A.: The complexity of theorem-proving procedures. In: Proc. Third Annual ACM Symposium on Theory of Computing, STOC 1971, pp. 151–158. ACM, New York (1971), http://doi.acm.org/10.1145/800157.805047
11. Cooper, G.F.: The computational complexity of probabilistic inference using bayesian belief networks (research note). Artif. Intel. 42(2-3), 393–405 (1990)
12. da Costa, P.C.G., Laskey, K.B., Laskey, K.J.: Pr-owl: A bayesian ontology language for the semantic web. In: da Costa, P.C.G., d'Amato, C., Fanizzi, N., Laskey, K.B., Laskey, K.J., Lukasiewicz, T., Nickles, M., Pool, M. (eds.) URSW 2005 - 2007. LNCS (LNAI), vol. 5327, pp. 88–107. Springer, Heidelberg (2008)
13. d'Amato, C., Fanizzi, N., Lukasiewicz, T.: Tractable reasoning with bayesian description logics. In: Greco, S., Lukasiewicz, T. (eds.) SUM 2008. LNCS (LNAI), vol. 5291, pp. 146–159. Springer, Heidelberg (2008)
14. Darwiche, A.: Modeling and Reasoning with Bayesian Networks. Cambridge University Press (2009)
15. Downey, R.G., Fellows, M.: Parameterized Complexity. Monographs in Computer Science. Springer (1999)
16. Koller, D., Levy, A.Y., Pfeffer, A.: P-classic: A tractable probablistic description logic. In: Proc. 14th National Conference on Artificial Intelligence (AAAI 1997), pp. 390–397. AAAI Press (1997)
17. Lukasiewicz, T., Straccia, U.: Managing uncertainty and vagueness in description logics for the semantic web. J. of Web Semantics 6(4), 291–308 (2008)
18. Riguzzi, F., Bellodi, E., Lamma, E., Zese, R.: Epistemic and statistical probabilistic ontologies. In: Proc. 8th Int. Workshop on Uncertainty Reasoning for the Semantic Web (URSW 2012), vol. 900, pp. 3–14. CEUR-WS (2012)
19. Roth, D.: On the hardness of approximate reasoning. Artif. Intel. 82(1-2), 273–302 (1996)

OTTER Proofs in Tarskian Geometry

Michael Beeson[1] and Larry Wos[2]

[1] San José State University
[2] Argonne National Laboratory

Abstract. We report on a project to use OTTER to find proofs of the theorems in Tarskian geometry proved in Szmielew's part (Part I) of [9]. These theorems start with fundamental properties of betweenness, and end with the development of geometric definitions of addition and multiplication that permit the representation of models of geometry as planes over Euclidean fields, or over real-closed fields in the case of full continuity. They include the four challenge problems left unsolved by Quaife, who two decades ago found some OTTER proofs in Tarskian geometry (solving challenges issued in [15]).

Quaife's four challenge problems were: every line segment has a midpoint; every segment is the base of some isosceles triangle; the outer Pasch axiom (assuming inner Pasch as an axiom); and the first outer connectivity property of betweenness. These are to be proved without any parallel axiom and without even line-circle continuity. These are difficult theorems, the first proofs of which were the heart of Gupta's Ph. D. thesis under Tarski. OTTER proved them all in 2012. Our success, we argue, is due to improvements in techniques of automated deduction, rather than to increases in computer speed and memory.

The theory of Hilbert (1899) can be translated into Tarski's language, interpreting lines as pairs of distinct points, and angles as ordered triples of non-collinear points. Under this interpretation, the axioms of Hilbert either occur among, or are easily deduced from, theorems in the first 11 (of 16) chapters of Szmielew. We have found Otter proofs of all of Hilbert's axioms from Tarski's axioms (i.e. through Satz 11.49 of Szmielew, plus Satz 12.11). Narboux and Braun have recently checked these same proofs in Coq.

1 Introduction

Geometry has been a test bed for automated deduction almost as long as computers have existed; the first experiments were done in the 1950s. In the nineteenth century, geometry was the test bed for the development of the axiomatic method in mathematics, spurred on by the efforts to prove Euclid's parallel postulate from his other postulates and ultimately the development of non-Euclidean geometry. This effort culminated in Hilbert's seminal 1899 book [5]. In the period 1927–1965, Tarski developed his simple and short axiom system (described below). Some 35 years ago, the second author experimented with finding proofs from Tarski's axioms, reporting success with simple theorems, but leaving several unsolved challenge problems. The subject was revisited by Art Quaife, who

S. Demri, D. Kapur, and C. Weidenbach (Eds.): IJCAR 2014, LNAI 8562, pp. 495–510, 2014.

in his 1992 book [8] reported on the successful solution of some of those challenge problems using an early version of McCune's theorem-prover, OTTER. But there were several theorems that Quaife was not able to get OTTER to prove, and he stated them as "challenge problems" in his book. As far as we know, nobody took up the subject again until 2012, when the present authors set out to see whether automated reasoning techniques, and/or computer hardware, had improved enough to let us progress beyond Quaife's achievements.

The immediate stimulus was the existence of the almost-formal development of many theorems in Tarskian geometry in Part I of [9]. This Part I is essentially the manuscript developed by Wanda Szmielew for her 1965 Berkeley lectures on the foundations of geometry, with "inessential modifications" by Schwäbhauser. There are 16 chapters. Quaife's challenge problems occur in the first nine chapters. The rest contain other important geometrical theorems (described below). We set ourselves the goal to find OTTER proofs of each of the theorems in Szmielew's 16 chapters. Our methodology was to make a separate problem of each theorem, supplying OTTER with the axioms and the previously-proved theorems, as well as the (negated) goal expressing the theorem to be proved. We were sometimes forced to supply OTTER with more information than that, as we describe below.

We did succeed in 2012 to find OTTER proofs of Quaife's challenge problems, the last of which is Satz 9.6 in Szmielew. Verification of chapters 1 through 11 (which we completed with OTTER in 2013) suffices to formally prove Hilbert's axioms from those of Tarski. In this paper, we give Tarski's axioms, explain the challenge problems of Quaife and some of the axioms of Hilbert, discuss the difficulties of finding OTTER proofs of these theorems, and explain what techniques we used to find those proofs.

2 Tarski's Axioms

In about 1927, Tarski first lectured on his axiom system for geometry, which was an improvement on Hilbert's 1899 axioms in several ways: first, the language had only one sort of variables (for points), instead of having three primitive notions (point, line, and angle). Second, it was a first-order theory (Hilbert's axioms mentioned sets, though not in an essential way). Third, the axioms were short, elegant, and few in number (only about fifteen). They could be expressed comprehensibly in the primitive syntax, without abbreviations.

2.1 History

The development of Tarski's theory, started in 1927 or before, was delayed, first by Tarski's involvement in other projects, and then by World War II (galley proofs of Tarski's article about it were destroyed by bombs). The first publication of Tarski's axioms came in 1948 [10], and contained little more than a list of the axioms and a statement of the important metamathematical theorems about the theory (completeness, representation of models as $\mathbb{F}^2$ for $\mathbb{F}$ a real-closed field, quantifier-elimination and decidability). Tarksi then lectured on the subject at Berkeley in 1956–58, and published a reduced set of axioms in 1959 [11]. In the sixties, Tarski, Szmielew, Gupta, and Schwäbhauser (and some students) reduced the number of axioms still further. The manuscript that Szmielew prepared for her 1965 course became Part I of [9]. More details of the history of these axioms can be found in [12] (our main source) and the foreword to (the Ishi Press edition of) [9]. For our present purposes, the relevance of the history is mainly that there are three versions of Tarski's theory, the 1948 version, the 1959 version, and the 1965 version (published in 1983). The earlier experiments of Wos used the 1959 axioms, but Quaife used the 1965 version, as we do. The exact differences are explained below.

2.2   Syntax

The fundamental relations in the theory (first introduced by Pasch in 1852) are "betweenness", which we here write $\mathbf{T}(a, b, c)$, and "equidistance", or "segment congruence", which is officially written $E(a, b, c, d)$, and unofficially as $ab = cd$, segment ab is congruent to segment cd. The intuitive meaning of $\mathbf{T}(a, b, c)$ is that b lies between a and c on the line connecting a and c; Tarski used non-strict betweenness, so we do have $\mathbf{T}(a, a, c)$ and $\mathbf{T}(a, c, c)$ and even $\mathbf{T}(a, a, a)$. Hilbert used strict betweenness. Both of them wrote $\mathbf{B}(a, b, c)$, which is a potential source of confusion. We therefore reserve $\mathbf{B}$ for strict betweenness, and use $\mathbf{T}$ for Tarski's non-strict betweenness. The fact that Tarski's initial is 'T' should serve as a mnemonic device. Of course the equality relation between points is also part of the language.

2.3   Betweenness and Congruence Axioms

We write $ab = cd$ instead of $E(a, b, c, d)$ to enhance human readability. In OTTER files of course we use $E(a, b, c, d)$. The following are five axioms from the 1965 system.

$ab = ba$	(A1) reflexivity for equidistance
$ab = pq \land ab = rs \ \rightarrow \ pq = rs$	(A2) transitivity for equidistance
$ab = cc \ \rightarrow \ a = b$	(A3) identity for equidistance
$\exists x \, (\mathbf{T}(q, a, x) \land ax = bc)$	(A4) segment extension
$\mathbf{T}(a, b, a) \ \rightarrow \ a = b$	(A6) identity for betweenness

When using (A4) in OTTER, we Skolemize it:

$\mathbf{T}(q, a, ext(q, a, b, c)) \land E(a, ext(q, a, b, c), b, c)$ (A4) Skolemized

The original (1948) theory had the following additional fundamental properties of betweenness listed as axioms. (We follow the numbering of [12].)

$\mathbf{T}(a, b, b)$ (A12) Reflexivity for betweenness

$\mathbf{T}(a, b, c) \;\rightarrow\; \mathbf{T}(c, b, a)$ (A14) Symmetry for betweenness

$\mathbf{T}(a, b, d) \wedge \mathbf{T}(b, c, d) \;\rightarrow\; \mathbf{T}(a, b, c)$ (A15) Inner transitivity

$\mathbf{T}(a, b, c) \wedge \mathbf{T}(b, c, d) \wedge b \neq c \;\rightarrow\; \mathbf{T}(a, b, d)$ (A16) Outer transitivity

$\mathbf{T}(a, b, d) \wedge \mathbf{T}(a, c, d) \;\rightarrow\; \mathbf{T}(a, b, c) \vee \mathbf{T}(a, c, b)$ (A17) Inner connectivity

$\mathbf{T}(a, b, c) \wedge \mathbf{T}(a, b, d) \wedge a \neq b$ (A18) Outer connectivity
$\;\rightarrow\; \mathbf{T}(a, c, d) \vee \mathbf{T}(a, d, c)$

Of these only (A15) and (A18) appear in the 1959 version, because in 1956–57 Tarski and his students Kallin and Taylor showed that the other four are dependent (derivable from the remaining axioms). H. N. Gupta showed in his 1965 Ph. D. thesis [4] that (A18) is also dependent. The proof of (A18) is one of Quaife's challenge problems. Gupta also showed that (A15) implies (A6) using the other axioms of the 1959 system. Then one could have dropped (A6) as an axiom; but instead, Szmielew dropped (A15), keeping (A6) instead; then (A15) becomes a theorem.

All six of these axioms occur as theorems in [9]: (A12) is Satz 3.1, (A14) is Satz 3.2, (A15) is Satz 3.5, (A16) is Satz 3.7, (A18) is Satz 5.1, and (A17) is Satz 5.3. Hence, our research program of proving all the theorems in Szmielew's development using OTTER automatically captured these results as soon as we reached Satz 5.3.

2.4   The Five-Segment Axiom

Hilbert [5] treated angles as primitive objects and angle congruence as a primitive relation, and took SAS (the side-angle-side triangle congruence principle) as an axiom. In Tarski's theory, angles are treated as ordered triples of points, and angle congruence is a defined notion, so a points-only formulation of the SAS principle is required. The key idea is Tarski's "five-segment axiom" (A5), shown in Fig. 1.

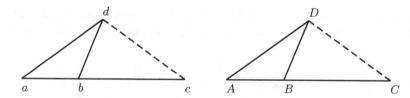

Fig. 1. The five-segment axiom (A5)

If the four solid segments in Fig. 1 are pairwise congruent, then the fifth (dotted) segments are congruent too. This is essentially SAS for triangles dbc and DBC. The triangles abd and ABD are surrogates, used to express the congruence of angles dbc and DBC. By using Axiom A5, we can avoid all mention of angles.

2.5   Pasch's Axiom

Moritz Pasch [6] (See also [7], with an historical appendix by Max Dehn) supplied (in 1882) an axiom that repaired many of the defects that nineteenth-century rigor found in Euclid. Roughly, a line that enters a triangle must exit that triangle. As Pasch formulated it, it is not in $\forall\exists$ form. There are two $\forall\exists$ versions, illustrated in Fig. 2.5. These formulations of Pasch's axiom go back to Veblen [13], who proved outer Pasch implies inner Pasch. Tarski took outer Pasch as an axiom in [11].

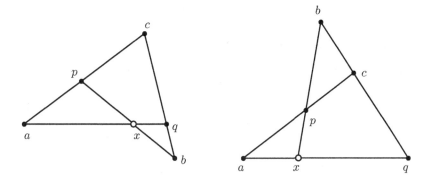

Fig. 2. Inner Pasch (left) and Outer Pasch (right). Line pb meets triangle acq in one side. The open circles show the points asserted to exist on the other side.

$$\mathbf{T}(a,p,c) \wedge \mathbf{T}(b,q,c) \;\rightarrow\; \exists x\,(\mathbf{T}(p,x,b) \wedge \mathbf{T}(q,x,a))\; \text{(A7) inner Pasch}$$
$$\mathbf{T}(a,p,c) \wedge \mathbf{T}(q,c,b) \;\rightarrow\; \exists x\,(\mathbf{T}(a,x,q) \wedge \mathbf{T}(b,p,x))\; \qquad \text{outer Pasch}$$

For use in OTTER, we introduce Skolem symbols $ip(a,p,c,b,q)$ and $op(p,a,b,c,q)$ for the point x asserted to exist.

Tarski originally took outer Pasch as an axiom. In [4], Gupta proved both that inner Pasch implies outer Pasch, and that outer Pasch implies inner Pasch, using the other axioms of the 1959 system. The proof of outer Pasch from inner Pasch is one of Quaife's four challenge problems.

2.6   Dimension Axioms

With no dimension axioms, Tarski's geometry axiomatizes theorems that are true in n-dimensional geometries for all n. For each positive integer n, we can specify that the dimension of space is at least n (with a lower dimension axiom $A8^{(n)}$), or at most n (with an upper dimension axiom $A9^{(n)}$). The upper dimension axiom says (in a first-order way) that the set of points equidistant from n given points is at most a line. The lower dimension axiom for n is the negation of the upper dimension axiom for $n - 1$. For the exact statements of these axioms see [12].

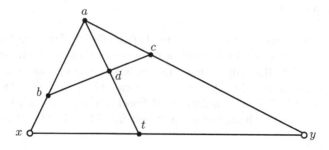

Fig. 3. Tarski's parallel axiom

2.7 Tarski's Parallel Axiom (A10)

In the diagram (Fig. 3), open circles indicate points asserted to exist.

$$\mathbf{T}(a,d,t) \wedge \mathbf{T}(b,d,c) \wedge a \neq d \rightarrow \exists x \exists y \, (\mathbf{T}(a,b,x) \wedge \mathbf{T}(a,c,y) \wedge \mathbf{T}(x,t,y)) \text{ (A10)}$$

The hypothesis says that t lies in the interior of angle a, as witnessed by b, c, and d. The conclusion says that some line through t meets both sides of the angle. Of course this fails in non-Euclidean geometry when both ab and ac are parallel to xy.

According to [12], Szmielew preferred to use the "triangle circumscription principle $(A10_2)$ as the parallel axiom. Substituting $(A10_2)$ was apparently one of the "inessential changes" made by Schwabhäuser. This principle says that if a, b, and c are not collinear, then there exists a point equidistant from all three.

2.8 Continuity Axioms

Axiom schema (A11) is not a single axiom, but an axiom schema, essentially asserting that first-order Dedekind cuts are filled. Models of A1-A11 are all isomorphic to planes $\mathbb{F}^2$ where $\mathbb{F}$ is a real-closed field. One can also consider instead of (A11), the axioms of line-circle continuity and/or circle-circle continuity, which assert the existence of intersection points of lines and circles, or circles and circles, under appropriate hypotheses. None of the continuity axioms are used in the work reported on in this paper. Szmielew's development proceeds strictly on the basis of A1-A10.

3 Methodology

In this section we describe, with illustrative examples, the techniques we used in this project.

3.1 How OTTER Works

Readers familiar with OTTER can skip this section. It is not an introduction to OTTER, but an attempt to make the subsequent information about our methodology comprehensible to those who do not have expertise with OTTER; at least, it should enable such readers to make sense of the input files and proofs we exhibit below and on the project's website [1]. For more information about OTTER, see [17].

OTTER is a clause-based resolution theorem prover. One writes -A for the negation of A. One writes A | B for disjunction ("or"). One does not write "and" at all, but instead one enters the two clauses separately. One writes $A \rightarrow B$ as -A | B. Similarly one writes $P \wedge Q \rightarrow R$ as -P | -Q | R.

Variables begin with x,y,z,w,u,v. Names beginning with any other letter are constants. A resolution theorem-prover requires the goal to be negated and entered as clauses. For example, to prove $A(x) \rightarrow \exists y\, B(x,y)$, we would enter the following clauses:

```
A(c).
-B(c,y).
```

After proving this theorem, if we want to use it to prove the next theorem, we invent a new Skolem symbol f and enter the theorem as -A(x) | B(x,f(x)).

The input to OTTER is contained in two lists, the "set of support" (sos) and "usable". The fundamental run-time loop of OTTER moves a clause from sos to usable, and then tries to use one of the specified inference rules to generate new clauses from that clause and other clauses in usable. If conclusions are generated, OTTER has to decide whether to keep them or not. If it decides to keep them, they are placed on sos, where they can eventually be used to generate yet more new clauses. If the empty clause is generated, that means a proof has been found, and it will be output.

The fundamental problem of automated deduction is to avoid drowning in a sea of useless conclusions before finding the desired proof. One tries to get control over this by assigning "weights" to clauses, adjusting those weights in various ways, and using them to control both which clauses are kept, and which clause is selected from sos for the next iteration of the loop. By default: the weight of a clause is the number of its symbols; the next clause selected is the lightest one in sos; and clauses are kept if their weight does not exceed a parameter max_weight. More sophisticated ways of setting the weights have been developed over the past decades and are discussed below. The idea is to get the weights of the important clauses to be small, and then to squeeze down max_weight to prevent drowning.

In addition to techniques involving weighting, there are other ways to control OTTER's search:

- Use a propitious combination of rules of inference. For an introduction to these rules please refer to [17].
- You have some control over which clause will be selected from sos at the next iteration, using OTTER's pick_given_ratio.
- You have some control over how the search starts and what kind of proof you want to look for (forward, backward, or bi-directional) by choosing which of your clauses to put in sos and which to put in usable.

3.2   Hints

Putting a clause into list(hints) causes OTTER to give that clause, if deduced, a low weight, causing it to be retained, even if its default weight would have been so large as to cause it to be discarded. One has options (specified at the top of an OTTER file) to cause this weight adjustment to apply to clauses that match the hints, or subsume the hints, or are subsumed by the hints. The way we use hints is describe in Section 3.5. The technique of hints was invented by Veroff [14] and later incorporated into OTTER. As a technical note: when using hints, you should always include these lines, without which your hints will not have the desired effect.

```
assign(bsub_hint_wt,-1).
set(keep_hint_subsumers).
```

Another similar technique is known as *resonators*. This is more useful when one has a proof in hand, and wishes to find a shorter proof. For the exact differences between hints and resonators, see [16], p. 259.

3.3   Giving OTTER the Diagram

This refers to a technique we learned from Quaife [8]. Suppose we are trying to find an x such that $A(a, b, x)$. If we let OTTER search, without making any attempt to guide the search, essentially OTTER will generate all the points that can be constructed from a and b (and other points in the problem or previously constructed in clauses that were kept) by the Skolem functions for segment extension and inner Pasch. Imagine trying to prove a geometry theorem that way: just take your ruler and draw all the lines you can between the points of the diagram, and label all the new points formed by their intersections (that is using Pasch), and construct every point that can be constructed by extending a segment you have by another segment you have, and see if any of the new points is the point you are trying to construct, and if not, repeat. You will generate a sea of useless points, even if you discard those with too many construction steps.

In order to guide OTTER to construct the right points, we "give it the diagram" by defining the points to be constructed, using the Skolem functions. For example, consider the "crossbar theorem" (Satz 3.17). See Fig. 4, which shows the diagram and the Otter input (not shown are the axioms A1-A6 and the previous theorems, which are placed in list(usable)). The two lines defining r and q are what we call "giving OTTER the diagram." With those lines present, OTTER finds a proof instantly. Remove them, and OTTER does not find a proof (at least, not in ten minutes).

The reason this works is that q, being a single letter, has weight 1, so clauses involving q have much lower weight than would the same clauses with q replaced by $ip(c, s, a, t, r)$, which has weight 6. A clause involving the terms defining both q and r would have weight at least 14, so it would not be considered very soon, and meantime, many other points will be constructed.

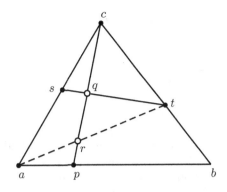

```
list(sos).
T(a,s,c).
T(b,t,c).
T(a,p,b).
-T(p,x,c)|-T(s,x,t).
r = ip(c,t,b,a,p).
q = ip(c,s,a,t,r).
end_of_list.
```

Fig. 4. The crossbar theorem asserts that q exists, given the other points except r. To prove it we construct first r and then q, using inner Pasch twice.

In this simple example, if you put all clauses into sos, and nothing in usable, OTTER does eventually find a proof without being given the diagram, but it is certainly much slower than with the diagram. In other, more complex examples, we think this technique is essential. Here, for example, are the lines from the input file for Satz 5.1, the inner connectivity of betweenness, describing the two rather complicated diagrams on pp. 39–40 of [9]:

```
c1=ext(a,d,c,d).
d1=ext(a,c,c,d).
p=ext(c1,c,c,d).
r=ext(d1,c,c,e).
q=ext(p,r,r,p).
b1 = ext(a,c1,c,b).
b2 = ext(a,d1,d,b).
e = ip(c1,d,b,d1,c).
```

3.4   Assistance with Proof by Cases

We found that OTTER often "got stuck" when Szmielew's proof required an argument by cases. We could sometimes get around this difficulty by simply adding a clause A | -A, where the proof needs to proceed by cases on A. It seems that OTTER prefers constructive proofs! This technique is called "tautology adjunction" by the second author, who used it decades ago in proving that subgroups of index 2 are normal. We used this in many input files. Here we discuss just one example. The inner connectivity of betweenness (A17 above, Satz 5.3 in Szmielew) is derived as an easy corollary of Satz 5.1, which is

$$a \neq b \wedge \mathbf{T}(a,b,c) \wedge \mathbf{T}(a,b,d) \;\rightarrow\; \mathbf{T}(a,c,d) \vee \mathbf{T}(a,d,c).$$

The natural way to formulate `list(sos)` for this problem would be

```
a != b.
T(a,b,c).
T(a,b,d).
-T(a,c,d).
-T(a,d,c).
```

Of course, that does not suffice to find a proof. So, we added the description of the diagram, as given in the previous section. Unfortunately OTTER could still not find a proof, even with hints.

The proof in Szmielew proceeds by cases. The methodology we followed in such cases was this:

- Add one case to sos, e.g. `c=c1`. (`c1` is a constant from the diagram, above.)
- If we find a proof: add the steps of that proof as hints.
- Remove that case, and add its negation, e.g. `c != c1`
- If we find a proof: add its steps also as hints.
- Now remove both cases and add their disjunction: `c = c1 | c != c1`.

Often we could find a proof. The amazing thing, to us, was that the same file with the tautology deleted would often *not* find a proof. We do not understand exactly why tautology adjunction works, but in practice, it often helps.

The example at hand required two divisions into cases (so tautology disjunction was applied recursively). The first is the case whether `c=c1` or not, and the second, whether `d1=e` or not. In the input file, one can see in the two commented lines in `list(sos)` the traces of completing the last argument by cases.

```
% d1 = e.
% d1!= e.
d1=e | d1!=e.
c = c1 | c != c1.
```

We do not mean to imply that this was all there was to proving Satz 5.1. This was just the last difficulty. By that time, the input file already contained a long list of hints obtained by methods described in the next section. The final proof has 131 steps.

3.5   Supplying Proof Steps

Our methodology was as follows: when trying to prove a theorem, we prepared an input file with the negated goal in `list(sos)`, and the axioms and previously proved theorems in list(usable), and the choice of inference rules as described below. If this did not find a proof, and the proof in Szmielew had a diagram, we supplied the diagram. If that still did not work, then we tried supplying some intermediate goals. We would list some important steps of the proof from

Szmielew (in negated form) in `list(passive)`, with answer literals having numbers for those goals. When this file is run, one sees which of the intermediate goals are being proved. Among other things, this showed us where OTTER was "getting stuck." But even though we found no proof of the main goal, we had the proofs of some intermediate goals. We converted these proof steps to hints, and ran OTTER again. Sometimes more intermediate goals would be reached. One can tinker with `max_weight`: sometimes a smaller `max_weight` may keep one from drowning, and sometimes a larger `max_weight` may allow a vital clause to be kept. With luck, this process would converge.

We note a technical point: Exactly how does one extract a list of hints from the proofs in an OTTER output file? The lines of a proof contain extraneous material, such as line numbers of the ancestors and justifications of the inferencea, that must be stripped out. Since this has to be done iteratively and often, and sometimes the list of clauses to be converted is long, one needs an automated method to do this. The first author used a PHP script, and the second author used a Unix shell script, for this purpose.

3.6 Divide and Conquer

If the methods described above still did not produce a proof, we tried to cut the task in half as follows: we picked one of the intermediate steps (already in `list(passive)`, but not being proved), and added its positive form to `list(sos)`. In the abstract, say we are trying to prove conclusion C from hypothesis A. The proof, we think, should proceed by first proving B from A, and then proving C from A and B. If we can't succeed in proving C from A in one run, it makes sense to assume B as well as A, and try to prove C. In a separate run, we try to prove B from A. That may be easier with the negated form of B in `list(sos)` instead of `list(passive)`.

3.7 Choice of Inference Rules and Settings

We mentioned above that one of the ways OTTER can be controlled is through a propitious choice of inference rules and settings. We tinkered with our choices often, in the hope that a different choice would be better. We did not find that one choice was always best. We always used hyperresolution, and we sometimes used unit resolution or binary resolution. Sometimes it helped to turn binary resolution on. Sometimes it helped to turn binary resolution off. Generally, we think it worked better with both hyperresolution and binary resolution allowed, but not always. Occasionally we could "break through" using `sos_queue` or `set(input_sos_first)`. We *often* changed the values of the parameters `max_weight`, `max_distinct_vars`, `pick_given_ratio`, and `max_proofs`. We used paramodulation for equality reasoning. The second author believes that failing to use paramodulation was an important reason for the limited success of the experiments he made in the 1980s in Tarskian geometry; but we also note that Quaife writes that paramodulation was available for use but seldom actually used in his proofs.

3.8   What about Prover9? Or E, Spass, Vampire?

The question inevitably arises, why did we use OTTER instead of the newer provers mentioned? The short answer is, we did use Prover9, and others tried E, Spass, and Vampire. Whenever we could not get a proof with OTTER, the first author would try it in Prover9. But in every case, Prover9 also could not find a proof; and sometimes Prover9 would not succeed when OTTER would. In so-far unpublished work, Narboux *et. al.* tried E, Spass, and Vampire on the theorems from Szmielew, with a 14% success rate. This was in a completely automated mode, so not directly comparable to the 100% success rate we achieved with a human-OTTER team.

4   Results

All the input files and resulting OTTER proofs that we found are posted on the web at [1]. Here we list some of the more difficult proofs we found, with the length of the proofs. It is not meaningful to list the execution times, as the posted input files (for the difficult proofs) were developed iteratively, and contain many hints based on previous runs, so most of the posted input files produce a proof more or less immediately.

4.1   Properties of Betweenness

In section 2.3, we listed six difficult theorems (A12-A18), each of which Tarski originally took as an axiom, and their numbers as theorems in [9]. There is another theorem about betweenness that occurs as a challenge problem in [15], namely the "five-point theorem":

$$\mathbf{T}(z, w, v) \wedge \mathbf{T}(z, y, v) \wedge \mathbf{T}(w, x, y) \;\to\; \mathbf{T}(z, x, v).$$

We found OTTER proofs of all those theorems. The following table gives the length of these proofs, which is perhaps some indication of the relative difficulty of finding them.

A12	Satz 3.1	Reflexivity for betweenness	4 steps
A14	Satz 3.2	Symmetry for betweenness	4 steps
A15	Satz 3.5	Inner transitivity	4 steps
A16	Satz 3.7	Outer transitivity	16 steps
A17	Satz 5.3	Inner connectivity	131 steps
A18	Satz 5.1	Outer connectivity	4 steps

4.2   Midpoints, Perpendiculars, and Isosceles Triangles

The "midpoint theorem" asserts that every segment has a midpoint. The traditional Euclidean construction involves the intersection points of two circles, but we are required to prove the theorem from A1-A9. (Not even the parallel axiom

A10 is to be used.) This is a difficult problem, and was apparently not solved until Gupta's 1965 thesis [4]. Two important preliminary steps are the erection of a perpendicular to a line at a given point, and the "Lotsatz", which says we can drop a perpendicular to a line from a point not on the line. Remember this must be done without circles! A very clever observation of Gupta was that it is easy to construct the midpoint of ab if ab is the base of an isosceles triangle (only two applications of inner Pasch are needed). This plays a key role in the proof of the Lotsatz. The two theorems on perpendiculars are used to construct the midpoint. Of course, once we have midpoints and perpendiculars, it is trivial to show that every segment is the base of an isosceles triangle; that theorem does not even occur explicitly in [9]. An important lemma used in the proofs of these theorems is the "Krippenlemma" (too long to state here). Here are the lengths of our proofs of these theorems:

Satz 7.22	Krippenlemma	132 steps
Satz 7.25	Base of isosceles triangle has a midpoint	123 steps
Satz 8.18	Lotsatz: there is a perpendicular to a line from a point not on the line	332 steps
Satz 8.21a	There is a perpendicular to a line through a point on the line on the opposite side from a given point not on the line.	108 steps
Satz 8.22b	Given segment ab and perpendiculars ap and qb, and point t on line ab between p and q, with $ap \leq qb$, then segment ab has a midpoint.	233 steps
Satz 8.22	Every segment has a midpoint	23 steps

4.3 The Diagonals of a Rhombus Bisect Each Other

One of the challenges in [15], solved by Quaife, was to prove that the diagonals of a rectangle bisect each other. That problem assumed that opposite sides are equal and the diagonals meet in the midpoint of one diagonal; you are to prove that the other diagonal is also bisected. A more general problem is found in Satz 7.21, which asserts that in a quadrilateral with opposite sides equal, the diagonals meet, and bisect each other. Our proof of this theorem has 31 steps.

4.4 Inner and Outer Pasch

The proof that inner Pasch implies outer Pasch (using A1-A6 and A8) was one of the major results of Gupta's thesis, and enabled Szmielew to replace outer Pasch by inner Pasch as an axiom. This theorem was one of Quaife's four challenges. It is Satz 9.6 in [9]. The proof posted on our archive is 111 steps, preceded by proofs Satz 9.4 and Satz 9.5 of 57 and 45 steps. Satz 9.5 is the "plane separation theorem", important in its own right.

4.5 Hilbert's Axioms

Hilbert's theory can be interpreted in Tarski's, using pairs of points for lines and ordered triples of points for angles and planes. His axioms (so interpreted) all turn out to be either axioms, theorems proved in [9], or extremely elementary consequences of theorems proved in [9]. The theorems of [9] needed are 2.3,2.4,2.5,2.8; 3.2,3.13;6.16, 6.18; 8.21, 8.22; 9.8, 9.25, 9.26, 11.15, 11.49, and finally Hilbert's parallel axiom is Satz 12.11. We have posted OTTER proofs of all these theorems. Narboux and Braun have proof-checked Hilbert's axioms in Tarskian geometry, using Coq [2].

5 Discussion

5.1 Proof Checking *vs.* Proof Finding

"Proof checking" refers to obtaining computer-verified proofs, starting with human-written proofs. "Proof finding" refers to the traditional task of automated deduction, finding a proof by searching a large space of possible proofs, either without possessing a proof or without making use of a known proof. The use of hints blurs this distinction, as we shall now explain. If we have a proof in hand (whether generated by human or machine), and we enter its steps as hints, with a low `max_weight`, we force OTTER to find a proof containing mostly the same formulas as the proof in the hints. (The order of deductions might be different.) Thus we can almost always ensure that Otter finds a proof, if we have a proof in hand. One could plausibly claim that this is proof-checking, not proof-finding.

What if, instead of entering all the steps as hints, we enter some key steps in `list(passive)`, and generate some proofs of some of those steps, and put those steps in as hints? Now are we doing proof-checking, or proof-finding? What may have appeared to be a clearcut distinction turns out to be a continuum of possibilities.

5.2 1992 *vs.* 2014

The research reported here shows how much progress has occurred in automated reasoning in that time period. Indeed, approximately thirty years ago, almost all of the theorems cited in this article were out of reach. The question arises, whether this advance might be due simply to the increased memory capacity and speed of modern computers. Perhaps Quaife, equipped with one of our computers, would have found these proofs? Perhaps we, constrained to run on a computer from 1990, might not have found them? We argue that this is not the case: the improvements are due not to faster hardware but to the techniques described above, namely generating partial proofs (of intermediate steps) and using their steps as hints; using the right combination of inference rules and settings; using tautology adjunction to help with proofs by cases; and divide-and-conquer. We note that Quaife did have (in fact, invented) the technique of giving OTTER the diagram. We did not actually try to run on a 1990 computer, and we do not doubt that it would have been painful and discouraging; but we think the main credit should go to Veroff's invention of hints, and the uses of hints developed by the second author and applied here.

6   Conclusions

We used OTTER to find proofs of the theorems in Tarskian geometry in the first nine chapters of Szmielew's development in Part I of [9]. Those theorems include the four unsolved challenge problems from Quaife's book[8], and the verification of Hilbert's axioms.

Input files and the resulting proofs for all the theorems that we have proved from Szmielew are archived at [1], where they can be viewed or downloaded. The second author has also conducted many experiments aimed at shortening some of these proofs or finding forward proofs instead of backwards or bidirectional proofs; we have not reported on those experiments here.

References

1. Beeson, M.: The Tarski formalization project,
   http://www.michaelbeeson.com/research/FormalTarski/index.php
2. Braun, G., Narboux, J.: From Tarski to Hilbert. In: Ida, T., Fleuriot, J. (eds.) Automated Deduction in Geometry 2012 (2012)
3. Caviness, B.F., Johnson, J.R. (eds.): Quantifier Elimination and Cylindrical Algebraic Decomposition. Springer, Wien/New York (1998)
4. Gupta, H.N.: Contributions to the Axiomatic Foundations of Geometry. Ph.D. thesis, University of California, Berkeley (1965)
5. Hilbert, D.: Foundations of Geometry (Grundlagen der Geometrie), 2nd English edn. Open Court, La Salle (1960), translated from the tenth German edition by Leo Unger. Original publication date (1899)
6. Pasch, M.: Vorlesung über Neuere Geometrie. Teubner, Leipzig (1882)
7. Pasch, M., Dehn, M.: Vorlesung über Neuere Geometrie. B. G. Teubner, Leipzig (1926), The 1st edn. (1882), which is the one digitized by Google Scholar, does not contain the appendix by Dehn
8. Quaife, A.: Automated Development of Fundamental Mathematical Theories. Springer, Heidelberg (1992)
9. Schwabhäuser, W., Szmielew, W., Tarski, A.: Metamathematische Methoden in der Geometrie: Teil I: Ein axiomatischer Aufbau der euklidischen Geometrie. In: Teil II: Metamathematische Betrachtungen (Hochschultext). Springer (1983); reprinted 2012 by Ishi Press, with a new foreword by Michael Beeson
10. Tarski, A.: A decision method for elementary algebra and geometry. Tech. Rep. R-109, 2nd revised edn., reprinted in [3], pp. 24–84. Rand Corporation (1951)
11. Tarski, A.: What is elementary geometry? In: Henkin, L., Suppes, P., Tarksi, A. (eds.) The Axiomatic Method, with Special Reference to Geometry and Physics. Proceedings of an International Symposium held at the Univ. of Calif., Berkeley, December 26, 1957-January 4, 1958. Studies in Logic and the Foundations of Mathematics, pp. 16–29. North-Holland, Amsterdam (1959); available as a 2007 reprint, Brouwer Press, ISBN 1-443-72812-8
12. Tarski, A., Givant, S.: Tarski's system of geometry. The Bulletin of Symbolic Logic 5(2), 175–214 (1999)
13. Veblen, O.: A system of axioms for geometry. Transactions of the American Mathematical Society 5, 343–384 (1904)
14. Veroff, R.: Using hints to increase the effectiveness of an automated reasoning program. Journal of Automated Reasoning 16(3), 223–239 (1996)

15. Wos, L.: Automated reasoning: 33 basic research problems. Prentice Hall, Englewood Cliffs (1988)
16. Wos, L.: Automated reasoning and the discovery of missing and elegant proofs. Rinton Press, Paramus (2003)
17. Wos, L., Pieper, G.W.: A fascinating country in the world of computing. World Scientific (1999)

NESCOND: An Implementation of Nested Sequent Calculi for Conditional Logics

Nicola Olivetti[1] and Gian Luca Pozzato[2]

[1] Aix-Marseille Université, CNRS, LSIS UMR 7296 - France
nicola.olivetti@univ-amu.fr
[2] Dipartimento di Informatica - Universitá di Torino - Italy
gianluca.pozzato@unito.it

Abstract. We present NESCOND, a theorem prover for normal conditional logics. NESCOND implements some recently introduced NESted sequent calculi for propositional CONDitional logics CK and some of its significant extensions with axioms ID, MP and CEM. It also deals with the *flat* fragment of CK+CSO+ID, which corresponds to the logic C introduced by Kraus, Lehmann and Magidor. NESCOND is inspired by the methodology of leanT^AP and it is implemented in Prolog. The paper shows some experimental results, witnessing that the performances of NESCOND are promising. The program NESCOND, as well as all the Prolog source files, are available at http://www.di.unito.it/~pozzato/nescond/

1 Introduction

Conditional logics are extensions of classical logic by a *conditional* operator $\Rightarrow$. They have a long history [10, 11], and recently they have found an interest in several fields of AI and knowledge representation. Just to mention a few (see [2] for a complete bibliography), they have been used to reason about prototypical properties, to model belief change, to reason about access control policies, to formalize epistemic change in a multi-agent setting. Conditional logics can also provide an axiomatic foundation of nonmonotonic reasoning [9]: here a conditional $A \Rightarrow B$ is read "normally, if A then B".

In previous works [1, 2] we have introduced *nested sequent calculi*, called $\mathcal{NS}$, for propositional conditional logics. Nested sequent calculi [4–6, 8] are a natural generalization of ordinary sequent calculi where sequents are allowed to occur within sequents. However, a nested sequent always corresponds to a formula of the language, so that we can think of the rules as operating "inside a formula", combining subformulas rather than just combining outer occurrences of formulas as in ordinary sequent calculi. The basic normal conditional logic CK and its extensions with ID, MP and CEM are considered, as well as the cumulative logic C introduced in [9] which corresponds to the *flat* fragment (i.e., without nested conditionals) of the conditional logic CK+CSO+ID.

Here we describe an implementation of $\mathcal{NS}$ in Prolog. The program, called NESCOND, gives a PSPACE decision procedure for the respective logics, and it is inspired by the methodology of leanT^AP [3]. The idea is that each axiom or rule of

S. Demri, D. Kapur, and C. Weidenbach (Eds.): IJCAR 2014, LNAI 8562, pp. 511–518, 2014.
© Springer International Publishing Switzerland 2014

the nested sequent calculi is implemented by a Prolog clause of the program. The resulting code is therefore simple and compact: the implementation of NESCOND for CK consists of only 6 predicates, 24 clauses and 34 lines of code. We provide experimental results by comparing NESCOND with CondLean [12] and GOALD$\mathcal{U}$CK [13]. Performances of NESCOND are promising, and show that nested sequent calculi are not only a proof theoretical tool, but they can be the basis of efficient theorem proving for conditional logics.

2   Conditional Logics and Their Nested Sequent Calculi

We consider a propositional conditional language $\mathcal{L}$ over a set ATM of propositional variables. Formulas of $\mathcal{L}$ are built as usual: $\bot$, $\top$ and the propositional variables of ATM are *atomic formulas*; if A and B are formulas, then $\neg A$ and $A \circ B$ are *complex formulas*, where $\circ \in \{\wedge, \vee, \rightarrow, \Rightarrow\}$. We adopt the *selection function semantics*. We consider a non-empty set of possible worlds $\mathcal{W}$. Intuitively, the selection function f selects, for a world w and a formula A, the set of worlds of $\mathcal{W}$ which are *closest* to w given the information A. A conditional formula $A \Rightarrow B$ holds in a world w if the formula B holds in *all the worlds selected by f for w and A*.

Definition 1 (Selection function semantics). *A model is a triple* $\mathcal{M} = \langle \mathcal{W}, f, [\ ] \rangle$ *where* $\mathcal{W}$ *is a non empty set of worlds,* $f : \mathcal{W} \times 2^{\mathcal{W}} \longmapsto 2^{\mathcal{W}}$ *is the* selection function, *and* $[\ ]$ *is the* evaluation function, *which assigns to an atom* $P \in ATM$ *the set of worlds where* P *is true, and is extended to boolean formulas as follows:* $[\top] = \mathcal{W}$; $[\bot] = \emptyset$; $[\neg A] = \mathcal{W} - [A]$; $[A \wedge B] = [A] \cap [B]$; $[A \vee B] = [A] \cup [B]$; $[A \rightarrow B] = [B] \cup (\mathcal{W} - [A])$; $[A \Rightarrow B] = \{w \in \mathcal{W} \mid f(w, [A]) \subseteq [B]\}$. *A formula* $F \in \mathcal{L}$ *is valid in a model* $\mathcal{M} = \langle \mathcal{W}, f, [\ ] \rangle$, *and we write* $\mathcal{M} \models F$, *if* $[F] = \mathcal{W}$. *A formula* $F \in \mathcal{L}$ *is valid, and we write* $\models F$, *if it is valid in every model, that is to say* $\mathcal{M} \models F$ *for every* $\mathcal{M}$.

The semantics above characterizes the *basic conditional system*, called CK, where no specific properties of the selection function are assumed. An axiomatization of CK is given by ($\vdash$ denotes provability in the axiom system):

- any axiomatization of the classical propositional calculus (prop)
- If $\vdash A$ and $\vdash A \rightarrow B$, then $\vdash B$ (Modus Ponens)
- If $\vdash A \leftrightarrow B$ then $\vdash (A \Rightarrow C) \leftrightarrow (B \Rightarrow C)$ (RCEA)
- If $\vdash (A_1 \wedge \cdots \wedge A_n) \rightarrow B$ then $\vdash (C \Rightarrow A_1 \wedge \cdots \wedge C \Rightarrow A_n) \rightarrow (C \Rightarrow B)$ (RCK)

Moreover, we consider the following standard extensions of the basic system CK:

SYSTEM	AXIOM	MODEL CONDITION
ID	$A \Rightarrow A$	$f(w, [A]) \subseteq [A]$
CEM	$(A \Rightarrow B) \vee (A \Rightarrow \neg B)$	$\mid f(w, [A]) \mid \leq 1$
MP	$(A \Rightarrow B) \rightarrow (A \rightarrow B)$	$w \in [A]$ implies $w \in f(w, [A])$
CSO	$(A \Rightarrow B) \wedge (B \Rightarrow A) \rightarrow ((A \Rightarrow C) \rightarrow (B \Rightarrow C))$	$f(w, [A]) \subseteq [B]$ and $f(w, [B]) \subseteq [A]$ implies $f(w, [A]) = f(w, [B])$

In Figure 1 we present nested sequent calculi $\mathcal{N}S$, where S is an abbreviation for CK$\{+X\}$, and X $\in \{$CEM, ID, MP, ID+MP, CEM+ID$\}$. A nested sequent Γ is defined inductively as follows: a formula of $\mathcal{L}$ is a nested sequent; if A is a formula and Γ is a nested sequent, then $[A : \Gamma]$ is a nested sequent; a finite multiset of nested sequents is a nested sequent. A nested sequent can be displayed as

$$\Gamma(P, \neg P) \quad (AX) \qquad\qquad \Gamma(\top) \quad (AX_\top) \qquad\qquad \Gamma(\neg\bot) \quad (AX_\bot) \qquad\qquad \frac{\Gamma(A)}{\Gamma(\neg\neg A)}(\neg)$$
$${}_{P \in ATM}$$

$$\frac{\Gamma(A) \quad \Gamma(B)}{\Gamma(A \wedge B)}(\wedge^+) \qquad \frac{\Gamma(\neg A, \neg B)}{\Gamma(\neg(A \wedge B))}(\wedge^-) \qquad \frac{\Gamma(A, B)}{\Gamma(A \vee B)}(\vee^+) \qquad \frac{\Gamma(\neg A) \quad \Gamma(\neg B)}{\Gamma(\neg(A \vee B))}(\vee^-)$$

$$\frac{\Gamma(\neg A, B)}{\Gamma(A \to B)}(\to^+) \qquad \frac{\Gamma(A) \quad \Gamma(\neg B)}{\Gamma(\neg(A \to B))}(\to^-) \qquad \frac{\Gamma(\neg(A \Rightarrow B), [C : \Delta, \neg B]) \quad A, \neg C \quad C, \neg A}{\Gamma(\neg(C \Rightarrow B), [A' : \Delta])}(\Rightarrow^-)$$

$$\frac{\Gamma([A : B])}{\Gamma(A \Rightarrow B)}(\Rightarrow^+) \quad \frac{\Gamma(\neg(A \Rightarrow B), A) \quad \Gamma(\neg(A \Rightarrow B), \neg B)}{\Gamma(\neg(A \Rightarrow B))}(MP) \quad \frac{\Gamma([A : \Delta, \Sigma], [B : \Sigma]) \quad A, \neg B \quad B, \neg A}{\Gamma([A : \Delta], [B : \Sigma])}(CEM)$$

$$\frac{\Gamma([A : \Delta, \neg A])}{\Gamma([A : \Delta])}(ID) \qquad \frac{\Gamma, \neg(A \Rightarrow B), [A' : \Delta, \neg B] \quad \Gamma, \neg(A \Rightarrow B), [A : A'] \quad \Gamma, \neg(A \Rightarrow B), [A' : A]}{\Gamma, \neg(A \Rightarrow B), [A' : \Delta]}(CSO)$$

Fig. 1. The nested sequent calculi $\mathcal{N}S$

$$A_1, \ldots, A_m, [B_1 : \Gamma_1], \ldots, [B_n : \Gamma_n],$$

where $n, m \geq 0$, $A_1, \ldots, A_m, B_1, \ldots, B_n$ are formulas and $\Gamma_1, \ldots, \Gamma_n$ are nested sequents. A nested sequent can be directly interpreted as a formula by replacing "," by $\vee$ and ":" by $\Rightarrow$, i.e. the interpretation of $A_1, \ldots, A_m, [B_1 : \Gamma_1], \ldots, [B_n : \Gamma_n]$ is inductively defined by the formula $\mathcal{F}(\Gamma) = A_1 \vee \ldots \vee A_m \vee (B_1 \Rightarrow \mathcal{F}(\Gamma_1)) \vee \ldots \vee (B_n \Rightarrow \mathcal{F}(\Gamma_n))$.

We have also provided nested sequent calculi for the flat fragment, i.e. without nested conditionals, of CK+CSO+ID, corresponding to KLM logic **C** [9]. The rules of the calculus, called $\mathcal{N}\mathbf{C}_{KLM}$, are those ones of $\mathcal{N}$CK+ID (restricted to the flat fragment) where the rule $(\Rightarrow^-)$ is replaced by the rule (CSO).

In order to present the rules of the calculus, we need the notion of context. Intuitively a context denotes a "hole", a *unique* empty position, within a sequent that can be filled by a sequent. We use the symbol ( ) to denote the empty context. A context is defined inductively as follows: $\Gamma( ) = \Delta, ( )$ is a context; if $\Sigma( )$ is a context $\Gamma( ) = \Delta, [A : \Sigma( )]$ is a context. Finally, we define the result of filling "the hole" of a context by a sequent. Let $\Gamma( )$ be a context and Δ be a sequent, then the sequent obtained by filling the context by Δ, denoted by $\Gamma(\Delta)$ is defined as follows: if $\Gamma( ) = \Lambda, ( )$ then $\Gamma(\Delta) = \Lambda, \Delta$; if $\Gamma( ) = \Lambda, [A : \Sigma( )]$ then $\Gamma(\Delta) = \Lambda, [A : \Sigma(\Delta)]$. The notions of derivation and of derivable sequent are defined as usual. In [1] we have shown that:

Theorem 1. *The nested sequent calculi $\mathcal{N}S$ are sound and complete for the respective logics, i.e. a formula F of $\mathcal{L}$ is valid in CK+X if and only if it is derivable in $\mathcal{N}S$.*

As usual, in order to obtain a decision procedure for the conditional logics under consideration, we have to control the application of the rules $(\Rightarrow^-)/(CSO)$, (MP), (CEM), and (ID) that otherwise may be applied infinitely often in a backward proof search, since their principal formula is copied into the respective premise(s). We obtain a sound, complete and terminating calculus if we restrict the applications of these rules as follows [1, 2]: $(\Rightarrow^-)$ can be applied only once to each formula $\neg(A \Rightarrow B)$ with a context $[C : \Delta]$ in each branch, the same for (CSO) in the system CK+CSO+ID; (ID) can be applied only once to each context $[A : \Delta]$ in each branch; (MP) can be applied only once to each formula $\neg(A \Rightarrow B)$ in each branch. For systems with (CEM), we need a more complicated mechanism: due to space limitations, we refer to [1] for this case. These results give a PSPACE decision procedure for their respective logics.

3 Design of NESCOND

In this section we present a Prolog implementation of the nested sequent calculi $\mathcal{NS}$. The program, called NESCOND (NESted sequent calculi for CONDitional logics), is inspired by the "lean" methodology of lean$T^A P$, even if it does not follow its style in a rigorous manner. The program comprises a set of clauses, each one of them implements a sequent rule or axiom of $\mathcal{NS}$. The proof search is provided for free by the mere depth-first search mechanism of Prolog, without any additional ad hoc mechanism.

NESCOND represents a nested sequent with a Prolog list, whose elements can be either formulas F or pairs [Context, AppliedConditionals] where:

- Context is also a pair of the form [F, Gamma], where F is a formula of $\mathcal{L}$ and Gamma is a Prolog list representing a nested sequent;
- AppliedConditionals is a Prolog list [A_1=>B_1, A_2=>B_2, ..., A_k=> B_k], keeping track of the negated conditionals to which the rule $(\Rightarrow^-)$ has been already applied by using Context in the current branch. This is used in order to implement the restriction on the application of the rule $(\Rightarrow^-)$ in order to ensure termination.

Symbols $\top$ and $\bot$ are represented by constants true and false, respectively, whereas connectives $\neg, \wedge, \vee, \rightarrow,$ and $\Rightarrow$ are represented by !, ^, v, ->, and =>.

As an example, the Prolog list [p, q, !(p => q), [[p, [q v !p, [[p,[p => r]],[]], !r]],[p => q]], [[q, [p, !p]],[]]] represents the nested sequent $P, Q, \neg(P \Rightarrow Q), [P : Q \vee \neg P, [P : P \Rightarrow R], \neg R], [Q : P, \neg P]$. Furthermore, the list [p => q] in the leftmost context is used to represent the fact that, in a backward proof search, the rule $(\Rightarrow^-)$ has already been applied to $\neg(P \Rightarrow Q)$ by using $[P : Q \vee \neg P, [P : P \Rightarrow R], \neg R]$.

Auxiliary Predicates. In order to manipulate formulas "inside" a sequent, NESCOND makes use of the three following auxiliary predicates:

- deepMember(+Formulas, +NS) succeeds if and only if either (i) the nested sequent NS representing a nested sequent Γ contains all the formulas in the list Formulas or (ii) there exists a context [[A, Delta], AppliedConditionals] in NS such that deepMember(Formulas, Delta) succeeds, that is to say there is a nested sequent occurring in NS containing all the formulas of Formulas.
- deepSelect(+Formulas, +NS, -NewNS) operates exactly as deepMember, however it removes the formulas of the list Formulas by replacing them with a place-holder hole; the output term NewNS matches the resulting sequent.
- fillTheHole(+NewNS, +Formulas, -DefNS) replaces hole in NewNS with the formulas in the list Formulas. DefNS is the output term matching the result.

NESCOND for CK. The calculi $\mathcal{NS}$ are implemented by the predicate prove(+NS, -ProofTree). This predicate succeeds if and only if the nested sequent represented by the list NS is derivable. When it succeeds, the output term ProofTree matches with a representation of the derivation found by the prover. For instance, in order to prove that the formula $(A \Rightarrow (B \wedge C)) \rightarrow (A \Rightarrow B)$ is valid in CK, one queries NESCOND with the goal: `prove([[(a => b ^ c) -> (a => b)],ProofTree)`. Each clause of `prove` implements an axiom or rule of $\mathcal{NS}$. To search a derivation of a nested sequent Γ, NESCOND proceeds as follows. First of all, if Γ is an axiom, the goal will succeed immediately by using one of the following clauses for the axioms:

```
prove(NS,tree(ax)):-deepMember([P,!P],NS),!.
prove(NS,tree(axt)):-deepMember([top],NS),!.
prove(NS,tree(axb)):-deepMember([!bot],NS),!.
```

implementing (AX), $(AX_\top)$ and $(AX_\perp)$, respectively. If Γ is not an instance of the axioms, then the first applicable rule will be chosen, e.g. if a nested sequent in Γ contains a formula A v B, then the clause implementing the $(\vee^+)$ rule will be chosen, and NESCOND will be recursively invoked on the unique premise of $(\vee^+)$. NESCOND proceeds in a similar way for the other rules. The ordering of the clauses is such that the application of the branching rules is postponed as much as possible.

As an example, the clause implementing $(\Rightarrow^-)$ is as follows:

```
1.   prove(NS,tree(condn,A,B,Sub1,Sub2,Sub3)):-
2.       deepSelect([!(A => B),[[C,Delta],AppliedConditionals]],
                                              NS,NewNS),
3.       \+member(!(A => B),AppliedConditionals),
4.       prove([A,!C],Sub2),
5.       prove([C,!A],Sub3),!,
6.       fillTheHole(NewNS,[!(A => B),[[C,[!B|Delta]]
                      ,[!(A => B)|AppliedConditionals]]],DefNS),
7.       prove(DefNS,Sub1).
```

In line 2, the auxiliary predicate `deepSelect` is invoked in order to find both a negated conditional $\neg(A \Rightarrow B)$ and a context $[C : \Delta]$ in the sequent (even in a nested subsequent). In this case, such formulas are replaced by the placeholder `hole`. Line 3 implements the restriction on the application of $(\Rightarrow^-)$ in order to guarantee termination: the rule is applied only if $\neg(A \Rightarrow B)$ does not belong to the list `AppliedConditionals` of the selected context. In lines 4, 5 and 7, NESCOND is recursively invoked on the three premises of the rule. In line 7, NESCOND is invoked on the premise in which the context $[C : \Delta]$ is replaced by $[C : \Delta, \neg B]$. To this aim, in line 6 the auxiliary predicate `fillTheHole(+NewNS,+Formulas,-DefNS)` is invoked to replace the `hole` in `NewNS`, introduced by `deepSelect`, with the negated conditional $\neg(A \Rightarrow B)$, which is copied into the premise, and the context $[C : \Delta, \neg B]$, whose list of `AppliedConditionals` is updated by adding the formula $\neg(A \Rightarrow B)$ itself.

NESCOND for Extensions of CK. The implementation of the calculi for extensions of CK with axioms ID and MP are very similar. For systems allowing ID, contexts are triples `[Context, AppliedConditionals, AllowID]`. The third element `AllowID` is a flag used in order to implement the restriction on the application of the rule (ID), namely the rule is applied to a context only if `AllowID=true`, as follows:

```
prove(NS,tree(id,A,SubTree)):-
    deepSelect([[[A,Delta],AppliedConditionals,true]]],
                                      NS,NewNS),!,
    fillTheHole(NewNS,[[[A,[!A|Delta]],AppliedConditionals,
                             false]]],DefNS),
    prove(DefNS,SubTree).
```

When (ID) is applied to [Context, AppliedConditionals, true], then the predicate `prove` is invoked on the unique premise of the rule `DefNS`, and the flag is set to `false` in order to avoid multiple applications in a backward proof search.

The restriction on the application of the rule (MP) is implemented by equipping the predicate `prove` by a third argument, `AppliedMP`, keeping track of the negated conditionals to which the rule has already been applied in the current branch. The clause of `prove` implementing (MP) is:

```
1.  prove(NS,AppliedMP,tree(mp,A,B,Sub1,Sub2)):-
2.      deepSelect([[!(A => B)],NS,NewNS),
3.      \+member(A => B,AppliedMP),!,
4.      fillTheHole(NewNS,[A,!(A => B)],NS1),
5.      fillTheHole(NewNS,[!B,!(A => B)],NS2),
6.      prove(NS1,[A => B|AppliedMP],Sub1),
7.      prove(NS2,[A => B|AppliedMP],Sub2).
```

The rule is applicable to a formula $\neg(A \Rightarrow B)$ only if [A => B] does not belong to `AppliedMP` (line 3). When (MP) is applied, then [A => B] is added to `AppliedMP` in the recursive calls of `prove` on the premises of the rule (lines 6 and 7).

The implementation of the calculus for the flat fragment of CK+CSO+ID, corresponding to KLM cumulative logic **C**, is similar to that for CK+ID; the only difference is that $(\Rightarrow^-)$ is replaced by (CSO). This does not make use of the predicate `deepSelect` to "look inside" a sequent to find the principal formulas $\neg(A \Rightarrow B)$ and $[C : \Delta]$: since the calculus only deals with the *flat* fragment of the logic under consideration, such principal formulas are directly selected from the current sequent by easy membership tests (standard Prolog predicates `member` and `select`), without searching inside other contexts. Due tu space limitations, we omit details for extensions with CEM.

4   Performance of NESCOND

The performances of NESCOND are promising. We have tested it by running SICStus Prolog 4.0.2 on an Apple MacBook Pro, 2.7 GHz Intel Core i7, 8GB RAM machine. We have compared the performances of NESCOND with the ones of two other provers for conditional logics: CondLean 3.2, implementing labelled sequent calculi [12], and the goal-directed procedure GOALD$\mathcal{U}$CK [13]. We have tested the three provers (i) on randomly generated sequents, obtaining the results shown in Figure 2 and (ii) over a set of valid formulas. Concerning CK, we have considered 88 valid formulas obtained by translating K valid formulas ($\Box A$ is replaced by $\top \Rightarrow A$, whereas $\Diamond A$ is replaced by $\neg(\top \Rightarrow \neg A)$) provided by Heuerding, obtaining the results in Figure 3.

Concerning (i), we have tested the three provers over 2000 random sequents, whose formulas are built from 15 different atomic variables and have a high level of nesting (10): NESCOND is not able to answer only in 0.05% of cases (1 sequent over 2000) within 10 seconds, whereas both GOALD$\mathcal{U}$CK and CondLean are not able to conclude anything in more than 3% of cases (60 tests over 2000). If the time limit is extended to 2 minutes, NESCOND answers in 100% of cases, whereas its two competitors have still more than 1.30% of timeouts. The difference is much more significant when considering sequents with a lower level of nesting (3) and whose formulas contain only 3 different atomic variables: with a time limit of 10 seconds, NESCOND is not able to answer only in 9.09% of cases, whereas both CondLean and GOALD$\mathcal{U}$CK are not able to conclude in 16.55% and in 51.15% of cases, respectively: this is explained by the fact that NESCOND is faster than the other provers to find 355 not valid sequents (against 17 of CondLean and 34 of GOALD$\mathcal{U}$CK) within the fixed time limit.

Concerning (ii) the performances of NESCOND are also encouraging. Considering CK, NESCOND is not able to give an answer in less than 10 seconds only in 5 cases over 88, against the 8 of CondLean and the 12 of GOALD$\mathcal{U}$CK; the number of timeouts drops to 4 if we extend the time limit to 1 minute, whereas this extension has no effect on the competitors (still 8 and 12 timeouts). We have similar results also for extensions of CK as shown in Figure 3: here we have not included GOALD$\mathcal{U}$CK since the most formulas adopted do not belong to the fragments admitting goal-directed proofs [13].

number of tests: 2000	Prop. vars: 15, Depth: 10, Timeout: 10 s			number of tests: 2000	Prop. vars: 15, Depth: 10, Timeout: 2 min		
	yes	no	timeout		yes	no	timeout
GoalDUCK	75,65%	21,35%	3,00%	GoalDUCK	74,50%	23,65%	1,85%
CondLean	77,75%	19,10%	3,15%	CondLean	74,90%	23,80%	1,30%
NESCOND	77,75%	22,20%	0,05%	NESCOND	75,40%	24,65%	0,00%

number of tests: 2000	Prop. vars: 3, Depth: 3, Timeout: 10 s		
	yes	no	timeout
GoalDUCK	47,15%	1,7%	51,15%
CondLean	77,6%	0,85%	16,55%
NESCOND	73,2%	17,75%	9,05%

Fig. 2. NESCOND vs CondLean vs GOALD$\mathcal{U}$CK over 2000 random sequents

Timeouts for CK

	100ms	1s	10s	1m
CondLean	12,50%	12,50%	9,09%	9,09%
GoalDUCK	21,71%	18,26%	13,88%	13,88%
NESCOND	12,50%	10,23%	5,68%	4,55%

Timeouts for extensions of CK

	1ms	1s	10s
CondLean	48,08%	36,54%	28,85%
NESCOND	38,46%	28,85%	26,92%

Timeouts for systems without CEM

	1ms	1s	10s
CondLean	19,23%	15,38%	13,46%
NESCOND	5,77%	0,00%	0,00%

Timeouts for systems CEM and CEM+ID

	1ms	1s	10s
CondLean	51,72%	37,93%	27,59%
NESCOND	58,62%	51,72%	48,28%

Fig. 3. NESCOND vs CondLean vs GOALD$\mathcal{U}$CK: timeouts over valid formulas

These results show that the performances of NESCOND are encouraging, probably better than the ones of the other existing provers for conditional logics (notice that there is no set of acknowledged benchmarks for them). Figure 3 shows that this also holds for extensions of CK: for systems not allowing CEM, NESCOND gives an answer in 95% of the tests (all of them are valid formulas) in less than 1ms. The performances worsen in systems with CEM because of the overhead of the termination mechanism.

5   Conclusions and Future Issues

We have presented NESCOND, a theorem prover for conditional logics implementing nested sequent calculi introduced in [1]. Statistics in section 4 show that nested sequent calculi do not only provide elegant and natural calculi for conditional logics, but they are also significant for developing efficient theorem provers for them. In future research we aim to extend NESCOND to other systems of conditional logics. To this regard, we strongly conjecture that adding a rule for the axiom (CS) $(A \wedge B) \rightarrow (A \Rightarrow B)$ will be enough to cover the whole cube of the extensions of CK generated by axioms (ID), (MP), (CEM) and (CS). This will be object of subsequent research. We also aim at comparing the performances of NESCOND with those of CoLoSS [7], a generic-purpose theorem prover for coalgebraic modal logics which can handle also basic conditional logics.

References

1. Alenda, R., Olivetti, N., Pozzato, G.L.: Nested Sequent Calculi for Conditional Logics. In: del Cerro, L.F., Herzig, A., Mengin, J. (eds.) JELIA 2012. LNCS, vol. 7519, pp. 14–27. Springer, Heidelberg (2012)
2. Alenda, R., Olivetti, N., Pozzato, G.L.: Nested Sequents Calculi for Normal Conditional Logics. Journal of Logic and Computation (to appear)
3. Beckert, B., Posegga, J.: leantap: Lean tableau-based deduction. JAR 15(3), 339–358 (1995)
4. Brünnler, K., Studer, T.: Syntactic cut-elimination for common knowledge. Annals of Pure and Applied Logic 160(1), 82–95 (2009)
5. Fitting, M.: Prefixed tableaus and nested sequents. A. Pure App. Log. 163(3), 291–313 (2012)
6. Goré, R., Postniece, L., Tiu, A.: Cut-elimination and proof-search for bi-intuitionistic logic using nested sequents. In: Advances in Modal Logic, vol. 7, pp. 43–66 (2008)
7. Hausmann, D., Schröder, L.: Optimizing Conditional Logic Reasoning within CoLoSS. Electronic Notes in Theoretical Computer Science 262, 157–171 (2010)
8. Kashima, R.: Cut-free sequent calculi for some tense logics. St. Logica 53(1), 119–136 (1994)
9. Kraus, S., Lehmann, D., Magidor, M.: Nonmonotonic reasoning, preferential models and cumulative logics. Artificial Intelligence 44(1-2), 167–207 (1990)
10. Lewis, D.: Counterfactuals. Basil Blackwell Ltd. (1973)
11. Nute, D.: Topics in conditional logic. Reidel, Dordrecht (1980)
12. Olivetti, N., Pozzato, G.L.: CondLean 3.0: Improving Condlean for Stronger Conditional Logics. In: Beckert, B. (ed.) TABLEAUX 2005. LNCS (LNAI), vol. 3702, pp. 328–332. Springer, Heidelberg (2005)
13. Olivetti, N., Pozzato, G.L.: Theorem Proving for Conditional Logics: CondLean and Goal-Duck. Journal of Applied Non-Classical Logics (JANCL) 18(4), 427–473 (2008)

Knowledge Engineering for Large Ontologies with Sigma KEE 3.0

Adam Pease[1] and Stephan Schulz[2]

[1] Articulate Software
apease@articulatesoftware.com
[2] Institut für Informatik, Technische Universität München
schulz@eprover.org

Abstract. The Suggested Upper Merged Ontology (SUMO) is a large, comprehensive ontology stated in higher-order logic. It has co-evolved with a development environment called the Sigma Knowledge Engineering Environment (SigmaKEE). A large and important subset of SUMO can be expressed in first-order logic with equality. SigmaKEE has integrated different reasoning systems in the past, but they either had to be significantly modified, or integrated in a way that multiple queries to the same theory required expensive full re-processing of the full knowledge base.

To overcome this problem, to create a simpler system configuration that is easier for users to install and manage, and to integrate a state-of-the-art theorem prover we have now integrated Sigma with the E theorem prover. The E distribution includes a simple server version that loads and indexes the full knowledge base, and supports interactive queries via a simple interface based on text streams. No special modifications to E were necessary for the integration, so SigmaKEE can be easily upgraded to future versions.

1 Introduction

The Suggested Upper Merged Ontology (SUMO) [6,7] is a large, comprehensive ontology stated in higher-order logic [2]. It has co-evolved with a development environment called the *Sigma Knowledge Engineering Environment* (SigmaKEE) [8].

SUMO and Sigma have been employed in applications for natural language understanding [9], database modeling [11] and sentiment analysis [10], among others.

A large and important subset of SUMO can be expressed in first-order logic with equality. This subset has been used in several instances of of the LTB division of the yearly CADE ATP System Competition (CASC)[13,18,14] The LTB (*Large Theory Batch*) division of CASC is concerned with reasoning in large theories, and in particular with the task of answering a series of queries over a large, relatively static background theory.

Since a large part of SUMO can be expressed in first-order logic, SigmaKEE provides first-order reasoning capabilities to support the user in interacting with

S. Demri, D. Kapur, and C. Weidenbach (Eds.): IJCAR 2014, LNAI 8562, pp. 519–525, 2014.

the knowledge base. Earlier versions of SigmaKEE have been integrated with a customized special purpose version of Vampire [15]. However, later versions of Vampire were not compatible with the customized version and the integration interface with SigmaKEE. As a consequence, Sigma did not have access to the latest deduction technologies.

An interim measure has been integration with the TPTPworld environment [22] that supports remote access to the entire suite of theorem provers competing in CASC, also adding facilities for generating explicit proof objects even with provers that do not natively have that capability. In recent years, an initial integration has also been done for the LEO-II higher-order logic prover [1]. However, both approaches require re-processing of the full ontology for each query, and thus result in significant overhead, resulting in noticeable delays even for simple user queries.

To overcome this problem, to use the latest first-order theorem proving technology, and to create a simpler system configuration that is easier for users to install and manage, SigmaKEE 3.0 now integrates the well-known theorem prover E [16,17]. The E distribution includes a simple server version that loads and indexes the full knowledge base, and supports interactive queries via a simple interface based on text streams. No special modifications to E were necessary for the integration, so Sigma users will be able to upgrade to successive versions of E as they are released.

2   Architecture and User Interface

Sigma is a Java and JSP system that typically will run under the Apache Tomcat web server. It consists of a set of tools integrated at the user level by an HTML interface. These include:

- *translation* of theories to and from THF, TPTP, SUO-KIF, OWL and Prolog formats
- *mapping* theories to other theories based on similarity of terms names and definitions
- structured *browsing* of hyperlinked and sorted theory content, including tree-structured presentation of hierarchies
- *natural language paraphrases* of theories in many different languages
- structured browsing of *WordNet* [4] and Open Multilingual Wordnet [3] and their links to SUMO
- various kinds of *static analysis* tools for theories, as well as structured inference using theorem proving as a client that attempts to find theory contradictions

The core of the system consists of data structures to manage knowledge bases, their constituent files, and the statements contained in the files. Analysis, display, natural language processing and inference components have been added incrementally around the common data structures. In particular, the facilities to handle TPTP input and output, and integration with the TPTPworld environment were added to support development for the CASC competition. Sigma is an Integrated Development Environment for ontologies in the same sense Eclipse is an IDE for Java programming. While actual development of theories is performed in a text editor, a suite of tools assists in the process of developing text files of SUMO-based theories and a typical development process involves having both Sigma and a text editor running, and the developer frequently switching attention between the two.

By integrating E into Sigma, SUMO developers can more rapidly test new theories for consistency, as well as applying them in applications involving automated deduction. Sigma supports a hyperlinked proof output that links steps in each proof to display of the statements from which they are derived.

Users interact with Sigma via a web browser. The user interface is straightforward, consisting of a window in which queries or statements are made in SUO-KIF format, and hyperlinked proof results, which are similar to a standard textbook proof. Steps in the proof are presented along with a brief justification of how they were derived, whether directly asserted to the knowledge base, or derived via rules of inference from previous steps.

In a typical workflow, a user writes theory content in a text editor and periodically loads the file into Sigma. He will frequently use the Sigma browsing tools to inspect other portions of SUMO, for example, to find useful relations and classes to help model the knowledge of the task at hand, or simply to check relation argument orders or argument types. The user can pose queries to E to test the coverage of the new theory content. Debugging a query that doesn't yield an answer often involves breaking down a complex chain of reasoning into smaller and simpler steps, adding knowledge to complete elements of the chain and gradually building up to the full line of reasoning desired. This will involve running queries, checking proofs of lemmas, checking existing portions of SUMO to see if knowledge that might be assumed present is actually present, and in the desired form. Based on iterations of this process, the user will keep adding new statements to a theory in a text editor until the modeling or application task is complete. In typical usage, this development process is broadly quite similar to modern programming, just that the language is strictly declarative, rather than functional or procedural, and therefore the analysis and development tools themselves are different. But the cycle of development in a text editor, use of analysis, browsing and debugging tools, then further editing or development of the "program" is a familiar one.

3   SigmaKEE/E Integration

SigmaKEE has been integrated with E 1.8[16,17]. E is a powerful theorem prover for full first order logic with equality. It has a number of features that make it an attractive choice for this integration:

- E supports the TPTP standards for input and output. In particular, it reads specifications and writes proof objects in the TPTP-3 language [20] and uses

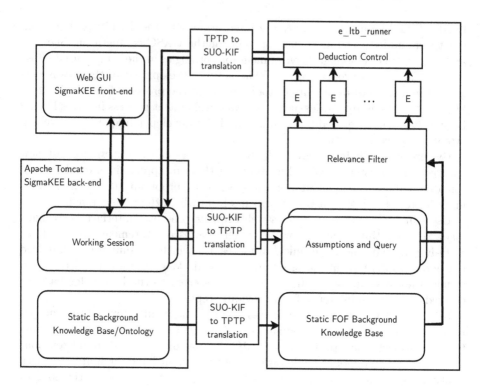

Fig. 1. Main architectural components and data flows of SigmaKEE with respect to the deduction component

the SZS result ontology [19] to signal success or failure of a proof attempt. Thus, the main interfaces are well-defined and results are easy to parse.

- As of version 1.7, E supports the proposed TPTP answer standard [21]. It can report one or multiple answers to queries, i.e. instantiations for top-level existentially quantified variables in the conjecture that make it true. This is particularly valuable for query-answering.
- With version 1.8, E can generate and print checkable proof objects with barely measurable overhead [17]. The proof objects make logical dependencies obvious, and can also help to debug the ontology, in particular by identifying inconsistencies.

In addition to support for well defined I/O standards and fast proof generation, E also provides a prototypical implementation of *deduction as a service*. The e_ltb_runner control program in the E distribution supports the efficient execution of multiple queries over a static background theory, both in batch mode and in interactive mode. This interactive mode forms the base of the interface between the SigmaKEE core and the deduction system.

On start-up, e_ltb_runner reads a specification file that describes the constant background theory, given in the form of files of clauses and formulas in TPTP-3 syntax. This initial knowledge base is indexed with a bi-directional index from formulas to function symbols and vice versa. Moreover, various statistics on the distribution of function symbols in the knowledge base are computed. These pre-computed indices and values allow the efficient application of a parameterized variant of the SInE algorithm [5].

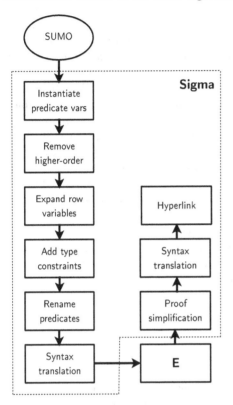

Fig. 2. Language transforms

The program then enters interactive mode and reads input from the user via `stdin`. Users can provide additional formulas, either directly or by specifying TPTP input directives to load axioms from files. Typically, a query consists of a number of additional assumptions and a conjecture (or *question*, if answer instantiations are desired). These formulas are temporarily integrated into the knowledge base and the indices are updated in a way that allows for the efficient retraction of the formulas and resetting of the indices. When the user indicates that the current specification is complete, `e_ltb_runner` runs various different relevancy filters over the extended knowledge base and extracts a number of individual proof problems, each of which contains the conjecture or query and a number of potentially useful axioms. These are handed to different instances of E in automatic mode. If one of the instances finds a proof (or a counter-saturation), all provers are stopped, and the result, along with the proof or the derivation of the saturation, is provided back on the standard output channel. If all instances of E time out or hit other resource limits, the proof attempt fails.

Once the job is processed, the additional formulas are removed from the knowledge base and the indices, and the system waits for the next user command.

Figure 1 provides an overview of the integration with SigmaKEE with E. On start-up, the SigmaKEE back-end translates the static ontology into TPTP format. It starts up `e_ltb_runner` in interactive mode, passing the translated ontology as as the background theory. The back-end connects to the deduction component via `stdin` and `stdout`.

New knowledge entered by the user is kept in a separate working session. When the user wants to query the knowledge base, the formulas of the current working session and the query are translated to TPTP syntax and provided as a job to `e_ltb_runner`. There, they are added to the background knowledge, the relevance filters are applied, and different instances of E try to find an answer to the query. If successful, the proof is handed to the SigmaKEE back-end, where it is translated back to SUO-KIF.

Several transformations are required to convert SUO-KIF into TPTP, and to restore the content in the TPTP3 format proofs to their authored SUO-KIF versions, as shown in Figure 2. While these transforms are described in more detail in [7] and [12] a brief overview here may be helpful. First, variables that are in the predicate position in a rule are removed by instantiating every such rule with every predicate from the knowledge base that is applicable. Next, the remaining higher-order logic content that is not expressible in TPTP FOF syntax is removed. Then row-variables, which stand for multiple arguments in variable-arity relations are expanded, treating this construct as a macro. SUMO requires type constraints for the arguments to all relations. To fully implement this in a non-sorted logic such as TPTP FOF, we prefix all rules with type constraints. While in TPTP-3 implementations any symbol identifier can be used either a function symbol of a given arity or a predicate symbol of a given arity, SUO-KIF does not share this restriction. Hence, in cases where a symbol is used in more than one role, occurrences of one type are renamed.

Lastly, the actual syntactic transformation of SUO-KIF, which conforms to LISP S-Expressions is converted to the Prolog syntax of TPTP. Upon return from E, the SZS ontology tags are extracted to provide the overall status of the result. The proof is simplified to removed repeated appearances of the same statement. Answer variables are removed. The syntactic transform is now run in reverse, converting TPTP statements to SUO-KIF. Finally, predicates are returned to their originally authored names.

4   Conclusion

Sigma KEE 3.0 brings together a practical development environment for creating expressive logical theories and a leading first order theorem prover. It is a start at providing the same sort of powerful development approach for logical theories that programmers have long enjoyed for procedural and object-oriented development.

SigmaKEE and SUMO offer a development tool suite and a reusable library of content on which to build new theories. All the tools are open source, in hopes of inviting collaboration. E is available from http://eprover.org and as part of the Sigma distribution from http://sigmakee.sourceforge.net.

References

1. Benzmüller, C., Pease, A.: Progress in automating higher-order ontology reasoning. In: Konev, B., Schmidt, R., Schulz, S. (eds.) Workshop on Practical Aspects of Automated Reasoning (PAAR 2010). CEUR Workshop Proceedings, Edinburgh, UK (2010)
2. Benzmüller, C., Pease, A.: Reasoning with Embedded Formulas and Modalities in SUMO. In: The ECAI 2010 Workshop on Automated Reasoning about Context and Ontology Evolution (August 2010)

3. Bond, F., Paik, K.: A survey of wordnets and their licenses. In: Proceedings of the 6th Global WordNet Conference (GWC, Matsue, pp, 64–71 (2012)
4. Fellbaum, C.: WordNet: An Electronic Lexical Database. Language, Speech, and Communication. MIT Press (1998), http://books.google.com.hk/books?id=Rehu80OzMIMC
5. Hoder, K., Voronkov, A.: Sine Qua Non for Large Theory Reasoning. In: Bjørner, N., Sofronie-Stokkermans, V. (eds.) CADE 2011. LNCS, vol. 6803, pp. 299–314. Springer, Heidelberg (2011)
6. Niles, I., Pease, A.: Toward a Standard Upper Ontology. In: Welty, C., Smith, B. (eds.) Proceedings of the 2nd International Conference on Formal Ontology in Information Systems, FOIS 2001 (2001)
7. Pease, A.: Ontology: A Practical Guide. Articulate Software Press, Angwin (2011)
8. Pease, A., Benzmller, C.: Sigma: An Integrated Development Environment for Logical Theories. AI Comm. 26, 9–97 (2013)
9. Pease, A., Li, J.: Controlled English to Logic Translation. In: Poli, R., Healy, M., Kameas, A. (eds.) Theory and Applications of Ontology. Springer (2010)
10. Pease, A., Li, J., Nomorosa, K.: WordNet and SUMO for Sentiment Analysis. In: Proceedings of the 6th International Global Wordnet Conference (GWC 2012), Matsue, Japan (2012)
11. Pease, A., Rust, G.: Formal Ontology for Media Rights Transactions. In: Garcia, R. (ed.) Semantic Web Methodologies for E-Business Applications. IGI publishing (2008)
12. Pease, A., Sutcliffe, G.: First Order Reasoning on a Large Ontology. In: Proceedings of the CADE-21 Workshop on Empirically Successful Automated Reasoning on Large Theories, ESARLT (2007)
13. Pease, A., Sutcliffe, G., Siegel, N., Trac, S.: Large Theory Reasoning with SUMO at CASC. AI Comm. 23(2-3), 137–144 (2010); Special issue on Practical Aspects of Automated Reasoning
14. Pelletier, F.J., Sutcliffe, G., Suttner, C.: The Development of CASC. AI Communications 15(2), 79–90 (2002)
15. Riazanov, A., Voronkov, A.: The Design and Implementation of VAMPIRE. Journal of AI Communications 15(2/3), 91–110 (2002)
16. Schulz, S.: E – A Brainiac Theorem Prover. AI Comm 15(2/3), 111–126 (2002)
17. Schulz, S.: System Description: E 1.8. In: McMillan, K., Middeldorp, A., Voronkov, A. (eds.) LPAR-19 2013. LNCS, vol. 8312, pp. 735–743. Springer, Heidelberg (2013)
18. Sutcliffe, G., Suttner, C.: The state of CASC. AI Comm 19(1), 35–48 (2006)
19. Sutcliffe, G., Zimmer, J., Schulz, S.: TSTP Data-Exchange Formats for Automated Theorem Proving Tools. In: Sorge, V., Zhang, W. (eds.) Distributed Constraint Problem Solving and Reasoning in Multi-Agent Systems. Frontiers in Artificial Intelligence and Applications, pp. 201–215. IOS Press (2004)
20. Sutcliffe, G., Schulz, S., Claessen, K., Van Gelder, A.: Using the TPTP Language for Writing Derivations and Finite Interpretations. In: Furbach, U., Shankar, N. (eds.) IJCAR 2006. LNCS (LNAI), vol. 4130, pp. 67–81. Springer, Heidelberg (2006)
21. Sutcliffe, G., Stickel, M., Schulz, S., Urban, J.: Answer Extraction for TPTP, http://www.cs.miami.edu/~tptp/TPTP/Proposals/AnswerExtraction.html (acccessed August 7, 2013)
22. Trac, S., Sutcliffe, G., Pease, A.: Integration of the TPTPWorld into SigmaKEE. In: Proceedings of IJCAR 2008 Workshop on Practical Aspects of Automated Reasoning (PAAR 2008). CEUR Workshop Proceedings (2008)

Author Index